FORBIDDEN ARCHEOLOGY

The Hidden History of the Human Race

FORBIDDEN ARCHEOLOGY

The Hidden History of the Human Race

MICHAEL A. CREMO
RICHARD L. THOMPSON

BHAKTIVEDANTA BOOK PUBLISHING, INC.
Los Angeles · Sydney · Stockholm · Bombay

Permission Credits:

Figure 2.4. "Patterns of grooves and ridges produced by a serrated shark tooth moving along the surface of a whale bone," is from *Journal of Paleontology* (1982, 56:6). Used with permission.

Figure 5.11. "A Folsom blade embedded in the lower surface of a travertine crust from Sandia Cave, New Mexico," is reprinted by permission of the Smithsonian Institution Press from *Smithsonian Miscellaneous Collections;* vol. 99, no. 23. c Smithsonian Institution, Washington, D. C. October 15, 1941. plate 7.

Figures 3.5, 3.29, and 4.12. The drawings of stone tools from Olduvai Gorge, Tanzania, are from *Olduvai Gorge* by Mary Leakey (1971) and are reprinted by permission of the Cambridge University Press.

Figure 5.15 (left). "Stone bowl from Nakura, Kenya," is from *The Stone Age Cultures of Kenya Colony,* by Louis Leakey (1931) and is used by permission of the Cambridge University Press.

Figures 5.5–7, and 5.9. The drawings of stone tools from Sheguiandah, Canada, are from *The Canadian Field-Naturalist* (1957, vol. 71). Used with permission.

Readers interested in the subject matter of this book are invited to correspond with the authors at:

Torchlight Publishing, Inc.
P.O. Box 52, Badger,
CA 93603, USA.
www.torchlight.com

First edition, 1993
First edition, revised, 1996, 1998

Published by Bhaktivedanta Book Publishing, Inc.

Cataloging-in-Publication Data
Cremo, Michael A.
    Forbidden Archeology: the hidden history of the human race / by
Michael A. Cremo and Richard L. Thompson.
    p.   cm.
    Includes bibliographical references and index.
    Preassigned LCCN: 92-76168
    ISBN 0-89213-294-9

    1. Man, Prehistoric. 2. Human evolution. I. Thompson, Richard L., 1947-
II. Title.
GN720.C74   1993   573.3
                    QBI93-573

Printed in India at Indira Printers, New Delhi-110 020.

Dedicated to

His Divine Grace
A. C. Bhaktivedanta Swami Prabhupāda

oṁ ajñāna-timirāndhasya jñānāñjnana-śalākayā
cakṣur unmīlitaṁ yena tasmai śrī-gurave namaḥ

Contents

PART II: ACCEPTED EVIDENCE

APPENDICES

Foreword

I perceive in *Forbidden Archeology* an important work of thoroughgoing scholarship and intellectual adventure. *Forbidden Archeology* ascends and descends into the realms of the human construction of scientific "fact" and theory: postmodern territories that historians, philosophers, and sociologists of scientific knowledge are investigating with increasing frequency.

Recent studies of the emergence of Western scientific knowledge accentuate that "credible" knowledge is situated at an intersection between physical locales and social distinctions. Historical, sociological, and ethnomethodological studies of science by scholars such as Harry Collins, Michael Mulkay, Steven Shapin, Thomas Kuhn, Harold Garfinkel, Michael Lynch, Steve Woolgar, Andrew Pickering, Bruno Latour, Karin Knorr-Cetina, Donna Haraway, Allucquere Stone, and Malcolm Ashmore all point to the observation that scientific disciplines, be they paleoanthropology or astronomy, "manufacture knowledge" through locally constructed representational systems and practical devices for making their discovered phenomenon visible, accountable, and consensual to a larger disciplinary body of tradition. As Michael Lynch reminds us, "scientists construct and use instruments, modify specimen materials, write articles, make pictures and build organizations."

With exacting research into the history of anthropological discovery, Cremo and Thompson zoom in on the epistemological crisis of the human fossil record, the process of disciplinary suppression, and the situated scientific handling of "anomalous evidence" to build persuasive theory and local institutions of knowledge and power.

In Cremo and Thompson's words, archeological and paleoanthropological "'facts' turn out to be networks of arguments and observational claims" that assemble a discipline's "truth" regardless, at times, of whether there is any agreed upon connection to the physical evidence or to the actual work done at the physical site of discovery. This perspective, albeit radical, accords with what I see as the best of the new work being done in studies of scientific knowledge.

Forbidden Archeology does not conceal its own positioning on a relativist spectrum of knowledge production. The authors admit to their own sense of place in a knowledge universe with contours derived from personal experience with Vedic philosophy, religious perception, and Indian cosmology. Their intriguing discourse on the "Evidence for Advanced Culture in Distant Ages" is light-years from "normal" Western science, and yet provokes a cohesion of probative thought.

In my view, it is just this openness of subjective positioning that makes *Forbidden Archeology* an original and important contribution to postmodern scholarly studies now being done in sociology, anthropology, archeology, and the history of science and ideas. The authors' unique perspective provides postmodern scholars with an invaluable parallax view of historical scientific praxis, debate, and development.

Pierce J. Flynn, Ph.D.
Department of Arts and Sciences
California State University, San Marcos,
Calif., U.S.A.

Introduction and Acknowledgments

In 1979, researchers at the Laetoli, Tanzania, site in East Africa discovered footprints in volcanic ash deposits over 3.6 million years old. Mary Leakey and others said the prints were indistinguishable from those of modern humans. To these scientists, this meant only that the human ancestors of 3.6 million years ago had remarkably modern feet. But according to other scientists, such as physical anthropologist R. H. Tuttle of the University of Chicago, fossil bones of the known australopithecines of 3.6 million years ago show they had feet that were distinctly apelike. Hence they were incompatible with the Laetoli prints. In an article in the March 1990 issue of *Natural History,* Tuttle confessed that "we are left with somewhat of a mystery." It seems permissible, therefore, to consider a possibility neither Tuttle nor Leakey mentioned—that creatures with anatomically modern human bodies to match their anatomically modern human feet existed some 3.6 million years ago in East Africa. Perhaps, as suggested in the illustration on the opposite page, they coexisted with more apelike creatures. As intriguing as this archeological possibility may be, current ideas about human evolution forbid it.

Knowledgeable persons will warn against positing the existence of anatomically modern humans millions of years ago on the slim basis of the Laetoli footprints. But there is further evidence. Over the past few decades, scientists in Africa have uncovered fossil bones that look remarkably human. In 1965, Bryan Patterson and W. W. Howells found a surprisingly modern humerus (upper arm bone) at Kanapoi, Kenya. Scientists judged the humerus to be over 4 million years old. Henry M. McHenry and Robert S. Corruccini of the University of California said the Kanapoi humerus was "barely distinguishable from modern *Homo.*" Similarly, Richard Leakey said the ER 1481 femur (thighbone) from Lake Turkana, Kenya, found in 1972, was indistinguishable from that of modern humans. Scientists normally assign the ER 1481 femur, which is about 2 million years old, to prehuman *Homo habilis.* But since the ER 1481 femur was found by itself, one cannot rule out the possibility that the rest of the skeleton was also anatomically modern. Interestingly enough, in 1913 the German scientist Hans Reck found at Olduvai Gorge, Tanzania, a complete anatomically modern human skeleton in strata over 1 million years old, inspiring decades of controversy.

Here again, some will caution us not to set a few isolated and controversial examples against the overwhelming amount of noncontroversial evidence showing that anatomically modern humans evolved from more apelike creatures fairly recently—about 100,000 years ago, in Africa, and, in the view of some, in other parts of the world as well.

But it turns out we have not exhausted our resources with the Laetoli footprints, the Kanapoi humerus, and the ER 1481 femur. Over the past eight years, Richard Thompson and I, with the assistance of our researcher Stephen Bernath, have amassed an extensive body of evidence that calls into question current theories of human evolution. Some of this evidence, like the Laetoli footprints, is fairly recent. But much of it was reported by scientists in the nineteenth and early twentieth centuries. And as you can see, our discussion of this evidence fills up quite a large book.

Without even looking at this older body of evidence, some will assume that there must be something wrong with it—that it was properly disposed of by scientists long ago, for very good reasons. Richard and I have looked rather deeply into that possibility. We have concluded, however, that the quality of this controversial evidence is no better or worse than the supposedly noncontroversial evidence usually cited in favor of current views about human evolution.

But *Forbidden Archeology* is more than a well-documented catalog of unusual facts. It is also a sociological, philosophical, and historical critique of the scientific method, as applied to the question of human origins and antiquity.

We are not sociologists, but our approach in some ways resembles that taken by practitioners of the sociology of scientific knowledge (SSK), such as Steve Woolgar, Trevor Pinch, Michael Mulkay, Harry Collins, Bruno Latour, and Michael Lynch.

Each of these scholars has a unique perspective on SSK, but they would all probably agree with the following programmatic statement. Scientists' conclusions do not identically correspond to states and processes of an objective natural reality. Instead, such conclusions reflect the real social processes of scientists as much as, more than, or even rather than what goes on in nature.

The critical approach we take in *Forbidden Archeology* also resembles that taken by philosophers of science such as Paul Feyerabend, who holds that science has attained too privileged a position in the intellectual field, and by historians of science such as J. S. Rudwick, who has explored in detail the nature of scientific controversy. As does Rudwick in *The Great Devonian Controversy*, we use narrative to present our material, which encompasses not one controversy but many controversies—controversies long resolved, controversies as yet unresolved, and controversies now in the making. This has necessitated extensive quoting from primary and secondary sources, and giving rather detailed accounts of the twists and turns of complex paleoanthropological debates.

For those working in disciplines connected with human origins and antiquity, *Forbidden Archeology* provides a well-documented compendium of reports absent from many current references and not otherwise easily obtainable.

One of the last authors to discuss the kind of reports found in *Forbidden Archeology* was Marcellin Boule. In his book *Fossil Men* (1957), Boule gave a decidedly negative review. But upon examining the original reports, we found Boule's total skepticism unjustified. In *Forbidden Archeology,* we provide primary source material that will allow modern readers to form their own opinions about the evidence Boule dismissed. We also introduce a great many cases that Boule neglected to mention.

From the evidence we have gathered, we conclude, sometimes in language devoid of ritual tentativeness, that the now-dominant assumptions about human origins are in need of drastic revision. We also find that a process of knowledge filtration has left current workers with a radically incomplete collection of facts.

We anticipate that many workers will take *Forbidden Archeology* as an invitation to productive discourse on (1) the nature and treatment of evidence in the field of human origins and (2) the conclusions that can most reasonably drawn from this evidence.

In the first chapter of Part I of *Forbidden Archeology,* we survey the history and current state of scientific ideas about human evolution. We also discuss some of the epistemological principles we employ in our study of this field. Principally, we are concerned with a double standard in the treatment of evidence.

We identify two main bodies of evidence. The first is a body of controversial evidence (A), which shows the existence of anatomically modern humans in the uncomfortably distant past. The second is a body of evidence (B), which can be interpreted as supporting the currently dominant views that anatomically modern humans evolved fairly recently, about 100,000 years ago in Africa, and perhaps elsewhere.

We also identify standards employed in the evaluation of paleoanthropological evidence. After detailed study, we found that if these standards are applied equally to A and B, then we must accept both A and B or reject both A and B. If we accept both A and B, then we have evidence placing anatomically modern humans millions of years ago, coexisting with more apelike hominids. If we reject both A and B, then we deprive ourselves of the evidential foundation for making any pronouncements whatsoever about human origins and antiquity.

Historically, a significant number of professional scientists once accepted the evidence in category A. But a more influential group of scientists, who applied standards of evidence more strictly to A than to B, later caused A to be rejected and B to be preserved. This differential application of standards for the acceptance and rejection of evidence constitutes a knowledge filter that obscures the real picture of human origins and antiquity.

In the main body of Part I (Chapters 2– 6), we look closely at the vast amount of controversial evidence that contradicts current ideas about human evolution.

We recount in detail how this evidence has been systematically suppressed, ignored, or forgotten, even though it is qualitatively (and quantitatively) equivalent to evidence favoring currently accepted views on human origins. When we speak of suppression of evidence, we are not referring to scientific conspirators carrying out a satanic plot to deceive the public. Instead, we are talking about an ongoing social process of knowledge filtration that appears quite innocuous but has a substantial cumulative effect. Certain categories of evidence simply disappear from view, in our opinion unjustifiably.

Chapter 2 deals with anomalously old bones and shells showing cut marks and signs of intentional breakage. To this day, scientists regard such bones and shells as an important category of evidence, and many archeological sites have been established on this kind of evidence alone.

In the decades after Darwin introduced his theory, numerous scientists discovered incised and broken animal bones and shells suggesting that tool-using humans or human precursors existed in the Pliocene (2–5 million years ago), the Miocene (5–25 million years ago), and even earlier. In analyzing cut and broken bones and shells, the discoverers carefully considered and ruled out alternative explanations—such as the action of animals or geological pressure—before concluding that humans were responsible. In some cases, stone tools were found along with the cut and broken bones or shells.

A particularly striking example in this category is a shell displaying a crude yet recognizably human face carved on its outer surface. Reported by geologist H. Stopes to the British Association for the Advancement of Science in 1881, this shell, from the Pliocene Red Crag formation in England, is over 2 million years old. According to standard views, humans capable of this level of artistry did not arrive in Europe until about 30,000 or 40,000 years ago. Furthermore, they supposedly did not arise in their African homeland until about 100,000 years ago.

Concerning evidence of the kind reported by Stopes, Armand de Quatrefages wrote in his book *Hommes Fossiles et Hommes Sauvages* (1884): "The objections made to the existence of man in the Pliocene and Miocene seem to habitually be more related to theoretical considerations than direct observation."

The most rudimentary stone tools, the eoliths ("dawn stones") are the subject of Chapter 3. These implements, found in unexpectedly old geological contexts, inspired protracted debate in the late nineteenth and early twentieth centuries.

For some, eoliths were not always easily recognizable as tools. Eoliths were not shaped into symmetrical implemental forms. Instead, an edge of a natural stone flake was chipped to make it suitable for a particular task, such as scraping, cutting, or chopping. Often, the working edge bore signs of use.

Critics said eoliths resulted from natural forces, like tumbling in stream beds. But defenders of eoliths offered convincing counterarguments that natural forces could not have made unidirectional chipping on just one side of a working edge.

In the late nineteenth century, Benjamin Harrison, an amateur archeologist, found eoliths on the Kent Plateau in southeastern England. Geological evidence suggests that the eoliths were manufactured in the Middle or Late Pliocene, about 2–4 million ago. Among the supporters of Harrison's eoliths were Alfred Russell Wallace, cofounder with Darwin of the theory of evolution by natural selection; Sir John Prestwich, one of England's most eminent geologists; and Ray E. Lankester, a director of the British Museum (Natural History).

Although Harrision found most of his eoliths in surface deposits of Pliocene gravel, he also found many below ground level during an excavation financed and directed by the British Association for the Advancement of Science. In addition to eoliths, Harrison found at various places on the Kent Plateau more advanced stone tools (paleoliths) of similar Pliocene antiquity.

In the early part of the twentieth century, J. Reid Moir, a fellow of the Royal Anthropological Institute and president of the Prehistoric Society of East Anglia, found eoliths (and more advanced stone tools) in England's Red Crag formation. The tools were about 2.0–2.5 million years old. Some of Moir's tools were discovered in the detritus beds beneath the Red Crag and could be anywhere from 2.5 to 55 million years old.

Moir's finds won support from one of the most vocal critics of eoliths, Henri Breuil, then regarded as one of the world's preeminent authorities on stone tools. Another supporter was paleontologist Henry Fairfield Osborn, of the American Museum of Natural History in New York. And in 1923, an international commission of scientists journeyed to England to investigate Moir's principal discoveries and pronounced them genuine.

But in 1939, A. S. Barnes published an influential paper, in which he analyzed the eoliths found by Moir and others in terms of the angle of flaking observed on them. Barnes claimed his method could distinguish human flaking from flaking by natural causes. On this basis, he dismissed all the eoliths he studied, including Moir's, as the product of natural forces. Since then, scientists have used Barnes's method to deny the human manufacture of other stone tool industries. But in recent years, authorities on stone tools such as George F. Carter, Leland W. Patterson, and A. L. Bryan have disputed Barnes's methodology and its blanket application. This suggests the need for a reexamination of the European eoliths.

Significantly, early stone tools from Africa, such as those from the lower levels of Olduvai Gorge, appear identical to the rejected European eoliths. Yet they are accepted by the scientific community without question. This is probably because they fall within, and help support, the conventional spatio-temporal framework of human evolution.

But other Eolithic industries of unexpected antiquity continue to encounter strong opposition. For example, in the 1950s, Louis Leakey found stone tools over 200,000 years old at Calico in southern California. According to standard views, humans did not enter the subarctic regions of the New World until about 12,000 years ago. Mainstream scientists responded to Calico with predictable claims that the objects found there were natural products or that they were not really 200,000 years old. But there is sufficient reason to conclude that the Calico finds are genuinely old human artifacts. Although most of the Calico implements are crude, some, including a beaked graver, are more advanced.

In Chapter 4, we discuss a category of implements that we call crude paleoliths. In the case of eoliths, chipping is confined to the working edge of a naturally broken piece of stone. But the makers of the crude paleoliths deliberately struck flakes from stone cores and then shaped them into more recognizable types of tools. In some cases, the cores themselves were shaped into tools. As we have seen, crude paleoliths also turn up along with eoliths. But at the sites discussed in Chapter 4, the paleoliths are more dominant in the assemblages.

In the category of crude paleoliths, we include Miocene tools (5–25 million years old) found in the late nineteenth century by Carlos Ribeiro, head of the Geological Survey of Portugal. At an international conference of archeologists and anthropologists held in Lisbon, a committee of scientists investigated one of the sites where Ribeiro had found implements. One of the scientists found a stone tool even more advanced than the better of Ribeiro's specimens. Comparable to accepted Late Pleistocene tools of the Mousterian type, it was firmly embedded in a Miocene conglomerate, in circumstances confirming its Miocene antiquity.

Crude paleoliths were also found in Miocene formations at Thenay, France. S. Laing, an English science writer, noted: "On the whole, the evidence for these Miocene implements seems to be very conclusive, and the objections to have hardly any other ground than the reluctance to admit the great antiquity of man."

Scientists also found crude paleoliths of Miocene age at Aurillac, France. And at Boncelles, Belgium, A. Rutot uncovered an extensive collection of paleoliths of Oligocene age (25 to 38 million years old).

In Chapter 5, we examine very advanced stone implements found in unexpectedly old geological contexts. Whereas the implements discussed in Chapters 3 and 4 could conceivably be the work of human precursors such as *Homo erectus* or *Homo habilis,* given current estimates of their capabilities, the implements of Chapter 5 are unquestionably the work of anatomically modern humans.

Florentino Ameghino, a respected Argentine paleontologist, found stone tools, signs of fire, broken mammal bones, and a human vertebra in a Pliocene formation at Monte Hermoso, Argentina. Ameghino made numerous similar discoveries in Argentina, attracting the attention of scientists around the world. Despite Ameghino's unique theories about a South American origin for the hominids, his actual discoveries are still worth considering.

In 1912, Ales Hrdlicka, of the Smithsonian Institution, published a lengthy, but not very reasonable, attack on Ameghino's work. Hrdlicka asserted that all of Ameghino's finds were from recent Indian settlements.

In response, Carlos Ameghino, brother of Florentino Ameghino, carried out new investigations at Miramar, on the Argentine coast south of Buenos Aires. There he found a series of stone implements, including bolas, and signs of fire. A commission of geologists confirmed the implements' position in the Chapadmalalan formation, which modern geologists say is 3–5 million years old. Carlos Ameghino also found at Miramar a stone arrowhead firmly embedded in the femur of a Pliocene species of *Toxodon,* an extinct South American mammal.

Ethnographer Eric Boman disputed Carlos Ameghino's discoveries but also unintentionally helped confirm them. In 1920, Carlos Ameghino's collector, Lorenzo Parodi, found a stone implement in the Pliocene seaside *barranca* (cliff) at Miramar and left it in place. Boman was one of several scientists invited by Ameghino to witness the implement's extraction. After the implement (a bola stone) was photographed and removed, another discovery was made. "At my direction," wrote Boman, "Parodi continued to attack the *barranca* with a pick at the same point where the bola stone was discovered, when suddenly and unex-pectedly, there appeared a second stone ball. . . . It is more like a grinding stone than a bola." Boman found yet another implement 200 yards away. Confounded, Boman could only hint in his written report that the implements had been planted by Parodi. While this might conceivably have been true of the first implement, it is hard to explain the other two in this way. In any case, Boman produced no evidence whatsoever that Parodi, a longtime employee of the Buenos Aires Museum of Natural History, had ever behaved fraudulently.

The kinds of implements found by Carlos Ameghino at Miramar (arrowheads and bolas) are usually considered the work of *Homo sapiens sapiens.* Taken at face value, the Miramar finds therefore demonstrate the presence of anatomically modern humans in South America over 3 million years ago. Interestingly enough, in 1921 M. A. Vignati discovered in the Late Pliocene Chapadmalalan formation at Miramar a fully human fossil jaw fragment.

In the early 1950s, Thomas E. Lee of the National Museum of Canada found advanced stone tools in glacial deposits at Sheguiandah, on Manitoulin Island in northern Lake Huron. Geologist John Sanford of Wayne State University argued that the oldest Sheguiandah tools were at least 65,000 years old and might be as much as 125,000 years old. For those adhering to standard views on North American prehistory, such ages were unacceptable.

Thomas E. Lee complained: "The sites discoverer [Lee] was hounded from his Civil Service position into prolonged unemployment; publication outlets were cut off; the evidence was misrepresented by several prominent authors . . . ; the tons of artifacts vanished into storage bins of the National Museum of Canada; for refusing to fire the discoverer, the Director of the National Museum, who had proposed having a monograph on the site published, was himself fired and driven into exile; official positions of prestige and power were exercised in an effort to gain control over just six Sheguiandah specimens that had not gone under cover; and the site has been turned into a tourist resort. . . . Sheguiandah would have forced embarrassing admissions that the Brahmins did not know everything. It would have forced the rewriting of almost every book in the business. It had to be killed. It was killed."

The treatment received by Lee is not an isolated case. In the 1960s, anthropologists uncovered advanced stone tools at Hueyatlaco, Mexico. Geologist Virginia Steen-McIntyre and other members of a U.S. Geological Survey team obtained an age of about 250,000 years for the site's implement-bearing layers. This challenged not only standard views of New World anthropology but also the whole standard picture of human origins. Humans capable of making the kind of tools found at Hueyatlaco are not thought to have come into existence until around 100,000 years ago in Africa.

Virginia Steen-McIntyre experienced difficulty in getting her dating study on Hueyatlaco published. "The problem as I see it is much bigger than Hueyatlaco," she wrote to Estella Leopold, associate editor of *Quaternary Research.* "It concerns the manipulation of scientific thought through the suppression of 'Enigmatic Data,' data that challenges the prevailing mode of thinking. Hueyatlaco certainly does that! Not being an anthropologist, I didn't realize the full significance of our dates back in 1973, nor how deeply woven into our thought the current theory of human evolution has become. Our work at Hueyatlaco has been rejected by most archaeologists because it contradicts that theory, period."

This pattern of data suppression has a long history. In 1880, J. D. Whitney, the state geologist of California, published a lengthy review of advanced stone tools found in California gold mines. The implements, including spear points and stone mortars and pestles, were found deep in mine shafts, underneath thick, undisturbed layers of lava, in formations that geologists now say are from 9 million to over 55 million years old. W. H. Holmes of the Smithsonian Institution, one of the most vocal nineteenth-century critics of the California finds, wrote: "Perhaps if Professor Whitney had fully appreciated the story of human evolution as it is understood today, he would have hesitated to announce the conclusions formulated [that humans existed in very ancient times in North America], notwithstanding the imposing array of testimony with which he was confronted." In other words, if the facts do not agree with the favored theory, then such facts, even an imposing array of them, must be discarded.

In Chapter 6, we review discoveries of anomalously old skeletal remains of the anatomically modern human type. Perhaps the most interesting case is that of Castenedolo, Italy, where in the 1880s, G. Ragazzoni, a geologist, found fossil bones of several *Homo sapiens sapiens* individuals in layers of Pliocene sediment 3 to 4 million years old. Critics typically respond that the bones must have been placed into these Pliocene layers fairly recently by human burial. But Ragazzoni was alert to this possibility and carefully inspected the overlying layers. He found them undisturbed, with absolutely no sign of burial.

Modern scientists have used radiometric and chemical tests to attach recent ages to the Castenedolo bones and other anomalously old human skeletal remains. But, as we show in Appendix 1, these tests can be quite unreliable. The carbon 14 test is especially unreliable when applied to bones (such as the Castenedolo bones) that have lain in museums for decades. Under these circumstances, bones are exposed to contamination that could cause the carbon 14 test to yield abnormally young dates. Rigorous purification techniques are required to remove such contamination. Scientists did not employ these techniques in the 1969 carbon 14 testing of some of the Castenedolo bones, which yielded an age of less than a thousand years.

Although the carbon 14 date for the Castenedolo material is suspect, it must still be considered as relevant evidence. But it should be weighed along with the other evidence, including the original stratigraphic observations of Ragazzoni, a professional geologist. In this case, the stratigraphic evidence appears to be more conclusive.

Opposition, on theoretical grounds, to a human presence in the Pliocene is not a new phenomenon. Speaking of the Castenedolo finds and others of similar antiquity, the Italian scientist G. Sergi wrote in 1884: "By means of a despotic scientific prejudice, call it what you will, every discovery of human remains in the Pliocene has been discredited."

A good example of such prejudice is provided by R. A. S. Macalister, who in 1921 wrote about the Castenedolo finds in a textbook on archeology: "There must be something wrong somewhere." Noting that the Castenedolo bones were anatomically modern, Macalister concluded: "If they really belonged to the stratum in which they were found, this would imply an extraordinarily long standstill for evolution. It is much more likely that there is something amiss with the observations." He further stated: "The acceptance of a Pliocene date for the Castenedolo skeletons would create so many insoluble problems that we can hardly hesitate in choosing between the alternatives of adopting or rejecting their authenticity." This supports the primary point we are trying to make in *Forbidden Archeology,* namely, that there exists in the scientific community a knowledge filter that screens out unwelcome evidence. This process of knowledge filtration has been going on for well over a century and continues right up to the present day.

Our discussion of anomalously old human skeletal remains brings us to the end of Part I, our catalog of controversial evidence. In Part II of *Forbidden Archeology,* we survey the body of accepted evidence that is generally used to support the now-dominant ideas about human evolution.

Chapter 7 focuses on the discovery of *Pithecanthropus erectus* by Eugene Dubois in Java during the last decade of the nineteenth century. Historically, the Java man discovery marks a turning point. Until then, there was no clear picture of human evolution to be upheld and defended. Therefore, a good number of scientists, most of them evolutionists, were actively considering a substantial body of evidence (cataloged in Part I) indicating that anatomically modern humans existed in the Pliocene and earlier. With the discovery of Java man, now classified as *Homo erectus,* the long-awaited missing link turned up in the Middle Pleistocene. As the Java man find won acceptance among evolutionists, the body of evidence for a human presence in more ancient times gradually slid into disrepute.

This evidence was not conclusively invalidated. Instead, at a certain point, scientists stopped talking and writing about it. It was incompatible with the idea that apelike Java man was a genuine human ancestor.

.As an example of how the Java man discovery was used to suppress evidence for a human presence in the Pliocene and earlier, the following statement made by W. H. Holmes about the California finds reported by J. D. Whitney is instructive. After asserting that Whitney's evidence "stands absolutely alone," Holmes complained that "it implies a human race older by at least one-half than *Pithecanthropus erectus,* which may be regarded as an incipient form of human creature only." Therefore, despite the good quality of Whitney's evidence, it had to be dismissed.

Interestingly enough, modern researchers have reinterpreted the original Java *Homo erectus* fossils. The famous bones reported by Dubois were a skullcap and femur. Although the two bones were found over 45 feet apart, in a deposit filled with bones of many other species, Dubois said they belonged to the same individual. But in 1973, M. H. Day and T. I. Molleson determined that the femur found by Dubois is different from other *Homo erectus* femurs and is in fact indistinguishable from anatomically modern human femurs. This caused Day and Molleson to propose that the femur was not connected with the Java man skull.

As far as we can see, this means that we now have an anatomically modern human femur and a *Homo erectus* skull in a Middle Pleistocene stratum that is considered to be 800,000 years old. This provides further evidence that anatomically modern humans coexisted with more apelike creatures in unexpectedly remote times. According to standard views, anatomically modern humans arose just 100,000 years ago in Africa. Of course, one can always propose that the anatomically modern human femur somehow got buried quite recently into the Middle Pleistocene beds at Trinil. But the same could also be said of the skull.

In Chapter 7, we also consider the many Java *Homo erectus* discoveries reported by G. H. R. von Koenigswald and other researchers. Almost all of these bones were surface finds, the true age of which is doubtful. Nevertheless, scientists have assigned them Middle and Early Pleistocene dates obtained by the potassium-argon method. The potassium-argon method is used to date layers of volcanic material, not bones. Because the Java *Homo erectus* fossils were found on the surface and not below the intact volcanic layers, it is misleading to assign them potassium-argon dates obtained from the volcanic layers.

The infamous Piltdown hoax is the subject of Chapter 8. Early in this century, Charles Dawson, an amateur collector, found pieces of a human skull near Piltdown. Subsequently, scientists such as Sir Arthur Smith Woodward of the British Museum and Pierre Teilhard de Chardin participated with Dawson in excavations that uncovered an apelike jaw, along with several mammalian fossils of appropriate antiquity. Dawson and Woodward, believing the combination of humanlike skull and apelike jaw represented a human ancestor from the Early Pleistocene or Late Pliocene, announced their discovery to the scientific world. For the next four decades, Piltdown man was accepted as a genuine discovery and was integrated into the human evolutionary lineage.

In the 1950s, J. S. Weiner, K. P. Oakley, and other British scientists exposed Piltdown man as an exceedingly clever hoax, carried out by someone with great scientific expertise. Some blamed Dawson or Teilhard de Chardin, but others have accused Sir Arthur Smith Woodward of the British Museum, Sir Arthur Keith of the Hunterian Museum of the Royal College of Surgeons, William Sollas of the geology department at Cambridge, and Sir Grafton Eliot Smith, a famous anatomist.

J. S. Weiner himself noted: "Behind it all we sense, therefore, a strong and impelling motive. . . . There could have been a mad desire to assist the doctrine of human evolution by furnishing the 'requisite' 'missing link.' . . . Piltdown might have offered irresistible attraction to some fanatical biologist."

Piltdown is significant in that it shows that there are instances of deliberate fraud in paleoanthropology, in addition to the general process of knowledge filtration.

Finally, there is substantial, though not incontrovertible, evidence that the Piltdown skull, at least, was a genuine fossil. The Piltdown gravels in which it was found are now thought to be 75,000 to 125,000 years old. An anatomically modern human skull of this age in England would be considered anomalous.

Chapter 9 takes us to China, where in 1929 Davidson Black reported the discovery of Peking man fossils at Zhoukoudian (formerly Choukoutien). Now classified as *Homo erectus*, the Peking man specimens were lost to science during the Second World War. Traditionally, Peking man has been depicted as a cave dweller who had mastered the arts of stone tool manufacturing, hunting, and building fires. But a certain number of influential researchers regarded this view as mistaken. They saw Peking man as the prey of a more advanced hominid, whose skeletal remains have not yet been discovered.

In 1983, Wu Rukang and Lin Shenglong published an article in *Scientific American* purporting to show an evolutionary increase in brain size during the 230,000 years of the *Homo erectus* occupation of the Zhoukoudian cave. But we show that this proposal was based on a misleading statistical presentation of the cranial evidence.

In addition to the famous Peking man discoveries, many more hominid finds have been made in China. These include, say Chinese workers, australopithecines, various grades of *Homo erectus,* Neanderthaloids, early *Homo sapiens,* and anatomically modern *Homo sapiens.* The dating of these hominids is problematic. They occur at sites along with fossils of mammals broadly characteristic of the Pleistocene. In reading various reports, we noticed that scientists routinely used the morphology of the hominid remains to date these sites more precisely.

For example, at Tongzi, South China, *Homo sapiens* fossils were found along with mammalian fossils. Qiu Zhonglang said: "The fauna suggests a Middle-Upper Pleistocene range, but the archeological [i.e., human] evidence is consistent with an Upper Pleistocene age." Qiu, using what we call morphological dating, therefore assigned the site, and hence the human fossils, to the Upper Pleistocene. A more reasonable conclusion would be that the *Homo sapiens* fossils could be as old as the Middle Pleistocene. Indeed, our examination of the Tongzi faunal evidence shows mammalian species that became extinct at the end of the Middle Pleistocene. This indicates that the Tongzi site, and the *Homo sapiens* fossils, are at least 100,000 years old. Additional faunal evidence suggests a maximum age of about 600,000 years.

The practice of morphological dating substantially distorts the hominid fossil record. In effect, scientists simply arrange the hominid fossils according to a favored evolutionary sequence, although the accompanying faunal evidence does not dictate this. If one considers the true probable date ranges for the Chinese hominids, one finds that various grades of *Homo erectus* and various grades of early *Homo sapiens* (including Neanderthaloids) may have coexisted with anatomically modern *Homo sapiens* in the middle Middle Pleistocene, during the time of the Zhoukoudian *Homo erectus* occupation.

In Chapter 10, we consider the possible coexistence of primitive hominids and anatomically modern humans not only in the distant past but in the present. Over the past century, scientists have accumulated evidence suggesting that human-like creatures resembling *Gigantopithecus, Australopithecus, Homo erectus,* and the Neanderthals are living in various wilderness areas of the world. In North America, these creatures are known as Sasquatch. In Central Asia, they are called Almas. In Africa, China, Southeast Asia, Central America, and South America, they are known by other names. Some researchers use the general term "wildmen" to include them all. Scientists and physicians have reported seeing live wildmen, dead wildmen, and footprints. They have also catalogued thousands of reports from ordinary people who have seen wildmen, as well as similar reports from historical records.

Myra Shackley, a British anthropologist, wrote to us: "Opinions vary, but I guess the commonest would be that there is indeed sufficient evidence to suggest at least the possibility of the existence of various unclassified manlike creatures, but that in the present state of our knowledge it is impossible to comment on their significance in any more detail. The position is further complicated by misquotes, hoaxing, and lunatic fringe activities, but a surprising number of hard core anthropologists seem to be of the opinion that the matter is very worthwhile investigating."

Chapter 11 takes us to Africa. We describe in detail the cases mentioned in the first part of this introduction (Reck's skeleton, the Laetoli footprints, etc.). These provide evidence for anatomically modern humans in the Early Pleistocene and Late Pliocene.

We also examine the status of *Australopithecus*. Most anthropologists say *Australopithecus* was a human ancestor with an apelike head, a humanlike body, and a humanlike bipedal stance and gait. But other researchers make a convincing case for a radically different view of *Australopithecus*. Physical anthropologist C. E. Oxnard wrote in his book *Uniqueness and Diversity in Human Evolution* (1975): "Pending further evidence we are left with the vision of intermediately sized animals, at home in the trees, capable of climbing, performing degrees of acrobatics, and perhaps of arm suspension." In a 1975 article in *Nature,* Oxnard found the australopithecines to be anatomically similar to orangutans and said "it is rather unlikely that any of the Australopithecines . . . can have any direct phylogenetic link with the genus *Homo*."

Oxnard's view is not new. Earlier in this century, when the first australopithecines were discovered, many anthropologists, such as Sir Arthur Keith, declined to characterize them as human ancestors. But they were later overruled. In his book *The Order of Man* (1984), Oxnard noted: "In the uproar, at the time, as to whether or not these creatures were near ape or human, the *opinion* that they were human won the day. This may well have resulted not only in the defeat of the contrary *opinion* but also the burying of *that part of the evidence* upon which the contrary opinion was based. If this is so, it should be possible to unearth this *other part of the evidence*." And that, in a more general way, is what we have done in *Forbidden Archeology*. We have unearthed buried evidence, evidence which supports a view of human origins and antiquity quite different from that currently held.

In Appendix 1, we review chemical and radiometric dating techniques and their application to human fossil remains, including some of those discussed in

Chapter 6. In Appendix 2, we provide a limited selection of evidence for ancient humans displaying a level of culture beyond that indicated by the stone tools discussed in Chapters 3–5. And in Appendix 3, we provide a table listing almost all of the discoveries contained in *Forbidden Archeology.*

Some might question why we would put together a book like *Forbidden Archeology,* unless we had some underlying purpose. Indeed, there is some underlying purpose.

Richard Thompson and I are members of the Bhaktivedanta Institute, a branch of the International Society for Krishna Consciousness that studies the relationship between modern science and the world view expressed in the Vedic literature. This institute was founded by our spiritual master, His Divine Grace A. C. Bhaktivedanta Swami Prabhupada, who encouraged us to critically examine the prevailing account of human origins and the methods by which it was established. From the Vedic literature, we derive the idea that the human race is of great antiquity. To conduct systematic research into the existing scientific literature on human antiquity, we expressed the Vedic idea in the form of a theory that various humanlike and apelike beings have coexisted for a long time.

That our theoretical outlook is derived from the Vedic literature should not disqualify it. Theory selection can come from many sources—a private inspiration, previous theories, a suggestion from a friend, a movie, and so on. What really matters is not a theory's source but its ability to account for observations.

Our research program led to results we did not anticipate, and hence a book much larger than originally envisioned. Because of this, we have not been able to develop in this volume our ideas about an alternative to current theories of human origins. We are therefore planning a second volume relating our extensive research results in this area to our Vedic source material.

Given their underlying purpose, *Forbidden Archeology* and its forthcoming companion volume may therefore be of interest to cultural and cognitive anthropologists, scholars of religion, and others concerned with the interactions of cultures in time and space.

At this point, I would like to say something about my collaboration with Richard Thompson. Richard is a scientist by training, a mathematician who has published refereed articles and books in the fields of mathematical biology, remote sensing from satellites, geology, and physics. I am not a scientist by training. Since 1977, I have been a writer and editor for books and magazines published by the Bhaktivedanta Book Trust.

In 1984, Richard asked his assistant Stephen Bernath to begin collecting material on human origins and antiquity. In 1986, Richard asked me to take that material and organize it into a book.

As I reviewed the material provided to me by Stephen, I was struck by the very small number of reports from 1859, when Darwin published *The Origin of*

Species, until 1894, when Dubois published his report on Java man. Curious about this, I asked Stephen to obtain some anthropology books from the late nineteenth and early twentieth centuries. In these books, including an early edition of Boule's *Fossil Men,* I found highly negative reviews of numerous reports from the period in question. By tracing out footnotes, we dug up a few samples of these reports. Most of them, by nineteenth-century scientists, described incised bones, stone tools, and anatomically modern skeletal remains encountered in unexpectedly old geological contexts. The reports were of high quality, answering many possible objections. This encouraged me to make a more systematic search. Digging up this buried literary evidence required another three years. Stephen Bernath and I obtained rare conference volumes and journals from around the world, and together we translated the material into English. The results of this labor provided the basis for Chapters 2–6 in *Forbidden Archeology.*

After I reviewed the material Stephen gave me about the Peking man discoveries, I decided we should also look at recent hominid finds in China. While going through dozens of technical books and papers, I noticed the phenomenon of morphological dating. And when I reviewed our African material, I encountered hints of the dissenting view regarding *Australopithecus.* My curiosity about these two areas also led to a fruitful extension of our original research program.

Writing the manuscript from the assembled material took another couple of years. Throughout the entire period of research and writing, I had almost daily discussions with Richard about the significance of the material and how best to present it. Richard himself contributed most of Appendix 1, the discussion of the uranium series dating of the Hueyatlaco tools in Chapter 5, and the discussion of epistemological considerations in Chapter 1. The remainder of the book was written by me, although I relied heavily on research reports supplied by Stephen Bernath for Chapter 7 and the first part of Chapter 9, as well as Appendix 2. Stephen obtained much of the material in Appendix 2 from Ron Calais, who kindly sent us many Xeroxes of original reports from his archives.

In this second printing of the first edition of *Forbidden Archeology,* we have corrected several small errors in the original text, mostly typographical. The account of a wildman sighting by Anthony B. Wooldridge, originally included in Chapter 10, has been deleted because we have since learned that the author has retracted his statements.

Richard and I are grateful to our Bhaktivedanta Institute colleagues and the other reviewers who read all or part of the manuscript of *Forbidden Archeology.* We have incorporated many, but not all, of their suggestions. Full responsibility for the content and manner of presentation lies with us.

Virginia Steen-McIntyre was kind enough to supply us with her correspondence on the dating of the Hueyatlaco, Mexico, site. We also had useful discussions about stone tools with Ruth D. Simpson of the San Bernardino County

Museum and about shark teeth marks on bone with Thomas A. Deméré of the San Diego Natural History Museum.

I am indebted to my friend Pierce Julius Flynn for the continuing interest he has displayed in the writing and publication of *Forbidden Archeology*. It is through him that I have learned much of what I know about current developments in the social sciences, particularly semiotics, the sociology of knowledge, and postmodern anthropology.

This book could not have been completed without the varied services of Christopher Beetle, a computer science graduate of Brown University, who came to the Bhaktivedanta Institute in San Diego in 1988. He typeset almost all of the book, going through several revisions. He also made most of the tables, processed most of the illustrations, and served as a proofreader. He made many helpful suggestions on the text and illustrations, and he also helped arranged the printing.

For overseeing the design and layout, Richard and I thank Robert Wintermute. The illustrations opposite the first page of the introduction and in Figure 11.11 are the much-appreciated work of Miles Triplett. The cover painting is by Hans Olson. David Smith, Sigalit Binyaminy, Susan Fritz, Barbara Cantatore, and Michael Best also helped in the production of this book.

Richard and I would especially like to thank the international trustees of the Bhaktivedanta Book Trust, past and present, for their generous support for the research, writing, and publication of this book. Michael Crabtree also contributed toward the printing cost of this book.

Finally, we encourage readers to bring to our attention any additional evidence that may be of interest, especially for inclusion in future editions of this book. We are also available for interviews and speaking engagements. Correspondence may be addressed to us at Bhaktivedanta Book Publishing, Inc., 3764 Watseka Avenue, Los Angeles, CA 90034.

Michael A. Cremo
Alachua, Florida
April 24, 1995

Part I
ANOMALOUS EVIDENCE

1

The Song of the Red Lion

One evening in 1871, an association of learned British gentlemen, the Red Lions, gathered in Edinburgh, Scotland, to feed happily together and entertain each other with humorous songs and speeches. Lord Neaves, known well for his witty lyrics, stood up before the assembled Lions and sang twelve stanzas he had composed on "The Origin of Species a la Darwin." Among them:

> *An Ape with a pliable thumb and big brain,*
> *When the gift of gab he had managed to gain,*
> *As Lord of Creation established his reign*
> *Which Nobody can Deny!*

His listeners responded, as customary among the Red Lions, by gently roaring and wagging their coattails (Wallace 1905, p. 48).

1.1 DARWIN HESITATES

Just a dozen years after Charles Darwin published *The Origin of Species* in 1859, growing numbers of scientists and other educated persons considered it impossible, indeed laughable, to suppose that humans were anything other than the modified descendants of an ancestral line of apelike creatures. In *The Origin of Species* itself, Darwin touched but briefly on the question of human beginnings, noting in the final pages only that "Light will be thrown on the origin of man and his history." Yet despite Darwin's caution, it was clear that he did not see humanity as an exception to his theory that one species evolves from another.

Other scientists were not as hesitant as Darwin to directly apply evolutionary theory to the origin of the human species. For these scientists, Darwinism helped explain the remarkable similarity between humans and apes. Even before Darwin published *The Origin of Species*, Thomas Huxley had been investigating anatomical similarities between apes and humans. Huxley clashed with Richard Owen, who insisted that human brains had a unique feature—the hippocampus major. At a meeting of the British Association for the Advancement of Science

3

in 1860, Huxley presented evidence showing that brains of apes also had the hippocampus major, thus nullifying a potential objection to the idea that humans had evolved from apelike ancestors. Exuding his usual self-confidence, Huxley (Wendt 1972, p. 71) had written his wife before the British Association meeting: "By next Friday evening they will all be convinced that they are monkeys!"

Huxley did not limit himself to convincing scientists of this proposition. He delivered to working men a series of lectures on the evolutionary connection between humans and lower animals, and in 1863 he published *Man's Place in Nature,* in which he summarized in popular form his arguments for human descent from an apelike creature by the mechanism of Darwinian evolution. In his book, Huxley presented detailed evidence showing the similarity of the human anatomy to that of the chimpanzees and gorillas. The book, intended for general readership, inspired violent criticism but sold well. Scientists continue to use the similarity between humans and apes as an argument in favor of the evolution of humans from apelike ancestors.

Scientists have extended the argument to the molecular level, and have presented evidence showing that there is 99 percent agreement between the DNA sequences of human genes and the corresponding genes of chimpanzees. This certainly suggests a close relationship between humans and chimpanzees, and on a broader scale the shared biochemical mechanisms of living cells indicate a relationship between all living organisms. However, the mere existence of patterns of similarity does not tell us what this relationship is. From an *a priori* standpoint, it could be a relationship of descent by Darwinian evolution, or it could be something quite different. To actually show evolutionary descent, it is necessary to find physical evidence of transforming sequences of ancestors.

In a companion volume to this book, we will fully discuss the argument that the genealogical tree of human descent can be traced out using biomolecular studies involving mitochondrial DNA and other genetic material. For now, we shall simply point out that interpretation of patterns of molecular similarity in terms of genealogical trees presupposes (rather than proves) that the patterns came about by evolutionary processes. In addition, the assignment of ages to such patterns of relationships depends on archeological and paleoanthropological studies of ancient human or near human populations. Thus, in the end, all attempts to show the evolution of species (the human species in particular) must rely on the interpretation of fossils and other remains found in the earth's strata.

By the time Darwin published *The Origin of Species* in 1859, some key finds relevant to human origins had already been made. About 15 years previously, Edouard Lartet had found in Miocene strata at Sansan in southern France the first fossils of *Pliopithecus,* an extinct primate thought to be ancestral to the modern gibbons. About this discovery Lartet wrote in 1845: "This corner of ground once supported a population of mammals of much higher degree than those here today. . . . Here are represented various degrees in the scale of animal life, up to and including the Ape. A higher type, that of the human kind, has not been found here; but we must not hastily conclude from its absence from these ancient formations that it did not exist" (Boule and Vallois 1957, pp. 17–18). Lartet was hinting that human beings might have existed in Miocene times, over 5 million years ago, an idea that would not win any support from today's scientists.

In 1856, Lartet reported on *Dryopithecus,* a fossil ape discovered by Alfred Fontan near Sansan. This Miocene ape is thought to be anatomically related to the modern chimpanzees and gorillas. Although *Pliopithecus* and *Dryopithecus* provided Darwinists with possible distant ancestors for humans and modern apes, there were no fossils of intermediate beings connecting humans with these Miocene primates. However, in the same year Lartet reported on *Dryopithecus,* the first evidence that intermediate prehuman forms may have existed was found in the Neander valley in Germany.

1.2 THE NEANDERTHALS

In the latter part of the seventeenth century, a minor German religious poet and composer named Joachim Neumann sometimes wandered through the Dussel River valley, in solitary communion with nature. He used the pseudonym Neander, and after his death the local people called the valley the Neanderthal. Two centuries later, others came to the pleasant little valley of the Dussel not for peace of mind but to quarry limestone for the Prussian construction industry. One day in August of 1856, while excavating the Feldhofer cave high on a steep slope of the valley, some workmen discovered human fossils and gave them to Herr Beckershoff. Beckershoff later dispatched a skullcap and some other large bones to J. Carl Fuhlrott, a local schoolteacher with a well-known interest in natural history. Recognizing the fossils as possible evidence of humanity's great antiquity, Fuhlrott in turn gave them to Herman Schaaffhausen, a professor of anatomy at the University of Bonn.

At this time, most of the scientists considering the question of human antiquity believed that Europe had once been inhabited by a roundheaded primitive race who used tools of stone and bronze. This race had later been been replaced by an invading longheaded race who knew how to use iron. The two races were not, however, regarded as being linked by evolution. In 1857, Professor Schaaffhausen delivered reports to scientific gatherings in Germany, calling the newly discovered Neanderthal man a representative of a "barbarous aboriginal race," perhaps descended from the wildmen of northwestern Europe mentioned in the works of various Roman authors such as Virgil and Ovid. Schaaffhausen called special attention to the Neanderthal skull's primitive features—its thick bone structure and its pronounced brow ridges—as evidence of its antiquity and difference from the modern racial type. Others suggested it was simply the skull of a modern man, heavily deformed by disease. And there the matter rested until 1859, when Darwin published *The Origin of Species,* setting off intense speculation about humanity's possible descent from more primitive apelike creatures.

The Neanderthal discovery was then no longer a topic for discussion only among the members of the Natural History Society of the Prussian Rhineland and Westphalia. The heavyweights of European science moved in to pass judgement. Charles Lyell, then recognized as the world's preeminent geologist, came to Germany and personally investigated both the fossils and the cave in which they had been found. He felt nothing conclusive could be deduced from the Neanderthal skeleton. For one thing it was "too isolated and exceptional" (Lyell 1863, p. 375). How could generalizations about human prehistory be drawn from just one set of bones which happened to have some "abnormal and ape-like" features? Lyell also felt that its age was "too uncertain." The unstratified cave deposits in which it had been found could not be assigned a place in the sequence of geological periods. Accompanying animal fossils might have helped establish the age of the Neanderthal man, but none had been found.

Many scientists, especially those opposed to evolutionary doctrines, thought the skeleton was that of a pathologically deformed individual of the recent era. The German anatomist Rudolf Virchow, for example, believed the crude features of the Neanderthal specimen could be explained by deformities resulting from rickets and arthritis. Thirty years after first expressing this opinion in 1857, Virchow still held it, and also continued to dismiss the idea that the Neanderthal bones represented a stage in human evolution from lower species. "The idea that men arose from animals," said Virchow, "is entirely unacceptable in my view, for if such transitional men had lived there would be evidence of it, and such evidence does not exist. The creature preliminary to man has just not been found" (Wendt 1972, pp. 57–58).

A British scientist argued that the "skull belonged to some poor idiotic hermit whose remains were found in the cave where he died" (Goodman 1982, p. 75). Dr. F. Mayer, an anatomist at Bonn University suggested, like Virchow, that the Neanderthal man's bent leg bones had been caused by childhood rickets, or perhaps many years of horse riding. In 1814, Cossack cavalry had moved through the area in pursuit of Napoleon's army. Was the Neanderthal man a wounded Cossack who had crawled into the cave and died? Mayer saw this as a distinct possibility. But Thomas Huxley, writing in *Natural History Review* (1864), asked how a dying soldier got in a cave 60 feet up a steep valley wall and buried himself. And where was his uniform?

An old skull dug up at Forbe's Quarry, during the building of fortifications at Gibraltar in 1848, entered the discussion. On investigation, the fossil skull had

turned out to be quite similar to the Feldhofer cave specimen, prompting George
Busk, professor of anatomy at the Royal College of Surgeons, to write in 1863:
"the Gibraltar skull adds immensely to the scientific value of the Neanderthal
specimen, showing that the latter does not represent . . . a mere individual
peculiarity, but that it may have been characteristic of a race extending from the
Rhine to the Pillars of Hercules. . . . even Professor Mayer will hardly suppose
a rickety Cossack engaged in the campaign of 1814 had crept into a sealed fissure
in the Rock of Gibraltar" (Goodman 1982, p. 77).

In 1865, Hugh Falconer said the Gibraltar skull represented "a very low type
of humanity—very low and savage, and of extreme antiquity—but still a man and
not halfway between a man and a monkey and certainly not the missing link"
(Millar 1972, p. 62). In similar fashion, Huxley concluded, after examining the
detailed drawings of the Neanderthal skull sent to him by Lyell, that the
Neanderthals were not the missing link sought by scientists. Despite the skull's
somewhat primitive features and its apparent great age, it was in Huxley's
opinion quite close to the modern type, close enough to be classified as simply
a variation. "In no sense," he said, "can the Neanderthal bones be regarded as the
remains of a human being intermediate between Men and Apes" (Huxley 1911,
p. 205). Most modern scientists agree with Huxley's analysis and see the
Neanderthals as a recent offshoot from the main line of human evolution.
The Neanderthals are sometimes designated *Homo sapiens neanderthalensis,* indicating a close relationship with the modern human type.

Huxley (1911, pp. 207–208) then went on to ask, "Where then, must we look
for primaeval Man? Was the oldest *Homo sapiens* pliocene or miocene, or yet
more ancient? In still older strata do the fossilized bones of an ape more
anthropoid, or a man more pithecoid, than any yet known await the researches
of some unborn paleontologist? Time will show."

1.3 HAECKEL AND DARWINISM

Possible intermediate forms between humans and apes were of great concern
to the German anatomist Ernst Haeckel. Haeckel, whose specialty was embry-
ology, was an avid advocate of Darwin's theory of evolution by natural selection.
He was also famous for his own theory that ontogeny, the step-by-step growth of
an animal (or human) embryo, faithfully represents the creature's phylogeny, or
evolutionary development over millions of years from a simple, one-celled
organism. However, this theory, which is summed up by the slogan "ontogeny
recapitulates phylogeny," has long been rejected by twentieth-century scientists.

Haeckel had illustrated his theory with drawings of embryos of different kinds
of animals. Unfortunately, some of his drawings turned out to be fakes, and he
was tried before the court of Jena University on charges of fraud. In his defense

he declared: "A small percent of my embryonic drawings are forgeries: those namely, for which the observed material is so incomplete or insufficient as to compel us to fill in and reconstruct the missing links by hypothesis and comparative synthesis. I should feel utterly condemned . . . were it not that hundreds of the best observers and biologists lie under the same charge" (Meldau 1964, p. 217). If Haeckel's sweeping accusation is correct, this may have important bearing on the mode of anatomical reconstruction employed for the many "missing links" we will discuss in this book.

Haeckel's enthusiasm for Darwinism was boundless, and he showed no hesitation in proclaiming the essence of the theory, the survival of the fittest, as the foundation of his whole view of reality. An early advocate of social Darwinism, he said: "A grim and ceaseless struggle for life is the real mainspring of the purposeless drama of the world's history. We can only see a 'moral order' and 'design' in it when we ignore the triumph of immoral force and the aimless features of the organism. Might goes before right as long as the organism exists" (Haeckel 1905, p. 88).

In *Descent of Man*, Darwin himself (1871, p. 501) wrote: "With savages, the weak in body and mind are soon eliminated and those that survive commonly exhibit a vigorous state of health. We civilized men, on the other hand, do our utmost to check the process of elimination; we build asylums for the imbecile, the maimed, and the sick; we institute poor-laws; and our medical men exert their utmost skill to save the life of every one to the last moment. . . . Thus the weak members of civilized societies propagate their kind. No one who has attended to the breeding of domestic animals will doubt that this must be highly injurious to the race of man. . . . hardly anyone is so ignorant as to allow his worst animals to breed." Modern supporters of Darwin's theory routinely downplay such unsettling statements.

Haeckel was one of the first to compose the familiar phylogenetic tree, showing different groups of living beings related to each other like branches and limbs coming from a central trunk. At the top of Haeckel's tree is found *Homo sapiens*. His immediate predecessor was *Homo stupidus*, "true but ignorant man." And before him came *Pithecanthropus alalus*, the "apeman without speech"—the missing link. Haeckel scored another first by commissioning a highly realistic painting of *Pithecanthropus alalus*, thus starting the longstanding tradition of presenting hypothetical human ancestors to the general public through the medium of lifelike pictures and statues.

Haeckel published his view of human evolution in 1866, in *General Morphology of Organisms*, and in 1868, in *Natural History of Creation*. These books appeared several years before Darwin came out with *Descent of Man*, in which Darwin acknowledged Haeckel's work. Haeckel believed humans had arisen from a primate ancestor in South Asia or Africa: "Considering the extraordinary resemblance between the lowest woolly-haired men, and the highest man-like apes . . . it requires but a slight imagination to conceive an intermediate form connecting the two" (Spencer 1984, p. 9).

1.4 THE SEARCH BEGINS

In his book *The Antiquity of Man,* first published in 1863, Charles Lyell, like Huxley and Haeckel, expressed the belief that fossils of a creature intermediate between the apes and humans would someday be found. The most likely places were "the countries of the anthropomorphous apes . . . the tropical regions of Africa, and the islands of Borneo and Sumatra" (Lyell 1863, p. 498).

Of course, it should be kept in mind that the missing link was not expected to connect modern humans with modern monkeys, but instead with the fossil apes. The first human ancestor, it was thought, must have branched off from the Old World monkeys sometime before the Miocene period. As Darwin himself stated (1871, p. 520): "We are far from knowing how long ago it was when man first diverged from the Catarrhine [Old World monkey] stock; but it may have occurred at an epoch as remote as the Eocene period; for that higher apes had diverged from the lower apes as early as the Upper Miocene period is shewn by the existence of *Dryopithecus.*"

Dryopithecus is still recognized as an early precursor of the anthropoid or humanlike apes, which include gorillas, chimpanzees, gibbons, and orangutans. As previously noted, *Dryopithecus* was discovered by Alfred Fontan, near Sansan in the Pyrenees region of southern France. In 1856, the find was reported to the scientific world by Edouard Lartet, who also gave it its name, which means "forest ape." In 1868, Louis Lartet, the son of Edouard Lartet, reported on fossils of the earliest fully modern humans, discovered near Cro-Magnon in south-western France. Recently, Cro-Magnon man has been assigned a date of 30,000–40,000 years. At the time, no fossils intermediate between *Dryopithecus* and Cro-Magnon man, except the Neanderthal man bones from Germany and Gibraltar, had been found (or so it appears from today's accounts).

In general, Lyell wanted to see the presence of anatomically modern humans pushed far back in time—but not too far. There were limits: "we cannot expect to meet with human bones in the Miocene formations, where all the species and nearly all the genera of mammalia belong to types widely differing from those now living; and had some other rational being, representing man, then flourished, some signs of his existence could hardly have escaped unnoticed, in the shape of implements of stone or metal" (Lyell 1863, p. 399).

This idea links the origin of humans directly with the succession in time of mammalian species, and it would be seen today as implicitly evolutionary. However, Lyell (1863, p. 499) proposed withholding final judgement regarding human evolution until a great many fossils confirming modern humanity's link with *Dryopithecus* were discovered: "At some future day, when many hundred species of extinct quadrumana [primates] may have been brought to light, the naturalist may speculate with advantage on this subject."

Still, Lyell clearly felt we should not let the lack of such evidence prejudice us against the idea of evolution. "The opponents of the theory of transmutation sometimes argue," he wrote, "that, if there had been a passage by variation from the lower Primates to Man, the geologist ought ere to this have detected some fossil remains of the intermediate links of the chain" (Lyell 1863, p. 435). But Lyell went on to suggest that "what we have said respecting the absence of gradational forms between the recent and pliocene mammalia . . . may serve to show the weakness in the present state of science of any argument based on such negative evidence, especially in the case of man, since we have not yet searched those pages of the great book of nature, in which alone we have any right to expect to find records of the missing links alluded to" (1863, pp. 435–436). He believed the proper paleontological pages were to be found in Africa and the East Indies. It is there, he felt, that "the discovery, in a fossil state, of extinct forms allied to the human, could be looked for" (Lyell 1863, p. 498).

Lyell's approach was reasonable, since he advocated withholding judgement until enough evidence was gathered. However, while rejecting arguments based on a lack of evidence, he was perhaps implicitly assuming that the discovery of semihuman forms would confirm modern humanity's descent from those forms. This is an error (and a perennial one), for the presence of a semihuman form does not preclude the contemporary or prior existence of fully human forms.

1.5 DARWIN SPEAKS

We have now seen that Huxley, Haeckel, and Lyell all wrote major works dealing with the question of human origins and that they did so before Darwin, who had deliberately held back from treating the question in *The Origin of Species*. Finally, in 1871 Darwin came out with his own book, *Descent of Man*. Explaining his delay, Darwin (1871, p. 389) wrote: "During many years I collected notes on the origin or descent of man, without any intention of publishing on the subject, but rather with the determination not to publish, as I thought that I should thus only add to the prejudices against my views. It seemed to me sufficient to indicate, in the first edition of my 'Origin of Species,' that by this work 'light would be thrown on the origin of man and his history;' and this implies that man must be included with other organic beings in any general conclusion respecting his manner of appearance on this earth."

In *Descent of Man*, Darwin was remarkably explicit in denying any special status for the human species. "We thus learn that man is descended from a hairy, tailed quadruped, probably arboreal in its habits, and an inhabitant of the Old World. . . . the higher mammals are probably derived from an ancient marsupial animal, and this through a long series of diversified forms, from some amphibian-like creature, and this again from some fish-like animal. In the dim obscurity of the past we can see that the early progenitor of all the Vertebrata must have been an aquatic animal. . . . more like the larvae of the existing marine Ascidians than any other known form" (Darwin 1871, p. 911). It was a bold statement, yet one lacking the most convincing kind of proof—fossils of species transitional between the ancient dryopithecine apes and modern humans.

The absence of evidence of possible transitional forms may not provide a proper disproof of evolution, but one can argue that such forms are required in order to positively prove the theory. Yet aside from the Neanderthal skulls and a few other little-reported finds of modern morphology, there were no discoveries of hominid fossil remains. This fact soon became ammunition to those who were revolted by Darwin's suggestion that humans had apelike ancestors. Where, they asked, were the fossils to prove it?

1.6 THE INCOMPLETENESS OF THE FOSSIL RECORD

Darwin himself (1871, p. 521) felt forced to reply and sought to defend himself by appealing to the imperfection of the fossil record: "With respect to the absence of fossil remains, serving to connect man with his ape-like progenitors, no one will lay much stress on this fact who reads Sir C. Lyell's discussion (*Elements of Geology* 1865, pp. 583–585 and *Antiquity of Man* 1863, p. 145), where he shews that in all the vertebrate classes the discovery of fossil remains has been a very slow and fortuitous process. Nor should it be forgotten that those regions which are the most likely to afford remains connecting man with some extinct ape-like creatures, have as yet not been searched by geologists."

Lyell (1863, p. 146) had argued that it was not "part of the plan of nature to store up enduring records of a large number of individual plants and animals which have lived." Rather nature tends to regularly clear her files, employing "the heat and moisture of the sun and atmosphere, the dissolving power of carbonic and other acids, the grinding teeth and gastric juices of quadrupeds, birds, reptiles, and fish, and the agency of many other invertebrata" (Lyell 1863, p. 146). Lyell also pointed out that researchers who had attempted to dredge human fossils from the sediments on the sea bottom had also been unsuccessful. He cited the attempt of the team of MacAndrew and Forbes who "failed utterly in drawing up from the deep a single human bone" and found no human artifacts "on a coast line of several hundred miles in extent, where they approached within less than half a mile of a land peopled by millions of human beings" (Lyell 1863, pp. 146–147).

To the present day, the drastic incompleteness of the fossil record has remained a critical factor in paleontology. Most popular presentations of evolution give the idea that the layers of sedimentary rock offer a complete and incontrovertible record of the progressive development of life on earth. But geologists who have studied the matter have come up with some astounding findings. For example, Tjeerd H. van Andel looked at a series of sandstone and shale deposits in Wyoming, parts of which apparently were submerged in a body of water resembling our present Gulf of Mexico. The rates at which sediment is deposited in the Gulf of Mexico are known. Applying these rates to the Wyoming strata, van Andel calculated they could have been deposited in 100,000 years. Yet geologists and paleontologists agreed that the series spans a time of 6 million years. That means that 5.9 million years of strata are missing. Van Andel (1981, p. 397) stated: "We may repeat the experiment elsewhere; invariably we find that the rock record requires only a small fraction, usually 1 to 10 percent, of the available time. . . . thus it appears that the geological record is exceedingly incomplete."

What about the sea bottom? Shouldn't the lack of erosional forces present on continental land masses result in a more complete record there? Van Andel (1981, p. 397) answered: "This turned out to be far from true. In the South Atlantic, for example, barely half of the history of the last 125 Myr is recorded in the sediment. It is no better in other oceans and surely worse for shallow marine and continental environments."

This has definite implications regarding the fossil record. Van Andel (1981, p. 398) warned that "key elements of the evolutionary record may be forever out of reach." J. Wyatt Durham, a past president of the Paleontological Society, pointed out that according to theory, about 4.1 million fossilizable marine species have existed since the Cambrian period some 600 million years ago. Yet only 93,000 fossil species have been catalogued. Durham (1967, p. 564) concluded: "Thus conservatively we now know about one out of every 44 species of invertebrates with hard parts that has existed in the marine environment since the beginning of the Cambrian. I think this ratio is unrealistically conservative; probably one out of every 100 is closer to reality."

When we turn from marine organisms to the totality of living organisms, the situation only gets worse. David M. Raup, curator of Chicago's Field Museum, and Steven Stanley, a paleontologist at Johns Hopkins University, estimated that 982 million species have existed during the earth's history, compared with the 130,000 known fossil species. They concluded that "only about .013 of one percent of the species that have lived during this 600 million year period have been recognized in the fossil record" (Raup and Stanley 1971, p. 11).

What does this have to do with human evolution? The standard idea is that the fossil record reveals a basic history, true in outline even though not known in every detail. But this might not at all be the case. Can we really say with complete certainty that humans of the modern type did not exist in distant bygone ages? Consider van Andel's point that out of 6 million years, only 100,000 may be represented by surviving strata. In the unrecorded 5.9 million years there is time for even advanced civilizations to have come and gone leaving hardly a trace.

Darwin's appeal to the incompleteness of the fossil record served to explain the absence of evidence supporting his theory. It was, nevertheless, basically a weak argument. Admittedly, many key events in the history of life probably have gone unrecorded in the surviving strata of the earth. But although these unrecorded events might support the theory of human evolution, they might radically contradict it.

Today, however, almost without exception, modern paleoanthropologists believe that they have fulfilled the expectations of Darwin, Huxley, and Haeckel by positive discoveries of fossil human ancestors in Africa, Asia, and elsewhere. We will now give a brief summary of these discoveries, placing them within the framework of the history of life on the earth as reconstructed by paleontologists. In this summary, we shall introduce the standard system of geological dates and time divisions that we will use throughout the book.

1.7 THE GEOLOGICAL TIMETABLE

The story of life on earth now accepted by paleontologists can be outlined as follows. About 4.6 billion years ago the earth came into being as part of the formation of the solar system. The earliest evidences of life are fossils reputed to be of single-celled organisms. These date to 3.5 billion years ago. It is said that only single-celled organisms inhabited the earth until about 630 million years ago, when simple multicellular creatures first make their appearance in the fossil record.

Then, some 590 million years ago, there was an explosive proliferation of invertebrate marine life forms, such as trilobites. This marks the beginning of the Paleozoic era and its first subdivision, the Cambrian period. The first fish are often said to have appeared in the Ordovician period, beginning 505 million years ago, but Cambrian fish have now been reported. In the Silurian period, beginning some 438 million years ago, the first land plants entered the fossil record. We note, however, that spores and pollen from such plants have been reported from Cambrian and even Precambrian marine strata (Jacob *et al.* 1953, Stainforth 1966, McDougall *et al.* 1963, Snelling 1963). In the Devonian period, which began 408 million years ago, the first amphibians came on the scene, followed by early reptiles in the Carboniferous period, the beginning of which is set at about 360 million years ago. Next is the Permian period, which began some 286 million years ago and marks the end of the Paleozoic era.

The next period is the Triassic, which began some 248 million years ago and is marked by the appearance of the first mammals. In the succeeding Jurassic period, which extends from 213 million years to 144 million years ago, paleontologists note the appearance of the first birds. The Jurassic and Triassic periods, along with the following Cretaceous period, are famous as the Age of the Dinosaurs and are known collectively as the Mesozoic era. At the end of the Cretaceous period, some 65 million years ago, the dinosaurs mysteriously died out.

Then comes the Cenozoic era. The name Cenozoic is made of two Greek words meaning "recent" and "life." The Cenozoic is divided into seven periods: Paleocene, Eocene, Oligocene, Miocene, Pliocene, Pleistocene, and finally the Holocene or most recent period, dating back 10,000–12,000 years. The dates for these periods, and the periods comprising the Paleozoic and Mesozoic eras, are given in Table 1.1. These dates are taken from *A Geologic Time Scale*, a recent text on radiometric dating (Harland *et al.* 1982).

The geological time divisions were largely formulated in the nineteenth century, on the basis of stratigraphic considerations. Initially, there was no way to assign quantitative dates to these divisions, and thus geologists referred to them qualitatively—a particular period was simply said to be earlier or later than another. In the twentieth century, scientists began to assign quantitative dates by means of radiometric methods, and they have continued to revise these dates periodically up to the present time. Thus today many roughly equivalent systems of dates are used by different geologists and paleontologists.

In general, we will use the dates in Table 1.1 throughout this book. When authors from the nineteenth century or early twentieth century assign a fossil to, say, the Miocene period, we will state that the fossil is from 5 to 25 million years old. The author in question may have had no quantitative estimate of the age of his fossil, or he may have had an estimate quite different from 5 to 25 million years. However, if the modern dates from Table 1.1 are correct for the Miocene, and the early author correctly assigned his fossil to the Miocene on the basis of stratigraphy, then it is valid for us to use the modern dates. We will do this since it helps us compare the old discovery with modern discoveries, which are generally given quantitative radiometric dates.

In some cases, the geological periods assigned to certain strata in the nineteenth century have been revised by modern geologists. For example, some Miocene strata have been reassigned to the Pliocene period. In general, whenever strata in a given locality have been identified, we have tried to look up the periods assigned to them in current geological literature. We have then given dates to these strata on the basis of the modern period assignments.

However, this method is often inadequate for assigning dates to nineteenth-century Pliocene and Pleistocene sites. In recent years, dates ranging from 2.7 to 15.0 million years have been assigned to the start of the Pliocene, with many

TABLE 1.1

Geological Eras and Periods

Era	Period	Start in Millions of Years Ago
Cenozoic	Holocene	.01
	Pleistocene	2
	Pliocene	5
	Miocene	25
	Oligocene	38
	Eocene	55
	Paleocene	65
Mesozoic	Cretaceous	144
	Jurassic	213
	Triassic	248
Paleozoic	Permian	286
	Carboniferous	360
	Devonian	408
	Silurian	438
	Ordovician	505
	Cambrian	590

vertebrate paleontologists favoring 10–12 million years. Other scientists have used the potassium-argon method to assign a date of 4.5–6.0 million years to the start of the Pliocene, and in Table 1.1 this date is listed as 5 million years (Berggren and Van Couvering 1974).

The Pliocene-Pleistocene boundary is defined as the base of the Calabrian, a marine stratigraphic subdivision from Italy, and this is now thought to be approximately 1.8 million years old. However, for this book the terrestrial mammalian fauna associated with the Pliocene and Pleistocene are of primary importance, since evidence pertaining to ancient human beings is typically dated on the basis of associated mammalian bones. A key faunal subdivision associated with the Pliocene and Pleistocene is the Villafranchian, which is divided into early, middle, and late sections, with dates ranging from 3.5–4.0 million years to 1.0–1.3 million years. Since many vertebrate paleontologists assigned the Villafranchian entirely to the Pleistocene, the starting date of the Pleistocene was sometimes given as 3.5–4.0 million years. At present, however, the Villa-franchian is divided between the Pleistocene and Pliocene, and the basal Calabrian date of 1.8–2.0 million years is assigned to the beginning of the Pleistocene (Berggren and Van Couvering 1974).

As a result, the best way to arrive at a quantitative date for a nineteenth-century site with Villafranchian (or later) fauna is to refer to modern estimates for the age of that site in years, and we have tried to do this as much as possible.

For sites with pre-Villafranchian fauna, the period will be Early Pliocene or earlier, and it is adequate for the purposes of this book to arrive at a date using Table 1.1 and the period presently assigned to the site.

In this book, we will take the modern system for granted, accepting it, for the sake of argument, as a fixed reference frame to use in studying the history of ancient humans and near humans. However, it is clear on closer examination that this reference frame is by no means fixed, and it may be that further study will reveal as much ambiguity in the evidence for its different time divisions and fossil markers as we have found in the evidence for ancient humans.

Certainly, experts in geology have sometimes expressed dissatisfaction with the established geological time divisions. For example, Edmund M. Spieker (1956, p. 1803) made the following remarks in a lecture delivered to the American Association of Petroleum Geologists: "I wonder how many of us realize that the time scale was frozen in its present form by 1840. . . . How much world geology was known in 1840? A bit of western Europe, none too well, and a lesser fringe of eastern North America. All of Asia, Africa, South America, and most of North America were virtually unknown. How dared the pioneers assume that their scale would fit the rocks in these vast areas, by far most of the world? Only in dogmatic assumption. . . . And in many parts of the world, notably India and South America, it does not fit. But even there it was applied! The founding fathers went forth across the earth and in Procrustean fashion made it fit the sections they found, even in places where the actual evidence literally proclaimed denial. So flexible and accommodating are the 'facts' of geology."

1.8 THE APPEARANCE OF THE HOMINIDS

The first apelike beings appeared in the Oligocene period, which began about 38 million years ago. The first apes thought to be on the line to humans appeared in the Miocene, which extends from 5 to 25 million years ago. These include the Dryopithecine ape *Proconsul africanus* and *Ramapithecus,* which is now thought to be an ancestor of the orangutan.

Then came the Pliocene period. During the Pliocene, the first hominids, or erect-walking humanlike primates, are said to appear in the fossil record. The term hominid should be distinguished from hominoid, which designates the taxonomic superfamily including apes and humans. The earliest known hominid is *Australopithecus*, the "southern ape," and is dated back as far as 4 million years, in the Pliocene.

This near human, say scientists, stood between 4 and 5 feet tall and had a cranial capacity of between 300 and 600 cubic centimeters (cc). From the neck down, *Australopithecus* is said to have been very similar to modern humans, whereas the head displayed some apelike and some human features.

One branch of *Australopithecus,* known as the "gracile" or lighter branch, is thought to have given rise to *Homo habilis* around 2 million years ago, at the beginning of the Pleistocene period. *Homo habilis* appears similar to *Australopithecus* except that his cranial capacity is said to have been larger, between 600 and 750 cc.

Homo habilis is thought to have given rise to *Homo erectus* (the species that includes Java man and Peking man) around 1.5 million years ago. *Homo erectus* is said to have stood between 5 and 6 feet tall and had a cranial capacity varying between 700 and 1,300 cc. Most paleoanthropologists now believe that from the neck down, *Homo erectus* was, like *Australopithecus* and *Homo habilis,* almost the same as modern humans. The forehead, however, still sloped back from behind massive brow ridges, the jaws and teeth were large, and the lower jaw lacked a chin. It is believed that *Homo erectus* lived in Africa, Asia, and Europe until about 200,000 years ago.

Paleoanthropologists believe that anatomically modern humans (*Homo sapiens sapiens*) emerged gradually from *Homo erectus.* Somewhere around 300,000 or 400,000 years ago the first early *Homo sapiens* or archaic *Homo sapiens* are said to have appeared. They are described as having a cranial capacity almost as large as that of modern humans, yet still manifesting to a lesser degree some of the characteristics of *Homo erectus,* such as the thick skull, receding forehead, and large brow ridges. Examples of this category are the finds from Swanscombe in England, Steinheim in Germany, and Fontechevade and Arago in France. Because these skulls also possess, to some degree, Neanderthal characteristics (Gowlett 1984, p. 85; Bräuer 1984, p. 328; Stringer *et al.* 1984, p. 90), they are also classified as pre-Neanderthal types. Most authorities now postulate that both anatomically modern humans and the classic Western European Neanderthals evolved from the pre-Neanderthal or early *Homo sapiens* types of hominids (Spencer 1984, pp. 1–49).

In the early part of the twentieth century, some scientists advocated the view that the Neanderthals of the last glacial period, known as the classic Western European Neanderthals, were the direct ancestors of modern human beings. They had brains larger than those of *Homo sapiens sapiens.* Their faces and jaws were much larger, and their foreheads were lower, sloping back from behind large brow ridges. Neanderthal remains are found in Pleistocene deposits ranging from

30,000 to 150,000 years old. However, the discovery of early *Homo sapiens* in deposits far older than 150,000 years effectively removed the classic Western European Neanderthals from the direct line of descent leading from *Homo erectus* to modern humans.

The type of human known as Cro-Magnon appeared in Europe approximately 30,000 years ago (Gowlett 1984, p. 118), and they were anatomically modern. Scientists used to say that anatomically modern *Homo sapiens sapiens* first appeared around 40,000 years ago, but now many authorities, in light of the Border Cave discoveries in South Africa, say that they appeared around 100,000 years ago (Rightmire 1984, pp. 320–321).

The cranial capacity of modern humans varies from 1,000 cc to 2,000 cc, the average being around 1,350 cc. As can be readily observed today among modern humans, there is no correlation between brain size and intelligence. There are highly intelligent people with 1,000 cc brains and morons with 2,000 cc brains.

Exactly where, when, or how *Australopithecus* gave rise to *Homo habilis,* or *Homo habilis* gave rise to *Homo erectus,* or *Homo erectus* gave rise to modern humans is not explained in present accounts of human origins. However, one thing paleoanthropologists do say is that only anatomically modern humans came to the New World. The earlier stages of evolution, from *Australopithecus* on up, are all said to have taken place in the Old World. The first arrival of human beings in the New World is generally said to have occurred some 12,000 years ago, with some scientists willing to grant a Late Pleistocene date of 25,000 years.

Even today there are many gaps in the presumed record of human descent. For example, there is an almost total absence of fossils linking the Miocene apes with the Pliocene ancestors of modern apes and ancestral humans, especially within the span of time between 4 and 8 million years ago.

Perhaps it is true that fossils will someday be found that fill in the gaps. Yet, and this is extremely important, there is no reason to suppose that the fossils that turn up will be supportive of evolutionary theory. What if, for example, fossils of anatomically modern humans turned up in strata older than those in which the dryopithecine apes were found? Even if anatomically modern humans were found to have lived contemporaneously with *Dryopithecus* (or even a million years ago, 4 million years after the Late Miocene disappearance of *Dryopithecus*), that would be enough to throw the current accounts of the origin of humankind completely out the window.

In fact, such evidence has already been found, but it has since been suppressed or conveniently forgotten. Much of it came to light immediately after Darwin published *The Origin of Species*, before which there had been no notable finds except Neanderthal man. In the first years of Darwinism, there was no clearly established story of human descent to be defended, and professional scientists made and reported many discoveries that now would never make it into the pages of any journal more academically respectable than the *National Enquirer.* Most of these fossils and artifacts were unearthed before the discovery by Eugene Dubois of Java man, the first protohuman hominid between *Dryopithecus* and modern humans.

Java man was found in Middle Pleistocene deposits generally given an age of 800,000 years. The discovery became a benchmark. Henceforth, scientists would not expect to find fossils or artifacts of anatomically modern humans in deposits of equal or greater age. If they did, they (or someone wiser) concluded that this was impossible and found some way to discredit the find as a mistake, an illusion, or a hoax. Before Java man, however, reputable nineteenth-century scientists found a number of examples of anatomically modern human skeletal remains in very ancient strata. And they also found large numbers of stone tools of various types, as well as animal bones bearing signs of human action.

1.9 SOME PRINCIPLES OF EPISTEMOLOGY

Before beginning our survey of rejected and accepted paleoanthropological evidence, we shall outline a few epistemological rules that we have tried to follow. Epistemology is defined in *Webster's New World Dictionary* (1978) as "the study or theory of the origin, nature, methods, and limits of knowledge." When engaged in the study of scientific evidence, it is important to keep the "nature, methods, and limits of knowledge" in mind; otherwise one is prone to fall into a number of illusions.

One important illusion, sometimes called the illusion of "misplaced concreteness," is that a scientific study deals directly with facts, and that scientific arguments appealing to the facts can prove statements about reality. For example, one might suppose that an argument involving facts in the form of fossil bones can prove that anatomically modern humans really did arise in Africa 100,000 years ago. Thinking this, one might strongly argue, on the basis of certain facts, that the statement "anatomically modern humans arose in Africa 100,000 years ago" represents the truth. If the facts are part of reality, and the arguments are sound, then surely the conclusion must be true. Or, at least, granting our human fallibility, we can be reasonably confident that it is true.

The problem here is that in the field of paleoanthropology the facts being considered are not directly part of reality. Indeed, if a "fact" is examined closely it is found to resolve into (1) arguments based on further "facts," or (2) claims that someone has witnessed something at a particular time and place. Thus "facts" turn out to be networks of arguments and observational claims.

To some extent, this is true of the facts discussed in any field of science. But the facts of paleoanthropology have certain key limitations that should be pointed out. First, the observations that go into paleoanthropological facts tend to involve rare discoveries that cannot be duplicated at will. For example, some scientists in this field have built great reputations on the basis of a few famous discoveries, and others, the vast majority, have spent their whole careers without making a single significant find.

Second, once a discovery is made, key elements of the evidence are destroyed, and knowledge of these elements depends solely on the testimony of the discoverers. For example, one of the most important aspects of a fossil is its stratigraphic position. However, once the fossil is removed from the earth, the direct evidence indicating its position is destroyed, and we simply have to depend on the excavator's testimony as to where he or she found it. Of course, one may argue that chemical or other features of the fossil may indicate its place of origin. This is true in some cases but not in others. And in making such judgements, we also have to depend on reports concerning the chemical and other physical properties of the strata in which the fossil was allegedly found.

Persons making important discoveries sometimes cannot find their way back to the sites of those discoveries. After a few years, the sites are almost inevitably destroyed, perhaps by erosion, by complete paleoanthropological excavation, or by commercial developments (involving quarrying, building construction, and so forth). Even modern excavations involving meticulous recording of details destroy the very evidence they are recording, and leave one with nothing but written testimony to back up many key assertions. And many important discoveries, even today, involve very scanty recording of key details.

Thus a person desiring to verify paleoanthropological reports will find it very difficult to gain access to the "real facts," even if he or she is able to travel to the site of a discovery. And, of course, limitations of time and money make it impossible to personally examine more than a small percentage of the totality of important paleoanthropological sites.

A third problem is that the facts of paleoanthropology are seldom (if ever) simple. A scientist may testify that "the fossils were clearly weathering out of a certain Early Pleistocene layer." But this apparently simple statement may depend on many observations and arguments involving geological faulting, the possibility of slumping, the presence or absence of a layer of hillwash, the presence of a refilled gully, and so on. If one consults the testimony of another person present at the site, one may find that he or she discusses many important details not mentioned by the first witness.

Different observers sometimes contradict one another, and their senses and memories are imperfect. Thus, an observer at a given site may see certain things, but miss other important things. Some of these things might be seen by other observers, but this could turn out to be impossible because the site has become inaccessible.

Then there is the problem of cheating. This can occur on the level of systematic fraud, as in the Piltdown case. As we shall see, to get to the bottom of this kind of cheating one requires the investigative abilities of a super Sherlock Holmes plus all the facilities of a modern forensic laboratory. Unfortunately, there are always strong motives for deliberate or unconscious fraud, since fame and glory await the person who succeeds in finding a human ancestor.

Cheating can also occur on the level of simply omitting to report observations that do not agree with one's desired conclusions. As we will see in the course of this book, investigators have sometimes admitted that they have observed artifacts in certain strata, but never reported this because they did not believe the artifacts could possibly be of that age. It is very difficult to avoid this, because our senses are imperfect, and if we see something that seems impossible, then it is natural to suppose that we may be mistaken. Indeed, this may very well be the case. Thus, cheating by omitting to mention important observations can have an important effect on paleoanthropological conclusions, but it cannot be eliminated. It is simply a limitation of human nature that, unfortunately, can have a considerably deleterious impact on the empirical process.

The drawbacks of paleoanthropological facts are not limited to excavations of objects. Similar drawbacks are also found in modern chemical or radiometric dating studies. For example, a carbon 14 date might seem to involve a straightforward procedure that reliably yields a number—the age of an object. But actual dating studies often turn out to involve complex considerations regarding the identity of samples, and their history and possible contamination. They may involve the rejection of some preliminary calculated dates and the acceptance of others on the basis of complex arguments that are seldom explicitly published. Here also the facts can be complex, incomplete, and largely inaccessible.

The conclusion we draw from these limitations of paleoanthropological facts is that in this field of study we are largely limited to the comparative study of reports. Although "hard evidence" does exist in the form of fossils and artifacts in museums, most of the key evidence that gives importance to these objects exists only in written form.

Since the information conveyed by paleoanthropological reports tends to be incomplete, and since even the simplest paleoanthropological facts tend to involve complex, unresolvable issues, it is difficult to arrive at solid conclusions about reality in this field. What then can we do? We suggest that one important thing we can do is compare the quality of different reports. Although we do not have access to the real facts, we can directly study different reports and objectively compare them.

A collection of reports dealing with certain discoveries can be evaluated on the basis of the thoroughness of the reported investigation and the logic and consistency of the arguments presented. One can consider whether or not various skeptical counterarguments to a given theory have been raised and answered. Since reported observations must always be taken on faith in some respect, one can also inquire into the qualifications of the observers.

We propose that if two collections of reports appear to be equally reliable on the basis of these criteria, then they should be treated equally. Both sets might be accepted, both might be rejected, or both might be regarded as having an uncertain status. It would be wrong, however, to accept one set of reports while rejecting the other, and it would be especially wrong to accept one set as proof of a given theory while suppressing the other set, and thus rendering it inaccessible to future students.

We apply this approach to two particular sets of reports. The first set consists of reports of anomalously old artifacts and human skeletal remains, most of which were discovered in the late nineteenth and early twentieth centuries. These reports are discussed in Part I of this book. The second set consists of reports of artifacts and skeletal remains that are accepted as evidence in support of current theories of human evolution. These reports range in date from the late nineteenth century (the *Pithecanthropus* of Dubois) to the 1980s, and they are discussed in Part II. Due to the natural interconnections between different discoveries, some anomalous discoveries are also discussed in Part II.

Our thesis is that in spite of the various advances in paleoanthropological science in the twentieth century there is an essential equivalence in quality between these two sets of reports. We therefore suggest that it is not appropriate to accept one set and reject the other. This has serious implications for the modern theory of human evolution. If we reject the first set of reports (the anomalies) and, to be consistent, also reject the second set (evidence currently accepted), then the theory of human evolution is deprived of a good part of its observational foundation. But if we accept the first set of reports, then we must accept the existence of intelligent, toolmaking beings in geological periods as remote as the Miocene, or even the Eocene. If we accept the skeletal evidence presented in these reports, we must go further and accept the existence of anatomically modern human beings in these remote periods. This not only contradicts the modern theory of human evolution, but it also casts grave doubt on our whole picture of the evolution of mammalian life in the Cenozoic era.

In general, if A contradicts B it is not necessary to prove that A is right in order to prove that B is wrong. To discredit B, all that is required is to show that A and B are both equally well supported by arguments and evidence. Then they cancel each other out. That is the case with our two sets of reports.

In making this study, there are a number of basic features of modern geology and paleontology that we are accepting as a fixed reference framework. These are the system of geological time divisions, the modern radiometric dates for these divisions, the succession of faunal types in successive time divisions of the

Cenozoic era, and the basic principles of stratigraphy.

It might be argued that if we are going to advocate a conclusion as radical as the one we just mentioned, then we might as well challenge these items as well. After all, if scientists can be completely wrong about the geological time range of human beings, why should we expect them to be right about the time ranges of various mammals?

The answer to this objection is that the various elements in our fixed reference frame may well be in need of reevaluation. However, in this study it would be impractical to delve into these matters in sufficient detail to demonstrate the specific defects that may exist in this geological and paleontological framework. Given the total body of available paleoanthropological evidence, we can only conclude that something must be seriously wrong with our current scientific picture of human evolution.

The point could be made that even if human beings existed in much earlier periods than is currently believed possible, this still does not contradict the theory of evolution. The evolution of humans could simply have taken place at earlier times. Our answer is that the material we are presenting can be interpreted in that way, and indeed it was so interpreted by most of the scientists who originally presented it. In fact, no matter what evidence is presented for the existence of human beings at a particular date, it is always possible to suppose that they evolved from lower forms at an earlier time.

It can also be said, however, that if the empirical basis for the current view of human evolution proves faulty, then the credibility of evolutionary theory in general is brought into question. After all, if the imposing empirical edifice of evolution from *Australopithecus* to *Homo sapiens* is just a house of cards, then how quick should one be to accept another elaborate evolutionary scheme?

1.10 THEORIES AND ANOMALOUS EVIDENCE

We have spoken of "anomalous evidence" and "evidence accepted in support of modern theory." In general, a piece of evidence is anomalous only in relation to a particular theory. If one could look at the world without any theoretical presuppositions (conscious or unconscious), one would see nothing anomalous. Unfortunately, one would probably experience little but a welter of meaningless sense perceptions, since it is through theoretical understanding that we give meaning to what we perceive.

In this connection a famous remark by Einstein is worth considering: "It may be heuristically useful to keep in mind what one has observed. But on principle it is quite wrong to try grounding a theory on observable quantities alone. In reality the opposite happens. It is the theory which determines what we can observe" (Brush 1974, p. 1167).

If Einstein is right, then as theories change, observations should also change. And this is indeed what we find in paleoanthropology. As we shall see, large amounts of paleoanthropological evidence were amassed in the late nineteenth and early twentieth centuries in support of a theory that humans or near humans were living in the Pliocene, Miocene, or earlier periods. This evidence was not regarded as anomalous by the scientists who introduced it, since they were contemplating theories of human origins (mainly along the lines of Darwinian evolution) that were compatible with this evidence. Then, with the development of the modern theory that humans like ourselves evolved in the Pleistocene, this evidence became highly unacceptable, and it vanished from sight.

One prominent feature in the treatment of anomalous evidence is what we could call the double standard. All paleoanthropological evidence tends to be complex and uncertain. Practically any evidence in this field can be challenged, for if nothing else, one can always raise charges of fraud. What happens in practice is that evidence agreeing with a prevailing theory tends to be treated very leniently. Even if it has grave defects, these tend to be overlooked. In contrast, evidence that goes against an accepted theory tends to be subjected to intense critical scrutiny, and it is expected to meet very high standards of proof.

This double standard is described in the following way by the archeologist George Carter (1980, p. 318): "When a new idea is advanced, it necessarily challenges the previous idea. This disturbs the holders of the previous idea and threatens their security. The normal reaction is anger. The new idea is then attacked, and support of it is required to be of a high order of certainty. The greater the departure from the previous idea, the greater the degree of certainty required, so it is said. I have never been able to accept this. It assumes that the old order was established on high orders of proof, and on examination this is seldom found to be true."

Of course, in this study the "new" ideas that we are bringing forward are actually older than the established ideas they contradict. One might say that these old ideas were properly repudiated many years ago, and it is absurd for us to resurrect them today. After all, science has advanced, and the methods we use today are far superior to those used a hundred years ago. For example, today we can date samples using nuclear physics, and the science of taphonomy has been developed to explain how materials are transformed when they are buried.

The answer to this objection is that we cannot accept *a priori* that the paleoanthropological studies of today are so superior in thoroughness, concept, and methodology to those of a hundred years ago. The existence of new dating methods does not rule out the validity of old stratigraphic studies. Indeed stratigraphy remains an essential tool in paleoanthropology. New methods can also create new sources of error, and some apparently new fields of study (such as taphonomy) were studied extensively in the past using different nomenclature.

The only way to really be sure of the relative value of new and old paleoanthropological reports is to undertake an actual comparative study of these reports, and that is what we attempt to do in this book. Another point, of course, is that anomalous findings are also being made today, and as we shall see, some of these involve the latest paleoanthropological techniques.

In discussing the anomalous and accepted reports in Parts I and II, we have tended to stress the merits of the anomalous reports, and we have tended to point out the deficiencies of the accepted reports. It could be argued that this indicates bias on our part. Actually, however, our objective is to show the qualitative equivalence of the two bodies of material by demonstrating that there are good reasons to accept much of the rejected material, and also good reasons to reject much of the accepted material. It should also be pointed out that we have not suppressed evidence indicating weaknesses in the anomalous findings. In fact, we extensively discuss reports that are highly critical of these findings, and give our readers the opportunity to form their own opinions.

1.11 THE PHENOMENON OF SUPPRESSION

As George Carter pointed out, some ideas or observations deviate more than others from an accepted theoretical viewpoint. If a finding is slightly anomalous, it may win acceptance after a period of controversy. If it is more anomalous, it may be studied for some time by a few scientists, while being rejected by the majority. For example, today we see that some scientists, such as Robert Jahn of Princeton University, publish parapsychological studies, while most scientists completely disregard this subject. Finally, there are some observations that so violently contradict accepted theories that they are never accepted by any scientists. These tend to be reported by scientifically uneducated people in popular books, magazines, and newspapers.

As time passes and theories change, the status of anomalous observations also changes. In some cases (as shown, for example, by the theory of continental drift), evidence once considered anomalous may later attain scientific acceptability. In other cases, evidence which was acceptable, or marginally acceptable, may become so anomalous that professional scientists will completely reject it.

This process of rejection does not usually involve careful scrutiny of the evidence by the scientists who reject it. Human time and energy are limited, and most scientists prefer to focus on positive research goals, rather than spend time scrutinizing unpopular claims. In the scientific community, the word will go out that certain findings are bogus, and this is enough to induce most scientists to avoid the rejected material.

When theories change, and a certain body of ideas and discoveries becomes unacceptable, there is generally a period of time during which prominent scientists will publish systematic attacks against the unwanted findings. (In the parlance of some scientists at the British Museum, these attacks are known as "demolition jobs.")

If the attacks are successful, then after some last attempts at rebuttal by diehard supporters, scientists will realize it is not in their best interest to defend the unwanted material or be associated with it. A shroud of silence descends over the rejected evidence, and it continues to exist only in fossilized form in the moldering pages of old scientific journals. As time passes, a few dismissive mentions may be made in occasional footnotes, and then a new generation of scientists grows up, largely unaware that the earlier evidence ever existed.

This process of suppression of evidence is illustrated by many of the anomalous paleoanthropological findings discussed in this book. This evidence now tends to be extremely obscure, and it also tends to be surrounded by a neutralizing nimbus of negative reports, themselves obscure and dating from the time when the evidence was being actively rejected. Since these reports are generally quite derogatory, they may discourage those who read them from examining the rejected evidence further.

However, the negative reports generally provide many references to earlier positive reports. When these are examined in detail, it is often found that they contain a wealth of detailed information and reasoning not adequately dealt with in the later negative critiques. Thus to properly evaluate anomalous evidence, there is no alternative to examining in detail the arguments and evidence presented in the original reports. And that is what we now propose to do.

2

Incised and Broken Bones: The Dawn of Deception

Intentionally cut and broken bones of animals comprise a substantial part of the evidence for human antiquity. They came under serious study in the middle of the nineteenth century and have remained the object of extensive research and analysis up to the present.

In the decades following the publication of Darwin's *The Origin of Species,* many scientists found incised and broken bones indicating a human presence in the Pliocene, Miocene, and earlier periods. Opponents suggested that the marks and breaks observed on the fossil bones were caused by the action of carnivores, sharks, or geological pressure. But supporters of the discoveries offered impressive counterarguments. For example, stone tools were sometimes found along with incised bones, and experiments with these implements produced marks on fresh bone exactly resembling those found on the fossils. Scientists also employed microscopes in order to distinguish the cuts on fossil bones from those that might be made by animal or shark teeth. In many instances, the marks were located in places on the bone appropriate for specific butchering operations.

Nonetheless, reports of incised and broken bones indicating a human presence in the Pliocene and earlier are absent from the currently accepted stock of evidence. This exclusion may not, however, be warranted. From the incomplete evidence now under active consideration, scientists have concluded that humans of the modern type appeared fairly recently. But in light of the evidence covered in this chapter, the soundness of that conclusion is somewhat deceptive.

2.1 ST. PREST, FRANCE (EARLY PLEISTOCENE OR LATE PLIOCENE)

Just above the famous cathedral town of Chartres in northwestern France, at St. Prest, in the valley of the Eure River, there are gravel pits, where, in the early nineteenth century, workmen occasionally turned up fossils. These were first reported to the scientific world in 1848 by Monsieur de Boisvillette, the engineer in charge of the local bridges and causeways. The numerous fossils, including many extinct animals such as *Elephas meridionalis, Rhinoceros leptorhinus, Rhinoceros etruscus, Hippopotamus major,* and a giant beaver called *Trogontherium cuvieri,* were judged to be characteristic of the Late Pliocene (de Mortillet 1883, pp. 28–29).

A further indication of the fossils' great antiquity was the fact that the gravels in which they were found lay at an elevation of 25 to 30 meters [82 to 98 feet] above the present level of the Eure, where an ancient river once ran in a different bed. The geological reasoning is as follows. When rivers cut valleys into a plain, the most recent gravels will normally be found near the bottom of the valley. Gravels found further up on the sides of the valley were deposited earlier by the same river, or other rivers, before the valley reached its present depth. The higher the gravels, the greater their age.

In April of 1863, Monsieur J. Desnoyers, of the French National Museum, came to St. Prest to gather fossils. From the sandy gravels he recovered part of a rhinoceros tibia, upon which he noticed a series of narrow grooves, longer and deeper than could have resulted from minor fracturing or weathering. To Desnoyers, some of the grooves appeared to have been produced by a sharp knife or blade of flint. He also observed small circular marks that could well have been made by a pointed implement (de Mortillet 1883, p. 43). Later, upon examining collections of St. Prest fossils at the museums of Chartres and the School of Mines in Paris, Desnoyers recognized upon a diverse assortment of bones the same types of marks. He then reported his findings to the French Academy of Sciences, maintaining that while some of the marks could possibly be attributed to glacial action others were definitely the work of humans.

If Desnoyers concluded correctly that the marks on many of the bones had been made by flint implements, then it would appear that human beings had been present in France before the end of the Pliocene period. One might ask, "What's wrong with that?" In terms of our modern understanding of paleoanthropology, quite a bit is wrong. The presence at that time in Europe of beings using stone tools in a sophisticated manner would seem almost impossible. It is believed that at the end of the Pliocene, about 2 million years ago, the modern human species had not yet come into being. Only in Africa should one find primitive human ancestors, and these were limited to *Australopithecus* and *Homo habilis* considered the first toolmaker.

At this point, some will inevitably question whether the nineteenth-century scientists were correct in assigning the St. Prest site to the Late Pliocene. The short answer to this question is a qualified yes.

As we mentioned in our discussion of the geological time periods in the previous chapter (Section 1.7), the dating of sites at the Pliocene-Pleistocene boundary remains a matter of intense controversy. Since the St. Prest site lies roughly in this period, one might expect various authorities to place it differently. And it turns out that this is in fact the case.

The American paleontologist Henry Fairfield Osborn (1910, p. 391) placed St. Prest in the Early Pleistocene. In times closer to our own, Claude Klein (1973, pp. 692–693) reviewed French opinion regarding the age of the St. Prest fauna. In 1927, Charles Deperet characterized St. Prest as Late Pliocene. G. Denizot placed St. Prest in the Cromerian interglacial stage of the Middle Pleistocene, a view he consistently maintained into the late 1960s. In 1950, P. Pinchemel referred St. Prest to the Late Pliocene. More recently, F. Boudier, in 1965, placed St. Prest in the Waalian temperate stage of the late Early Pleistocene, with a quantitative date of about 1 million years (Klein 1973, p. 736).

Others have arrived at different quantitative dates for St. Prest. Tage Nilsson (1983, p. 158) stated that two sites in the Central Massif region of France, Sainzelles and Le Coupet, yielded potassium-argon dates of 1.3–1.9 million years. Nilsson (1983, p. 158) then said: "St. Prest, near Chartres in northern France, is held to be closely related." Nilsson considered the three sites Late Villafranchian, or Early Pleistocene.

Let us now consider some of the species that were listed as present at St. Prest. *Elephas meridionalis* (sometimes called *Mammuthus meridionalis*) is said by modern authorities (Maglio 1973, p. 79) to have existed in Europe from about 1.2 million to 3.5 million years ago. Osborn (1910, p. 313) places *Rhinoceros (Dicerorhinus) leptorhinus* in the Plaisancian (or Piacenzian) age of the Pliocene. Osborn placed the Plaisancian age in the Early Pliocene, but Romer (1966, p. 334) places the Plaisancian in the Late Pliocene. *Rhinoceros (Dicerorhinus) etruscus,* according to Nilsson (1983, p. 475), occurs in Europe from the Villafranchian, which begins in the Late Pliocene, to the early Middle Pleistocene. But Savage and Russell (1983, p. 339) list occurrences of *Dicerorhinus etruscus* as early as the Ruscinian age of the Early Pliocene. According to Osborn (1910, p. 313), *Hippopotamus major,* a larger version of the modern hippopotamus, is found in the Late Pliocene and throughout the Pleistocene in Europe. *Hippopotamus major* is sometimes referred to as *Hippopotamus amphibius antiquus.* This species is listed by Savage and Russell (1983, p. 351) as part of the Pliocene Villafranchian fauna of Europe. *Trogontherium cuvieri,* the giant extinct beaver, is found in Pliocene faunal lists (Savage and Russell 1983, p. 352) and persisted until the Mosbachian age of the early Middle Pleistocene (Osborn 1910, p. 403). Thus all the above species were in existence during the Pliocene period.

Add it all up, and it can be seen that a Late Pliocene date for St. Prest is not out of the question. And, as noted previously, some twentieth-century scientists (Pinchemel and Deperet) have in fact assigned St. Prest to this period. That would place toolmaking hominids in Europe at over 2 million years ago.

How recent could St. Prest possibly be? The presence of *Elephas meridionalis*, which survived in Europe until 1.2 million years ago (Maglio 1973, p. 79) would appear to impose a late Early Pleistocene limit. The potassium-argon dates of 1.3–1.9 million years for French sites having a fauna similar to that of St. Prest (Nilsson 1983, p. 158) offer another guidepost. Kurtén (1968, p. 24), like Boudier (1965), assigns St. Prest to the Waalian temperate stage of the Early Pleistocene. Some authorities place the Waalian stage at about 1.1–1.2 million years (Nilsson 1983, p. 144). But Senéze, a French site tentatively attributed to the Waalian temperate stage, is estimated to be about 1.6 million years old (Nilsson 1983, p. 158). From all this, one could conclude that the St. Prest site, at the more recent end of its probable date range, might be just 1.2–1.6 million years old. Even at this date, incised bones would still be anomalous. The oldest undisputed evidence for the presence of *Homo erectus* in Europe dates back only about 700,000 years (Gowlett 1984, p. 76). Also, the oldest occurrences of *Homo erectus* in Africa have dates of about 1.5 million years.

Even in the nineteenth century, Desnoyers's discoveries of incised bones at St. Prest provoked controversy. Professor Bayle, a paleontologist at the School of Mines, responded to Desnoyers's report by claiming that it was he, with his own instruments, who had incised and otherwise marked the bones of St. Prest during the process of cleaning them. Dr. Eugene Robert accepted this explanation and communicated it to the French Academy of Sciences.

In response, Desnoyers (1863) protested that his careful scientific presentation had been attacked by means of a brief rumorlike report, submitted without any credible evidence. To his accusers, Desnoyers went on to reply, in a paper published in the proceedings of the French Academy of Sciences, that the bones of St. Prest, found in sand, did not require metal instruments in order to be cleaned. Furthermore, the grooves and other markings were visible on bones that had not needed any kind of cleaning whatsoever. Perhaps the professor of paleontology at the School of Mines, Dr. Bayle, truly had been sufficiently clumsy to have extensively damaged the valuable bones under his care. But Desnoyers did not believe anyone could say the same of the many capable and careful collectors who also had specimens of fossil bones from St. Prest bearing the exact same striations and incisions. In the words of Desnoyers (1863, p. 1201): "Let us admit, against all probability, that the memoir of the preparator and conservator of the collection is true, and all the bones of St. Prest in his possession have been subjected to the kind of alteration to which he pleads guilty. Very well. That assertion itself serves to demonstrate the action of the hand of man on all the other bones from the same locality, which, fortunately, have been preserved in other collections, from dangerous influences. The marks on them are incontestably primitive, and are completely identical to those produced by the chisels and burins of the functionary of the School of Mines."

Desnoyers (1863, p. 1201) was further annoyed that persons who had never even seen the bones claimed that the impressions on them were made by the tools of the workmen in the St. Prest sand pits. He pointed out that this supposition is clearly disproved by the fact that the grooves were covered with the same magnesium deposits and dendrites found on other sections of the bone. Dendrites are crystalline mineral deposits that form branching treelike patterns. If the cuts on the fossil bones had been made by the tools of modern excavators or museum employees, the dendrites would have been scraped away. In some cases, the grooves and marks were still tightly filled with compacted sand from the deposits in which they were discovered.

Desnoyers (1863, p. 1201) suggested that doubters examine the actual specimens: "One would see that the incisions, which furrow the bones across their width and cut their edges, are frequently crossed by the longitudinal cracks resulting from dessication. These cracks were unquestionably produced *after* the marks made when the bone was fresh; they were produced during the course of fossilization. The distinct characteristics of these two kinds of markings are proof that the one is older than the other."

Recent tool marks probably would have cut through the dessication markings in recognizable fashion, erasing the lighter and shallower cracks. Desnoyers's careful analysis foreshadows the modern discipline of taphonomy, the scientific study of the changes undergone by bone and other objects in the course of entombment and fossilization.

About one of his finds, Desnoyers (1863, p. 1201) noted, "One would see on the horn of a giant deer a large incision at the base, an incision difficult to distinguish from those found on the horns of deer from caverns of later geological eras." In other words, the incision on the deer horn was placed appropriately for a human cut mark.

The prominent British geologist Charles Lyell agreed that the St. Prest gravel beds were of Pliocene age. He observed, however, that among the fauna was the large extinct beaver, *Trogontherium,* and asked how one could be certain it was not the teeth of this animal that produced the marks on the fossil bones (Lyell 1863, appendix p. 4). Gabriel de Mortillet, professor of prehistoric anthropology at the École d'Anthropologie in Paris, stated in his book *Le Préhistorique* (1883, p. 45) that Lyell's supposition was inadmissible because the marks on the bones of St. Prest were not at all of the character of those of a rodent's teeth. In particular, they were too narrow to have been made by the strong and powerful incisors of *Trogontherium.*

De Mortillet had his own ideas about the cause of the marks on the fossil bones of St. Prest. Some authorities had suggested glaciers had been responsible for the markings. But de Mortillet said that glaciers had not reached that particular region of France. Modern authorities (Nilsson 1983, p. 169) agree on this point—the extreme southern limit of the North European glaciation passed through the Netherlands and Central Germany. De Mortillet also rejected human action as the cause of the marks on the bones.

The key to understanding the marks, according to de Mortillet, could be found in the statement by Desnoyers that they appeared to have been made by a sharp blade of flint. According to de Mortillet (1883, pp. 45–46), that was true, only the flint, instead of being moved by the hand of man, had been moved by natural force—a very strong underground pressure that caused the sharp flints to slide across the bones with force sufficient to cut them. As evidence, de Mortillet cited the fact that he had observed flints from the St. Prest gravels and elsewhere that displayed on their surfaces deep scratches. At this point it should be mentioned that in *Le Préhistorique* de Mortillet rejected every single one of the many discoveries of incised bones made up to that time, almost always offering the same explanation—that the marks were caused by sharp stones moved by subterranean geological pressures.

But in the case of the St. Prest bones, Desnoyers (1863, p. 1201) responded to de Mortillet's objections, observing: "many of the incisions have been worn by later rubbing, resulting from transport or movement of the bones in the midst of the sands and gravels. The resulting markings are of an essentially different character than the original marks and striations, and offer superabundant proof of their different ages." In other words, marks from subterranean pressure may indeed be found upon the bones, but, according to Desnoyers, they can be clearly distinguished from the earlier marks attributed to human action.

So who was right, Desnoyers or de Mortillet? Some authorities believed the question could be settled if it could be demonstrated that the gravels of St. Prest contained flint tools that were definitely of human manufacture. This same demand—for the tools that made the marks—is often made today in cases of anomalous discoveries of incised bones (Section 2.3). The Abbé Bourgeois, a clergyman who had also earned a reputation as a distinguished paleontologist, carefully searched the strata at St. Prest for such evidence. By his patient research he eventually found a number of flints that he believed were genuine tools and made them the subject of a report to the Academy of Sciences in January, 1867 (de Mortillet 1883, p. 46).

Even this did not satisfy de Mortillet (1883, pp. 46–47), who said of the flints discovered by Bourgeois at St. Prest: "Many others that he found there, and which are now deposited in the collection of the School of Anthropology, do not have conclusive traces of human work. The slidings and pressures that resulted in striations on the surfaces of the flints have also left on their sharp edges a number of chips that greatly resemble retouching by humans. This is what deceived Bourgeois. In effect, of the flints discovered at St. Prest, many present a false appearance of having been worked."

It appears that in our attempt to answer one question, the nature of cut marks on bones, we have stumbled upon another, the question of how to recognize human workmanship on flints and other stone objects. This latter question shall be fully treated in the next chapter. For now we shall simply note that judgements about what constitutes a stone tool are a matter of considerable controversy even to this day. It is, therefore, quite definitely possible to find reasons to question de Mortillet's rejection of the flints found by Bourgeois. Certainly, the bare observation that some of the flints collected by Bourgeois did not, in de Mortillet's opinion, show signs of human work does not change the fact that others, however few, did in fact show such signs. And the presence of stone tools at St. Prest would satisfy a key demand for the verification of intentional cuts on fossil bones found there.

The famous American paleontologist Henry Fairfield Osborn (1910, p. 399) made these interesting remarks in connection with the presence of stone tools at St. Prest: "the earliest traces of man in beds of this age [Early Pleistocene by his estimation] were the incised bones discovered by Desnoyers at St. Prest near Chartres in 1863. Doubt as to the artificial character of these incisions has been removed by the recent explorations of Laville and Rutot, which resulted in the discovery of eolithic flints, fully confirming the discoveries of the Abbé Bourgeois in these deposits in 1867."

So as far as the discoveries at St. Prest are concerned, it should now be apparent that we are dealing with paleontological problems that cannot be quickly or easily resolved. Certainly there is not sufficient reason to categorically reject these bones as evidence for a human presence in the Pliocene. This might lead one to wonder why the St. Prest fossils, and others like them, are almost never mentioned in textbooks on human evolution, except in rare cases of brief mocking footnotes of dismissal. Is it really because the evidence is clearly inadmissible?

Or is, perhaps, the omission or summary rejection more related to the fact that the potential Late Pliocene antiquity of the objects is so much at odds with the standard account of human origins? In theory, scientists proclaim themselves ready to follow the facts wherever they might lead. But in practice, the social mechanisms of the scientific community set limits beyond which its members in good standing may cross only at their peril. When eminent authorities announce their rejection of certain categories of evidence, others hesitate to mention similar evidence out of fear of ridicule. Thus anomalous evidence gradually slides from disrepute into complete oblivion.

Along these lines, Armand de Quatrefages, a member of the French Academy of Sciences and a professor at the Museum of Natural History in Paris, wrote in his book *Hommes Fossiles et Hommes Sauvages* (1884, p. 90): "The objections made to the existence of humans in the Pliocene and Miocene periods seem to habitually be more related to theoretical considerations than to direct observation." De Quatrefages (1884, p. 91) further stated: "The existence of man in the Secondary epoch is not at all contrary to the principles of science, and the same is true of Tertiary man."

This is quite a shocking statement, considering that the most recent Secondary period is the Cretaceous, which ended approximately 65 million years ago. Supposedly, only very small and primitive mammals existed in the Cretaceous, dodging the last of the dinosaurs. Evidence for human beings in the Cretaceous would most certainly cast a great thundering cloud of doubt over Darwin's seemingly invincible hypothesis. But for now, our focus is on the more recent Tertiary epoch. Even if anatomically modern human beings were found to have existed in the latest Pliocene, at a mere 2 million years ago, that would still call into question the evolutionary picture of human origins.

In *Hommes Fossiles et Hommes Sauvages,* de Quatrefages gave a summary of the evidence for his assertions about humans existing in the very distant past and then stated (1884, p. 96): "The preceding historical samples are incomplete and abbreviated. But they suffice, I believe, to make comprehensible that the conviction, agreed upon by many modern scientists of diverse disciplines, relative to the existence of Tertiary man, is not formed lightly but is the result of serious and repeated study."

Concerning the presence of ancient man at St. Prest, de Quatrefages (1884, pp. 89–90) wrote: "Mr. Desnoyers has affirmed his existence, based on the examination of incisions manifestly intentional found on the bones of *Elephas meridionalis* and other great mammals of the same age. This discovery was greatly contested, by among others Lyell, who declared he was not able to accept that the incisions on the bones were demonstrably the work of man until he could be shown the instruments that did it. The Abbé Bourgeois responded to this desire. But 20 years later, de Mortillet, in opposing all the results of this research, simply raises objections which when made the object of attentive study turn out to have little foundation."

Elsewhere in *Hommes Fossiles et Hommes Sauvages,* de Quatrefages (1884, p. 17) succinctly reaffirmed the evidence for the presence of humans in the Pliocene at St. Prest: "The researches of Mr. Desnoyers and the Abbé Bourgeois do not leave any doubt in this regard. Mr. Desnoyers first discovered in 1863, on bones found in the gravel pits of St. Prest, near Chartres, imprints that he did not hesitate to report as being made by the action of flint implements in the hands of human beings. A little later, the Abbé Bourgeois confirmed and completed this important discovery when he found in the same place the worked flints that had made the incisions on the bones of *Elephas meridionalis, Rhinoceros leptorhinus,* and other animals. I have examined at leisure the bones studied by Desnoyers, as well as the scrapers, borers, lance points, and arrowheads collected by the Abbé Bourgeois. From the start, I have had little doubt, and everything has been confirming that first impression. Thus man lived on the globe at the end of the Tertiary era. And he left traces of his industry; he had at this time both arms and tools. The honor of the first recognition of this fact, so little in accord with all that was believed only a short time ago, goes incontestably to Mr. Desnoyers."

Here it should be noted that it would of course be possible to more briefly summarize and paraphrase reports such as these. There are two reasons for not doing so. The first is that paleoanthropological evidence mainly exists in the form of reports, some primary and others secondary. Very few individuals, even experts in the field, have the opportunity to engage in firsthand inspection of the fossils themselves, scattered in collections around the world. Even if one is able to do so, one is still not able to be sure about the exact circumstances of the discovery. This is critical, because the interpretation of the significance of a fossil depends as much on the exact position in which it was found as on the fossil itself. In most cases, for all investigators except the original discoverers, the real evidence is the reports themselves, which give the details of the discovery, and we shall therefore take the trouble to include many selections from such reports, the exact wording of which reveals much. Contemporary discussions of these original reports, both those which are positive and those which are negative, are also illuminating.

A second consideration is that the particular reports referred to in this chapter are extremely difficult to obtain. Almost no reference to them will be found in modern textbooks. Most of them come from rare nineteenth-century paleontological and anthropological books and journals, the majority in languages other than English. This being the case, translated excerpts of the original reports have been judged preferable to paraphrases and footnotes, and will serve as a unique introduction to a vast store of buried evidence.

A final consideration is that proponents of evolutionary theory often accuse authors who arrive at nonevolutionary conclusions of "quoting out of context." It therefore becomes necessary to quote at length, in order to supply the needed context.

The controversy over the St. Prest finds was noted by S. Laing, a popular British nonfiction author of the late nineteenth century, whose well-researched books on scientific subjects, intended for the general public, reached a wide audience. After discussing the site at St. Prest, Laing (1893, p. 113) stated: "In these older gravels have been found stone implements, and bones of the *Elephas Meridionalis* with incisions evidently made by a flint knife worked by a human hand. This was disputed as long as possible, but Quatrefages, a very cautious and competent authority, states in his latest work, published in 1887, that it is now established beyond the possibility of doubt."

2.2 A MODERN EXAMPLE: OLD CROW
RIVER, CANADA (LATE PLEISTOCENE)

Before moving on to further examples of nineteenth-century discoveries that challenge modern ideas about human origins, let us consider a more recent investigation of intentionally modified bones. One of the most controversial questions confronting New World paleoanthropology is determining the time at which humans entered North America. The standard view is that bands of Asian hunter-gatherers crossed over the Bering land bridge about 12,000 years ago. Some authorities are willing to extend the date to about 30,000 years ago, while an increasing minority are reporting evidence for a human presence in the Americas at far earlier dates in the Pleistocene. We shall examine this question in greater detail in coming chapters (Sections 3.8, 4.8, 5.1, 5.2, 5.4, 5.5). For now, however, we want only to consider the fossil bones uncovered at Old Crow River in the northern Yukon territory as a contemporary example of the type of evidence dealt with in this chapter.

In the 1970s, Richard E. Morlan of the Archeological Survey of Canada and the Canadian National Museum of Man, conducted studies of modified bones from the Old Crow River sites. Morlan concluded that many bones and antlers exhibited signs of intentional human work executed before the bones had become fossilized. The bones, which had undergone river transport, were recovered from an Early Wisconsin glacial floodplain dated at 80,000 years B.P. (before present).

But R. M. Thorson and R. D. Guthrie (1984) published a taphonomic study showing that the action of river ice could have caused the alterations that suggested human work to Morlan. Thorson and Guthrie performed experiments in which large blocks of ice containing bones frozen within them were dragged behind trucks over various surfaces, reproducing the effect of river ice scraping against rocks and gravels. In a 1986 reappraisal of his previous work, Morlan, considering the taphonomic experiments of Thorson and Guthrie, admitted "the observed effects are impressive for the hazards they might pose to recognition of artificial alterations among redeposited fossils." He went on to note: "However some critical variables probably were not simulated adequately (e.g., texture and hardness of the substrate, buoyancy of the ice block), and it is noteworthy that many of the experimental bones are more profoundly altered than those recovered from natural environments. Certainly these experiments have not shown that all the altered fossils from Old Crow Basin can be attributed to river icing and breakup" (Morlan 1986, p. 29).

Nevertheless, Morlan did in fact back away, in almost all cases, from his earlier assertions that the bones he had collected had been modified by human agency. He gave alternate explanations, such as the river ice hypothesis, but cautioned: "The alternate interpretations do not prove that humans were not present in Early Wisconsinan time, but they show that such ancient presence of people cannot be demonstrated on the basis of evidence gathered thus far" (Morlan 1986, p. 27). He went on to say: "This conclusion differs from earlier statements, but it is not necessarily a retraction of those statements. I have definitely changed my mind about some of my earlier interpretations, but in most cases I am simply trying to enlarge our conceptual framework and to stimulate further observations and discussions" (Morlan 1986, pp. 28–29).

But even though Morlan recanted his previous assertions of human work on 30 bone specimens, he believed four others still bore signs of being definite human artifacts. At Johnson Creek, near Old Crow valley, he found a "fresh-fractured *Bison* sp. radius" *in situ*. The radius is one of the long bones of the lower forelimb. "Although it is not out of the question that the bison bone was broken by carnivores," stated Morlan (1986, p. 36), "its massive size and micro-relief features indicative of dynamic fracture suggest that it was broken by man. The enclosing matrix of organic silt is suggestive of a thaw-lake deposit and yields a date of >37,000 B.P."

At another locality, Morlan found two large mammal long bones and a bison rib, all three bearing incisions. Morlan (1986, p. 36) stated about these three bones and the bison radius discussed in the previous paragraph: "The cuts and scrapes . . . are indistinguishable from those made by stone tools during butchering and defleshing of an animal carcass. These four specimens comprise the most formidable barrier to a global dismissal of our supposed Early Wisconsinan archaeological record."

Morlan (1986, p. 36) then added: "While this paper was in press . . . two cut bones . . . were sent to Dr. Pat Shipman, Johns Hopkins University, for examination under the scanning electron microscope. The marks were examined with reference to a collection of more than 1000 documented marks on bones, and the provenience [source] of the specimens was not made known until after the marks had been identified. The surface of the large mammal long bone fragment is damaged and difficult to evaluate, but Dr. Shipman positively identified the mark on the Bison rib as a tool mark." Morlan (1986, p. 28) noted that stone implements have been found in the Old Crow River area and in nearby uplands, but not in direct association with bones.

What this all means is that the bones of St. Prest, and others like them, cannot be so easily dismissed. Evidence of the same type is still considered important today, and the methods of analysis are almost identical to those practiced in the nineteenth century. De Quatrefages and other scientists of that era compared specimens of cut bone with bones bearing undisputed signs of human workmanship. They also performed experiments on fresh bone. Like modern students of taphonomy, they gave detailed consideration to the changes that bones would undergo during the process of entombment and fossilization. They examined bones with a microscope. It should be noted that an electron microscope is not required for such study. A modern authority, John Gowlett (1984, p. 53), said: "Under a microscope, marks made by man are distinguishable in various ways from those made by carnivores. Dr. Henry Bunn (University of California) observed through an optical microscope at low magnification that stone tools leave V-shaped cuts, which are much narrower than rodent gnawing marks."

As the Old Crow River case clearly shows, modern scientists use methods not much different from those practiced in the nineteenth century. We can just picture Thorson and Guthrie, in previous nineteenth-century incarnations, driving a horsedrawn cart, rather than a truck, and dragging behind them a big block of ice filled with bones over a rough gravel road in northern France, trying to prove the bones of St. Prest were marked by natural forces. Amusing as the image may appear, this is the type of technologically unsophisticated yet important work that still goes into resolving questions about incised bones. But as Morlan's study shows, all questions about the Old Crow bones have not been clearly decided one way or another. He changed his mind about some of his specimens, but remained convinced about others. This ambiguity and inconclusiveness is typical of the empirical approach to such evidence.

In addition to debating whether or not the cut marks on the Old Crow bones were made by stone tools or natural forces, scientists were concerned about the age of the bones. If the bones were seen as bearing signs of human work and if they were also dated to the Early Wisconsin period, that would challenge the date for the earliest entry of humans into North America. The view now dominant is that Siberian hunters crossed the Bering Strait land bridge in the latest Pleistocene and passed through an ice-free corridor into what is now the United States about 12,000 years ago. Nevertheless, as we shall see throughout this book, there is a lot of controversial, hotly debated evidence showing that human beings were present in the Americas far before 12,000 years ago. Those scientists favoring the 12,000-year date tend to believe the marks on the Old Crow bones were caused by geological action of some kind, even though the marks have in some cases been judged identical to those caused by stone tools. This is something we shall encounter again and again. Similarly, preconceptions about the relatively recent origin of anatomically modern humans often influence scientists to reject evidence that they would otherwise take as proof of a human presence.

2.3 THE ANZA-BORREGO DESERT, CALIFORNIA (MIDDLE PLEISTOCENE)

Another recent example of incised bones like those found at St. Prest, again related to the presence of humans in the New World, is a discovery made by

George Miller, curator of the Imperial Valley College Museum in El Centro, California. Miller, who died in 1989, reported that six mammoth bones excavated from the Anza-Borrego Desert bear scratches of the kind produced by stone tools. Uranium isotope dating carried out by the U.S. Geological Survey indicated that the bones are at least 300,000 years old, and paleomagnetic dating and volcanic ash samples indicated an age of some 750,000 years (Graham 1988).

One established scholar said that Miller's claim is "as reasonable as the Loch Ness Monster or a living mammoth in Siberia," while Miller countered that "these people don't want to see man here because their careers would go down the drain" (Graham 1988). Here, perhaps, we see preconceptions influencing the established scholar to reject evidence which, if given a more suitably recent date, he might have accepted.

The incised mammoth bones from the Anza-Borrego Desert came up in a conversation we had with Thomas Deméré, a paleontologist at the San Diego Natural History Museum (May 31, 1990). Deméré said he was by nature skeptical of claims such as those made by Miller. He called into question the professionalism with which the bones had been excavated, and pointed out that no stone tools had been found along with the fossils. Furthermore, Deméré suggested that it was very unlikely that anything about the find would ever be published in a scientific journal, because the referees who review articles probably would not pass it. We later learned from Julie Parks, the present curator of George Miller's specimens, that Deméré had never inspected the fossils or visited the site of discovery, although he had been invited to do so (Parks, personal communication, June 1, 1990).

As of June 1990, the Anza-Borrego mammoth bones were still under study. Deposits of sandy matrix were being painstakingly removed from the incisions on the bones, so that the incisions could be examined by a scanning electron microscope. Hopefully, inspection of the minute striations on the surfaces of the cuts under high magnification will confirm whether or not they are characteristic of stone tools. Parks (personal communication, June 1, 1990) said that one incision apparently continues from one of the fossil bones to another bone that would have been located next to it when the mammoth skeleton was articulated. This is suggestive of a butchering mark. Accidental marks resulting from movement of the bones in the earth after the skeleton had broken up probably would not continue from one bone to another in this fashion.

The lesson to be learned from the marked bones found at Old Crow River and in the Anza-Borrego Desert is this: the marked bones of St. Prest and others like them discovered in the nineteenth century should be kept in the active file of paleoanthropological evidence. Even today, scientists are not always able to immediately determine whether or not marks on bones were made by natural forces, animals, or humans. Much careful study and analysis is required to arrive at a conclusion, and even then not all experts will agree. Therefore the marked bones discussed in this chapter and the reports about them should be seriously examined, and be available for reexamination. If fossils do not pass the test of a certain investigator or school of investigators at a particular point in time, they should not be cast into the outer darkness, so that later researchers will not even know they exist. Rather they should be placed in a category of disputed evidence. In that way, in the event of improvements in the methods of analysis or changes in theoretical constructs of human prehistory, the evidence will be available for further study. Who knows? In the future, new pieces to the puzzle of human origins may give new meaning to old pieces that previously did not quite fit.

2.4 VAL D'ARNO, ITALY (EARLY PLEISTOCENE OR LATE PLIOCENE)

Specimens incised in a manner similar to those of St. Prest were found by Desnoyers in a collection of bones gathered from the valley of the Arno River (Val d'Arno) in Italy. The grooved bones were from the same types of animals found at St. Prest—including *Elephas meridionalis* and *Rhinoceros etruscus*. They were attributed to the Late Pliocene stage called the Astian (de Mortillet 1883, p. 47). This would yield a date of 2.0–2.5 million years. Some authorities (Harland *et al.* 1982, p. 110) put the Astian in the Middle Pliocene, at 3–4 million years ago.

Modern scientists divide the fauna from the Val d'Arno into two groups—the Upper Valdarno and Lower Valdarno. The Upper Valdarno is assigned to the Late Villafranchian, which is given a quantitative date of 1.0–1.7 million years (Nilsson 1983, pp. 308–309). The Lower Valdarno is placed in the Early Villafranchian, or Late Pliocene, at around 2.0–2.5 million years ago (Nilsson 1983, pp. 308–309).

It is not clear to which group the incised bones reported by Desnoyers belong. But the fact that de Mortillet referred them to the Astian stage of the Late Pliocene seems to indicate that they might be assigned to the Lower Valdarno. On faunal grounds this would not be out of the question. We know that *Elephas meridionalis* occurs in the Lower Valdarno (Maglio 1973, p. 56). As mentioned in our discussion of St. Prest, *Rhinoceros (Dicerorhinus) etruscus* is reported in the Late Pliocene (Nilsson 1983, p. 475) in Europe, and even as far back as the Early Pliocene (Savage and Russell 1983, p. 339). De Mortillet listed *Equus arnensis* as present at Val d'Arno. *Equus* is typical of Pleistocene faunal assemblages, but examples of *Equus* are known from the Early Villafranchian (Kurtén 1968, p. 147), which is generally thought to extend into the Late Pliocene.

2.5 SAN GIOVANNI, ITALY (LATE PLIOCENE)

In addition, grooved bones also were discovered in other parts of Italy. On September 20, 1865, at the meeting of the Italian Society of Natural Sciences at Spezzia, Professor Ramorino presented bones of extinct species of red deer and rhinoceros bearing what he believed were human incisions (de Mortillet 1883, pp. 47–48). These specimens were found at San Giovanni, in the vicinity of Siena, and like the Val d'Arno bones were said to be from the Astian stage of the Pliocene period. De Mortillet (1883, p. 48), not deviating from his standard negative opinion, stated that he thought the marks were most probably made by the tools of the workers who extracted the bones.

2.6 RHINOCEROS OF BILLY, FRANCE (MIDDLE MIOCENE)

On April 13, 1868, A. Laussedat informed the French Academy of Sciences that P. Bertrand had sent him two fragments of a lower jaw of a rhinoceros. They were from a pit near Billy, France. One of the fragments had four very deep grooves on it. These grooves, situated on the lower part of the bone, were approximately parallel and inclined at a 40-degree angle to the longitudinal axis of the bone. They were 1–2 centimeters (a half inch or so) in length, and the deepest was 6 mm (a quarter inch) in depth (Laussedat 1868, p. 752). According to Laussedat, the cut marks appeared in cross section like those made by a hatchet on a piece of hard wood. And so he thought the marks had been made in the same way, that is, with a handheld stone chopping instrument, when the bone was fresh. That indicated to Laussedat (1868, p. 753) that humans had been contemporary with the fossil rhino in a geologically remote time.

Just how remote is shown by the fact that the jawbone was found in a calcareous sand stratum at a depth of 8 meters (26 feet), in between other strata of the Mayencian age of the Middle Miocene. Furthermore, the incised jawbone was from a species, *Rhinoceros pleuroceros,* judged by Laussedat to be characteristic of the Early Miocene. According to modern authorities (Savage and Russell 1983, p. 214), *Rhinoceros (Dicerorhinus) pleuroceros* occurs in the Agenian land mammal age of the Early Miocene.

At the meeting of the Academy of Sciences, Mr. Hebert asked if one could be sure of the authenticity of the incisions on the fossil. Edouard Lartet responded with a demonstration that the marks, the surfaces of which had the same appearance as the other parts of the bone, indeed dated from the time of burial (de Mortillet 1883, p. 49).

By what agency were the marks produced? De Mortillet (1883, p. 50) rejected straightaway the idea of gnawing by carnivores, because the incisions did not display the appropriate characteristics. Animal gnawing tends to be accompanied by significant destruction of the bone, whereas the rhinoceros jawbone from Billy bore only the four rather clear incisions. Were they produced by human beings? De Mortillet thought not. The imprints of a stone edge used as a saw are easily recognizable, and there were no traces of sawing on the bone. Because of their irregular edges, cutting instruments of stone generally leave small striations along the longitudinal axis of the V-shaped groove produced. But on the markings of the Billy fossil the striations were said to be transverse to this axis, i.e., running from the top of the cut, vertically down to the bottom of the groove. Furthermore, the marks on the jawbone were wider and deeper than might be expected from the action of a thin stone blade drawn across the bone.

De Mortillet thought the marks were not produced by a stone chopping instrument as proposed by Laussedat. The blow of a stone handaxe, according to de Mortillet, leaves an imprint with rounded sides. The marks on the jawbone of Billy, however, were straight-sided, and could not, in the opinion of de Mortillet, have been the result of a stone hatchet blow. Furthermore, he noted that the mark of the blow of a hatchet is distinguished by a surface clean and sharp on the side hit by the blade, and abrupt and rough on the side from which the splinter of bone separates. In the imprints on the jaw of Billy, this feature was, said de Mortillet, absent (1883, p. 50).

What then had been the cause? De Mortillet, sticking to his usual explanation, wrote in *Le Préhistorique* (1883, pp. 50–51): "They are simply geological impressions. All geologists know that there exist in many terrains, especially Miocene, rocks that have profound impressions on them. The cause is not easily recognized, but the fact that it has been observed is incontestable. There is a great similarity between the marks on some of these rocks and those on the jaw of Billy. I have collected at Tavel (Gard), and given to the museum of Saint-Germain, a quartzite rock, a very hard rock, bearing marks completely analogous to those on the specimen presented by Mr. Laussedat. On examining with care and at length this bone, one notices on one of the extremities a small impression produced by crushing. There is no removal of material, simply compression. This impression, which is of the same aspect as the other marks on the bone, is their contemporary and serves to explain them."

About marks on stones from Miocene formations, de Mortillet, as mentioned above, admitted that "the cause is not easily recognized." It is known that glaciers can groove bedrock, but this phenomenon is not applicable to grooved stones (or fossil bones) from preglacial Miocene formations. De Mortillet mentioned a grooved piece of quartzite. But quartzite is a very hard rock (7 on the Mohs scale of hardness, with talc at 1 and diamond at 10). It would thus require a harder mineral, which de Mortillet did not name, and extreme pressure, which de Mortillet did not explain, to mark quartzite with deep grooves. One must also consider the possibility that grooves in quartzite might be caused by chemical corrosion and recrystallization rather than cutting.

It is apparent that neither we nor de Mortillet know for certain what produced the grooves in the quartzite rock he found at Tavel. But it is probably not the same agency that would produce grooves on bone, a very different material, found in a freshwater deposit of calcified sand (de Mortillet 1883, p. 49). In essence, we find de Mortillet proposing that we should accept a completely unknown geological mechanism to explain the marks on the rhinoceros jaw of Billy, in preference to the known mechanism of human action. Although de Mortillet may be right, he offers insufficient evidence to justify his view.

Another factor to consider is the character and placement of the marks on the rhinoceros jaw of Billy. A highly regarded modern authority on cut bones is Lewis R. Binford, an anthropologist from the University of New Mexico at Albuquerque. In *Bones: Ancient Men and Modern Myths,* a comprehensive study of incised faunal remains, Binford pointed out that a key element in distinguishing human incisions from others is the exact placement of the marks. Extensive research has shown that in almost all cultures, ancient and modern, butchering marks tend to occur, though with some degree of variation, on specific bones and in specific locations on those bones, as dictated by the anatomy of the animal. For example, Binford (1981, p. 101) stated: "Marks on the mandible [lower jaw] tend to be slightly oblique incised marks on the inside of the mandible generally opposite the M2 tooth [second molar]. The marks are believed to originate from the underside of the mandible and to be related to the severing of the mylohyoid muscle during the removal of the tongue." The marks described by Laussedat appear to conform to this general description, but because no drawing or photo accompanied the available reports on the Billy jawbone, this remains to be more exactly confirmed.

The marks on the jawbone of Billy, which Laussedat described as a group of short parallel cuts, also appear to be consistent with the type of pattern that might be made by stone implements. According to Binford (1981, p. 105): "Most of the cut marks made on bones with metal tools are almost hairline in size. . . . the marks are generally long, resulting from cuts running across tissue for considerable distances. Cutting with stone tools requires a much less continuous action, more of a series of short parallel strokes. . . . Marks from stone tools tend to be short, occurring in groups of parallel marks, and to have a more open cross section."

It seems difficult to categorically reject human action on the rhinoceros jawbone of Billy, at least on the basis of the available published information. The action of carnivores can be safely ruled out. The geological explanation proposed by de Mortillet appears unlikely. The cut marks are on a bone that typically would be cut in butchering operations, and they appear to be in an appropriate location on the bone. In addition, the short length and parallel grouping of the marks resembles the pattern to be expected from the use of stone tools. So despite de Mortillet's objections, it does not seem impossible that a stone instrument pressed forcefully on a bone could make the kind of marks found on the Miocene rhinoceros fossil from Billy, France.

2.7 COLLINE DE SANSAN, FRANCE (MIDDLE MIOCENE)

The report of the rhinoceros jaw of Billy led to the opening, at the meeting of the French Academy of Sciences on April 20, 1868, of a sealed packet deposited at the Academy on May 16, 1864 by the researchers F. Garrigou and H. Filhol. These gentlemen wrote on that date: "We now have sufficient evidence to permit us to suppose that the contemporaneity of human beings and Miocene mammals is demonstrated" (Garrigou and Filhol 1868, p. 819). This evidence was a collection of bones, apparently intentionally broken, from Sansan (Gers), France. Especially noteworthy were broken bones of the small deer *Dicrocerus elegans*. The bone beds of Sansan were judged to be of Middle Miocene age (Mayencian). One may consider the devastating effect that the presence of human beings about 15 million years ago would have on current evolutionary doctrines.

Were the nineteenth-century scientists correct in their determination of the age of the site? Once more, the answer to this question is yes. Modern authorities (Romer 1966, p. 334) still place Sansan in the Middle Miocene, and *Dicrocerus elegans* is assigned to the Helvetian land mammal stage, which is considered Middle Miocene (Klein 1973, p. 566; Romer 1966, p. 334).

According to de Mortillet, Edouard Lartet, who also excavated fossils from Sansan and himself sent to Garrigou some of the bones on which Garrigou and Filhol founded their assertions, did not believe in human action on the bones. There were many broken bones at Sansan, and de Mortillet (1883, pp. 64–65), in his usual fashion, said that some were broken at the time of fossilization, perhaps by dessication, and others afterward by movement of the strata.

Garrigou, however, maintained his conviction that the bones of Sansan had been broken by humans, in the course of extracting marrow. He made his case in 1871 at the meeting in Bologna, Italy, of the International Congress of Prehistoric Anthropology and Archeology. Garrigou (1873) first presented to the Congress a series of recent bones with undisputed marks of butchering and breaking. For comparison, he then presented bones of the small deer (*Dicrocerus elegans*) collected from Sansan. Among them was a humerus (the long bone of the upper forelimb) with a set of breaks exactly resembling those on a cow humerus from the Neolithic age. On its inner surface, the deer bone bore a profound incision, filled up with material from the stratum in which it was found.

Garrigou also displayed a radius (one of the bones of the lower forelimb) presenting a longitudinal fracture terminating at a right angle to the end of the bone. The fracture had the same patina as the rest of the bone, indicating the break was made when the bone was fresh, and the broken part had a surface so clean and sharp that it was impossible to see it as a natural geological effect. Subterranean pressure and shifting, if it had occurred, would have almost certainly damaged the perfectly intact edges and joint surfaces of the fractured long bone. In making these observations, Garrigou showed a good grasp of taphonomic principles. He also pointed out that the longitudinal fracture on the specimen he showed was identical to those encountered on hundreds of similar bones at Sansan.

Here we may note that longitudinal fracturing is characteristic of breaking bone for the purpose of obtaining marrow. Binford (1981, p. 162) stated: "Marrow is primarily contained in the medullary cavity of the body or shaft of long bones. This shaft is shaped like a cylinder, so access to the medullary cavity and hence the marrow is facilitated by collapsing or fracturing the cylinder longitudinally. Transversal fractures in the center of long-bone shafts do not provide ready access to the marrow."

Garrigou also showed that many of the bone fragments had very fine and delicate striations such as found on broken bones of the Late Pleistocene. The marks could be indications of processing the bone for marrow breaking, as described by Binford: "The secret of controlled breakage of marrow bones is the removal of the periosteum [the sheath of connective tissue covering bone surfaces] in the area to be impacted. The Nunamiut invariably do this by scraping it back with the edge of a knife, a rough surface on a hammerstone, or almost any handy crude scraping tool. This means that longitudinal scratches and striations along the shafts of long bones are commonly produced when bones are prepared for cracking during marrow processing. Such marks are noted in Mousterian [Neanderthal] assemblages" (Binford 1981, p. 134).

Garrigou also displayed two metacarpals (foot bones), each with the smaller end removed by a direct blow. He pointed out that since flint tools had been found in the Miocene, one should not be astonished to find the effects of their usage. Food is the primary human need, so one should expect to observe signs of human attempts to secure it (Garrigou 1873, p. 137). In the next three chapters, we shall consider in detail the evidence for flint tools in the Miocene and Pliocene, but for now we should keep in mind that reports of such discoveries were very common at this time, and were accepted by many reputable scientists.

Garrigou did, however, meet with strong opposition at the Congress, from, among others, Professor Japetus Steenstrup, secretary of the Danish Royal Society of Science and director of the Museum of Zoology in Copenhagen. Steenstrup argued that a broken bone should have a percussion mark

(Garrigou 1873, p. 140). The fractured edges of a bone fragment should converge at this point, where a blow had been struck. According to Steenstrup, the bones displayed by Garrigou did not show percussion marks and converging fractures. Steenstrup therefore believed that the bones had been broken by the gnawing of carnivores.

Garrigou disagreed that fragments must show a percussion mark; its absence would not, in the case of any particular fragment, rule out direct impact as the cause of fracturing. In experiments, Garrigou had seen fresh bones broken into many flakes by a blow, and only one or two flakes would have the percussion marks. And if the instrument used happened to be sharply pointed, the bone would split immediately like a piece of wood, with no percussion imprint whatsoever (Garrigou 1873, p. 141).

The observations of both Steenstrup and Garrigou are in line with modern test data. In support of Steenstrup, we find that Binford stated (1981, p. 163): "Impact scars from hitting the bone during marrow cracking are quite distinctive. First, they are almost always at a single impact point, which results in driving off short but rapidly expanding flakes inside the bone cylinder. At the point of impact the bone may be notched, in that a crescent-shaped notch is produced in the fracture edge of the bone." But Binford's surveys showed that only about 14–17 percent of bone splinters in marrow cracking assemblages will have impact notches on them, indicating human action; this lines up with Garrigou's assertion that the vast majority of fragments will not have the impact marks. It would seem appropriate to analyze some Sansan bone splinter assemblages in terms of Binford's impact notch frequency criterion to test for human or animal action.

Garrigou also pointed out that Steenstrup's assertion that the bone breakage was caused by animal gnawing was incorrect, because the bones should then have displayed the marks of their canines and molars, and such was not the case. Animal gnawing results in extensive bone destruction, and the clean edges of the longitudinal fractures described by Garrigou contradicted that hypothesis.

Binford (1981, pp. 179–180) advised: "If one observes a pattern of bone destruction and knows that destruction is the normal consequence of animal behavior, one should view one's task as disproof of the proposition that animals were responsible for the observed patterns. . . . One might suspect that the reverse strategy might prove helpful when a pattern of bone breakage or modification by percussion is noted. Namely, knowing that breakage is a normal consequence of human behavior, one should view one's task the disproof of the proposal that man was responsible." The bones of Sansan seem to fit in the category of breakage rather than destruction.

What sort of tests might be applied to disprove human action? Binford pointed out that animals typically destroy the articulator (or joint) ends of long bones during gnawing, whereas human breakage normally does not result in articulator destruction. Binford (1981, p. 173) suggested that it should therefore be possible to examine ratios of articulator ends to shaft pieces in broken bone assemblages as a method of discriminating between animal and human action. In the case of animal action, one would expect a low ratio of articulator ends to be present. Of course, the possibility that animals might scavenge bones left by humans introduces a complicating factor.

So in the case of the broken bones of Sansan we once more encounter evidence for a human presence in very ancient times. This evidence certainly cannot be ruled out in the absence of further study. Garrigou's methodology and analysis appear to be quite rigorous, relying on sound taphonomic principles, extensive comparison with bones indisputably broken by human action, and evidence gathered from direct experiments in bone breakage patterns. We can only wonder why this report has remained buried. Whatever the reason, it would appear that the present data collection upon which ideas about human origins are based may be quite incomplete.

2.8 PIKERMI, GREECE (LATE MIOCENE)

At a place called Pikermi, near the plain of Marathon in Greece, there is a fossil-rich stratum of Late Miocene (Tortonian) age, explored and described by the prominent French scientist Albert Gaudry. During the meeting in 1872 at Brussels of the International Congress of Prehistoric Anthropology and Archeology, Baron von Dücker reported that broken bones from Pikermi proved the existence of humans in the Miocene (von Dücker 1873, pp. 104–107). Modern authorities still place the Pikermi site in the Late Miocene (Nilsson 1983, p. 476; Jacobshagen 1986, pp. 213, 221).

Von Dücker first examined numerous bones from the Pikermi site in the Museum of Athens. He found 34 jaw parts of *Hipparion* (an extinct three-toed horse) and antelope as well as 19 fragments of tibia and 22 other fragments of bones from large mammals such as rhinoceros. All showed traces of methodical fracturing for the purpose of extracting marrow. According to von Dücker (1873, p. 104), they all bore "more or less distinct traces of blows from hard objects." He also noted many hundreds of bone flakes broken in the same manner. It would thus appear that these fractured bones would satisfy the requirements of nineteenth-century authorities such as Steenstrup as well as modern authorities such as Binford with regard to impact notches as a sign of intentional breakage.

In addition, von Dücker observed many dozens of crania of *Hipparion* and antelope showing methodical removal of the upper jaw in order to extract the brain. The edges of the fractures were very sharp, which may generally be taken as a sign of human breakage, rather than breakage by gnawing carnivores or geological pressures. One might question whether the bones in the museum collection actually belonged to the Miocene stratum of Pikermi, but many of them had a matrix of red clay clearly confirming the layer from which they were recovered. The museum personnel said, however, that no stone tools or traces of fire had been found with the bones.

Von Dücker then journeyed to the Pikermi site itself to continue his investigation. During the course of his first excavation, he found dozens of bone fragments of *Hipparion* and antelope and reported that about one quarter of them bore signs of intentional breakage. In this regard, one may keep in mind Binford's finding that in assemblages of bones broken in the course of human marrow extraction about 14–17 percent have signs of impact notches. "I also found," stated von Dücker (1873, p. 105), "among the bones a stone of a size that could readily be held in the hand. It is pointed on one side and is perfectly adapted to making the kinds of marks observed on the bones."

Von Dücker's second excavation was made in the presence of one of the founders of the International Congress of Prehistoric Anthropology and Archeology, Professor G. Capellini of Bologna, Italy. Capellini, who believed that broken bones were by themselves insufficient to demonstrate the presence of human beings at a site, did not attach as much significance to the Pikermi finds as did von Dücker. Nevertheless, he thought the bones had been fractured before the time of deposit.

Capellini reported that he had visited the museum and found the majority of bones were not broken by humans, as believed by von Dücker. Capellini pointed out that in fact there were many bones and skulls on display that remained whole and in good condition. Von Dücker replied that the fact that some bones were not broken did not change the fact that others were broken, and these in a way that suggested intentional work. He noted that Gaudry had naturally selected the best bones for his museum displays (von Dücker 1873, p. 106). Von Dücker stated that Capellini's very brief examination could hardly compare with his own lengthy and careful study, lasting for a period of several months, both in the museum and at the site.

De Mortillet stated that von Dücker's report was submitted to Gaudry, who found no evidence of human work. De Mortillet also examined the bones, and agreed with Gaudry and Capellini that the breakage was "accidental." It is, however, interesting to note that von Dücker, after communicating his observations to Gaudry, received the following statement from Gaudry: "I find every now and then breaks in bones that resemble those made by the hand of man. But it is difficult for me to admit this" (von Dücker 1873, p. 107). In Gaudry's remark surfaces one of the central questions confronting us in our examination of the treatment of paleoanthropological evidence. The evidence appears in general to be quite ambiguous. So on what basis can one draw conclusions? Gaudry hinted that his preconceptions were in subtle conflict with his perceptions. Humans in the Miocene? It was too difficult for him to admit. Preconception triumphed, however quietly, over perception.

In the final analysis, what are we to make of the fractured bones of Pikermi? Any clear answer to that question shall have to wait until such time as the final analysis is made. And it remains doubtful whether any totally "final" analysis ever can be made. Ambiguity is inherent in the enterprise. Surely, we cannot yet conclude, on the basis of the available reports, that humans were not responsible for the breakage observed on *Hipparion* bones from the Miocene formations at Pikermi, Greece.

Another thing to keep in mind is that some modern researchers believe that in general evidence for human breaking of bone has been neglected or gone unrecognized. Robert J. Blumenschine and Marie M. Selvaggio, anthropologists at Rutgers University, conducted experiments in which they used pieces of sandstone to break African mammal (gazelle, impalla, wildebeest) longbones in order to extract marrow. According to *Science News* of July 2, 1988: "The resulting pits and grooves or 'percussion marks' on the bones, usually found near the notches created by the impact of stone, look much like carnivore tooth marks at first glance, the researchers report in the June 24 *Nature*." But the scanning electron microscope revealed "patches of distinctive parallel lines" different from those made by hyaena teeth. Blumenschine and Selvaggio maintained, stated *Science News,* that "researchers probably have underestimated or over-looked the breaking of bones by early humans to obtain marrow."

2.9 PIERCED SHARK TEETH FROM THE RED CRAG, ENGLAND (LATE PLIOCENE)

At a meeting of the Royal Anthropological Institute of Great Britain and Ireland, held on April 8, 1872, Edward Charlesworth, a Fellow of the Geological Society, showed many specimens of shark (*Carcharodon*) teeth, each with a hole bored through the center, as is done by South Seas islanders for the purpose of making weapons and necklaces. The teeth were recovered from the Red Crag formation, indicating an age of approximately 2.0–2.5 million years (Nilsson 1983, p. 106).

The record of the meeting, published in the journal of the Anthropological Institute, informs us: "Mr. Charlesworth pointed out the conditions under which boring molluscs, as *Pholas* and *Saxicava,* perforate the texture of stones or other solid substances, and glanced at the perforating action of burrowing sponges (*Cliona*) and destructive annelides (*Teredo*). Reasons were given at length why these could not have produced such perforations as those now exhibited. The most searching and cautious examination was also bestowed to demonstrate that the perforating body, whatever it was, was coeval with the crag period; *i.e.,* that specimens existed in which the true crag matrix filled up the hole from end to end, thus showing that it had been immersed in the crag sea after the period of its perforation" (Charlesworth 1873, p. 91).

Charlesworth (1873, pp. 91–92) did not personally suggest human agency, but did show a letter from Professor Owen, who had carefully examined the specimens and stated: "the ascription of the perforations to human mechanical agency seemed the most probable explanation of the facts."

During the ensuing discussion, Mr. Whitaker suggested tooth decay as the cause, noting one specimen with holes in various stages, from slight indentation to perforation (Charlesworth 1873, p. 92). Then Dr. Spencer Cobbold, an expert on parasites, suggested parasites as the agent of perforation but admitted, according to the summary report: "it might be said with truth, perhaps, that no entozoon [internal animal parasite] had hitherto been known to take up its abode in the bones or teeth of fishes" (Charlesworth 1873, p. 92).

At that point Dr. Collyer gave his opinion in favor of human action. The record of the meeting summarized his remarks as follows: "He had carefully examined by aid of a powerful magnifying glass the perforated shark's teeth. . . . The perforations, to his mind, were the work of man. His reasons were—First, the bevelled conditions of the edges of the perforations. Secondly, the irregularity of the borings. Thirdly, the central position of the holes in the teeth. Fourthly, the choice of the thin portions of the tooth where it would be most easily perforated. Fifthly, the marks of artificial means employed in making the borings. Sixthly, they are at the very place in the tooth that would be chosen in making an instrument of defence or offence, or for ornament in the form of a necklace. Seventhly, the fact that rude races—as the Sandwich Islanders or New Zealanders—have from time immemorial used sharks' teeth and bored them identically with those exhibited. His reasons for supposing the perforations not to have been produced by molluscs, or boring-worms, or any parasitic animal, were—First, those creatures invariably had a purpose in making a hole for lodgement; it was therefore evident they would not choose the thin portion of the tooth, which would be totally unadapted for the object sought. Secondly, there was not a case on record of any parasite or mollusc or worm boring a fish's tooth. Thirdly, those animals had no idea that the exact centre of the tooth would be preferable to the lateral portion. Fourthly, had the holes been the result of animal borings, they would have presented a uniform appearance. As to the tooth being perforated by decay, that seemed to him the most extraordinary proposition. The appearance of a decayed tooth had no analogy whatever to the borings presented. Moreover, sharks were not subject to decayed teeth" (Charlesworth 1873, p. 93).

Mr. T. McKenny Hughes then argued against human boring, pointing out that in some cases the holes on the front and back sides of the tooth are not perfectly lined up with each other. It is not, however, obvious how this would preclude human action. Just to consider one possibility, one could easily imagine a worker partially boring the tooth on one side, turning it over, and completing the perforation by boring in from a slightly different angle starting on the other side.

Hughes then offered another curious objection. He observed that the same types of perforation are found on fossils not only in the Crag, a formation on the Plio-Pleistocene boundary, but also on shells in other deposits more ancient, such as the green sandstone strata of Secondary age. He asserted that it was clearly impossible for humans to have existed at this remote time; therefore the perforations in fossils in the green sandstone were clearly natural in origin. And, by analogy, so were the perforations in the shark teeth from the Red Crag. Here is yet another very typical example of preconceptions determining what kind of evidence for human antiquity can be accepted. Another possible way to look at the perforated shells found in the older green sandstone strata is that they also could be the result of the action of human beings. As previously mentioned, the most recent Secondary period is the Cretaceous, which ended about 65 million years ago.

In any case, Hughes suggested that the perforations in the Red Crag shark teeth were caused by a combination of wear, decay, and parasites (Charlesworth 1873, p. 93). Mr. G. Busk presented the same conclusion at the 1872 meeting of the International Congress of Prehistoric Anthropology and Archeology in Brussels. In *Le Préhistorique*, de Mortillet (1883, p. 68) sarcastically remarked that it was really curious how some people searched so obstinately for proof of the existence of Tertiary humans in marine deposits.

But in looking at the arguments presented in this case, both those in favor of human work and those opposed, it would seem that obstinacy is more clearly evident in those who refused to accept the possibility of human action. What are the alternatives that were presented? Some suggested tooth decay, although sharks are not known to have cavities; others suggested parasites, although one of Britain's leading experts admitted there was no known instance of a parasite inhabiting the teeth of fish or sharks. Others suggested wear had a role to play, though one would be hard pressed to find examples in nature of wear causing clean round holes through the centers of teeth.

2.10 CARVED BONE FROM THE DARDANELLES, TURKEY (MIOCENE)

In the *Journal of the Royal Anthropological Institute of Great Britain and Ireland*, Frank Calvert (1874, p. 127) reported: "I have had the good fortune to discover, in the vicinity of the Dardanelles, conclusive proofs of the existence of man during the Miocene period of the tertiary age. From the face of a cliff composed of strata of that period, at a geological depth of eight hundred feet, I have extracted a fragment of the joint of a bone of either a dinotherium [*Deinotherium*] or a mastodon, on the convex side of which is deeply incised the unmistakable figure of a horned quadruped, with arched neck, lozenge-shaped chest, long body, straight fore-legs, and broad feet. There are also traces of seven or eight other figures which, together with the hind quarters of the first, are nearly obliterated. The whole design encircles the exterior portion of the fragment, which measures nine inches in diameter and five in thickness. I have found in different parts of the same cliff, not far from the site of the engraved bone, a flint flake and some bones of animals, fractured longitudinally, obviously by the hand of man for the purpose of extracting the marrow, according to the practice of all primitive races."

Calvert (1874, p. 127) added: "There can be no doubt as to the geological character of the formation from which I disinterred these interesting relics. The well known writer on the geology of Asia Minor, M. de Tchihatcheff, who visited this region, determined it to be of the miocene period; and the fact is further confirmed by the fossil bones, teeth, and shells of the epoch found there. I sent drawings of some of these fossils to Sir John Lubbock, who obligingly informs me that having submitted them to Messrs. G. Busk and Jeffreys, those eminent authorities have identified amongst them the remains of dinotherium, and the shell of a species of melania, both of which strictly appertain to the miocene epoch."

The *Deinotherium* is said by modern authorities to have existed from the Late Pliocene to the Early Miocene in Europe (Romer 1966, p. 386). It is thus quite possible that Calvert's dating of the Dardanelles site as Miocene was correct. The Miocene is now said to extend from 5 to 25 million years before the present. According to the current dominant view, only exceedingly apelike hominids are supposed to have existed during that period. Even a Late Pliocene date of 2.5–3.0 million years for the Dardanelles site would predate the first toolmaking hominid (*Homo habilis*).

Calvert appears to have been sufficiently qualified to estimate the date of the Dardanelles site. David A. Traill (1986a, pp. 53–54), a professor of classics at the University of California at Davis, gives this information about him: "Calvert was the most distinguished of a family of British expatriates that was prominent in the Dardanelles . . . . he had a good knowledge of geology and paleontology." Calvert conducted several important excavations in the Dardanelles region.

Calvert also played a very important role in finding the site of the famous city of Troy. Scholars usually give the credit for this to Heinrich Schliemann. But Traill (1986a, pp. 52–53) said of Calvert: "After excavating the 'Tumulus of Priam' on the Balli Dağ (1863) and reading Charles Maclaren's *A Dissertation on the Topography of the Plain of Troy* (Edinburgh 1822), he decided that Hissarlick must be the site of Troy. He purchased part of the mound and started to excavate in 1865, but lack of funds and the pressure of other commitments caused him to abandon the task. . . . After Schliemann's unsuccessful diggings at Bunarbashi in 1868, Calvert persuaded him . . . that Hissarlick, not Bunarbashi, was the true site of Troy. Schliemann later downplayed both the significance of Calvert's excavations and his role in awakening his interest in Hissarlick and successfully appropriated all the glory for himself. Calvert, however, was much the better scholar."

During his excavations, Schliemann came upon a group of weapons, utensils, and ornaments that he called "Priam's Treasure." Calvert reviewed this find and Schliemann's excavations in general. Traill (1986b, p. 120) stated: "He pointed out, with remarkable acuity, that the excavated material should be dated before 1800 B.C. and after 700 B.C. but that nothing was attributable to the period between these dates. Since the missing period included the time of the Trojan War, these findings enraged Schliemann. His response was to ridicule Calvert's views and misrepresent his role in the excavation of Hissarlick. . . . Calvert was, as far as I have been able to determine from extensive reading of his correspondence, scrupulously truthful." The so-called treasure of Priam, thought Calvert, was genuine, but not of the classical Trojan era, and this view conforms with the opinion of modern scholars.

Altogether, Calvert seems to have been a quite competent field investigator, with a reputation for truthful and careful reporting. It thus seems that in the case of his Miocene discoveries, he would not have missed any obvious sign that the carved bone, broken bones, and stone implements he discovered had been recently cemented into the deposits. It should be noted that the carved bone from the Dardanelles was no less securely positioned stratigraphically than a great many thoroughly accepted discoveries. Most of the Java *Homo erectus* finds and most of the East African *Australopithecus*, *Homo habilis,* and *Homo erectus* finds occurred on the surface and are presumed to have washed out from underlying formations varying from Middle Pleistocene to Late Pliocene in age.

In *Le Préhistorique,* de Mortillet did not dispute the age of the Dardanelles formation. Instead he commented that the simultaneous presence of a carved bone, intentionally broken bones, and a flint flake tool was almost too perfect, so perfect as to raise doubts about the finds (de Mortillet 1883, p. 69). This is quite remarkable. In the case of the incised bones of St. Prest, de Mortillet complained that no stone tools or other signs of a human presence were to be found at the site. But here, with the requisite items discovered along with the carved bone, de Mortillet said the ensemble was "too perfect," hinting at cheating.

De Mortillet then alluded to the well-publicized disputes between Calvert and Schliemann, which he claimed had discredited both men. In addition to Calvert's disagreements with Schliemann about the dates of his archeological discoveries at Hissarlick and their relation to the classical Troy of Homer, there were also some financial bickerings. Calvert and Schliemann had an agreement that they would share the proceeds from the sale of any discoveries at Hissarlick. A particularly fine statue was the source of some controversy, with Calvert charging that Schliemann paid him far less than it was actually worth (Traill 1986a, pp. 53–54). But it seems that Calvert emerges from all this as an honorable and truthful person, who had a better grasp of the archeology of the Hissarlick site than Schliemann. This tends to increase, rather than decrease, the credibility of Calvert's reporting about his Miocene discoveries.

Finally, de Mortillet (1883, p. 69) stated that because no further reports of a serious nature or new discoveries of human artifacts had emerged from the Dardanelles site, the original Miocene finds reported by Calvert should be considered unconfirmed. But perhaps if new finds had been made, de Mortillet would have reacted as he had to the first ones—by calling them "too perfect," questioning the character of the discoverer, and demanding more discoveries.

2.11 BALAENOTUS OF MONTE APERTO, ITALY (PLIOCENE)

During the latter part of the nineteenth century, fossil whale bones bearing curious marks turned up in Italy. On November 25, 1875, G. Capellini, professor of geology at the University of Bologna, reported to the Institute of Bologna: "Recently as I was cleaning a bone that I myself extracted from the blue Pliocene clay, synchronous with that of the Grey Crag of Anvers, of Astian age, I saw to my great surprise on the dorsal surface a notch and an incision. The former, especially, was so clean cut and deep as to indicate it was made by a very sharp instrument. I am able to say that the bone found is so completely petrified as to preserve all the most delicate details of its microscopic structure; furthermore, it has acquired such hardness that it is not possible to scratch it with a steel point. This circumstance enables us to completely reject suggestions that tend to attribute the marks to modern action" (de Mortillet 1883, p. 56). During further cleaning Capellini discovered three other lighter marks on the bone. He announced this discovery and others that followed at the Academy of Lynxes at Rome and the International Congress of Prehistoric Anthropology and Archeology meetings at Budapest in 1876 and Paris in 1878. Capellini, a founding member of the Congress, was a prominent member of the European scientific community.

The whale bones studied by Capellini were from the extinct small baleen whale *Balaenotus,* which is characteristic of the Late Pliocene of Europe (Romer 1966, p. 393). This confirms Capellini's assignment of his discoveries to the Pliocene.

In 1876, Capellini showed his principal specimens at the Congress at Budapest, where he told the members (1877, pp. 46–47): "For fifteen years I have been researching and studying cetacean fossils. After my work on the *Balaenopteridae* in the province of Bologna, I decided to undertake researches into the baleen whales of Tuscany. . . . By the kindness of Professor D'Ancona, I was able to examine at my leisure the remains of fossil baleen whales at the Museum of Natural History of Florence. I then became convinced of the great importance of extending my researches beyond the specimens in the glass cases and dusty vaults of the museums. I was certain that direct investigations in the strata that had already yielded much precious material would be extremely fruitful for further progress in the study of fossil whales."

We shall now consider Capellini's extensive report in detail, making liberal use of direct quotations, translated from the original French. This procedure is being followed for the two reasons previously mentioned: (1) a report, in this case a very important one, is itself, for all practical purposes, the evidence; and (2) readers could not otherwise obtain the original report except by referring to a rare nineteenth-century volume of conference proceedings.

"In October of 1875," continued Capellini (1877, p. 47), "I journeyed to Siena to continue my stratigraphic studies of that region's Tertiary terrains and at the same time examined the remains of fossil cetaceans in the museum of the Académie des Fisiocritici. On the advice of Dr. Brandini, I also began excavations at Poggiarone, in the neighborhood of Monte Aperto. I was greatly fortunate to make a double discovery: first, I recovered numerous remains of a skeleton of *Balaenotus*, a fossil cetacean first recognized by van Beneden, and heretofore found only in the Grey Crag of Anvers; and secondly on these very same bones I noticed the first traces of the hand of man, demonstrating the coexistence of human beings with the Pliocene whales of Tuscany."

Capellini went on to display some samples of his discoveries. "I have the honor," he said, "of presenting remarkable specimens that bear marks which, by their form and placement on the fossil bones, demonstrate in an irrefutable manner the action of a being manipulating an instrument. This is the opinion of all the most experienced naturalists and anatomists, not only in Italy, but from all over Europe, who have examined these specimens, judging them without preconceived ideas" (Capellini 1877, p. 47). It may be noted that by considering the "form and placement" of the cuts, Capellini was adhering to modern criteria for distinguishing human workmanship from animal gnawing on bone. His reference to scientists tending to have "preconceived ideas" is particularly relevant to our discussion.

Regarding the geological age of the strata in which the *Balaenotus* fossils had been discovered, Capellini observed in his report: "The geological position of the strata in which the *Balaenotus* was found in the neighborhood of Monte Aperto and the shells that were found in the same bed do not permit us to doubt their Pliocene age and their resemblance to the Grey Crag of Anvers. The alternation of beds entirely of sand with others of clay and sand, give evidence that the animal was beached in the shallows along the shore of an island of the Pliocene archipelago that occupied what is now central Italy during the last part of the Tertiary epoch."

Capellini (1877, p. 48) then described the placement of the cut marks on the fossil bones: "The marks on the skeleton of the *Balaenotus* are found on the lower extremities, the exteriors of the ribs, and on the apophyses [*spines*] of the vertebrae."

The presence of cuts on the vertebral spines, or apophyses, conforms with the observations of Binford (1981, p. 111), who stated that in flesh removal, cuts are made to free flesh from the dorsal spines of the thoracic and lumbar vertebrae, producing "cut marks . . . commonly oriented transversely or slightly obliquely to the dorsal spines of the thoracic vertebrae." As far as the ribs are concerned, Binford (1981, p. 113) stated that in the most common butchering operation "transverse marks, derived from the removal of the tenderloin, occur along the dorsal surface of the rib just to the side of the proximal end of the rib." The marks observed by Capellini, all on the dorsal (exterior) surface of the rib, correspond to this description.

Applying principles of taphonomic analysis, Capellini (1877, p. 49) then stated: "On the dorsal apophysis of an almost complete lumbar vertebra, I have moreover marked the presence of intersecting cutmarks and next to them one sees tiny oysters, evidence that indicates the deposition took place in very shallow water not far from the shore. One should not forget that the entire region formerly occupied by the sea in the environs of Siena has been raised and lowered many times, which accounts for the alternation of marine, brackish, and freshwater deposits one is able to observe and study at Siena." These alternations are indications of a littoral, or shoreline, area, which is important. Some critics believed the marks had been made by the teeth of sharks, and according to their analysis this would necessitate deep water.

For example, in his book *Le Préhistorique,* de Mortillet (1883, p. 59) stated that some Italian naturalists (Strobel and de Stefani) were of the opinion that the beds yielding bones of *Balaenotus* were not littoral but deep ocean. This seems to be at variance with the firsthand observations of Capellini, who was himself an experienced geologist. In his review, de Mortillet does not mention the evidence that Capellini cited in support of his conclusion that the location where the *Balaenotus* bones were found represented the shallows along the beach of the Pliocene sea.

"Having surveyed the excavations of the remains of the skeletons of *Balaenotus* in the environs of Siena," Capellini (1877, pp. 49–50) went on to say, "I was able to easily account for the existence of the marks on only one side, and always the same side. In effect, it is evident that for the specimen in question the marks were made by a human being that came upon the animal beached in shallow waters, and by means of a flint knife or with the aid of other instruments attempted to detach pieces of flesh." Capellini (1877, p. 50) added: "From the position of the remains of the *Balaenotus* of Poggiarone, I am convinced that the animal ran aground in the sand and rested on its left side and that the right side was thus exposed to the direct attack of humans, as is demonstrated by the places in which marks are found on the bones." The fact that only the bones on one side of the whale were marked would tend to rule out any purely geological explanation as well as the action of sharks in deep water.

Capellini (1877, p. 50) noted: "That which happens at present to the *Balaenopteridae* and cachalots [sperm whales] that from time to time become beached on our shores also happened to the *Balaenotus* of Poggiarone and to other small whales on the shores of the islands of the Pliocene sea." Capellini (1877, p. 50) then made an important observation: "After an attentive examination of skeletons found in the majority of Europe's museums of natural history, it is very easy to convince oneself that all of these, which were prepared by humans, present the same kinds of markings as those on the bones you have seen and others which I will show you." Comparison with examples of undoubted human work is still one of the main methods scientists use in determining whether incisions on bones are of human origin.

Capellini (1877, p. 51) then reported that he had found examples of the kind of tool that might have made the cuts on the bones: "In the vicinity of the remains of the *Balaenotus* of Poggiarone, I collected some flint blades, lost in the actual beach deposits." He added: "with those same flint implements I was able to reproduce on fresh cetacean bones the exact same marks found on the fossil whale bones" (Capellini 1877, p. 51).

"Before leaving the environs of Siena," Capellini (1877, p. 51) went on to explain, "I should point out that the remains of a human being found in 1856 by the Abbé Deo Gratias in the marine Pliocene clays of Savona in Liguria can be referred to approximately the same geological horizon as Poggiarone and other locales in Tuscany where I have found numerous cetacean remains." The details of the discovery of human skeletal remains in the Pliocene at Savona will be discussed at length in Chapter 6, which also contains many other such reports. For now, it will be sufficient to note that the discoveries of incised bones in the middle and late nineteenth century were accompanied by a great many simultaneous discoveries of flint implements and actual human skeletal remains in Pliocene and Miocene strata. These discoveries are practically never mentioned in modern textbooks. It bears repeating that the existence of human beings of the modern type in the Pliocene period would completely demolish the presently accepted evolutionary picture of human origins.

Capellini then discussed another find of human skeletal remains that he believed to be contemporary with the incised whale bones he had discovered in Pliocene strata. "In my first notice on Pliocene man in Tuscany (Nov. 1875) I mentioned the human cranium discovered by Professor Cocchi in the upper valley of the Arno, in Tuscany, and for the moment I accepted the conclusions given by my associate concerning the age of the strata in which the cranium was found." Cocchi had given them a Pleistocene date.

"Dr. F. Major, however," said Capellini (1877, pp. 51–52) to his colleagues at the Congress of Budapest, "has for many years been particularly interested in studying the fossil vertebrates of the upper valley of the Arno, and after new researches into the geological position of the human skull found at Olmo has reached an opinion contrary to that of Professor Cocchi. According to Dr. Major, the fossils of the strata in which the cranium of Olmo was found and those collected with the cranium itself by Professor Cocchi prove the Pliocene age of the stratum and that it is contemporary with the marine deposits containing incised bones of small whales." Modern authorities, however, assign a Pleistocene date to the Olmo skull (Appendix 1.2.1).

"Some months after the discovery of the *Balaenotus* of Poggiarone," continued Capellini (1877, p. 52), "I was, by means of similar discoveries, able to conclude that Pliocene man was present on other islands in the Tuscan archipelago. In examining the numerous remains of fossil cetaceans which Sir R. Lawley recently contributed to the museum of Florence, I discovered a fragment of a humerus and three fragments of cubitus with marks just as well-defined and instructive as those in question. Among the remains of *Balaenotus* from La Collinella, near Castelnuovo della Misericordia in the valley of the Fine, there have been recovered a good number with incrustations of gypsum. It was in the course of removing these incrustations, aided by the preparator E. Bercigli, that I noticed the markings. Shortly thereafter, the specimens were examined and the marks confirmed by M. d'Ancona, professor of paleontology, M. Giglioli, professor of zoology and comparative anatomy, Dr. Cavanna, Dr. Ch. Major, and others."

Many of Italy's leading scientists concurred with Capellini's judgement that the markings were caused by sharp instruments manipulated by human beings. Capellini (1877, p. 53) said: "The unanimous opinion of the naturalists of Florence, confirmed by that of the anatomists and naturalists of Bologna, all of whom examined the specimens with great care, was also supported by the academicians of the Rome Society of Lynxes, the names of whom may be found at the end of my published memoir."

Returning to consideration of the actual specimens, Capellini (1877, p. 53) said: "The Museum of Florence has allowed me to present these precious specimens for the inspection of the members of the Congress. I am very pleased to present them to the assembly because all of you interested in this question can verify that drawings alone do not allow one to appreciate all the fine details that permit us to exclude explanations other than that of a human being or other animal, who operated with the aid of instruments, and who by means of cuts in several directions, mostly deep and confined to a very limited area, was often able to facilitate the breaking of the bone."

"On one of the fragments of cubitus," said Capellini (1877, p. 53), "I left intact a portion of the gypsum incrustation that covered a deep incision, a section of which is visible. If one removed the gypsum one would see that the entire mark had been made on the bone while it was fresh, and then conserved by fossilization and incrustation." This was good proof that the cut marks were not made in recent times.

Capellini (1877, pp. 53–54) also found similar cut marks on the apophyses of vertebrae he saw in the whale bone collection of Lawley. "The fragment of the dorsal apophysis of a lumbar vertebra, in the space of a few centimeters," stated Capellini (1877, p. 54), "presents on the right side nine different incisions oriented in different directions. In examining the original with the aid of a lens, one can assure oneself that these marks, and the other marks that you will see, were made when the bone was fresh. One may also note that one side of the cut is smooth while the other is rippled, as occurs when one, with a knife or other instrument, marks a bone, either by a direct blow or by manipulation of the instrument in the manner of ordinary cutting [Figure 2.1]. It is to be remarked that the side of the bone opposite that bearing the marks is intact, and whatever incisions have been inflicted on the bone are so profound as to have been able to break it off. Two fragments of the apophyses of vertebrae broken at the place where they were cut or grooved are represented . . . in my memoir." The marks on the spine of the lumbar vertebra are in a location that according to Binford typically displays cut marks from butchering operations.

Capellini then returned to geological considerations, describing the location at which several of his specimens were found. "The pieces . . . come from San Murino,

Figure 2.1. Magnified cross section of a cut on a fossil whale bone from a Pliocene formation at Monte Aperto, Italy (de Quatrefages 1887, p. 97).

near Pieve Santa Luce on the coast of the ancient Pliocene island of Monte Vaso, on La Collinella, in the valley of the Fine. Some meters from where M. Paco, a fossil hunter, found bone fragments of small whales, the ancient limestone rocks, which formed the shore of the Pliocene sea, are regularly pierced by lithophages. Because the depth at which these creatures establish their residences and leave their traces is well known, it is, in the valley of the Fine near Santa Luce, quite easy to establish the ancient level of the sea frequented by the small whales that human beings came upon in the Pliocene period, just as in our own day we come upon small whales beached on the shores of the Mediterranean." Here is more evidence that the whale bones were most probably deposited in shallows by the shore. It is surprising that de Mortillet neglected to mention this in his review, where he gave the impression that scientific opinion is decidely in favor of a deep water interpretation.

Returning to the question of the age of the strata in which the fossil whale bones were found, Capellini, himself a professor of geology, then stated (1877, pp. 55–56): "Among those who recognize without difficulty the work of humans in the markings on the whale bones, are some who are not persuaded that they are ancient, and who have demanded to know if there is perhaps not some doubt about the judgement that the beds bearing the bones of *Balaenotus* are really Pliocene in age. This question has been discussed by me in my memoir presented to the Rome Society of Lynxes in the presence of eminent geologists and paleontogists from Central Italy, such as Messrs. Sella, Meneghini, Ponzi, and others, who confirmed all that I had said. Their exact knowledge of the locality sufficed to allow them to appreciate the geological drawings by which I sought to decipher and record the stratigraphic series of the ancient fjord (presently the valley of the Fine) where the cetaceans perished in the Pliocene. After the publication of my memoir, complete with geological notices, I believe it useless to here repeat all the facts about the age of the strata of the small whales and the circumstances favoring the opinion that the whales were captured by human beings."

After Capellini's presentation, the members of the Congress engaged in discussion. Sir John Evans accepted the geological age of the fossils, but said he thought the bones had some marks that appeared to have been made by the teeth of fish. This suggested to him that the bones had lain on the bottom of the sea, where the other more prominent marks were perhaps made by the teeth of sharks. He believed that proof for the strata being on the shoreline was lacking. Thus, questioned Evans, if humans did exist in the Pliocene, how could it be that they were getting food from the deep sea? Furthermore the marks were so sharp that if it were an instrument that made them, it would seem to have been one of metal rather than stone. He also maintained that marks made accidentally by humans in detaching flesh would be of a different nature (Capellini 1877, pp. 56–57).

These appear to be fairly weak objections. Capellini gave adequate geological reasons to suggest at least the strong probability that the strata in which the fossils were found were littoral. Capellini had also examined, in museums, many skeletons of whales from which the flesh had been detached by humans, and had found the markings practically identical to those on the fossil bones of the Tuscan *Balaenotus*. Capellini (1877, p. 51) had in at least one case found flint implements near fossil whale bones and demonstrated that the flint blades could make marks identical to those found on the bones. Evans simply seems to have had some strong bias against the presence of humans in the Pliocene.

Next to speak was Paul Broca, a surgeon and secretary general of the Anthropological Society, headquartered in Paris. Broca was famous as an expert on the physiology of bones, particularly the skull. He lined up on the side of Capellini. Interestingly enough, Broca was a Darwinist, but the evidence he supported at the Congress of Budapest in 1876 would, if accepted now, completely destroy the modern Darwinian picture of human evolution.

"The discovery of Quaternary [Pleistocene] man was the greatest event in modern anthropology," said Broca. "It opened a great field of investigation, and none here can fail to recognize its importance, because, it was this event, one could say, that was most responsible for the grand movement of ideas that resulted in the founding of our Congress. The discovery of Tertiary man could be an even greater event, because the period it could add to the life of humanity is incomparably greater than that we know at present" (Capellini 1877, p. 57). The Tertiary includes the Pliocene, Miocene, Oligocene, Eocene, and Paleocene periods.

"This is not the first time this question has arisen in our discussions," continued Broca. "Already in 1874, at the Congress at Brussels, Abbé Bourgeois showed a series of flints from Tertiary strata and in which he believed he could see proof of human work, but few shared his opinion. For my part, I examined many times the flints of Abbé Bourgeois, and remained among those not accepting his demonstrations. The other facts relative to Tertiary man that have been put forward, from Europe and America, have not been conclusive enough for me. To this day I remain doubtful about the stratigraphic location and about the work attributed to human hands" (Capellini 1877, p. 57). In the next few chapters of this book, one will have the chance to draw one's own conclusions about the many discoveries of flint implements and human skeletal remains referred to here by Broca.

"But today," confessed Broca, "for the first time, I sense my doubts disappearing. I would declare myself entirely convinced, if I were relying totally on my own judgement. But I should also take into account the judgement of my colleagues. I should fear that I might be mistaken when I find myself opposed by such competent men as Franks and Evans. With these reservations, I shall explain the evidence that leads me to admit the interpretation of Capellini" (Capellini 1877, pp. 57–58).

Broca then proceeded to present arguments against the hypothesis that the marks on the fossil bones of *Balaenotus* had been produced by the teeth of sharks. "In the first place," he said, "it is evident that the marks shown to us have been produced by cutting. All the world agrees on this point. We are only discussing the question of whether these cuts were made by the sharp pointed teeth of sharks or by the human hand armed with sharp flint. There is another point which seems to me incontestable. That is that all the incisions, in their diverse forms, those perpendicular as well as oblique, can be easily reproduced, with all their characteristics, with a flint implement on fresh whale bones. The hypothesis of Capellini explains very well the observed facts, while the other hypothesis encounters very strong objections. Capellini has remarked with reason that every bite should produce two imprints corresponding to the two jaws that seize the bone at two opposite points. But without exception all the incisions are on the convex surface of the ribs, with the concave surface totally exempt from all markings. I do not believe that one can respond to this argument" (Capellini 1877, p. 58).

Here Broca seems to be thinking that the shark would completely devour the whale carcass, thus breaking apart the rib cage. Given the feeding frenzies of sharks, especially the great white shark, present in the Pliocene as *Cacharodon megalodon,* one might expect this to happen. Otherwise, it is difficult to see how the shark could place bite marks on both sides of the rib.

Some years later, de Mortillet (1883, p. 62) suggested, in *Le Préhistorique,* that the particular nature of a shark's jaw and method of biting would result in tooth marks being placed on only one side of a bone subjected to its attack. As usual, however, de Mortillet only painted speculative scenarios and did not present any hard experimental evidence.

Broca continued: "Among the incisions, the majority penetrate obliquely into the bone. One of the sides of the V-shaped incision slices into the bone at a small angle, departing only slightly from the horizontal plane of the surface of the bone; while the opposite side, shorter than the first, is abrupt, almost vertical. The incision shows breakage. That is to say, the cutting action results in the separation of a small shaving of bone, broken at its base [Figure 2.1]. The cutting action of a sharp edge produces marks of this type. I don't believe that the teeth of any animal could produce the same effect" (Capellini 1877, p. 58). The same thing was admitted by de Mortillet himself, who raised the point in his discussion of the bones of St. Prest (Section 2.1).

"Finally," said Broca, "—and I insist on this point, which Capellini touched upon only lightly—the direction of certain of the marks is incompatible with the idea of a bite. The jaws do not execute such a movement. They open and they close.

The sort of curve described by a tooth rests always on the same plane. The incision produced by a pointed tooth on a hard surface, convex and immobile, is of determinate form. It is that of a plain curve, from one point to another by the shortest path, like a meridian on the surface of a sphere. The majority of incisions before our eyes do not present such a character. Here is one among others in which the direction changes many times [Figure 2.2]. . . . the whole incision is made up, first, of a path perpendicular to the axis of the rib, then another longitudinal path, and finally an oblique one. It is a turning movement that a jaw could not make. The human hand, on the contrary, is capable, because of its multiple articulations, of perfect mobility, of guiding and inclining in every direction over the surface the instruments with which it is armed" (Capellini 1877, pp. 58–59).

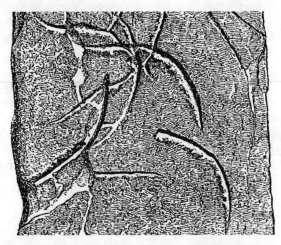

Figure 2.2. A Pliocene whale scapula from Monte Aperto, Italy, with cut marks similar to those described by Broca (de Quatrefages 1887, p. 97).

Even though there may be some justification for pursuing the shark hypothesis with regard to the markings on the Pliocene whale bones of Italy, there is no reason to immediately abandon the hypothesis of human action, for which there is a great deal of evidence.

It is interesting that Broca, one of the foremost authorities on bone physiology of his time, favored Capellini's view that the marks on the fossil whale bones were the product of intentional human work. Perhaps not all of Broca's observations about the action of teeth on bone are correct. But this does not detract from Capellini's conclusions, which were founded on years of painstaking research, and not on Broca's extemporaneous statements.

After Broca's remarks, Capellini (1877, p. 60) himself offered some concluding words: "I have of course taken into consideration bones gnawed by different animals. At the same time, I have not neglected to examine all the kinds of fish teeth found in the same strata as the small whales, of which Mr. Lawley possesses a truly extraordinary collection. If one comes to tell me that with such teeth (using them as tools) he has been able to make such marks as you see on the fossil bones, I am ready to admit this, but if he pretends that the fish itself made the marks, that is another thing. In that case I would invite my illustrious contradictor to bring to my consideration the species of fish to which he would attribute marks identical to those we know as the work of man." Capellini (1877, p. 61) pointed out that such objections had not been raised by the naturalists who were knowledgeable about fish, but rather by archeologists.

One naturalist suggested the marks had been made by a swordfish, and to demonstrate this had taken a swordfish beak in hand, delivering thrusts that left some impressive marks on pieces of fresh whalebone. But even de Mortillet (1883, p. 61), on seeing them and comparing them with the incisions on the Tuscany fossils, rejected this view.

De Quatrefages was among the scientists accepting the Monte Aperto *Balaenotus* bones as being cut by sharp flint instruments held by a human hand. He wrote: "However one may try, using various methods and implements of other materials, one will fail to duplicate the marks. Only a sharp flint instrument, moved at an angle and with a lot of pressure, could do it" (de Quatrefages 1884, pp. 93–94). De Quatrefages believed a band of Pliocene hunters found the whale beached and set upon it with stone knives of the type used by the present-day Australian aboriginals.

The whole issue was nicely summarized in English by S. Laing, who wrote in 1893 (pp. 115–116): "An Italian geologist, M. Capellini, has found in the Pliocene strata of Monte Aperto, near Siena, bones of the *Balaeonotus,* a well-known species of a sort of Pliocene whale, which are scored by incisions obviously made by a sharp-cutting instrument, such as a flint knife guided by design, and by a human hand. At first it was contended that these incisions might have been made by the teeth of fishes, but as specimens multiplied, and were carefully examined, it became evident that no such explanation was possible. The cuts are in regular curves, and sometimes almost semi-circular, such as a sweep of the hand could alone have caused, and they invariably show a clean cut surface on the outer or convex side, to which the pressure of a sharp edge was applied, with a rough or abraided surface on the inner side of the cut. Microscopic examination of the cuts confirms this conclusion, and leaves no doubt that they must have been made by such an instrument as a flint knife, held obliquely and pressed against the bone while in a fresh state, with considerable force, just as a savage would do in hacking the flesh off a stranded whale. Cuts exactly similar can now be made on fresh bone by such flint knives, and in no other known or conceivable way. It seems, therefore, more like obstinate prepossession, than scientific skepticism, to deny the existence of Tertiary man, if it rested only on this single instance."

Continuing his commentary, Laing (1893, p. 116) stated: "As regards the evidence from cut bones it is very conclusive, for experienced observers, with the aid of the microscope, have no difficulty in distinguishing between cuts which may have been made accidentally or by the teeth of fishes, and those which can only have been made in fresh bone by a sharp cutting instrument, such as a flint knife."

A modern authority, Binford, stated (1981, p. 169): "There is little chance that an observer of modified bone would confuse cut marks inflicted during dismembering or filleting by man using tools with the action of animals." Binford (1981, p. 169) further noted: "The marks of animals' teeth are somewhat different. They follow the contours of the bone's surface. . . . Tooth marks may frequently take the form of depressed or mashed lines. . . . On many of the wolf specimens, the tooth mark under magnification appears as a 'cracked' surface scar rather than as a cut or incision in the bone."

But the teeth of sharks are sharper than those of terrestrial mammalian carnivores such as wolves and might produce marks on bone that more closely resemble those that might be made by cutting implements. After inspecting fossil whale bones in the paleontology collection of the San Diego Natural History Museum, we concluded that shark's teeth can in fact make marks closely resembling those that might be made by implements. However, we also concluded that it is nevertheless possible, in some cases, to distinguish marks made by implements from those made by shark teeth.

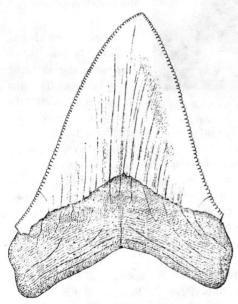

Figure 2.3. Tooth of *Carcharodon megalodon,* a Pliocene great white shark (G. de Mortillet and A. de Mortillet 1881, plate 4, figure 19).

The bones we saw were from a small Pliocene species of baleen whale. The marks on one bone, a jaw fragment, were the subject of a report by Thomas A. Deméré and Richard A. Cerutti (1982) of the San Diego Natural History Museum. The ventral margin of the jaw fragment showed a pair of V-shaped grooves that ran transversely to that surface (Deméré and Cerutti 1982, p. 1480). One of the marks measured 16 mm (0.63 inch) long, and slightly curved. The other one ran 11 mm (0.43 inch) in a straight line. Our inspection of the incisions through a magnifying lens showed evenly spaced parallel longitudinal striations such as one would expect from the serrated edge of a shark's tooth (Figure 2.3).

Even so, Deméré, who showed us the marked fossil at the San Diego Natural History Museum paleontology collection on May 31, 1990, stated that as far as he was concerned these V-shaped incisions alone were inconclusive. That is to say, they might have been caused by something other than shark teeth.

More useful for diagnostic purposes was another mark on the bone. Deméré and Cerutti (1982, p. 1480) described this as a beveled surface "characterized by 12 sinuous but parallel small-scale ridges and grooves." Deméré and Cerutti (1982, p. 1480) went on to state: "This very distinctive pattern has been duplicated by us using a piece of paraffin and a tooth from the Pliocene great white shark, *Carcharodon sulcidens* Agassiz, 1843. . . . The teeth of *Carcharodon* are characterized by serrated margins." The pattern of grooves and ridges observed on the fossil whale bone (Figure 2.4) could have been produced by a glancing blow, with the edge of the tooth scraping along the surface of the bone rather than cutting into it.

Figure 2.4. Pattern of grooves and ridges produced by a serrated shark tooth moving across the surface of a whale bone (Deméré and Cerutti 1982, p. 1481).

With this knowledge, it should be possible to reexamine the Pliocene whale bones of Italy and arrive at some fairly definite conclusions as to whether or not the marks on them were made by shark teeth. Patterns of parallel ridges and grooves on the surfaces of the fossils, such as those described by Deméré and Cerutti, would be an almost certain sign of shark predation or scavenging. And if close examination of deep V-shaped cuts also revealed evenly spaced, parallel longitudinal striations, that, too, would have to be taken as evidence that shark teeth made the cuts. One would not expect the surfaces of marks made by flint blades to display evenly spaced striations.

Even so, care would have to be taken to examine each and every cut on the fossil whale bones. Deméré and Cerutti (1982, p. 1480) reported that carcasses of sea otters, with the bones marked by shark teeth, have been found washed up on the California coast. One can imagine that in the past a whale carcass, partially devoured by sharks, might similarly have washed ashore, and then been butchered by humans. Therefore fossil whale bones might bear both the marks of shark teeth and human implements.

The following statement by Deméré and Cerutti (1982, p. 1480) calls attention to one of the drawbacks of the way anomalous evidence is treated by the scientific community: "It appears then that our fossil specimen preserves a late Pliocene scavenging and/or predator event by *Carcharodon* on cetaceans. To our knowledge this represents the first well-documented report of such activity." It is significant that two working paleontologists, with a special interest in shark teeth and whale bones, were unaware of the extensive debate that occurred in the nineteenth century on the topic of possible *Carcharadon* (versus human) markings on Pliocene cetaceans. Therefore, rather than casting controversial evidence into oblivion, it would be wiser, perhaps, to somehow keep it readily available for further study. That is one purpose of this book.

2.12 HALITHERIUM OF POUANCÉ, FRANCE (MIDDLE MIOCENE)

In 1867, L. Bourgeois caused a great sensation when he presented to the members of the International Congress of Prehistoric Anthropology and Archeology, meeting in Paris, a *Halitherium* bone bearing marks that appeared to be human incisions (de Mortillet 1883, p. 53). *Halitherium* is a kind of extinct sea cow, an aquatic marine mammal of the order Sirenia.

The fossilized bones of *Halitherium* had been discovered by the Abbé Delaunay in the shell beds at Barriére, near Pouancé in northwestern France (Maine-et-Loire). Delaunay was surprised to see on a fragment of the humerus, a bone from the upper forelimb, a number of cut marks (Figure 2.5). The surfaces of the cuts were of the same appearance as the rest of the bone and were easily distinguished from recent breaks, indicating that the cuts were quite ancient. The bone itself, which was fossilized, was firmly situated in an undisturbed stratum, making it clear that the marks on the bone were of the same geological age. Furthermore, the depth and sharpness of the incisions showed that they had been made before the bones had fossilized. Some of the incisions appeared to have been made by two separate intersecting strokes. Even de Mortillet (1883, pp. 53–55) admitted that

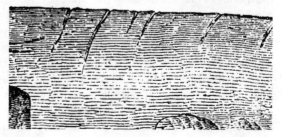

Figure 2.5. Cut marks on *Halitherium* bone from the Miocene at Pouancé, France (de Mortillet 1883, p. 54).

they did not appear to be the products of subterranean scraping or compression. But he would not admit they could be the product of human work, mainly because of the age of the stratum in which the bones were found. The shell beds of this region were said to date to the period represented by the Mayencian formation of the Middle Miocene. But they could be somewhat older. The marine layers in which the *Halitherium* bone was discovered, known as the Faluns of Anjou, are assigned by modern authorities to the Early Miocene (Klein 1973, table 6). *Halitherium* is generally thought to have existed in Europe from the Early Miocene to the Early Oligocene (Romer 1966, p. 386).

De Mortillet (1883, p. 55) wrote in his book *Le Préhistorique*, "This is much too old for man." It is easy enough to see how a scientist who was committed to the evolutionary hypothesis would think so—the Middle Miocene dates as far back as 15 million years, and the Early Miocene to somewhere around 25 million years.

Here again, we have a clear case of theoretical preconceptions dictating how one will interpret a set of facts. De Mortillet (1883, p. 55) attributed the marks on the bones to large sharks of the requin family: "It is a fact that the shell beds of Anjou contain an abundance of sharp pointed teeth of fish of this family. These fish, encountering *Halitherium* beached on the coast, then ate them and left on their bones the numerous marks of their voracity and the strength of their teeth." De Mortillet (1883, p. 55) also stated that on May 5, 1879 Mr. Tournouër presented to the Geological Society of France an incised *Halitherium* bone, attributing the marks to shark teeth. However, in light of the foregoing discussion, it seems the case of the *Halitherium* bone of Pouancé should remain open for further investigation.

On the general subject of cut bones as a category of viable evidence, Laing (1894, pp. 353–354) wrote in his book *Human Origins*, which went through five reprintings: "cut bones afford one of the most certain tests of the presence of man. The bones tell their own tale, and their geological age can be certainly identified. Sharp cuts could only be made on them while the bones were fresh, and the state of fossilization, and presence of dendrites or minute crystals alike on the side of the cuts and on the bone, negate any idea of forgery. The cuts can be compared with thousands of undoubted human cuts on bones from the reindeer and other later periods, and with cuts now made with old flint knives on fresh bones. All these tests have been applied by some of the best anthropologists of the day, who have made a special study of the subject, and who have shown their caution and good faith by rejecting numerous specimens which did not fully meet the most rigorous requirements. . . . The only possible alternative suggested is, that they might have been made by gnawing animals or fishes. But as Quatrefages observes, even an ordinary carpenter would have no difficulty in distinguishing between a clean cut made by a sharp knife, and a groove cut by repeated strokes of a narrow chisel; and how much more would it be impossible for a Professor trained to scientific investigation, and armed with a microscope, to mistake a groove gnawed out by a shark or rodent for a cut made by a flint knife."

Laing's observations are significant in that they counter certain modern prejudices about the caliber of scientific work at that time. On first encountering reports like those concerning the cut bones of St. Prest, Monte Aperto, or Pouancé, one might think something like this: "How quaint these nineteenth-century scientists were, in those old days of the infancy of paleoanthropological investigation. How quick they were to accept questionable evidence upon cursory inspection." But from Laing's statements we can see that scientists like de Quatrefages, Desnoyers, and Capellini were carefully applying standards of investigation and evaluation comparable to those of the present day. In particular, they displayed a considerable grasp of the principles of the modern discipline of taphonomy. One might also postulate something like the following: "Well, perhaps in the nineteenth century, before there were many actual human fossils uncovered, these naturalists focused undue attention on these cut bones, reading too much into them, because they had nothing else to concern themselves with." But even today, many researchers are investigating the presence of humans at certain sites solely on the basis of animal bones bearing signs of intentional workmanship. And, as we shall see in coming chapters, it is not true that nineteenth-century naturalists interested in human antiquity had nothing but cut bones to study. They also extensively investigated many finds of stone tools and human skeletal remains that have since slipped into near total obscurity.

2.13 SAN VALENTINO, ITALY (LATE PLIOCENE)

In 1876, at a meeting of the Geological Committee of Italy, M. A. Ferretti showed a fossil animal bone bearing "traces of work of the hand of man, so evident as to exclude all doubt to the contrary" (de Mortillet 1883, p. 73). This bone, of elephant or rhinoceros, was found firmly in place in Astian (Late Pliocene) strata in San Valentino (Reggio d'Emilie), Italy. The bone's dimensions are 70 mm (2.8 inches) by 40 mm (1.6 inches). Of special interest is the fact that the fossil bone has an almost perfectly round hole at the place of its greatest width. According to Ferretti, the hole in the bone was not the work of molluscs or crustaceans. The next year Ferretti showed to the Committee another bone bearing traces of human work. It was found in blue Pliocene clay, of Astian age, at San Ruffino. This bone appeared to have been partially sawn through at one end, and then broken. De Mortillet, who included the above-mentioned information in his book, stated (1883, p. 77) that he had not seen the bones nor heard any further discussion about them. This indicated to him that they had not been (and thus should not be) taken seriously. It would perhaps have been more appropriate, and scientific, for de Mortillet to have inspected the bones before concluding they were of little scientific value. Many modern scientists react in a similar fashion when confronted with unfamiliar, little-discussed anomalous evidence. They assume it is not of any importance; otherwise they would have seen it discussed in the published works of scientists committed to the established views.

At a scientific conference held in 1880, G. Bellucci, of the Italian Society for Anthropology and Geography, called attention to recent discoveries in San Valentino and Castello delle Forme, near Perugia. Found there were bones of different animals bearing incisions, both straight and intersecting, and with imprints probably made with rocks employed for the purpose of breaking the bones. Bellucci said there were also two specimens of carbonized bones, and finally flint flakes. All were recovered from lacustrine Pliocene clays, characterized by a fauna like that of the classic Val d'Arno. According to Bellucci, these objects proved the existence of man in the Tertiary period in Umbria (Bellucci and Capellini 1884).

2.14 CLERMONT-FERRAND, FRANCE (MIDDLE MIOCENE)

Turning once more to France, we note that in the late nineteenth century the museum of natural history at Clermont-Ferrand had in its collection a femur of *Rhinoceros paradoxus* with grooves on its surface. The specimen was found in a freshwater limestone at Gannat, in a quarry said to be dated by fossils to the Mayencian age of the Middle Miocene (de Mortillet 1883, p. 52). M. Pomel presented this piece to the anthropological section of the French Association for the Advancement of Science meeting of 1876 in Clermont. Pomel said the marks were from carnivores, which were numerous in the French Middle Miocene. But de Mortillet disagreed that an animal could have been responsible. He pointed out that the grooves on the Miocene rhinoceros femur could not have been made by a rodent, because rodent incisors usually leave pairs of parallel marks. The grooves on the rhinoceros femur were not arranged in pairs. De Mortillet also believed that the marks were not caused by larger carnivores, because, as noted by Binford (1981, p. 169) in modern times, carnivore teeth leave many irregular impressions and cause distinctive patterns of bone destruction. Binford stated that "association of scoring with patterns of destruction is not expected when man dismembers an animal with tools." According to this standard, the Miocene rhinoceros femur, which displayed scoring but no pattern of destruction, might very well have been cut by ancient humans using stone tools.

For de Mortillet, however, the marks were a purely geological phenomenon. He concluded that the grooves on the rhinoceros femur of Clermont-Ferrand were probably produced by the same subterranean pressures responsible for the marks on the Billy specimen (de Mortillet 1883, p. 52). But de Mortillet's own description (1883, p. 52) of the markings on the bone leaves this interpretation open to question: "The impressions occupy a portion of the inner surface near the condyles. They are parallel grooves, somewhat irregular, transverse to the axis of the bone." The condyles are the rounded prominences on the articulator, or joint, surfaces at the end of the femur, or thighbone. The orientation and position of the marks on the fossil were identical to those of incisions made in the course of butchering operations on a long bone such as the femur. Binford's studies (1981, p. 169) revealed: "cut marks are concentrated on articulator surfaces and are relatively rare as transverse marks on long bone surfaces. . . . cut marks from stone tools are most commonly made with a sawing motion resulting in short and frequently multiple but roughly parallel marks. Such marks are generally characterized by an open cross section. Another characteristic of cut marks derived from the use of stone tools is that they rarely follow the contours of the bone on which they appear. That is, the cut does not show equal pressure in depressions and along prominent ridges or across the arc of a cylinder." As described by de Mortillet, the short parallel grooves found on the Miocene rhinoceros femur conform to these criteria, leaving one to wonder how it is possible that chance geological pressures could so closely duplicate, in terms of position and character, the distinctive marks of human butchering.

The Miocene dating of the Clermont-Ferrand site is confirmed by the presence of *Anthracotherium magnum,* an extinct mammal of the hippopotamus family. In fact, the site could be older than Middle Miocene. According to one modern authority, *Anthracotherium* existed in Europe from the Late Miocene to the Early Eocene (Romer 1966, p. 389). Savage and Russell (1983, p. 245) last report *Anthracotherium* in the Orleanian land mammal stage of the Early Miocene.

2.15 CARVED SHELL FROM THE RED CRAG, ENGLAND (LATE PLIOCENE)

In a report delivered to the British Association for the Advancement of Science in 1881, H. Stopes, F.G.S. (Fellow of the Geological Society), described a shell, the surface of which bore a carving of a crude but unmistakably human face. The carved shell was found in the stratified deposits of the Red Crag (Stopes 1881, p. 700). The Red Crag, part of which is called the Walton Crag, is thought to be of Late Pliocene age. According to Nilsson (1983, p. 308), the Red (Walton) Crag is between 2.0 and 2.5 million years old.

Just how the discovery (Figure 2.6) was received was detailed by Marie C. Stopes, the discoverer's daughter, in an article in *The Geological Magazine* (1912, p. 285): "in 1881, when it was brought forward by Mr. Henry Stopes at a British Association meeting, it was considered *wrong* to suggest that man could have been alive at so early a date." Arguing against forgery, Marie Stopes (1912, p. 285) stated: "It should be noted that the excavated features are as deeply coloured red-brown as the rest of the surface. This is an important point, because when the surface of Red Crag shells are scratched they show white below the colour. It should also be noticed that the shell is so delicate that any attempt to carve it would merely shatter it." It is therefore quite possible that this shell was carved and deposited in the Red Crag strata during the Late Pliocene. If true, this would place intelligent human beings in England as far back as 2.0 million and maybe as much as 2.5 million years ago. One should keep in mind that in terms of conventional paleoanthropological opinion, one does not encounter such works of art until the time of fully modern Cro-Magnon man in the Late Pleistocene, about 30,000 years ago.

Figure 2.6. Carved shell from the Late Pliocene Red Crag formation, England (M. Stopes 1912, p. 285).

Discoveries of incised bones dating back to the Pliocene or earlier persisted into the early part of the twentieth century. Opposition to them also persisted, and eventually prevailed. For example, Hugo Obermaier, professor of prehistoric archeology at the University of Madrid, wrote (1924, pp. 2–3): "traces (chiefly fluted, engraved, or grooved) have been observed on the bones of animals and shells of molluscs in Tertiary deposits at Saint-Prest, Sansan, Pouancé, and Billy, France; in the Tertiary basin of Antwerp, Holland; at Monte Aperto near Siena, Italy; in North and South America; and in several other places. . . . it is easy to explain supposed traces of human activity as the result of natural causes—such, for example, as the gnawing or biting of animals, earth pressure, or the friction of coarse sand." But can we say for certain that this "easy" explanation is the correct one?

2.16 BONE IMPLEMENTS FROM BELOW THE RED CRAG, ENGLAND (PLIOCENE TO EOCENE)

In the early twentieth century, J. Reid Moir, the discoverer of many anomalously old flint implements (Section 3.3), described "a series of mineralised bone implements of a primitive type from below the base of the Red and Coralline Crags of Suffolk" (1917a, pp. 116–131). The top of the Red Crag in East Anglia is now considered to mark the boundary of the Pliocene and Pleistocene, and would thus date back about 2.0–2.5 million years (Romer 1966, p. 334; Nilsson 1983, p. 106). The older Coralline Crag is Late Pliocene and would thus be at least 2.5–3.0 million years old. The beds below the Red and Coralline Crags, the detritus beds (Table 2.1, p. 78), contain materials ranging from Pliocene to Eocene in age (Section 3.3.2). Objects found there could thus be anywhere from 2 million to 55 million years old. One group of Moir's specimens is of triangular shape (Figure 2.7). In his report, Moir (1917a, p. 122) stated: "These have all been formed from wide, flat, thin pieces of bone, probably portions of large ribs, which have been so fractured as to now present a definite form. This triangular form has, in every case, been produced by fractures *across* the natural 'grain' of the bone." Moir (1917a, p. 116) then began to describe some of his attempts to reproduce the specimens: "having conducted a number of experiments in which

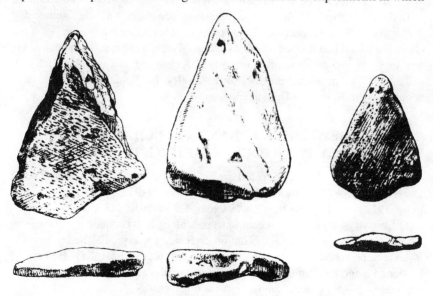

Figure 2.7. Three bone tools from the detritus bed beneath the Coralline Crag, which contains materials ranging from Pliocene to Eocene in age. These implements could thus be anywhere from 2 to 55 million years old (Moir 1917a, plate 26).

mineralised and unmineralised bones were subjected to the effects of fortuitous blows and pressure, and after having fractured numerous modern shank bones of the bullock by striking and cutting them with flints and other stones held in the hand with a view of thus shaping them to the forms of the sub-Crag examples, he [the author] is compelled to regard these latter specimens as undoubted works of man." According to Moir, the triangular pieces of fossilized whale bone discovered in the strata below the Coralline Crag might have once been used as spear points.

Moir had himself collected most of the specimens, but he also described one discovered by another naturalist, a Mr. Whincopp, of Woodbridge in Suffolk, who had in his private collection a "piece of fossil rib partially sawn across at both ends" (Moir 1917a, p. 117). This object came from the detritus bed below the Red Crag and was "regarded by both the discoverer and the late Rev. Osmond Fisher as affording evidence of human handiwork" (Moir 1917a, p. 117). Indications of sawing would be quite unexpected on a fossil bone of this age. A piece of sawn wood was recovered from the more recent Cromer Forest Bed in the same region (Section 2.20).

Osmond Fisher, who was a Fellow of the Geological Society, made some interesting discoveries of his own. In a review published in *The Geological Magazine,* Fisher (1912, p. 218) wrote: "When digging for fossils in the Eocene of Barton Cliff I found a piece of jet-like substance about 9½ inches square and 2¼ inches thick. . . . It bore on at least one side what seemed to me marks of the chopping which had formed it into its accurately square shape. The specimen is now in the Sedgwick Museum, Cambridge." Jet is a compact velvety-black coal that takes a good polish and is often used as jewelry. The Eocene period dates back about 38–55 million years from the present.

2.17 DEWLISH ELEPHANT TRENCH, ENGLAND (EARLY PLEISTOCENE TO LATE PLIOCENE)

Osmond Fisher also discovered an interesting feature in the landscape of Dorsetshire—the elephant trench at Dewlish. Fisher (1912, pp. 918–919) stated in his 1912 review: "This trench was excavated in chalk and was 12 feet deep, and of such a width that a man could just pass along it. It is not on the line of any natural fracture, and the beds of flint on each side correspond. The bottom was of undisturbed chalk, and one end, like the sides, was vertical. At the other end it opened diagonally on to the steep side of a valley. It has yielded substantial remains of *Elephas meridionalis*, but no other fossils. . . . This trench, in my opinion, was excavated by man in the later Pliocene age as a pitfall to catch elephants; and if so, it proves that he was already an intelligent and social being." *Elephas meridionalis,* or "southern elephant," was in existence in Europe from

1.2 to 3.5 million years ago (Maglio 1973, p. 79). Thus, while the bones found in the trench at Dewlish could conceivably be Early Pleistocene in age, they might also date to the Late Pliocene.

In Fisher's original reports in the *Quarterly Journal of the Geological Society of London*, we find the following more detailed description: "The trench was . . . followed for about 103 feet, until it suddenly terminated in a smooth 'apse-like' end. . . . It was a deep, narrow trench, with nearly vertical sides of undisturbed chalk. Mr. [Clement] Reid says: 'The fissure (or rather trough) ended abruptly, without any trace of a continuing join; it was not a fault, for the lines of flint-nodules corresponded on each side'" (O. Fisher 1905, p. 35). The base of the trench was reported to be a smooth surface of chalk, twelve feet down (O. Fisher 1905, p. 36). Photographs accompanying the report show the vertical walls of the trench, carefully chipped as if with a large chisel.

In response to suggestions that flowing water might have excavated the trench, Fisher (1905, p. 36) stated: "A stream in such a locality would be unlikely to excavate a deep and narrow channel, much less, if it did so, would it come to an abrupt ending. And, even if we could account for the natural formation of such a trench, how came it that the remains of so many elephants were found in it, and (so far as appears) no other animals?"

Fisher (1905, p. 36) referred to reports showing that primitive hunters of modern times made use of similar trenches: "Sir Samuel Baker describes this method of taking elephants by natives of Africa. He says that an elephant cannot cross a ditch with hard perpendicular sides, which will not crumble nor yield to pressure. Pitfalls 12 to 14 feet deep are dug in the animals' routes towards drinking-places, and covered with boughs and grass. The pits are made of different shapes, according to the individual opinions of the trappers. When caught, the animals are attacked with spears while in their helpless position, until they at last succumb through loss of blood. . . . If the stream which now runs at the bottom of the hill, despite subsequent changes in the contour of the country already existed, then this trench would have been made in a position suitable to intercept the route to the drinking place."

Some critics pointed out that the trench appeared too narrow to accommodate a fully grown elephant, but evidently the deep trench was simply meant to incapacitate an adult animal by injuring its legs or to capture a young animal. Also, further excavation of the trench by the Dorset Field Club, as reported in a brief note in *Nature* (October 16, 1914; p. 511), revealed that "instead of ending below in a definite floor it divides downward into a chain of deep narrow pipes in the chalk." But it is not unlikely that ancient humans might have made use of small fissures to open a larger trench in the chalk. It would be worthwhile to examine the elephant bones found in the trench for signs of cut marks or selective preservation.

2.18 MORE ON IMPLEMENTS FROM BELOW THE RED CRAG (PLIOCENE TO EOCENE)

Ten years after his first report (Section 2.16), J. Reid Moir (1927, pp. 31–32) again described fossilized bone implements taken from below the Red Crag formation (Figure 2.7): "In the sub-Red Crag Bone Bed where these flint implements are found, there are a number of bones comprising, chiefly, pieces of whale rib, very highly mineralised. Among these I have found certain specimens that have every appearance of having been shaped by man. Such pieces are of great rarity and assume, usually, a definite pointed form which cannot well have been produced by any natural, non-human means. The 'worked' portions of these bones show the same deep and ancient coloration of the other parts of the specimens, and experiments which I have carried out demonstrate that, in the present mineralised state of the bones, it is not possible to shape them to the forms they have assumed. In order to produce such forms from bone I found it necessary to operate on fresh specimens, and that these, by 'flaking' and rubbing with a hard quartzite pebble, could be made into shapes quite comparable with those found below the Red Crag. I have little doubt, therefore, that these latter specimens have been shaped by man and represent the most ancient bone implements yet discovered."

Bone implements, like incised bones, remain a major category of paleoanthropological evidence. For example, Mary Leakey (1971, p. 235) has reported from Olduvai Gorge in Africa: "It is probable that the majority of the broken mammalian bones found on living sites in Bed I and II at Olduvai merely represent food debris. Some may also have been further broken by carnivores after the sites were abandoned. There is, however, a relatively small number which appear to have been artificially flaked and abraded."

Leakey (1971, p. 235) then gave the following example: "Part of an equid [horse family] first rib showing evidence of polishing and smoothing at the fractured end. . . . There is an oblique fracture of the shaft of the rib, towards the proximal end, which runs transversely from the lower to the upper margin. One edge of the fracture is abraded and smooth, showing that the bone was used after it had been broken."

She also described a series of humeri (the bones of the upper forelimb): "A proportion of these specimens appears to represent the ends of bones in which the shafts were shattered to extract the marrow and which have been subsequently utilised, but others, including the pointed series and those split longitudinally, seem to have been expressly shaped" (M. Leakey 1971, p. 236).

Leakey qualified her apparent acceptance of these implements with only this statement: "At the time of this writing there is, as yet, no general agreement regarding the extent to which bone was worked and used in Lower and Middle Pleistocene times. It is evident that more basic research on the effect of artificial fracture and use of bone, as distinct from damage caused by natural means, is required before bone debris from early living sites can be satisfactorily interpreted" (M. Leakey 1971, p. 235).

Despite this cautionary remark, Mary Leakey's statements about the bone implements of Olduvai Gorge seemed positive. The question is this: will scientists show the same openmindedness in the case of the sub-Crag bone tools reported by J. Reid Moir? If the answer is yes, then paleoanthropologists will have to rework their ideas about human origins to include toolmaking humans over 2 million years ago, and maybe as much as 55 million years ago, in England.

2.19 IMPLEMENTS FROM CROMER FOREST BED, ENGLAND (MIDDLE TO EARLY PLEISTOCENE)

J. Reid Moir (1927, pp. 49–50) also wrote of bone tool finds from the Cromer Forest Bed: "During this year (1926) Mr. J. E. Sainty found upon the beach at Overstrand a piece of heavily mineralized bone which is evidently referable to the Cromer Forest Bed. . . . the bone is of a markedly implemental form; in fact, on the surface figured and at the butt-end, it exhibits flaking and hacking, which, judging from the experiments I carried out in shaping this material, I think has been intentionally produced. . . . Sir Arthur Keith, F.R.S. [Fellow of the Royal Society], who examined the specimen, has kindly given me the following opinion upon it: 'There can be no doubt, I think, that your implement has been fashioned out of the lower jaw of the larger whalebone whales. None of the original surface of the bone is left; it has been removed by flaking.' From the extreme fossilization of this specimen, I judge it to belong to the earliest Cromer Forest Bed deposit, and to be contemporary with the great flint implements found at that horizon. Remains of whales have been discovered in the Forest Bed and it was doubtless the skeleton of one of these that supplied the material from which this implement was made by one of the earliest Cromerian men."

The most comprehensive recent study of the Cromer Forest Bed formation is by R. G. West. According to West (1980, p. 201), the oldest part of the Cromer Forest Bed is the Sheringham member. West identified the lower part of the Sheringham member, representing the base of the Cromer Forest Bed, with the Pre-Pastonian cold stage of East Anglia (Table 2.1, p. 78).

Even after much study, West was not able to give a conclusive date for the Pre-Pastonian. He suggested that the lowest level of the Pre-Pastonian, might be equivalent to the basal part of the northwestern European cold stage called the Erburonian. This would give the Pre-Pastonian cold stage a maximum age of about 1.75 million years (West 1980, fig. 54). But Nilsson (1983, p. 308) puts the base of the Erburonian at 1.5 million years.

TABLE 2.1

Stratigraphy of East Anglia

Est. Age (Years B.P.)	Traditional Divisions	Stages of West (1980)	Northwest Europe
.4 million	Cromer Till (g)	Anglian (g)	Elster (g)
.8 million / 1.0 million / 1.5 million / 2.0 million	Cromer Forest Bed *lower limit (Nilsson)* Weybourne Crag Norwich Crag *(West)*	Cromerian (t) Beestonian (c) Pastonian (t) Pre-Pastonian (c)	Cromer complex (i/g) Menapian (g) Waalian (t) Erburonian (c) Tiglian (t)
2.5 million	Red Crag	Waltonian (c)	Praetiglian (c)
	Detritus Bed (Cretaceous ® Pliocene)		
	Coralline Crag (Pliocene)		
38.0 million	Detritus Bed (Cretaceous ® Pliocene)		
55.0 million	London Clay (Eocene)		
	Chalk (Cretaceous)		

Cold (c), temperate (t), glacial (g), and interglacial (i).

According to West (1980, fig. 54), the Pre-Pastonian cold stage of East Anglia might also be identified, on paleomagnetic grounds, with the Menapian glaciation of northwestern Europe at .8–.9 million years. The Pre-Pastonian might also be identified with the early part of the northwestern European Cromer complex, a series of alternating glacials and interglacials extending from about .4 million to .8 million years ago (West 1980, p. 120; Nilsson 1983, p. 308). The early part of the Cromer complex of glacials and interglacials can be estimated at about .6–.8 million years according to the correlation table of Nilsson (1983, p. 308).

Therefore, according to West, the Cromer Forest Bed series might be as old as 1.75 million years or as young as .6–.8 million years. Nilsson (1983, p. 308) shows the Cromer Forest Bed series beginning at about .8 million years ago.

So if the heavily mineralized bone implement reported by Moir actually did come from the lowest levels of the Cromer Forest Bed, as he surmised, it might be as much as 1.75 million years old. The oldest *Homo erectus* fossils from Africa only date back about 1.6 million years.

If, however, we take the younger of the possible dates for the oldest levels of the Cromer Forest Bed (about .6 million years) that would still be quite anomalous for England. According to Nilsson (1983, p. 111), the oldest stone tools from England come from Westbury-sub-Mendip deposits equivalent to the terminal phase of the Cromer Forest Bed, at about .4 million years ago.

Of course, Moir could have been wrong about the source of the mineralized bone implement. The beds at Overstrand cover almost the entire span of Cromer Forest Bed time (West 1980, p. 159). Thus the implement from Overstrand might have come not from the earliest but from the latest part of the Cromer Forest Bed sequence, making it the same age as the stone tools from Westbury-sub-Mendip, about .4 million years old—quite within the range of conventional acceptability. This possibility makes it all the more remarkable that the bone tool reported by Moir is not given serious attention by modern paleoanthropologists.

In some additional remarks on the Cromer Forest Bed discoveries, Moir (1927, p. 50) went on to describe incised bones rather than bones modified as tools: "The discovery of flint implements in the Forest Bed induced me to make a close examination of the mammalian bones from this deposit, in the possession of Mr. A. C. Savin of Cromer. This examination revealed three specimens, all found in the peat, representing the upper part of the Forest Bed at West Runton, by Mr. Savin, which show on their surface clearly defined cuts which, I think, can only have been produced by flint knives in removing flesh . . . the Cromer examples are quite comparable with others exhibiting cuts which I have discovered in various later prehistoric epochs. The lines are fine, and straight, and were evidently produced by a sharp-edged flint. Some of the smaller mammals might cut a bone with their teeth in a similar way, but they could not produce such long cuts as are present on the bones from West Runton. Nor is it possible to regard these markings as due to glacial action."

The part of the Cromer Forest Bed sequence represented especially well at West Runton is the Upper Freshwater Bed. According to West, the Upper Freshwater Bed, as defined during Moir's time, contained elements as old as the Pastonian temperate stage. The Pastonian stage of East Anglia was thought by West (1980, fig. 54) to be equivalent to the latter part of the Waalian temperate stage of northwestern Europe, dated at 1 million years (Nilsson 1983, p. 308).

Alternatively, the Pastonian temperate stage might correlate with an inter-glacial within the Cromer glacial complex, at about .5 million years. In any case, West (1980, p. 116) believed most of the Upper Freshwater Bed was within the time range of the Cromer complex of northwestern Europe, giving it an age of .4–.8 million years (Nilsson 1983, p. 308).

Taken together, the different estimates of the age of the Upper Freshwater Bed would give the cut bones from West Runton a possible date range of between 0.4 and 1.0 million years. At the older end of the date range, the cut bones would be extremely anomalous; at the younger end, less so.

Moir observed that the marks on the West Runton bones were not of the kind produced by glaciers and further noted that the bed in which the specimens were found contained many fragile, unbroken shells and thus appeared undisturbed. "The bones comprise part of the humerus of a large bison, and portions of the lower jaws, with teeth in place, of deer," stated Moir (1927, p. 50). The cuts, he also observed, ran under thick ferruginous deposits, indicating their great age.

"I have recently carried out some experiments in scraping modern bones with a sharp flake of flint," continued Moir (1927, p. 51), "and find that the cuts so produced are in every way comparable with those upon the Cromer examples. It was noticed that these latter specimens, in addition to the easily recognised cuts, exhibited a large number of minute incisions which could only be examined adequately by means of a lens. Upon the experimental bones I found that a precisely similar assemblage of small cuts was present, and I have no doubt that these are due to the microscopic projections present on the cutting-edge of the flint which I used."

The specific identifying characteristics of incisions made with flint flakes on bone have been confirmed by modern investigators such as Rick Potts and Pat Shipman. John Gowlett (1984, p. 53) stated: "Their work involved use of the electron microscope, at a very high magnification. They found that many bones from Olduvai preserved carnivore gnawing marks, as well as stone tool cut-marks. Very close parallel striations were indisputable evidence of the stone tools, for no edge of a flake is perfectly straight, and each protruding sharp piece leaves its mark." It is apparent that Moir's methods of identification compare favorably with those employed by modern professional paleoanthropologists.

2.20 SAWN WOOD FROM CROMER FOREST BED, ENGLAND (MIDDLE TO EARLY PLEISTOCENE)

J. Reid Moir (1927, p. 47) also described a piece of cut wood from the Cromer Forest Bed (Figure 2.8) that suggested human action: "the late Mr. S. A. Notcutt of Ipswich dug out of this deposit, at the foot of the cliff near Mundesley, a piece

Figure 2.8. Piece of wood from the Cromer Forest Bed, England. The piece of wood, apparently sawn at the right end, is between 0.5 and 1.75 million years old (Moir 1917b).

of wood which, in my opinion, was shaped by man. The bed in which the wood was found consisted of undisturbed sand and gravel, and was overlain by Lower Glacial Clay *in situ*."

The beds at Mundesley extend from the lattermost Cromer Forest Bed times, at about .4 million years, to the lower part of the Pre-Pastonian cold stage, estimated variously at 0.8 or 1.75 million years (West 1980, p. 182; Nilsson 1983, p. 308). But most of the Mundesley strata are identified with the Cromerian temperate stage of East Anglia (West 1980, p. 201). One should note that the Cromerian temperate stage of East Anglia, dated roughly at .4–.5 million years, is not the same as the Cromer complex of northwestern European glacials and interglacials, dated at .4–.8 million years (Nilsson 1983, p. 308).

In the course of his comments about the piece of cut wood, Moir (1927, p. 47) made these observations: "The specimen, which is quite comparable with other wood found in the Forest Bed, is . . . slightly curved, four-sided, and is flat at one end and pointed at the other. . . . The flat end appears to have been produced by sawing with a sharp flint, and at one spot it seems that the line of cutting has been corrected [Figure 2.9], as is often necessary when starting to cut wood with a modern steel saw. The present form of the specimen is due to the original round

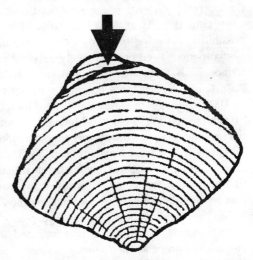

Figure 2.9. Cross section of a piece of cut wood from the Cromer Forest Bed. The arrow indicates a groove, possibly from an initial cut by a sawing implement (Moir 1917b).

piece of wood—which has been identified by Dr. A. B. Rendle, F.R.S., as yew—having been split four times longitudinally in the direction of its natural grain. The pointed end is somewhat blackened as if by fire, and it is possible that the specimen represents a primitive digging stick used for grubbing up roots."

While there is an outside chance that beings of the *Homo erectus* type might have been present in England during the time of the Cromer Forest Bed, the level of technological sophistication implied by this sawn wood tool is extraordinary and suggestive of *sapiens*-like capabilities. In fact, it is hard to see how this kind of sawing could have been produced even by stone implements. Small flint chips mounted in a wooden holder, for example, would not have produced the clean cut evident on the specimen because the wooden holder would have been wider than the flint teeth. Hence one could not have cut a narrow groove with such a device. A saw blade made only of stone would have been extremely brittle and would not have lasted long enough to perform the operation. Furthermore it would have been quite an accomplishment to make such a stone blade. Thus it seems that only a metal saw could produce the observed sawing. Of course, a metal saw at .4–.5 million years is quite anomalous.

It is remarkable that the incised bones, bone implements, and other artifacts from the Red Crag and Cromer Forest Beds are hardly mentioned at all in today's standard textbooks and references. This is especially true in the case of the Cromer Forest Bed finds, most of which are, in terms of their age, bordering on the acceptable, in terms of the modern paleoanthropological sequence of events.

In Gowlett's *Ascent to Civilization* (1984, p. 88), we read: "There is a possibility that some finds from Britain are older than the Hoxnian [an inter-glacial period dated approximately 330,000 years ago]: for example the high terrace finds from Fordwich and from Kent's Cavern near Torquay. The importance of such finds lies in the demonstration that perhaps as much as 500,000 years ago, man was able at least for a time to colonize Europe out to its extremities. At Westbury-sub-Mendip, in south-west England, remains of ex-tinct animals associated with very few stone tools suggest contemporaneity with the Cromerian phase, estimated at c. 0.7–0.5 million years, and named after beds in eastern England, where there are faunal remains but no archaeological traces." Elsewhere Gowlett stated "it is safest to assume that the first occupation of Europe would have been by tool-making men in the earlier Pleistocene." This would "imply a date about 1.5 million years ago" (Gowlett 1984, p. 76).

Considering that Gowlett was prepared to find evidence of toolmaking humans in Europe at 1.5 million years ago, it is odd to find him stating that the Cromer Forest Bed contains "no archaeological traces" (Gowlett 1984, p. 88). Gowlett, a professor at Oxford University, should have been knowledgeable about the recent history of paleoanthropology in England. Was he unaware that in the early twentieth century Moir and others found bone tools, incised bones, and other artifacts (including a whole flint industry) in the Cromer Forest Bed? That would seem unlikely. Did he think the finds to be not genuine? Perhaps he was aware of the discoveries and considered them genuine but deliberately avoided including them in his discussion, even though they would have helped his case. Why? It could be that mentioning them would have implied his acceptance of the still older sub-Red Crag discoveries of Moir and others, which pose a strong challenge to the whole scenario of human origins and antiquity.

2.21 CONCLUDING WORDS ABOUT INTENTIONALLY MODIFIED BONE

It is really quite curious that so many serious scientific investigators in the nineteenth century and early twentieth century independently and repeatedly reported that marks on bones from Miocene, Pliocene, and Early Pleistocene formations were indicative of human work. Among the researchers making such claims were Desnoyers, de Quatrefages, Ramorino, Bourgeois, Delaunay, Bertrand, Laussedat, Garrigou, Filhol, von Dücker, Owen, Collyer, Calvert, Capellini, Broca, Ferretti, Bellucci, Stopes, Moir, Fisher, and Keith.

Were these scientists deluded? Perhaps so. But cut marks on fossil bones are an odd thing about which to develop delusions—hardly romantic or inspiring. Were the above-mentioned researchers victims of a unique mental aberration of the last century and the early part of this one? Or does evidence of primitive hunters really abound in the faunal remains of the Tertiary and early Quaternary?

Assuming such evidence is there, one might ask why it is not being found today. One very good reason is that no one is looking for it. Evidence for intentional human work on bone might easily escape the attention of a scientist not actively searching for it. If a paleoanthropologist is convinced that tool-making human beings did not exist in the Middle Pliocene, he is not likely to give much thought to the exact nature of markings on fossil bones from that period.

Even for those prepared to find signs of human work, the interpretation of marks on fossil bones is a difficult matter. This led Binford (1981, p. 181) to write: "One might reasonably ask at this point that if we cannot establish a pattern of bone modification unambiguously referable to man, why study the faunal products of man and seek greater understanding of his highly variable behavior? The answer to this is simply that the basic task of anthropology—of which archaeology is a part—is to seek an understanding of man's variable cultural behavior." Binford clearly defined the dilemma inherent in the empirical approach to such questions—it is imperfect, yet there appears to be no other choice. So it seems that great caution is required. In fact, our study of the empirical methods used by paleoanthropologists suggests these methods cannot give a completely reliable picture of the past, and of human origins in particular.

3
Eoliths

3.1 ANOMALOUSLY OLD STONE TOOLS

Even when considered alone, the evidence gathered from incised and broken bones, as detailed in the preceding chapter, inflicts heavy damage on the conception that toolmaking hominids emerged only in the Pleistocene. But we now turn to a more extensive and significant category of evidence—ancient stone implements.

Nineteenth-century scientists turned up large quantities of what they presumed to be stone tools and weapons in Early Pleistocene, Pliocene, Miocene, and older strata. These were not marginal discoveries. They were reported by leading anthropologists and paleontologists in well-established journals, and were thoroughly discussed at scientific congresses. But today hardly anyone has heard of them. One wonders why. As in the case of the bones discussed in the previous chapter, the hard facts of these discoveries, though disputed, were never conclusively invalidated. Instead, reports of these ancient stone implements were, as time passed, simply put aside and forgotten as different theoretical scenarios of human evolution came into vogue.

Here is what appears to have taken place. In the 1890s, Eugene Dubois discovered and promoted the famous, yet dubious, Java ape-man (Section 7.1). Many scientists accepted Java man, found unaccompanied by stone tools, as a genuine human ancestor. But because Java man was found in Middle Pleistocene strata, the extensive evidence for toolmaking hominids in the far earlier Pliocene and Miocene periods no longer received much serious attention. How could such toolmaking hominids have appeared long before their supposed ape-man ancestors? Such a thing would be impossible; so better to ignore and forget any discoveries that fell outside the bounds of theoretical expectations.

And that is exactly what happened—whole categories of facts were interred beneath the surface layers of scientific cognition. By patient research we have, however, managed to locate and recover a vast hoard of such buried evidence, and our review of it shall take us from the hills of Kent in England to the valley of the Irrawady in Burma. We shall also give consideration to anomalously old crude stone tool industries discovered by researchers in the late twentieth century.

The anomalous stone tool industries we shall consider fall into three basic divisions: (1) eoliths, (2) crude paleoliths, and (3) advanced paleoliths and neoliths.

According to some nineteenth-century authorities, eoliths (or "dawn stones"), were stones with edges naturally suited for certain kinds of uses. These, it was said, were selected by humans and used as tools with little or no further modification. Often one or more of the natural edges of the stone would be chipped to make it more suitable for a desired function. To the untrained eye, Eolithic stone implements were often indistinguishable from ordinary broken rocks, but specialists in lithic technology developed criteria for identifying upon them signs of human modification and usage.

In the case of more sophisticated stone tools, called paleoliths, the signs of human manufacture were more obvious, involving an attempt to form the whole of the stone into a recognizable tool shape. Questions about such implements centered mainly upon the determination of their correct age. Some Paleolithic implements, such as those used in Europe during the Late Stone Age and in recent historical times by the American Indians, display a high degree of artistry and craftsmanship, with very fine and elaborate chipping and graceful, symmetrical shapes. Most of the implements we shall be examining, however, are far more rudimentary. In fact, some researchers of the nineteenth and twentieth centuries have categorized them among the eoliths. But we have chosen to make a rough distinction between eoliths and crude paleoliths. While the eoliths are formed from naturally broken pieces of stone, perhaps with some slight chipping on a working edge, the crude Paleolithic industries include some specimens that have been deliberately flaked from stone cores and then modified by more extensive chipping into definite tool shapes. In distinguishing crude paleoliths from eoliths, we have also relied on experts who have testified that anomalously old paleoliths from the Pliocene, Miocene, and earlier periods are identical to accepted Paleolithic implements of the Late Pleistocene.

Our third division, advanced paleoliths and neoliths, refers to anomalously old stone tools that resemble the very finely chipped or smoothly polished stone industries of the standard Late Paleolithic and Neolithic periods.

Over the years, the terms eolith, paleolith, and neolith have been used in various ways. For most researchers, they have denoted not only levels of technical development but also a definite temporal sequence. Eoliths would be the oldest implements, followed in turn by the paleoliths and neoliths. But in the course of our discussion we will mainly use these terms to indicate degrees of workmanship. The evidence, we propose, makes it impossible to assign dates to stone tools simply on the basis of their form.

In this chapter, we shall discuss anomalous eoliths. In Chapter 4, we shall discuss anomalous crude paleoliths, and in Chapter 5, we shall discuss anomalous advanced paleoliths and neoliths. This threefold division is not perfect. We were confronted with borderline cases in which assignment to one chapter or another was difficult. Within the cruder stone tool industries are often found individual implements and groups of implements that might be classified as more sophisticated; and similarly, among the more sophisticated industries are found examples of implements that might be classified among the most crude. Also, some individual researchers discovered a number of industries, of varying levels of complexity, and for the sake of convenience, these have been grouped together. Because of this, it has not been possible, or practical, to achieve a complete segregation of tool types in different chapters. Still, we have found it useful to attempt to make a rough division between (1) the Eolithic, (2) the crude Paleolithic, (3) and the advanced Paleolithic and Neolithic types.

Having expressed these cautions, we can now embark upon our examination of the Eolithic stone tools, beginning with those found by Benjamin Harrison in England and proceeding to tools found in other countries during the latter part of the nineteenth century. We shall then consider the discoveries of J. Reid Moir in England. In the last sections of this chapter, we shall examine attempts by H. Breuil and A. S. Barnes to discredit Eolithic industries, and finally we shall review modern examples of Eolithic industries.

3.2 B. HARRISON AND THE EOLITHS OF THE KENT PLATEAU, ENGLAND (PLIOCENE)

3.2.1 Young Harrison

The small town of Ightham, in Kent, is situated about twenty-seven miles southeast of London. Nearby one finds the home of the unfortunate second wife of Henry VIII, Anne Boleyn, who lost her head to the executioner's blade. In the more sedate years of the Victorian era, a respectable small businessman named Benjamin Harrison kept a grocery shop in Ightham. On holidays he roamed the nearby hills and valleys, collecting flint implements which, though now long forgotten, were for decades the center of protracted controversy in the scientific community.

Even as a boy, Harrison was interested in geology and read Lyell's *Principles of Geology* at age thirteen. In the course of his walks, he grew well acquainted with the landscape around Ightham. This region of southeastern England, known as the Weald of Kent and Sussex, had a complex geological history. In the past, it was a broad rise. In later times, the central part of the rise was eroded away by

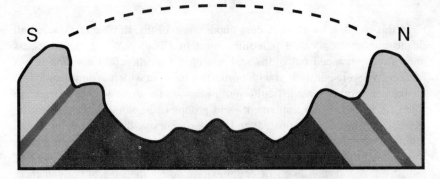

Figure 3.1. The Weald region of Kent and Sussex, England. The dotted line shows the ancient land surface, now eroded away, leaving the present North Downs (N) and South Downs (S) (Moir 1924, p. 638). The Kent Plateau is in the North Downs region.

the forces of nature (Figure 3.1), leaving hills to the north (the North Downs) and south (the South Downs). The North Downs rise to the Kent Plateau near Ightham, and it is on the Kent Plateau that Harrison made some of his most significant discoveries.

Young Harrison developed into an accomplished amateur paleoanthropologist. Perhaps semi-professional would be a better word than amateur, for Harrison did much of his work in close consultation with, and sometimes under the direct supervision of, Sir John Prestwich, the famous English geologist, who lived in the vicinity. Harrison also corresponded regularly with other scientists involved in paleoanthropological research and carefully catalogued and mapped his finds, according to standard procedures.

A room over Harrison's shop served as a museum where he kept his flint tools. On the walls he displayed geological maps of the Weald region of Kent and Sussex, water colors of implements he had found, and portraits of Charles Darwin, Sir John Prestwich, and Sir John Evans.

3.2.2 Neoliths and Paleoliths

Harrison's first finds were not of the very crude Eolithic variety. They were Neolithic implements. Neoliths are smooth-surfaced, polished stone artifacts, displaying highly sophisticated craftsmanship. According to modern opinion, Neolithic cultures date back only about 10,000 years, and are associated with agriculture and pottery. Harrison found neoliths scattered over the present land surfaces around Ightham.

In the early 1860s, the discoveries of Boucher des Perthes in France were attracting the attention of British scientists. Boucher des Perthes had found paleoliths in the gravels of the Somme River valley. These implements were older and somewhat cruder than the neoliths Harrison was collecting. Having learned of the finds of Boucher des Perthes, Harrison himself began to search for similar specimens. These Paleolithic implements, although cruder than Neolithic implements, are still easily recognized as objects of human manufacture. They are thus distinct from Eolithic implements. Modern authorities would assign European Paleolithic tools to the Middle and Late Pleistocene. Harrison looked for paleoliths in ancient deposits of gravel on river terraces, and in 1863 discovered his first paleolith in a gravel pit near Ightham (E. Harrison 1928, p. 46). In addition to searching himself, Harrison trained local workmen to recognize flint implements and collect them for him. Over the years, he amassed a substantial collection of paleoliths.

In 1878, William Davies, a geologist of the British Museum, saw some of Harrison's flint implements and agreed that some of them were paleoliths. Harrison sent a report and some specimens to Sir John Lubbock, who also stated that some of the implements were definitely Paleolithic. G. Worthington Smith, of the Royal Anthropological Institute, visited Ightham and after inspecting the flints initially agreed that some were paleoliths but then later changed his mind (E. Harrison 1928, p. 81).

In 1879, Harrison first met Sir John Prestwich, an eminent geologist, who had a country house eight miles away, at Shoreham. Harrison asked Prestwich some questions about the geological position of the discoveries of Boucher des Perthes in relation to the present level of the Somme River. From Prestwich's window, they could see the Darent River valley. Prestwich said: "If we take the Darent to be the Somme, the gravels would lie at about the level of the railway station." The author of Benjamin Harrison's biography, Sir Edward R. Harrison, wrote (1928, p. 84): "As this remark was made, it flashed through Harrison's mind that some of his own palaeoliths had been found in gravels that were higher, in relation to the level of the streams to which they belonged, than was the level of the railway station in respect to the Darent. Broadly speaking, greater relative height meant greater antiquity, and, consequently, amongst his finds were implements that might be older than those found by Boucher des Perthes in the Somme valley."

To further clarify the matter, let us suppose we have a river running on a level plain a million years ago. As it excavates a channel, it will deposit gravel on the terraces of its banks. As the river descends through the strata, it will deposit more gravels at successively lower levels. In this way, it may be seen that the oldest river gravels, about one million years old, would be found at the higher levels of the valley, while the most recent ones would be found at the lowest levels, on the banks of the present river. The ages of the different levels of gravel are therefore the reverse of the ages of a typical sequence of geological strata, in which the higher strata are the youngest and the lower strata are the oldest. It should, however, be kept in mind, that in actual practice, the assignment of ages to river terraces and gravels is rarely so simple as in this hypothetical illustration.

On September 11, 1880, Harrison made a typical discovery. Sir Edward R. Harrison (1928, p. 87) informs us: "He walked to examine a bed of gravel lying on High Field, at the head of the gorge of the Shode. In this gravel, far above the present level of the stream, he found a palaeolithic implement. His thoughts, on making this discovery, must have been somewhat as follows. The gravel was a very ancient gravel, even in a geological sense, and in it was an implement that had been made by man and carried down afterwards by a stream running at a much higher level than the present stream, to the position in which it was found. So man was older than the very old gravel. Harrison sent news of his find to Prestwich, who came at once to Ightham to see for himself the geological position in which the implement had been found." Prestwich pronounced it a very old bed and advised further research. Prestwich himself and workers under his direction made similar finds.

As word of the newly discovered stone implements spread, James Geikie, one of England's leading geologists, wrote about them on May 2, 1881 to G. Worthington Smith: "They will yet be found in such deposits and at such elevations as will cause the hairs of cautious archaeologists to rise on end. I hope other observers will take a hint from you and search for paleolithic implements in places which have hitherto been looked upon as barren of such relics" (E. Harrison 1928, p. 91).

Geikie's remarks about searching for stone tools "in places which have hitherto been looked upon as barren of such relics" help clarify why modern scientists do not often report finding evidence for a human presence in very ancient times. Because of their preconceptions, they do not look for such evidence in all the places where it might be found. For example, since modern scientists do not accept a fully human presence in the Pliocene, they do not look for advanced stone tools in Pliocene deposits. And if they do find such tools in unexpectedly old deposits, they explain them away. But in the nineteenth century, it was not clear to scientists that they should not be looking for evidence of a human presence in the Pliocene and earlier. So they looked for it, and when they found it, reported it straightforwardly.

In 1887, Harrison read an article by Alfred Russell Wallace on human antiquity in America and then wrote Wallace a letter. Wallace, famous for publishing a scientific paper on evolution by natural selection before Darwin, wrote to Harrison: "I am glad you find my article on 'The Antiquity of Man in America' interesting. It is astonishing the amount of incredulity that still prevails among geologists as to any possible extension of the evidence as to greater antiquity than the paleolithic gravels. The wonderful 'Calaveras skull' has been so persistently ridiculed, from Bret Harte upwards, by persons who know nothing of the real facts, that many American geologists even seem afraid to accept it" (E. Harrison 1928, p. 130).

The Paleolithic gravels referred to by Wallace are equivalent to those of the Somme region, in which Boucher des Perthes found stone tools. These belong to the Middle Pleistocene period of the Quaternary. The Calaveras skull as well as many stone tools were found in far older Tertiary strata in California. The Tertiary includes the Pliocene, Miocene, Oligocene, Eocene, and Paleocene periods. We shall discuss the Calaveras skull and several related discoveries later in this book (Sections 6.2.6, 5.5). The tactic of persistent ridicule mentioned by Wallace was, however, so effective that a good many modern students of paleoanthropology have never even heard of the California finds.

Prestwich and Harrison considered some of the stone implements found near Ightham to be Tertiary in age. The geological reasons for this opinion were discussed by Prestwich in a paper presented to the Geological Society of London in 1889. In preparation for his report, Prestwich asked Harrison to catalog and map his finds. Harrison did so, with the following results: 22 flint implements had been found at elevations over 500 feet, 199 at elevations between 400 and 500 feet, and 184 at elevations under 400 feet, amounting to a total of 405 implements found since 1880 (E. Harrison 1928, p. 129).

In his presentation to the Geological Society, with Harrison sitting in the audience, Prestwich first demonstrated that the higher formations of gravel around Ightham could not have been deposited by the present streams, at any point in their history. He gave evidence showing that the Shode could not have flowed any higher than the 340-foot level (Prestwich 1889, p. 273). Thus the tools in the gravels at elevations over 400 feet must have been quite old, having been deposited by ancient rivers.

This analysis is confirmed by modern authorities. Francis H. Edmunds, in a study published by the Geological Survey of Great Britain, wrote (1954, p. 59): "Occasional patches of gravel, unassociated with any present river system, have been recorded at various localities in the Wealden District. . . . they cap hilly ground and occur usually about 300 ft. above sea level. They consist of a few feet of roughly-bedded flint or chert gravel in a clayey matrix."

Prestwich, having discussed the geological history of the high-level gravels, which he called hill drifts, then dealt with an important question regarding the implements found in them. Could these implements, perhaps of recent origin, have been dropped into the very old hill drift gravels in an age not long past? Prestwich believed that this was true of some of the implements, the Neolithic ones. But along with the Neolithic tools, dropped in the ancient hill drift gravels within the last few thousand years, there were, according to Prestwich, far older Paleolithic tools. These could be distinguished from the Neolithic tools by their deeply stained surfaces and the wear on their edges. Prestwich (1889, p. 283) stated that the paleoliths "exhibit generally the deep uniform staining of brown, yellow, or white, together with the bright patina, resulting from long imbedment in drift-deposits of different characters." In addition, he said that some of the paleoliths were "more or less rolled and worn at the edges by drift-action—some very much so" (Prestwich 1889, p. 283). The neoliths were relatively unstained and unworn.

Sir John Prestwich (1889, p. 286) went on to say about the paleoliths found by Harrison near Ightham: "It is clear from the condition of the implements that, although now occurring on the surface of the ground, they, unlike the neolithic flints, which are unstained and unaltered except by atmospheric agencies, have been imbedded in some matrix which has produced an external change of structure and colour; while the matrix itself, which has been removed by denudation, has nevertheless in several instances left traces on the implements sufficient to indicate its nature."

Describing the remnants of one kind of matrix, Prestwich (1889, p. 289) stated: "a considerable portion of these paleolithic implements are studded on one side with small dark-brown concretionary incrustations of iron peroxide and sand. . . . From this we may infer that both the flint implements and the flints have at one time been imbedded in a sandy, ferruginous matrix, just as the film of calcite on the under side of some of the St. Acheul specimens shows them to come from one of the seams of calcareous sand or chalky gravel common in the drift there, or as the ferruginous concretions on the Dunks Green specimens indicate their origin in that drift."

The identity of the matrix is hinted at by Edmunds (1954, p. 47): "At intervals along the higher parts of the North Downs, and near the crest of the Chalk escarpment, patches of rusty brown sand are present." The hill drifts of the North Downs and the plateau drifts of the Chalk Escarpment are the locations where Harrison found most of his implements. Edmunds (1954, p. 47) further noted: "similar blocks of fossiliferous ironstone or ferruginous sandstone occur on the South Downs near Beachy Head. The fossils have been proved to be of Pliocene age."

"Unfortunately," stated Edmunds (1954, p. 47), "no fossils have been found in the sand resting on the top of the Downs, but their general resemblance to the fossiliferous sandstones . . . leads to the conclusion that they are the remains of an extensive sheet of sands laid down during a marine transgression which is thought to have taken place subsequent to the Miocene." Ferruginous sandstone like that of the South Downs also occurs in the Lenham Beds of the Weald region. Some modern authors (Klein 1973, table 6) date the Lenham Beds to the Early Pliocene or Late Miocene. According to Edmunds, the sandy deposits on the North Downs, the Lenham Beds, and the ferruginous sandstone of the South Downs would all three be of the same Pliocene age.

Granting Edmunds's explanation of the history and age of the iron-stained sands found on the North Downs and Chalk Escarpment, we can consider two hypothetical accounts about how stone implements might have come to be present in them.

The first account involves a Miocene origin for the implements. In the Late Miocene, toolmakers might have left implements on a land surface in the Weald region of southern England, which was later submerged by rising sea levels in the Early Pliocene. The implements were then embedded in marine deposits. Later in the Pliocene, the region again became a land surface, the central portion of which was uplifted (Figure 3.1). Rivers flowing down from the central uplands, in a northerly direction, eroded the ferruginous marine sands. The flint implements and ferruginous sands were deposited in the places where they are now found—as hilltop drifts at very high elevations on the North Downs and as plateau drifts on the Chalk Escarpment (Figure 3.2). During the Pleistocene glacial periods that followed, a different river system carved out valleys and deposited valley drift gravels on terraces below the North Downs hilltops and the Chalk Plateau, with their deposits of sands and gravels from the Pliocene.

Our second account involves a Pliocene origin for the tools. As above, a marine transgression took place in the Early Pliocene, depositing layers of

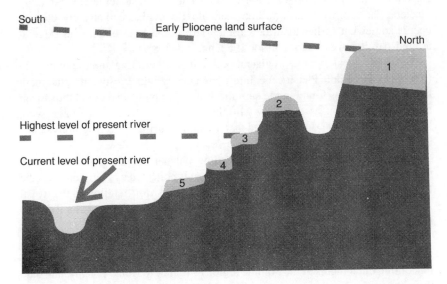

Figure 3.2. The relationships of gravel deposits (drifts) to generalized Weald landscape. (1) Plateau drift deposited by rivers flowing north over the Early Pliocene land surface. (2) Hilltop drifts deposited by a now vanished Late Pliocene river. (3–5) Progressively younger valley drifts deposited by the present river in the Middle and Late Pleistocene.

sediment. Later in the Pliocene, the region again became a land surface, drained by rivers. People living along the banks of these rivers left stone tools, which were transported by the river to their present locations on the North Downs hilltops and the Chalk Plateau. This took place before the present river systems came into being. Embedded in the gravel deposits for long periods of time, the flint implements acquired their coloration and patina. These implements, their edges worn by transport, could not be any younger than the now-vanished northward-flowing rivers. Any implements more recently dropped into these gravels would have remained unrolled and unworn because no water was flowing at that high level. The new rivers were flowing at much lower levels.

How old were the Paleolithic flint implements on the Kent Plateau and in the hilltop drifts? Prestwich (1889, p. 292) concluded: "physiographical changes and the great height of the old chalk plateau, with its 'red clay with flints' and 'southern drift' high above the valleys containing the Postglacial deposits, point to the great antiquity—possibly Preglacial—of the palaeolithic implements found in association with these summit drifts."

According to current opinion, glaciers approached, but did not actually cover the Kent Plateau. The Cromer Till of East Anglia, north of the Kent Plateau, represents the earliest definite geological evidence of glaciation in southern England (Nilsson 1983, pp. 112, 308). A till is a deposit of stones left by retreating glaciers. The Cromer till is .4 million years old. But evidence of an arctic climate occurs somewhat earlier than the Cromer Till, in the Beestonian cold stage at around .6 million years ago (Nilsson 1983, pp. 108, 308).

So strictly speaking, the preglacial period in southern England might be said to begin in the Middle Pleistocene. Interpreted in this light, Prestwich's statement that the implements found in the summit drifts were preglacial could thus mean they were as recent as the early Middle Pleistocene. But, as we have seen, Edmunds (1954, p. 47) has proposed that the summit drifts, the ferruginous sands, are in fact Pliocene in age.

Hugo Obermaier (1924, p. 8), a leading paleoanthropologist of the early twentieth century, stated that the flint implements collected by Harrison from the Kent Plateau "belong to the Middle Pliocene." J. Reid Moir, a fellow of the Royal Anthropological Institute, also referred Harrison's discoveries to the Tertiary (Section 3.3.1).

A Late or Middle Pliocene date for the implements of the Kent Plateau would give them an age of 2–4 million years. Modern paleoanthropologists attribute the Paleolithic implements of the Somme region of France to *Homo erectus,* and date them at just .5–.7 million years ago. The oldest currently recognized implements in England are about .4 million years old (Nilsson 1983, p. 111). So the Paleolithic implements of the Kent Plateau pose a number of difficulties for modern paleoanthropology.

3.2.3 Eoliths

Among the Paleolithic implements collected by Benjamin Harrison from the Kent Plateau were some that appeared to belong to an even more primitive level of culture. These were the eoliths, or dawn stones (Figure 3.3). This name eventually came to be used for a wide variety of very crude stone tool industries from England and other countries.

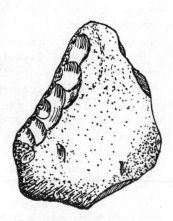

The Paleolithic implements discovered by Harrison, although somewhat crude in appearance, had been extensively worked in order to bring them into definite tool and weapon shapes (Figure 3.4). The Eolithic implements, however, were, as defined by Harrison, natural flint flakes displaying only retouching along the

Figure 3.3. An eolith from the Kent Plateau (Moir 1924, p. 639).

edges. Such tools are still used today by primitive tribal people in various parts of the world, who pick up a stone flake, chip one of the edges, and then use it for a scraper or cutter.

The question then arises as to how such eoliths could be distinguished from broken pieces of flint unmodified by human action. There were, of course, difficulties in making such distinctions, but even modern experts accept lithic assemblages resembling the eoliths collected by Harrison as genuine human artifacts. We shall consider this subject in greater detail in the course of this chapter, but for now we shall mention as an example the crude cobble and flake

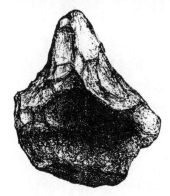

Figure 3.4. These implements from the Kent Chalk Plateau were characterized as paleoliths by Sir John Prestwich (1889, plate 11). Prestwich (1889, p. 294) called the one on the left, from Bower Lane, "a roughly made implement of the spear-head type."

tools of the lower levels of Olduvai Gorge (Figure 3.5). The Olduvai Gorge implements are extremely crude, but to our knowledge, no paleoanthropologists have ever challenged their status as intentionally manufactured objects.

Harrison believed that the Kent eoliths belonged to an older period than that represented by his paleoliths. But in his 1889 report, Sir John Prestwich did not make a distinction between the two forms. Of the eoliths, Sir Edward R. Harrison stated: "Prestwich in his paper made no attempt to claim for them a higher antiquity than that of the Plateau paleoliths, with which they seemed to be associated" (E. Harrison 1928, p. 145). As we have seen, the nature of the drift gravels on the Kent Plateau and the hilltops of the North Downs suggested a Late Pliocene age for the implements.

In the aftermath of Prestwich's presentation, Harrison found himself somewhat of a celebrity. His name appeared in newspapers, and scientists from all parts of the world began to make the pilgrimage to his museum above his grocery shop in Ightham. In June of 1889, the members of the Geological Society of

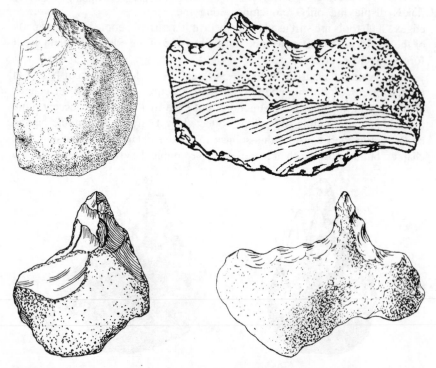

Figure 3.5. Top: Stone implements from Olduvai Gorge (M. Leakey 1971, pp. 45, 113). Bottom: Implements found by Benjamin Harrison on the Kent Plateau, England (Moir 1924, p. 639; E. Harrison 1928, p. 342).

London visited Ightham for a tour of the sites from which the stone implements had been recovered.

Even the considerable authority of Prestwich was, however, not enough to end all controversy regarding Harrison's discoveries, particularly the eoliths. Many scientists still saw in the eoliths nothing but the result of purely natural, rather than artificial forces. Nevertheless, Harrison was gradually winning converts. On September 18, 1889, A. M. Bell, a Fellow of the Geological Society, wrote to Harrison: "I am glad that you saw the veteran Professor [Prestwich], and that his verdict on these unbulbed scrapers coincides with our own. I have looked again and again at the edges of those which I selected, and with an increasing feeling that there is a human purpose dimly visible in the working. There seems to be something more in the uniform though rude chipping than mere accidental attrition would have produced. I have come to this conclusion with diffidence: first, because I had hitherto regarded the bulb or trace of artificial blow as a *sine qua non;* second, and more important, because I feel and have all along felt that the real enemy to such a story as ours is the too enthusiastic friend who *sees what is not there;* but having made my conclusion, I hold it with all firmness. Until I see flints carefully and uniformly chipped all round their edges, and only in one direction of blow, by natural action, I shall believe that these are artificial" (E. Harrison 1928, p. 151).

A modern expert in lithic technology, Leland W. Patterson, also believes it is possible to distinguish even very crude intentional work from natural action. Considering "a typical example of a flake that has damage to its edge as a result of natural causes in a seasonally active stream bed," Patterson (1983, p. 303) stated: "Fractures occur randomly in a bifacial manner. The facets are short, uneven, and steeply transverse across the flake edge. It would be difficult to visualize how random applications of force could create uniform, unidirectional retouch along a significant length of a flake edge. Fortuitous, unifacial damage to an edge generally has no uniform pattern of retouch." Unifacial tools, those with regular chipping confined to one side of a surface, formed a large part of the Eolithic assemblages gathered by Harrison and others.

Prestwich, however, was at first very cautious about the eoliths, feeling more comfortable with the more readily identifiable paleoliths. But gradually he began to change his mind. On September 10, 1890, Harrison and Prestwich were searching the West Yoke ocherous gravels, which were stained red (ocher) by iron compounds. Harrison wrote: "Professor Prestwich was impressed by the great spread of worn gravel, and remarked that it was a 'capital exhibition of ochreous drift in an important position.' At his request I filled my satchel with the water-worn flints, which were scattered over the field in abundance. It was the dawn of the era of the eoliths, for on this day he pressed me to take home specimens that only a few months earlier he would have regarded as too doubtful to be preserved" (E. Harrison 1928, pp. 155–156).

In 1891, Prestwich presented at the Geological Society of London another paper, titled "On the Age, Formation, and Successive Drift-Stages of the Valley of the Darent; with Remarks on the Palaeolithic Implements of the District and on the Origin of its Chalk Escarpment." In this paper, Prestwich (1891, p. 163) described a paleolith found by Harrison in a hole dug for the planting of a tree: "I have now seen the fine specimen. . . . It is 6 inches long by 3¾ in. wide, very flat and round pointed, and shows no wear. It more resembles one of the large St. Acheul types. It was found on the top of the soil last thrown out of the hole." It is not clear what kind of sediments the tool was found in, but the manner in which Prestwich related the find suggests that he regarded it as a demonstration that the paleoliths were to be found not only on the surface, but *in situ.*

In addition to the paleoliths, Prestwich mentioned some of the cruder Eolithic implements. This brought some inquiries from William Topley, a fellow of the Geological Society and the author of a Geological Survey memoir on the Weald region. Harrison wrote in his diaries: "Mr. William Topley at the reading of the Darent paper said that he wished to know if there was any clear case of the flints being found in place. He added that the antiquity of the gravels in such an elevation [on the Plateau] was beyond question and certainly preceded the excavation of the great Chalk valleys and the present features of the Weald. In consequence of these remarks I went to the Vigo inn, and searched in and near the post holes dug for a fence. I found worked stones and thus recorded my first finds *in situ*" (E. Harrison 1928, p. 161). Thus the eoliths as well as paleoliths were to be found within the earth, and not just on the surface.

Harrison also noted that in most cases his eoliths occurred in places where there were no paleoliths. To him, this indicated a different age for the two types of implements.

A. R. Wallace, who was greatly interested in Harrison's finds, asked him for a copy of Prestwich's Darent paper. Harrison forwarded the paper to Wallace, who later replied: "I read Mr. Prestwich's paper with great interest, especially with regard to the rude type of implements, which I had never seen represented before. They are certainly very distinct from the well-formed palaeolithic weapons, and their having a separate area of distribution is strong proof of their belonging to a different and earlier period" (E. Harrison 1928, p. 370).

3.2.4 More on The Geology of the Kent Plateau

In 1891, Sir John Prestwich presented a third major paper on the stone implements of the Kent Plateau. In this paper, delivered to the Royal Anthropological Institute, Prestwich pointed out that the Chalk Plateau of Kent, where Harrison found paleoliths and eoliths, is bounded by a large valley running across its southern border. According to Prestwich, this valley was scooped out by water action during the glacial period. The Kent Plateau, however, contained drift gravels like those present on the South Downs, the hills that still exist on the other side of the southern valley. Prestwich (1892, p. 250) stated: "as the flint implements are closely associated with this plateau drift, and are limited to the area over which it extends, we are led to infer the pre-glacial or early glacial age of the men by whom they were fabricated." Just to clarify the reasoning, let us imagine ourselves in the Late Pliocene, looking south from the present North Downs and Kent Plateau. Instead of the valley now there, we would see the rising surface of the Weald dome (Figure 3.1, p. 88). At this time, according to Prestwich, the now-vanished dome uplands would have been inhabited by humans who made crude stone tools. Rivers and streams running down from the uplands flow north, depositing their gravels and sediments, along with stone tools, on the surface of the region now occupied by the North Downs and Kent Plateau. The rivers also flow south from the divide of the central dome uplands, to the South Downs.

This process continues until the Pleistocene, a time of increased precipitation. Torrents of water flowing along an east-west axis, carve out a large valley where the Weald uplands once rose. Now the landscape is considerably changed, leaving the Kent Plateau and hills in the north separated by a deep, wide valley from hills to the south. At this point, the rivers no longer flow onto the plateau, but rather empty into the valley. But the old gravels and sediments, containing eoliths, remain on the Kent Plateau surface. They could only have been deposited there before the excavation of the valley. The proof of the accuracy of this scenario: the gravels and sediments found today on the Kent Plateau surface greatly resemble those found on the South Downs, now separated from the Kent Plateau by the great transverse valley. As we have seen, Edmunds (1954, p. 47) has identified the ferruginous deposits topping the North Downs with those now found in the South Downs. Since certain kinds of tools were found only in the ferruginous gravels and other such deposits on the North Downs and Kent Plateau, Prestwich concluded that these tools were made by the humans who lived on the central dome uplands, before the glacial period.

Modern authorities relate the geological history of the rivers of the Weald region and their gravel deposits in much the same way as outlined above. For example, Francis H. Edmunds, in a study published by the Geological Survey of Great Britain, wrote (1954, p. 69): "The original rivers of the Wealden district . . . flowed either northward or southward from an east-to-west watershed along the main axial line of the Weald." These rivers left north-south gaps in the Weald landscape, some of which are not used by the present river systems. Edmunds (1954, p. 63) stated: "Certain physical features, notably the position of the river gaps through the North and South Downs, connect modern topography with that of the pre-Pliocene epoch." A map by Edmunds (1954, p. 71) shows the Plateau gravels as having been deposited by the rivers flowing from south to north. This tends to confirm the views of Prestwich, who believed the Plateau gravels were laid down by rivers flowing north from the central dome uplands during the Pliocene and perhaps the preglacial Pleistocene.

Concerning the Plateau deposits (Clay-with-flints), Edmunds thought some were produced locally by dissolution of the underlying chalk formations, which contain flint. But Edmunds (1954, p. 56) added: "The Clay-with-flints in several Wealden localities, however, contains a major proportion of material which could not have been so derived, but which represents *remainié* Tertiary beds, of Eocene and Pliocene ages."

This suggests that the worn and patinated eoliths (and paleoliths) found in the Plateau deposits could very well be of Tertiary age.

Maps supplied by Edmunds (1954, p. 71) show that the north-south river systems, which laid down the Tertiary Plateau gravels and the hill drifts, were later diverted into their present east-west channels. These east-west rivers deposited the Pleistocene gravels on terraces below the hill drifts, the higher terraces being the oldest (Figure 3.2, p. 93). This process of gravel deposition began during the glacial period.

The stone implements found in the higher terrace gravels of the present rivers were, according to Prestwich, similar to the Paleolithic implements encountered in the Somme region of France, where Boucher des Perthes conducted his investigations. In his address to the Anthropological Institute, Prestwich explained that in the Kent Plateau region Neolithic implements were mainly found in the lower, more recent, river beds along with fossil remains of mammoth, woolly rhinoceros, reindeer, and other Ice Age mammals.

To summarize, the eoliths were found mainly in the Pliocene drift gravels on the top of the Plateau, crude paleoliths mainly in the hilltop drifts of Pliocene rivers, better paleoliths mainly in the Pleistocene higher gravels of the present rivers, and polished neoliths in the lower more recent river gravels.

Most of the high Plateau discoveries were surface finds. But Prestwich (1892, p. 251) noted that "from the deep staining of the implements, and their occasional incrustations with iron oxide, we have reason to believe that they have been imbedded in a deposit beneath the surface." This is significant. If the implements were embedded beneath the surface of the now-vanished dome uplands for a long time before they were transported to the Plateau, that would indicate an indefinitely great age for them. In other words, they were at least Late Pliocene in age, and perhaps far older.

Some of the Plateau implements were found not on the surface but *in situ* deep within the preglacial Plateau drift gravels. This would tend to rule out the supposition that the implements were of fairly recent origin and had been dropped on the drift gravels by the later inhabitants of the Plateau region. Prestwich (1892, p. 251) stated: "A fine specimen was found at South Ash in making a hole two feet deep for planting a tree, but as it was picked up on the thrown-out soil, its exact position beneath the surface remains of course uncertain. It was the same with the one obtained in a post-hole at Kingsdown. For two others we have, however, the personal testimony of Mr. Harrison. One he took out of a bank of the red-clay-with-flints on the side of a pond and at the depth of two and a half feet, and the other from a bed of 'deep red clay,' two feet in depth, at the Vigo."

In a footnote to the above passage, Prestwich (1892, p. 251) went on to say: "Mr. Bullen has just had a trench dug on the top of Preston Hill. It was nearly five feet deep, through surface soil (one foot); and the red-clay-with-flints in which, at a depth of three feet ten inches from the surface, he found an unworn white flint—apparently the broken point of a small implement." As we have seen, Edmunds (1954, p. 56) characterized major portions of the clay-with-flints deposits as *remainié* beds of Tertiary age, some Pliocene, some even Eocene.

3.2.5 The Relative Antiquity of Eoliths and Paleoliths

Returning to the eoliths found on the surface of the Plateau, Prestwich (1892, p. 252) asked: "could these implements, like the neolithic implements which occur on the same ground, have been dropped on the surface where they are now found, at some later date?" Although most of the Neolithic implements were found in the lower river terraces, some did occur on the Plateau. Prestwich (1892, p. 252) went on to state, in response to his own question: "The answer to this is, that these neolithic implements show only weathering by exposure on the surface, and are found at all levels, whereas the plateau implements, besides their wear and colour, present all the physical characteristics due to having been *imbedded in a special drift*, and *are confined to a special area*. The two sorts, although found on the same ground, remain perfectly distinguishable."

Prestwich (1892, p. 252) then gave an extensive answer to an objection raised by Sir John Evans: "Then again, is it not possible that similar rude specimens occur in the valley drifts, and have been overlooked owing to the prevalence of the better finished implements to which attention had been exclusively given." If eoliths were found in connection with the paleoliths or neoliths in the valleys, that might weaken Prestwich's argument for their great age, which was based on the fact that they tended to be found only in the very ancient Plateau drifts. Prestwich (1892, p. 252) answered as follows: "A large number of rude and badly finished specimens have been collected in the valley drifts, but they all belong to one set of types, and though I have seen and handled many hundreds of these, I question whether, with the exception of the *derived* specimens [those washed down from the Plateau] to be named presently, there were any like the ruder and most primitive of the plateau types. The distinction is as well marked as that between the ruder specimens of Roman pottery and rude early British pottery."

Prestwich (1892, p. 252) went on to state: "Boucher de Perthes collected everything in the Somme district, which showed any traces of workmanship, howsoever indistinct, or even of similitude, yet I do not remember that in his great collection there were specimens of the peculiar character of these plateau implements." In other words, the evidence from the Somme region confirmed Prestwich's hypothesis that the Kent Plateau eoliths were of a distinct type, different from superficially similar crude implements of later periods. In a footnote, Prestwich (1892, p. 252) added: "I do have one specimen given me by M. Boucher de Perthes, from near St. Riquier, five miles north-east of Abbeville, which may belong to this group. It is said to have been found at a depth of four metres [about 13 feet], and evidently comes from the red clay drift, which there caps, as it does here, the higher chalk hills. It is four inches long by one and half inches wide, rod-shaped, very roughly chipped all around and at ends, and has a white patina, to which some of the red clay as yet adheres." This discovery would appear to be well worth looking into, and is representative of the intriguing items one comes across in old journal articles. It might represent a stone implement far older than the others discovered by Boucher de Perthes in the river gravels of the Somme valley at Abbeville, now dated to the Middle Pleistocene, about a half million years old.

After giving testimony about not finding specimens like the Plateau eoliths in Boucher des Perthes's collection, Prestwich (1892, pp. 252–253) stated: "Nor had Mr. Harrison, during his rigorous examination of the Shode Valley, discovered any specimens in the valley drifts of the Ightham district to correspond with the group of plateau implements. At my request, he has re-examined several of these localities, as well as the large pit at Aylesford in the Medway Valley, and the pits at Milton Street (Swanscombe) in the Thames Valley, with this special object in view. He reports to me that he finds no contemporary specimens of the plateau type, and very few derived specimens of that type."

Prestwich (1892, p. 268) then cited evidence from De Barri Crawshay, who stated: "I find that on examination of my collection of over 200 specimens of implements and scrapers from the 100 foot level around Swanscombe, Kent, I have but one . . . which is a plateau specimen undoubtedly derived. . . . I have always made specially careful search for all these ochreous flints in the low level gravels, and have rarely found one at all."

Derived specimens are those washed down from the Plateau and left in the lower level gravels. Prestwich (1892, p. 253) stated: "The derived plateau specimens are easily distinguished, by their greater wear, distinct colour, and peculiar shapes, from the implements contemporary with these valley drifts."

The valley Paleolithic specimens were very extensively worked, with fine, regular chipping, and generally took the form of points meant, perhaps, to be used as spear heads. There were some crude, unfinished specimens among them, but they were obviously of the same type as the finished paleoliths, and not of the Plateau type (Prestwich 1892, p. 255).

About the Plateau eoliths, Prestwich (1892, p. 256) stated: "The trimming slight though it may be, is to be recognised by its being at angles or in places incompatible with river drift agencies, and such as could not have been produced by natural causes." Prestwich admitted that some specimens resembling the more advanced valley paleoliths were found along with the Plateau eoliths, and stated (1892, p. 257): "It is not easy to account for the presence of these abnormal specimens. If contemporaneous with the others, we might assume that there were then some workmen more skilled than their neighbors in the fabrication of flint implements." Working against this hypothesis, according to Prestwich, was the fact that the rude Eolithic specimens were heavily patinated and were very worn, whereas the finished Paleolithic specimens were unpatinated and had perfectly sharp edges. Prestwich surmised the latter might have been left on the Plateau by Paleolithic men in more recent times, long after the eoliths had been deposited. Prestwich (1892, p. 258) then made a very important observation: "Though the work on the plateau implements is often so slight as scarcely to be recognisable, even modern savage work, such as exhibited for example by the stone implements of the Australian natives, show, when divested of their mounting, an amount of work no greater or more distinct, than do these early palaeolithic specimens." This implies that it is not necessary to attribute the Plateau eoliths to a primitive race of ape-men. Since the eoliths are practically identical to stone tools made by *Homo sapiens sapiens,* there is no reason to rule out, *a priori,* the possibility that the eoliths (and the paleoliths) may have been made by humans of the fully modern type in England during the Late Pliocene. As we shall demonstrate later on (Section 6.2), scientists of the nineteenth century made several discoveries of skeletal remains of anatomically modern human beings in strata of Pliocene age.

In the discussion that followed Prestwich's presentation of his report, Sir John Evans repeated his point that the presence in the Plateau drift gravels of paleoliths made it possible that eoliths were contemporary with them and thus more recent than Prestwich and Harrison believed (Prestwich 1892, p. 271). Years later Harrison wrote in a letter, dated June 3, 1908, to W. M. Newton: "At the meeting of the Anthropological Institute in 1891, Dr. Evans closed his observations with the following sentence, 'Before we accept these' [the Eolithic implements]—looking at Prestwich—'we must think twice,'—looking at me—'we must think thrice, and'—looking round the whole meeting—'we must think again'" (E. Harrison 1928, p. 165).

Other members of the Anthropological Institute also commented. General Pitt-Rivers maintained that stones resembling the eoliths were to be found in all gravels, insinuating that eoliths were simply a product of purely natural forces (Prestwich 1892, p. 272). In support of Prestwich, J. Allen Brown reported that some flints from the upper terraces of the Thames River resembled the Ightham ones, and might be of the same age and origin (Prestwich 1892, p. 275). The journal of the Anthropological Institute recorded a summary of Prestwich's concluding remarks: "In reply, Professor Prestwich said that he had looked forward to the possibility of there being some substantial objections to his views which might have escaped him. He had, however, heard nothing but an amplified repetition of the very same difficulties which had occurred to him, and had been discussed and explained in the paper" (Prestwich 1892, p. 275).

Careful study of the report bears out Prestwich's statement. With regard to the doubts of General Pitt-Rivers, Prestwich had already demonstrated that the chipping on the eoliths was quite different from that produced by purely natural forces on river gravels. He had also offered explanations for the presence of both paleoliths and eoliths in the Plateau gravels, explaining that some of the paleoliths, which were sharp and unworn, had probably been introduced into the Plateau gravels at a much later period than the deeply stained and much worn eoliths.

Sir Edward R. Harrison (1928, p. 166) gave a summary of the three papers presented by Prestwich: "The first paper opened up the subject of Harrison's discoveries by describing the palaeolithic implements found around Ightham in the post-glacial valley gravels, in the glacial high-level gravels, and in the very ancient, pre-glacial gravels of the high Chalk Plateau. . . . The second paper, on the drift stages of the Darent valley, added to the evidence contained in the Ightham paper. . . . The third paper was directed to the character of the rude implements, the nature of the chipping upon their edges, the classification of the specimens in groups representing different kinds of tools, and the other reasons that existed for attributing them to the hand of man." In light of Prestwich's testimony, it is remarkable that most modern studies of stone implements generally do not mention Harrison's eoliths, and those few that do give only brief, highly critical, and often sarcastic notices of dismissal.

3.2.6 A. R. Wallace Visits Harrison

On November 2, 1891, Alfred Russell Wallace, who was at that time one of the world's most famous scientists, paid an unannounced visit to Benjamin Harrison at his grocery shop in Ightham. Harrison recorded the incident in his notebooks: "Dr. A. R. Wallace, accompanied by Mr. Swinton of Sevenoaks, dropped in unexpectedly at 10.30. I had previously purchased Dr. Wallace's

Travels on the Amazon, and from his portrait, which forms the frontispiece to this work, I recognized him before he entered my shop. I therefore greeted him with 'Dr. Wallace, I presume,' a recognition which puzzled him until I explained that I had many times studied his portrait. This evidently pleased him. A long and patient examination was made of the old types of implement and of some later paleoliths" (E. Harrison 1928, p. 169). Harrison then took Wallace on a walking tour of the sites where the implements had been found.

Harrison also noted: "When I was showing him my rude implements and placing them in groups, he asked, 'Was it not a pleasure to you to find such agreement in form and work when first you became certain of them?' I answered that it was a supreme time. . . . Our conversation turned to the subject of the new and startling find of implements in the auriferous gravels of North America, startling in the fact that although their positions indicated a high antiquity, yet their forms were similar to those of implements in use by the Indians at the time of the discovery of the continent in the fifteenth century" (E. Harrison 1928, pp. 169–170). The stone implements from the auriferous, or gold bearing, gravels were of Neolithic type (Section 5.5). As we shall show, they provided evidence for the presence of humans of the modern type in the very early Pliocene, or perhaps even as far back as the Eocene.

The day following his visit to Ightham, Wallace wrote in a letter to Harrison: "I was very greatly interested in your collection of the *oldest* paleoliths. Could you not write a popular article giving an account of your discovery of them, with all the main features of their form and peculiarities, and the special areas in which they are found, illustrated by outline sketches of all the chief types of form, and laying particular stress on the fact that each of these *types,* however made, is illustrated by numbers of specimens showing how natural flint pebbles of suitable form have been selected, and by being chipped on one side only, have been brought to the required shape and edge? If you could write as you speak, I think such a paper would be published by one of the good reviews" (E. Harrison 1928, p. 171). Harrison did not write such an article immediately, but, according to Sir Edward Harrison, in 1904 he published a pamphlet along the lines suggested by Wallace.

On March 14, 1892, the noted Scottish geologist Sir Archibald Geikie wrote to Benjamin Harrison about the paper presented by Prestwich at the Anthropological Institute: "I was delighted to receive a copy of Mr. Prestwich's paper [on eoliths] a few days ago, and to read his account of your very successful investigations. It is a strange tale which these implements tell, and you may be congratulated on the successful result of your long and laborious, but, no doubt, very interesting quest. Yes, paleolithic man is *old.* . . . I am at present preparing a work the object of which is to show the results of glacial and archaeological researches into the antiquity of man which have been obtained up to the present time. The more one investigates the question, the further into the past does paleolithic man seem to recede" (E. Harrison 1928, p. 175).

3.2.7 More Objections

Worthington G. Smith, repeating a common objection, wrote to Harrison on March 26, 1892: "It appears to me that the importance of your discovery of implements rests on your lighting on *genuine undoubted* examples on the high levels. I don't attach much importance myself to the dubious and disputed forms [the eoliths], because such forms occur with genuine implements in all paleolithic gravels. The very rudest forms can never mean anything, unless such forms are exclusive, and pertain only to certain deposits" (E. Harrison 1928, p. 175). Here Smith appears to have ignored all the evidence amassed by Prestwich for the greater antiquity of the Plateau eoliths, even when found in association with more advanced Paleolithic types. Among other things, Prestwich repeatedly emphasized that the eoliths, and some of the paleoliths, are very much worn and patinated whereas other paleoliths and neoliths retain the original color of the flint and have sharp edges.

All that aside, however, it appears that Harrison did find locations in which the eoliths occur by themselves. Sir Edward R. Harrison (1928, p. 176) has stated of the eoliths: "Harrison was influenced principally by their rude character, and he thought it likely that they were, for that reason, the tools of a race older than paleolithic man. Subsequently, when excavations had been made in the drifts, he found confirmation of his views in the fact that whilst certain drifts produced occasional paleoliths in apparent association with rude implements, there was also on Parsonage Farm and elsewhere, an older drift or 'buried channel' which, in his experience, contained rude implements alone."

Of course, the fact that the eoliths are sometimes found by themselves had already been reported by Prestwich. All this reveals much about scientific discussion concerning anomalous evidence. Scientists whose preconceptions dispose them to reject certain evidence often tend to repeat their objections even after they have been met with apparently adequate responses, as if the response had never been made. Doctrinaire scientists also set conditions they believe should be met, even when such conditions have already been met. All of this makes for an Alice-in-Wonderland type of discourse: "My dear sir, I have found crudely chipped stone tools alone." "Well sir, I really think you should find these chipped stone tools alone." "But I have sir." "Then you very well should do so, or I shall never believe you." Or: "Dear sir, let me demonstrate how this set of stone tools is older than this other set." "Very well, but I really think you should now demonstrate that this set of tools is older than the other set." "But I already have." "Yes, but you should do it, and until you do so, I shall never believe you."

Sir John B. Evans provides a good example of this kind of interchange. Evans wrote to Harrison on October 29, 1892: "A certain number of flints, such, for instance, as several from Ash, are to my mind undoubtedly fashioned by man; there are others which probably have been worked, and others again which possibly have had their edges retouched. The great majority, however, seem to me to have assumed their present forms by natural agency. . . . When the more perfect implements are found with these ruder forms, there is no reason for regarding them as otherwise than contemporary . . . everyone will accept the ordinary forms of paleolithic implements as having been found at the high levels, and I am doubtful as to the desirability of complicating the question with a second race of men and a set of implements of extremely questionable character" (E. Harrison 1928, p. 184). Here Evans admitted that some of the rude implements display signs of human work. If he admitted that some, however few, were the result of human work, this conclusion was not nullified by the fact that the "great majority" appeared to have been the result of natural action. As for the relative ages of the eoliths and the paleoliths, he appears to have either missed or deliberately ignored all the evidence suggesting that the Eolithic implements could have been more ancient.

A troubled Harrison wrote to Prestwich, who replied on November 15, 1892: "No explanation necessary. Your collection stands on its merits. Differences of opinion there will always be. All you have to say is that Sir John Evans accepts some specimens and rejects others. Let everyone judge for himself" (E. Harrison 1928, p. 185).

Despite the continuing controversy, the British Museum still thought enough of the eoliths to purchase, in 1893, a set of representative specimens (E. Harrison 1928, p. 186). Harrison, meanwhile, continued his investigations, with the special intention of proving that the eoliths occurred not in all gravels, as some critics asserted, but only in special locations, in the very old Pliocene drift. In many gravel deposits around Ightham, Harrison noted the complete absence of any stones resembling his Eolithic implements. For example, Harrison's notebook entry for September 3, 1893 read: "To Fane Hill—a long search, but not a single specimen of old old work." Sir Edward R. Harrison (1928, p. 188) stated: "This negative evidence confirmed Harrison in his opinion that the eoliths had been artificially chipped. Had they been merely the work of natural forces it was to be expected that they would be found in large numbers in all flint-bearing gravels alike."

For years, Harrison's eoliths continued to be a topic of serious discussion in scientific societies, including the British Association for the Advancement of Science. Sir Edward R. Harrison (1928, p. 192) wrote: "A. M. Bell championed the cause of the rude implements at the meeting of the British Association at Edinburgh in 1892. It fell to Professor T. Rupert Jones to undertake a like service in 1894, when the meeting was held at Oxford." The 1894 meeting was, according to A. M. Bell, who wrote to Harrison on August 10, 1894, "not a triumph . . . not a defeat, but leaves things much as they were" (E. Harrison 1928, p. 193).

3.2.8 The British Association Sponsors Excavations

In order to resolve the controversy over the age of the eoliths, the British Association, a prestigious scientific society, financed excavations in the high-level Plateau drift and other localities in close proximity to Ightham (E. Harrison 1928, p. 194). The purpose was to show definitively that eoliths were to be found not only on the surface but *in situ*, deep within the Pliocene preglacial gravels. Alfred R. Wallace had also expressed a desire for such proof, having written to Harrison on November 8, 1893: "I suppose you have not found any of your old flints yet, *in situ* by digging, or in the undisturbed gravel at some distance below the surface. When you do that you will have more converts" (E. Harrison 1928, p. 189). It would appear that Harrison had already found some eoliths *in situ* (such as the ones from the post holes dug near Vigo Inn, see Section 3.2.4), but this excavation, financed by the respected British Association, would be more conclusive.

It should be noted that many accepted flint industries were initially discovered on the surface. For example, John Gowlett (1984, p. 72) described the finds at Olorgesailie, in Kenya: "Hand axes were found weathering out on the surface by Louis and Mary Leakey, and it soon became evident that this was one of the major Middle Pleistocene localities of East Africa." Today there is an open-air museum at Olorgesailie, where visitors may walk on catwalks above a land surface covered with stone implements. A similar situation is found at Kilombe in the Kenya rift valley. Gowlett (1984, p. 68) stated: "Kilombe is a massive Acheulean site in Kenya. Artifacts on this site were first noticed in 1972 by geologist Dr. W. B. Jones as an extensive scatter on the surface, evidently weathered out from nearby Pleistocene beds." Describing the Kilombe hand axes, which were made from flakes of stone, Gowlett (1984, p. 70) stated: "many of these large flakes were only gently retrimmed in the final shaping and the original form is quite apparent." The Kilombe flake implements, with only slight human modification, conform to the description of eoliths. At both Kilombe and Olorgesailie, stone implements were later found *in situ*. The same was true of the sites on the Kent Plateau.

The British Association selected Harrison himself to supervise the Plateau excavations, under the direction of a committee of scientists. Harrison recorded in his notebooks that he found many examples of eoliths *in situ*, including "thirty convincers" (E. Harrison 1928, p. 189).

3.2.9 The Royal Society Exhibition

In 1895, the same year that the Geological Society of London awarded him part of the Lyell Fund (E. Harrison 1928, p. 196), Harrison was invited to exhibit his eoliths at a meeting of the Royal Society. He was quite pleased to have the chance to show his specimens to this scientific elite (E. Harrison 1928, p. 197). Sir Edward Harrison (1928, p. 197) stated: "This was an opportunity not to be missed, and he informed Prestwich of his intention to send for exhibition the specimens found in situ in the excavation in the drift at Parsonage Farm. Prestwich did not dissent from this proposal, but he advised the exhibition also of carefully selected surface specimens, arranged in groups. Harrison followed this counsel in the main, but he included too large a proportion of specimens from the pit, and amongst them specimens which did not impress those who saw them so much as he had hoped."

Some scientists, however, were quite impressed, among them E. T. Newton, a Fellow of the Royal Society and paleontologist of the Geological Survey of Great Britain, who wrote to Harrison on December 24, 1895: "I hope you will not mind your specimens remaining with me until after the Christmas holidays. I feel satisfied that most of them, to say the least, show human work, and some of these are definitely from one of the pits. . . . Some of the specimens I should be very doubtful about, but there are others that I cannot bring myself to believe are accidental; they have been done intentionally, and, therefore, by the only intellectual being we know of, *Man*" (E. Harrison 1928, p. 202). Here we have an example of a qualified scientist fully accepting as genuine human artifacts some of the eoliths excavated from the Pliocene Plateau drifts. Modern authorities, who have never examined the specimens in question, might thus be cautious of prematurely dismissing them.

3.2.10 The Problem of Forgery

Of course, recognizing intentional human work is always beset with many difficulties, and in his notebooks Harrison mentioned one of the most vexing—forgery. On March 26, 1896, Harrison was visited by William Cunnington, a Fellow of the Geological Society. Harrison wrote: "He was well acquainted with Flint Jack, the notorious forger of flint implements. Flint Jack's first appearance was characteristic. He entered Mr. Cunnington's office, and, taking from his pocket some flints wrapped in paper, said, 'I hear you buy flint arrowheads.' 'You are Flint Jack.' 'Yes,' he replied, 'I am, and as I was passing I thought you would like to see some arrow-heads made!' On one occasion Mr. Cunnington set him up in life and gave him decent clothing, hoping to reform him, but in vain. Mr. Cunnington sent him to Farmingham to get some fossils. On his return he produced a stone which he said he had bought for a shilling from a shepherd. Recognizing at once that the stone was a forgery, Mr. Cunnington accused him of making it, and refused to have anything more to do with him. The forged implement was made of sandstone. Flint Jack had shaped it with a pick, and had afterwards rubbed it over with earth to disguise its new appearance" (E. Harrison 1928, p. 205).

Harrison was not without his own experience of forgers. In his notebook entry for May 29, 1894, he stated that Smith, an Ightham laborer, had told him: "When Seldon and I were working on the railway he said to me, 'I wonder whether we shall find any flints for Mr. Harrison.' We did not find any of the right sort, not your sort, you know. He said, 'Here's a big 'un. I'll take him home and hammer him up a bit, file him, and make him look like one of the right sort.' When he brought it to you, he thought you would not know it, but would think it was one of the right sort. He asked you if it was one of the right sort, and you said, 'This is one of your own make, Seldon.' Seldon said, 'I thought he would not know, but I was too tricky, he know'd it. Its no use taking home-made ones to him, he knows too much. But he give me some tobacco for being tricky'" (E. Harrison 1928, p. 195). It should, however, be noted that it is not only laborers who are responsible for forgery. As we shall see, in the case of Piltdown man the finger of guilt points in the direction of the scientists themselves.

3.2.11 "The Greater Antiquity of Man"

In 1895, Sir John Prestwich published in *Nineteenth Century*, a popular magazine directed at the intelligent public, a review of the Ightham implements titled "The Greater Antiquity of Man." Since it gives, in layman's terms, an excellent summary of the scientific issues involved in the eolith question, we shall here reproduce some sections of the article.

Prestwich (1895, p. 621) first described the Kent Plateau, as it existed in the Pliocene epoch: "It then, was a high level plain of chalk covered by argillaceous [clayey] and drift beds, which thus became furrowed by the escaping rainfall; and as the furrows gradually deepened they ended in the formation of the existing chalk valleys. It will, therefore, be seen that these valleys must be newer than the hills through which they are cut, and consequently that the beds of sand and gravel, with the remains of extinct mammalia, together with the flint implements of Palaeolithic man, found in these valleys, must also be newer than the drift scattered on the summit of those hills."

It was in the Plateau drift that the eoliths were to be found. Prestwich distinguished them from Paleolithic implements. Paleolithic implements were very elaborately worked into recognizable tool and weapon forms. Describing the much more crudely fashioned eoliths, Prestwich (1895, p. 622) wrote: "Other scrapers have been formed out of split Tertiary flint-pebbles, sometimes split naturally, and at other times artificially. The edges are trimmed generally all round, so as to act as a rough scraper in whatsoever position the pebble may best be held. At the present day a similar practice prevails among some North American Indians, who, whenever in want of a scraper, select a pebble, which they split and then trim the edges. They rarely keep the old scraper, fresh ones being so easily obtained. This tool is called a *pashoa,* or scraper, and is used by the Shoshone Indians to dress skins."

Prestwich then pointed out that these rude Eolithic implements from the Pliocene Plateau drifts had features that distinguished them from rude implements that might be found in more recent deposits. "But says one critic, rudeness of form is no test of age, and leaves it to be inferred that these specimens are no older than other rude forms of later ages. Who of the advocates of the plateau implements ever said that it was? I know of none. We particularly remarked in 1892 that rudeness of form alone was no proof of antiquity, and that there were plenty of very rude specimens of the valley types. We would again emphasise the fact that there are rude implements not only of the valley gravels, but also of neolithic times, whilst among the stone implements of living savages there are many as rude as those of the plateau group" (Prestwich 1895, p. 624).

Prestwich (1895, p. 624) went on to say: "Each epoch had, however, its typical forms, and these are broadly persistent, howsoever rude the specimens may be. In the neolithic period axe and chisel shapes predominate; in the valley gravels the long pointed and spatula-shaped implements are characteristic of the period; and in the plateau group various forms for scraping and hammering prevail. There are, no doubt, pointed forms in the plateau group, but they have a different *cachet* from those of the valley group, as these again differ from those of the subsequent Stone period. There are, besides, certain generalised forms which persist throughout all the periods, though perhaps varying a little in some minor details. Simple flakes likewise, more or less worked, are found in all three periods."

Prestwich then pointed out that many Eolithic implements had been found not on the surface but in excavations into the drift deposits. Of these drift deposits on the Plateau, Prestwich (1895, p. 624) stated: "The drift on that surface is certainly not of local origin, as is shown by the presence in it of fragments of strata derived from the hills some miles distant to the south." As previously noted, the drift could only have arrived in its present position on the Plateau before the chalk valleys, which now intervene between the Plateau and the southern hills, were excavated.

Answering the charge that the eoliths were perhaps naturefacts rather than artifacts, Prestwich (1895, p. 625) stated: "It has also been frequently asserted that these implements are natural forms produced by the friction of the shingle on the shore or in the beds of rivers. Challenged to show any such natural specimens, those who have made the assertion have been unable, although nearly three years have elapsed since the challenge was given, to bring forward a single such specimen. If, moreover, implements were formed in that manner, they should be found in gravel beds of all ages and origins. So far from running water having this constructive power, the tendency of it is to wear off all angles, and reduce the flint to a more or less rounded pebble."

So here one of Britain's foremost geologists, a Fellow of the Geological Society, and a Fellow of the Royal Society, made quite a coherent case for the human origin and Pliocene date of the Eolithic implements collected by Benjamin Harrison. He answered in a convincing manner all possible objections to his interpretation. Of course, some scientists maintained their opposition, as might be expected of persons with strongly held beliefs. Nevertheless, we must still wonder why, as far as modern paleoanthropology is concerned, the Plateau eoliths have completely disappeared from view. Apparently there is no place in the modern views on human origins for toolmaking hominids in England at least 2–4 million years ago in the Pliocene period.

3.2.12 On the Treatment of Anomalous Evidence

In 1896, Prestwich died, but Harrison, in his prominent patron's absence, continued with the Plateau excavations and answered the doubters. On May 18, 1898, Harrison wrote to W. J. Lewis Abbott, reproducing in his letter a poem called "That Chocolate Stone," written by his son (E. Harrison 1928, p. 219):

> *If only that chocolate stone could explain what the dickens it did in the past,*
> *That those sages might cease from exciting the brain, and the hatchet be buried at*
>   *last,*
> *Whether eolith, neolith, nature, or man, could they but of that question dispose,*
> *Why, those eminent men might relinquish the pen—till a new controversy arose.*

This verse, light and humorous though it may be, strikes at the very heart of an important epistemological consideration. In the absence of direct knowledge of the past, any discussion of paleoanthropological evidence, which is always somewhat ambiguous, is certain to involve controversy, because of the differing preconceptions and methods of analysis of the participants in the debate. Empiricism thus becomes inextricably entangled with speculative modes of thought and deeply held emotional biases and prejudices. In most cases, the speculation and bias are carefully masked with a thin veneer of fact. But as imperfect as this process may be, it is, for scientists, the only one that can be applied; therefore, one can at least insist on consistent application of principles and close reasoning from the observed facts. This granted, the case made by Prestwich and Harrison held up quite well against the arguments thrown by their opponents, who simply seemed to be searching for ways to reject something they were *a priori* not prepared to accept.

An interesting example of this may be found in G. Worthington Smith's continued opposition to Harrison's eoliths. On March 22, 1899, Benjamin Harrison wrote in a letter to Sir Edward R. Harrison (1928, p. 224): "After I became acquainted with Mr. Worthington Smith in 1878, he from time to time sent me interesting trifles, which were duly marked and placed in a drawer. In going through this lot yesterday, I came upon some interesting rude specimens from Basuto Land. These are about as rude as can be, and are facsimiles of those now found in Bushmen's caves in Central Africa. They feature [resemble] my rude implements. Strange that Smith classes all my Plateau finds [eoliths] as cretins, make-beliefs, casuals, travesties—anything but human made. And yet, as long ago as 1880, he sent me those then-acknowledged stones, as if to encourage me to look for similar specimens. When I find them, he scouts [rejects] them!" Here we have an apparent instance of inconsistent application of principles on the part of Smith.

Harrison wrote to Smith about this, who replied, in a somewhat humorous tone, on March 23, 1899, that although he vaguely recalled perhaps having sent some flakes and stones, he failed to see what bearing they had on the present question: "I don't quite see what . . . modern flakes have to do with high-level implements." Smith then stated that he himself had found stones resembling eoliths but never took them home. He then concluded his letter to Harrison with more humor: "Now I hope you are quite well and blessed with a happy and peaceful mind, without pre-glacial nightmares . . . and palaeolithic tailless apes" (E. Harrison 1928, pp. 224–225). The not so subtle ridicule of the very idea of *Homo sapiens* existing in the Tertiary is typical of the unscientific methods used by scientists to dismiss evidence that falls outside their particular circle of comprehension. Smith's admission that he himself had deliberately avoided collecting specimens of eoliths is also somewhat damaging to the notion of evenhanded scientific treatment of controversial questions. It often happens that anomalous evidence is ignored. Smith's statement that he failed to see any connection between modern flakes and ancient ones is also quite curious, for such comparative studies of lithic technologies were, and presently are, recognized as an appropriate method for evaluating intentional human work on stone objects.

Smith once wrote to Harrison, who had asked him to consider certain points bearing on the eolith question: "As for answering questions and giving opinions about *dubious subjects*, it is not always easy, and silence, philosophic doubt, or no settled convictions are better, especially in face of a high priest like you. It is like a Salvation Army captain full of zeal, coming here and asking me about Noah and his ark, Balaam and his ass, and Jonah and his whale. The better plan, according to my view, is to bolt and say nothing" (E. Harrison 1928, p. 187). When one considers the support given to Harrison's discoveries by reputable scientists such as Sir John Prestwich, Smith's characterization of Harrison seems a bit unfair. As we shall see, the put-offs and put-downs from Smith's repertoire are, for a good many scientists, still the favored methods for dealing with evidence that has uncomfortable implications for established views on human evolution. They avoid acknowledging anomalous evidence, never discuss it on its merits, and if pressed, simply ridicule it and those who support it.

3.2.13 More Honors for Harrison

As time passed, however, Harrison continued to receive more honors and his eoliths more attention. In 1899, upon recommendation by Prime Minister Balfour, Queen Victoria awarded him a prestigious Civil List pension "in consideration of your researches on the subject of prehistoric flint implements" (E. Harrison 1928, p. 230). The Royal Society also granted him an annuity. That same year, T. Rupert Jones made a presentation about eoliths at the British Association meeting in Dover, exhibiting some small implements that attracted much attention (E. Harrison 1928, p. 231). In August of 1900, Arthur Smith Woodward of the British Museum and Professor Packard of Brown University paid Harrison a visit. Packard accepted all of Harrison's finds as genuine and Woodward agreed that the Plateau drift in which the eoliths were found was probably Pliocene in age (E. Harrison 1928, p. 237). On August 21, 1900, Harrison received a letter from Dr. H. P. Blackmore, who stated that he accepted the eoliths because of "the fairly uniform heights of deposits in which eoliths are found: differing greatly in age of deposit from the more recent river drift or paleolithic gravels" (E. Harrison 1928, pp. 237–238). In 1902, at the British Association meeting in Belfast, W. J. Knowles and F. J. Bennett came out in favor of the eoliths, while Boyd Dawkins was opposed. Some of Harrison's eoliths were placed on exhibition in the British Museum.

Ray E. Lankester, who was a director of the British Museum (Natural History), became a supporter of Harrison's Kent Plateau eoliths. On April 15, 1904, Lankester wrote to Harrison: "Good health and happiness to you—courageous and indomitable discoverer of pre-paleolithic man" (E. Harrison 1928, p. 271). Sir Edward R. Harrison stated: "Professor Ray Lankester, who expressed publicly his belief that the eoliths were artificial, and in the Romanes lecture in Oxford, in 1905, declared that they carried 'the antiquity of man at least as far back beyond the paleoliths as these are from the present day', desired to emphasize the value, as evidence of purpose, of similarity of shape of certain eoliths, and wrote to Harrison for specimens to illustrate a book that he had in course of preparation. He was impressed by the large number of implements with a 'tooth-like prominence rendering the flint fit for use as a "borer"' and also by a group he called trinacrial, from their resemblance in shape to the island of Sicily" (E. Harrison 1928, p. 270). In his presidential address to the British Association in 1906, Lankester affirmed his belief in "the human authorship" of Harrison's eoliths (E. Harrison 1928, p. 270).

As time passed, Benjamin Harrison continued to win more and more converts. Sir Edward R. Harrison (1928, pp. 287–288) wrote: "A visit from Professor Max Verworn of Göttingen, who had come to England in connexion with the centenary of Charles Darwin's birth, gave Harrison great pleasure. Professor Verworn, who stated that he did not at first believe in eoliths or in any of the supposed evidence of Tertiary man, but had modified his views after personal investigation of the Miocene deposits of the Cantal [Section 4.3.3], spent five days at Ightham. The fullest use was made of the time available, both in Harrison's museum and in the field. Professor Verworn found an interesting old paleolith in situ in the Plateau gravel at the Vigo, an implement that from its position near the crest of the Chalk escarpment, and its rolled condition, could only have come from the vanished Wealden hills. . . . Harrison could not have wished for a more striking discovery to have been made by his visitor in order to satisfy him of the great antiquity of man in Kent." If Sir Edward Harrison is using the word paleolith in its then accepted sense, we have here an account of an implement more technically advanced than the Eolithic type being found in the very old gravels of the Plateau, and having the worn appearance of implements belonging to those gravels. This gives added support to the possibility that humans of the modern type may have existed in later Tertiary times in England, perhaps 2–4 million years ago.

On July 25, 1909, Professor Verworn wrote to Harrison from Göttingen: "If up to then I had the slightest doubt of the artificial nature of the eoliths of Kent, my visit on the spot and your splendid collection would have quite converted me" (E. Harrison 1928, p. 288).

3.2.14 More Opposition

The controversy over the eoliths continued well into the twentieth century. On April 28, 1911, Lord Avebury (Sir John Lubbock) wrote to Harrison: "I am satisfied that many, if not most of your eoliths are worked, though the numbers are staggering. I am not satisfied, however, that palaeolithic implements are in all cases younger" (E. Harrison 1928, pp. 294–295). In his last edition of his book *Prehistoric Times,* Lord Avebury fully accepted the eoliths of Harrison, as well as the implements of J. Reid Moir, which we shall discuss in the next section of this chapter (E. Harrison 1928, p. 305). The opposition, however, continued to criticize the eoliths. In 1911, F. N. Haward published a paper purporting to show that natural forces were able to chip flints in a way that gives the impression of human work. We shall discuss Haward's objections in connection with the flint implements of J. Reid Moir.

At this point, one may question the necessity of giving such a detailed treatment of the Harrison eoliths. There are several good reasons for doing so. The authors have discovered that modern students of paleoanthropology are generally not at all acquainted with many nineteenth-century discoveries demonstrating the presence of humans of the modern type in Tertiary times. And when these discoveries are brought to the attention of modern students, they tend to categorize them as "crackpot" or "oddball" cases that somehow gained some public notoriety and were quickly dismissed when brought to the attention of scientific authorities. We have also noted a strong prejudice against anomalous evidence that is "old." Old accepted evidence is honored—for example, Java man, highlighted in all modern textbooks, was a nineteenth-century discovery. But the less familiar nineteenth-century evidence, which goes against the theories presented in modern textbooks, is tainted with suspicion, more so if one has never even heard of it before. In such cases, one often encounters in modern students a very strong assumption that if one has not even heard of some anomalous evidence, then it must have been completely rejected on purely scientific grounds long ago. One reason for presenting a detailed account of anomalous evidence is to show that it was not always of a marginal, crackpot nature. Rather anomalous evidence was quite often the center of serious, longstanding controversy within the very heart of elite scientific circles, with advocates holding scientific credentials and positions just as prestigious as those of the opponents. By presenting detailed accounts of the interplay of conflicting opinion, we hope to give the reader a chance to answer for himself or herself the crucial question—was the evidence actually rejected on purely scientific grounds, or was it dropped from consideration and forgotten simply because it did not lie within the parameters of certain circumscribed theories?

In his book *Ancient Hunters and their Modern Representatives,* W. J. Sollas of Oxford rejected Harrison's finds (E. Harrison 1928, p. 298). In response, Harrison sent him an eolith. On February 1, 1912, Sollas wrote to Harrison: "The specimen you send for my inspection is one of the most interesting of your finds that I have seen. I read its history as follows: (1) Natural agencies detached it as an irregular flake from a flint nodule. . . . (2) It lay in the bed of a stream with the rough side uppermost and was battered on the exposed surface by pebbles, which have left percussion cones as their mark. . . . (3) Still later, it was chipped in a remarkable manner over a portion of its margin" (E. Harrison 1928, p. 298). Here Sollas attributed a remarkable sequence of manufacturing steps to purely natural forces. The end result was a sharp-edged flint implement, something not usually to be expected from the movement of stones in a stream, the random battering of which, as modern authorities point out and anyone can see, tends to produced rounded pebbles.

Sollas then observed: "It is the chipping which is of especial interest to both of us. Two explanations may be given: (1) That the chipping is the result of superincumbent pressure acting upon a yielding substratum. In favour of this it may be pointed out that the chipping is confined to the margin, which we might judge from the general shape of the stone to have thinned off a blunt edge. (2) That the chipping was done by man. In favour of this is the fact that over one part of the specimen the chipping is such as to remove all sharp edges, as if it had been intended for a comfortable hold for the hand . . . while on the opposite side the chipping has produced a projecting point which would be very effective if the flint were used as a weapon for striking a blow. In fact this flint would make a splendid 'knuckle duster.' I should not wonder if this was its true nature. But I should not like to commit myself to the assertion that it was" (E. Harrison 1928, pp. 298–299). One wonders why he should not like to commit himself. The points raised here by Sollas himself seem to run very much in favor of the hypothesis that the stone object was of human manufacture.

Sollas then stated (E. Harrison 1928, p. 299): "Granting that it was, however, what does it prove? The patina of the latest chipping is not deep, it looks to my eyes remarkably fresh, and, since palaeolithic implements are found in your deposits, what evidence have you to show that this was not also palaeolithic?" Here the same old question, to which Prestwich long ago had given a detailed and convincing scientific response, came up again. To repeat Prestwich's basic points, the Eolithic implements, being quite well worn, were distinctly different in appearance from the paleoliths; furthermore, they were sometimes found by themselves in specific deposits. Despite his doubts, Sollas did, however, request more samples for the Oxford museum and Harrison sent six.

At the beginning of the First World War, the British Army, perhaps fearing a German invasion, dug trenches on the hills around Ightham, creating more exposures of gravel for Benjamin Harrison to search. Sir Edward R. Harrison (1928, p. 317) wrote that one of the local flint hunters trained by Benjamin Harrison "joined up at the outbreak of war in 1914, was stationed in the Somme valley, found a palaeolith when digging a trench, carried it with him 'over the top', and finally brought it safely to Ightham, and to Harrison, when he came home on leave."

Harrison died in 1921, and his body was buried on the grounds of the parish church, St. Peter's, in Ightham. On his gravestone one finds the words: "He found in life, 'books in the running brooks, sermons in stones, and good in everything'" (E. Harrison 1928, p. 331). A memorial tablet, set in the north wall of St. Peter's on July 10, 1926, bears this inscription: "IN MEMORIAM.—Benjamin Harrison of Ightham, 1837–1921, the village grocer and archaeologist whose discoveries of eolithic flint implements around Ightham opened a fruitful field of scientific investigation into the greater antiquity of man. A man of great mind and of kindly disposition" (E. Harrison 1928, p. 332). Factually speaking, however, the "fruitful field of scientific investigation into the greater antiquity of man" opened by the eoliths of the Kent Plateau was buried along with Harrison.

3.3 DISCOVERIES BY J. REID MOIR IN EAST ANGLIA

Our journey of exploration now takes us to the southeast coast of England and the discoveries of J. Reid Moir. Starting in 1909, Moir found flint implements in and beneath the Red and Coralline Crags of East Anglia (Suffolk). We shall first give an overview of Moir's discoveries and then discuss in detail the scientific controversies they sparked, concluding with a survey of recent opinion.

3.3.1 Moir and Harrison

J. Reid Moir, a fellow of the Royal Anthropological Institute and president of the Prehistoric Society of East Anglia, was acquainted with Benjamin Harrison's eoliths. Moir (1927, p. 17) believed the gravels on the Kent Plateau, from which Harrison had recovered his eoliths, were the remnants of an old Tertiary land surface, perhaps as old as the Eocene. But, as we have seen, some authorities would assign the gravels of the Kent Plateau to the Pliocene (Sections 3.2.2, 3.2.4). Moir wrote: "It is probable that these flints were shaped by a race of ape-like people who lived on a land surface which existed at one time over what is now the Weald of Kent, which was then enjoying a tropical climate. . . . They were probably small, squat men, with very ape-like skulls and projecting jaws, and in many ways more like animals than men" (1927, pp. 17–18, 19).

Moir was an evolutionist. He believed that the degree of primitiveness shown by a very old stone tool industry was indicative of the correspondingly primitive physiological character of the toolmaker. But even today tribal people, physiologically identical to MIT computer scientists, make implements just like the crudest ever found in ancient strata. Furthermore, skeletal remains of fully human character have been found in strata dating back to the Pliocene and even further (Sections 6.2, 6.3). It is therefore possible that the eoliths discovered by Harrison were made by human beings of the type *Homo sapiens sapiens*.

Harrison found many eoliths during excavations sponsored by the British Association for the Advancement of Science. But he found most on the surface and although geologist Sir John Prestwich argued strongly for their Tertiary age, stubborn critics remained doubtful. The geological position of Moir's finds was more secure, for most of them were found *in situ*, deep below the land surface in various locations in East Anglia.

3.3.2 The Age of the Crag Formations

The Red Crag formation (Table 2.1, p. 78), in which Moir made some of his most significant discoveries, is composed of the shelly sands of a sea that once washed the shores of East Anglia. At some places beneath the Red Crag is found a similar formation called the Coralline Crag. Some authorities have placed the Red Crag wholly within the Early Pleistocene. For example, J. M. Coles (1968, p. 19) proposed that the boundary between the Red and Coralline Crags represents the boundary between the Pleistocene and the Pliocene. Others have said that the Red Crag spans the Pleistocene-Pliocene boundary. W. H. Zaguin (1974), for example, placed the lower part of the Red Crag in the Pliocene (Nilsson 1983, p. 108). And still others, such as A. S. Romer (1966, p. 334), put the Red Crag entirely within the Pliocene. Claude Klein (1973, table 6) also placed the Red Crag in the Pliocene and gave it a date of 2.5–4.0 million years.

Tage Nilsson (1983) called attention to potassium-argon dates for Icelandic formations that some experts correlate with those of East Anglia. Nilsson (1983, p. 106) stated: "If the correlation of the uppermost Tjörnes Beds with the Butley Red Crag is justified, this would imply a probable age of 2.5–3.0 million years for the youngest Red Crag in Britain." According to Nilsson, this view is supported by paleomagnetic data, which suggest a date of over 2.5 million years for the Red Crag. Paleomagnetic dating relies on the fact that the magnetic field of the earth periodically reverses. Signs of this can be detected in various formations, which are thus labeled normal and reversed in terms of their magnetic polarity. Nilsson (1983, p. 106) stated: "the Red Crag in East Anglia is normally magnetized and probably referable to the later part of the Gauss Normal Epoch, and [is] thus more than 2.5 million years old."

After studying the range of geological opinion, we have arrived at a conservative age estimate of at least 2.0–2.5 million years for the Red Crag. The range of dates assigned to the Red Crag raises an important question. The conventionally accepted evidence for human evolution comes from sites representing only the last 2 or 3 million years of the earth's history. Much depends upon being able to arrange fossils from these sites in an accurate temporal sequence. But if the quantitative age determinations of fairly recent formations can vary by hundreds of thousands of years, or even a million or more, years, then the integrity of proposed evolutionary sequences, at least insofar as they are founded on stratigraphic evidence, becomes problematic.

Below the Crags of East Anglia are found detritus beds, sometimes called bone beds, composed of a mixture of loose materials—sands, gravels, shells, and bones derived from a variety of older formations. According to Moir, the detritus beds also contain stone implements.

It is certain that these stone implements are older than the Late Pliocene Red Crag. But how much older they actually are depends upon how one interprets the detritus bed below the Red Crag. J. Reid Moir (1924, pp. 642–643) wrote: "The sub-Red Crag detritus bed, which is sometimes as much as three feet in thickness, is, as its name implies, composed of materials of different periods occurring prior to the time when the deposit was laid down. Sir Ray Lankester has shown that these varying materials have been derived from the following sources:—(a) the chalk, [Cretaceous] (b) the London Clay [Eocene], (c) a Miocene land surface, (d) a marine Pliocene deposit (the Diestian Sand), (e) the earlier sweepings of a land surface which submerged after the Diestian deposit, and (f) later sweepings of the same land surface. It will thus be seen that the flint implements, now to be described, that were found in the detritus bed, may be referable to any of the periods represented by c, e, or f of the above list. We have no reason to think that at the epochs when the chalk and the London Clay were being laid down, man was present upon this planet, nor can he well be associated with the marine accumulation (d)."

Modern authorities give similar accounts of the detritus bed below the Red Crag. Tage Nilsson (1983, p. 105) stated: "At the bottom of the Red Crag deposits there is often a stony layer, constituting a kind of basal conglomerate, the Red Crag Nodule Bed. This mainly consists of flint pebbles and phosphorite nodules, washed out from older bedrock. It contains usually densely mineralized and often well-rounded and polished mammal fossils, which must in part be reworked from Eocene and other pre-Quaternary deposits."

According to Nilsson (1983, p. 105), some fossils of Villafranchian species (such as *Mammuthus meriodionalis*) were found in the detritus bed. The Villafranchian land mammal stage spans the Pliocene-Pleistocene boundary. This might suggest that the Red Crag detritus bed contains materials from the Early Pleistocene.

Arguing against this is the fact that the detritus beds are often found *in situ* beneath the intact Red and Coralline Crags (Moir 1924, p. 641), which can be safely referred to the Pliocene. Thus the Villafranchian component of the detritus bed fauna can be assigned to the Pliocene (rather than Pleistocene) part of the Villafranchian stage.

We note that potassium-argon dates obtained for a Villafranchian site in France reached 2.5–3.0 million years (Nilsson 1983, pp. 24, 158). We can therefore conclude that the age of the materials in the detritus beds at the base of the Crags range from Late Pliocene to perhaps Cretaceous in age. The Cretaceous chalk is, however, a marine formation, making the Eocene London Clay the earliest habitable land surface in the stratigraphic sequence of East Anglia.

3.3.3 Tools from Below the Red Crag
(Pliocene to Eocene)

J. Reid Moir found in the sub-Crag detritus beds many types of stone tools, showing varying degrees of intentional work (Figure 3.6). He concluded that the cruder tools were older. Since the detritus beds, according to this scheme, appeared to contain a succession of stone tools from different periods, perhaps as far back as the Eocene, Moir (1935, pp. 360–361) wrote: "then it becomes necessary to recognize a much higher antiquity for the human race than has hitherto been supposed. I am fully aware of the implication of such a conclusion and the responsibility attaching to those who support it. Nevertheless, after a very careful and painstaking examination of all the available facts, I have been compelled to accept this conclusion as true, and have no hesitation in stating that such is the case."

Moir connected the crudest tools, resembling the Harrison eoliths, with the Miocene elements of the detritus bed below the Red Crag. He considered them to be contemporary with the flint implements discovered in French Miocene formations at Aurillac (Section 4.3). But further than that he would not go, having stated, as above mentioned: "We have no reason to think that at the epochs when the chalk and the London Clay were being laid down, man was present upon this planet" (Moir 1924, p. 643).

J. Reid Moir may have thought in that way, but, as we shall demonstrate later in this book, there is evidence that humans of the fully modern type were in fact present throughout the Tertiary, including the Eocene period, during the time when the London Clay was being deposited upon the underlying Cretaceous chalk (Sections 3.4.2, 5.5, 6.2, 6.3; also Appendix 2).

So the stone implements collected by Moir from beds below the Red Crag formation could be of that age. In fact, it is quite possible that any of the stone implements, from the crudest to the most sophisticated, could be referred to any period from the Late Pliocene to the Eocene.

At the very least, then, the implements are Late Pliocene in age. But according to present evolutionary theory one should not expect to

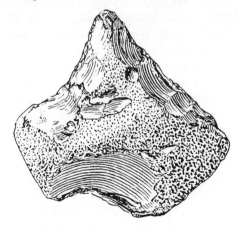

Figure 3.6. Pointed implement from below the Red Crag (Moir 1935, p. 364). This specimen is over 2.5 million years old.

find signs of toolmaking humans in England at 2–3 million years ago. Two million years before the present, our toolmaking hominid ancestors (of the *Homo habilis* type) should still have been confined to their homeland in Africa. Three million years before the present, we should expect to find in Africa only the apelike australopithecines, who are not generally recognized as makers of stone tools.

Of his Miocene inhabitants of England, Moir (1927, p. 31) wrote: "Unfortunately, no actual bones of the people who made these implements have yet been discovered, but, judging from these specimens, we conclude that their makers were possessed of considerable strength, and represent an early and brutal stage in human evolution."

We do not deny the possibility that ape-man-like creatures might have been responsible for the implements reported by Moir. But even today, modern humans are known to manufacture very primitive stone tools. It is thus possible that beings very much like *Homo sapiens sapiens* could have made the crudest of the implements recovered by Moir from below the Red Crag. In the absence of skeletal evidence directly connected with the stone tools, it is impossible to say with certainty what kind of creature manufactured them. All may have been made by humanlike creatures, all may have been made by ape-man-like creatures, or perhaps some were manufactured by humanlike creatures and others by ape-man-like creatures.

The implements themselves were a matter of extreme controversy. Many scientists thought them to be products of natural forces rather than of human work. Nevertheless, Moir had many influential supporters. These included Henri Breuil, who personally investigated the sites (Section 3.4.7). He found in Moir's collection an apparent sling stone from below the Red Crag (Section 5.3.1). Another supporter was Archibald Geikie, a respected geologist and president of the Royal Society (Millar 1972, p. 100). Yet another was Sir Ray Lankester, a director of the British Museum. Lankester identified from among Moir's specimens a representative type of implement he named rostro-carinate. This word calls attention to two prominent characteristics of the tools. "Rostro" refers to the beaklike shape of the working portion of the implements, and "carinate" refers to the sharp keellike prominence running along part of their dorsal surface (Moir 1927, p. 26).

Lankester presented a detailed analysis of the "Norwich test specimen" (Figure 3.7). A particularly good example of the rostro-carinate type of implement, it was discovered beneath the Red Crag at Whitlingham, near Norwich (Moir 1927, p. 28; Osborn 1921, p. 576). If the Norwich test specimen is from below the Red Crag, it would be over 2.5 million years old. If it is from below the Norwich Crag as suggested by Sparks and West (1972, p. 234), it would be over 2.0 million years old (Table 2.1, p. 78). The Norwich test specimen combined a good demonstration of intentional work with clear stratigraphic position. Sir Ray Lankester wrote in a Royal Antropological Institute report in 1914: "it is not possible for anyone acquainted with flint-workmanship and also with the non-human fracture of flint to maintain that it is even in a remote degree possible that the sculpturing of this Norwich test flint was produced by other than human agency" (Coles 1968, p. 27).

Professor J. M. Coles of Cambridge University (1968, p. 27) later noted: "His description was full and was accompanied by drawings and photographs showing that approximately 40 flakes had been removed from various angles and positions around the flint, consisting of two cleaving

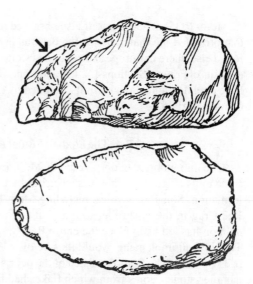

Figure 3.7. The Norwich test specimen. J. Reid Moir (1927, p. 28) said it was found beneath the Red Crag at Whitlingham, England. The beak (arrow) forms the working portion of the implement, which, if from below the Red Crag, would be over 2.5 million years old.

fractures, a group of large conchoidal flakings, and a third group of smaller flakings directed upon specific parts, particularly the beaked portion."

At a lecture before the Royal Society in London, Lankester said he hoped "that no one would venture to waste the time of the society by suggesting that sub-crag flints had been flaked by natural causes, as by so doing it would be plain that they had a very scanty knowledge of such matters." Someone present did, however, venture to suggest exactly that. Lankester said it was "the sort of thing I would expect to hear from a savage." Another time G. Worthington Smith, known to us for his skeptical exchanges with Benjamin Harrison of Ightham, said of the eoliths and pebble tools: "We have here choppers that do not chop and borers that do not bore." Lankester retorted: "You, sir, are a bore who does bore" (Millar 1972, p. 100).

About the age of the rostro-carinate tools, Lankester stated in 1941: "I do not intend to proceed without caution to any conclusion on this subject, but it seems to me quite possible that there is a close relationship between the men who made the Upper Miocene rostro-carinate implements of Aurillac [Section 4.3] and those who made similar implements in Suffolk before the deposit of the Red Crag" (Moir 1935, p. 359).

Moir (1935, p. 360) himself also observed that intact counterparts of the beds that provided the materials for the detritus layer below the Red Crag could be found elsewhere in Europe and contained stone implements: "the Upper Miocene deposits of France [Sections 4.2, 4.3] and some older beds in Belgium [Section 4.4] have already yielded flaked flints, claimed by certain competent investigators as of human origin."

3.3.4 The Foxhall Finds (Late Pliocene)

One important set of discoveries by Moir occurred at Foxhall, where he found stone tools (Figure 3.8) not in the detritus bed but in the middle of the Red Crag formation. Some authorities, including Moir, have placed the upper part of the Red Crag in the Early Pleistocene, but our review of the range of geological opinion has led us to place the entire Red Crag formation in the Late Pliocene. The Foxhall implements would thus be over 2.0 million years old. Moir (1927, p. 33) wrote: "The finds consisted of the debris of a flint workshop, and included hammer-stones, cores from which flakes had been struck, finished implements, numerous flakes, and several calcined stones showing that fires had been lighted at this spot. The Foxhall implements are, in the majority of cases, of a yellowish white colour, and more finely made than the still more ancient specimens found at the base of the Crag, and give us a very clear idea of the type of workmanship of which these ancient Suffolk people of Early Quaternary times were capable. While if the famous Foxhall human jaw-bone, which was apparently not very primitive in form, was, indeed, derived from the old land surface now buried deep beneath the Crag and a great thickness of Glacial Gravel, we can form the definite opinion that these ancient people were not very unlike ourselves in bodily

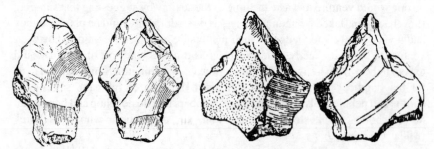

Figure 3.8. Front and rear views of two stone tools from the Red Crag at Foxhall, England. They are Late Pliocene in age. Henry Fairfield Osborn (1921, p. 572) said of the tool on the left: "Two views of pointed flint implement flaked on the upper and lower surfaces and with a constricted base, from sixteen-foot level of Foxhall pit. Primitive arrowhead type, which may have been used in the chase." Of the implement on the right, Osborn wrote: "Borer (*perçoir*) from sixteen-foot level of Foxhall."

characteristics." The jaw spoken of by Moir has an interesting history (Section 6.2.1). For now, we shall simply note that some scientists who examined it considered it like that of a modern human being.

It is unfortunate that the Foxhall jaw is not available for further study, for it might offer further confirmation that the flint implements from Foxhall were of human manufacture. But even without the presence of actual human skeletal remains, the tools themselves point strongly to a human presence in England during the Late Pliocene, perhaps 2.0–2.5 million years ago.

The American paleontologist Henry Fairfield Osborn (1921, p. 573) came out strongly in favor of the implements having been manufactured by human beings and argued for a Pliocene date: "Proofs which have rested hitherto on the doubtful testimony of irregular eoliths generally considered by archaeologists as not of human manufacture, now rest on the firm foundation of the Foxhall flints in which human handiwork cannot be challenged; these proofs have convinced the most learned and most conservative expert in flint industry in Europe today, namely Abbe Henri Breuil of the *Institut de Paleontologie Humaine.*" According to Osborn (1921, p. 573), the Foxhall specimens included borers, arrowheadlike pointed implements (some hafted), scrapers, and side scrapers very much like early Mousterian *racloirs.*

Osborn (1921, p. 566) concluded: "This discovery of man in Pliocene time delights the present writer for a personal reason, namely, because it tends to render somewhat more probable his prophecy made in April 1921, before the National Academy of Sciences at Washington that one of the great surprises in store for us in science is the future discovery of Pliocene man with a large brain." This sort of talk would not go down very well today.

Osborn (1921, p. 565) backed not only the Foxhall flints but the rest of Moir's work as well: "The discoveries of J. Reid Moir of evidences of the existence of Pliocene man in East Anglia open a new epoch in archaeology. . . . they bring indubitable evidence of the existence of man in southeast Britain, man of sufficient intelligence to fashion flints and to build a fire, before the close of the Pliocene time and before the advent of the First Glaciation."

But whether one accepts Osborn's Pliocene date or Moir's Early Pleistocene estimate, neither is to be expected if one accepts the standard version of hominid evolution in an African homeland. This is especially true if, as the Foxhall jaw indicates, the maker of the Foxhall flint tools was fully human. The first *Homo sapiens* are thought to have come into existence only a couple of hundred thousand years ago at most, and the standard textbook version is that fully modern *Homo sapiens sapiens* is only about 100,000 years old.

Another scientist won over by the Foxhall finds was Hugo Obermaier, previously a consistent and vocal opponent of Eolithic discoveries. Obermaier was one of those scientists who believed that eoliths were produced by natural

forces similar to the forces operating in cement and chalk mills
(Section 3.5). But Obermaier
(1924, p. 41) wrote: "Very recently a large bed of flints with
evidences of fire has been found
on the eastern coast of England
near Norwich and beneath the
Late Pliocene deposits known
as the 'Red Crag' and the 'Norwich Crag.' The authenticity of
the flints as of human origins is
disputed by some archaeologists,
but is accepted by others, including Louis Capitan, the veteran archaeologist of France, and
Henri Breuil, who is frequently
quoted in these pages. This discovery of Foxhall is the first
evidence we have of the existence of Tertiary man."

Someone might have asked
Obermaier if, having accepted
the Foxhall flint tools as proof of
human existence in the Tertiary, he might reevaluate any of
the Tertiary Eolithic industries
he had once rejected.

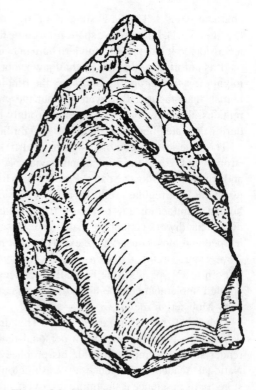

Figure 3.9. Pointed tool from the Cromer Forest Bed, East Runton, England (Moir 1927, p. 45). It could be from about .8 million to 1.75 million years old, depending upon how one dates the Cromer Forest Bed.

3.3.5 Cromer Forest Bed (Middle or Early Pleistocene)

Thus far we have considered Moir's discoveries in the Tertiary bone bed below the Late Pliocene Red Crag and his finds in the Red Crag itself at Foxhall. We shall now turn to some discoveries in the more recent Cromer Forest Bed of Norfolk. As we have seen (Section 2.19; Table 2.1, p. 78), the Cromer Forest Bed dates from about .4 million years to about .8 million years ago, or perhaps even as much as 1.75 million years ago. During this period, according to Moir, the delta of the Rhine extended to East Anglia.

Moir found specimens of a stone industry (Figure 3.9), including large handaxes, lying on the beach at Cromer and East Runton in Norfolk. He stated that they originated from a stone bed exposed in the base of the cliffs along the shore. Moir (1924, p. 649) wrote: "The Cromer specimens are found chiefly upon the foreshore.... They lie upon the chalk, and have evidently been derived from a formation at the very base of the Cromer Forest Bed series of deposits, which form the lowermost strata of the high bluffs of the Norfolk coast.... In some places, as at East Runton, about two miles northwestward of Cromer, large areas of the implementiferous bed can be seen *in situ* upon the chalk, and from this deposit have been recovered several very definite examples of Early Paleolithic hand axes." If the implements are, as Moir stated, from the lowest part of the Cromer Forest Bed formation, they would, according to modern estimates, be at least .8 million and perhaps as much as 1.75 million years old.

Moir (1924, p. 652) went on to describe the implements: "There is no doubt that the Cromer industry shows an advance from the sub-Crag culture, but it is nevertheless closely related to it. The ancient Cromerians, using probably large hammerstones of flint, were able to detach in some cases enormous flakes of flint, and the whole industry is on a large and massive scale. On the foreshore at Cromer the contents of a workshop site were found, comprising hand axes, choppers, side scrapers, points, and numerous flakes.... Their skill in flint-flaking is evidenced by the immense flake scars produced by the primary quartering blows, the well-formed striking platforms, and the regular and accurate secondary flaking." Critics of anomalous stone tools often ask for just the type of evidence reported by Moir—a variety of finished tool types and flakes in close association, indicating a workshop site.

3.3.6 Moir Versus Haward

Having briefly reviewed Moir's discoveries beneath the Red and Coralline Crags, in the Red Crag at Foxhall, and from the Cromer Forest Bed, we shall now examine the history of the scientific controversies surrounding them. J. M. Coles (1968) of Cambridge gave a rare modern summary of the disputes.

In 1919, F. N. Haward attacked Moir's discoveries, claiming that they were the product of geological pressure acting on flint. Moir and A. S. Barnes replied to Haward in articles published in the *Proceedings of the Prehistoric Society of East Anglia.* Moir (1919, p. 158) made the following comments: "It appears that Mr. Haward has found in the Norwich Stone Bed a flint, or flints, which exhibit a flake detached, but not removed from the parent block, and he concludes, and rightly concludes, that such flakes have become so detached since the bed in which they occur was laid down. He draws attention to the well-known fact that flints in the chalk, and, I may add, in other deposits as well, break up into pieces of varying size, and that such breakage is of natural origin. And once more I am in agreement. But here, I fear, we take widely different paths."

Haward believed the cause of breakage to be pressure. Moir agreed that this was indeed one possible cause, and pointed out that he had himself published a paper on this topic ("The Fractured Flints of the Eocene 'Bull-Head' at Coe's Pit, Bramford, near Ipswich") in the *Journal of the Prehistoric Society of East Anglia*. Moir (1919, pp. 158–159) went on to state: "But I know also that pressure flakes exhibit certain peculiarities of their surfaces which differentiate them markedly from other flakes which have been removed by percussion, and so far as I can ascertain, Mr. Haward has not yet demonstrated, scientifically, that the few flakes upon which he bases his portentous argument have without doubt been detached by pressure. It may also be recalled that the Norwich Stone Bed, as I can testify from actual observation, contains very often fragile bones of mammals, and the sands above it, at Whitlingham, where a large proportion of the sub-crag implements described by Mr. Clarke have been found, have embedded in them even more fragile shells. And it is legitimate to ask why, if pressure is fracturing the hard, resistant flints in the Stone Bed, the easily-broken organic remains mentioned are quite frequently found intact." Rejecting the pressure hypothesis, Moir suggested another explanation. Before being embedded in the deposit a flint nodule might have been subjected to blows strong enough to produce incipient bulbs of percussion. Later, under the influence of heat, for example, the flakes might have come off. Moir (1919, p. 159) added that Haward himself had noted that some flaked flints he studied bore signs of percussion.

Moir (1919, p. 159) then stated: "But whatever the exact cause of such fracturing may be, it is clear that such cases are very rare, and moreover, when they are found, only one or two flakes are seen to be in contact with the parent block. Yet Mr. Haward does not hesitate to infer that all the other flints exhibiting *numerous* flake-scars upon their surfaces, and a definite implemental form, have been produced by this same natural fracturing. When also it is remembered that many, if not most, of these latter specimens show by their colouration and condition that they are definitely more ancient than the bed in which they now occur, it will be seen that this inference rests upon a very attenuated and shaky basis. But if this is the case in regard to the Norwich Stone Bed flints, what is one to say about the extension of Mr. Haward's inference to the specimens found under totally different conditions beneath the Red Crag of Suffolk, and where, up to the present, no evidence of any fracturing *in situ* has been seen?" Moir (1919, pp. 160–161) pointed out that the specimens found below the Red Crag displayed only signs of flake removal by percussion, with no sign of pressure fracturing.

Moir (1919, p. 161) concluded his remarks with this affirmation: "students of human and animal bones have regarded the existence of man in the Pliocene as almost a necessity, and from my later researches I incline to the belief that not only was man present on this earth at that period, but that he was then culturally much more advanced than has hitherto been imagined."

3.3.7 Warren's Attack on Moir

Still the opposition to Moir continued, with scientists clinging with remarkable tenacity to variations of the natural pressure-flaking hypothesis. Coles (1968, p. 27) stated: "A more scholarly attack on the authenticity of the 'industries' was made by S. Hazzledine Warren in 1921, who claimed that mechanical movement of flint upon flint under pressure produced flaking comparable to that seen on not only the Kentish eoliths but also the rostro-carinates and other Crag assemblages. Warren based his argument upon his observations of fractured flints in Eocene deposits in Essex, and upon experiments. Moir and Barnes defended themselves vigorously, and claimed that natural pressure flaking could easily be distinguished from the edge-flaking on the Kentish eoliths and on the Crag series. The naturally-produced specimens claimed by Warren to be of rostro-carinate form from the Essex gravels were said to be entirely different." In 1923, an international commission of scientists concluded that the flaking on the specimens collected by Warren was in fact different from that on Moir's implements (Section 3.3.8).

Warren's report was delivered in an address to the Geological Society of London, and was later published in the Society's journal. In the Eocene location studied by Warren, the flints were lying beneath layers of sediment, upon a chalk surface, where he claimed that they had been subjected to pressure and differential movement by "solution of the chalk surface." In other words, the flints had been crushed by the pressure of overlying layers as they slipped into holes eroded in the chalk by the action of ground water. Warren claimed to have found, in locations where such crushing had occurred, many specimens resembling not only eoliths but Mousterian implements as well.

Of one such specimen (Figure 3.10), Warren (1920, p. 248) stated: "This, a

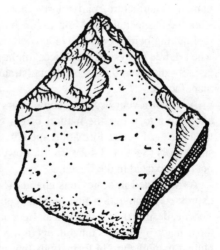

Figure 3.10. S. Hazzledine Warren said that this object, which he believed to be the product of natural pressure flaking, almost exactly resembled a Mousterian trimmed point implement (MacCurdy 1924b, p. 657). But although found in an Eocene formation, it could in fact be of human manufacture.

good example of a trimmed-flake point, is the most remarkable specimen of the group. If considered by itself, upon its own apparent merits, and away from its associates and the circumstances of its discovery, its Mousterian affinities could scarcely be questioned. But, like all the other specimens illustrated, I dug it out of the Bullhead bed myself in circumstances which preclude the possibility of mistake." In this connection George Grant MacCurdy, director of the American School of Prehistoric Research in Europe, wrote: "Warren states that if the best selected flakes from the Bullhead Bed were mingled with flakes from a prehistoric workshop floor, they could never be separated again unless it were by their mineral condition" (1924b, pp. 657–658).

Much depends upon whether or not the flaking on Warren's specimens actually resembled that of Paleolithic humans. If the flaking was different, then Warren's argument against Moir becomes irrelevant. If, on the other hand, the flaking was similar, then what are we to make of specimens, such as the one depicted in Figure 3.10, which are so very much like accepted Paleolithic stone tools?

Warren appeared to take for granted, in a fashion typical of those who shared his prejudices, that it was impossible to find in Eocene strata implements of human manufacture, particularly those displaying a relatively high level of stoneflaking technique. Moir, as we have seen, expressed the same view—no toolmaking beings could have existed in the Eocene (Section 3.3.3). But those who are free from such prejudices might justifiably wonder whether Warren had actually discovered, in the Eocene strata of Essex, a genuine object of human manufacture.

A similar event occurred some years earlier in France, where H. Breuil, in attempting to prove the natural origin of eoliths by geological pressure, also found in an Eocene formation specimens exactly resembling Late Paleolithic stone tools (Section 3.4.2). Breuil, however, was convinced that humans could not have existed in the Eocene.

As we noted in a previous chapter (Section 2.9), T. McKenny Hughes also expressed a conviction that humans could not have existed as far back as the Eocene, despite the presence in an Eocene formation of pierced shark's teeth like those made by today's inhabitants of the South Pacific. Other finds of objects of human manufacture in formations that might be as old as the Eocene occurred in California (Section 5.5). In this context, Warren's Bullhead Bed discoveries, if regarded as genuine implements, do not seem so out of place.

In the discussion that followed Warren's report, Mr. Dewey, one of the scientists present, pointed out that in some cases the Kent eoliths and Moir's rostro-carinates are found in the middle of Tertiary sedimentary beds and not directly on the hard chalk. This circumstance, said Dewey, would rule out the particular pressure explanation given by Warren.

Warren had displayed some specimens during his talk. But Reginald Smith complained that Warren (and Breuil in France) had compared their natural productions with only a few of the very poorest eolith specimens. Smith accused Warren of discouraging research in early deposits.

The record stated: "Mr. H. Bury thought it unfortunate that such a discussion should have been raised without a fair representation of both sides of the case among the exhibits. The author [Warren] and Mr. Haward had brought forward the best specimens that they could find in support of their case; but for comparison they only produced some half dozen very inferior Kentish eoliths, and no sub-Crag implements at all. It was a mistake to suppose that believers in Pliocene man had ignored these pressure-flaked flints from the Eocene beds; on the contrary, the differences in detail which they observed between the two categories formed an essential factor in their argument" (Warren 1920, p. 251).

Bury's point is well worth noting, for one often encounters something like the following in discussions of eoliths by their detractors. The skeptical authority will point out that such and such scientist found in Tertiary strata stone objects he incautiously believed to be of human manufacture and that the discovery was a matter of controversy for some years until such and such scientist delivered his definitive report that conclusively demonstrated that the stone objects had been produced by the pressure of the overlying layers. But in recounting this history the skeptical authority ignores the fact that the original discoverer had carefully considered and dismissed that very possibility. In considering the eolith question with an open mind, one learns to be suspicious of definitive disproofs, which often turn out to be quite rickety intellectual contraptions.

The notes of the discussion also recorded the following ironic remarks by one of the members of the Geological Society: "Mr. A. S. Kennard congratulated the author [Warren] on an important discovery, and considered that the paper strongly supported the claim for the human origin of the Kentish eoliths. He agreed with the author that it was unfair to decide from a few examples, and that the proper test was the whole group. Judged by this standard, neither of the series shown [by Warren] resembled the Kentish eoliths, since the more numerous and characteristic specimens [shown by Warren] were quite unknown on the Plateau" (Warren 1920, p. 251). Kennard thus turned the tables on Warren, taking his attempt to dismiss the eoliths as proof of their genuineness.

3.3.8 An International Commission of Scientists Decides in Favor of Moir

At this point, the controversy over Moir's discoveries was submitted to an international commission of scientists for resolution. Coles (1968, p. 27) related that this group "was overwhelmingly in support of Moir's conclusions, that the flints from the base of the Red Crag near Ipswich were in undisturbed strata, and that some of the flaking was indubitably of artificial origin." In the words of the commission report: "The flints are found in a stratigraphic position, without trace of resorting, at the base of the Red Crag. A certain number of the flints do not appear to have been made by anything other than voluntary human action" (Lohest et al. 1923, p. 44).

The commission, formed at the request of the International Institute of Anthropology, was composed of Dr. L. Capitan, professor at the College of France and the School of Anthropology; Paul Fourmarier, professor of applied geology at the University of Liége and the School of Anthropology; Charles Fraipont, professor of paleontology at the University of Liége and the School of Anthropology; J. Hamal-Nandrin, professor of the School of Anthropology at Liége; Max Lohest, professor of geology at the University of Liége and the School of Anthropology; George Grant MacCurdy, professor at Harvard University; Mr. Nelson, archeologist of the National Museum of Natural History of New York; and Miles Burkitt, professor of prehistory at the University of Cambridge (Lohest et al. 1923, p. 54).

The commission wanted to settle the following questions (Lohest et al. 1923, p. 53): "(1) At the point where the flints considered worked were discovered, is it established that the strata in which they were found are definitely Pliocene and that no action of resorting or intrusive deposition is responsible for the introduction into ancient beds of modern objects? (2) Are the flints found among rocks or other conditions that could have produced pseudo-retouching by impact or pressure?" Concerning the flints themselves, the commission was to answer the following questions: "(1) Are the flints of the Crag worked, retouched, or utilized? (2) Can the retouching be compared to that produced by natural physical action? (3) Can one affirm that the flaking and retouching are due to intelligent and voluntary work?"

To answer these questions, the commission visited the principal sites where Moir had collected his specimens, including locations at Ipswich, Thorington Hall, Bramford, and Foxhall Road. They also examined the collection at the Ipswich Museum, the personal collection of Moir, and Warren's collection of pressure-flaked flints from the Bullhead Eocene beds. Also visited were the collections at the Cambridge Museum and the British Museum at South Kensington, as well as the collection of Mr. Westlake at Fordingbridge near Salisbury, which included his enormous collection of flints from Puy Courny and Puy de Boudieu near Aurillac, France (Lohest et al. 1923, p. 54).

The geologists Max Lohest and Paul Fourmarier reported on the stratigraphy of Moir's discoveries. Lohest and Fourmarier stated: "The purpose of our mission to Ipswich was to verify whether flints showing indisputable signs of intentional work are in fact encountered in undisturbed Tertiary strata" (Lohest et al.

1923, p. 54). These two experts confirmed, at Thorington Hall, that the Red Crag lies upon the Eocene London Clay, and that at the bottom of the Red Crag there is a "detritus bed," which contains flints (not rolled), flint pebbles, phosphate nodules, fossil remains of deer, and also flints showing signs of intentional work.

Lohest and Fourmarier reported: "After minute examination, we believe we can affirm that the Red Crag, because of its cross-bedded stratification and numerous fossils at the pit at Thorington Hall, constitutes incontestably a primary deposit in place, not reformed, and that the deposit is Pliocene and formed in the immediate vicinity of the seashore. If the flints of this deposit are really the work of an intelligent being, then there is no doubt, according to us, that this being existed in England before the great marine invasion of *Trophon antiquum,* considered by all geologists as dating to the late Tertiary epoch" (Lohest *et al.* 1923, pp. 55–56).

J. Hamal-Nandrin and Charles Fraipont also reported on the geological considerations: "The detritus bed from which the flints are recovered is surmounted by several meters of Red Crag deposits containing Pliocene shells. The Red Crag is apparently an ancient shore, and the shells accumulated in the sand on the actual shore. There are very delicate shells, such as bivalves; many are found whole, and the least pressure, the least touch, causes them to break. A deposit of this type is primary, not composite or resorted (*remanié*). It is in the underlying detritus bed that the flints are found. At Thorington Hall the detritus bed lacks many rocks. It contains coprolites, phosphate nodules, and only some small flint pebbles. The superimposed Red Crag is also almost without rocks" (Lohest *et al.* 1923, p. 57).

Hamal-Nandrin and Fraipont then stated: "The rarity of rocks does not permit us to suppose that the flints may have been retouched by shocks or pressure *in situ.* It had to be done, either naturally or artificially, before their incorporation into the beds. Below the detritus bed is the London clay, from which some rolled blocks have been incorporated into the detritus bed. The detritus bed contains, along with bones of whales, fossils of terrestrial mammals certainly characteristic of the Pliocene. This gives evidence that it was upon an ancient land surface that the sea of the Late Pliocene deposited the Red Crag, a shoreline formation at this point. If the flints from below the Red Crag at Thorington Hall, in undisturbed strata, give signs of intelligent work, the being that used them is Pliocene" (Lohest *et al.* 1923, p. 57).

Hamal-Nandrin and Fraipont then turned their expertise to determining the presence of signs of intentional work on the sub-Crag flints: "A certain number of the pieces collected from below the Red Crag, and now found in the collections of Mr. Reid Moir and the Ipswich Museum, present, in our opinion, the characteristics that distinguish worked flints: a striking platform, clear bulb of percussion, and edges with series of small flakes removed, indicating intentional retouching and utilization as a tool. If you were to find these in strata of the Mousterian period, you would not hesitate to say that they are tools showing intentional work and utilization. . . . In our present state of knowledge, we cannot see that anything other than intelligent action could be capable of producing such effects. . . . At Thorington Hall, the rarity of stones and their dispersal does not permit us to suppose that the flints have been naturally retouched by impact or pressure. One can observe that in the level where the specimens are found one does not find any worn and fractured flints other than the ones appearing to be the result of intentional work. The worked flints are not only rare, but extremely rare, according to prehistorians who have studied the strata" (Lohest et al. 1923, p. 58).

After studying several of the collections of flints previously mentioned, Hamal-Nandrin and Fraipont declared themselves in favor of Moir's view that the sub-Crag flints were implements of human manufacture. They further stated: "The chipped edges of the flints collected by Mr. Warren from the Eocene Bullhead beds, along with those produced artificially by him, are very different from the edges of those belonging to the detritus beds below the Crag at Ipswich" (Lohest et al. 1923, p. 58).

Capitan's report also supported Moir's position, both on the sub-Crag finds and those from the Cromer Forest Bed and related formations on the Norfolk coast. Capitan noted that the Pleistocene Boulder Clay had yielded to Moir and others some rare specimens of Mousterian type. But the middle glacial gravels below the Boulder Clay, according to Capitan, contained an enormous number of flints modified by glacial action. The flakes and their pseudoretouching from purely natural causes became an object of special study and consideration, for the precise purpose of comparison with the flints recovered from below the Red Crag. Certain pieces from the glacial gravels did, however, appear to be clearly worked, resembling the Chellean and pre-Chellean types of tools. They were chipped in simple fashion and had a bright characteristic patina (Lohest et al. 1923, p. 59).

Capitan described the Red Crag as a sandy clay, colored red by oxides of iron, containing isolated siliceous stones, phosphate concretions of round small size, fragments of shells, rare shark teeth and even more rare whale bones, and also relatively small pieces of fractured flint. These elements, he noted, were concentrated in a layer at the base of the Crag. Capitan stated: "This is the detritus bed. It is here exclusively (except at Foxhall where there is a second bed almost the same as this) that one finds, only after great trouble, isolated in the midst of the sands, and never in contact with other flints, some flakes and pieces of broken flint, and even more rarely the typical Red Crag specimens" (Lohest et al. 1923, p. 60).

Members of the commission carried out four excavations into the detritus bed over the course of four days and found five or six typical specimens. Capitan stated: "I will not neglect to say that the flints were absolutely in place in compact terrain; two reposed at Thorington Hall on the underlying clay. . . . at Thorington Hall you have a detritus bed covered by marine sands. So everything there is from either before or contemporaneous with the sea that deposited the Crag" (Lohest *et al.* 1923, p. 60).

Studying the specimens of Moir and those at the Ipswich Museum, Capitan categorized them as doubtful, probable, and definite. About half the total specimens were in the doubtful category, with almost another half in the probable category. In the probable group were all those flakes that showed traces of adaptation or retouching identical to that on accepted tools. Capitan stated: "We consider that the greatest number of these pieces are genuine tools bearing diverse traces of intentional work which one can distinguish, with practice, from natural fracturing and flaking. But if someone wants to express doubts, then we leave the discussion to them and will not seek to demonstrate the intentional work" (Lohest *et al.* 1923, pp. 61–62).

But Capitan stated that in addition to the many specimens in the probable category the commission recognized twenty pieces as indisputably worked: "They are of definite form, exactly like accepted Mousterian pieces. These are not freaks of nature or naturally broken stones used without modification as tools—they were products of volition, and show signs of a definite intent to construct a particular kind of tool" (Lohest *et al.* 1923, p. 62). The commission selected eleven pieces for reproduction in their report: two Mousterian-like side scrapers (*racloirs*), two discoidal end scrapers (*grattoirs*), two points, two blades (one with much retouching), an actual handaxe, a sort of big chisel, and a big retouched piece of the *grattoir* form.

Capitan, praising the rigorous scientific procedures applied by Moir and his collaborators, then stated: "One might object that the small number of definite specimens is not sufficient, but this is due to the extremely rigorous process of selection. We are persuaded that a great many of the ones not selected are also worked" (Lohest *et al.* 1923, p. 62). Capitan added: "The small number selected for this demonstration is deliberate because their legitimacy as products of human industry cannot in the least be challenged even by technical experts" (Lohest *et al.* 1923, p. 62).

Capitan concluded: "We need not uselessly continue the discussion about whether these pieces are worked or not, giving undue attention to explanations from incompetents. For any person who has any real acquaintance with the characteristics of worked flints, such questions will not come up" (Lohest *et al.* 1923, pp. 62–63). If one rejected Moir's finds, stated Capitan, then one would have to reject about 80 percent of the generally accepted Mousterian pieces

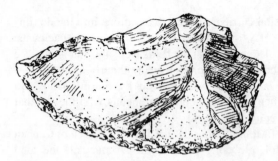

Figure 3.11. A side scraper (*racloir*) discovered beneath the Red Crag at Thorington Hall, England (Lohest *et al.* 1923, p. 63).

(Lohest *et al.* 1923, p. 63).

Capitan next described some of the undisputed specimens. These came from Thorington Hall, Bramford, and the Bolton Company brickfield. From Moir's reports (1924), it appears that the primary tool-bearing layer at each of these sites is the detritus bed below the Red Crag. This would make the flint tools Capitan described at least 2.5 million years old. And because the detritus bed contains materials from ancient Eocene land surfaces, the tools might be up to 55 million years old.

Concerning an implement from below the Red Crag at Thorington Hall (Figure 3.11), Capitan said: "The very best piece . . . is a great and thick *racloir* (side scraper) fashioned from an irregular oval flake, with numerous bulbs of percussion. It is of the same form as many of the most typical Mousterian *racloirs,* and like them it is retouched on all sides. On the outer surface, near the point of the instrument . . . a carefully retouched depression accommodates a finger for gripping the implement. In truth, this is a piece that can just as much be said to have been manufactured by humans as the best Mousterian *racloirs.* On the plane surface, on the other end of the implement . . . is an enormous bulb of percussion" (Lohest *et al.* 1923, p. 63).

Of two discoidal *grattoirs* (end scrapers) recovered from Thorington Hall (Figure 3.12), Capitan stated: "Made from thick flakes, and carefully retouched

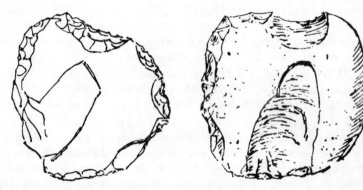

Figure 3.12. Two discoidal scrapers from below the Red Crag at Thorington Hall, England (Lohest *et al.* 1923, p. 64).

all around, they both have in the middle of the upper surface a long deep flake removed. On the other side of each, which is smooth, there is a bulb of percussion" (Lohest *et al.* 1923, p. 63).

In using a *grattoir,* or end scraper, the scraping edge of the implement is held lengthwise along the line of force (or end first). In using a *racloir,* or side scraper, the tool's scraping edge is held perpendicular to the line of force (or sideways).

In addition, Capitan drew attention to a particular implement (Figure 3.13) that he described as being "well retouched on every side and having an extremity terminating in a bevelled edge carefully made by regular retouching" (Lohest *et al.* 1923, p. 64).

Capitan also noted "a big *racloir,* with the cortex partially removed and with the cutting edge carefully dressed and adapted by a series of regular and multiple retouchings. This edge is so perfectly rectilinear as to give clear indication it is of human origin" (Lohest *et al.* 1923, p. 65).

Another implement (Figure 3.14) was retouched on two of its edges and displayed on its face three long flake scars. The fact that the three flake scars on the implement were parallel was, according to Capitan, a certain sign they were deliberately removed in succession. He believed this specimen from below the Red Crag appeared to be a handaxe absolutely identical to the best pre-Chellean types from the Somme region of France.

Capitan (Lohest *et al.* 1923, p. 66) described another specimen as follows:

Figure 3.13. An implement from below the Red Crag (Lohest *et al.* 1923, p. 65).

Figure 3.14. An implement from below the Red Crag at Bramford, England (Lohest *et al.* 1923, p. 66).

"A thin blade with a bulb on the inferior surface, and a very precise imprint of a second blade removed from the upper part. This work is absolutely human" (Figure 3.15). Yet another object illustrated in Capitan's report was a pointed implement, with an apparent bulb of percussion visible at the base (Figure 3.16).

In concluding his analysis, Capitan definitively stated that "there exist at the base of the Crag, in undisturbed strata, worked flints (we have observed them ourselves). These are not made by anything other than a human or hominid which existed in the Tertiary epoch. This fact is found by us prehistorians to be absolutely demonstrated" (Lohest *et al.* 1923, p. 67).

Surprisingly, even after the commission report, Moir's opponents, such as Warren, persisted in attempting to show that the flint implements from beneath the Red Crag and elsewhere were the product of some kind of natural pressure flaking.

Moir and Barnes kept defending their position and picked up supporters. Coles (1968, p. 29) stated: "In 1932 T. D. Kendrick outlined some of the different viewpoints, and came down strongly in support of Moir, not so much on the geological problems involved as on the character of some of the flints." About Moir's flints and other Eolithic industries, Kendrick said that "many of them are to be regarded as 'probably artifacts,' while there are one or two (in the British Museum) . . . that I feel certain are man's handiwork" (Coles 1968, p. 29).

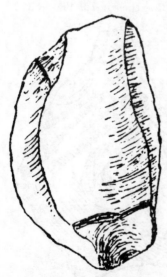

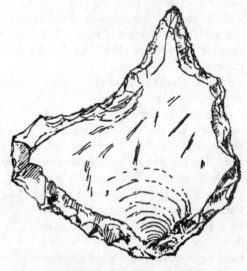

Figure 3.15. A blade implement found beneath the Red Crag formation at Bramford, England (Lohest *et al.* 1923, p. 66).

Figure 3.16. A pointed implement from below the Red Crag formation, England, thought to be from Late Pliocene to Eocene in age (Lohest *et al.* 1923, p. 65).

3.3.9 Continued Opposition

As far as the opposition was concerned, their attempted counterexplanations became more strained; indeed, it seems no proposal was too extreme to win the support of those who for one reason or another could not find room for Moir's discoveries within the bounds of their paleoanthropological parameters.

Coles (1968, p. 29) informs us: "One of the final statements was made by Warren in 1948 in an address to the geological section of the Southwestern Union of Scientific Societies. . . . He agreed with Moir in considering that, at the present day, wave action was not an effective process in the fracturing of flint in a way comparable to that seen on Moir's Crag specimens, but tried to find some other natural process that could have flaked the submarine flints exposed by erosion of that Chalk. Warren concluded that during the formation of the Crag deposits, the area must have been subject to the arrival of icebergs from the north. Such ice, grounding near the shores of the Crag sea, might well have caused the pressure-crushing and striation of the flints exposed on the sea bed. These arguments, apart from being practically the last word in the controversy, also neatly disposed of many of the points made by Moir about the differences between sea action fractures and his 'implements,' and allowed the exposure and deposition of fragile marine shells amidst the ice-fractured stone beds."

We do not yet have in our possession a copy of Warren's 1948 address, but one gets the impression, from Coles's account, that the iceberg hypothesis was a somewhat desperate exercise in pure speculation. One wonders whether icebergs move onto shorelines in the manner suggested by Warren; and granting that they may, is there at present any hard evidence, anywhere in the Arctic or Antarctic regions, suggesting they have produced implementlike objects in the manner suggested by Warren? To our knowledge, no one has given any proof that icebergs can produce the numerous bulbs of percussion and elaborate retouching reported above by Capitan. Furthermore, as pointed out previously, many of the Red Crag specimens are lying in the middle of sediments and not on hard rock surfaces against which an iceberg might have crushed them. In addition, Coles (1968, p. 29) reported that at Foxhall implements occur in layers of sediment that appear to represent land surfaces and not beach deposits. This would also rule out the iceberg action imagined by Warren.

3.3.10 Silence Ends the Debate

After Warren put forward his iceberg explanation, the controversy faded. Coles (1968, p. 28) wrote: "That . . . the scientific world did not see fit to accept either side without considerable uncertainty must account for the quite remarkable inattention that this East Anglian problem has received since the days of active controversy." This may be in part true, but there is another possible explanation—that the scientific community decided silence was a better way to bury Moir's discoveries than active and vocal dissent. By the 1950s, with scientific opinion lining up solidly behind an Early Pleistocene African center for human evolution, there would have been little point, and perhaps some embarrassment and harm, in continually trying to disprove evidence for a theoretically impossible Pliocene habitation of England. That would have kept both sides of the controversy too much alive. The policy of silence, deliberate or not, did in fact prove highly successful in removing Moir's evidence from view. There was no need to defeat something that was beneath notice, and little to gain from defending or supporting it either.

3.3.11 Recent Negative Evaluations of Moir's Discoveries

Although most modern authorities do not even mention Moir's discoveries, a rare notice of dismissal may be found in *The Ice Age in Britain,* by B. W. Sparks and R. G. West (1972, p. 234): "The beginnings of tool manufacture are shrouded in doubt by the similarity of primitive tools to naturally-occurring flaked pebbles. The earliest dated tools identified are found in Africa (Lower Pleistocene, 1.75 million years) and are of the so-called chopper tool or pebble tool type, made by striking a few flakes from the side of a pebble in one or two directions. Such an industry has been associated with *Homo habilis* and *Homo erectus*. In Britain such Lower Pleistocene industries have not been found. But early in this century many flints from the Lower Pleistocene Crags were described as being artifacts, such as the flints, some flaked bifacially, in the Red Crag near Ipswich, and the so-called rostrocarinates from the base of the Norwich Crag near Norwich. All are now thought to be natural products. They do not satisfy the requirements for identification as a tool, namely, that the object conforms to a set and regular pattern, that it is found in a geologically possible habitation site, preferably with other signs of man's activities (e.g. chipping, killing, or burial site), and that it shows signs of flaking from two or three directions at right angles." Sparks and West, of Cambridge University, are experts on the Pleistocene in Britain.

Briefly responding to Sparks and West, we may note that Moir and other authorities, such as Osborn and Capitan, were able to classify the Crag specimens into definite tool types (handaxes, borers, scrapers, etc.) comparable to those included in accepted Paleolithic industries, including the Mousterian. The Foxhall site, with the Foxhall jaw, was taken by many authorities to represent a geologically possible habitation site. Moir (1927, p. 33) considered it to be a workshop area and noted signs of fire having been used there. As far as flaking from several directions at right angles is concerned, this is not the only criterion that might be applied for judging human workmanship upon stone objects. Even so, M. C. Burkitt of Cambridge (1956, p. 104) did find flaking from several different directions at right angles on some of the implements that were collected by J. Reid Moir.

Among other scientists who opposed Moir's discoveries and saw fit to say so in print was K. P. Oakley. Coles (1968, p. 29) stated: "Although Oakley (1961) goes so far as to say that 'the chipping in some cases suggests intelligent design,' he believes that none can be accepted without some reserve." As have many other opponents of crude stone tool industries, Oakley included in his book some illustrations of natural products that supposedly resembled objects thought to be implements. Leland W. Patterson (1983, p. 303) has responded: "As an example of superficial observation, the author has received comments that the edge damage on natural flakes illustrated by Oakley resembles retouch patterns of unifacial tools. A careful examination of Oakley's illustrations shows that the flake scars do not form a uniform pattern as is characteristic of the results of perpendicular force applications in making unifacial tools. In Oakley's illustrations, flake scars at the edge go at a variety of angles from the plane of the ventral face of the specimens, instead of being parallel flake scars mainly perpendicular to the plane of the ventral face. Flake scars also vary widely in size."

Yet another late-twentieth-century opponent of eoliths was F. Clark Howell. Coles (1968, pp. 27-30) stated: "Howell [1966, p. 89] dismisses all of this material by stating that 'the angles of fracture and the nature of the flake removal . . . fall outside the range of variation of specimens known otherwise to be of human manufacture,' but this is surely not a valid basis for rejection, particularly in view of the variability of known industries of the Lower and early Middle Pleistocene throughout the Old World." We fully agree with Coles on this point and shall more fully discuss the important matter of angles of fracture later in this chapter.

3.3.12 A Slightly Favorable Modern Review of Moir's Finds

Coles himself provides an exception to the usual instinctive rejection of Moir's discoveries (or complete silence about them). He felt it "unjust to dismiss all this material without some consideration" (Coles 1968, p. 22). But as we shall see, Coles did, after some consideration, dismiss almost all of it.

Concerning the Forest Bed discoveries, Coles (1968, pp. 24, 27) stated: "Of the immense quantity of flints available, only a small proportion were flaked, and Moir believed, rightly it seemed, that wave-action could not have caused this fracturing. Most of the flaked pieces were irregular, but a few straight-edged retouched flakes occurred. Moir examined other areas of foreshore, to serve as a check on natural flaking in exposures, and claimed that there were no struck flakes outside his 'workshop-sites,' which had yielded both fractured flakes and cores."

But then Coles (1968, p. 27) shifted to negative expressions: "These sites, however, are not generally accepted as showing any sign of man's activity." In fact, there is no "general acceptance" of any of Moir's sites. But to what extent does general acceptance reflect the actual truth regarding the human manufacture of Moir's implements? Coles himself admitted that the flaking on the Forest Bed specimens had probably not been accomplished by the action of waves but did not himself propose any specific alternative explanation.

Coles (1968, p. 24) went on to say: "The sites lay on the foreshore, and Moir believed that the occupation had taken place on the Stone Bed, and that it should therefore extend under the cliffs at Cromer." But Coles (1968, p. 27) asserted that "the flint deposit is believed to occur only on the foreshore, and not to extend under the Cromer Forest Bed in the cliffs at Cromer."

Coles appears to have been wrong about this. West (1980), who conducted extensive geological research on the Cromer Forest Bed Formation, made several references to the Cromer Stone Bed underlying the Cromer Forest Bed formations. He identified it as the source of the flints found on the foreshore at various locations and said it was of the same general age as the top part of the Norwich Crag (Table 2.1, p. 78). The Stone Bed, and any implements from it, would thus be about 1.0 to 1.5 million years old.

And about the Foxhall implements, from bands of black sediment in the middle of the Red Crag, Coles (1968, p. 29) had this to say: "they, and they alone, were stratified in such a position as to make their presence and fracturing *in situ* under the conditions envisaged by Warren most unlikely."

This statement is not completely accurate. Capitan reported that the implements from the detritus bed below the Red Crag were also found in conditions that ruled out natural fracture by either pressure or impact (Section 3.3.8). As at Foxhall, the implements were found in sandy deposits, distant from other pieces of flint. Burkitt made similar observations (Section 3.3.13).

In any case, Coles (1968, p. 24) made this favorable comment about the implement-bearing layers at Foxhall: "Above and below were horizontally stratified clean sand deposits, showing no evidence of natural agencies sufficient to flake the flints found sporadically in the two dark layers." The flints were also unrolled. Their sharp edges indicated to Coles that the flaking was human in origin. The random battering of natural forces tends not to preserve sharp edges.

Coles (1968, p. 29) further explained that the dark layers in which the flints were found "may represent temporary periods of land exposure during a general marine phase in this area." In other words, the layers represent a probable habitation site. Coles (1968, p. 29) added that the relative rarity of the flints, as well as the fact that it was hard to account for their presence by natural means, indicated that they arrived at their positions in the dark layers in the Red Crag by artificial (that is to say, human) agencies.

"Unfortunately, however," said Coles (1968, p. 29), "few of the flints found by Moir are convincing; a number are small flakes little over one inch in length, others are larger with edge flaking. One or two are bifacially retouched."

The presence of bifacial retouch (retouching on both sides of an edge) is an extremely good indication of human manufacture. Leland W. Patterson, an expert on lithic technology, stated (1983, p. 304): "random forces could seldom produce a long interval of bifacially retouched edge that is sharp. Natural fractures tend to produce blunt and rounded bifacial edges, because of the steep transverse nature of most natural fracture." That even one bifacially retouched implement was found at Foxhall is highly significant. It means that the other flakes cannot be so easily dismissed.

The fact that many of the flaked objects found at Foxhall are small does not rule out human manufacture. At many sites, small flakes are regarded as byproducts of the tool manufacturing process. Another possibility is that the small flakes themselves might have been used as implements. John Gowlett (1984, p. 144) wrote in *Ascent to Civilization:* "Microliths are very small stone tools, generally 3 cm long [about 1.2 inches] or less, made from small flakes or segments of blades. Usually one side has been blunted by the 'backing' technique, a form of retouching in which tiny flakes are struck off the edge. . . . flakes which lack retouch are just as likely to have been used as tools."

In fact, microliths, which occur principally in the Middle and Late Stone Ages, are regarded as a technological advance upon the earlier large handaxe industries. They are typical of *Homo sapiens,* and are identified with highly evolved cultural activities such as agriculture and bow-and-arrow hunting. For example, Gowlett (1984, p. 145) stated that "the tools were sometimes fitted end-to-end, in a row, into a curved blade desirable in a sickle." Therefore, small size alone should not lead one to label stone flakes "unconvincing" as tools.

Coles himself (1968, p. 29) noted that one should be careful in ruling out human workmanship simply because stone objects do not appear convincing: "it must be born in mind that a number of the flakes from a site such as Vértesszölös . . . might also not have been accepted as demonstrating human workmanship if they had not been found on an undoubted working floor, in association with other human activities."

Is there any evidence at Foxhall, in addition to the flaked flints, that might lend support to a human presence? The answer to this question is yes. First of all, the variety of flints and flakes found at Foxhall suggested a workshop location. Second, Moir noted the presence of burned stones, a sign that fire had been used at the site. And, finally, as previously noted, a fully human jawbone was recovered at Foxhall, from the same levels that contained the stone implements.

Confronted with this uncomfortable fact, Coles (1968, p. 28) lapsed into the reflexlike response typical of scientists with strong preconceptions about what might and might not be found in strata of certain ages: "As far as Foxhall is concerned, the presence of the jawbone, quite clearly *Homo sapiens,* suggests disturbance of some sort. Perhaps local landslip has occurred, bringing an upper Crag deposit on top of a recent land surface, which itself overlay Crag sands *in situ.*" But Coles did not provide any actual geological evidence that such a "landslip" had actually occurred. Coles's proposal adds nicely to our collection of examples demonstrating how scientists adhering to the particular view of human evolution now in vogue must often engage in speculative mental exercises in order to bring anomalous evidence within the bounds of an acceptable time frame.

But despite his generally negative opinion about Moir's discoveries, Coles nevertheless felt that three particular implements were worthy of further study. These were: (1) "the undoubted handaxe apparently from the Cromer Till at Sidestrand in Norfolk" (Coles 1968, p. 29); (2) a handaxe from the Stone Bed at Whitlingham; and (3) a handaxelike implement from the detritus bed below the Red Crag at Bramford. According to Coles (1968, p. 29), these three objects were the "one positive source of support for Moir's views." Otherwise, Coles felt that Warren's iceberg hypothesis was essentially correct.

Here we would like to emphasize that we do not share Coles's suspicion that Warren's highly speculative iceberg hypothesis is preferable to the findings of the international commission of geologists and anthropologists, who held that Moir's implements were definitely made by humans. Therefore, we do not believe that the final decision about Moir's discoveries must rest solely on the interpretation of the three test specimens mentioned by Coles. Nevertheless, they are significant, and we shall now examine them, beginning with the Sidestrand find.

Moir (1923, p. 135) gave this description of the Sidestrand handaxe discovery (Figure 3.17): "The specimen was discovered lying upon its flat under-surface, and firmly embedded in Boulder Clay at the foot of the cliff, which passed directly into, and was apparently part of, the underlying mass."

Coles (1968, p. 27) mentioned that the Boulder Clay at Sidestrand, Norfolk, in which the "undoubted handaxe" was found, was apparently the Cromer Till. The Cromer Till is from the Anglian glacial period (Table 2.1, p. 78), which began about .4 million years ago. But the handaxe "is believed to have been transported by glacial action from the upper part of the Cromer Forest Bed" (Coles 1968, p. 27).

In this regard, Moir (1923, pp. 136–137) stated: "The occurrence of this specimen in Boulder Clay, a deposit composed solely of *derived* material, makes it certain that the land surface, upon which the implement originally lay, must be looked for in some deposit more ancient than the Till at Sidestrand. Examination of the Cromer Forest Bed immediately underlying the Lower Glacial deposits of the Norfolk Coast has shown me that flints, exhibiting flake-scars of the same colour as those of Mr. Sainty's specimen, occur freely in the Upper Freshwater Bed (the highest division of the Cromer Forest Bed series), and it seems to me very probable that the implement originally belonged to this deposit."

The implement was quite unworn. Moir (1927, p. 47) explained this as follows: "the glacial clay . . . very frequently contains portions of the Fresh-water Bed, which were torn up by the glacier in its advance." The sharp edges of the implement could thus have been preserved by the surrounding Fresh-water Bed materials.

According to West (1980, p. 116), the Upper Freshwater Bed, as defined by J. Reid Moir and his contemporaries, includes materials ranging from the last part of the Cromerian temperate stage, at .4–.5 million years B.P., to the beginning part of the Pre-Pastonian cold stage, at 1.50–1.75 million years B.P. (Table 2.1, p. 78).

At 1.5 million years ago, the Sidestrand specimen, accepted by Coles as a definite handaxe, would be quite anomalous. Handaxes of this sort are usually attributed to *Homo erectus,* but according to the standard human evolutionary theory, at 1.5 million years ago, *Homo erectus* should still have been confined to Africa, where he should only recently have come into being. At .4–.5 million years, however, the Sidestrand specimen would be barely within the range of conventionally accepted stone implements in England.

Let us now consider the two remaining test specimens mentioned by Coles. The first is an implement (Figure 3.18) "apparently from the Stone Bed at the base of the Norwich Crag, at Whitlingham." The Whitlingham site

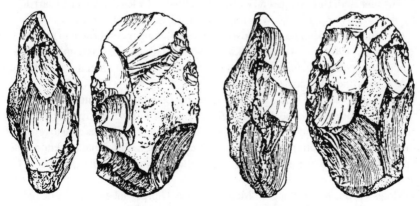

Figure 3.17. Four views of a stone implement from the Cromer Till at Sidestrand (Moir 1927, p. 46). Coles (1968, p. 29) called it an "undoubted handaxe."

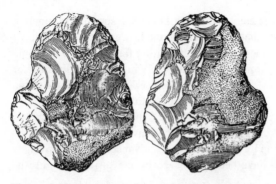

Figure 3.18. Implement from Whitlingham, England (Coles 1968, p. 26, after Moir). Coles (1968, p. 29) called it "convincing as a handaxe." J. Reid Moir said it came from the Stone Bed bed beneath the Norwich Crag, giving it an age of about 2 million years.

is at the foot of a cliff near Thorpe, on the Norfolk coast. Coles stated (1968, p. 29): "On the face of it, this object is convincing as a handaxe. Unfortunately, it was not discovered *in situ* but lay with fallen material at the foot of a tall section. . . . It is possible that this object came from the till and not from the Stone Bed although Sainty and Moir claimed it was definitely from the latter."

But Coles (1968, p. 24) also said: "This implement is in fresh condition, and it is unlikely that it could have survived transport in this condition." This observation suggests that the implement might have come not from the glacial till but from the much older Stone Bed. An implement crushed beneath a moving glacier would probably have had its sharp edges removed. In our discussion of the Sidestrand specimen earlier in this section, we noted that Moir offered an explanation why the handaxe found there was not worn by glacial action—it might have been incorporated within a large piece of Forest Bed sediment taken up by the advancing glacier. Moir backed up this assertion by stating that the glacial clay at Sidestrand does in fact contain intact pieces of the Upper Freshwater Bed. But this special explanation (which Coles did not mention) does not necessarily apply at the Whitlingham site. Therefore, the unworn condition of the Whitlingham handaxe is consistent with its being incorporated in the Stone Bed. Coles (1968, p. 24) noted that the Stone Bed at Whitlingham contained "abundant shells, *in situ,* and unbroken," as well as "many slender nodules of flint . . . also undamaged."

In addition to the handaxe, a good many other flaked flint objects were recovered from the Stone Bed at Whitlingham, England. In regard to these discoveries, Breuil said (1922, pp. 228–229): "Mr. Reid Moir was able to retrieve some pieces in a stratigraphic position at the base of a cliff. That the enormous flakes found there were made by very violent human percussion cannot be doubted."

Coles (1968, p. 24) stated: "Many of these Thorpe flakes were believed to exhibit deliberate flaking. The flakes include irregular forms with even retouch along one or two edges." The presence of these other flaked implements "in a stratigraphic position" at the base of the cliff at Whitlingham tends to confirm the Stone Bed as the source of the handaxe.

The Whitlingham handaxe, if from the glacial gravels that make up the Cromer Till, would be not much more than .4 million years old. But if, as is most likely, the handaxe is from the Stone Bed underlying the Early Pleistocene Norwich Crag, it would be about 2 million years old (Table 2.1, p. 78).

Coles (1968, p. 29) said that his last test specimen (Figure 3.19) "was found at the base of the Red Crag at Bramford in Suffolk and its stratigraphical horizon is not in doubt." He added: "It lay in the Detritus Bed in London Clay and was sealed by Crag sands. It is reminiscent of Chellean axes with triangular sections, but is considerably rolled; although it bears some 25 flake scars, and has lost all its cortex, the irregular nature of the object itself is not convincing."

In another description of the same piece, Coles (1968, p. 24) stated that it was "superficially of handaxe form, with thirteen flake scars upon one face, and twelve upon the other." He added: "These scars appear to have been directed from a multiplicity of positions on the edges, and are sufficiently elongated to overlap at the center of one face, producing thereby a triangular sectioned 'tool'" (Coles 1968, p. 24).

The position of this specimen in the detritus bed beneath the Red Crag means that it is at least Pliocene in age (2–5 million years old). But because the detritus bed contains materials from land surfaces dating back as far as the Eocene, the handaxe could be as much as 55 million years old.

All in all, Coles, in spite of his negative conclusions, can be commended for his willingness to discuss Moir's discoveries. At the end of his review, he stated: "A fair comment on the East Anglian material would, I think, be concerned to point out that the typology of the claimed implements was not necessarily outside the range of variation known from humanly worked industries in Europe and Africa, but that we have very little information about the natural flaking processes available in East Anglia in early Pleistocene times, some of which might well have been

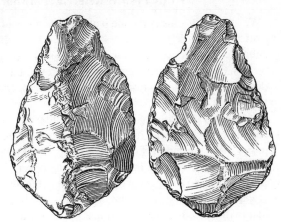

Figure 3.19. Handaxe from below the Red Crag at Bramford, England (Moir 1935, p. 364). It could be anywhere from 2 million to 55 million years old.

capable of producing flaked flints including bifacially-worked 'handaxes'; no natural sources are known today which could do this under observation. Our greatly augmented evidence about the chronology of early tool-making in other parts of the world continues, however, to suggest how extraordinary it would be if the East Anglian Crag industries were of human manufacture" (Coles 1968, p. 30). This is an incredible line of reasoning. No natural forces known to today's scientists can account for the production of the handaxes and other flaked implements. Nevertheless, Coles hesitates to accept them as the product of intentional human work.

It may be that in terms of the "greatly augmented evidence" available to Coles, human manufacture of the East Anglia specimens would seem extraordinary in terms of "chronology," that is to say, their unexpected age. But in terms of the even more greatly augmented evidence presented in this book, human manufacture of the East Anglia implements during the late Tertiary and earliest Pleistocene would seem quite within the bounds of the ordinary.

In this regard, a modern authority, Gowlett (1984, p. 76), reported that four flakes and five pebble choppers were found at Le Vallonet, southern France, in old beach sediments dated 1–2 million years old. If we assign these eolithlike stone tools to the oldest part of their probable date range, they would be roughly contemporary with some of the East Anglia specimens, such as those from Foxhall. Gowlett called the Le Vallonet specimens doubtful, yet he mentioned them in his book. He did not, however, mention Moir's discoveries.

3.3.13 Positive References to Moir's Finds

We shall now consider some isolated examples of positive scientific reporting on J. Reid Moir's discoveries from the latter half of the twentieth century. Cambridge University archeologist and anthropologist, M. C. Burkitt, who served on the international commission that examined Moir's implements in the 1920s, gave favorable treatment to them in his book *The Old Stone Age*, published in 1956.

Burkitt was particularly impressed with the site at Thorington Hall, 2 miles south of Ipswich, where flint implements had been collected from the Crag deposits. "At Thorington Hall bivalve shells with the hinges still intact have been collected from just above the artifacts. This is very important evidence for the prehistorian, as no subsequent differential movement of the gravel, such as might have caused fracturing of the contained flints, can have taken place, since it would certainly have led to the smashing of the delicate hinges of these shells. Incidentally, too, at this site, as well as at Foxhall, the deposit in which the specimens occur is of a sandy nature and not packed with pebbles. So even if differential movement had occurred no fracturing due to the pressure of one stone against another could have resulted" (Burkitt 1956, p. 108). That the implements were found isolated in apparently undisturbed sandy deposits also appears to rule out Warren's suggestion (Section 3.3.9) that they were formed by icebergs crushing flint against the underlying chalk.

As far as Foxhall was concerned, Burkitt (1956, pp. 108, 110) stated: "At Foxhall the chipped flints were found at two different levels only, and this can be best explained if we consider that these levels were actually old land surfaces on which man lived, in other words that we are dealing with 'floors' or actual occupation sites."

Burkitt (1956, p. 110) further stated: "The argument that the flints were chipped elsewhere by natural forces and later incorporated in these late pliocene gravels cannot always be maintained. Small flakes, as well as large specimens, occur together and this would not happen under such circumstances, as the selective action of flowing water would cause the smaller and lighter specimens to be collected together at one site and the larger and heavier objects at another." Burkitt's strong arguments in favor of actual living floors at Foxhall help resolve the doubts expressed by Coles and others about human manufacture of the flint objects found there.

Regarding Moir's discoveries from the Cromer Forest Bed formations, Burkitt (1956, pp. 112–113) wrote: "For the most part these consist of large flakes carefully struck off from a core, the striking platform being unfaceted and frequently inclined at a high angle to the main flake surface. Although there is not always any further trimming, a sharp cutting edge has often been obtained. . . . Occasionally more finished tools are found and rarely specimens of a core-tool type such as choppers, etc. have been collected. Essentially, however, it is a flake industry with which we are dealing. . . . It would appear that these chipped specimens were made by men who lived at a time when the earlier beds of the Cromer Forest series were being laid down, for a few undoubted artifacts have been discovered in them, and the horizon at which they occur probably represents the ancient land surface on which these makers of the Cromerian industries wandered, collecting the raw material for their tools from exposures of the stone bed below. Actually most of the Cromer Forest Bed is now also masked by talus."

Burkitt (1956, p. 112) then delivered a striking conclusion about the implements discovered in and below the Red Crag: "the eoliths themselves are mostly much older than the late pliocene deposits in which they were found. Some of them might actually date back to pre-pliocene times." In other words, he was prepared to accept the existence of intelligent toolmaking hominids in England over 5 million years ago. Because there is much evidence, including skeletal remains (as we shall show in our coming chapters), that humans of the fully modern type existed in pre-Pliocene times, there is no reason to rule out the possibility that Moir's implements from the below the Crag formations were made by *Homo sapiens* over 5 million years ago.

Another supporter of Moir's finds was Louis Leakey (1960d, pp. 66, 68), who wrote: "It is more than likely that primitive humans were present in Europe during the Lower Pleistocene, just as they were in Africa, and certainly a proportion of the specimens from the sub-crag deposits appear to be humanly flaked and cannot be regarded merely as the result of natural forces." Implements from below the Crags would, however, be not Early (Lower) Pleistocene but at least Late Pliocene in age.

Leakey (1960d, p. 68) then made an important point: "It must be constantly borne in mind that although simple pebble chopping tools without any more elaborate forms are typical of the Kafuan and Oldowan, similar tools continued to be made and used by the makers of much more advanced cultures, just as we ourselves still use candles although we also have electric light." This observation is essential to understanding lithic remains. There is no reason to suppose that crude stone tools, found in Early Pleistocene or Tertiary beds, must have been made by correspondingly primitive hominids. This is especially true when we consider that examples of much more sophisticated tools, of kinds universally attributed to *Homo sapiens,* occur in beds of the same Early Pleistocene and Tertiary ages (Chapter 5), as do skeletal remains indistinguishable from those of modern human beings (Chapter 6).

These discoveries are not well known, having been forgotten by science over the course of many decades or in many cases eliminated by a biased process of knowledge filtration. The result is that modern students of paleoanthropology are not in possession of the complete range of scientific evidence concerning human origins and antiquity. Rather most people, including professional scientists, are exposed to only a carefully edited selection of evidence supporting the currently accepted theory that protohuman hominids evolved from apelike predecessors in Africa during the Late Pliocene and Early Pleistocene, and that modern humans subsequently evolved from the protohuman hominids in the Late Pleistocene, in Africa or elsewhere. This book is intended to supply those concerned with paleoanthropological studies access to the full range of evidence. Objectively reviewed, the totality of evidence, in the form of incised bones (Chapter 2), stone implements (Chapters 3–5), and human skeletal remains (Chapter 6), suggests that the current theory of an African evolution is erroneous. It appears that toolmaking hominids indistinguishable from *Homo sapiens sapiens* were present in habitable areas all over the planet far back into the Tertiary epoch. This does not, however, rule out the simultaneous presence of more apelike hominids, some of whom may have manufactured some of the most primitive stone implements. In Appendix 2, we catalog selected radical evidence suggesting higher cultural levels in the Tertiary and even earlier.

3.4 BREUIL AND BARNES: TWO FAMOUS DEBUNKERS OF EOLITHS

In paleoanthropology, we sometimes encounter the definitive debunking report—a report that is repeatedly cited as having decisively invalidated a particular discovery or general category of evidence. In the case of European eoliths, two papers are good examples of definitive debunking reports. These are H. Breuil's paper claiming that pseudoeoliths were formed by geological pressure in the French Eocene formations at Clermont (Oise), and A. S. Barnes's paper claiming to demonstrate, by statistical analysis of platform striking angles, the natural origin of Eolithic industries. We shall now review these two papers.

3.4.1 Breuil's Attempt to End the Eolith Controversy

In 1910, Abbé Henri Breuil conducted investigations he thought would put an end to the eolith controversy. In his often cited report ("Sur La Présence d'Éolithes a la Base de l'Éocene Parisien"), Breuil said that for several years his attention had been drawn to the gravel pits of Belle-Assise, near Clermont, in the department of Oise, northeast of Paris. Excavations there had exposed a bed of chalk, which formed the stratigraphic base for the overlying formations. Above the chalk was a bed of clay containing layers of angular pieces of flint, interspersed with layers of gravel and sand. Above the flint-bearing clay was a very thick deposit of greenish Bracheux sands, which belong to the Thanetian formation, at the base of the Eocene (Obermaier 1924, p. 12). Breuil concluded that the flint-bearing beds below the sands must therefore belong to the very beginning of the Eocene. They would thus be about 50–55 million years old according to modern dating. Some modern authorities put the Thanetian formation as far back as the Late Paleocene, at 55–60 million years (Marshall *et al.* 1977, p. 1326). Above the Bracheux sands were gravel deposits from the Pliocene and Pleistocene.

"With the onset of the discussions concerning the question of how the eoliths were produced, " wrote Breuil (1910, p. 386), "I frequently thought that an examination of the broken flints at the base of the Bracheux sands of the Thanetian at Belle-Assise would yield some interesting observations." Breuil gathered specimens over the course of three years, carefully observing patterns of breakage. "I always avoided using metal tools to extract the flints, and also took care to reject those that had been subjected to contact from the picks of the workers. It is somewhat easy, at the moment one extracts a flint, to examine its surfaces, to see if the fracturing has been produced recently, or if the breakage took place before the excavation. The surfaces of ancient fractures always have thin deposits of iron or manganese" (Breuil 1910, p. 386).

Breuil (1910, pp. 386–387) then stated: "Having noted, without any possibility of doubt, the presence of flints with fractures indicative of intentional work and retouching, and thus resembling what are called eoliths, I invited many persons to come and confirm the fact. Capitan, Cartailhac, and Obermaier were able, along with me, to collect these characteristic flint objects with their own hands. Mr. Commont, with whom I had the pleasure of making an inspection of the flint-bearing strata, also collected some specimens. Furthermore, Commont found flints with features resembling intentional work and retouching in various Eocene exposures in Picardy. The stratigraphic position of the discoveries was the same as at Belle-Assise."

Breuil then described specimens that displayed retouching, bulbs of percussion, and striking platforms. Some showed regular bifacial flaking, typical of Late Paleolithic implements. Others had chipping confined to the side of the flake opposite the bulb of percussion, another characteristic of human work. But Breuil (1910, p. 388) warned: "If in our descriptions we use terminology that normally is applied to proper tools of human manufacture, that is nothing more than a convention, a manner of expression, and does not at all signify that we suppose for an instant we are dealing with ancient implements made by people of Eocene or pre-Eocene times."

Breuil felt that human action could be ruled out with complete certainty because the flints were found in an Eocene formation. Like many other scientists, he could not imagine human beings existed in the Eocene, when the mammals known from fossils were, apparently, quite different from those of today. Breuil (1910, p. 406) wrote of "the absolute unlikelihood of the presence, before the deposit of the Bracheux sands or during their deposit, of an intelligent being, a worker of flint."

But if human action were to be excluded, how, then, had the flint objects been produced? Searching for a natural explanation, Breuil (1910, pp. 387–388) wrote: "It is easy to observe that the flints have not been subjected to transport, for their sharpest edges remain intact. From among the processes that could have resulted in their fracture, one can therefore eliminate the mechanical action of water, either of oceans or rivers. Further examination of the fracturing gives evidence of a different kind of mechanical action, which was able to produce facets and impressions analogous to those produced by intentional human work, or by energetic localized force. A bulb of percussion, more or less clearly present, is often found at the point where a flake was taken off from the surface of the parent block. One can totally eliminate a thermal origin of the fractures, because fractures produced by heat, in the form of surface flaking or cracking of the entire flint block, are completely different."

Breuil (1910, p. 403) then presented specimens that he believed shed a very clear light on the mode of production of the "pseudotools" he had reviewed:

"They are pieces of flint which were flaked while in their positions in the interior of the beds, the fragments remaining in contact with each other. It is easy to see that these fragments present conchoidal fracturing, with the production of positive and negative bulbs of percussion" (Figure 3.20).

Conchoidal fracturing is fracturing that results in elevations or depressions shaped like the curved inner surface of a shell. A positive (raised) bulb of percussion is found on the surface of a flake detached from a flint core. The core retains a negative impression of the bulb. Breuil held that the fracturing that produced these bulbs of percussion was the result of geological pressure. But what about the further signs of modification that normally are present on even the crudest eoliths?

To account for this, Breuil also described a few flakes, found adjacent to parent blocks of flint, that had some chips removed from an edge. According to Breuil (1910, p. 403), geological pressure caused this apparent retouching. He proposed that as a flake was detached, it rotated, causing chips to be removed from its thinner edge as it scraped over the surface of the parent block of flint (Figure 3.21).

We shall give careful attention to Breuil's arguments, because similar reasoning has been used in attempts to discredit many of the discoveries discussed in this book.

For example, Hugo Obermaier (1924, p. 4) observed in his book *Fossil Man in Spain*: "The controversy concerning Thenay [France] did not subside until the year 1901,

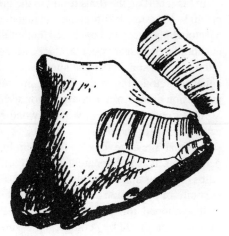

Figure 3.20. Henri Breuil (1910, p. 405) found examples of flakes removed from parent blocks of flint by geological pressure in an Eocene formation in Clermont (Oise), France. Such specimens, he believed, showed that eoliths were not made by human beings.

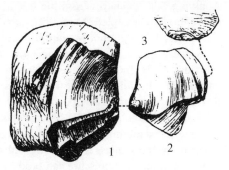

Figure 3.21. (1) Parent block of flint, found in an Eocene formation at Clermont (Oise), France. (2) Flake, apparently removed by geological pressure, found in contact with parent block of flint. (3) Opposite side of flake, with one edge chipped, apparently by geological pressure (Breuil 1910, p. 406).

when L. Capitan and G. d'Ault du Mesnil showed how purely natural agencies might produce effects very similar to human handiwork, one of the most important being earth pressure above the brittle flint."

In his report on the flints found in the gravel pit at Belle-Assise, Breuil (1910, pp. 403–404) stated: "From the fact that the flakes found in connection bore signs of retouching, it can be concluded that retouching, bulbed flakes, and blocks with conchoidal flake scars were produced here exclusively by compression within the interior of the soil. . . . If one attempted to reproduce, on an intact block of flint, either the retouching or the flaking, one would have to employ the processes of percussion and vigorous compression used in working stone."

This might lead one to wonder whether geological pressure was in fact the actual cause for the observed effects. A modern authority (L. Patterson 1983) stated that pressure flaking very rarely produces clearly marked bulbs of percussion. It is not apparent from Breuil's drawings how well developed the bulbs of percussion are on his specimens. Breuil (1910, p. 388) himself described the bulbs of percussion as only "more or less" clearly present. But if the bulbs are well developed, this would, according to Patterson's view, make it unlikely that they were produced by geological pressure.

In general, the bulb of percussion, as the name itself indicates, is taken as a sign of intentional percussive fracturing. But perhaps Breuil was correct in his supposition that geological pressure flaking could produce clear bulbs and retouching, like those found on implements made by humans. In that case, no crudely chipped stone object should be recognized as a genuine tool unless found directly in contact with other unambiguous evidence of human involvement. Applying this standard across the board, one would have to reject numerous conventionally accepted stone tools, such as the many crude Oldowan tools of East Africa that were not found in the immediate vicinity of hominid fossils.

As we shall see, Breuil (Section 3.4.2), like S. Hazzeldine Warren in England (Section 3.3.7), found Eocene objects resembling not only crude eoliths but advanced tools of the Late Stone Age. Breuil and Warren nevertheless believed that all of these toollike specimens—the most sophisticated as well as the crudest—were the product of natural geological forces. This implies that even specimens resembling very good Paleolithic implements should not be securely identified as tools unless found along with definite signs of human habitation. Of course, if geological pressure can produce very good "tools," then even if such "tools" were found along with signs of human habitation, one could not tell if they were produced by nature or by humans. In order to satisfy skeptics like Breuil, it seems one would have to find even the best sort of implement clutched in the fossil fingers of a human hand.

But perhaps Breuil was wrong to suppose that geological pressure caused the bulbs of percussion on the many specimens he found in the Eocene at Belle-

Assise. His only evidence was the few bulbed flakes he found directly in contact with parent blocks of flint. Here we can refer to J. Reid Moir's explanation of the same phenomenon (Section 3.3.6). F. N. Haward had found flakes in contact with parent blocks of flint in the stone bed below the Norwich Crag. Haward said they were removed by geological pressure alone, but Moir suggested the following. Before the flints were covered by the deposit, intentional (presumably human) percussion caused the formation of incipient bulbed flakes, which were later completely removed from the parent blocks by geological pressure or heat.

In any case, taking Breuil's specimens as examples of pressure flaking, there is yet another problem to consider. It can be safely assumed that the specimens pictured by Breuil are among the better examples of flints found with flakes in contact with the parent block. But in studying the illustrations (Figures 3.20, 3.21), it is readily apparent that the flaking and retouching are extremely crude, far more so than that manifest on the other specimens of cores and flakes selected by Breuil as examples of pseudoeoliths (Figure 3.22).

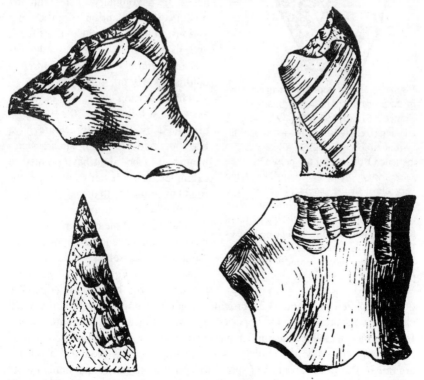

Figure 3.22. These objects, from an Eocene formation at Clermont (Oise), France, were characterized by H. Breuil as "pseudoeoliths" (Breuil 1910, pp. 389, 392, 400, 401).

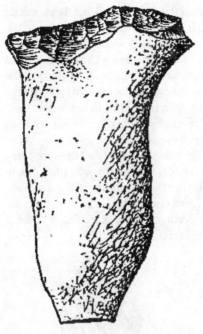

Figure 3.23. A stone object discovered in the Eocene strata at Clermont (Oise), France (Breuil 1910, p. 394). It was characterized by Breuil as a pseudoeolith, produced by geological pressure. As evidence Breuil cited the presence in the same formation of detached flakes lying very close to the parent blocks of flint (Figures 3.20, 3.21). But implementlike objects as sophisticated as the one pictured here were not found with detached flakes lying nearby. This raises serious doubts about the viability of Breuil's geological pressure hypothesis.

It seems, therefore, unfair to insist that the numerous better looking "pseudoeoliths" from the Eocene at Clermont, such as those shown in Figure 3.22, must have been formed by the same process of natural geological pressure flaking that had produced the extremely crude flakes.

But that is just what Breuil did in his report: "By means of this simple mechanical process, which one is able to perceive quite literally, there have nevertheless resulted the fractures, cleavages, terminal and marginal retouchings that simulate with extreme perfection the action of a voluntary agent with the preconceived intention of producing various elementary industrial artifacts, and, in exceptional cases, pseudomorphs of definite implements, not only eoliths" (1910, pp. 403–404).

This assertion does not, however, very easily follow from the examples presented by Breuil. He would have been justified in making such a statement only if he could have pointed to examples of the better looking eoliths found in contact with the parent blocks. And this he did not do.

Also, some of the implementlike objects from the Eocene formation at Clermont were themselves whole pieces of flint, from which chips had been removed to form the working edge. The object depicted in Figure 3.23 provides a good example. The unidirectional chipping concentrated on the upper edge is typical of intentional human work. If Breuil had discovered the implement shown in Figure 3.23 with a dozen or more chips lying alongside the chipped edge, we might be less doubtful about his argument. But in the absence of such a demonstration, intentional human work remains a more viable explanation.

3.4.2 "Two Truly Exceptional Objects" (Eocene)

The unsatisfactory nature of Breuil's geological pressure hypothesis becomes even clearer when we turn our consideration to what Breuil (1910, p. 402) called "two truly exceptional objects, of which the site of discovery, in the interior of the beds, is absolutely certain."

Describing the first object (Figure 3.24), which he characterized as a *grattoir*, or end scraper, Breuil (1910, p. 402) wrote: "The *grattoir* presents a blackish green patina, extremely brilliant, which is present on only a small number of small pieces of flint found in the sands."

The formation of patina occurs where the cortex, or rough outer surface of the flint, is chipped away, exposing the glassy interior to the atmosphere. Breuil (1910, p. 403) observed: "The great majority of the flints are without patination, and their fracturing occurred in the interior of the soil at undetermined times and places."

Breuil believed the presence of a brilliant patina on a small number of the flaked flints in the Eocene formation at Clermont meant they were fractured before they were incorporated into that formation. "Consequently," said Breuil (1910, p. 403), "it can be concluded that the fracturing of these flints occurred in pre-Eocene times." Therefore the pressure fracturing mechanism that Breuil used to explain the eolithlike objects at Clermont would not necessarily apply to the *grattoir* now under discussion.

In further describing the *grattoir* from the Eocene of Clermont, France, Breuil (1910, p. 402) observed: "Its plane of fracture shows a clear bulb of percussion; the other face shows fine and regular retouching, principally on the working edge, with a point at the apex, and on the left border. The chipping is less well-defined on the right side. This object is a veritable pseudomorph of an Azilio-Tardenoisian *grattoir.*" Scientists generally attribute the Azilio-Tardenoisian stone implements to *Homo sapiens sapiens* in the Late Pleistocene of Europe.

Figure 3.24. This flint object was found by H. Breuil and H. Obermaier in an Eocene formation at Clermont (Oise), France (Breuil 1910, p. 402). Breuil said it was identical in form to certain Late Pleistocene implements, but he nevertheless considered it the product of natural geological pressure.

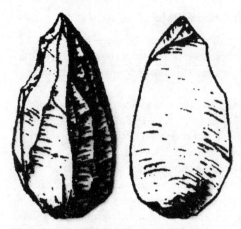

Figure 3.25. A flint object found in an Eocene formation at Clermont (Oise), France (Breuil 1910, p. 402). Although H. Breuil said it resembled a Late Pleistocene pointed tool, he claimed it was formed by geological pressure.

Breuil (1910, p. 402) then stated about the *grattoir*: "That it was discovered in place, at the base of the Eocene sands of Bracheux at Belle-Assise, is a cause of profound stupefaction." Indeed it is. We can see no justification for attributing the highly sophisticated flaking on this piece to the kind of crude pressure flaking exemplified by the few specimens cited above by Breuil. It thus appears that we are confronted with yet another example of a stone object displaying definite signs of intentional human work being found in very ancient strata, in this case over 50 million years old. Significantly, it was found by Breuil and Obermaier in person. So it seems that even these two stalwart eolith debunkers may have unwittingly discovered an anomalously old implement of advanced type.

Describing the second exceptional object (Figure 3.25), which he characterized as "another very curious pseudomorph," Breuil (1910, p. 402) wrote: "It is a very fine lamellar, or scalelike, flake, a little short, with, on its dorsal surface, multiple traces of longitudinal flaking, equally lamellar. At the point, the left side has some fine flaking on the dorsal surface; the other side shows fine chipping, like that produced by a burin. This object itself could be a micro-burin of Eyzies." Les Eyzies is a Late Pleistocene site in France. It would have been quite remarkable to find a piece like this as a flake in contact with the parent block, and with the chips taken from it lying next to it. But nothing remotely approaching this was reported by Breuil. The examples he did cite and illustrate were of the crudest sort possible, being essentially nothing more than randomly fractured pieces of stone.

It is quite remarkable that Breuil should have included two technologically sophisticated specimens, of Late Paleolithic type, in his report without recognizing they were sufficient to demolish his entire argument. He skipped right over them, apparently genuinely unaware of their significance. But objects exactly resembling implements of the Late Paleolithic type, especially when found in an undisputed Eocene stratum, should not be skipped over. We can only request the reader to carefully consider what damage the demonstrated presence of toolmaking human beings over 50 million years ago in France would do to all current evolutionary explanations of human origins and antiquity.

Of course, one can always insist that the two remarkable objects reported by Breuil were products of nature. In that case, one could dismiss any stone tools, including conventionally accepted Late Pleistocene tools, for the same reason.

3.4.3 An Attempt to Trap Rutot

After describing the finds he had made at Clermont, France, Breuil launched an attack on the Belgian scientist A. Rutot, who had found a series of crude stone tool industries during the first decade of the twentieth century (Section 4.4). Breuil (1910, pp. 404–406) wrote: "Is it possible to distinguish the real eoliths from those produced by nature? We have read, from the pen of Mr. Rutot [1906], that 'the recognition and appreciation of eoliths is not simple or elementary, as many persons believe. . . . It can be in certain cases very difficult to distinguish a pseudoeolith from a real one, just as the task of determining the difference between the closely related Cerithes and Pleurotomes is not easy to accomplish at first glance.' If Mr. Rutot were confronted with our flints from Belle-Assise, would he judge them the work of an intelligent being, or simply curious and troubling pseudomorphs? Shown by Mr. Capitan a choice selection of our best specimens, Mr. Rutot, in the absence of information about their stratigraphic position, was willing to formulate his judgement. He considered them to be so well fashioned as to belong to the transition from the Eolithic to the Paleolithic, the Strepyan, according to his system, the primitive Chellean in French usage. According to Rutot, certain specimens 'bear rudimentary traces of intentional work, as might be found in trial attempts.' In others 'the intentional work is of a much better character.' Another 'has been utilized as a scraper, of which it has the character.' Another long piece 'bears on its end attempts at work, for making a dagger or piercer.' Another is 'a very good *racloir* [side scraper], very well worn from use and retouched.' Another is 'a very good *grattoir* [end scraper], equally well worn from use and retouched.' Finally there is a very good 'throwing stone.' Mr. Rutot considers the morphology of the flints of Belle-Assise as characteristic of intentional work, surpassing the simple retouching of natural flakes found in eoliths, and marking the appearance of real intentional manufacture of definite tool types in the dawn of the Paleolithic. Shown the series collected by Mr. Commont, from both Belle-Assise and Picardy, Rutot gave the same diagnosis, though honestly acknowledging he had trouble with the Eocene age of such objects."

If one accepts Breuil's explanation that all of the specimens from Belle-Assise were formed by geological pressures, as demonstrated by a few examples of crudely chipped flakes found in contact with parent blocks of flint, then, of course, Rutot comes off very badly. One can only conclude that the unwitting Belgian geologist foolishly accepted naturally flaked flints as objects of human manufacture. But, as we have shown, Breuil's attempted explanation does not adequately account for all of the implementlike objects found in the early Eocene beds at Belle-Assise and elsewhere. Breuil (1910, p. 287) wrote: "Although parts of broken blocks of flint are frequently found still lying in close connection, this is not the rule, and one does not often find such cases, especially in the sand which is less compacted." It would thus appear that examples of flakes lying next to their parent blocks (Figures 3.20, 3.21) were not all that numerous. Furthermore, the flakes found in contact with the parent blocks did not very closely resemble the many other specimens that Breuil called "pseudotools" (Figures 3.22, 3.23). In particular, the flakes in contact with parent blocks did not at all resemble the two Late Paleolithic type implements found at Belle-Assise (Figures 3.24, 3.25).

Therefore the assumption that all the specimens shown to Rutot were produced by natural forces is unwarranted. The presence of a few naturally broken flints at Belle-Assise does not rule out the possibility that many others, resembling implements, were in fact made by humans, especially since the latter category display more elaborate patterns of chipping than visible in the few specimens demonstrably broken by geological pressure. It is, therefore, quite possible that Rutot's judgements about the specimens shown to him by Capitan were entirely correct, and that Breuil had inadvertently been the discoverer of a new Eolithic industry in the Eocene. Worthy of note is the fact that Rutot found signs of utilization on the edges of many of the specimens. The hypothesis that implemental shapes with signs of wear on the appropriate working edges could have been produced by blind natural forces will induce in at least some unprejudiced minds a sense of improbability.

3.4.4 The Role of Preconception in the Treatment of Eolith Evidence

It can thus be seen that Breuil's main support was simply his unfounded belief that humans or protohumans capable of manufacturing even the crudest stone tools could not have existed in the Eocene. His view was shared by Hugo Obermaier. Many supporters of eoliths have pointed out that modern tribal people, such as the Australian aboriginals, make eolithlike implements. But Obermaier (1924, p. 16) protested: "If, then, from the actual [modern] eoliths we should draw the conclusion that, for the sake of consistency, similar forms from the Tertiary must also be considered as artefacts, we should find ourselves forced to admit the existence of man in Oligocene and perhaps even Eocene times. For these Tertiary products are in no way less 'human' than the corresponding modern forms, and must therefore presuppose similar cultural demands. Both Rutot in regard to Boncelles [Section 4.4], and Verworn in regard to Cantal [Section 4.3], urge the point that the flints from these sites—which really do conform most admirably to the human hand—'appear to have been expressly made for it.' Well, the same is true of Belle-Assise!" It is obvious that Obermaier, like Breuil, was a prisoner of a belief that humans could not have existed in the Eocene. But this belief appears to have been arrived at independently of the available evidence.

Obermaier, citing the work of Max Schlosser, who studied fossil apes at Fayum in Egypt, further stated: "Viewed from the standpoint of palaeontology all this is untenable. The forms most closely related to the Eocene man of Clermont would be the Pachylemurae [lemurs]! The oldest known fossil anthromorph, the Oligocene *Propliopithecus,* was probably no larger than a baby. No one can seriously believe [wrote Schlosser] 'that so small a creature could use such large stones as the eoliths. Neither could this be said of *Anthropodus,* which certainly did not attain the size of a twelve-year-old child. According to this, the theory of Pliocene eoliths must also be abandoned'" (Schlosser 1911, p. 56; Obermaier 1924, pp. 16–17). It should, however, be kept in mind that these statements were founded upon a carefully edited version of the fossil record that deliberately excluded discoveries of fully human fossil skeletal remains in Pliocene, Miocene, Eocene, and even more ancient strata (Sections 6.2, 6.3). But even taking Obermaier's statements as they stand, they exhibit a questionable logic. Obermaier should not have absolutely ruled out the existence of humanlike primates in the Tertiary simply because the only primate fossils recovered up till that time were nonhumanlike.

3.4.5 The Double Standard in Operation

Seeing the eolith question from another point of view, Breuil (1910, p. 406) stated: "It is established that the criterion for distinguishing these natural productions from flints truly used by man, or flints rudimentarily worked by him, has not yet been discovered, and probably does not exist." Many authorities, from the nineteenth century up to the present, would disagree with this observation. The works of Leland W. Patterson (L. Patterson 1983, L. Patterson *et al.* 1987), outline a combination of criteria (including bulbs of percussion, retouching, striking platform geometry, repetition of particular forms, etc.) for judging human workmanship in even the crudest assemblages. Patterson (1983, p. 303) has stated: "Any experienced lithic analyst with a 10-power magnifier can distinguish fortuitously shaped flakes from unifacial tools."

Breuil (1910, p. 407) did, however, admit: "One is not able to conclude from the discoveries at Belle-Assise that there is no such thing as an Eolithic industry, no intentional work on natural stone flakes, no first manifestation of rudimentary tool types." He then stated that "in order to determine the presence of an intelligent being something more than calling attention to signs of adaptation is required, because the work of nature and that of human beings can be easily confounded. The objects should possess a degree of intentional work that is particularly clear, or should occur in an assembly of circumstances that rule out natural causes, or demonstrate, by the association of food debris or signs of fire, that human beings lived there" (Breuil 1910, p. 407).

But in many cases supporting evidence of the type specified by Breuil has been found in connection with stone implements. The stone tools discovered by Florentino Ameghino in an Early Pliocene formation at Monte Hermoso, Argentina, were accompanied by burned earth, remnants of hearths, burned and broken animal bones, and even human fossil remains, yet these implements were not generally recognized by the scientific community (Section 5.1.1).

Summarizing his case, Breuil (1910, p. 407) stated: "It is clear that we have here many pseudomorphs that show extreme signs of 'wear,' not only eoliths, but types truly recognized as Paleolithic, such as the marvelous small scraper in figure 67 [our Figure 3.24]. If nature, in exceptional circumstances without doubt, is able to produce objects that resemble advanced industrial types, perfectly defined and discovered in their normal geological position outside all possibility of error, there is thus very good reason to show caution regarding manifestations of the most elementary type of human activity, and to show great care before basing overambitious theories on such problematic findings. All this has been established in a definite manner and with all clarity."

This statement hinges on accepting Breuil's opinion that forces of nature are actually responsible for "types truly recognized as Paleolithic." Nothing in his report demonstrated that this is in fact true. As we have seen, the examples he gave of flints obviously broken in place (Figures 3.20, 3.21) do not compare very well with even the cruder "pseudomorphs" he collected at the Belle-Assise site (Figure 3.22). He also gave no real explanation for the highly organized chipping on the more advanced "pseudomorphs" (Figure 3.25). It would thus seem that Breuil himself was the one who was guilty of constructing overambitious theories on the shaky foundations of problematic findings.

3.4.6 How Scientists Cooperated in Propagating Untruths about Eoliths

Breuil's paper was quite influential and is still cited today as proof that eoliths are natural rather than artificial productions. As an example of how Breuil's study was used shortly after it appeared, we can point to *The Origin and Antiquity of Man* (1912) by G. F. Wright, an American geologist. In a discussion of eoliths,

Wright (1912, pp. 338–339) recounted how S. Hazzeldine Warren had shown that cart wheels rolling on gravel roads produced chipped flints like eoliths, and how Marcellin Boule had collected chipped flints resembling eoliths from machinery used for the production of cement. Wright, after lamenting that some scientists, like Rutot, were still promoting eoliths, wrote: "Within the year past, however, Abbé Breuil has apparently been able to give a finishing touch to the evidence discrediting the artificial character of the eoliths. We will content ourselves with quoting the summary of this evidence given by Professor Sollas" (Wright 1912, p. 340).

Wright then quoted from *Ancient Hunters* by W. J. Sollas (1911, pp. 67–69): "These [eoliths] were found by the Abbé Breuil in Lower Eocene sands (Thanétien) at Belle-Assize, Clermont (Oise). M. Breuil shows in the most convincing manner that they all owe their formation to one and the same process, *i.e.* to movements of the strata while settling under pressure of the soil. The flint nodules crowded together in a single layer are thus squeezed forcibly one against the other, and flaking is the inevitable result. . . . In many cases the flakes are still to be found in connection with the parent nodule, lying apposed to the surface from which they have been detached." Wright published a reproduction of Breuil's drawing of some very crude flakes lying next to parent blocks of flint. Sollas had used the same drawing in his book. As we noted in our previous discussion, the degree of "workmanship" on the flakes pictured in these drawings (Figures 3.20, 3.21) hardly approaches that of even the crudest of eoliths.

The quotation from Sollas (1911) about Breuil's pseudoeoliths continued: "They display just the same forms as other Tertiary 'eoliths,' ranging from the obviously purposeless to those which simulate design and bear bulbs of percussion and marginal retouches. Among the most artificial looking are a few which present an astonishing degree of resemblance to special forms of genuine implements; attention may be directed to two in particular, which are compared by the Abbe Breuil, the one to Azilio-Tardenoisian flakes, and the other to the small burins of Les Eyzies; in their resemblance to artificial forms these simulacra far transcend any 'eoliths' which have been found on other horizons of the Tertiary series" (Wright 1912, p. 341). Sollas implied that Breuil found at Clermont examples such as these last two, with flakes in place. There is, however, a little dishonesty in this presentation. Sollas should have mentioned that although some pieces of flint were found with flakes lying nearby, these were, although displaying, in some cases, bulbs of percussion and secondary chipping, decidedly nonimplemental in character. Of course, most of the blame lies with Breuil, who wrote the original report.

Sollas concluded: "On the important question of man's first arrival on this planet we may for the present possess our minds in peace, not a trace of unquestionable evidence of his existence having been found in strata admittedly older than the Pleistocene" (Wright 1912, pp. 341–342). This view is still prominent today, although there are hundreds of discoveries, a good many of which are discussed in this book, that invalidate it.

The case of Wright and Sollas shows how researchers who share a certain bias (in this case a prejudice against evidence for Tertiary humans) cooperate by citing a poorly constructed "definitive debunking report" (in this case by Breuil) as absolute truth in the pages of authoritative books and articles in scientific journals. It is a very effective propaganda technique. After all, how many people will bother to dig up Breuil's original article, in French, and, applying critical intelligence, see for themselves if what he had to say really made sense?

3.4.7 Breuil Supports Moir

It is interesting to note that Breuil's "definitive" 1910 report came before most of J. Reid Moir's discoveries in East Anglia. Eventually, when Moir's finds began to attract considerable attention, Breuil, and other scientists, went to England to conduct firsthand evaluations. Surprisingly enough, Breuil backed Moir.

M. C. Burkitt (1956, p. 107) wrote: "Messrs Breuil and Boule, who came over to see the finds, still maintained their skeptical attitude. Mr. Moir, however, was undaunted and continued his researches at new sites until finally at Foxhall, a few miles from Ipswich, he collected a series of specimens of such a nature that an examination of them by M. Breuil caused him to change his ideas completely and to join the ever-growing company of those prehistorians who believed in the existence of man as early as late tertiary times."

It is noteworthy that such a conservative and cautious researcher as Breuil should have come out in favor of Moir. During his visit to England, Breuil had specifically searched Moir's sites for any evidence of soil movement and pressure. But he found none. George Grant MacCurdy, director of the American School of Prehistoric Research in Europe, wrote in *Natural History:* "Breuil is authority for the statement that conditions favoring the play of natural forces do not exist in certain . . . deposits of East Anglia, where J. Reid Moir has found worked flints" (MacCurdy 1924b, p. 658).

Some of these deposits are found in the middle of the Red Crag at Foxhall. About Foxhall, Breuil (1922, p. 228) stated: "There is a twin layer in the superior part of the Red Crag, representing without doubt land surfaces that temporarily emerged shortly before the final retreat of the sea during the upper Pliocene." As we have seen, modern authorities still place the Red Crag in the Late Pliocene (Section 3.3.2). Breuil (1922, p. 228) added: "Here there are no causes of natural mechanical fracturing—no rolling, no scraping, no contusion, no flints found in great quantities of stone. The flints are scattered, not numerous, have sharp angles, and are small in size, just as occurs in a level where the products and byproducts of lithic industry are present. The signs of intentional flaking are very well defined, and one also finds waste products of such flaking. One finds flint cores. Bulbs of percussion are very certain. One finds the same types as at the base of the Red Crag [in the sub-Crag detritus beds]. Furthermore I have noted instances of parallel successive flake removal."

Moir himself (1924, p. 647) informs us that Breuil "definitely accepted the view that the sub-Crag implements were made by man." In 1922, after visiting sub-Crag sites at Thorington Hall and Bramford, Breuil (1922, p. 228) wrote: "The level in which the flints are found represents a land surface that existed prior to the invasion of the Red Crag seas, which occurred in the upper Pliocene, bringing in a fauna adapted to the cold. There certainly does exist cause for mistaken identification of implements, such as intense compression of the soil, which, by means of mechanical action, many times produced examples of flaking and fracturing, including bulbed flakes, with edges showing chipping resembling retouching and signs of utilization. Nevertheless, there are some flint specimens that bear very well-defined bulbs of percussion, manifesting patterns of flaking that could only be obtained by removing successive flakes by repeated blows in the same direction. This flaking oftentimes gives the appearance of retouching, and absolutely resembles flaking of human origin. I am not aware of any action of compression that could produce these results. The mechanical action of rivers or the sea can also be eliminated as causes, as can thermal action. There are some flints that show evidence of having been burned. I reject the majority of rostrocarinates [a type of eolith] as not being the product of intentional work, but I do accept as the true product of intentional work an important number of specimens. These are not simply eoliths but are absolutely indistinguishable from classic flint implements."

Breuil's statement that some of the objects from below the Red Crag were "absolutely indistinguishable from classic flint implements" is highly significant. The sub-Crag formations, which lie between the Late Pliocene Red Crag and the Eocene London Clay, could be anywhere from 2 to 55 million years old. We thus have a situation analogous to that at the Belle-Assise site in France, where Breuil found in Eocene formations two "pseudomorphs" resembling classic Paleolithic implements of the Late Pleistocene. In the case of the sub-Crag implements Breuil stated he was "not aware of any action of compression that could produce these results." This differed from the position he took regarding the two specimens from the Eocene of Belle-Assise, namely, that they were produced by geological compression. Breuil's views about the authenticity of some of Moir's implements nevertheless add considerable weight to the conclusion that the objects found at Belle-Assise were also the product of intentional human work rather than geological compression. One wonders why, if Breuil was prepared to accept the sub-Crag objects were manufactured by humans, he did not change his views about the two objects found at Belle-Assise.

Breuil, once an avid supporter of Moir's finds, apparently became noncommittal later on. In a late edition of *Men of the Old Stone Age*, published posthumously, Breuil and Raymond Lantier (1965, p. 56), in considering the Crag specimens, stated only that "traces of fire and a certain number of flakes might be accepted, though their angle of cut is generally against it." One wonders why there is no mention of the objects Breuil (1922, p. 28) previously said were "not simply eoliths but are absolutely indistinguishable from classic flint implements."

3.4.8 Barnes and the Platform Angle Controversy

Another important element in the eolith controversy was the platform angle test, promoted by Alfred S. Barnes. Barnes, who defended Moir against attacks by Haward and Warren in the 1920s, later became opposed. In 1939, he delivered what many authorities still regard as the death blow to the Red Crag and Cromer Forest Bed tools. But Barnes did not limit his attention to East Anglia. In his study, titled "The Differences Between Natural and Human Flaking on Prehistoric Flint Implements," Barnes (1939, p. 99) considered stone tool industries from France, Portugal, Belgium, and Argentina, as well as those of Moir.

Supporters of the view that implements from the above sites were of human manufacture generally argued that natural forces could not produce the kinds of chipping observed on the objects in question. Barnes admitted that random concussion would not produce effects such as regular, unidirectional chipping along a single edge. He also felt that simple pressure from overlying beds, as proposed by Breuil (Section 3.4.1), was also not a very satisfactory agent, because it did not produce specimens with good striking platforms or clearly marked bulbs of percussion (Barnes 1939, pp. 106–107). But Barnes went on to give some examples of natural forces that, in his opinion, were capable of producing objects resembling eoliths. He called attention to some flints collected from the Blackheath Eocene marine beds at Stanstead in Surrey. At this site, by a process called foundering, flint nodules had descended 20 to 40 feet into cavities eroded in the chalk, where they were crushed by masses of large pebbles from the overlying beds. Some chipped flints were found lying in contact with the parent blocks (Barnes 1939, p. 103).

Besides foundering, another natural force that could, according to Barnes (1939, p. 106) and others, produce eolithlike specimens was solifluction, in which a large mass of frozen gravel thaws and then flows rapidly down a slope.

Barnes admitted that judgements based on simple visual inspection of chipping thought to have been caused by foundering or solifluction were liable to be very subjective. So he proposed that attention should be focused on some measurable feature of the implements that could be objectively evaluated. For this purpose, Barnes chose what he called the "angle platform-scar."

Barnes (1939, p. 107) explained: "It may be said of natural fractures in general that some really good pseudomorphs of human work may be found, but when a number of specimens are examined, examples of aberrant flaking will be present. These aberrant flakes either serve no useful purpose in connection with the supposed tool or occur in positions where they would not be found in human work, or present angles platform-scar which are obtuse. The angle platform-scar is the angle between the platform or surface on which the blow was struck or the pressure was applied which detached the flake, and the scar left on the tool where the flake has been detached."

We find Barnes's description of the angle to be measured somewhat ambiguous. We have spoken with experts in lithic technology at the San Bernardino County Museum, including Ruth D. Simpson, and they have also been unable to specify exactly what angle Barnes was measuring.

In any case, in the angle platform-scar, Barnes believed he had found the objectively measurable feature by which one could distinguish natural chipping from human work. However, as noted later in this section, modern authorities such as Leland W. Patterson have extensively critiqued Barnes's methodology.

Barnes (1939, p. 109) made these observations: "When we examine the tools of Paleolithic man we find that they are furnished with acute edges (less than 90 degrees) for cutting and scraping, for such edges are more effective for these purposes than edges with obtuse angles (90 degrees and over). There is a further reason why on humanly made tools we find that the majority of angles platform scar are acute and that is because the tool maker must be able to control the flakes he removes. . . . In the author's experience of making flint implements he finds that for satisfactory control of the flaking the angles platform-scar lie between 20 degrees and 88 degrees."

In order to be effective, the measurement had to be applied not to a single specimen, but to a large sample of specimens from the industry in question. Barnes (1939, p. 111) stated that a sample "may be considered of human origin if not more than 25% of the angles platform-scar are obtuse (90 degrees and over)." Having established this, Barnes (1939, p. 111) delivered a devastating conclusion: "None of the eoliths examined by the author . . . (Pre-Crag Suffolk, Kent, Puy Courny, Belgium, etc.) . . . comply with the criterion and therefore they cannot be considered to be of human origin."

Interestingly enough, it appears that Moir himself was aware of the Barnes criterion and believed his specimens were within the required range. In 1935, four years before Barnes came out with his report, Moir analyzed his own specimens in terms of angles. He first noted that flint implements "are all, of necessity, made upon the same general plan," utilizing "a more or less flat striking-platform in the production of the implements" (Moir 1935, p. 355). He then decided to examine "the angle of the secondary edge-flaking exhibited by a series of pre-Crag implements, a factor largely under the control of the flint flaker" (Moir 1935, p. 355).

The term "secondary edge-flaking" appears to refer to flakes removed from the edge of a selected piece of naturally broken flint in order to fashion it into an implement. Although one cannot say so with absolute certainty, the angle of this secondary edge flaking apparently corresponds to the "angle platform-scar" of Barnes. Moir (1935, p. 355) noted "Professor A. S. Barnes was the first to draw attention to the significance of such measurements of flint implements."

Moir (1935, pp. 355–356) then gave the results of his study: "A quantity of pre-Crag implements to the number of 181, composed of 55 specimens of Group No. 1, 55 specimens of Group No. 2, 13 specimens of Group No. 3, 55 specimens of Group No. 4, and 3 specimens of Group No. 5, were measured with the following results. It was found that the average angle of edge-flaking of Group No. 1 was 88½ degrees, of Group No. 2, 75½ degrees, of Group No. 3, 82 degrees, of Group No. 4, 79 degrees and of Group No. 5, 69 degrees."

From these average figures alone we cannot verify that Moir's samples met Barnes's statistical requirement that at most 25 percent of the measured angles in each group exceed 90 degrees. But the angles Moir measured clearly tended to be acute, and he believed his tools satisfied Barnes's requirement.

Nevertheless, Barnes believed he had demolished, in his brief 1939 report, every anomalously old stone tool industry found by scientists over the previous 75 years. For Barnes, and almost everyone else in the scientific community, the controversy was over. But factually speaking, Barnes was beating a dead horse, because the controversy about the eoliths and other Tertiary stone tool industries had long since ceased to be a burning issue. With the discoveries of Java man and Peking man, the scientific community had become increasingly convinced that the key transition from apelike precursors to toolmaking humans (or proto-humans) had taken place in the Early to Middle Pleistocene, thus making the lithic evidence for Tertiary humans a sideshow topic of little serious concern. Barnes, however, could be seen as performing the valuable, if menial task, of sweeping away some useless remnants of irrelevant evidence. Thereafter, whenever the topic of very old stone tool industries happened to come up, as it still does from time to time, scientists could conveniently cite Barnes's report. Even today scientists studying stone tools apply the Barnes method.

Barnes's 1939 paper is typical of the definitive debunking report, which can be conveniently cited again and again to completely resolve a controversial question, making any further consideration of the matter superfluous. But on close examination, it appears that Barnes's definitive debunking report may be in need of some debunking itself.

Alan Lyle Bryan, a Canadian anthropologist, recently wrote (1986, p. 6): "The question of how to distinguish naturefacts from artifacts is far from being resolved and demands more research. The way the problem was resolved in England, by application of the Barnes' statistical method of measuring the angles of platform scar, is not generally applicable to all problems of differentiating naturefacts from artifacts." During a phone conversation with one of us on May 28, 1987, Bryan stated that application of the Barnes criterion would, for example, eliminate any blade tools struck from polyhedral cores. He also expressed a cautious belief that Barnes may have gone too far in trying to eliminate all of the anomalous European stone tool industries. Giving attention to more recent discoveries, Bryan said that Peter White has shown there are Late Pleistocene Australian tools that do not conform to Barnes's specifications.

An example of an industry that apparently does not conform with the Barnes criterion is the Oldowan, from the lower levels of the Olduvai Gorge. At site DK at the bottom of Bed I, 242 whole flakes were recovered. A striking platform angle could be measured on 132 of these. Mary Leakey (1971, p. 39) recorded the following results:

70–89°	90–109°	110–129°	130°+
4.6%	47.7%	46.2%	1.5%

As can be seen, over 95 percent of the angles are obtuse. However, it is not clear from Leakey's report exactly which angle was being measured. We discussed this with Ruth D. Simpson and her colleagues at the San Bernardino County Museum of Natural History, near Redlands, California. They were also unable to tell from Mary Leakey's report exactly what angle was being measured. This is a general problem that we have encountered in our review of angle studies on stone tool industries. The vagueness of the descriptions of the angles being measured by various investigators makes it difficult to compare findings and calls into question the scientific usefulness of such reporting.

As far as the implements from Olduvai are concerned, if the angle being measured was the angle used by Barnes, or an equivalent angle, then the Oldowan industry, although universally accepted, does not meet the Barnes criterion. Considering the extremely crude nature of the objects, which Louis Leakey said were comparable to Moir's implements, it is remarkable that they have never been subjected to the slightest challenge by the scientific community. This is probably because the Oldowan industry offers support to the African evolution hypothesis of human origins, which is accepted as dogma.

During the 1950s, the Barnes method was criticized by George F. Carter, who had discovered crude stone implements at various sites in the San Diego area, principally at Texas Street. The tools, mostly pebble choppers and quartzite flakes, were referred to the last interglacial. They were assigned dates of about 100,000 years, which violates the currently accepted idea that humans entered the Americas no more than 30,000 years ago, with most authorities adhering to a more conservative figure of approximately 12,000 years.

Reacting to attempts to dismiss the tools by the same methods used to reject the European eoliths, Carter (1957, p. 323) stated: "Comparison of the San Diego County material with that of Europe has severe limitations placed upon it that seem to have been missed by some people. The lithic materials are extremely different—quartzite and porphyries in California versus glassy rocks of the flint family in Europe. There is no frost action of solifluction or any related phenomenon in the San Diego area now nor was there any during the Pleistocene. There is no limestone area to founder and produce pressures."

Specifically referring to the Barnes method, Carter (1957, p. 329) noted: "Clearly, many of the usual criteria for judging the human authorship of stonework do not apply to such a tradition. Regrettably this seems to apply especially to the platform-angles method of testing which was so useful in distinguishing between human and natural work in England. Barnes' (1939) platform-angles on a bifacially flaked tool are much lower than 90 degrees. Those on flakes and cores of an industry such as that of Texas Street are normally about 90 degrees. It should not be overlooked that plano-convex tools normally have high platform angles." Plano-convex tools are those that are flat on one side and convex on the other. So here we have another example of an industry that was accepted (at least by Carter and his supporters) as being of human manufacture and that does not conform to the Barnes criterion.

In the preceding paragraphs, we have reviewed a number of stone tool industries that appear to be exceptions to the criterion proposed by Barnes. If these industries can be considered exceptions, then why not any or all of the various Eolithic industries that Barnes rejected?

Leland W. Patterson, the principal author of a recent study on the stone implements discovered at the Calico site in California, has also examined the application of the Barnes method. At Calico, stone objects believed to be of human manufacture have been found in strata dated by uranium series analysis to about 200,000 years before the present. They are, therefore, like the Texas Street implements, highly anomalous. We shall discuss these and similar finds relating to the human settlement of the Americas more fully in Section 3.8. For now, we shall confine ourselves to studying the application of the Barnes method to the Calico specimens, which are quite similar to Eolithic implements.

Barnes angle measurements were used by L. A. Payen (1982) to dismiss the Calico specimens. But L. Patterson and his coauthors (1987, p. 92) believed that measurement of Barnes's angle was not suitable for this purpose. Patterson defined the Barnes angle, or beta angle (Figure 3.26), as "the angle between the ventral surface and the platform plane" (L. Patterson *et al.* 1987, p. 92). Patterson, however, preferred to measure the striking platform angle, which he defined as the angle between the dorsal surface of the flake and the platform plane (Figure 3.26).

Patterson observed: "For general lithic analysis, the striking platform angle is a better attribute than the 'beta' angle . . . because prominent bulbs of force on ventral surfaces of flakes can frequently interfere with 'beta' angle measurement" (L. Patterson 1983, p. 301).

When Patterson and his coworkers measured striking platform angles rather than beta angles, their results differed from Payen's: "Acute platform angles were found on 94.3% of the Calico flakes with intact platforms as compared with 95.5% of the experimental sample. The average platform angle of the

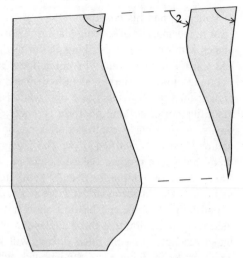

Figure 3.26. (1) The Barnes, or beta, angle, measured on a stone core. (2) The Barnes, or beta, angle, measured on a flake detached from the stone core. (3) L. Patterson's striking platform angle, also measured on a detached flake.

Calico flakes was 78.7%, with a standard deviation of 8.3%. This is consistent with the usual products of intentional flaking" (L. Patterson *et al.* 1987, p. 97).

Why such a difference from Payen's findings? Patterson and his coauthors stated: "A question can be raised as to the nature of Payen's sample. Only specimens that are candidates for representation as products of controlled flaking should be subject to analysis of platform geometry. A large amount of analytical 'noise' can be introduced by analyzing miscellaneous specimens of broken stone that possibly are not the result of controlled flaking. It is common in many lithic industries to find large quantities of non-diagnostic broken stone that are not the products of controlled flaking" (L. Patterson *et al.* 1987, p. 92). This might be true of some of the anomalously old European stone tool sites.

From Barnes's report, it appears that he was measuring mainly secondary flake scars on possible implements. He said that he would measure 100 angles, from about 30 tools. Thus he would measure an average of 3.33 angles per object. As far as eoliths are concerned, they are mostly natural flint flakes or blocks that have been subjected to some limited intentional retouching. So they should have both intentional and natural flake scars. If Barnes randomly picked 3 flake scars per eolith for his measurements, it is quite possible that this would introduce enough obtuse angles to violate his requirement that no more than 25 percent of the measured angles should exceed 90 degrees.

Patterson and his coauthors (1987, p. 92) then stated: "Another source of error in the analysis of striking platform geometry is the confusion of secondary planes with true residual striking platforms on flakes." Patterson (1983, p. 301) had earlier pointed out: "In collections both of man-made and naturally fractured stone . . . Barnes identified many specimens with flake-scar angles greater than 90 degrees. These observations must result either from incorrect identification of striking platform geometry or from incorrect angle measurements, if man-made controlled flaking or simulated controlled flaking by nature is being identified. Core flake-scar edge angles, and corresponding 'beta' angles on product flakes, cannot be obtuse in controlled flaking. On a flake, the striking platform and 'beta' angles are most often incorrectly identified when a secondary fracture has removed the true residual surface of the striking platform and has left another flake scar surface which gives the incorrect impression that these angles are obtuse. It must be emphasized that intact examples of *controlled flaking* will have striking platform and 'beta' angles under 90 degrees. . . . Studies such as that published by R. E. Taylor and L. A. Payen that use 'beta' angles on flakes as the basis for concluding that the sites of Calico and Texas Street do not have man-made specimens are questionable for the reasons given here."

Further emphasizing this fundamental flaw in the Barnes method, Patterson (1983, pp. 301–302) stated: "Previous investigators have obtained the impression that collections of naturally produced lithic flakes have many striking platforms with obtuse angles, but this appears mainly to be a case of incorrect identification of striking platform geometry. . . . Collections of naturally fractured rock often superficially appear to have a high percentage of flakes with striking platforms that have obtuse angles simply because so many residual striking platforms are missing and secondary fracture planes are incorrectly identified as remnant striking platforms."

Thus even collections of naturally broken stone should satisfy the Barnes criterion, if the original striking platform angles can be properly identified. It would thus appear that the method devised by Barnes is not appropriate for distinguishing between the effects of natural forces and intentional human work on pieces of stone.

"Probably the greatest problem with the Barnes method," observed Patterson, "is that it considers only a single attribute, and it is very difficult to conclusively demonstrate the presence or absence of human workmanship in that manner" (L. Patterson *et al.* 1987, p. 92). In another paper, Patterson gave some guidelines for a more suitable method of determining whether or not flaked stone objects are of human manufacture: "Demonstrate the likelihood of human manufacture by *combinations* of key attributes. Studies of single attributes will always remain unconvincing" (L. Patterson 1983, pp. 298–299).

Among the key attributes that Patterson suggested were the presence of clearly marked striking platforms (especially those modified for better flaking), multiple examples of tool types, platform angle measurements, the presence of bulbs of percussion and associated ripple lines, and the geological context. Other attributes that might be considered are the presence of regular retouching, sharp edges (nature tends to produce rounded edges), and signs of parallel flake removal. This balanced approach is typical of the methodology applied by the original discoverers of the stone tool industries discussed in the preceding pages.

Let us now consider in greater detail some of the key attributes identified by Leland W. Patterson and others. Patterson considered the bulb of percussion to be the single most important identifying factor. With regard to Calico, Patterson and his coauthors stated: "Of the 3,336 flakes from five Calico units, 26.1% had force bulbs and were classified as diagnostic flakes. In the experimental knapping project, using hard percussion, 24.3% of the 473 flakes possessed force bulbs and were classified as diagnostic flakes. By comparison, flakes produced by mechanical crushing (pressure force) usually have a very low percentage of distinguishable force bulbs, as shown by samples of flakes from mechanical gravel crushers" (L. Patterson *et al.* 1987, p. 95).

Patterson (1983, p. 300) also pointed out that percussive fracturing tends to produce prominent ripple lines radiating from the impact point, whereas pressure fracturing produces finer ripple lines. In addition, percussion fracturing can result in the presence of eraillures, small chips removed from the ventral surface of the force bulbs.

Patterson stressed the bulb of percussion, force ripples, and eraillures as very important in making an identification because stone flaking by humans almost always involves percussive techniques, whereas naturally broken stone is generally the result of pressure flaking. Patterson and his coauthors stated: "To date there is no documented situation where natural forces have produced large concentrations of percussion made flakes" (L. Patterson *et al.* 1987, p. 96).

In some controversial cases, Patterson has suggested that "the geological context of a lithic collection becomes important in determining if nature would have had the probable capability of fracturing rock, especially in a percussive manner" (L. Patterson 1983, p. 299). He added: "The only published manner that nature can do much percussive fracturing is under high-energy, ocean-beach storm conditions. . . . viscous liquids and slurries inhibit high-velocity percussive interactions of rocks. Pressure fracturing gives different lithic attributes than the percussive-type flaking used by early man. . . . Another condition in which nature can break rock is when flint nodules are held in a secure limestone matrix and there is a shift in the mass. Here, it is common to see shear fractures that have none of the key attributes of percussive fracture patterns" (L. Patterson 1983, p. 299). Here we see that Patterson, in common with the original proponents of many early stone tool industries, believed that it is possible to clearly distinguish natural pressure fracturing from that caused by intentional percussion flaking techniques.

In regard to tool type analysis and distribution patterns, Patterson commented: "Even if nature can produce lithic objects resembling simple man-made items, nature is not likely to do this often. Therefore, the frequency of occurrence at a given location of specimens with similar morphologies is important in demonstrating probable manufacturing patterns. Production of numerous lithic specimens with consistent morphology is certainly not a habit of nature. Quantitative data on amounts of each specimen type should therefore always be presented" (L. Patterson 1983, p. 298).

Patterson warned against the type of purely speculative interpretation often encountered in the writings of critics of anomalous lithic industries. An example would be Warren's suggestion that grounding icebergs were responsible for Moir's specimens. Patterson stated: "Even the personal opinion of a lithic expert is of little value if explicit technological reasons cannot be given to explain an opinion, either positive or negative. . . . The comments of [C. V.] Haynes on the Calico site lithic collection are a good example of subjective comments, without consideration of specific lithic attributes that could distinguish man-made manufacturing patterns. A list is given of ways that stone *could* fracture from natural causes, and then an opinion is given that the Calico lithics are the result of natural fractures, without presenting any detailed specific qualitative and quantitative studies of the attributes of the lithic materials in question. This type of subjective discussion should be avoided, as it unduly influences general opinion without any real basis" (L. Patterson 1983, p. 298).

In light of the views presented by Bryan, Carter, and Patterson, it is clear that wholesale rejection of the Eolithic and other early stone tool industries by application of the Barnes criterion is unwarranted. As a rule, the proponents of the anomalously old industries appear to have reached their conclusions by sounder analytic techniques than the opponents of such industries, whose objections mainly take the form of suggesting, with inadequate supporting evidence, various ways in which natural forces, principally pressure flaking, could have produced the objects in question.

So what are we left with? At this stage in our review of ancient stone implements, we find that we have some very credible reports, by reputable scientists, of stone tool industries dating well back into the Tertiary epoch. We should, however, point out that our investigations, although thorough, are by no means complete. In the course of our research, which we can only characterize as a preliminary survey, we have had to leave many leads unpursued (Eolithic industries from Tunisia, Egypt, etc.). We fully expect that future editions of this book will contain increasing numbers of authenticated examples of very ancient stone tool industries, as they come to our attention either in the course of our own investigations or through submissions by others.

3.5 CEMENT MILL EOLITHS?

From the late 1800s to the present, some scientists have challenged the human manufacture of eoliths and other crude stone implements, claiming that flakes of flint just like them are produced by machinery at cement factories. Alfred Russell Wallace wrote to Benjamin Harrison on June 8, 1907: "I suppose you know that a considerable number of eoliths have been found recently on the high gravels of the New Forest, near Fordingbridge, by Mr. Westlake and others. But the most important thing recently is the attack on the human origin of eoliths by the production during some process of crushing flints on the Continent of forms which are alleged to be *identical* with those of the eoliths in every detail. Opinion seems to be strongly divided, but I have seen no really careful judgement after close comparison. Have you seen them? Can you not get a set of them in exchange for yours, and give us a careful comparison? That would be worth while" (E. Harrison 1928, p. 278).

"Harrison was alive to the challenge to the eoliths arising out of the alleged resemblances of battered mill-made specimens of rude implements," observed Sir Edward R. Harrison (1928, p. 278). "He visited several brickyards and cement works in order to examine the stones that had been struck by the revolving rakes of the machines, and came away convinced that the chipped stones so produced were distinguishable from the typical Kent eoliths."

In one of his notebook entries for the year 1907, Harrison wrote of a visit to a cement mill: "Had over an hour's search on the waste heap, but could find no 'eoliths.' Two bulbed flakes found. One or two stones, having been accidentally rehit near the same place, bore some resemblance to poor eoliths, but still with a difference" (E. Harrison 1928, p. 275).

The charge that stones randomly chipped in mills resembled crude tools was also made in regard to other Eolithic industries in England and elsewhere in Europe. In 1905, Hugo Obermaier, with A. Laville, M. Boule, and E. Cartailhac, visited a chalk mill at Guerville, near Mantes, close to the Seine. Obermaier (1924, p. 11) wrote: "These mills consist of tanks filled with water, in which lumps of chalk with flint nodules embedded in them are rapidly rotated. In order to separate these nodules from the chalk and to pulverize the latter, chalk-lumps and water are subjected by means of turbines to a centrifugal motion of four meters [thirteen feet] per second. . . . The eoliths produced by the chalk mills, equally with those found in river deposits, showed forms with either partial or entire retouch around the edges, notched edges more or less deeply incurved, specimens that might be classed as scrapers, burins, and even planing tools."

But a modern expert, Leland W. Patterson, has pointed out a method for distinguishing between random natural chipping on edges of stone and intentional human chipping. Patterson (1983, p. 304) stated: "Lithic objects in nature are generally free to move or are loosely held by surrounding materials. Randomly applied forces under this condition will tend to be very oblique to the edge of the flake. Fractures then occur transversely to flake edges in the direction of least mass resistance." This kind of random chipping quickly removes sharp edges from flakes of stone. Furthermore, the chip scars tend to be of various sizes, rather than uniform in size, and tend to be oriented in many directions, rather than in a single direction.

Patterson's own studies of crushed gravel from cement factories demonstrated: "In crushed gravel there are few objects that resemble man-made cores. There are also no long sections of flake edges with uniform, unifacial retouch" (L. Patterson 1983, p. 306). Eoliths and other early stone implements, it may be recalled, are characterized by unifacial retouch—chipping confined to one side of a sharp edge.

It thus seems that a careful student of lithic technology would be able to offer a response to the challenge by Obermaier, who believed running water was a better explanation of eoliths than human action. One might ask if any pieces with sharp edges were found at the chalk mill at Mantes? Obermaier said he saw "sharp edged types, and others in which the edge had been completely worn away." He observed, "The sharp-edged types resulted after remaining in the mill from eight to ten hours, the others after a longer time in the water" (Obermaier 1924, p. 11). This evidence supports Patterson's observation that random natural action tends to quickly wear away sharp edges, making it probable that sharp-edged Eolithic specimens with regular unifacial retouch were manufactured by human beings. Rapidly running water does not produce such effects.

Obermaier, however, tried to overcome this difficulty by proposing mechanisms that would result in only brief random percussive action on flints, a few hours over the course of perhaps millions of years. Here, as many times previously, we find a scientist eager to discredit unwelcome discoveries moving into the realm of extremely improbable special explanations. Obermaier referred to a deposit of Quaternary eoliths discovered by P. Wernert and R. R. Schmidt at Steinheim in the valley of the Stuben, near Württemberg, Germany. Wernert and Schmidt stated: "We were able to show at the site itself how the fragments of flint were borne along by the stream in the principal valley and suddenly drawn into whirlpools caused by the inflow of a tributary stream. By this means the flints were subjected to a strong rotary movement which, however, was limited and intermittent in action, and therefore did not result in such continuous wearing away as would transform the flints into rounded pebbles" (Obermaier 1924, pp. 11–12).

Even if the whirlpool explanation is granted, application of Patterson's method of analyzing edge damage should result in identification of these specimens as the product of random natural forces rather than intentional human work. In fact, Obermaier himself (1924, p. 12), reported that A. Rutot, who discovered a famous series of crude stone implements in Belgium (Section 4.4), visited the German site in 1911 and pronounced the objects found there to be "pseudoeoliths." Even a supporter of eoliths was apparently not as eager to see a human implement in every piece of broken stone as his opponents might have believed. He was able to distinguish a pseudoeolith produced by natural forces from eoliths of human manufacture.

Rutot's own specimens were more sophisticated than Harrison's eoliths. They were, nevertheless, sometimes called eoliths by authors who applied the term to almost any anomalously old and relatively unrefined tools. In the course of the debate about whether or not Rutot's specimens were made by humans, the German scientist H. Hahne concluded they were distinct from machine-chipped rocks. In his book *Human Origins: A Manual of Prehistory,* George Grant MacCurdy, a professor of prehistoric anthropology at Yale University, wrote (1924a, pp. 91-92): "After a careful comparison of machine-made eoliths from both Mantes and Sassnitz with eoliths from Belgium, Hahne's conclusions are as follows: (1) the chalk-mill flints are all scratched and otherwise marked by the iron teeth of the mill; (2) the sides of all the larger pieces are bedecked with scars from blows that were not properly placed to remove a flake; (3) almost every piece shows more or less of the original chalky crust of the nodule; (4) anything like a systematic chipping of an edge or margin is never found, except for a very short stretch, where one would expect it to be carried along the entire margin; this is quite different from the long retouched margins of most eoliths; (5) the same edge is often rechipped first on one side and then on the other, absolutely without meaning or purpose (the 'reverse working' of true eoliths is quite another thing); (6) in the mill product, coarse chipping alternates with fine retouches along the same margin, while on the eolith there is a regularity and orderly sequence of chipping; (7) the repeated rechipping of the same edge, while others are left untouched, does not occur in machine-made eoliths; (8) the chief difference is between the haphazard and meaningless on one hand, and the purposeful on the other. The most prominent and easily breakable parts suffer most in passing through the mill. They are often retained intact, or only slightly altered to serve as a handhold on the eolith, and there is a logical relationship between the worked and unworked portion."

During the early decades of the twentieth century, the recurring cement-mill accusations were also leveled against the finds of J. Reid Moir in England. But M. C. Burkitt, a Cambridge archeologist and anthropologist, rejected the various attempts to account for the chipping on crude stone implements by reference to mechanical agencies. In 1905, Marcellin Boule had published a long article about cement-machine chipping that produced pieces of stone resembling eoliths. In his book *The Old Stone Age,* Burkitt (1956, p. 104) noted: "It is certainly true that specimens showing a remarkable series of chippings are produced by such machines, but no mechanical machine or natural force can chip a flint, dealing the blows from only two or three directions, more or less at right angles to one another." Burkitt (1956, p. 104) believed that some of Moir's flint specimens met that criterion and pointed out that "a number of serious students believe that the Kent specimens [of B. Harrison] are really the result of human workmanship."

It therefore appears that in no case were opponents of anomalously old crude stone tool industries able to conclusively demonstrate that implements representative of these industries could be duplicated by the action of cement and chalk mills. Thus they failed to show that the implements were in fact the product of purely natural forces rather than intentional human work. Instead, various researchers, from the late nineteenth century to the present, have presented criteria by which crude stone implements can be distinguished from the products of random battering of lithic materials, and have shown that the stone tool industries under consideration satisfied these criteria.

3.6 IMPACT OF THE ENGLISH EOLITHIC INDUSTRIES ON MODERN IDEAS OF HUMAN EVOLUTION

If scientists were to resurrect the eoliths of the Kent Plateau and East Anglia, at least granting them some serious consideration, then how would they fit into the current scenario of human evolution?

3.6.1 Eoliths of the Kent Plateau

First let us consider the implements discovered by Benjamin Harrison on the Kent Plateau. For the sake of the discussion that follows, let us set aside all the evidence for stone tool industries in the Miocene and earlier geological periods (see for example, Sections 4.1–4, 4.7, 5.5), and let us consider just the Kent Plateau implements. The reasoning behind this approach is as follows. If we take seriously the evidence for the presence of toolmaking beings in Europe during the Miocene period, then the whole story of human evolution currently accepted, with the *Homo* line originating in Africa and migrating to Europe and Asia during the Early Pleistocene, must be completely wrong. For the present, we just want to consider why the Eolithic implements of England, by themselves, present problems for advocates of the currently accepted doctrines of human evolution.

We have seen that the eoliths of the Kent Plateau may be referred to the Pliocene period in England. The end of the Pliocene is generally placed at about 2 million years ago, although some place it at about 1.6 million years ago (Gowlett 1984, p. 200). Hugo Obermaier, one of the important scientists working in the field of paleoanthropology during the early twentieth century, wrote of "the eoliths from the chalk plateau of Kent in southern England, which belong to the Middle Pliocene" (1924, p. 8).

A Middle Pliocene date would make the eoliths of Kent 3–4 million years old. Most paleoanthropologists now put the origin of *Homo sapiens* of the fully modern type (technically known as *Homo sapiens sapiens*) at a maximum of 100,000 years before the present. The immediate forerunner of *Homo sapiens sapiens*, technically known as archaic *Homo sapiens* or early *Homo sapiens*, would date back only 200,000–300,000 years. *Homo erectus*, the supposed ancestor of early *Homo sapiens*, dates back roughly 1.5 million years in Africa (Johanson and Edey 1981, p. 283), and *Homo habilis*, the supposed ancestor of *Homo erectus*, dates back only 2 million years. According to the standard account, the hominids of the Late and Middle Pliocene would have been very primitive australopithecines, none of which are thought to have been makers of stone tools.

Just for the sake of argument, let us suppose that the eoliths of the Kent Plateau can be referred to the very latest Pliocene, at about 2 million years B.P. This is, of course, too early for *Homo sapiens*. It is also too early for *Homo erectus*. Even if we push the first appearance of *Homo erectus* back further than 1.5 million years, the 2-million-year minimum age for the Eolithic implements of the Kent Plateau still causes some problems. According to the most widely accepted scenario of human evolution, *Homo erectus* was the first hominid to leave Africa, and did not do so any earlier than about a million years ago. Thus even an Early Pleistocene date for the Harrison implements from the Kent Plateau would be problematic.

Up to now, we have been speaking of the standard evolutionary account of human origins, with the major transitions taking place in Africa. But there is a second, less widely held version of the human evolutionary process. According to this account, the transition to *Homo erectus* and *Homo sapiens* took place not in Africa alone but across a wider geographical range (Gowlett 1984). This means that the precursors of *Homo erectus*, creatures like *Homo habilis*, must have already been existing outside Africa, perhaps as much as 2 million years ago. According to some scientists, *Homo habilis* made the very primitive stone tools found in the lower levels of Olduvai Gorge, tools very much like eoliths. It is therefore within the realm of theoretical possibility (for some paleoanthropologists) that a creature like *Homo habilis* may have made the eoliths found by Benjamin Harrison in England.

One would thus have to make relatively few changes in current theory to accommodate the Harrison eoliths. But once such evidence has been condemned, it must apparently remain so perpetually, with no chance of rehabilitation. Even scientists whose theories the tainted evidence might support ignore it. Why? Perhaps because if some relatively benign evidence of this kind were to be resurrected, then more threatening evidence might also emerge from the crypt.

3.6.2 East Anglian Tools and the African Origins Hypothesis

The implements discovered by J. Reid Moir pose a similar set of problems, which were, interestingly enough, recognized by a modern researcher (Coles 1968). Some of Moir's discoveries in the Cromer Forest Bed were referred to the Middle Pleistocene. Others, from the Red Crag, were referred to the Early Pleistocene or Late Pliocene. For the purposes of this discussion, we shall set aside implements from the detritus bed below the Red Crag, which could be dated anywhere from the Pliocene to the Eocene.

J. M. Coles (1968, p. 30) summarized his review of Moir's East Anglian discoveries by stating: "in view of the evidence of early man in North Africa and in Southern Europe, there is nothing basically startling about the presence of human industries in East Anglia at the beginning of the Middle Pleistocene. The axe from Sidestrand, if it is, in fact, a paleolithic tool and not a neolithic rough-out in an erosion pocket, suggests that man was present during the Cromerian interglacial period, or early in Mindel times. This would not be out of step with the evidence of man's presence in Europe . . . during these periods, but the character of the handaxe is rather surprising. But even more surprising would be the existence of a handaxe tradition encompassing the Whitlingham axe in the Norwich Crag phase, or pre-Günzian age, which at the moment would seem radically out of step with our evidence for early man and early industries, in both Africa and Europe. The evidence for humanly-struck flints at Foxhall, certainly the most puzzling of all the East Anglian sites, if accepted, would extend back to the earliest Villafranchian, and would indicate that an enormous gap in our evidence for early man existed, if we were to maintain our belief in an African origin."

In suggesting, however obliquely, that the belief in an African origin might be open to question, Coles is, in the light of the most widely accepted view, verging on heresy. The early Villafranchian stage, in which Coles placed the Foxhall tools, belongs to the Late Pliocene, extending from 2.0 to 3.5 million years ago (Section 1.7). According to our conservative estimate, the Foxhall site would most likely fall toward the latter part of the early Villafranchian stage, between 2.0 and 2.5 million years before the present (Section 3.3.4). One would not expect to find toolmaking humans present in England at that time. According to the African origins story, one should find during that period, in Africa alone, just apelike *Australopithecus,* who is not thought to have been a toolmaker.

In Coles we see a modern establishment scientist approaching the point of giving serious consideration to one of the conclusions warranted by the evidence presented in this book, namely, that an African origin for the *Homo* line is a myth. Coles found himself confronted with a spectrum of anomalous evidence. Some of it was mildly surprising to him, some of it more surprising. This is to be expected—that there should be, in the range of evidence ignored by the scientific establishment, a certain number of cases that approach the borderlines of acceptability. However, in light of the evidence we have thus far considered, and evidence we shall consider in coming chapters, it is clear that the Late Pliocene discoveries Coles found most surprising are just the tip of an iceberg of anomalous evidence that extends into the depths of the Tertiary and beyond.

We suggest it is the threatening nature of this vast body of anomalous evidence that might cause establishment science to steadfastly refuse to consider even the borderline evidence. One thing leads to another. If the borderline evidence is admitted, then the more surprising evidence comes one step closer to acceptance. And then what very quickly happens, as Coles hinted, is that the African origins hypothesis evaporates. And then where would paleoanthropology be? Lost in a raging sea of evidence suggesting all kinds of impossible things. A strong sense of vertigo is bound to arise, because there is a lot of evidence, every bit as good as Moir's discoveries, that puts human beings back as far as the Miocene (Sections 4.1–3), Oligocene (Section 4.4), and Eocene (Section 5.5). At that point, not only the idea of an African origin but also the whole concept of an evolutionary origin of the human species becomes untenable. And if scientists are forced to give up an evolutionary explanation of human origins, what does that say about the whole theory of evolution?

Those who have staked their prestige on the slogan "evolution is a fact not a theory" might counter that the evidence for evolution in general is "overwhelming." There are, of course, millions of species that might be considered, but here we are focusing on one, the human species, and testing the hypothesis of its evolutionary origin. In this defined area of investigation, we have documented overwhelming evidence contradicting the proposal that the modern human type evolved from more apelike predecessors. Trying to avoid the implications of this thought-provoking evidence by bringing in *ex cathedra* claims of evolutionary progressions in the fossil histories of myriad other species is inappropriate.

3.6.3 Recent Pakistan Finds (Plio-Pleistocene Boundary)

Resistance to the idea that representatives of the *Homo* line may have been present outside Africa around 2 million years ago is apparent in reactions to some recent discoveries in Pakistan. These were reported in a New York Times News Service story appearing in the *San Diego Union* edition of August 30, 1987. The story told of "reports from British archaeologists working in northern Pakistan that they have found 2-million-year-old chopping tools believed to have been made by early humans." The reports were from the British journal *New Scientist*. The news article continued: "If such a significantly earlier time of migration is established, it would presumably mean that a more primitive species in the human lineage, *Homo habilis,* was the first to leave Africa and did so soon after learning to make stone tools. The prevailing view now is that the later *Homo erectus,* which had a considerably larger brain capacity, initiated the human migration about a million years ago." To those accepting the prevailing view, the English eoliths, discovered in the nineteenth century, and the new Pakistani stone tools, both at least 2 million years old, present a problem.

The article went on to explain how mainstream scientists considering the Pakistan tools dealt with this problem—they tried to discredit the discovery. "Sally McBrearty, an anthropologist at William and Mary College who has done research in Pakistan, complains that the discoverers 'have not supplied enough evidence that the specimens are that old and that they are of human manufacture.'" Our review of anomalous stone implements should make us suspicious of this sort of claim. As we have seen, it is fairly typical procedure for scientists to demand higher levels of proof for anomalous finds than for evidence that fits within the established ideas about human evolution.

The New York Times News Service article then stated: "Like many experts, McBrearty was skeptical of the 2-million-year date because the discovery was made in a river plain, which is 'not a good stratigraphical context.' The sediment layers there have been so mixed up by flowing water over time that geologists have a hard time determining whether artifacts are embedded in their original sediments." As previously noted, if this standard were to be applied uniformly, then there should be similar skepticism regarding many important paleoanthropological finds, which were also made in river plains and other places, such as caves, with poor stratigraphy. One good example is the famous Java man, the first bones of which were taken from a flood plain directly on the edge of a river.

Finally, the news service article stated: "Anthropologists also noted that pebbles fracture easily as they roll through flowing water, resulting in shapes that can be mistaken for artifacts." Do these anthropologists think that the British scientists who discovered the implements in Pakistan were unaware of this problem, which has been the object of serious study for over a century? As we have seen earlier in this chapter (Sections 3.2.3, 3.2.5, 3.2.11), authorities ranging from Sir John Prestwich (1892, p. 256; 1895, p. 625) to Leland W. Patterson (1983, p. 108) have pointed out that fortuitous damage to stones in stream beds can be clearly distinguished from intentional human work.

Now let us look at the report on the discovery of the Pakistani tools published in *New Scientist,* and see how it matches up with the newspaper statements of scientists critical of the find. In the New York Times New Service story, Sally McBrearty strongly suggested that the reported 2-million-year date for the Pakistani implements was very uncertain, but *New Scientist* stated: "These artefacts are surprisingly old, but the date is convincing" (Bunney 1987, p. 36). McBrearty also claimed that the stratigraphic context was not good, hinting that if the objects were tools, they did not belong to the beds where they were found.

But the *New Scientist* stated: "Such doubts do not apply in the case of the stone pieces from the Soan Valley southeast of Rawalpindi, argues Robin Dennell, the field director of the Paleolithic Project of the British Archaeological Mission and the University of Sheffield. He and his colleague Helen Rendell, a geologist at the University of Sussex, report that the stone pieces, all of quartzite, were so firmly embedded in a deposit of conglomerate and gritstone called the Upper Siwalik series, that they had to chisel them out" (Bunney 1987, p. 36). According to the *New Scientist,* the dating was accomplished using a combination of paleomagnetic and stratigraphic studies.

The New York Times News Service article left the reader with the strong impression that the objects in question were quite probably formed by random concussion in stream beds, and it did not mention any of the evidence in favor of their human manufacture. However, the *New Scientist* gave its readers with a more balanced treatment: "Of the pieces that they extracted, eight, Dennell believes are 'definite artefacts.' In Dennell's view, the least equivocal artefact is a piece of quartzite that a hominid individual supposedly struck in three directions with a hammer stone, removing seven flakes from it [Figure 3.27]. This multifaceted flaking together with the fresh appearance of the scars left on the remaining 'core' make a 'very convincing' case for human involvement, Dennell told *New Scientist*" (Bunney 1987, p. 36).

So what is going on with the find in Pakistan? It appears we may have a recent example of scientists being

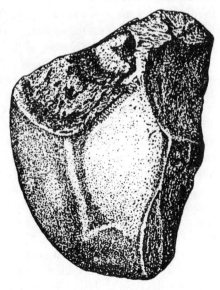

Figure 3.27. A stone tool discovered in the Upper Siwalik formation in Pakistan (Bunney 1987, p. 36). British scientists estimated its age at about 2 million years.

unable to objectively evaluate evidence that contradicts their preconceptions about the progress of human evolution. In this case, we find that scientists holding the view that *Homo erectus* was the first representative of the *Homo* line to leave Africa, and did so about a million years ago, were apparently quite determined to discredit stone tools found in Pakistan, about 2 million years old, rather than modify their ideas. We can just imagine how such scientists would react to stone tools found in Miocene contexts.

3.6.4 Siberia and India (Early Pleistocene to Late Pliocene)

Many other discoveries of stone implements around 2 million years old have been made at other Asian sites, in Siberia and northwestern India. Turning first to Siberia, let us consider what A. P. Okladinov and L. A. Ragozin called the riddle of Ulalinka. These two scientists reported in 1984: "Quite recently it was thought that the Siberia Paleolithic was not more than 20–25,000 years ago. Everything changed after a Paleolithic site, bearing no similarities with any site known before, was discovered in 1961 on the slopes of the steep bank of the Ulalinka River, at the edge of the city of Gorno-Altaisk, the capital of the autonomous oblast. Stone tools of primeval man were found here in the form of cobble stones only partially worked over by a coarse chipping. Half or even two-thirds of such a stone retained its original pebbly surface, a kind of scale, which had been removed only at the working end of the tool, at its cutting edge. A person not acquainted with the technology of those remote times would have tossed this stone away, seeing nothing striking in it. But the stone from Ulalinka can tell an archaeologist, a specialist in such things, a great deal" (Okladinov and Ragozin 1984, p. 5). Six hundred such tools were found at Ulalinka.

After the discovery of the implements, geologists dated the Ulalinka site at 40,000 years. This dating poses no particular problems for modern ideas about human evolution. The tools could have been made by anatomically modern *Homo sapiens*, or perhaps by some late survivals of a Neanderthal population in Siberia. But subsequent studies put the Ulalinka site in the late Middle Pleistocene, giving ages that range from 150,000 to 400,000 years (Okladinov and Ragozin 1984, pp. 5–6). Then, in 1977, Okladinov and Ragozin conducted new excavations and determined that the implement-bearing stratum was much older than scientists previously thought. They stated: "the pebble tools belong to the middle part of the Kochkov horizon, the Podpusk-Lebiazh'e layers, formed roughly 2.5 million to 1.5 million years ago. This conclusion was confirmed by thermoluminescent analysis done by A. I. Shliukov, Director of the Geochronology Group of the Faculty of Geography of the Moscow State University. . . . it was found that the cultural layer with the Ulalinka pebble tools was more than 1.5 million years old" (Okladinov and Ragozin 1984, pp. 11–12). The faunal remains at the site were comparable to the middle Villafranchian (Early Pleistocene or Late Pliocene) of Europe (Okladinov and Ragozin 1984, p. 12).

Okladinov and Ragozin (1984, p. 12) also reported: "Similar pebble tools were found in China, together with two knives made of hominid incisor teeth. This is the so-called Yuanmou man. His age, according to paleomagnetic data, is from 1.5 to 3.1 million years; the accepted date is 1.7 million years."

Okladinov and Ragozin (1984, p. 14) then posed a question: "was the Ulalinka man an aborigine or did he come in from somewhere else?" It was possible, they stated, that the ancestors of Ulalinka man had migrated from Africa. If so, the migration must have occurred well over 1.5 million years ago, and the being that migrated would therefore have been *Homo habilis*.

But the Russian scientists apparently had some patriotic impulse, and favored the idea that the ancestors of the Ulalinka hominid had not migrated from elsewhere. Okladinov and Ragozin (1984, pp. 15–19) therefore proposed an extensive search for skeletal remains of a possible ancestor of Ulalinka man in Siberia, hinting that Siberia, not Africa, might very well have been the cradle of humanity. In a paleoanthropological reflection of the wider Sino-Soviet conflict, Okladinov and Ragozin (1984, p. 18) proposed: "It is not impossible that Sinanthropus [Peking man] stems from the Ulalinka hominids." In other words, China man came from Russia man. The Chinese, however, believed the reverse to be true.

Okladinov and Ragozin were not the first scientists to broach the idea that human beings evolved within the borders of the former Soviet Union. Alexander Mongait, an archeologist, wrote (1959, p. 64): "today it may be surmised that Transcaucasia was within the vast zone where man first appeared. . . . In 1939, the remains of an anthropoid ape, which lived at the end of the Tertiary period, was found in Eastern Georgia in a locality called Udabno. It was named Udabnopithec. This find confirmed the possibility that mankind originated in Trans-Caucasia (in addition to other regions embracing South Asia, South Europe, and Northeast Africa). But in order to substantiate this hypothesis, science needed the chief link—if not the remains of primitive man himself, then at least the most ancient implements of labor. In 1946–48, S. M. Sardaryan and M. Z. Panichkina, while surveying Satani-dar (Mount Satan), which is situated close to Mount Bogutlu in Armenia, found crude obsidian implements of the most ancient forms dating from the Chellean period; to date, these implements are the most ancient of the archaeological finds in the U.S.S.R. and make up yet another link in the chain of facts proving that the southern areas of the Soviet Union were part of the region where man grew out of the animal state."

Another scientist, Yuri Mochanov, discovered stone tools resembling the European eoliths at a site overlooking the Lena River at Diring Yurlakh, Siberia. The formations from which these implements were recovered were dated by potassium-argon and magnetic methods to 1.8 million years before the present. Mochanov, leaving aside the standard African origins concept, proposed the simultaneous emergence of man in Siberia and Africa during the very early Pleistocene. Mochanov stated: "I couldn't believe my eye, at first. After all, I had always argued against finding such primitive pebble tools in this part of Siberia" (Daniloff and Kopf 1986). Some have argued that Siberia was too cold for human habitation. But Pavel Melnikov, director of the Permafrost Institute at Yakutsk, stated that "paleobotanists, studying pollens and seeds in ancient layers, have concluded that the Siberian climate a million years ago was much like today and could have supported people" (Daniloff and Kopf 1986). There is no reason to rule out the possibility that these toolmaking people might have been very much like modern *Homo sapiens.* And here is something else to consider—if the climate was like that of today these ancient Siberians surely would have needed clothing, indicating an advanced level of culture.

Recent evidence from India also takes us back about 2 million years. Many discoveries of stone tools have been made in the Siwalik Hills region of northwestern India. The Siwaliks derive their name from the demigod Shiva (Sanskrit *Siva*), the lord of the forces of universal destruction. Roop Narain Vasishat, an anthropologist at Punjab University, objected strongly to the idea that "the Siwalik hominoids did not evolve into hominids and the prehistoric stone tool making man in this region was an intruder from outside" (1985, pp. xiv–xv). Some Indian scientists, like the Russians and Chinese, believe that the key steps in human evolution took place within their nation.

In 1981, Anek Ram Sankhyan, of the Anthropological Survey of India, North Western Region, reported: "the author recovered a Palaeolithic implement from the Upper Siwalik horizon, about 8 kms [5 miles] east of Haritalyangar village" (1981, p. 358). Sankhyan offered this description of the implement: "The stone tool under reference is a typical Bifacial Chopper made on a large dark-coloured quartzite cobble, 12.5 cm [4.9 inches] in length, 9.3 cm [3.7 inches] in breadth, and 6.5 cm [2.6 inches] in maximum thickness at the butt end. The core exhibits multiple flaking scars on nearly half of its surface on both sides forming a sharp and broad cutting edge. One surface is smoothly flaked and tapering whereas the other carried a large and deep flake scar, besides other smaller flake scars near the edge. The butt end is unworked and rounded for a comfortable grasp" (1981, pp. 358–359).

On the age of the implement, Sankhyan (1981, p. 358) reported: "The stone tool was recovered from a thin band of pebbles distributed in patches over a grey shale horizon. . . . Prasad (1971) assigns these beds to the Tatrot Formation (Upper Pliocene)." Sankhyan subsequently discovered many more stone tools apparently from the same Tatrot horizon (1983, pp. 126–127). Other Indian researchers have made similar finds in the same area.

The above-mentioned Siberian and Indian discoveries, at 1.5–2.5 million years old, do not agree very well with the standard view that *Homo erectus* was the first representative of the *Homo* line to emigrate from Africa, doing so about a million years ago. But, as previously mentioned, they might agree with the view that creatures like *Homo habilis* migrated from Africa about 2 million years ago. One prominent scientist expressing this view is John Gowlett of Oxford. Gowlett wrote (1984 p. 59): "Although it is sometimes suggested that human occupation of the East only started with the migration of *Homo* [*erectus*] from Africa at the beginning of the Pleistocene, this seems unlikely. Some of the very first fossil hominid remains ever found are those of *Homo erectus* from Java, which can hardly have been the first stop on a migration route. In addition to these historic finds made by Eugene Dubois in 1891 near the Solo River, other more primitive specimens have since been discovered in the older Djetis beds." The Djetis beds were given a potassium-argon date of 1.9 million years (Jacob 1972; Gowlett 1984, p. 59). But subsequent tests (Bartstra 1978; Nilsson 1983, p. 329) gave the Djetis beds a far younger date of less than 1 million years. In Chapter 7, we shall see that the Java *Homo erectus* discoveries are, however, all highly questionable, because they are practically all surface discoveries. This means that the stratigraphic context, and consequently the dates, are not firmly established.

In any case, Gowlett (1984, p. 58) proposed: "Human evolution is likely to have taken place across a continuous band of the tropics and subtropics. . . . our only certain evidence comes from a thin scattering of archaeological sites and human remains. These testify directly to the early occupation of large areas, including southern Africa and the Far East, from 2 or 3 million years ago." Gowlett did not offer very much further in the way of detail, but from the whole of his discussion it would appear he was suggesting that *Homo habilis* and perhaps even the australopithecines were spread widely throughout this region 2–3 million years ago. In this case, why did Gowlett not mention the Eolithic implements of England, also 2–3 million years old? It would seem they would have lent support to his hypothesis.

There come to mind at least three reasons why Gowlett did not mention the English eoliths in connection with his hypotheses about human evolution: (1) he was aware of the discoveries of Harrison, Moir, and others, but accepted the verdict of Barnes and others that they were products of natural forces; (2) he was aware of the Early Pleistocene and Late Pliocene eoliths of England but hesitated to mention them because of their embarrassing connection with older eoliths from the Early Pliocene, Miocene, and earlier periods; (3) he was unaware of the discoveries.

Many modern students of paleoanthropology are in fact completely unaware of reports of crude stone tool industries from the Tertiary and early Quaternary.

Why? The eolith evidence was buried decades ago by skeptical scientists, at a time when it did not fit in so well with then current theories of human evolution. During the 1930s, the oldest human ancestors completely accepted by science were the Java *Homo erectus* and Peking *Homo erectus*, which dated back to the Middle Pleistocene, about a half million years ago. This did not leave any place for a toolmaking being in England during the Early Pleistocene, 1–2 million years ago or Late Pliocene, 2–3 million years ago. Now, the understanding of human evolution has changed, and there are some versions with which the English eolith evidence seems somewhat compatible. But hardly any scientists are now familiar with the discoveries of Harrison or Moir. So here is a good argument for not burying controversial evidence so deeply that it is hardly remembered—it may become relevant in light of future developments.

Again, it should be kept clearly in mind that in discussing how the English eoliths relate to modern evolutionary scenarios centering on a Late Pleistocene origin of the human species, we are deliberately excluding from consideration the extensive evidence (in the form of incised and broken animal bones, stone implements, and modern human skeletal remains) that places humans of the modern type in the Early Pliocene, the Miocene, and even more distant geological periods. When this evidence is admitted into the discussion, as we believe it should be, the discovery of stone implements in the Pliocene in England or anywhere else poses no particular problems.

Where has all of the preceding discussion left us? The main conclusion is that most modern paleoanthropologists are unable to cope with stone tools from periods and places that even slightly deviate from entrenched ideas about the time for the migration of the *Homo* line out of its Africa homeland. Evidence is submitted to intense negative criticism for no other reason than that it conflicts with established views. If this is true of evidence that lies on the very borderline of acceptability, then what kind of treatment can one expect for otherwise good evidence that happens to lie completely beyond the range of current expectations, such as the Miocene implements discovered in France and Portugal (Sections 4.1–3)? Silence and ridicule are the receptions most likely to be encountered.

Of course, even after having heard all of the arguments for eoliths being of human manufacture, arguments which will certainly prove convincing to many, some might still legitimately maintain a degree of doubt. Could such a person, it might be asked, be forgiven for not accepting the eoliths? The answer to that question is a qualified yes. The qualification is that one should then reject other stone tool industries of a similar nature. This would mean the rejection of large amounts of currently accepted lithic evidence, including for example, the Oldowan industries of East Africa and the crude stone tool industry of Zhoukoudian (Choukoutien) in China.

3.7 ACCEPTABLE EOLITHS: THE STONE TOOLS OF ZHOUKOUDIAN AND OLDUVAI GORGE

We shall now examine some stone tools broadly similar to but in some cases even more primitive than European eoliths such as those found by Benjamin Harrison and J. Reid Moir. Unlike the European eoliths, these implements are unquestioningly accepted by modern paleoanthropologists. It would seem, however, that if tools comparable to the European eoliths are considered genuine, then to be consistent, the European eoliths should also be accepted as genuine.

3.7.1 Accepted Implements from Zhoukoudian (Middle Pleistocene)

One industry similar to the European Eolithic industries is that found at Zhoukoudian, the site of the Peking man discoveries. The Zhoukoudian tools, comprising natural flakes modified with unifacial chipping, compare favorably with the European eoliths. In fact, the crudeness of the tools at Zhoukoudian (Figure 3.28) was unexpected. Peking man was classified as *Homo erectus,* who in Europe and Africa was usually associated with the more advanced bifacially flaked Acheulean implements. Anthropologist Alan Lyle Bryan (1986, p. 7) stated: "less than 2% of the 100,000 artifacts recovered from the living floors at Zhoukoudian Locality I exhibit bifacial edge retouch."

Zhang Shensui of China described the implements from the lower levels of Locality 1 at Zhoukoudian: "Tools fashioned from cores, pebbles and small chunks of stone outnumber those made on flake blanks. This assemblage is typologically simple, consisting primarily of choppers and scrapers. Points and gravers occur only rarely and are very crudely retouched" (Zhang 1985, p. 168). When illustrations of the eoliths found on the Kent Plateau and in East Anglia (Figures 3.3, p. 95; 3.6, p. 121; 3.12, p. 136) are set alongside those of tools from Zhoukoudian, we do not notice much of a difference in workmanship.

Figure 3.28. These tools from the Zhoukoudian cave seem cruder than the anomalously old Pliocene and Miocene eoliths of Europe (Black *et. al.* 1933, pp. 115, 131, 132).

3.7.2 The Oldowan Industry (Early Pleistocene)

A second industry very much like the European eoliths is the Oldowan industry, initially discovered by Mary and Louis Leakey in Beds I and II of Olduvai Gorge, Tanzania, during the 1930s. Many of the Oldowan implements were described by Mary Leakey in the third volume of *Olduvai Gorge,* published by the Cambridge University Press in 1971.

From the published reports, which is all we really have to go on, it is not possible to easily distinguish European eoliths, such as those collected by Harrison on the Kent Plateau, from some of the Oldowan tools. This is readily seen in the illustrations in Mary Leakey's book, which show the apparent identity between the two types. Although made of different kinds of stone, they look remarkably alike. Furthermore, Leakey's verbal descriptions could just as well be applied to eoliths. One might say that there are subtle distinctions not revealed in the reports, but then what does that say about the quality of scientific reporting on stone tool industries?

Mary Leakey stated that the Oldowan industry was found in locations ranging from upper Bed I to the base of Bed II at Olduvai Gorge. Describing the primary Oldowan industry, she stated: "It is characterised by choppers of various forms, polyhedrons, discoids, scrapers, occasional hammer stones, utilised cobbles and light-duty utilised flakes" (M. Leakey 1971, p. 1). In Bed II, Leakey found an industry she called Developed Oldowan, which contained more spheroid types than the Oldowan. Bed II also yielded a second industry, Developed Oldowan B, which contained some bifacially flaked tools (less than 40 percent of the assemblage). Bifacially flaked tools are those with chipping on both surfaces of the edges. In the upper part of Middle Bed II, there occurred Acheulean assemblages, in which more than 40 percent of the tools were bifacially flaked. Even these were still quite crude. According to Leakey, "The Acheulean appears to be an early form in which the bifaces exhibit minimal flaking and considerable individual variation" (M. Leakey 1971, p. 2). The Acheulean type of Olduvai appears to correspond with the Paleolithic implements described by Harrison and Prestwich, while the Oldowan type, especially its unifacially flaked specimens, appears to roughly correspond with the flint implements described as eoliths. We shall mainly concern ourselves with the Oldowan industry.

The majority of the Oldowan tools were classified as "choppers," made of volcanic cobblestone and also of quartz and quartzite. Leakey stated: "These are essentially jagged and lack secondary trimming, although utilisation has often resulted in the edges having been chipped and blunted" (M. Leakey 1971, p. 1). In other words, these are even cruder than the eoliths of the Kent Plateau, most of which display some form of intentional secondary trimming. Careful searching, however, has failed to reveal a single published challenge to the authenticity of the Oldowan specimens as genuine human artifacts.

One might argue that hominid fossils have been recovered at Olduvai Gorge, while none were found on the Kent Plateau. It should, however, be noted that crude stone tools were being excavated at Olduvai Gorge by Louis and Mary Leakey for decades before any currently accepted hominid fossil remains were recovered. In 1959, the Leakeys discovered the first fossil bones of a new primitive apelike hominid, which they regarded as humanlike and named *Zinjanthropus* (Section 11.4.1). They initially attributed the stone tools of Olduvai Gorge to *Zinjanthropus*. Not long thereafter, however, the bones of a more advanced hominid, *Homo habilis,* were found nearby (Section 11.4.2). *Zinjanthropus* was demoted from his status as toolmaker, and *Homo habilis* replaced him.

But although the designation of the toolmaker was changed, the tools themselves remained unquestioned. The principal reason why the implements discovered in Olduvai Gorge have not been subjected to the same sorts of challenges directed at the eoliths discovered in Europe is hinted at in the following statement made by Mary Leakey (1971, p. 280): "evidence for the manufacture of tools by means of using one tool as an instrument to make another is one of the most important criteria in deciding whether any particular taxon has reached the status of man. . . . If evidence of toolmaking is not counted as a decisive factor for the human status it is difficult to see what alternative can be used for determining at what point it had been reached. Evolutionary changes must have been so gradual that it will never be possible for the threshold to be recognised on the evidence of fossil bones alone. This would be true even if a far more complete evolutionary sequence of material were available for study: with the scanty and often incomplete material that has survived it is clearly out of the question. An arbitrary definition based on cranial capacity is also of doubtful value, since the significance of cranial capacity is closely linked with stature or body size, of which we have little precise information in respect of early hominids."

Scientists almost unanimously accept the idea that the genus *Homo* arose in Africa, developing from the australopithecine hominids around 2 million years ago. The strong need for stone tools as corroborating evidence of humanlike status may thus explain, at least in part, the extremely lenient treatment of the Oldowan industry. If they were not accepted as tools, that would greatly detract from the status of the African hominids as human ancestors.

In her report on Olduvai Gorge, Mary Leakey identified, besides the choppers previously mentioned, several other types of implements, which, from her descriptions, appear to correspond to the eoliths found in Europe. She described "various fragments of no particular form but generally angular, which bear a minimum of flaking and some evidence of utilisation" (M. Leakey 1971, p. 6).

Another category of Oldowan tools was scrapers of various types. Leakey described the heavy-duty scrapers of Bed II, which were fashioned from quartzite flakes, as follows: "Many of the heavy-duty scrapers are impossible to assign to any particular type and consist merely of amorphous pieces of lava, quartz, or quartzite, with at least one flat surface from which steep trimming has been carried out along one edge" (M. Leakey 1971, p. 6). About "discoidal scrapers," Leakey wrote: "the tools are seldom entirely symmetrical and they are usually trimmed on only part of the circumference" (M. Leakey 1971, p. 6). These scrapers conform to the descriptions of the eoliths discovered on the Kent Plateau of England.

Another type similar to a common variety of eolith was the nosed scraper. About this type of tool, Leakey stated: "There is a median projection on the working edge, either bluntly pointed, rounded, or occasionally spatulate, flanked on either side by a trimmed notch or, more rarely, by straight convergent trimmed edges" (M. Leakey 1971, p. 6). Hollow scrapers, with a broad curved indentation on one side of the stone forming the working edge, are another type common both to the Eolithic and Oldowan assemblages. Leakey described this type as follows: "Specimens in which the notch is unquestionably prepared are relatively scarce in both the heavy- and light-duty groups, although light-duty flakes and other fragments with notches apparently caused by utilisation are common" (M. Leakey 1971, p. 6). In other words, on these Oldowan specimens, as in the case of eoliths, the working edge of the stone had simply been modified by slight chipping or use.

One of the more remarkable coincidences of form may be found in the presence of tools called awls or borers in both Eolithic and Oldowan assemblages. Of the awls in the Developed Oldowan, Mary Leakey (1971, p. 7) stated: "They are characterized by short, rather thick, pointed projections, generally at the distal ends of flakes, but sometimes on a lateral edge. In the majority, the points are formed by a trimmed notch, on either one or both sides, but occasionally by straight convergent trimmed edges. The points are often blunted by use and have sometimes been snapped off at the base." This description perfectly applies to the awls collected and displayed by both Harrison and Moir. The identity of the Oldowan and English specimens is very much evident in Figure 3.5 (p. 96).

About the above-mentioned light-duty flakes and fragments, Leakey wrote: "Flakes and other small fragments with chipping and blunting on the edges occur in both the Oldowan and developed Oldowan but are more common in the latter. They fall into three groups: (a) with straight edges; (b) with concave or notched edges; (c) with convex edges. There is also a miscellaneous group with indeterminate chipping. In specimens with straight edges, chipping is usually evident on both sides, while in the notched and convex series it is usually only

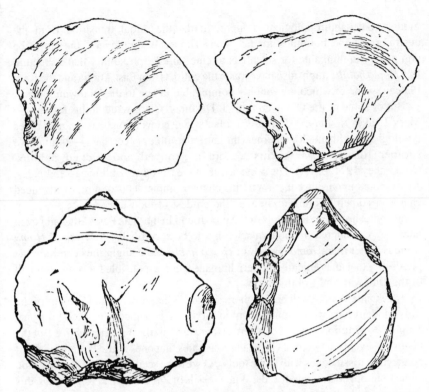

Figure 3.29. Top: Light-duty utilized flakes from Olduvai Gorge, Africa (M. Leakey 1971, p. 38). Bottom: Flaked flint implements from the Red Crag formation at Foxhall, England (Moir 1927, p. 34). The Olduvai specimens appear cruder and look less like implements than the specimens from England.

present on one face" (M. Leakey 1971, pp. 7–8). Leakey also described "light-duty utilised flakes" (Figure 3.29). Of these, she stated: "The utilised edges are sharp, with 'nibbled' one-directional flaking, which is sometimes present on two of the edges" (M. Leakey 1971, p. 37). The above descriptions could also apply to many of the European eoliths.

3.7.3 Who Made the Eolithic and Oldowan Implements?

Now comes a crucial question: to what sort of being should the manufacture of the quite similar Oldowan and Eolithic tool types be assigned? Most of the tools in both the Oldowan and Eolithic assemblages are very crude. Scientists are prepared to accept practically without question that the Oldowan implements were made by *Homo habilis,* a primitive hominid species which, according to modern paleoanthropological thought, marks the initial transition from the australopithecine hominids to the genus *Homo*. It should not, therefore, be completely unthinkable for scientists to entertain the possibility that a creature like *Homo habilis* might also have made the eoliths from East Anglia and the Kent Plateau, some of which are roughly comparable in age to the Oldowan tools.

But of some of the Oldowan tools J. Desmond Clark wrote in his forward to Mary Leakey's study: "Here are artefacts that conventional usage associates typologically with much later times (the late Paleolithic or even later)—diminutive scraper forms, awls, burins . . . and a grooved and pecked cobble" (M. Leakey 1971, p. xvi). The same is true of the European Eolithic assemblages. As we noted in our introduction to this chapter, implements of a more advanced character sometimes turn up in even the crudest of industries.

We note, however, that tools of the type found in the "late Paleolithic and even later" are considered by modern scientists to be specifically the work of *Homo sapiens* rather than *Homo erectus* or *Homo habilis*. We might thus entertain the possibility that anatomically modern humans were responsible for some if not all of the Oldowan and Eolithic tools.

The standard reply will be that there are no fossils showing that humans of the fully modern type were around then, in the Early Pleistocene or Late Pliocene, roughly 1–2 million years ago, whereas there are fossils of *Homo habilis*. But the history of events at Olduvai Gorge demonstrates that one should be careful about connecting fossil bones with stone tools. As we have seen, the Leakeys first found stone tools but no hominid fossils. When fossils of *Zinjanthropus* were found, this creature was designated as the toolmaker. But when additional fossils of the more advanced *Homo habilis* were found, *Homo habilis* replaced *Zinjanthropus* as the toolmaker. One cannot predict what further fossils might be found in the lower levels of Olduvai Gorge. Perhaps scientists might uncover fossils of *Homo sapiens*, who would then replace *Homo habilis* as the toolmaker.

Even in the absence of *Homo sapiens* remains, the advanced nature of some of the Oldowan tools raises questions about the correctness of attributing their manufacture to a creature as primitive as *Homo habilis*. The Leakeys found in Bed I of Olduvai Gorge bola stones and an apparent leather-working tool that might have been used to fashion leather cords for the bolas (Section 5.3.2). Using bola stones to capture game would seem to require a degree of intelligence and dexterity beyond that possessed by *Homo habilis*. This concern is heightened by the recent discovery of a relatively complete skeleton of *Homo habilis*, which shows this hominid to have been far more apelike than scientists previously imagined (Section 11.7).

It should be kept in mind that *Homo sapiens* fossils are quite rare even at Late Pleistocene sites where, according to conventional views, they should be expected to be found. Marcellin Boule (Boule and Vallois 1957, p. 145) noted that scientists searching for human fossils in the Prince's Cave at Grimaldi in southern Europe sifted through four thousand cubic yards of deposits without finding a single human bone. Nevertheless, stone tools and animal remains were both abundant in the cave. Thus the absence of *Homo sapiens* fossils at a particular site does not eliminate *Homo sapiens* as the maker of stone tools found there.

Furthermore, as described in Chapters 6 and 11, fossil skeletal remains of human beings of the fully modern type have been discovered by scientists in strata at least as old as the lower levels of Olduvai Gorge, Tanzania. Among them may be numbered the fossil human skeleton discovered in 1913 by Dr. Hans Reck in Bed II of Olduvai Gorge (Section 11.1), and some fossil human femurs discovered by Richard Leakey at Lake Turkana, Kenya, in a formation slightly older than Bed I at Olduvai (Section 11.3). Bed I is now dated at approximately 1.75 million years, and the top of bed II is dated at about 0.7–1.0 million years (M. Leakey 1971, pp. 14–15).

It is, therefore, not correct to say that there is no fossil evidence whatsoever for a fully human presence in the lower levels of Olduvai Gorge. In addition to fossil evidence, we have a report from Mary Leakey (1971, p. 24) about a controversial circular formation of stones at the DK site in lower Bed I: "On the north side, where the circle was best preserved, there were groups of stones piled up into small heaps. It is possible to identify six of these piles which rise to a height of 6–9 in. and are spaced at intervals of 2–2.5 ft., suggesting that they may have been placed as supports for branches or poles stuck into the ground to form a windbreak or rough shelter."

Leakey then continued: "In general appearance the circle resembles temporary structures often made by present-day nomadic peoples who build a low stone wall round their dwellings to serve either as a windbreak or as a base to support upright branches which are bent over and covered with either skins or grass" (M. Leakey 1971, p. 24). For the purpose of illustration, Mary Leakey provided a photograph of such a temporary shelter made by the Okombambi tribe of South West Africa (now Namibia).

Not everyone agreed with Leakey's interpretation of the stone circle. But accepting Leakey's version, the obvious question may be raised: if she believed the structure resembled those made by "present-day nomadic peoples" like the Okombambi, then why could she not assume that anatomically modern humans made the Olduvai stone circle 1.75 million years ago?

The same assumption might easily be made about even the crudest stone tools. Leakey stated in her book: "An interesting present-day example of unretouched flakes used as cutting tools has recently been recorded in South-West Africa and may be mentioned briefly. An expedition from the State Museum, Windhoek, discovered two stone-using groups of the Ova Tjimba people who not only make choppers for breaking open bones and for other heavy work, but also employ simple flakes, un-retouched and un-hafted, for cutting and skinning" (M. Leakey 1971, p. 269). Nothing, therefore, prevents one from entertaining the possibility that anatomically modern humans might have been responsible for even the crudest stone tools found at Olduvai Gorge and the European eolith sites.

At present, we find that humans manufacture stone tools of various levels of sophistication, from primitive to advanced. We also find evidence of the same variety of tools in the Pleistocene, Pliocene, Miocene, and even as far back as the Eocene. There are examples of relatively crude stone tools, such as those found by Ribeiro in Miocene formations in Portugal (Section 4.1). And there are also advanced stone tools, similar to those used by modern Indians in North America, from formations of Eocene antiquity in California (Section 5.5).

The simplest explanation is that anatomically modern humans, who make such a spectrum of tools today, also made them in the past. Continuity of tool types suggests continuity of toolmakers. We might call this the hypothesis of stasis. Alternatively, the evolutionary hypothesis requires us to reject all advanced stone tool industries from periods earlier than the Late Pleistocene. As for the remaining crude stone tools, we must reject the ones found in geological contexts older than the earliest Pleistocene or the latest Pliocene. We must then propose that various grades of subhumans made crude stone tools in the Late Pliocene and Early Pleistocene, and then when modern humans came along in the Middle and Late Pleistocene, they also made identical crude tools along with more advanced ones.

All in all, the hypothesis of stasis allows us to account for all the reported evidence in a more straightforward fashion. The only anomaly in this account of stasis is the absence of evidence for advanced civilization, with its intricate metallic productions and complex stone architecture, in very ancient times. Abundant evidence for such civilization appears to extend back only a few thousand years. There are, however, intriguing hints of the existence of advanced civilization millions of years ago. This evidence, reported in Appendix 2, is, however, not very extensive.

Granting the stasis hypothesis, we must therefore ask the following question. Why are there so many scientific reports of stone tools and cut bones indicating the presence of anatomically modern humans tens of millions of years ago yet so little evidence of more advanced civilization for the same time periods?

Here is one possible explanation. Although the scientists who reported much of the evidence contained in this book were prepared to find signs of a human presence in times far more ancient than allowed by current evolutionary theory, these scientists were themselves evolutionists. As such, they believed that in the past culture was more primitive than today. Therefore, they probably would not have given serious consideration to any evidence of advanced culture in very ancient times.

Did they encounter such evidence but refuse to report it? We cannot say for certain. What we do know is that evidence for advanced civilizations in very ancient times has been reported, but not often by scientists. Many of the reports have come from miners. Such reports are far more likely to turn up in old newspapers than scientific journals. We suspect that many finds suggestive of advanced civilizations in very ancient times have not been reported at all.

It is thus possible that our data base for the study of human origins and antiquity is quite incomplete. But what evidence we do have suggests that anatomically modern humans have been manufacturing stone tools of various degrees of sophistication since the Miocene and earlier.

To further complicate the picture, one could imagine *Homo sapiens* coexisting millions of years ago with species of humanlike apes, unrelated to human beings in any evolutionary sense. These humanlike creatures may have also been able to manufacture very crude stone implements. There are in fact reports from Central Asia of a living ape-man-like creature, the Almas, which is said to break stones for use as tools (Section 10.8), just like modern humans. Indeed, this is what the unedited record of skeletal remains and stone implements actually suggests—that human beings of modern type and more primitive creatures have been coexisting since time immemorial and manufacturing a whole array of tool types, from the crudest to the most advanced.

3.8 RECENT EXAMPLES OF EOLITHIC IMPLEMENTS FROM THE AMERICAS

Several anomalously old crude stone tool industries of Eolithic type have been discovered in the Americas. A careful study of the debates about these industries will add to our understanding of why and how the stone tools from Pliocene and Miocene sites in Europe have largely disappeared from view, as far as modern science is concerned.

3.8.1 Standard Views on the Entry of Humans into North America

The debates about various anomalous stone tool industries discovered in the Americas takes place in the context of the standard theory of the entry of humans into the New World. According to this theory, Siberian hunters crossed over the Bering Strait into Alaska on a land bridge that existed when the last glaciation lowered sea levels. During this glacial period, the Canadian ice sheet blocked southward migration until about 12,000 years ago, when the first American immigrants followed an ice free passage to what is now the United States. These people were the so-called Clovis hunters, famous for their characteristic doubly fluted spearpoints. These would correspond to the highly evolved stone implements of the later Paleolithic in Europe.

According to Jared Diamond (1987), these Clovis hunters quickly multiplied and peopled the entire habitable region of North and South America. Because a site in Patagonia, in the southernmost part of South America, is now dated at 10,500 years, the immigrants must have gone from the arctic, to the tropics, and on to the near antarctic regions of South America in little more than a thousand years. In their long march, these Clovis hunters exterminated over 70 percent of the large mammalian genera of the New World in an orgy of rapacious exploitation rivaled only by the European heirs of the territory they conquered (Diamond 1987, pp. 82–88).

The following arguments in favor of this theory were published in the popular science magazine, *Discover*, in June of 1987: "at excavated Clovis sites, conclusive evidence for artifacts made by other peoples has been found above but not below the level with Clovis tools; and there are no irrefutable human remains with irrefutable pre-Clovis dates anywhere in the New World south of the former Canadian ice sheet. Mind you, there are dozens of claims of sites with pre-Clovis human evidence, but all are marred by serious questions about whether the material used for radiocarbon dating was contaminated by older carbon, or whether the dated material was really associated with human remains, or whether the tools supposedly made by hand were just naturally shaped rocks. In contrast, the evidence for Clovis is undeniable, widely distributed, and accepted by archaeologists" (Diamond 1987, pp. 84, 86).

To put this theory into perspective, we should note that before World War II, anthropological authorities insisted that human beings first entered America just 4,000 years ago. Their initial reaction to the Clovis hunter theory was summed up by the anthropologist John Alsoszatai-Petheo (1986, pp. 18–19): "For . . . decades, American archaeologists would labor under the view of man's relative recency in the New World, while the mere mention of the possibility of greater antiquity was tantamount to professional suicide. Given this orientation, it is not surprising that when the evidence of the antiquity of man in America was finally reported from Folsom, Clovis, and other High Plains sites, it was rejected out of hand by established authorities despite the clear nature of the evidence at multiple locations, uncovered by different researchers, and seen and attested to by a large variety of professional visitor/observers. . . . The mind set of conservatives of the day left no room for acceptance."

Alsoszatai-Petheo argued that the history of the rejection of the Folsom and Clovis discoveries is now being repeated as conservative archeologists of the present day staunchly reject evidence for pre-Clovis man in America. Certainly, there are now many cases of archeological excavations using modern methods that have yielded dates as great as 30,000 years for humans in America.

For example, geological, archeological, and paleontological research at El Cedral, in the state of Sinaloa, northern Mexico, revealed human artifacts along with bones of extinct animals in "undisturbed stratified deposits on horizons radiocarbon-dated at 33,000 B.P., 31,850 B.P., 21,960±540 B.P., and older than 15,000 B.P." (Lorenzo and Mirambell 1986, p. 107). The date of 31,850 B.P. corresponds to a hearth found *in situ* and consisting of "a circle of proboscidean tarsal bones surrounding a zone of charcoal about 30 cm [a foot] in diameter and 2 cm [almost an inch] thick" (Lorenzo and Mirambell 1986, p. 111). Proboscideans are elephants of various kinds. Tarsal bones come from the ankle region.

Another case involves a fire pit found on California's Santa Rosa Island, off Santa Barbara, and investigated by archeologist Rainer Berger of UCLA. Laboratory testing showed that charcoal samples taken from the pit contained no measurable carbon 14. They are thus older than the 40,000-year limit imposed by the conventional radiocarbon dating method. The find is significant, since the fire pit contained crude chopping tools along with the bones of a bull-sized species of mammoth (*Science News* 1977a, p. 196).

Yet another interesting excavation took place in northeastern Brazil. At the rock-shelter of Boquierão do Sitio da Pedra Furada, a joint French-Brazilian team of archeologists dug through a stratified 3-meter [10-foot] deposit of sediment that was found to contain human occupational debris at all levels. The lowest levels included big circular hearths with large quantities of charcoal and ash. There were pebble tools, denticulates, burins, retouched flakes, and double-edged flakes, all made from local quartz and quartzite. There were also painted fragments of rock spalled or broken from the cave walls, which suggests that the tradition of rock painting well known in this part of Brazil may have existed during the earliest occupational period (Guidon and Delibrias 1986, pp. 769–771).

Charcoal from the lowest hearth in the deposit yielded carbon 14 dates of 31,700±830 years and 32,160±1,000 years. In addition, carbon 14 dates were obtained at a series of levels running throughout the entire deposit. These dates formed the following consistent series in years B.P.: 6,160, 7,750, 7,640, 8,050, 8,450, 11,000, 17,000, 21,400, 23,500, 25,000, 25,000, 25,200, 26,300, 26,400, 27,000, 29,860, 31,700, and 32,160 (Guidon and Delibrias 1986).

This excavation is of particular interest because it involved a controlled study of stratified cave deposits yielding hearths, artifacts, and a series of radiocarbon dates. These are some of the criteria often insisted upon by defenders of orthodox archeological theories. However, one can always point to flaws in unwanted evidence, and thereby adopt a double standard.

Of course, a small but increasing number of archeologists are now accepting that humans may have been living in South America as long as 30,000 years ago.

It might therefore be argued that the resistance to new findings exhibited by successive schools of archeologists is simply a healthy and unavoidable part of the scientific process. By applying the braking action of skepticism, science can make slow but steady progress, while avoiding wild, speculative excesses.

One answer to this is that by sticking to conservative viewpoints in anthropology one certainly does not avoid extreme speculation. The theory that Clovis hunters marched from northern Canada to the Tierra del Fuego in a few centuries is certainly speculative. And the sweeping denial of certain possibilities—such as the existence of humans in America at a certain date—can be just as much a speculative excess as their uncritical affirmation. In addition, it may happen that evidence suppressed as a result of such policies of denial is permanently lost, and important advancements in understanding will be delayed until similar evidence manages to surmount the barriers to acceptance in the future.

An alternative approach would be to recognize that in fields such as archeology, most empirical evidence is of a doubtful nature, whether it corroborates our views or contradicts them. Therefore, it would be best (though difficult in practice) to maintain all relevant evidence in a readily accessible form, without giving absolute credence to any current positive or negative interpretations. If this cannot be done, one should at least recognize that one may be aware of only a fraction of the evidence that has already been seriously studied—what to speak of the evidence that may be uncovered in the future.

The present method of rendering final judgement on controversial evidence by how well it fits with currently established theories does not seem to be scientifically healthy, and it can be argued that it may do irremediable damage not only to the progress of scientific knowledge, but also to the reputations of persons who happen to find controversial evidence. This is especially true when politics and intrigue enter into the scientific process. Such considerations appear to have played a major role in the negative treatment of evidence suggesting that human beings were living in the New World long before both the 12,000-year limit still favored by a majority of paleoanthropologists and the 30,000-year limit currently accepted by a growing minority. We shall now discuss a few recent examples of this evidence, in the form of anomalously old crude stone tool industries, with the aim of shedding more light on the social processes of acceptance and rejection of evidence in the scientific world.

3.8.2 Texas Street, San Diego (Early Late Pleistocene to Late Middle Pleistocene)

A good example of a controversial American early stone tool industry reminiscent of the European eoliths is the one discovered by George Carter (1957) in the 1950s at the Texas Street excavation in San Diego. At this site,

Carter (1957) claimed to have found hearths and crude stone tools at levels corresponding to the last interglacial period, some 80,000–90,000 years ago. Critics scoffed at these claims, referring to Carter's alleged tools as products of nature, or "cartifacts", and Carter was later publicly defamed in a Harvard course on "Fantastic Archeology" (Williams 1986, p. 41). However, Carter gave clear criteria for distinguishing between his tools and naturally broken rocks, and lithic experts such as John Witthoft (1955) have endorsed his claims.

In 1973, Carter conducted more extensive excavations at Texas Street and invited numerous archeologists to come and view the site firsthand. Almost none responded. Carter (1980, p. 63) stated: "San Diego State University adamantly refused to look at work in its own backyard."

Carter found evidence for a human presence during the last interglacial period at several other sites in San Diego and elsewhere in the southwestern United States. But he found it difficult to get his findings published in standard scientific journals. In 1960, an editor of *Science,* the journal of the American Academy for the Advancement of Science, asked Carter to submit an article about early humans in America. Carter did so, but the article was rejected. The editor wrote to Carter on February 1, 1960: "It was good of you to prepare a paper 'On the Antiquity of Man in America' for possible publication in our *Current Problems in Research* series in *Science.* In view of the fact that I invited you to prepare the paper for us, I especially regret to say that your paper, although it is interesting and deals with an important subject, is too controversial for publication in a general scientific magazine such as ours. I sought the advice of two highly competent advisers and they were in essential agreement with each other in their recommendations. They both thought that the paper was unsuitable for *Science*" (T. E. Lee 1977, p. 3).

Carter replied in a letter to the editor, dated February 2, 1960: "I must assume now that you had no idea of the intensity of feeling that reigns in the field. It is nearly hopeless to try to convey some idea of the status of the field of Early Man in America at the moment. But just for fun: I have a correspondent whose name I cannot use, for though he thinks that I am right, he could lose his job for saying so. I have another anonymous correspondent who as a graduate student found evidence that would tend to prove me right. He and his fellow student buried the evidence. They were certain that to bring it in would cost them their chance for their Ph.D's. At a meeting, a young professional approached me to say, 'I hope you really pour it on them. I would say it if I dared, but it would cost me my job.' At another meeting, a young man sidled up to say, 'In dig *x* they found core tools like yours at the bottom but just didn't publish them'" (T. E. Lee 1977, p. 4).

The inhibiting effect of negative propaganda on the evaluation of Carter's discoveries is suggested in the following statement by archeologist Brian Reeves: "Were actual artifacts uncovered at Texas Street, and is the site really

Last Interglacial in age? . . . Because of the weight of critical 'evidence' presented by established archaeologists, the senior author [Reeves], like most other archaeologists, accepted the position of the skeptics uncritically, dismissing the sites and the objects as natural phenomena" (Reeves *et al.* 1986, p. 66).

But when he took the trouble to look at the evidence himself, Reeves changed his mind. He wrote: "While visiting San Diego in 1976 the senior author had the opportunity to view some of George Carter's . . . collections from Texas Street . . . in Mission Valley. Among the fractured quartzite cobbles were many objects that appeared to Reeves and R. S. MacNeish to be culturally produced, modified, and utilized quartzite cobble artifacts" (Reeves *et al.* 1986, p. 66). Ten years later Reeves conducted several onsite investigations near Texas Street.

Many of the specimens he studied, although made from quartzite rather than flint, appear to be Eolithic: "In summary, the Mission Ridge quartzite cobble complex includes naturally produced sharp-pointed and edged bipolar cores, blocky quartzite pieces and irregular-shaped sharp-edged flakes. These fragments were not only utilized by man, but also modified into more formed flakes and tools (the horseshoe chopper, for example) as well as culturally manufactured, unifacially retouched and utilized flakes" (Reeves *et al.* 1986, p. 78).

Reeves concluded: "The bulk of the fractured quartzites recovered from Mission Ridge were, in our opinion, naturally broken but collected elsewhere and brought to the site by man for use primarily as ready-made expediency tools" (Reeves *et al.* 1986, p. 78). In light of Reeves's change of heart about Carter's tools, one wonders what would result from an openminded review of the European eoliths.

Reeves then made the following commentary on the unfair treatment professional scientists gave to the San Diego implements: "The fractured quartzite complex, as first claimed by Carter, is part of a Late to Middle Pleistocene quartzite cobble core/unifacial flake tradition of Pacific coastal-adapted people. . . . Had Carter's claims been taken seriously enough by professional archaeologists to undertake detailed field studies instead of simply dismissing them, we would have had a major body of data on Late Pleistocene North American coastal settlement" (Reeves *et al.* 1986, pp. 78–79). Reeves believed some of Carter's implements to be 120,000 years old.

Over several decades, many ancient human occupation sites were investigated around San Diego, and Carter (1957, pp. 370–371) constructed a tentative history of stone tool usage in this region over the last 90,000 years. After the Texas Street phase, characterized by crude stone tools, came the following developments: (1) The period of 55,000 to 80,000 years ago, represented by "strongly weathered manos and metates from basal positions in alluvium over interglacial beaches at Scripps campus and about La Jolla and Point Loma." There were also biface and plano-convex cobble core tools, and used flakes.

Manos and metates are grinding tools. Plano-convex tools are flat on one side and rounded, or convex, on the other. (2) The period of 30,000 to 55,000 years ago, with large, crude, percussively flaked, ovate knives, tending to be unifacial. (3) The period of 15,000 years to 30,000 ago with small, slender, leaf-shaped, double-convex knives, broad-stemmed knives, and abundant fine plano-convex tools. (4) Then came the recent San Dieguito and Yuman cultures.

According to standard views, practically all of the variegated lithic forms in this list would have to be either (1) incorrectly dated, or (2) products of human imagination applied to naturally broken stone. The manos and metates are especially interesting, since these grinding tools are generally associated with Neolithic, or very late Stone Age, culture. The oldest accepted examples, from Egypt, are thought to be only 17,000 years old (Gowlett 1984, p. 152).

3.8.3 Louis Leakey and the Calico Site in California (Middle Pleistocene)

As we have several times seen in previous chapters (and will see again in later chapters), some famous scientists have occasionally nurtured heretical ideas, despite the personal risks involved in opposing prevailing academic views. One example is Louis Leakey, world renowned for his discoveries in Africa. He began to have radical ideas about the antiquity of humans in America at a time when the entry date for the Siberian hunters was thought to be no greater than some 5,000 years ago. Eventually, Leakey journeyed to America and discovered a crude stone tool industry, of Eolithic type, at Calico, in southern California. The site was dated at over 200,000 years.

Leakey recalled: "Back in 1929–1930 when I was teaching students at the University of Cambridge, I began to look into the question of the antiquity of man in the Americas. Although there was no concrete evidence to indicate a remote age, I was so impressed by the circumstantial evidence that I began to tell my students that man must have been in the New World at least 15,000 years. I shall never forget when Ales Hrdlicka, that great man from the Smithsonian Institution, happened to be at Cambridge, and he was told by my professor (I was only a student supervisor) that Dr. Leakey was telling students that man must have been in America 15,000 or more years ago. He burst into my rooms—he didn't even wait to shake hands—and said, 'Leakey, what's this I hear? Are you preaching heresy?'" Leakey said, "No, Sir!" Hrdlicka replied, "You are! You are telling students that man was in America 15,000 years ago. What evidence have you?" Leakey replied, "No positive evidence. Purely circumstantial evidence. But with man from Alaska to Cape Horn, with many different languages and at least two civilizations, it is not possible that he was present only the few thousands of years that you at present allow" (L. Leakey 1979, p. 91).

Leakey continued to harbor unorthodox views on this matter, and in 1964 he made an effort to collect some definite evidence by initiating an excavation at a site known as Calico in the Mojave Desert of California. This site is situated near the shore of now-vanished Pleistocene Lake Manix, on the eroded remains of an alluvial fan of sediments washed down from the nearby Calico mountains. Over a period of eighteen years of excavation, some 11,400 artifacts were recovered from a number of levels. The oldest artifact-bearing level has been dated by the uranium series method to about 200,000 years B.P. (Budinger 1983).

There is general agreement among geologists about the great age of the Calico site, and ages as great as 500,000 years have been seriously proposed. However, as happened with Texas Street, mainstream archeologists have tended to reject the artifacts discovered at Calico as products of nature, and the Calico site tends to be passed over in silence in popular accounts of archeology. Indeed, it seemed that the iconoclastic Leakey, famous for so many revolutionary archeological discoveries, had committed a grave error in judgement in his foray into the New World. Leakey's biographer Sonia Cole (1975, p. 351) said, "For many colleagues who felt admiration and affection for Louis and his family, the Calico years were an embarrassment and a sadness."

Yet the artifacts of Calico also have their defenders, who give elaborate arguments showing that they were human artifacts, not "geofacts" resulting from natural processes. These archeologists include Phillip Tobias, the well-known associate of Raymond Dart, discoverer of *Australopithecus*. Tobias (1979, p. 97) declared: "when Dr. Leakey first showed me a small collection of pieces from Calico . . . I was at once convinced that some, though not all, of the small samples showed unequivocal signs of human authorship." Tobias went on to point out that the presence of naturally broken stones is to be expected, and does not detract from the validity of artifacts that are mixed in with them.

The arguments presented are reminiscent of the controversy over eoliths in Europe. Detractors such as archeologist C. Vance Haynes (1973, pp. 305–310) claimed that the natural banging together of stones in streams and shifting earth can simulate all the alleged Calico stone tools. On the other hand, defenders pointed out that these alleged natural processes did not occur at sites such as Calico, and could not have produced the observed, systematic patterns of lithic flaking even if they did occur (L. Patterson et al. 1987, pp. 91–105).

Geological evidence indicates that the Calico implements lie in an ancient mud flow context. In this regard, Ruth D. Simpson stated: "Natural forces in a mud flow would be expected to give mainly bidirectional random damage to flake edges. It would be difficult for nature to produce many specimens resembling man-made unifacial tools, with completely unidirectional edge retouch done in a uniform, directed manner. The Calico site has yielded many completely unifacial stone tools with uniform edge retouch. These include end scrapers, side scrapers, and gravers. Some gravers have bifacial retouch on points, which can be expected in even unifacial flake tool industries" (Simpson *et al.* 1986, p. 96). Flake tools with unifacial, unidirectional chipping, like those found at Calico, are typical of the European eoliths. Examples are also found among the Oldowan industries of East Africa. Among the best tools that turned up at Calico was an excellent beaked graver (Figure 3.30). Bola stones have also been reported (Minshall 1989, p. 110).

At an international archeological conference held in Mexico City, Mexico, in 1981, three of the defenders of Calico listed 17 criteria for human flaking which, according to them, were met by the artifacts discovered at the Calico site. Some of these criteria were (1) the presence of ripple lines and force bulbs with bulb scars, (2) striking platform angles under 90 degrees, (3) crushing of striking platforms, (4) no remaining cortex on either striking platforms or dorsal surfaces, (5) prismatic flakes and blades, (6) unifacial edge retouch, (7) flaking on certain edges and not others, (8) well-defined bifacial objects, and (9) specific workshop areas with evidence of stone working (Simpson *et al.* 1981).

Figure 3.30. A beaked graver—a stone tool from Calico in southern California, dated at about 200,000 years (Bryan 1979, p. 77).

Herbert L. Minshall stated that in 1985 several of the best small Calico implements were displayed at the annual meeting of the Society for American Archaeology in Denver, Colorado. Minshall wrote (1989, p. 111): "The tools were finally accepted as manmade, but now the objection was that they could not possibly have so great an age, even though 200,000 years was modest compared to many estimates for the age of the fan sediments. . . . One highly respected archaeologist actually suggested that the tools he was shown must have somehow fallen into the excavation from the surface."

In 1986, George Carter and Fred Budinger discovered an additional site at Calico. Minshall (1989, p. 111) stated that Carter and Budinger found "small stone specimens apparently worked by man and more than 20 feet below the dated volcanic ash stratum at the foot of the Calico/Mule Canyon fan" near the main Calico site. Fossils of typical Pleistocene mammals such as the sabertooth tiger, camel, horse, and mammoth were also found beneath the ash, which yielded a potassium-argon date of 185,000 years.

In general, however, the Calico discoveries have met with silence, ridicule, and opposition in the ranks of mainstream paleoanthropology. Ruth Simpson nevertheless stated: "The data base for very early man in the New World is growing rapidly, and can no longer simply be ignored, because it does not fit current models of prehistory in the New World. With the present data gaps that exist in our knowledge of the prehistory of man in the New World, any current proposed 'final' solutions to the early origins, migrations, and cultures of Pleistocene man in the New World are premature. At the present state of knowledge in early man research, there is a need for flexibility in thinking to assure unbiased peer reviews" (Simpson *et al.* 1986, p. 104). The same might also be said of the larger question of human evolution.

3.8.4 Toca da Esperança, Brazil (Middle Pleistocene)

Support for the authenticity of the Calico tools has come from a find in Brazil. In 1982, Maria Beltrao found a series of caves with wall paintings in the state of Bahia. In 1985, a trench was cut in the Toca da Esperança (Cave of Hope), and excavations in 1986 and 1987 "yielded stone tools associated with Quaternary fauna in a defined stratigraphic context" (de Lumley *et al.* 1988, p. 241).

There were four layers in the cave. The first layer was a hard carbonate crust, 20 to 60 centimeters (about 8 to 24 inches) thick. Beneath this were 3 layers of sand and sandy clay. In the lowest, Layer 4, stone implements were discovered along with abundant mammalian fossils. De Lumley *et al.* (1988, p. 241) commented: "Three bones . . . were dated by the uranium-thorium method using alpha and gamma-ray spectrometries, [giving] ages between 204,000 and 295,000 years." These tests were performed at three different laboratories—Gif-sur-Yvette, France; the University of California at Los Angeles; and the laboratory of the U.S. Geological Survey at Menlo Park, California (de Lumley *et al.* 1988, p. 243).

The tools were fashioned from quartz pebbles and were somewhat crude, like those from Olduvai Gorge. The implements included "a chopper with cutting-edge trimmed by three adjacent removals" (de Lumley *et al.* 1988, p. 243). The report pointed out that the nearest source of quartz pebbles is about 10 kilometers from the cave.

De Lumley *et al.* (1988, p. 242) stated: "the evidence seems to indicate that Early Man entered into the American continent much before previously thought." They went on to say: "In light of the discoveries at the Toca da Esperança, it is much easier to interpret the lithic industry of the Calico site, in the Mojave Desert, near Yermo, San Bernardino County, California, which is dated at between 150,000 and 200,000 years" (de Lumley *et al.* 1988, p. 245).

According to de Lumley and his associates, humans and protohumans entered the Americas from northern Asia several times during the Pleistocene. The early migrants, who manufactured the tools in the Brazilian cave, were *Homo erectus* (de Lumley *et al.* 1988, p. 242). While this view is in harmony with the consensus on human evolution, there is no reason why the tools in the Toca da Esperanca could not have been made by anatomically modern humans. As we have several times mentioned, such tools are still being manufactured by humans in various parts of the world.

Toca da Esperança provides a clear example of how the scientific community hesitates to change deeply held convictions. The discovery was made by a team headed by a famous French scientist, respected in his field. The site was systematically excavated according to strict principles. The implements were discovered *in situ,* in a defined stratigraphic context. They were clearly intentionally manufactured. They were found in conjunction with a typical Middle Pleistocene fauna, with many extinct species. The researchers admitted that it was not possible to assign a direct age to the cave on the basis of the biostratigraphic evidence (the Middle Pleistocene goes from about 100,000 to 1,000,000 years ago), but multiple uranium series tests gave ages of between 204,000 and 295,000 years. Of course, the uranium series dates could be wrong. But if these are wrong, then every uranium series date, including the ones used to buttress more acceptable finds, could also be wrong. Altogether, it is hard to see what more one could desire in the way of empirical evidence that would confirm the presence of intelligent toolmaking beings in the Americas in the Middle Pleistocene. Yet the consensus that humans entered the Americas fairly recently remains intact.

3.8.5 Alabama Pebble Tools

The crude stone tools of Bed I in Olduvai Gorge, Tanzania, are also paralleled, interestingly enough, by pebble tools from Alabama, U.S.A. that are almost identical in form (Figure 3.31, p. 208). These stone tools, reported by archeologist Daniel Josselyn (1966), can be found in great numbers in certain surface sites, where they are mixed in with artifacts from a variety of native American cultures.

Pebble tools are usually associated with very primitive levels of culture not thought to have ever existed in America. Thus, when Josselyn tried to acquaint other American archeologists with his finds, he did not receive an encouraging reaction. "Rather," as he put it, "to my horror, I learned that Pavlov could have studied 'conditioned reflexes' about as well in archaeologists as in dogs. Please,

Figure 3.31. Crude pebble chopper, from Alabama, U.S.A., undated (Josselyn 1966). Such tools usually imply very primitive cultures not thought to have existed in America.

please, believe that I say this with no critical rancor" (Josselyn 1966, p. 25). It was apparently "known" by some that no pebble tools were made in the New World.

Josselyn said that since the Alabama tools were not from stratified sites, they could not be dated, and he had no suggestion about their age. They could thus be quite recent, posing no threat to dominant views about the arrival of humans in the Americas. The problem here seems to be a fixation on the questionable idea that pebble tools must have been made by protohumans such as *Homo habilis* or *Homo erectus*. But human beings have used pebble tools in Asia and Africa in historic times.

3.8.6 Monte Verde, Chile (Late Pleistocene)

Another archeological site that has bearing on the evaluation of crude stone tools is the Monte Verde site in south central Chile. According to a report in *Mammoth Trumpet* (1984), this site was first surveyed by archeologist Tom Dillehay in 1976. Although the date of 12,500 to 13,500 years B.P. for the site is not highly anomalous, the archeological finds uncovered there challenge the standard Clovis hunter theory. The culture of the Monte Verde people was completely distinct from that of the Clovis hunters. Although these people made some bifacial implements, their lithic technology was based mainly on minimally modified pebble tools. Indeed, to a large extent, they obtained stone tools by selecting naturally occurring split pebbles. Some of these show signs of nothing more than usage; others show signs of deliberate retouching of a working edge. This is strongly reminiscent of the descriptions of the European eoliths.

In this case, the vexing question of artifacts versus geofacts was resolved by a fortunate circumstance: the site is located in a boggy area in which perishable plant and animal matter has been almost indefinitely preserved. Thus two pebble tools were found hafted to wooden handles. Twelve "architectural foundations" were found, made of cut wooden planks and small tree trunks staked in place. There were large communal hearths, as well as small charcoal ovens lined with clay. Some of the stored clay bore the footprint of a child 8 to 10 years old. Three crude wooden mortars were also found, held in place by wooden stakes. Grinding stones (metates) were uncovered, along with the remains of wild potatoes, medicinal plants, and sea coast plants with a high salt content. All in all, the Monte Verde site sheds an interesting light on the kind of creatures who might have made use of "crude pebble tools" during the Pliocene and Miocene in Europe or at the Plio-Pleistocene boundary in Africa. In this case, the culture was well equipped with domestic amenities made from perishable materials. Far from being subhuman, the cultural level was what we might expect of anatomically modern humans in a simple village setting even today.

By an accident of preservation, we thus see at Monte Verde artifacts representing an advanced culture accompanying the crudest kinds of stone tools. At sites millions of years older, we see only the stone tools, although perishable artifacts of the kind found at Monte Verde may have once accompanied them.

Finally, we note that Tom Dillehay found in the deepest stratum at Monte Verde a split basalt pebble, some wood fragments, two modified stones, and some charcoal dated at about 33,000 years B.P. (Bray 1986, p. 726).

3.8.7 Early Humans in America and the Eolith Question

The arguments about American sites tens and hundreds of thousands of years old are similar to those that took place among European scientists when the first evidence for prehistoric humans was coming to light. This was noted by anthropologist Alan Lyle Bryan, who wrote (1986, p. 5): "The present controversy over Early Man in America is analogous to that in Europe more than a century ago because the intellectual climate has been dominated for over 50 years by a particular paradigm which has seemed to fit most of the evidence but which fails to explain an increasing body of data. Rather than considering a new paradigm which might make the evidence sensible, skeptics have demanded that all evidence for 'pre-Clovis' be judged by more rigid standards of evidence and argument than are applied to later sites. . . . Arbitrary application of such rigid criteria to later sites, including Clovis sites, would relegate nearly all archaeological evidence to the 'not proven' category." It should, however, be noted that the European controversy of the nineteenth century, the full dimensions of which Bryan was probably unaware, is, like the debate on the antiquity of humans in the Americas, still very much an open question. The seriousness with which a modern paleoanthropologist might consider reports of stone tools apparently made by humans in the European Pliocene and earlier is likely to vary in inverse proportion to his commitment to the now-accepted views on human evolution.

Eolithic tools have been found not only in America and Europe but in Australia (R. A. Gould *et al.* 1971). They have been described as featuring "the casual use of available materials; the lack of emphasis on technological sophistication; the regular discarding of tools after a specific job had been completed; and an attitude which de-emphasizes symmetry, refinement, and systematic continuity in tool types, but instead focuses on the most convenient means of accomplishing the job at hand" (Alsoszatai-Petheo 1986, p. 22).

The human manufacture of the Australian specimens has been widely recognized in the scientific community. So why are not similar tools found in America granted equal recognition? Alan Lyle Bryan (1986, pp. 7-8) stated: "some definitely shaped tool (preferably something 'diagnostic') must be present in order to have acceptable 'proof' for the presence of Early Man. Anything less is now being labelled a 'myth,' and believers of myths cannot be scientific archaeologists. But if the Australian archaeologists had adhered to such strict criteria they would not have searched for and thereby recovered evidence for Pleistocene man on that continent. . . . It was realized that the only 'diagnostic' artifact categories may be simple flakes and cores. It was realized that simple retouched flakes are adequate to demonstrate the presence of early man, if they are recovered from datable stratigraphic contexts. . . . It is illogical to require the presence of diagnostic shaped tools in America and not to require their presence in Australia in order to prove that that continent was populated at least 40,000 years ago." But if simple retouched flakes are adequate to prove the existence of humans 40,000 years ago in Australia, and 200,000 years ago in America, why are they not adequate to prove the existence of toolmaking hominids 2 million years ago in England and even earlier elsewhere?

Obviously, the great mass of evidence for a human presence in the Pliocene and earlier, as presented in this book, does not fit within the narrow limits of current ideas on human evolution. Many will therefore hesitate to even consider such evidence. This being the case, it can be said that evolutionary preconceptions impose unreasonable constraints on what evidence may be introduced into discussions of human origins and antiquity. Evidence is excluded for no other reason than that it violates evolutionary expectations. If one were, however, to give even-handed treatment to all of the available evidence, then it would become impossible to coherently set forth any temporally sequential and physiologically progressive path of hominid development. Only a ruthlessly selective editing of the totality of paleoanthropological evidence allows an evolutionary picture of human origins to be sustained.

3.9 A RECENT EOLITHIC DISCOVERY FROM INDIA (MIOCENE)

We shall conclude our discussion of very crude stone tools, from as far back as the Eocene, with a recent example that shows the relevance of the issues raised in this chapter to modern paleoanthropological research.

K. N. Prasad (1982, p. 101) of the Geological Survey of India wrote in an abstract of his report: "A crude unifacial hand-axe pebble tool recovered from the late Mio-Pliocene (9–10 m.y. B.P.) at Haritalyangar, Himachal Pradesh, India is described. This crude flaked tool is assigned to *Ramapithecus*. The occurrence of this pebble tool in such ancient sediments indicates that early hominids such as *Ramapithecus* fashioned tools, were bipedal with erect posture, and probably utilized the implements for hunting." Prasad (1982, p. 102) added: "The implement was recovered *in situ,* during remeasuring of the geological succession to assess the thickness of the beds. Care was taken to confirm the exact provenance of the material, in order to rule out any possibility of its derivation from younger horizons." He also pointed out that *Ramapithecus* jaw fragments and teeth were found in the same horizon, the Nagri formation of the Middle Siwaliks.

Describing the tool itself, Prasad (1982, p. 102) stated: "The quartz artefact, heart-shaped (90 mm ´ 70 mm) [3.6 inches ´ 2.8 inches] was obviously fabricated from a rolled pebble, the dorsal side of which shows signs of rough flaking. . . . On the ventral side much of the marginal cortex is present at the distal end. Crude flaking has been attempted for fashioning a cutting edge. Marginal flaking at the lateral edge on the ventral side is visible." Prasad reminded his readers that another Indian scientist had recovered stone tools from the lower part of the Pinjor formation, corresponding to the Villafranchian stage of the European Late Pliocene. He then stated: "It is not improbable that fashioning tools commenced even as early as the later Miocene and evolved in a time-stratigraphic period embracing the Astian-Villafranchian" (Prasad 1982, p. 103). We agree, but the real question the identity of the toolmaker. As we shall see, *Ramapithecus* has not remained a viable candidate.

Ramapithecus first came to the attention of scientists in the 1930s. This creature, initially regarded as a fossil ape, was named after *Rama,* an incarnation of God described in the Vedas. In 1964 *Ramapithecus* achieved worldwide fame when Elwyn Simons and David Pilbeam reconstructed an upper jaw from two fragments, giving it a characteristically human parabolic shape. Simons and Pilbeam pronounced *Ramapithecus* to be a hominid, an erect, bipedal primate. In 1964, Elwyn Simons wrote in *Anthropology: "Ramapithecus punjabicus* is almost certainly man's forerunner of 15 million years ago. This determination increases tenfold the approximate time period during which human origins can now be traced with some confidence" (Fix 1984, p. 20). This was a bit of an overstatement, because between *Australopithecus* and *Ramapithecus* there was, and still is, a gap of several million years in the hominid fossil record.

In any case, *Ramapithecus* quickly received acclaim, in textbooks and journal articles, as the earliest human ancestor. As Richard Leakey and Roger Lewin wrote in 1977: *"Ramapithecus* . . . as far as one can say at the moment . . . is the first representative of the human family—the hominids" (Fix 1984, p. 20).

Others, however, maintained a more cautious attitude. In 1972, Maitland A. Edey wrote in *The Missing Link:* "On grounds of pure logic, it is tempting to regard *Ramapithecus* as a sort of proto-Australopithecine; after all, the Australopithecines had to start somewhere. But, however tempting such an idea may be, it is premature. We have no knowledge whatsoever of the nature of the rest of *Ramapithecus's* body. We do not know what its skull was shaped like or how large its brain was. We know nothing about its hand or foot. We do not know if it stood upright" (Fix 1984, p. 21). Herbert Wendt also expressed some doubts in *Ape to Adam:* "Whether *Ramapithecus,* which some experts think does not really belong to the race of hominids in the narrow sense of the term, was already a tool-maker we do not know" (Fix 1984, p. 21).

In 1979, information confirming the doubtfulness of *Ramapithecus* appeared in the journal *Natural History.* A. L. Zihlman and J. M. Lowenstein stated that a complete lower jaw of *Ramapithecus,* the first ever found, was V-shaped, unlike either the human jaw, which has a parabolic shape, or the ape jaw, which has parallel sides (Fix 1984, p. 21). In response, Pilbeam modified his position on *Ramapithecus,* placing it in a separate category related neither to apes nor humans. But three years later, *Ramapithecus's* status changed again. William R. Fix (1984, pp. 21-22) wrote: "the February 6, 1982 issue of *Science News* added a new twist to the *Ramapithecus* story. Compiling information from an article in *Nature* (January 21, 1982) and a telephone interview with Pilbeam, *Science News* now has *Ramapithecus* as 'part of the orangutan lineage.'" This newly defined *Ramapithecus* was definitely not a maker of stone tools.

As late as 1981, however, A. R. Sankhyan of the Anthropological Survey of India was writing (1981, pp. 358–359): "The Sivalik Group of rocks exposed in Haritalyangar area of district Bilaspur is famous for the well known Mio-Pliocene Hominoidea—*Dryopithecus, Gigantopithecus,* and *Ramapithecus,* the last of which is considered as the earliest hominoid ancestor of man and also believed to be an ad hoc toolmaker."

But a short time later this view was history. R. N. Vasishat, an anthropologist at Punjab University, wrote (1985, p. xiv): "Until the year 1982, scientists all over the world had unanimously been considering the genus *Ramapithecus* to be the earliest known hominid in the world and [*it*] was also presumed to be ancestral to *Australopithecus* and *Homo.* When this species was taken out of the family Hominidae, the Siwaliks became devoid of any evidence for the antecedents of Early Man. But the author is very sure, the void thus created is very temporary and there is no reason for us to believe that the Siwaliks will never yield fossil evidence or physical evidence of Early Man in the future."

It is interesting to note that with *Ramapithecus* demoted, the Siwaliks became "devoid of any evidence" for Early Man. But what about the above-mentioned stone tools reported by Prasad and Verma? Were these not still evidence? Here is yet another example of the curious manner in which scientists treat anomalous discoveries. Prasad's discovery of a Miocene implement is particularly significant in that it shows that evidence of the type reported by nineteenth-century scientists is still turning up and still being subjected to the same unfair treatment.

4

Crude Paleolithic Stone Tools

In the previous chapter, we considered anomalous stone tools of the crudest type, the eoliths. We shall now turn our attention to other stone tools, which, although also crude when compared with the sophisticated implements of the conventional Late Stone Age, represent an advance over the eoliths. These we have chosen to designate as crude paleoliths.

For some researchers, the terms eolith and paleolith represent a chronological succession, but we use these terms principally to make a distinction in the morphology of tool types. Eoliths, it may be recalled, are naturally broken pieces of stone that are used as tools with little or no further modification. A working edge might be retouched and show signs of wear. Paleoliths, however, are often deliberately flaked from stone cores and then more extensively modified.

As we have previously mentioned, arriving at clear-cut distinctions between eoliths and crude paleoliths is not always possible. Furthermore, a particular group of discoveries often includes implements of various levels of sophistication. In making decisions about what industries to put in this chapter, we have been guided by statements of scientists who favorably compared individual implements, and groups of implements, to recognized tools from much later periods. Anomalously old stone tool industries containing a good many implements comparable to the cruder kinds of classical Paleolithic implements have been selected for inclusion.

4.1 THE FINDS OF CARLOS RIBEIRO IN PORTUGAL (MIOCENE)

We first turn our attention to Carlos Ribeiro's discoveries in the Miocene of Portugal. The first hint of Ribeiro's work came to our attention quite accidentally. While going through the writings of the nineteenth-century American geologist J. D. Whitney, who reported evidence for Tertiary human beings in California, we encountered a sentence or two about Ribeiro having discovered flint implements in Miocene formations near Lisbon. We found more brief mentions

213

in the works of S. Laing, a popular English science writer of the late nineteenth century. Curious, we searched libraries, but turned up no works under Ribeiro's name and found ourselves at a dead end. Sometime later, Ribeiro's name turned up again, this time in the 1957 English edition of *Fossil Men* by Boule and Vallois, who rather curtly dismissed the work of the nineteenth-century Portuguese geologist. We were, however, led by Boule and Vallois to the 1883 edition of *Le Préhistorique,* by de Mortillet, who gave a favorable report of Ribeiro's discoveries, in French. By tracing out the references mentioned in de Mortillet's footnotes, we gradually uncovered a wealth of remarkably convincing original reports in French journals of archeology and anthropology from the latter part of the nineteenth century. The search for this buried evidence was very illuminating, demonstrating how the scientific establishment treats reports of facts that no longer conform to accepted views. Keep in mind that for most current students of paleoanthropology, Ribeiro and his discoveries simply do not exist. You have to go back to textbooks printed over 30 years ago to find even a mention of him. Did Ribeiro's work really deserve to be buried and forgotten? We shall present the facts and allow readers to form their own conclusions.

4.1.1 A Summary History of Ribeiro's Discoveries

Carlos Ribeiro was not an amateur. In 1857, he was named to head the Geological Survey of Portugal, and he would also be elected to the Portuguese Academy of Sciences. During the years 1860–63, he conducted studies of stone implements found in Portugal's Quaternary strata. Nineteenth-century geologists generally divided the geological periods into four main groups: (1) the Primary, encompassing the periods from the Precambrian through the Permian; (2) the Secondary, encompassing the periods from the Triassic through the Cretaceous; (3) the Tertiary, encompassing the periods from the Paleocene through the Pliocene; and (4) the Quaternary, encompassing the Pleistocene and Recent periods. During the course of his investigations, Ribeiro learned that flints bearing signs of human work were being found in Tertiary beds between Canergado and Alemquer, two villages in the basin of the Tagus River, about 35–40 kilometers (22–25 miles) northeast of Lisbon.

Ribeiro immediately began his own investigations, and in many localities found "flakes of worked flint and quartzite in the interior of the beds." Ribeiro (1873a, p. 97) said: "I was greatly surprised when I forcefully extracted, with my own hand, worked flints, from deep inside a bed of limestone which had been inclined at an angle of 30–50 degrees from the horizontal." The geology of the region indicated the limestone bed was of Tertiary age, yet the presence of the stone implements, so obviously the work of humans, placed Ribeiro in a dilemma. The discovery of the implements "deep inside" the beds seemed to rule out the possibility that they had been artificially introduced at some later period. So if he accepted the beds as Tertiary, then humans must have existed during that time. But Ribeiro felt he must submit to the prevailing scientific dogma that human beings were not older than the Quaternary. To this very day authorities hold that humans of the modern type did not appear until the very latest part of the Pleistocene. So Ribeiro looked for and found a way to designate the limestone formation as Quaternary. He remained troubled at heart, however, for the geological facts he himself had observed were leading him to the forbidden conclusion that humans had existed in times more ancient than the Quaternary (Ribeiro 1873a, p. 97).

In 1866, on the official geological maps of Portugal, Ribeiro reluctantly assigned Quaternary ages to certain of the implement-bearing strata. Upon seeing the maps, the French geologist de Verneuil took issue with Ribeiro's judgement, pointing out that the so-called Quaternary beds were, according to geological evidence, certainly Pliocene or Miocene.

Meanwhile, in France, the Abbé Louis Bourgeois, a reputable investigator, had reported finding stone implements in Tertiary beds, and some authorities had supported him. Thus, under the twin influences of de Verneuil's criticism and the discoveries of Bourgeois, Ribeiro resolved his inner conflict and decided that the geological and paleontological facts could no longer be ignored. He began openly reporting that implements of human manufacture were being found in Pliocene and Miocene formations in Portugal (Ribeiro 1873a, p. 98).

From the standpoint of modern geology, Ribeiro's assessment of the age of the formations in the Tagus River valley near Lisbon is generally correct. Modern authorities have observed seven Miocene cycles of sedimentation and one Pliocene cycle (Antunes *et al.* 1980, p. 136). The Late Tertiary (including the Pliocene and Miocene) is sometimes called the Neogene. In a study focusing on the Neogene formations of Europe, Ivan Chicha (1970, p. 50) said about Portugal: "The Neogene beds are known from the basin situated in the lower reach of the river Tejo [Tagus], in the environs of Lisbon. The Oligocene beds, prevalently of freshwater continental origin . . . are overlain by beds . . . which are placed in the oldest Miocene—Aquitanian." According to Chicha, these Aquitanian beds are surmounted by limestones and claystones that ascend to the Tortonian stage of the Late Miocene. Another recent study (Antunes *et al.* 1980, p. 138) included a chart showing the lithostratigraphic units in the Tagus basin. Limestones, such as those in which Ribeiro found stone tools, occur in the Middle and Early Miocene.

In considering stone implements, three questions must be answered: (1) is the specimen really of human manufacture? (2) has the age of the stratum in which it was discovered been properly determined? (3) was the implement incorporated into the stratum at the time the stratum was laid down, or was the implement introduced at a later date? As far as Ribeiro was concerned, he was convinced that he had satisfactorily answered all three questions. The toollike flint objects he studied were of human manufacture, they were found in strata mostly of Miocene age, and many appeared to be in primary position, although some of his specimens were found on the surface.

In 1871, Ribeiro presented to the members of the Portuguese Academy of Science at Lisbon a collection of flint and quartzite implements, including those gathered from the Tertiary formations of the Tagus valley. In 1872, at the International Congress of Prehistoric Anthropology and Archeology meeting in Brussels, Ribeiro gave a similar report on his discoveries and displayed more specimens, mostly pointed flakes. At that time, Bourgeois said that none appeared to be of human manufacture. Upon a new examination of Ribeiro's specimens, Bourgeois found one flint that he thought displayed signs of human work, but unfortunately it had not been found *in situ*. He therefore suspended judgement (de Mortillet 1883, p. 95). The English authority, A.W. Franks, who served as Conservator of National Antiquities and Ethnography at the British Museum, gave a more positive opinion. An expert in cultural remains, including tools, Franks stated that some of the specimens did appear to be the product of intentional work, but he reserved judgement on the age of the strata in which they had been found (Ribeiro 1873a, p. 99).

Ribeiro himself (1873b, p. 100) then addressed the Congress on the question of "the exact geological situation of the beds in which he had found worked flint flakes, the authenticity of which has been recognized by Mr. Franks and other members of the Congress." Ribeiro reported that one of the flints had been found in the reddish-yellow Pliocene sandstone on the left bank of the Tagus, to the south of Lisbon. He noted that these beds cover Miocene marine deposits (Ribeiro 1873b, p. 101). Modern authorities (Antunes *et al.* 1980, pp. 136–138) still show this basic sequence—Miocene marine deposits surmounted by Pliocene sandstone formations—in the Lisbon region.

"Concerning the other flints which Mr. Franks has declared bear evident traces of human workmanship," said Ribeiro (1873b, p. 102), "they were found in Miocene strata." He explained that on the way north from Lisbon to Caldas da Rainha, between the towns of Otta and Cercal, one comes to the steep hill of Espinhaço de Cão. According to Ribeiro (1873b, p. 102), it was in the sandstone beds of this hill, which lie under marine Miocene strata, that he found "flints worked by the hand of man before they were buried in the deposits." This would indicate the presence of human beings in Portugal at least 5 million years ago and perhaps as much as 25 million years ago. Figure 4.1 shows an implement from Espinhaço de Cão.

Ribeiro's Miocene flints made an impressive debut at Brussels, but remained controversial. At the Paris Exposition of 1878, Ribeiro displayed 95 specimens of Tertiary flint tools in the gallery of anthropological science. De Mortillet visited Ribeiro's exhibit and, in the course of examining the specimens carefully, found that 22 had indubitable signs of human work. This was quite an admission for de Mortillet, for, as described in Chapter 2, he habitually rejected all evidence for human work on incised and broken bones from the Tertiary. Gabriel de Mortillet, along with his friend and colleague Emile

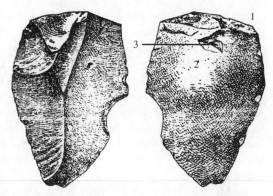

Figure 4.1. Implement found by Carlos Ribeiro, of the Geological Survey of Portugal, in a Miocene layer at Espinhaço de Cão (G. de Mortillet and A. de Mortillet 1881, plate 3). The ventral surface shows: (1) a striking platform, (2) bulb of percussion, and (3) eraillure.

Cartailhac, enthusiastically brought other paleoanthropologists to see Ribeiro's specimens, and they were all of the same opinion—a good many of the flints were definitely made by humans. Cartailhac then photographed the specimens, and de Mortillet later presented the pictures in his *Musée Préhistorique* (1881).

De Mortillet (1883, p. 99) wrote: "The intentional work is very well established, not only by the general shape, which can be deceptive, but much more conclusively by the presence of clearly evident striking platforms and strongly developed bulbs of percussion." The bulbs of percussion also sometimes had eraillures, small chips removed by the force of impact. In addition to the striking platform, bulb of percussion, and eraillure, some of Ribeiro's specimens had several long, vertical flakes removed in parallel, something not likely to occur in the course of random battering by the forces of nature.

De Mortillet's method of analysis is comparable to that employed by modern experts in lithic technology, who, like de Mortillet, emphasize that the toollike shape of a flint does not in itself establish human work. Leland W. Patterson, a contemporary expert in distinguishing artifacts from naturefacts, believes that the bulb of percussion is the most important sign of intentional work on a flint flake. If the flake also shows the remnants of a striking platform, then one can be even more certain that one is confronted with a flake struck deliberately from a flint core and not a piece of naturally broken flint resembling a tool or weapon.

"There can be no doubt," wrote de Mortillet (1883, p. 99) about Ribeiro's stone implements. "The diverse specimens are formed from big flakes, almost all of them triangular and without retouch, some in flint, some in quartzite. In looking at the collection, one believes oneself to be seeing Mousterian tools, only somewhat coarser than usual." Mousterian is the name given to the type of stone tool usually considered to have been made by the Neanderthals (*Homo sapiens neanderthalensis*), who are thought to have lived in the latter part of the Pleistocene. By making the comparison with the Late Pleistocene Mousterian implements, de Mortillet was pointing out that Ribeiro's specimens almost exactly resemble those that are universally acknowledged as being of human manufacture. Figure 4.2 shows one of Ribeiro's Miocene tools from Portugal and for comparison an accepted stone tool from the Mousterian cultural stage of the European Late Pleistocene. They share the typical features of intentional human work on stone: the striking platform, bulb of percussion, eraillure, and parallel removal of flakes.

De Mortillet (1883, pp. 99–100) further observed: "Many of the specimens, on the same side as the bulb of percussion, have hollows with traces and fragments of sandstone adhering to them, a fact which establishes their original position in the strata. The sandstone is inserted among strata of clays and limestones in the valley of the Tagus, together comprising a formation that attains in some places a depth of 400 meters [over 1,300 feet]. The beds have been dislocated and are in some places now resting almost in a vertical position. It is very evidently Tertiary terrain. Of the 22 worked specimens, 9 are indicated by Ribeiro to be Miocene. The others are Pliocene."

Plate 3 in de Mortillet's publication *Musée Préhistorique* (G. de Mortillet and A. de Mortillet 1881) featured illustrations of Ribeiro's Miocene and Pliocene discoveries. We have selected two for reproduction. Figure 4.3 depicts both sides of a flint flake recovered from a Tertiary formation at the base of Monte Redondo. This formation is said to belong to the Tortonian stage of the Late Miocene

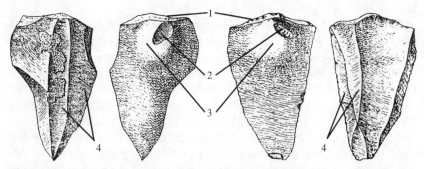

Figure 4.2. Left: Dorsal and ventral views of a stone tool recovered from a Tertiary formation in Portugal (de Mortillet 1883, p. 98). It would be over 2 million years old. Right: An accepted stone tool, less than 100,000 years old, from the Mousterian cultural stage of the European Late Pleistocene (de Mortillet 1883, p. 81). Both implements clearly display the following features of intentional human work: (1) striking platforms, (2) eraillures, (3) bulbs of percussion, and (4) parallel flake removal.

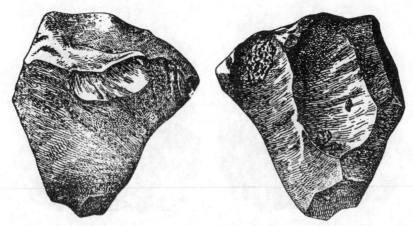

Figure 4.3. Ventral and dorsal surfaces of a flint tool found in a Late Miocene formation at Monte Redondo, Portugal (G. de Mortillet and A. de Mortillet 1881, plate 3).

(de Mortillet 1883, p. 102). The ventral surface of the flint flake shows "a large striking platform, bulb of percussion, and eraillure" (G. de Mortillet and A. de Mortillet 1881, plate 3). The dorsal surface of the flake bore proof that it was found in the Tertiary sandstones of Otta. Sandstone, just like that found at the base of Monte Redondo, adhered to the surface.

The quartzite flake shown in Figure 4.4 was found in a Pliocene formation at Barquinha, 103 kilometers (about 64 miles) northeast of Lisbon, Portugal. The ventral surface of the flake displays a striking platform, bulb of percussion, and eraillure (G. de Mortillet and A. de Mortillet 1881, plate 3). While this flake was still attached to the quartzite core, another flake was struck from it, as shown by a negative bulb of percussion on the dorsal surface of the flake.

In a report published in 1879, Cartailhac said about some of Ribeiro's specimens: "One would believe himself to be viewing a series of Mousterian stone implements, though somewhat cruder. The bulbs of percussion are generally quite prominent. . . . These pieces bear the proof that they were not found on the surface. On the faces of the

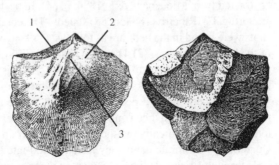

Figure 4.4. Quartzite tool found in a Pliocene formation at Barquinha, Portugal (G. de Mortillet and A. de Mortillet 1881, plate 3). The ventral surface (left), shows (1) a striking platform, (2) bulb of percussion, and (3) eraillure.

Figure 4.5. An implement found in a Miocene formation at Carregado, Portugal (Cartailhac 1879, plate 8).

flakes and in the hollows are found fragments of the sandstone which had encased them" (Cartailhac 1879, p. 439). One of the pieces (Figure 4.5) was found at Carregado in a Miocene formation and was described by Cartailhac as displaying "a bulb of percussion and retouch." Retouching, in the form of regular chipping along the edges of a flint flake, is a good indicator of intentional work.

4.1.2 An International Committee Vindicates Ribeiro

At the 1880 meeting of the International Congress of Prehistoric Anthropology and Archeology, which was held in Lisbon, Portugal, Ribeiro, now on his home ground, delivered another report and displayed more specimens that were "extracted from Miocene beds" (1884, p. 86). In his report ("L'homme Tertiaire en Portugal"), Ribeiro (1884, p. 88) stated: "The conditions in which the worked flints were found in the beds are as follows: (1) They were found as integral parts of the beds themselves. (2) They had sharp, well-preserved edges, showing that they had not been subject to transport for any great distance. (3) They had a patina similar in color to the rocks in the strata of which they formed a part."

The second point is especially important. Some geologists claimed that the flint implements had been introduced into Miocene beds by the floods and torrents that periodically washed over this terrain. According to this view, Quaternary flint implements may have entered into the interior of the Miocene beds through fissures and been cemented there, acquiring over a long period of time the coloration of the beds (de Quatrefages 1884, p. 95). But if the flints had been subjected to such transport, then the sharp edges would most probably have been damaged, and this was not the case.

The Congress assigned a special commission of scientists the task of directly inspecting the implements and the sites from which they had been gathered. In addition to Ribeiro himself, the commission included G. Bellucci of the Italian Society for Anthropology and Geography; G. Capellini, from the Royal University of Bologna, Italy, and known to us from Chapter 2 for his discoveries of incised Pliocene whale bones; E. Cartailhac, of the French Ministry of Public Instruction; Sir John Evans, an English geologist; Gabriel de Mortillet, professor of prehistoric anthropology at the College of Anthropology, Paris; and Rudolph Virchow, a German anthropologist. The other members were the scientists Choffat, Cotteau, Villanova, and Cazalis de Fondouce.

On September 22, 1880, at six in the morning, the gentlemen of the commission boarded a special train and proceeded north from Lisbon. During the rail journey, they gazed at the old forts topping the hilltops, and pointed out to each other the Jurassic, Cretaceous, and Tertiary terrains as they moved through the valley of the Tagus River. They stepped off the train at Carregado. It is on a line from Carregado north to Cercal that Ribeiro discovered most of his flints. They then proceeded to nearby Otta and two kilometers (just over a mile) from Otta arrived at the hill of Monte Redondo. At that point, the scientists dispersed into various ravines in search of flints.

Paul Choffat, a member of the commission and its secretary, later reported to the Congress: "Of the many flint flakes and apparent cores taken from the midst of the strata under the eyes of the commission members, one was judged as leaving no doubt about the intentional character of the work" (1884a, p. 63). This was the specimen found *in situ* by the Italian naturalist Bellucci (Figure 4.6). Choffat then noted that Bellucci had found on the surface other flints with incontestable signs of work. Some thought they were Miocene implements that had been removed from the Miocene conglomerates by atmospheric agencies, such as rain and wind, while others thought that the implements were of a much more recent date.

In his book *Le Préhistorique,* Gabriel de Mortillet gave an informative account of the events that took place at the Congress at Lisbon: "While the printer was preparing the first pages of this book," wrote

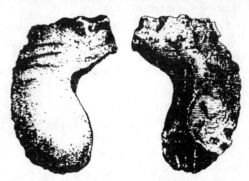

Figure 4.6. Flint implement found by G. Bellucci in an Early Miocene formation at Otta, Portugal (Choffat 1884b, figure 1). It was judged by a commission of scientists to be identical to Late Pleistocene implements of similar type.

de Mortillet (1883, p. 100), "I went to the meeting of the International Congress of Prehistoric Anthropology and Archeology in Lisbon, one of my purposes being to complete the table of strata containing evidence for the presence of humans. I was able to confirm in a very exact and positive manner the actuality of Ribeiro's discoveries, including the precise geological position of certain of his worked flints."

De Mortillet (1883, pp. 100–101) then proceeded to describe the scientists' excursion to Otta and Bellucci's remarkable discovery: "The members of the Congress arrived at Otta, in the middle of a great freshwater formation. It was the bottom of an ancient lake, with sand and clay in the center, and sand and rocks on the edges. It is on the shores that intelligent beings would have left their tools, and it is on the shores of the lake that once bathed Monte Redondo that the search was made. It was crowned with success. The able investigator of Umbria [Italy], Mr. Bellucci, discovered *in situ* a flint bearing incontestable signs of intentional work. Before detaching it, he showed it to a number of his colleagues. The flint was strongly encased in the rock. He had to use a hammer to extract it. It is definitely of the same age as the deposit. Instead of lying flat on a surface onto which it could have been secondarily recemented at a much later date, it was found firmly in place on the under side of a ledge extending over a region removed by erosion [Figure 4.7]. It is impossible to desire a more complete demonstration attesting to a flint's position in its strata." All that was needed was to determine the age of the strata. Study of the fauna and flora in the region around the Monte Redondo site showed that the formations present there can be assigned to the Tortonian stage of the Late Miocene period (de Mortillet 1883, p. 102).

"Therefore," con-
cluded de Mortillet
"during the Tortonian
epoch there existed
in Portugal an intel-
ligent being who
chipped flint just like
Quaternary humans"
(1883, p. 102). Some
modern authorities
consider the Otta con-
glomerates to be from
the Burdigalian stage
of the Early Miocene
(Antunes   *et   al.*
1980, p. 139).

Choffat   (1884b,
pp. 92–93) prsented,

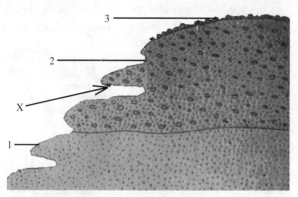

Figure 4.7. Stratigraphy of the site at the base of Monte Redondo hill in Otta, Portugal, where G. Bellucci found the implement pictured in Figure 4.6: (1) sandstone; (2) Miocene sandstone conglomerate with flints; (3) surface deposit of eroded flints. The arrow marked "X" indicates the position of the implement (de Mortillet 1883, p. 101).

in the form of answers to four questions, the conclusions of the commission members, who had not only examined the specimens Ribeiro exhibited at the Congress but also journeyed to Otta to conduct field investigations. The first two questions and answers dealt with the flints themselves: "(1) Are there bulbs of percussion on the flints on exhibition and on those found during the excursion? The commission declares unanimously that there are bulbs of percussion, and some pieces have several. (2) Are bulbs of percussion proof of intentional work? There are different opinions. They may be summarized as follows: de Mortillet considers that just one bulb of percussion is sufficient proof of intentional work, while Evans believes that even several bulbs on one piece do not give certitude of intentional chipping but only a great probability of such." Here it may once more be noted that modern authorities such as Leland W. Patterson (1983) consider one or more bulbs of percussion to be very good indicators of intentional work.

The remaining two questions concerned the positions in which the flints were found: "(3) Are the worked flints found at Otta from the interior of the beds or the surface? There are diverse opinions. Mr. Cotteau believes all are from the surface, and that those found embedded within the strata came down through crevasses in the beds. Mr. Capellini, however, believes that pieces found on the surface were eroded from the interior of the beds. De Mortillet, Evans, and Cartailhac believe there are two time periods to which the flints may be referred, the first being the Tertiary, the other being the Old and New Stone Ages of the Quaternary. The flints of the two periods are easy to distinguish by their form and patina. (4) What is the age of the strata of the worked flints? After only a moment's discussion the members declared they were in perfect accord with Ribeiro." In other words, the strata were Miocene, although some members of the commission believed that the flints found lying on the surface had not weathered out of the Miocene rock but instead had been dropped there in fairly recent times.

In the discussion that followed the presentation of Choffat's report, Capellini said: "I believe these flints to be the product of intentional work. If you do not admit that, then you must also doubt the flints of the later Stone Ages" (Choffat 1884b, pp. 97–98). According to Capellini, Ribeiro's Miocene specimens were almost identical to undoubted Quaternary flint implements. Capellini's remarks strike at one of the central issues in the treatment of scientific evidence—the application of a double standard in determining what evidence is to be accepted and what is to be rejected. If the standards used by the scientific establishment to reject finds such as Ribeiro's were applied in the same manner to conventionally accepted finds, then the accepted finds would also have to be rejected. And this would deprive the theory of human evolution of a substantial portion of its evidential foundation.

The next speaker, Villanova, provided a good example of the double standard treatment. Villanova was very doubtful, even about the Bellucci find. He said that in order to remove all cause for suspicion one would have to discover an unmistakably genuine implement firmly embedded not in a Miocene conglomerate but in the middle of undifferentiated Miocene limestone and alongside characteristic fossils (Choffat 1884b, p. 99). A conglomerate is a mass of rock composed of rounded stones of various sizes cemented together in a matrix of sandstone or hardened clay. Apparently, Villanova felt there was some reason to doubt the age of a stone tool found in a conglomerate—there was perhaps a chance it had entered recently and been cemented in with other stones. Or perhaps he doubted the age of the conglomerate at Otta, but the majority opinion was that this conglomerate was in fact Miocene.

Maybe it would have been better if the flint had been found in an undifferentiated stratum. The crucial point, however, is this: if Villanova's criterion were to be applied in all cases, this would wipe out most of the paleoanthropological evidence now accepted. The number of human fossils found in undifferentiated strata directly alongside characteristic fossils is rather small. For example, as we shall see in Chapter 9, the initial Java *Homo erectus* discovery was made in strata that had undergone considerable mixing, and almost all of the later Java *Homo erectus* finds were made on or quite near the surface. Beijing *Homo erectus* was found in cave deposits. Another point that will emerge in our discussion is this: sometimes anomalous finds are made in undifferentiated strata alongside characteristic fossils, and then some other means will be found to discredit them. Indeed, as previously mentioned, in his report to the International Congress of Prehistoric Anthropology and Archeology at Brussels in 1872, Ribeiro (1873a, p. 97) did tell of finding flint implements "deep inside" undifferentiated Miocene limestone beds.

Following Villanova, Cartailhac spoke. He said that if the question of the Miocene age of the implements were to be decided on the grounds of actual scientific evidence, the answer would have to be affirmative. Cartailhac believed that the coloration of many of the surface finds indicated they were eroded from Miocene beds, and he pointed out that some specimens had remnants of Miocene sediments adhering to them.

Cartailhac then asked the members to consider a particular specimen from Ribeiro's collection, which he had previously studied at the anthropological exposition in Paris. He stated: "I have seen on it two bulbs of percussion, and possibly a third, and a point that seems to truly be the result of intentional work. It has on its surface not a coloration that could be removed by washing but rather a surface incrustation of Miocene sandstone tightly adhering to it. A chemist would not permit us to say that such a deposit could form and attach itself to a flint lying, for whatever amount of time, on a sandstone surface" (Choffat 1884b, p. 100). In other words, the flint must have been lying within the Miocene bed itself, when it was formed. Cartailhac admitted that natural action might in rare occasions produce a bulb of percussion, but to have two on the same piece would be an absolute miracle. He believed that the many very good specimens discovered on the Miocene surface, where there was absolutely no trace of any other deposit, were really Miocene implements that had weathered out of the rock.

One may certainly disagree with Cartailhac. But then we may here note that in more recent times, the famous Lucy australopithecine fossils were found by D. C. Johanson on the surface of Pliocene deposits in Ethiopia, from which they were presumed to have weathered out (Johanson and Edey 1981, pp. 16–18). As we shall see in Chapter 9, the same is also true of many of the Java *Homo erectus* finds. Are these discoveries also to be doubted? Perhaps, but the real point is that the application of standards should be consistent. Unfortunately, as we shall see throughout this book, standards tend to be applied selectively, in conformity with the biases and expectations of the researcher.

After Cartailhac finished his remarks, Bellucci recounted his own noteworthy discovery of an implement (Figure 4.6, p. 221) in the Miocene conglomerate at Otta (Choffat 1884b, pp. 101–102). Before extracting it, he had shown it to many members of the commission, who saw that it was firmly integrated into the stratum (Figure 4.7, p. 222). It had been so firmly fixed in the Miocene sandstone conglomerate that he had not been able to remove it with his wooden tool, and had needed to use Cartailhac's iron pick to break the sandstone. Bellucci stated that the inner surface of the implement, the one adhering to the conglomerate, had not only the same reddish color as the conglomerate but also incrustations of tiny grains of quartzite that could not be detached even by vigorous washing.

Bellucci further pointed out that the elements composing the intact conglomerate corresponded perfectly with those found loose on the surface. This led Bellucci to conclude that the loose stones found on the surface at Otta were the result of weathering of the conglomerate. This indicated that flint implements found on the surface might also have come from the conglomerate, which was of Miocene age (Choffat 1884b, p. 103). By itself, this was, however, a fairly weak argument. Although the flints on the surface may have weathered out of the Miocene conglomerate, they also could have been dropped on the surface during the Late Pleistocene. But the fact that the implements had incrustations of Miocene sediments on them, and were the same color as the Miocene conglomerate, strongly supported the conclusion that the implements were themselves Miocene.

As for the signs of intentional work on the piece found *in situ*, Bellucci noted: "Mr. Evans says he believes in bulbs of percussion. Well look. This piece was detached from the surface of a flint core, and it not only has a magnificent bulb of percussion, but also one of its surfaces presents marks showing that another flake had been previously detached, in the same direction, when the implement had been still part of the flint nucleus" (Choffat 1884b, p. 104).

The last feature described by Bellucci, successive parallel flake removal from a core, is recognized today by experts in lithic technology as one of the surest signs of intentional work. The striking of two successive flakes from a flint core requires a considerable degree of expertise, and is quite beyond what might be expected from random shocks by purely natural forces. Patterson stated: "Humans will often strike multiple flakes in series from a single core, usually resulting in the production of some flakes with multiple facets on the dorsal face. In contrast, the removal of a few flakes from cores by random natural forces would not be expected to occur often by serial removals. . . . It is characteristic in human lithic manufacturing processes to use the same striking platform for multiple flake removals" (L. Patterson *et al.* 1987, p. 98).

When Cotteau's turn to speak came, he argued, like Villanova, that, in order to be accepted, finds of implements should be made only in undifferentiated, intact strata (Choffat 1884b, pp. 105–106). Cotteau observed that unless finds were made in undifferentiated, intact strata, the possibility always existed that the implements might have been washed in through fissures from the surface and cemented in place. In time, the fissure might be filled in, hiding its existence to researchers. It should, however, be noted, that Cotteau did not specifically address the conditions of Bellucci's discovery. Was there in fact a filled-in fissure near the place where Bellucci found the flint implement? Cotteau does not say. Furthermore, the position in which Bellucci found his implement, firmly in place on the underside of an overhanging section of the Miocene formation, argues against Cotteau's hypothesis. In general, Bellucci's opponents at the Congress offered only vague hypothetical objections.

Altogether, there seems little reason why Ribeiro's discoveries should not be receiving some serious attention, even today. Here we have a professional geologist, the head of Portugal's geological survey, making discoveries of flint implements in Miocene strata. In appearance the implements resembled accepted types, and they displayed characteristics that modern experts in lithic technology accept as signs of human manufacture. To resolve controversial questions, a congress of Europe's leading archeologists and anthropologists deputed a committee to conduct a firsthand investigation of one of the sites of Ribeiro's discoveries of Miocene implements. There a scientist discovered *in situ* an implement in a Miocene bed, a fact witnessed by several other members of the committee. Of course, objections were raised, but upon reviewing them, it does not appear to us that they were conclusive enough to cause an unbiased observer to categorically reject Bellucci's find in particular or Ribeiro's finds in general.

4.2 THE FINDS OF THE ABBÉ BOURGEOIS AT THENAY, FRANCE (MIOCENE)

We now turn our attention to the discoveries of the Abbé L. Bourgeois, rector of the seminary at Pontlevoy, Loire-et-Cher, France. On August 19, 1867, in Paris, Bourgeois presented to the International Congress for Prehistoric Anthropology and Archeology a report on flint implements he had found in Early Miocene beds at Thenay, in north central France, near Orleans (de Mortillet 1883, p. 85). Bourgeois, who had conducted research near Thenay for over twenty years, said that although the instruments were crudely made, they resembled the types of Quaternary implements (scrapers, borers, blades, etc.) he had found on the surface in the same region. He found on almost all of the Miocene specimens the standard indications of human work: fine retouching, symmetrical chipping, and traces of use. He also noted multiple examples of particular forms. Some of the flints, naturally translucent, were opaque, a sign that they had been burned. By performing experiments with fire and flint, Bourgeois had been able to reproduce the exact effect. The signs of fire on the flints were another strong indication that humans had made and used them.

The flint implements of Thenay were recovered from below the Calcaire de Beauce, a well-known Early Miocene limestone formation. Bourgeois recognized that the presence of stone tools in this geological position was indeed remarkable, having serious implications with regard to human antiquity. Yet, for him, the facts, uncomfortable though they might be to contemplate, spoke for themselves. De Mortillet (1883, p. 86) said that the layers of clay in which the flints were found were of Early Miocene or even Oligocene age. This would push back the presence of human beings in France to around 20–25 million years before the present. If this sounds impossible, one should ask oneself why. If the answer is that modern science's ideas about human evolution prevent one from seriously considering such a thing, one should honestly admit that one is allowing preconceived notions to unduly influence one's perception of facts and that this is unscientific. One with faith in the scientific method should maintain a willingness to change one's notions, even the most dearly held, in the face of facts that contradict them.

Modern geologists still agree with the determination that the deposits at Thenay are Miocene. As stated above, the implement-bearing layers lie below the Calcaire de Beauce. This limestone formation is now referred to the Aquitanian stage (Pomerol and Feugeur 1974, p. 142), which lies within the Early Miocene (Romer 1966, p. 334). Some French authorities (Klein 1973, p. 566) put the deposits of Thenay at the base of the Helvetian stage. The Helvetian stage is placed in the Middle Miocene (Romer 1966, p. 334). The base of the Helvetian would thus mark the boundary between the Middle and Early Miocene.

4.2.1 Debates About the Discoveries at Thenay

Bourgeois displayed his specimens at the house of the Marquis de Vibraye, and the members of the Paris congress of 1867 were allowed to examine them at their leisure. Although the form and appearance of the flints had been sufficient to convince Bourgeois they were of human manufacture, most of the visitors were hesitant to acknowledge this. De Mortillet (1883, p. 86) stated that "the ancient age of the strata in which they were found involuntarily indisposed the geologists and paleontologists." Here again we find a clear case of preconceptions (of what could and should be) dominating a decision whether or not to accept evidence.

Thus the flints from the Miocene of Thenay did not win much approval at their Paris debut. Only a few scientists, prominent among them the Danish naturalist Worsaae, admitted they were actual artifacts. Undeterred, Bourgeois continued his work, finding more and more specimens, and convincing individual paleontologists and geologists they were the result of intentional work. De Mortillet said he was one of the first to be so convinced. He and other scientists not only examined the collection of Bourgeois at Pontlevoy but also carefully studied the site at Thenay.

Some scientists questioned the stratigraphic position in which the flints had been found. The first specimens collected by Bourgeois, many of which showed signs of burning by fire, came from the slopes of rocky debris along the sides of a small valley cutting through the plateau at Thenay. Geologists such as Sir John Prestwich objected that these were essentially surface finds. In response, Bourgeois dug a trench in the valley and found flints showing the same signs of human work (de Mortillet 1883, p. 94).

Still unsatisfied, critics proposed that the flints found in the trench had come to their positions through fissures leading from the top of the plateau, where Quaternary implements were often found. To meet this objection, Bourgeois, in 1869, sank a pit into the top of the plateau (de Mortillet 1883, p. 95). In the course of the excavation, he came to a layer of limestone 32 centimeters (about one foot) thick, with no fissures through which Quaternary stone tools might have slipped to lower levels.

Deeper in his pit, at a depth of 4.23 meters (13.88 feet) in Early Miocene strata of the Aquitanian stage, Bourgeois discovered many flint tools. De Mortillet (1883, pp. 95–96) stated in *Le Préhistorique:* "There was no further doubt about their antiquity or their geological position." In the layer of Early Miocene clay containing the flint implements, Bourgeois found a hammer stone bearing evident signs of percussion. Hammer stones are primarily used to strike flakes from flint cores. In his collection, Bourgeois (1873, p. 90) had several other examples of hammer stones.

Despite the clear demonstration provided by the pit sunk in the middle of the plateau at Thenay, many scientists retained their doubts. A showdown came in Brussels, at the 1872 meeting of the International Congress of Prehistoric Anthropology and Archeology. There Bourgeois delivered a report

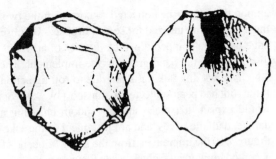

Figure 4.8. A pointed implement from a Miocene formation at Thenay, France (Bourgeois 1873, plate 1).

summarizing the history of his discoveries. In addition, he presented many specimens, figures of which were included in the published proceedings of the Congress. Describing a pointed specimen (Figure 4.8), Bourgeois (1873, p. 89) stated: "Here is an awllike specimen, on a broad base. The point in the middle has been obtained by regular retouching. This is a type common to all epochs. On the opposite side is a bulb of percussion, which although rare in the Tertiary flints of Thenay, here shows itself very well."

Bourgeois described another implement: "A very regularly shaped fragment of a flake that deserves the designation knife or cutter." He continued: "The edges have regular retouching, and the opposite side presents a bulb of percussion" (Bourgeois 1873, p. 49). On many of his specimens, noted Bourgeois, the edges on the part of the tool that might be grasped by the hand remained unworn, while those on the cutting surfaces showed extensive wear and polishing.

Another specimen (Figure 4.9), was characterized by Bourgeois (1873, p. 89) as a projectile point or an awl. He noted the presence of retouching on the edges, obviously intended to make a sharp point. Bourgeois (1873, p. 89) also saw

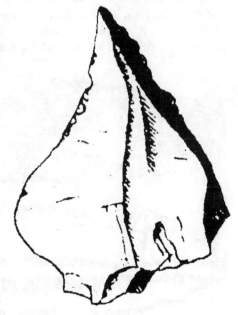

Figure 4.9. A pointed artifact from Miocene strata at Thenay, France, with retouching near the point (Bourgeois 1873, plate 2).

among the objects he collected "a core with the two extremities retouched with the aim of being utilized for some purpose." He observed: "The most prominent edge has been chipped down by a series of artificial blows, probably to prevent discomfort to the hand grasping the implement. The other edges remain sharp, which shows this flaking is not due to rolling action" (Bourgeois 1873, p. 89). If the flint had been subjected to transport by water or another natural agency, one would expect that the resultant random chipping and fracturing should have damaged all the edges, and not just one. For the sake of comparison, we show in Figure 4.10 the implement from the Early Miocene of Thenay alongside a similar accepted implement from the Late Pleistocene.

Then Bourgeois (1873, p. 90) described a final specimen: "A short scraper, with numerous and well-marked retouchings, in all respects resembling the Quaternary types found every day on the surface. On the other side, it presents . . . a bulb of percussion."

Bourgeois did not specify the exact places from which the above-mentioned specimens were taken—that is, from the exposed sections in the valley, from the valley trench, or from the pit sunk in the top of the plateau. But his reports suggest that implements recovered from all three places were quite similar.

In order to resolve any controversy, the Congress of Prehistoric Anthropology and Archeology nominated a fifteen-member commission to judge the discoveries of Bourgeois. A majority of eight members, including de Quatrefages and Capellini, voted that the flints were of human manufacture (de Mortillet 1883, p. 87).

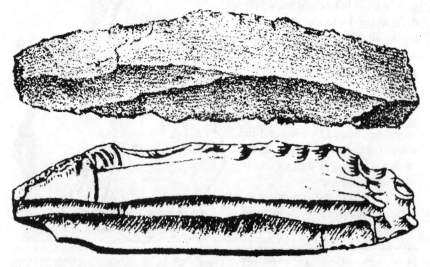

Figure 4.10. Top: A Late Pleistocene flint implement (Laing 1894, p. 366). Bottom: An implement from Early Miocene strata at Thenay, France (Bourgeois 1873, plate 2).

An additional member voted in favor of Bourgeois, but with some reservations. Only five of the fifteen found no trace of human work in the specimens from Thenay. One member expressed no opinion.

De Mortillet stated that if instead of considering just the numbers of votes that were cast, one considered their scientific merit, then the victory of Bourgeois was even greater. De Mortillet pointed out that among those voting in support of Bourgeois were the scientists who had especially devoted themselves to the study of flint tools, while among the dissenters were the scientists who had little or no experience in this area. Indeed, one of them, Dr. Fraas, of Germany, claimed at the Congress that the handaxes of the Quaternary gravels of the Somme region of France, accepted by almost all authorities as genuine human artifacts, were "an invention of French chauvinism" (de Mortillet 1883, p. 88).

Bourgeois gave a choice collection of flint tools from Thenay to the national museum of antiquities at St. Germain and also exhibited his best specimens at the exposition of anthropological science held in 1878. After his death, specimens were given to the museum of the School of Anthropology in Paris.

Many of the flints of Thenay have finely cracked surfaces indicating exposure to fire. Others, much more altered, have surfaces pitted with irregular holes. Was the cracking and pitting caused by weathering? De Mortillet (1883, p. 90) said that cracking resulting from fire and weathering could be very easily distinguished. Significantly, the normally translucent flints had become opaque. Experiments showed that it took a great deal of heat to discolor flints as much as those found at Thenay. The heat of the sun could not have done it. But if fire was the cause, was it fire used by humans or some kind of accidental fire?

In considering the possible causes of accidental fire, de Mortillet suggested that the three most likely possibilities were volcanic action, spontaneous vegetable combustion, or vegetation ignited by lightning. De Mortillet pointed out, however, that there were no volcanoes in the region and no layers of combustible plant material such as peat. Furthermore, the burned flints were found scattered at many locations throughout diverse levels in the same general area. This indicated to de Mortillet that the signs of burning were not the result of fires ignited by lightning. He appears to have reasoned as follows. The many localized signs of fire at numerous levels indicated continuous intentional use of small fires over a long period of time rather than occasional general conflagrations, such as might have occurred when grass, brush, or forest was ignited by lightning. The evidence strongly suggested that humans had regularly used fire to help fracture the flints.

Bulbs of percussion were rare on the Early Miocene flints of Thenay, but most displayed fine retouching of the edges. De Mortillet (1883, p. 92) stated that even though there were not many bulbs of percussion, retouching alone was a good sign of intentional work. The retouching tended to be concentrated on just one side of an edge, while the other side remained untouched; this is called unifacial flaking. De Mortillet, like modern authorities, believed that in almost all cases unifacial flaking is not the result of chance impacts but of deliberate work. Some researchers have suggested that in special instances unifacial flaking might result from natural forces that press one side of a flint against a hard surface, taking small chips off the edge (Section 3.4.1). De Mortillet (1883, pp. 92-93) admitted that this sometimes occurred; however the resultant chipping was generally very crude and irregular. In his book *Musée Préhistorique,* de Mortillet included reproductions of some Thenay flints that displayed very regular unifacial retouching—flakes removed in the same direction along one side of an edge (Figure 4.11). Some of the critics of Bourgeois commented that among all the Early Miocene flint pieces he collected at Thenay, there were only a very few good specimens, about thirty. But de Mortillet (1883, p. 93) stated: "Even one incontestable specimen would be enough, and they have thirty!"

A modern expert on stone implements, Leland W. Patterson, has stated (1983, p. 303): "Unifacially retouched stone tools are generally an important class of tools on archeological sites, and comprise a major portion of lithic artifacts of early man sites. This group can include well-known types of stone tools such as gravers, perforators, scrapers, notched tools, and some types of knives, choppers, and denticulates." The Thenay implements conformed to this description.

According to L. Patterson (1983, p. 303): "Completely unifacial tool shapes would be one of the most difficult items for nature to reproduce by random forces. It would be difficult for random forces unidirectionally to fracture flake edges

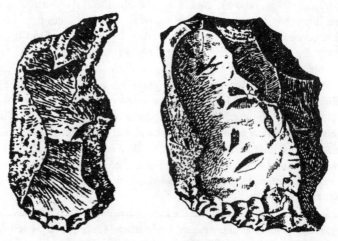

Figure 4.11. Unifacially retouched implements from the Early Miocene at Thenay, France (G. de Mortillet and A. de Mortillet 1881, plate 1).

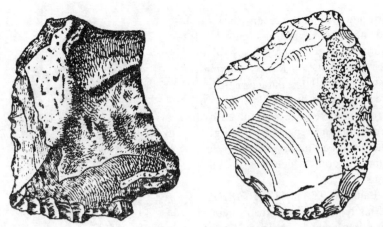

Figure 4.12. Left: A flint implement from an Early Miocene formation at Thenay, France (G. de Mortillet and A. de Mortillet 1881, plate 1). Right: An accepted implement from the lower middle part of Bed II, Olduvai Gorge, Africa (M. Leakey 1971, p. 113). The lower edges of both specimens show roughly parallel flake scars, satisfying the requirements of L. Patterson (1983) for recognition as objects of human manufacture.

only on one face. It would be even more difficult for fortuitous forces to create the long, uniform, parallel flake scars characteristic of purposefully made unifacial tools. . . . It follows, then, that it would be extremely difficult to conceive of nature fortuitously creating an entire group of various well-made unifacial tools, with multiple examples of each tool type, that is the usual demonstration of a kit of man-made stone tools." Patterson (1983, p. 303) added: "Any experienced lithic analyst with a 10-power magnifier can distinguish fortuitously shaped flakes from unifacial tools."

Illustrations of the flints from the Early Miocene of Thenay show the parallel flake scars of approximately the same size that, according to Patterson, are indicative of intentional human work. Figure 4.12 shows a unifacial implement from Thenay along with a similar accepted unifacial implement from Olduvai Gorge.

Through the writings of S. Laing, knowledge of the Thenay tools from the Early Miocene reached the intelligent reading public of the English-speaking countries. Because we desire to make this work a sourcebook of primary and contemporary secondary reports about anomalous evidence relating to human antiquity, we will include relevant passages from Laing's works.

Laing (1893, p. 113) wrote of the tools found at Thenay: "When these were first produced, the opinion of the best authorities was very equally divided as to their being the work of human hands, but subsequent discoveries have produced specimens as to which it is impossible to entertain any doubt, especially the flint knife and two small scrapers [the knife and one scraper appear in Figure 4.13] figured by M. Quatrefages at p. 92 of his recent work on *Races humaines*. They present all the characteristic features by which human design is inferred in other cases, viz.: the bulb of percussion and repeated chipping by small blows all in the same direction, round the edge which was intended for use."

Laing (1893, pp. 113–115) continued his review: "The human origin of these implements has been greatly confirmed by the discovery that the Mincopics of the Andaman Islands manufacture whet-stones or scrapers almost identical with those of Thenay, and by the same process of using fire to split the stones into the requisite size and shape. These Mincopics are not acquainted with the art of chipping stone into celts or arrow-heads, but use fragments of large shells, of which they have a great abundance, or of bone or hard wood, and the scrapers are employed in bringing these to a sharper point or finer edge. The main objection, therefore, at first raised to the authenticity of these relics of Miocene man, that they did not afford conclusive proof of design, may be considered as removed, and the objectors have to fall back on the assumption, either that the implements were fabricated by some exceptionally intelligent *Dryopithecus,* or that the Abbé Bourgeois may have been deceived by workmen, and mistaken in supposing that flints, which really came from overlying Quaternary strata, were found in the Miocene deposit. This hardly seems probable in

Figure 4.13. A scraper or borer (top) and a flint knife (bottom) from an Early Miocene formation at Thenay, France, reproduced by S. Laing (1894, pp. 364–365) from a book by A. de Quatrefages.

the case of such an experienced observer, and had it been so, the implements might have been expected to show the usual Quaternary types of celts, knives, and arrow-heads, fashioned by percussion, whereas the specimens found all bear a distinct type, being scrapers and borers of small size, and partly fashioned by fire. . . . On the whole, the evidence for these Miocene implements seems to be very conclusive, and the objections to have hardly any other ground than the reluctance to admit the great antiquity of man." Here we may note that collections of Quaternary implements often include scrapers and borers of the type found at Thenay, in addition to the more sophisticated projectile points and handaxes.

As an example of popular science writing, Laing's work is satisfactory. His mode of expression was reasonable and lucid. He did not oversimplify. The evidence he cited was faithfully reproduced from original scientific reports and was presented in an honest fashion. Especially strong was his report that the Andaman islanders made tools similar to those of Thenay by using fire to flake the stone. Modern authorities regard studies of present-day lithic technologies as useful in recognizing intentional human work on stone materials gathered from ancient sites.

In his book *Human Origins*, Laing (1894, p. 363) again wrote of the flint implements of Thenay: "The general form might be the result of accident, but fractures from frost or collisions simulating chipping could hardly be all in the same direction, and confined to one part of the stone. The inference is strengthened if the specimen shows bulbs of percussion, where the blows had been struck to fashion the implement, and if the microscope discloses parallel striae and other signs of use on the chipped edge, such as would be made by scraping bones or skins, while nothing of the sort is seen on the other natural edges." As we have seen, some of the flint objects from Thenay do have bulbs of percussion and signs of wear confined to working edges, in addition to regular unifacial retouching. Laing also mentioned that the Thenay specimens closely resembled later implements of undoubted human manufacture.

Laing (1894, p. 356) listed the flint implements found in the Early Miocene at Thenay as one of many cases "in which the preponderance of evidence and authority in support of Tertiary man seems so decisive, that nothing but a preconceived bias against the antiquity of the human race can refuse to accept it."

Laing (1894, pp. 363–364) told the history of the finds: "When specimens of the flints from Thenay were first submitted to the Anthropological Congress at Brussels in 1867, their human origin was admitted by MM. Worsae, de Vibraye, de Mortillet, and Schmidt, and rejected by MM. Nilson, Hebert, and others, while M. Quatrefages reserved his opinion, thinking a strong case made out, but not being entirely satisfied. M. Bourgeois himself was partly responsible for these doubts, for, like Boucher de Perthes, he had injured his case by overstating it, and including a number of small flints, which might have been, and probably were, merely natural specimens. But the whole collection having been transferred to the Archeological Museum at St. Germain, its director, M. Mortillet, selected those which appeared most demonstrative of human origin, and placed them in a glass case, side by side with similar types of undoubted Quaternary implements. This removed a great many doubts, and later discoveries of still better specimens of the type of scrapers have, in the words of Quatrefages, 'dispelled his last doubts,' while not a single instance has occurred of any convert in the opposite direction, or of any opponent who has adduced facts contradicting the conclusions of Quatrefages, Mortillet, and Hamy, after an equally careful and minute investigation."

Laing (1894, p. 370) then went on to say: "The scraper of the Esquimaux and the Andaman islanders is but an enlarged and improved edition of the Miocene scraper, and in the latter cases the stones seem to have been split by the same agency, viz. that of fire. The early knowledge of fire is also confirmed by the discovery, reported by M. Bourgeois in the Orleans Sand at Thenay, with bones of mastodon and dinotherium, of a stony fragment mixed with carbon, in a sort of hardened paste, which . . . must be the remnant of a hearth on which there had been a fire."

In any case, the evidence that an intelligent being of the human type produced the flints of Thenay around 20 million years ago in the Early Miocene seems overwhelming. But some authorities believed the being was not of the modern human type, but rather a more primitive ancestor, as required by evolutionary theory. The controversy was vehement. As this question will come up again and again in our review of evidence for the presence of humans in Tertiary times, we shall now give this matter some detailed consideration.

4.2.2 Evolution and the Nature of Tertiary Man

In his book *Hommes Fossiles et Hommes Sauvages,* A. de Quatrefages (1884, p. 80) noted: "The problem of Tertiary man is singularly obscured by the fact that solutions are too often dictated by opinions held *a priori*, deriving from extremely opposing theories." The opposing theories and opinions were those of the Darwinists and the Biblical creationists. Uncomfortable with the views of both these groups, de Quatrefages (1884, p. 80) went on to say: "The elements of a conviction based on purely scientific and rational grounds are not numerous. It is easy to see that men of equal intelligence and experience can have different opinions or hesitate to give any opinion whatsoever. But Darwinian doctrines and dogmatic religious convictions have obviously influenced scientific discussion on this matter."

As of the late nineteenth century, the only fossil remains relating to human origins yet discovered were those of the Neanderthals and Cro-Magnon man. As previously mentioned (Section 1.2), scientists favoring evolution thought that the Neanderthals, although somewhat primitive, were too humanlike to qualify as a missing link with the Miocene apes; and Cro-Magnon man, of course, was fully human. But Cro-Magnon man did put the fully human type well back into the Quaternary, contemporary with ice age mammals such as the mammoth and woolly rhinoceros.

This naturally led Darwinists to place the origin of the human species from apelike ancestors much further back in time. De Quatrefages (1884, pp. 80–81) noted: "Haeckel was the first to make a proposal. He put his *Homo alalus* (speechless man) and *Homo pithecanthropus* (ape-man) in the Pliocene, or late Tertiary. Darwin, taking after his German disciple, proposed that the initial transition from ancient apes to the precursors of modern humans, as signified by the loss of the ape's primitive coat of hair, occurred as early as the Eocene. Wallace cautiously suggested the middle Tertiary as the time during which an unspecified variety of ape attained the human form after a prolonged process of morphological evolution."

At this time, however, the visions of ape-men propounded by Darwin and Haeckel were purely hypothetical. No fossils of creatures truly transitional between the early Tertiary apes and Cro-Magnon man had been found. But what about the stone tools discovered in Miocene formations by Ribeiro in Portugal and by Bourgeois in France?

Anatole Roujou, a French evolutionist, reacted in an interesting fashion to the stone tools found at Thenay. Roujou said: "Being convinced of the transformation of species, I did not have to wait for the discovery of Miocene flints to demonstrate the existence of Tertiary man, because his existence is a necessary consequence of transformation, as currently understood, and an indispensable corollary to the ideas I hold about the morphological affinities of the mammals and their mode of descent" (de Quatrefages 1884, p. 81).

De Quatrefages (1884, p. 81) observed: "Roujou traced back to Tertiary man, whose existence he accepted on purely theoretical grounds, the several distinct present races of humans which, he believed, have existed since the Quaternary. Roujou saw no reason to suppose that humans like those presently existing could not have existed at the time the flint implements of Thenay were being made."

This is quite an interesting admission from an evolutionist. Today, evolutionists put the emergence of anatomically modern humans in the Late Pleistocene. Nevertheless, even from the standpoint of current evolutionary theory, there is, strictly speaking, no reason to rule out in advance the existence of modern human beings, or a closely related species, in the Miocene. After all, advocates of punctuated equilibrium no longer envision an uninterrupted process of gradual change from one species to another. The paleontological evidence, they say, shows that species remain static for long periods of time, millions of years, and that new species appear quite abruptly in the fossil record (Gould and Eldredge 1977). Accepting this point of view, we should not necessarily expect our ancestors to become progressively more primitive and apelike as we trace them back further and further. After all, there are many present-day creatures, turtles and alligators to name a couple, that have not changed substantially for tens of millions of years.

De Mortillet, also a Darwinist, took a somewhat different approach than Roujou. "He tries to accommodate the ideas of Darwin with the paleontological facts," wrote de Quatrefages (1884, p. 81). De Mortillet himself said: "The mammalian fauna has been replaced several times, at least thrice, since the implement-bearing deposits at Thenay were laid down. . . . Can human beings, who display one of the most complex levels of biological organization, have escaped from that law of transformation?" (de Quatrefages 1884, p. 81).

But from the standpoint of modern theory, species may change at different rates. Even if it is agreed that some mammalian species have been replaced several times since the Miocene, there is no reason to reject evidence that suggests the human species might not have been replaced. According to current thinking, speciation is a relatively abrupt and unpredictable occurrence rather than the result of an ongoing process of gradual, progressive change.

As can be seen from the different conclusions of Roujou and de Mortillet, evolutionary theory is quite flexible, perhaps too flexible. It seems almost any piece of paleoanthropological evidence can be accommodated within the elastic evolutionary framework.

De Mortillet went on to make the following observation. "If we see in the flint objects found at Thenay signs of intentional work, we can only conclude that it was the work not of anatomically modern human beings but of another human species, probably representative of a genus of human precursors that fills the gap between humans and animals" (de Quatrefages 1884, pp. 81–82).

De Mortillet called this precursor genus *Anthropopithecus,* existing in three species, the oldest, that of Thenay, being the link with the apes. The other two species were the makers of flint tools found by Ribeiro in Portugal (Section 4.1) and by Rames at Aurillac in southern France (Section 4.3.2).

"For de Mortillet," stated de Quatrefages (1884, pp. 82–83), "the existence of the *anthropopitheques* in Tertiary times is a necessary consequence of Darwinist doctrines. Their successive appearances and disappearances are equally indispensable for maintaining the accord between the progressive development of the human type and that of mammalian fauna. Encountering in the ancient layers of the earth flints bearing signs of intentional work, it was natural for him to interpret them as the first manifestations of primitive industry by a precursor of modern humans." De Mortillet's objections to anatomically modern humans in the Tertiary were, it seems, primarily theoretical, based on his Darwinian preconceptions.

Looking back on this formative era of modern paleoanthropology, one should carefully note the great strength of de Mortillet's faith in the existence of an apelike precursor of modern human beings. Darwinists were awaiting the appearance of the missing link just as expectantly as others awaited the coming of the Messiah. We may well ask: was it perhaps this strong faith and conviction, more than any other factor, that motivated later paleoanthropologists to designate certain apelike fossil creatures as the biological ancestors of the modern human type?

De Quatrefages (1884, p. 83) then continued: "De Mortillet is the first to admit that no one has as yet found the slightest remains of the *anthropopitheques;* and he combats the theory of Mr. Gaudry, who is disposed to attribute the worked flints of Thenay to the Miocene ape *Dryopithecus fontani.* But it remains for de Mortillet to reveal to us the exact character of that being, which evidently has, except in his own eyes, nothing but a completely theoretical existence. Others, however, are more daring. Haeckel and Darwin, on the basis of diverse considerations, have indicated some characteristics which would, in their opinion, enable us to recognize their ape-men. Finally Hovelacque, carrying to extremes the theory of transformationism, has compared point for point the corresponding traits of the highest anthropoid apes with those of the lowest forms of humanity; from this exercise, he has derived an intermediate form and believes he is able to trace a fairly complete portrait of the being that immediately preceded the first human of the modern type."

Such speculative visualization continues even today. Whereas Hovelacque had not a single fossil bone to work with, paleoanthropologists of later years had at least some starting point. But even so, the few fragments of bone they came to possess were, as we shall see in later chapters, quite insufficient to justify the countless elaborate technicolor visions of body types and lifestyles that to this day decorate museum exhibits and the pages of popular science publications. The main point to be gathered, however, is that the existence of apelike precursors of modern humans was, as de Quatrefages so perceptively noted, more a matter of dogmatic assertion than scientific fact. If this is kept in mind, the subsequent developments in paleoanthropology can be seen in a new light. Were the later "discoveries" of fossil apelike human ancestors the product of unbiased scientific investigation or of a fanciful prophetic quest that ended in true believers seeing in broken iron cups the holy grail?

"The majority of the authors responsible for the evolutionary views I have discussed speak very loudly in the name of free thought," stated de Quatrefages (1884, p. 83). The term "free thought," in this context, refers not to the modern constitutional guarantee of freedom of conscience but to the atheistic and deistic philosophies that arose in Europe during the eighteenth and nineteenth centuries, in opposition to established churches and their doctrines.

After commenting on the views of the Darwinist free thinkers, de Quatrefages (1884, p. 83) observed: "It is very curious to see how other authors arrive at the very similar conclusions starting from a quite different position, namely, the Mosaic tenets shared by the Christian faiths." De Quatrefages then went on to discuss the beliefs of Boucher de Perthes, the discoverer of the flints of Abbeville, who from Christianity derived the idea of pre-Flood humans, very different from present humans. Some Christian thinkers believed that the time before the Flood was of inestimable length and that the earth had once been inhabited by pre-Adamite humans, who were "rough sketches" of the present species. For such thinkers, including Boucher de Perthes, it was these primitive humans who made the crude stone tools of Tertiary times. Boucher de Perthes suggested that the fossil bones of the antediluvian race had already been found but had perhaps been mistaken for those of anthropoid apes. The pre-Adamite race of apelike humans, constitutionally incapable of understanding and worshiping God, was thought to have been destroyed by an inundation (not the Flood of Noah's time). After this catastrophe, and others, came the six days of the new creation during which the modern race of humans, capable of worshiping God, was brought into being, starting with Adam and Eve (de Quatrefages 1903, p. 31; 1884, pp. 84–88). The new human species was completely distinct from the old, with no connection by descent.

"On the other hand, for de Mortillet and Darwin and his disciples," observed de Quatrefages (1884, p. 89), "the successive creations are continuous. The present human being is connected to the ancient *anthropopitheque* by an uninterrupted line of descent. His form has been somewhat modified, the intelligence increased; but we are nothing else than, in the accepted physiological sense of the word, his great grandson. I will not here combat this last opinion. Everyone already knows the negative nature of my views toward the doctrine of transformationism. So likewise with the religious theories just reviewed." The question of Tertiary humans, in de Quatrefages's view (1884, p. 89), had become "as so much else which should have remained exclusively scientific, a theater of conflict between religious dogmatism and free thought." The same is still true today, as demonstrated by the ongoing debates between advocates of Darwinian evolution and Biblical creationism, particularly in the United States.

We share the views of de Quatrefages, in the sense that we are not satisfied with the dogmatic accounts of human origins given by either the Darwinian evolutionists or the Biblical creationists. The available empirical evidence appears to be at variance with both, which suggests that it would be advisable to seriously consider other theoretical systems. In a forthcoming book, we shall present an alternative account of human origins that agrees with all the facts more completely than the accounts given by either of the traditional opponents in the long-running debate on human origins.

4.2.3 Who Made the Flints of Thenay?

So the question remains: who made the flint implements of Thenay? Even if one assumes the presence of some primitive ape-man, how can one rule out the presence of human beings of the modern type in the same period? If you can bring *Homo habilis* or *Homo erectus* back to the Miocene, why not *Homo sapiens*?

Laing (1894, p. 370) said of the flints of Thenay: "their type continues, with no change except that of slight successive improvements, through the Pliocene, Quaternary, and even down to the present day. The scraper of the Esquimaux and the Andaman islanders is but an enlarged and improved edition of the Miocene scraper." If humans make such scrapers today, it is certainly possible, if not probable, that identical beings made similar scrapers back in the Miocene period. And, as we shall see in coming chapters, scientists did in fact uncover skeletal remains of human beings indistinguishable from *Homo sapiens* in Tertiary strata.

It thus becomes clearer why we no longer hear of the flints of Thenay. At one point in the history of paleoanthropology, several scientists who believed in evolution actually accepted the Thenay Miocene tools, but attributed them to a precursor of the human type. Evolutionary theory convinced them such a precursor existed, but no fossils had been found. When the expected fossils were found in 1891, in Java, they occurred in a formation now regarded as Middle Pleistocene. That certainly placed any supporters of Miocene ape-men in a dilemma. The human precursor, the creature transitional between fossil apes and modern humans, had been found not in the Early Miocene, 20 million years ago by current estimate, but in the Middle Pleistocene, less than 1 million years ago (Nilsson 1983, pp. 329–330). Therefore, the flints of Thenay, and all the other evidences for the existence of Tertiary humans (or toolmaking Tertiary ape-men), were quietly, and apparently quite thoroughly, removed from active consideration and then forgotten.

The alternative to burying the evidence from Thenay and elsewhere was uncomfortable—perhaps anatomically modern humans had coexisted with dryopithecine apes. This would have meant discarding the emerging evolutionary picture of human origins or revising it to such an extent as to make it appear far less credible. What to speak of anatomically modern humans, any kind of toolmaking hominids would have been, after the discovery of Java man, quite out of place in the Early Miocene of France.

Of course, this scenario about the treatment of evidence is somewhat hypothetical, but it would appear that something like this actually did occur within the scientific community, over the course of several decades in the late nineteenth and early twentieth centuries. The extensive evidence for the presence of toolmaking hominids in the Tertiary was in fact buried, and the stability of the entire edifice of modern paleoanthropology depends upon it remaining buried. If even one single piece of evidence for the existence of toolmakers in the Miocene or Early Pliocene were to be accepted, the whole picture of human evolution, built up so carefully in this century, would begin to disintegrate. Late Pliocene and Early Pleistocene tools found outside Africa also present difficulties. According to currently dominant ideas, *Homo erectus* was the first hominid to leave Africa and did so about one million years ago.

4.3 IMPLEMENTS FROM THE LATE MIOCENE OF AURILLAC, FRANCE

4.3.1 A Find by Tardy

Further discoveries of Tertiary stone tools were made at two principal sites (Puy Courny and Puy de Boudieu) near the town of Aurillac in the department of Cantal in south central France. In 1870, Anatole Roujou reported that Charles Tardy, a geologist well known for his Quaternary research, had removed a flint knife [Figure 4.14] from the exposed surface of a Late Miocene conglomerate at Aurillac. To describe the removal, Roujou (1870) used the word *arraché,* which means the flint had to be extracted with some force. According to Roujou,

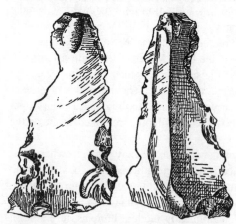

the stratum was proven to be Late Miocene in age by a characteristic fauna, including *Dinotherium giganteum* and *Machairodus latidens* (de Mortillet 1883, p. 97). De Mortillet, who thought the signs of intentional work on the flint were incontestable, declared that the object resembled undoubted Quaternary tools. Yet de Mortillet (1883, p. 97) believed Tardy's flint tool had only recently been cemented onto the surface of the Late Miocene conglomerate and therefore chose to assign it a Quaternary date.

Figure 4.14. The first stone tool found at Aurillac, France (Verworn 1905, p. 9).

4.3.2 Further Discoveries by Rames

The French geologist J. B. Rames was doubtful that the object found by Tardy was actually of human manufacture, but in 1877 Rames made his own discoveries of flint implements in the same region, at Puy Courny. De Mortillet stated that the flints collected by Rames were found in beds of white quartzite sand and whitish clay containing fossils of *Hipparion, Mastodon angustidens,* and other species of Late Miocene (Tortonian) age. Instead of being split by the action of fire, like the flints of Thenay, the specimens from Puy Courny were obviously chipped by percussion (de Mortillet 1883, p. 97).

S. Laing (1894, p. 357) provides a good review of the positive case for the implements found by Rames at Puy Courny: "The first question is as to the geological age of the deposits in which these chipped implements have been found. In the case of Puy Courny this is beyond dispute. In the central region of Auvergne there have been two series of volcanic eruptions, the latest towards the close of the Pliocene or commencement of the Quaternary period, and an older one, which from its position and fossils, is clearly of the Upper Miocene. The gravels in which the chipped flints were discovered by M. Rames, a very competent geologist, were interstratified with tuffs and lavas of these older volcanoes, and no doubt as to their geological age was raised by the Congress of French archaeologists to whom they were submitted. The whole question turns therefore on the sufficiency of the proofs of human origin, as to which the same Congress expressed themselves as fully satisfied."

Modern geologists still refer the fossiliferous sands of Puy Courny to the Miocene (Peterlongo 1972, pp. 134–135). The fauna (*Dinotherium giganteum, Mastodon longirostris, Rhinoceros schleiermacheri, Hipparion gracile,* etc.) is said to be reminiscent of that of Pikermi, Greece, and is judged to be characteristic of the end of the Pontian (Peterlongo 1972, p. 135). In the past, the Pontian was equated with the Early Pliocene, but Nilsson (1983, p. 19) stated that modern radiometric dating methods indicate that "the whole Pontian stage should be assigned to the latest Miocene time." According to French authorities also, the Pontian marks the end of the Miocene, and can be given a quantitative date of about 7–9 million years (Klein 1973, table 6).

Laing (1894, p. 358) then gave a detailed description of the signs of human manufacture that Rames had observed on the flints: "The specimens consist of several well-known palaeolithic types, celts, scrapers, arrow-heads, and flakes, only ruder and smaller than those of later periods. They were found at three different localities in the same stratum of gravel, and comply with all the tests by which the genuineness of Quaternary implements is ascertained, such as bulbs of percussion, conchoidal fractures, and above all, intentional chipping in a determinate direction. It is evident that a series of small parallel chips or trimmings, often confined to one side only of the flint and which have the effect of bringing it into a shape which is known from Quaternary and recent implements to be adapted for human use, imply intelligent design, and could not have been produced by the casual collisions of pebbles rolled down by an impetuous torrent."

According to Laing, de Quatrefages noted fine parallel scratches on the chipped edges of many specimens, indicating usage. These use marks were not present on other unchipped edges. The flint implements of Puy Courny were accepted as genuine at a congress of scientists in Grenoble (Laing 1893, p. 118).

In conclusion, Laing (1894, pp. 358–359) repeated another very important point that was made by de Quatrefages: "The chipped flints from Puy Courny also afford another very conclusive proof of intelligent design. The gravelly deposit in which they are found contains five different varieties of flints, and of these all that look like human implements are confined to one particular variety, which from its nature is peculiarly adapted for human use. As Quatrefages says, no torrents or other natural causes could have exercised such a discrimination, which could only have been made by an intelligent being, selecting the stones best adapted for his tools and weapons."

Leland W. Patterson (1983, pp. 305–306), a modern expert on lithic technology, has written: "The selective occurrence of certain types of raw material can be useful in identifying human activity at a specific location. The lack of a local source for a raw material is an argument in favor of transport by humans to a site. Another consideration is the selective occurrence of only certain types of raw materials for specimens proposed to be man-made. Man would tend to be selective in use of lithic raw materials, while nature would tend to fracture a wide variety of stone types in a random manner."

But Marcellin Boule gave a geological explanation for the fact that the objects thought to be tools were formed from only one of the many kinds of flint present at Puy Courny. As noted by Rames, the various kinds of flint all came from different layers of the underlying Oligocene formation. In 1889, Boule suggested that during the Late Miocene, only the layer containing the particular type of flint in question had been eroded. According to Verworn (1905, p. 10), that meant only this particular type of flint, lying loose on the surface, was available for toolmaking by intelligent beings in the Late Miocene.

But Boule completely rejected the idea that the flint objects of Aurillac were manufactured by humans or human evolutionary ancestors. His analysis of the erosion of flint at Aurillac was intended to demonstrate that in the Late Miocene, only a certain type of flint had been subjected to purely natural forces tending to create toollike forms.

Boule's account of the successive erosion of the various flint-bearing Oligocene layers may not, however, have been correct. Perhaps several layers eroded simultaneously. If so, this would preserve the point Rames made about intelligent selection of one kind of flint from among many for the purpose of toolmaking. But even if we do accept the sequence of geological events outlined by Boule, this still would not allow one to conclude that the chipped flint objects from the Late Miocene of Puy Courny were produced by purely natural forces. It would seem that all the other kinds of flint that later eroded from layers below the one described above should also have been shaped by natural forces into forms resembling tools. Considered in this way, Boule's explanation tends to explicitly confirm human rather than natural action.

Furthermore, Boule's geological explanation, if correct, merely accounts for the selection of a particular kind of flint. It does not explain the special character of the chipping on the flints. As previously mentioned, the chipping on the flints, confined to one side of one edge, with the chips removed consecutively and in parallel, was not of the type one would expect from random natural battering or geological pressures. In fact, the flint objects were, according to many authorities, identical to accepted unifacially flaked flint tools from the Late Pleistocene.

4.3.3 Verworn's Expedition to Aurillac

In the first part of the twentieth century, some professional scientists continued to recognize specimens from the sites near Aurillac as the work of human beings in the Late Miocene. Among them was Max Verworn of the University of Göttingen in Germany.

In his introduction to a lengthy report on the implements of Aurillac (Cantal), published in 1905, Verworn pointed out that the existence of human beings in the Pleistocene period had been established beyond doubt by skeletal remains, stone artifacts, and other objects of human manufacture. Verworn (1905, pp. 3–4) stated: "The fact that the skeletal remains so far discovered in our Pleistocene investigations can be recognized by their morphology as genuinely human should indicate, in the most lucid manner, to every modern researcher who stands upon the ground of the theory of descent, that the beginning of our race and its specific human characteristics must reach far beyond the Pleistocene, and, at very least, deep into the Tertiary. Yet despite this theoretical advancement in the investigation of natural history, science is very reluctant to enter fully into the question of the existence of Tertiary man, and any discussion of the evidence in this regard has been treated with utmost distrust and skepticism in the scientific community. Of course this is justifiable, because in all true science every provisional truth must pass the test of the critical fire of doubt before it can be granted full recognition."

In Verworn we have an excellent example of a scientist with Darwinian credentials accepting evidence (in this case, evidence for a human presence in the Miocene) that would completely contradict current Darwinian ideas about the origin of the human species. The present scientific establishment propagates the belief that only fundamentalist creationists and early scientists opposed to evolution have ever presented evidence contradicting the current evolutionary understanding of human origins. But this is far from the truth. Scientists who believed in evolution have been the main source of the information compiled in this book.

Scientific discussion of Tertiary humans peaked in the 1880s and decreased markedly in the final years of the nineteenth century. The question was reopened by Rutot's discoveries of flint implements in Belgium, which we shall consider later in this chapter (Section 4.4). Verworn, working in the very early years of the twentieth century, was himself at first quite doubtful about the human manufacture of eoliths, or "dawn stones," as the crudest of the early stone tools had come to be known.

Verworn (1905, pp. 4–5) wrote in his report on Aurillac: "I must confess that less than a year ago I was still skeptical about accepting the implemental nature of eoliths, and expressed my doubts at the meeting of the Göttingen Anthropological Society on July 22, 1904. Of course, I had seen with my own eyes only the finds of Dr. Hahne from the Pleistocene of the Magdeburg region, and I can say that regarding the greater part of Hahne's eoliths, in view of the strong inorganic influences upon them and the conditions of their occurrence, I still today maintain my skepticism, though I do recognize some isolated pieces that bear signs of human work. Meanwhile Herr Rutot was, in the course of the past year, kind enough to send to me as a gift a great collection of typical eoliths from the various levels of the Belgian Pleistocene, and after carefully analyzing them I could no longer maintain any doubts about their implemental nature. I was overcome with strong excitement. With these discoveries the traces of primitive culture extended far beyond all previous boundaries." Verworn, in these passages, is using the term eolith in a very broad sense. But as we shall see, he will later employ distinctions similar to the ones adopted in this book.

Verworn (1905, pp. 5–6) continued: "The question then arose for me, whether such evidence might extend back into the Tertiary. The evidence supporting this proposal gathered in earlier times, which in some cases had been introduced with great precision, had not been able to win general recognition. For me there was no doubt about the theoretical possibility of man existing in the Tertiary; the real question was whether or not the Tertiary ancestors of humankind had been capable of manufacturing stone tools, which would give evidence of their existence to those of us in a far removed time. I was still skeptical on this point. When Rutot and Klaatsch had become convinced of the existence of Tertiary eoliths and published some illustrations of such, I could not, from their descriptions and illustrations alone, reach any positive conclusion about their implemental nature. There is no alternative, for anyone who wants to come to his own decision, to having the objects in his own hands, to being able to turn them around and analyze all their features. Furthermore, it is necessary to understand the objects in terms of their circumstances of discovery by visiting the places from which they came, especially in order to come to firm conclusions about their geological age, which is required. So just as for years I had conducted my own experimental flint-flaking studies in order to understand flint objects bearing the characteristic signs of human work, I decided to conduct my own onsite excavations, and thus be in a position to be able to reach a definite decision, for or against the implemental nature of the Tertiary flints in question. I can honestly say that I entered upon my investigation without any preconceived opinions. I would have been just as happy to answer the question negatively as positively."

Verworn then had to decide where to conduct his search for implements. He was aware that France had furnished investigators with many examples of reputed Tertiary flint tools. The site at Thenay was a possibility, but two scientists, L. Capitan and P. Mahoudeau, had recently published an extremely negative report about the flint objects found there, so Verworn decided to look elsewhere.

Aurillac, in Cantal, where several discoveries of Late Miocene implements had been made over the course of many decades, seemed a more profitable place to conduct his study. Verworn also considered the valley of the Tagus at Lisbon, where Ribeiro had uncovered his Miocene specimens, but because no further discoveries had been made in that region, Verworn ruled out going there. At other sites, such as the Kent Plateau in southeast England and St. Prest in France, the geological context was thought to be Pliocene, not as suitable for Verworn's purposes as the older Miocene age of the implement-bearing formations at Aurillac. So Aurillac it would be (Verworn 1905, pp. 6–7).

On his way to France, Verworn visited Rutot in Brussels and examined specimens of stone implements in the Royal Museum of Natural History, including some from Aurillac. These had been forwarded to Brussels by the French geologists Pierre Marty and Charles Puech.

Verworn (1905, p. 7) noted: "Even these collections had pieces that I could not easily account for as being other than the product of human work, and the same was true of L. Capitan's large collection of flints from the same site that I soon thereafter had the opportunity to see. . . . Capitan has like Klaatsch personally conducted excavations at Aurillac, but has not yet published his findings. Despite the fact that my firsthand observation and testing of these discoveries was leading me to belief in a Miocene flint culture in the Auvergne, I must nevertheless state that my scientific skepticism, and my own previous negative convictions in this matter, were strong enough to inspire new doubts that brought my positive decision again into question. I knew that I had to see the things on the spot, that I must personally get to know the circumstances of discovery, that I must with my own hand remove specimens from the ground—otherwise, I would not be certain. So I traveled to Aurillac."

Verworn remained at Aurillac for six days. Pierre Marty, a local geologist who had written a monograph on the Late Miocene fauna of Joursac (in Cantal), explained to him the geology of the region. Marty also showed Verworn a site he had himself discovered at Puy de Boudieu, and Verworn's excavations there yielded him the majority of his specimens. Charles Puech, a geologist and engineer of roads for the department of Cantal, also gave Verworn extensive geological information.

Verworn (1905, p. 8) reported: "It happened that in the course of my very first excavation at Puy de Boudieu I had the luck to come upon a place where I found a great number of flint objects, whose indisputable implemental nature immediately staggered me. I had not expected this. Only slowly could I accustom myself to the thought that I had in my hand the tools of a human being that had lived in Tertiary times. I raised all the objections of which I could think. I questioned the geological age of the site, I questioned the implemental nature of the specimens, until I reluctantly admitted that all possible objections were not sufficient to explain away the facts. In what follows, I shall attempt to show all this in detail. At the same time, if anyone doubts the facts as presented, then let him, as I did, go and see."

Concerning de Mortillet's proposal that the maker of the implements of Aurillac was a small apelike human precursor called *anthropopithecus* (later *homosimius*), Verworn (1905, p. 11) said: "It hardly seems necessary to mention that these speculations, insofar as they are based on the flint tools, are completely arbitrary."

Describing his own discoveries at Aurillac, Verworn (1905, p. 16) wrote: "I especially noted at Puy de Boudieu, where I had the good fortune to come upon a very productive site, that the worked stones, sometimes 5, 10, or 15, would be grouped quite close to each other, separated only by a little tuff or clay, while for 50 to 80 centimeters [roughly 2 to 3 feet] around there would be no such nests or only a few single specimens. As far as appearance goes, the unworked stones appeared to be quite rolled. The worked specimens showed little or no evidence of rolling." Verworn (1905, p. 16) added: "at Puy de Boudieu I was almost exclusively excavating specimens with edges as sharp as when they had been made. All the quartz stones found among the flints are rolled until almost round." The presence of sharp-edged flint objects amidst rolled and rounded pebbles of other kinds of rock at Puy de Boudieu signified that the flint objects had not been subjected to much movement since their deposition and that the flaking upon them was therefore of human rather than geological origin. The fact that the sharp-edged implemental flints were found in groups suggested the presence of workshop sites.

Summarizing the geological context of the discoveries, Verworn gave the following account. The basal layers are Oligocene freshwater and brackish sedimentary deposits containing beds of flint. Above these are Miocene layers of fluviatile sands, stones, and eroded chalk containing fossils of *Dinotherium giganteum, Mastodon longirostris, Rhinoceros schleiermacheri, Hipparion gracile,* etc., along with flint implements. Layers of basalt from volcanic eruptions cover these Late Miocene implement-bearing layers and in some cases go under them. Above the basalt and the Miocene layers, there are some Pliocene layers, with *Elephas meridionalis* and other Pliocene mammals. Volcanic layers from Pliocene eruptions cover these. There was no further volcanic action, and the cold periods of the Pleistocene followed. Paleolithic and Neolithic implements of the standard types are found in the upper terraces (Verworn 1905, p. 17). The basic volcanic sequence outlined by Verworn is still accepted today (Autran and Peterlongo 1980, pp. 107–112).

Verworn pointed out that those who disputed the Miocene age of the Cantal flints had not visited the sites. Verworn (1905, p. 19) stated: "In fact, in connection with the age of the flints there is, among the geologists who have actually visited the sites, not the slightest degree of reservation. They are all in agreement, and outside of Noetling and Keilhack, I am not aware of any other who have expressed doubt."

Keilhack suggested that perhaps the volcanic eruptions, said by Verworn to have ended in the Pliocene, had in fact continued into the Quaternary. If this were true, then perhaps the implements, some of which were found between layers of lava, were more recent than the Pliocene or Miocene. But what about the fact that the implements were found together with Miocene fossils? Keilhack proposed that the action of streams had mixed in bones from older Miocene layers with more recent Quaternary flint implements.

To these objections Verworn replied as follows. First of all, in no case were fossils of mammals that lived only in the Pleistocene found together with flint implements beneath the lava at Aurillac. This indicated that there had been no Quaternary eruptions. Therefore, any flint-bearing beds found under the several layers of lava were definitely Pliocene or older. Furthermore, the layers of basalt and other volcanic rock were separated by freshwater sedimentary beds with sharply characteristic fossil remains. For example, one might find under a particular layer of basalt a sedimentary bed containing Pliocene fossils and under this another layer of basalt. Under this second bed of basalt, one might then find another sedimentary layer, this with fossil remains of Miocene plants and animals along with flint implements. And under a third layer of basalt one might find another Miocene sedimentary layer containing flint implements, this layer lying upon the Oligocene basement formation. From such evidence, Verworn concluded that the flint-bearing sedimentary beds below or directly above the lowest layer of basalt at Aurillac were Miocene rather than Pliocene in age.

Verworn (1905, p. 20) concluded: "So we find these implement-bearing layers always directly over the Oligocene or directly upon the basalt from the oldest eruptions, which directly cover the Oligocene layers. The fact that over these oldest eruptive masses one finds beds that contain a typical Late Miocene fauna, like that found at Joursac, with *Hipparion, Dinotherium,* etc., means that the underlying implement-bearing beds cannot be any more recent than the Late Miocene. Thus the second doubt of Keilhack, namely that the Miocene fauna has been secondarily introduced into the implement-bearing layers from below, is cleared away."

Verworn (1905, p. 21) then discussed at length various ways to identify human work on a flint object. He divided evidence of such work into two groups: (1) signs of percussion resulting from the primary blow that detached the flake from a flint core; (2) signs of percussion resulting from secondary edge-chipping on the flake itself.

On a flint flake, the principal signs of percussion from the main blow that detached the flake from a flint core would be a striking platform, bulb of percussion, and eraillure. According to de Mortillet, the presence of a striking platform, bulb of percussion, and eraillure together on a flake is a very good indicator of intentional work (Verworn 1905, pp. 21–22).

In addition to the three features mentioned above, Verworn (1905, pp. 22–23) described several more signs of percussion that can be observed on flint flakes (Figure 4.15). Concentrated near the point of impact on the top of the flake one can see a small formation of concentric circular cracks. Radiating from the point of impact and extending over the entire surface of the flake there is also visible a series of curved percussion marks, or force ripples. The stronger the blow that separated the flake from the flint core, the stronger the ripples. Raylike cracks, emanating from the point of impact, intersect the curved force ripples. Verworn also pointed out that in a flake made by percussion the plane of fracture is not straight. If one looks at the flake, edge on from the side, one sees that the ventral surface of the flake is convex at the bulb of percussion, near the top of the flake, and concave at the lower portion, giving an S-shaped contour. Sometimes one can also see on the striking platform a crush mark from a previous blow that failed to detach the flake from the flint core. Negative impressions of some of the above-mentioned features are sometimes visible on the core from which the flake was taken.

It would seem that the presence of combinations of these percussion signs would make it easy for one to identify human work on a flint object. But according to Verworn, this is not necessarily so. All the above-mentioned characteristics are symptoms of just one thing—a blow of sufficient force directed at a given point. But if nature could deliver the blow, then the presence of all the symptoms of percussion is not enough to establish human workmanship (Verworn 1905, p. 23).

The question as to whether nature can actually deliver such a blow has been much debated. Verworn (1905, p. 24) wrote: "It is generally recognized that extreme fluctuations of temperature and moisture, and the action of frost, do not result in fracturing that produces the above-mentioned features. It is otherwise with the question whether or not strongly agitated water, as in flooded mountain streams, waterfalls, or ocean shores, can throw stones together in such a way as to bring about the typical characteristics of percussion. I do not rule this out, but I would tend to believe that such things, if they occur, do so only in very isolated instances." In this respect, Verworn is in agreement with modern authorities on lithic technology such as Leland W. Patterson (1983) and George F. Carter (1957, 1979).

Verworn (1905, p. 24), willing to consider all possibilities, further stated: "I could also imagine that falling stones, loosened by erosion, could produce such effects, but again, very rarely. Finally it would appear to me that stones pressed against each other by glacial action could produce the characteristic symptoms. In summary, the possibility that purely inorganic factors could act on flint to produce the above-mentioned signs of percussion is something I do not wish to dispute. Therefore the bulb of percussion, eraillure, striking platform, force ripples, etc., are not, contrary to de Mortillet's view, definite criteria for

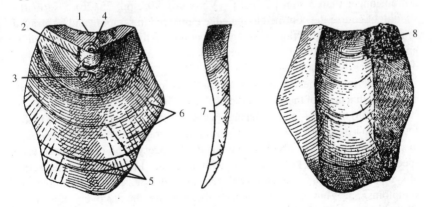

Figure 4.15. Diagnostic features of a struck flake (Verworn 1905, p. 22): (1) striking platform; (2) bulb of percussion; (3) eraillure; (4) point of impact, with concentric circular cracks; (5) curved force ripples; (6) cracks emanating from the point of impact; (7) S-shaped curve of the plane of fracture; (8) crush mark from a previous blow that failed to detach the flake from the flint core.

intentional flaking." Here Verworn was perhaps showing too much caution. Even in terms of his own analysis of evidence indicating percussion, it is not very likely that nature would, except in extremely rare circumstances, produce such combinations of effects.

Verworn believed that retouched edges on flint flakes were good, but again not absolutely certain, evidence of human manufacture. He recommended very careful study of the features of such retouching, including the depth and size of individual marks, the similarity of their planes of impact, and their arrangement in regular rows along the edges of the presume flint implement (Verworn 1905, pp. 24–25).

Unidirectional flaking on one side of an edge is generally taken as a very sure sign of human work, but Verworn (1905, pp. 27–28) stated he could "imagine circumstances in which you might have a sharp piece of flint sticking out from a wall of limestone, and then have pieces of rock falling from above, hitting the edge many times, producing unidirectional flaking."

Verworn recommended that special attention be given to signs of use on the edges of possible flint implements. One would expect that an implement used for scraping wood, bones, or skin, or for digging the earth, would display certain characteristic markings. Verworn conducted extensive experimental research in this area.

He concluded: "It is characteristic of use-patterns that there are only small marks on the edge, on the average no greater than 1–2 mm [.04-.08 inch]. Even when an edge is used with great pressure on the hardest materials, the use marks are no longer than 5 mm [.2 inch]" (Verworn 1905, pp. 25–26). Use marks should, of course, be confined to the edge employed in scraping and be arrayed in the appropriate direction, in a regular parallel fashion. Less pressure can be applied with a small flint, so use marks should be smaller on small pieces than on big pieces (Verworn 1905, p. 26).

Considering all the various characteristics of percussion and use, Verworn suggested that none of them are in themselves conclusive. Verworn (1905, p. 29) stated: "I propose that in each separate case a critical diagnosis must be made, founded on a deep and thorough analysis of the characteristics of each specimen in connection with the circumstances of its discovery. The diagnosis of each specimen should not be concerned with just one, but with a whole series of symptoms, just as a doctor analyzes internal diseases by a complex of symptoms. . . . What must concern us is therefore not the discovery of a single, all-embracing, universally applicable criterion for recognizing manufacture in stone implements; such a criterion does not exist in reality and every attempt to find one is fruitless. What we must concern ourselves with is the development of a critical diagnostic method, similar to that employed by physicians. The more carefully we develop this diagnostic method through observation and experiment, the more we shall be able to reduce the number of questionable factors. The critical analysis of a given combination of symptoms is the only thing that will put us in a position to make decisions."

This is the same methodology suggested by L. W. Patterson (1983). Patterson does, however, give more weight than Verworn to bulbs of percussion and unidirectional flaking along single edges of flakes, especially when numerous specimens are found at a site. Patterson's studies showed that natural forces almost never produce these effects in significant quantities.

Verworn (1905, p. 29) then provided an example to illustrate how his method of analysis might be applied: "Suppose I find in an interglacial stone bed a flint object that bears a clear bulb of percussion, but no other symptom of intentional work. In that case, I would be doubtful as to whether or not I had before me an object of human manufacture. But suppose I find there a flint which on one side shows all the typical signs of percussion, and which on the other side shows the negative impressions of two, three, four, or more flakes removed by blows in the same direction. Furthermore, let us suppose one edge of the piece shows numerous, successive parallel small flakes removed, all running in the same direction, and all, without exception, are located on the same side of the edge. Let us suppose that all the other edges are sharp, without a trace of impact or rolling. Then I can say with complete certainty—it is an implement of human manufacture."

Verworn, after conducting a number of excavations at sites near Aurillac, analyzed the many flint implements he found, employing the rigorously scientific methodology described above. He then came to the following conclusion: "With my own hands, I have personally extracted from the undisturbed strata at Puy de Boudieu many such unquestionable artifacts. That is unshakable proof for the existence of a flintworking being at the end of the Miocene" (Verworn 1905, pp. 29–30).

At his main excavation site at Puy de Boudieu, Verworn (1905, p. 30) found that the implements were sharp, showing no movement since they were deposited. Verworn (1905, p. 32) stated: "I find that in terms of size, shape, and adaptation to the human hand, these specimens are not different from Paleolithic implements. That, as is evident, rules out de Mortillet's supposition that the small size of the tools meant that the bodily size of the hypothetical *homosimius* was inferior to that of a human being. The tools do not give grounds for such a conclusion."

Verworn discovered in the Miocene formations at Aurillac 199 worked pieces of flint, 98 with bulbs of percussion. In reality, more should have been counted as having bulbs of percussion, for, in many cases, although the part of the flake with the bulb was broken off, the remainder of the flake showed all the usual signs of percussion. Most of the tools were 4–5 centimeters (about 2 inches) in size, although some went up to 10 centimeters (4 inches).

Verworn (1905, p. 33) wrote: "The typical signs of percussion, such as striking platform, bulb of percussion, eraillures, fissures of percussion, and curvature of the plane of fracture, were clearly evident. Only the force ripples on the plane of fracture were not very strongly developed, and the circular percussion marks near the point of impact were not to be seen very clearly, perhaps because of the opaqueness of the material and its strong, dark patination. The backs of the flakes sometimes bear upon them the cortex, but for the most part they display the scars of earlier flakes that always have been removed in the same direction. Sometimes four or five flake scars run over the back, and often the negative bulbs of percussion from these flakes are still well preserved. Next to them one often sees the strong crush marks of blows delivered in the same direction."

Verworn (1905, p. 33) performed his own experimental flint flaking and reported: "With hammer stones, I have struck from the flat pieces of flint from the Miocene beds a number of flint flakes, and these flakes closely resemble the old ones." Verworn stated that because of the cortex covering the flint, the blows had to be quite hard, resulting in well-marked bulbs of percussion like those on the Miocene flakes. The cushioning effect of the relatively soft cortex also accounted for the lightness of the rings of percussion on the flakes detached from the flint core.

In addition to flakes, Verworn also found many cores from which flakes had been struck. Verworn (1905, p. 34) analyzed the situation as follows: "In fact one finds a great number of slabs of flint, on the edges of which one finds characteristic flake scars with negative bulbs of percussion. . . . One might have taken a good slab and removed one or more flakes from the edge. One finds a number of flake scars next to each other on the edge, mostly removed by blows in the same direction, though there are some cases where they have been removed at different angles."

Most of the implements found by Verworn in the Miocene beds of Aurillac were scrapers of various kinds: "Some scrapers show only use marks on the scraping edge, while the other edges on the same piece are quite sharp and unmarked [Figure 4.16]. On other specimens the scraping edge displays a number of chips intentionally removed in the same direction. This chipping displays quite clearly all the usual signs of percussion. Even today the edges of the impact marks of previous blows on the upper part of some implements are perfectly sharp [Figure 4.17]. The goal of the work on the edges is clearly and without doubt recognizable as the removal of cortex or the giving of a definite form. On many pieces there are clearly visible handgrip areas, fashioned by the removal of sharp edges and points from places where they would injure or interfere" (Verworn 1905, pp. 37–38).

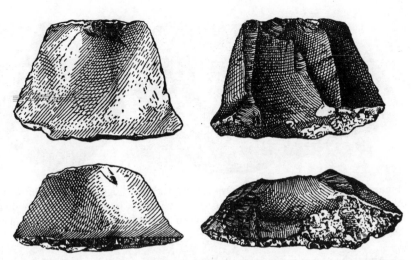

Figure 4.16. Four views of a flint scraper found in Late Miocene strata at Aurillac, France (Verworn 1905, p. 37). Top left: Ventral surface with large bulb of percussion. Bottom left: Ventral surface tilted to show the lower edge, with numerous small use marks. Top right: Dorsal surface of the scraper, showing removal of five large parallel flakes. Bottom right: Dorsal surface tilted to show the lower edge, with use marks on the left and remnants of cortex on the right.

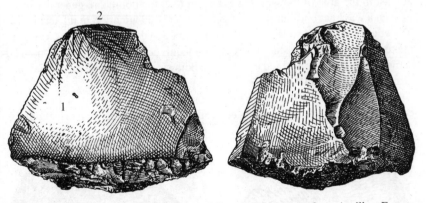

Figure 4.17. Left: Ventral surface of Late Miocene flint scraper from Aurillac, France, showing (1) bulb of percussion and (2) striking platform. The cortex of the flint has been removed from the lower edge by percussion, leaving numerous flake scars oriented in approximately the same direction. Right: Dorsal surface, showing five large parallel scars of flakes removed before the scraper itself was struck from the parent flint core. The upper left corner of the implement shows impact damage from one of the previous blows (Verworn 1905, p. 38).

Figure 4.18. Late Miocene flint scraper from Aurillac, France, with large flakes removed in parallel (Verworn 1905, p. 39). This feature reminded Verworn of Late Pleistocene examples.

Figure 4.19. A pointed flint implement from the Late Miocene at Aurillac, France (Verworn 1905, p. 40).

About the object in Figure 4.18, Verworn (1905, p. 39) said: "the flake scars on the scraper blade lie so regularly next to each other in parallel fashion that one is reminded of Paleolithic or even Neolithic examples." In the accepted sequence, Paleolithic and Neolithic tools are assigned to the later Pleistocene.

Verworn also found many pointed scrapers (Figure 4.19): "Among all the flint objects, these show most clearly the intentional fashioning of definite tool shapes, at least in the area of the working edges. In fact, the points are generally made in such a way that one can speak of genuine care and attention in the technique. The edges have been worked by many unidirectional blows in such a way as to make the intention of fashioning a point unequivocal. I characterize as pointed scrapers those tools on which the chips on both sides of the point run in the same direction" (Verworn 1905, p. 40).

Also found at Aurillac were notched scrapers (Figure 4.20), with rounded concave openings on the working edge suitable for scraping cylindrical objects

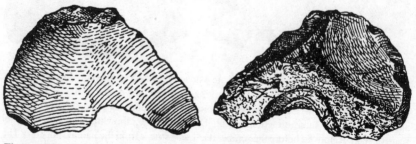

Figure 4.20. Left: Ventral surface of a notched scraper from the Late Miocene of Aurillac, France (Verworn 1905, p. 40). Right: Dorsal surface, showing removal of cortex on the working edge, upon which Verworn observed tiny use marks.

like bones or spear shafts. Verworn (1905, p. 41) observed: "In most cases the notched scrapers are made by chipping out one of the edges in a curved shape by unidirectional blows."

Figure 4.21. A Late Miocene flint tool from Aurillac, France. The point is formed by removal of many flakes in the same general direction (Verworn 1905, p. 41).

Verworn also uncovered several tools adapted for hammering, hacking, and digging. Describing the one in Figure 4.21, Verworn (1905, p. 41) wrote: "A large pointed tool for chopping or digging. It is formed from a natural slab of flint by the working of a point. One sees on the surfaces of the piece the cortex of the flint and at the top a point made from numerous flakes, mostly removed in the same direction." About another pointed tool, Verworn (1905, p. 41) stated: "This tool has on the side directly below the point a handgrip made by removing the sharp, cutting edges. It might have been a primitive handaxe used for hammering or chopping." Verworn also found tools he thought were adapted for stabbing, boring, and engraving.

Verworn (1905, pp. 44–45) concluded: "At the end of the Miocene there was here a culture, which was, as we can see from its flint tools, not in the very beginning phases but had already proceeded through a long period of development. . . . this Miocene population of Cantal knew how to flake and work flint."

The only visible signs of human work upon the Eolithic tools (Chapter 3) were use marks and perhaps slight chipping to improve the working edge. Verworn saw signs of more extensive intentional work on the tools of Aurillac (Cantal)—removal of cortex (the rough outer surface of the flint) to expose a sharp edge and the subsequent shaping of the edge for a particular purpose. But the modification was confined to the specific edge that was meant for use. Modification did not extend to the shaping of an entire implement, as in the Late Paleolithic and Neolithic. A third sign of intentional work on the tools from Aurillac was the removal of sharp edges to form a comfortable handgrip (Verworn 1905, pp. 44–47). For these reasons, we have placed the flint implements found by Verworn at Aurillac in the category of crude paleoliths.

Verworn (1905, p. 50) designated the implements of Aurillac as archaeoliths, placing them between eoliths and paleoliths. Eolithic industries, according to Verworn, are those in which the natural edges of pieces of stone are used as tools without any further modification. Use marks would be the only sign of human action upon them. In Archeolithic industries, the working edges of the tools are modified for specific purposes, and in Paleolithic industries the entire piece of stone is worked with some degree of artistry into a specific tool shape.

Verworn (1905, p. 50) believed that purely Eolithic cultures—with implements displaying no retouch, just use marks—had not yet been found. As can be seen, Verworn's definition of an eolith is somewhat different than the one we employ, which encompasses slight retouching as well as use marks on naturally produced stone flakes. Our category of crude paleoliths differs from the category of eoliths in that an industry of crude paleoliths would contain at least some tools deliberately struck from cores and subjected to more extensive retouching.

Verworn felt that geological considerations are primary in determining the age of stone tools, because different levels of culture exist at different times. Even today, he said, there are people who make and use the crudest sort of stone tools (Verworn 1905, p. 50). Verworn's methodology protects one from automatically assuming that a technologically advanced stone tool found in very old strata must in fact be recent or that a crude tool must necessarily be old.

Verworn (1905, p. 47) further stated: "Concerning the Miocene culture of Cantal, the facts teach us that we must guard against a mistake, often encountered in the field of prehistoric research when an ancient culture level is discovered. That mistake is forming too low an estimate of the culture in question. The Tertiary age of the culture in this case should in no circumstances force us into underestimating it." We fully agree with Verworn on this point.

Verworn (1905, pp. 48–49) went on to say: "Concerning the physiological status of the Miocene inhabitants of Cantal, I would like to make a few observations. I have already indicated that de Mortillet's conclusion from his study of the implements that the manufacturers were of small bodily size is fallacious, because the supposition that the tools are especially small is not supported by observation. I would, on the contrary, with a great deal of certainty say that the size of the implements points toward a being with a hand of the same size and shape as our own, and therefore a similar body. The existence of large scrapers and choppers that fill our own hands, and above all the perfect adaptation to the hand found in almost all the tools, seems to verify this conclusion in the highest degree. Tools of the most different sizes, which show with perfect clarity useful edges, use marks, and handgrips, lie for the most part so naturally and comfortably in our hands, with the original sharp points and edges intentionally removed from the places where a hand would grasp, that one would think the tools were made directly for our hands."

Of the manufacturers of the implements found at Aurillac, in Cantal, south central France, Verworn (1905, p. 49) stated: "While it is possible that this Tertiary form might possibly have stood closer to the animal ancestors of modern humans than do modern humans themselves, who can say to us that they were not already of the same basic physical character as modern humans, that the development of specifically human features did not extend back into the Late Miocene? Perhaps the Miocene inhabitants of Cantal were so highly developed that we could unquestionably give them the title of human being. Such a proposition is neither more nor less likely than de Mortillet's hypothesis of an intermediate form. On the other hand, what would prevent us from seeing in this Tertiary being a line of development parallel to the main line of human descent? All of these are simply possibilities that do not allow for proof or disproof, for the simple reason that we do not have any right to connect a specific culture level with a specific level of physiological development. So long as we have no bodily remains of the Tertiary inhabitants of Cantal, all we say will be speculation without meaning. On the same grounds, all attempts at linkage with *Pithecanthropus* of Trinil (Java man) are worthless. In one case we have cultural remains with no bodily remains, and in the other bodily remains with not a trace of cultural remains. We have simply a comparison of two unknowns. Nothing will come of it. We need patience and more material."

Verworn here makes an important point. From a viewpoint ranging from hundreds of thousands to several million years after the fact, it is very difficult to connect stone implements with particular sets of physiological remains from the same period, if such exist. As we explain in Chapter 6, fossil skeletal remains indistinguishable from those of fully modern humans have been found in Pliocene, Miocene, and even Eocene and earlier geological contexts. When we also consider that humans living today make implements not much different from those taken from Miocene beds in France and elsewhere, then the validity of the standard sequence of human evolution begins to seem tenuous. In fact, the standard sequence only makes sense when a lot of very good evidence is ignored. When all the available evidence, implemental and skeletal, is considered, it is quite difficult to construct any kind of evolutionary sequence. What we are left with is the supposition that there have been various types of human and humanlike beings, living at the same time and manufacturing stone tools of various levels of sophistication, for tens of millions of years into the past.

4.3.4 A Footnote on Aurillac

Shortly after Verworn's excavations at Aurillac (Cantal), the French researcher L. Mayet delivered a report about his own investigations, which led him to the conclusion that the objects found there were products of nature rather than the result of intentional human work. In a footnote to his famous report on the "pseudoeoliths" of Clermont (Section 3.4), Breuil referred to "Mayet's study of Cantal, where in the dislocated strata he found broken blocks of flint resembling eoliths." Breuil (1910, p. 407) stated: "There you also have some broken flints with the pieces still held in place by the sandy matrix." This was obviously to be taken as conclusive and final proof that the stone tools of Puy Courny, like those of Clermont, were produced by geological pressures rather than human action.

But not everyone responded as favorably as Breuil to Mayet's report, originally delivered at a meeting of the French Association for the Advancement of Science, held in Lyon in 1906. Dr. Hermann Klaatsch (1907, p. 765) later wrote: "At a time when the problem of primitive stone artifacts is in a phase permeated with complete lack of clarity, we must happily receive every work that without prejudiced views attempts a factual solution to the eolith puzzle, and we should also give due recognition to the courage of the author who attempts to deal with such troublesome material. In every genuine discussion, opposition is just as welcome as agreement. In this spirit, the authorities who, like myself, are in favor of the human manufacture of the Tertiary flint objects of Cantal, will find especially worthy of attention any work that attempts to demonstrate they were formed by purely natural causes." Klaatsch (1907, p. 765) added something Breuil neglected to mention: "It must be noted that L. Mayet in no way shares the radically negative standpoint of Boule, but instead fully recognizes the artifactual nature of the Belgian eoliths [Section 4.4]."

Mayet had twice visited the classic Cantal sites (Puy Boudieu and Puy Courny) and conducted excavations. Klaatsch (1907, p. 765) wrote: "After his introductory lecture, in which Mayet gave assurances that he could supply proofs of the natural process by which the flint objects had been formed, I was extremely disappointed by the way he sought to demonstrate his point. I had hoped that he would clearly inform me about the ways in which natural forces had acted so ingeniously as to transform the site at Puy Boudieu into 'a veritable eolith factory.' That significant shifting and partial resorting of the beds have occurred here is well known to anyone who has conducted excavations. But it remains for L. Mayet to make it plausible that these forces were responsible for the very sophisticated way in which pieces of flint have been broken and worked. Instead he puts off the knowledge-thirsty listener with the suggestion that one cannot precisely describe the action of these natural forces, among which he numbers 'atmospheric agents, variations in temperature, torrential waters, shifting of geological beds, and certainly other factors about which we remain ignorant.' It is as if he were trying to silence an unruly child by intimidating him with a multitude of hints of terrible future events, the consequences of which one could not even imagine." Breuil (1910, p. 407) had tried to do the same thing in his study: "It is clear that the observations made at Belle-Assise do not explain all the natural formations of the Eolithic type; the process that is observed can be juxtaposed with others, such as the action of torrents of water, periods of flooding, the trampling of animals and men, etc."

In discussing Mayet's conclusions about the Puy de Boudieu site, Klaatsch (1907, p. 765) made the following observation: "But about the fact that animal teeth in this frightful topsy-turvy have remained quite whole, as if that were possible, we hear nothing." In other words, if the geological pressures were sufficient to crush blocks of flint, why not the accompanying animal fossils? Klaatsch (1907, pp. 765–766) then stated: "I am therefore not satisfied by Mayet's concluding assertion that 'the action of the intense natural forces that have mixed together the sands and flints at this point are perfectly able to have produced the eoliths, eliminating the necessity of suggesting the intervention of human industry.' People who simply accept these closing words at face value will repeat them as wisdom, and it will afterwards appear that Mayet has proved the natural origin of the Tertiary flint implements. But no, we cannot proceed in this fashion. One should really demand that our adversaries in this debate should fight us on experimental grounds. This reasonable request to solve by experimentation the puzzle of how the flint objects could be produced by the 'intelligent' action of natural forces is not weakened by the fact that Mayet was unsuccessful in producing anything resembling a flint implement by the process of banging blocks of stone together."

Klaatsch (1907, p. 766) then turned his attention to Mayet's statements about the other site at Aurillac, Puy Courny: "Regarding Puy Courny, Mayet cannot call attention to any geological disturbances such as were present at the other site. Instead he seeks, by heaping up questions, to lead one around the complete lack of reasonable arguments and evidence in favor of his point of view. He simply states with utter complacency in his 'conclusions' that the eoliths of Puy Courny 'are in all likelihood the products of the same natural forces.' The fact that countless fossils found in the same beds remain completely unchanged by these forces is here also not mentioned."

Klaatsch (1907, p. 766) then answered one of Mayet's specific objections: "The great number of specimens at Puy de Boudieu startled him. But in another publication I have pointed out the great masses of artifacts that are to be found at stone workshops in Tasmania. Were such sites to be covered by a stream of lava and then again exposed, this would present much the same sort of scene that confronts one at Puy de Boudieu." In Africa also, there are sites with thousands of stone tools scattered about. "On the whole," stated Klaatsch (1907, p. 766), "I must sadly conclude that the work of Mayet has not brought us one step closer to solving the eolith problem."

4.3.5 A Final Report

As late as 1924, George Grant MacCurdy, director of the American School of Prehistoric Research in Europe, reported in *Natural History* about the flint implements of Puy Courny (Cantal). Finds similar to those of Rames at Puy Courny and Verworn at Puy de Boudieu had been made in England by J. Reid Moir. Some critics argued that natural forces, such as movements of the earth, had fractured flints by pressure, thus creating stone objects resembling tools. But scientists showed that in the particular locations where the flint tools were found, the geological evidence did not suggest the operation of such natural causes.

MacCurdy (1924b, p. 658) wrote: "Breuil is authority for the statement that conditions favoring the play of natural forces do not exist in certain Pliocene deposits of East Anglia, where J. Reid Moir has found worked flints. . . . Can the same be said of the chipped flints from Upper Miocene deposits near Aurillac (Cantal)? Sollas and Capitan have both recently answered in the affirmative. Capitan finds not only flint chips that suggest utilization but true types of instruments which would be considered as characteristic of certain Palaeolithic horizons. These not only occur but reoccur: punches, bulbed flakes, carefully retouched to form points and scrapers of the Mousterian type, disks with borders retouched in a regular manner, scratchers of various forms, and, finally, picks. He concludes that there is a complete similitude between many of the chipped flints from Cantal and the classic specimens from the best-known Palaeolithic sites." William Sollas held the chair of geology at Oxford, and Louis Capitan, a highly respected French anthropologist, was professor at the College of France.

4.4 DISCOVERIES BY A. RUTOT
IN BELGIUM (OLIGOCENE)

From France, let us now proceed to Belgium, where A. Rutot, conservator of the Royal Museum of Natural History in Brussels, made a series of discoveries that brought the question of anomalous stone tool industries into new prominence during the early twentieth century. Most of the industries identified by Rutot dated to the Early Pleistocene. The oldest of his Pleistocene industries, the Reutelian, was named after the small village of Reutel, east of Ypres. Then came the Mafflian and Mesvinian, named after the villages of Maffle and Mesvin. Last in the series was the more highly developed Strepyan industry, named after the town of Strépy. Rutot regarded the Strepyan as marking the transition to the true Paleolithic industries of the later Pleistocene (Obermaier 1924, p. 8).

But in 1907, Rutot's ongoing research resulted in much more startling finds, this time in the Oligocene, from 25 to 38 million years ago. Georg Schweinfurth gave an initial report in the *Zeitschrift für Ethnologie,* using the term eolith in its broadest sense to describe the new finds. But on the basis of Rutot's later published descriptions, we have classified the tools as crude paleoliths.

Schweinfurth (1907, pp. 958–959) stated: "The continuing search for eoliths in the high plateau of the Ardennes led to this discovery. . . . As Rutot searched a sand pit near Boncelles, 8 kilometers [5 miles] south from Lüttich, he found an eolith-bearing stone bed under the sands at a depth of 15 meters [49 feet]. The sand is generally regarded as Oligocene, but there were no fossils in it, and therefore the age of the bed is not certain. But in the course of further research Dr. Rutot found in another sand pit a well-developed marine fauna of the Late Oligocene, and at the bottom of this sand there was also a stone bed containing eoliths. Among them were choppers, anvil stones, knives, scrapers, borers, and throwing stones, all displaying clear signs of intentional work that produced forms exquisitely adapted for use by the human hand. Rutot has now brought together a complete series of these artifacts and is preparing for publication a comprehensive report, with illustrations, for the bulletin of the Geological Society of Belgium. On September 30, the fortunate discoverer had the pleasure to show the sites to 34 Belgian geologists and students of prehistory. They all agreed that there could be no doubt about the position of the finds."

Schweinfurth (1907, p. 959) then reproduced this preliminary statement by Rutot about the geology of the Boncelles region: "On the plateau (between the Maas and Ourthe rivers) the primary stone was covered with flint-bearing chalk, and during the Eocene period the chalk was eroded away, leaving behind heaps of flint that later formed the flint beds. At the beginning of the Late Oligocene a marine intrusion covered the flint beds, depositing 15 meters [49 feet] of fossil-bearing sands over them. Finally, during the Middle Pliocene, streams deposited an additional 3 meters [10 feet] of white quartz gravel (a formation now called the Kieselöolithe) along with beds of sand and clay. Then began the excavation of the present valleys." Rutot believed that human beings manufactured the Boncelles eoliths before the Oligocene marine intrusion, when the land surface was a flint-heaped lowland bordering the sea.

Rutot's complete report on the Boncelles finds appeared in the bulletin of the Belgian Society for Geology, Paleontology, and Hydrology and provided extensive verification of the preliminary reports cited above. Rutot (1907, p. 479) also supplied information that stone tools like those of Boncelles had been found in Oligocene contexts at Baraque Michel and the cavern at Bay Bonnet. At Rosart, on the left bank of the Meuse, stone tools had also been found in a Middle Pliocene context, thus making them as old as the eoliths of the Kent Plateau.

In his report on Boncelles, Rutot (1907, p. 442) stated that the initial discovery of implements had been made by E. de Munck, in a sand pit situated alongside the main roadway from Tilff to Boncelles, about 500 meters (1640 feet) from a crossroad at the place called "Les Gonhir." In the very bottom of the sand pit, workmen had excavated a hole about half a meter (a foot and a half) deep in order to extract flint to be used as gravel for roadbeds. This enabled de Munck to gather from the matrix of clayey yellow sand many flint flakes showing signs of fine retouching and utilization (Rutot 1907, p. 442). "It was these implements, including a scraper with a clear bulb of percussion and nicely retouched sharp edge, which convinced me that at the place pointed out by de Munck there existed a deposit of Tertiary eoliths that deserved to be explored and studied," said Rutot (1907, pp. 442–443). A bulb of percussion indicates the scraper was intentionally flaked from a flint core for the purpose of tool manufacture, which, according to our conventions, places such an implement in the category of the crude paleoliths, rather than the eoliths.

Rutot and de Munck worked together at Boncelles, enlarging and deepening the original excavation. The flint bed was about 1 meter (3 feet) thick and rested on a Devonian sandstone base, surmounted by 15 meters (49 feet) of Oligocene marine sands and clays (Rutot 1907, p. 443). Rutot and de Munck recovered over a hundred specimens, which Rutot (1907, p. 444) said represented "numerous examples of all the various Eolithic types, that is to say *percuteurs* (choppers), *enclumes* (anvils), *couteaux* (cutters), *racloirs* (side scrapers), *grattoirs* (end scrapers), and *percoirs* (awls)." Rutot (1907, p. 444) stated: "These tools display, in all their detailed features, the same characteristics as other well-known and authenticated Tertiary and Quaternary Eolithic industries." Rutot called the industry the Fagnian, after the name of the region, Hautes-Fagnes.

Another pit 500 meters (1640 feet) to the northwest of the first also yielded tools. Furthermore, this site provided confirmation of the Oligocene dating of the flint bed bearing the tools. Whereas the first site did not furnish any fossils, the layers of sediment above the flint bed at the second site contained many shell imprints. About a dozen species were recognized (Rutot 1907, p. 444). It was obvious that the shells represented a typical Oligocene assemblage. The most common species was *Cytherea beyrichi*. Rutot (1907, p. 447) stated: "This shell is characteristic of the Late Oligocene of Germany, notably the beds at Sternberg, Bünde, and Kassel. . . . The other recognizable species (*Cytherea incrassata, Petunculus obovatus, P. philippi, Cardium cingulatum, Isocardia subtransversa, Glycimeris augusta*, etc.) are all found in the Late Oligocene."

Rutot (1907, p. 448) concluded: "Therefore, the Eolithic industry found in the flint bed at the base of the Late Oligocene sands is at least Middle Oligocene in age." The Oligocene ranges from 25 million years ago to 38 million years ago. Rutot's interpretation of the stratigraphy at Boncelles is upheld by other authorities. Maurice Leriche (1922, p. 10) and Charles Pomerol (1982, p. 114) both characterize the sands of Boncelles as Chattian, or Late Oligocene.

"We are thus confronted with a grave problem, or rather a fact the importance of which one cannot escape," wrote Rutot (1907, p. 448). Referring to the controversies regarding the discoveries of some of the tools we discussed earlier, Rutot (1907, p. 448) observed: "In fact, it is not without a certain repugnance that some have been obliged to accept, in recent times, the idea of the existence of intelligent beings who made and used tools in the Late Miocene. And it is almost with a sense of relief that some have been able to decrease the importance once accorded to the site at Thenay, reported as Aquitanian [Early Miocene]."

"But now it appears," said Rutot (1907, p. 448), "that the notion of the existence of humanity in the Oligocene, at a time more ancient than that represented by Thenay, has been affirmed with such force and precision that one cannot detect the slightest fault. This is something that offends our old ideas, which have barely become habituated to the simple conception of humans in the Quaternary. But little by little the reality of Pliocene man of the Kent Plateau has been affirmed and accepted, which has in turn permitted the introduction of the idea of humanity in the Late Miocene, contemporary with *Mastodon, Hipparion,* and *Dryopithecus.*" The Late Miocene discoveries are probably those of Ribeiro in Portugal and of Tardy and others at Aurillac, in France.

"Of course," added Rutot (1907, p. 448), "passing abruptly from the Late Miocene to the Middle Oligocene may seem somewhat improbable; nevertheless it is proper to submit to the inevitable and accept the facts as they are, seeing that they are not susceptible to any different explanation."

"Moreover," continued Rutot (1907, pp. 448–449), "hesitation is no longer possible after the discovery of an industry fashioned by recently living Tasmanians, which has been brought to our attention through the research conducted by Dr. F. Noetling. The bringing to light of this industry is, as it were, providential, because it demonstrates quite positively that eoliths are a reality. The discovery shows that scarcely sixty years ago human beings were making and using implements that are, according to competent and impartial observers, absolutely of Eolithic type." Perhaps the Tasmanians would still have been making such implements during Rutot's time had they not been exterminated by European settlers in the middle of the nineteenth century.

Rutot then described in detail the various types of tools from the Oligocene of Boncelles, beginning with *percuteurs* (or choppers). "Concerning choppers," said Rutot (1907, pp. 451–452), "there exist almost always several distinct types, which are: plain choppers, sharpened choppers, pointed choppers, small choppers, and retouchers. Almost all of these are found at Boncelles. The plain chopper [Figure 4.22] is a pebble or block of stone that has been used to strike blows. Such choppers may or may

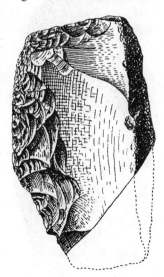

Figure 4.22. Plain chopper (*percuteur simple*) from below the Late Oligocene sands at Boncelles, Belgium (Rutot 1907, p. 452).

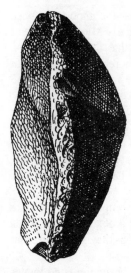

Figure 4.23. Sharpened chopper (*percuteur tranchant*). Rutot (1907, p. 452) noted use marks on the working edge.

not have retouching to facilitate gripping. These are rare at Boncelles, and the ones collected do not appear to have been used much. One notes on their surfaces relatively faint traces of the special and characteristic marks of percussion."

The sharpened chopper (Figure 4.23) was the most abundant type. The simple chopper described above could have been used as a hammer stone to strike flakes from blocks of flint, and these flakes could then have been fashioned into sharpened choppers. But at Boncelles, according to Rutot, many natural flakes, were scattered over the land surface, so it was not necessary to produce them artificially. After some retouching to enable them to be comfortably gripped in the hand, they could immediately be put to use. In contrast to the plain chopper, the sharpened chopper is fit for varieties of practical work (Rutot 1907, pp. 452–453).

"The sharpened choppers collected at Boncelles," wrote Rutot (1907, pp. 452–453), "are as fine and characteristic as possible. Clearly evident is the fact that most of the flaking from usage is angled to the left,

Figure 4.24. Small sharpened chopper (*tranchet*) from below the Late Oligocene sands at Boncelles, Belgium (Rutot 1907, p. 453). The sides show retouching to accommodate gripping by the hand, while the lower edge, said Rutot, shows use marks.

Figure 4.25. Pointed chopper (*percuteur pointu*) also from Boncelles, Belgium. Rutot (1907, p. 454) said it shows signs of use on both ends.

as always happens when an implement is gripped in the right hand. The opposite occurs when it is employed with the left hand."

The *tranchet* (Figure 4.24), according to Rutot, was a smaller version of the sharpened chopper. "The *tranchet,*" said Rutot (1907, pp. 453–454), "was certainly used for percussion, and the scratch marks of utilization on the edges are the same as those produced on the large sharpened choppers, though of much smaller size. It appears the *tranchet* rendered service analogous to that of a hatchet. This instrument is not rare at Boncelles, and we give an illustration of one. One notes on the vertical edges deliberate retouching, in the form of removal of sharp edges, for easy gripping, and on the lower horizontal edge one notes the irregular marks of utilization."

Rutot (1907, p. 454) noted: "The Oligocene of Boncelles also has pointed choppers [Figure 4.25], that is to say, elongated pieces of flint with one or two of the ends having been used to strike blows. They display on the utilized ends a characteristic star-shaped pattern of flaking, which one can see very well."

The final type of *percuteur* described by Rutot was the retoucher, which, as the name implies, is a small percussion implement used in the retouching of the edges of stone tools. He illustrated a retoucher (Figure 4.26) with very evident signs of use along the working edge (Rutot 1907, p. 454). Also found at the Boncelles sites were several anvil stones (Figure 4.27) characterized by a large flat surface showing definite signs of percussion (Rutot 1907, pp. 455–456).

Rutot then described some implements he called *couteaux,* best translated as cutters. "One can see that *couteaux* are made from relatively long flakes of flint,

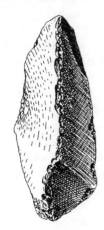

blunt on one side and sharp on the other. The blunt side generally retains the flint's cortex. Prolonged usage of the blade turns the rectilinear edge into a sawlike edge, with small irregular teeth. This is caused by chipping of the edge when the blade is pressed against the irregularities of the surface of the object being cut. The cutters were not retouched.

Figure 4.26. An Oligocene retoucher (*retouchoir*), with percussion marks on working edge (Rutot 1907, p. 454).

Figure 4.27. An Oligocene anvil (*enclume*) from the Boncelles, Belgium site showing signs of percussion around the circumference of the flat surface (Rutot 1907, p. 455).

Figure 4.28. Two views of a cutting implement (*couteau*) from below the Late Oligocene sands at Boncelles, Belgium (Rutot 1907, p. 456). The working edge shows use marks characteristic of cutting

They were used for a long time, until blunted by usage and polishing. It was rare that they were employed until completely unusable. At Boncelles one finds cutters [Figure 4.28] of a very characteristic type." (Rutot 1907, p. 456).

Rutot then described the *racloir,* or side scraper. The *racloir* was ordinarily made from an oval flake, produced either naturally or by deliberate flaking, with one of the longitudinal edges blunt and the opposite edge sharp (Figure 4.29). After retouching for a suitable grip, the blunt edge was held in the palm of the hand, and the sharp edge of the implement was moved along the length of the object to be scraped. During this operation, series of small splinters were detached from the cutting edge of the implement, thus dulling it. Rutot (1907, p. 458) stated: "The characteristic feature of the *racloir,* used as such, is the presence along the working edge of a series of small chip marks, all arranged in the same direction and located on the same side. When the implement became unusable, it was possible to restore its edge with the retoucher stone, allowing it to be further used."

"The special purpose of the retoucher," said Rutot (1907, p. 458), "was the striking upon a implement's working edge of a series of small regular blows in the same direction, detaching flakes from 2 millimeters to 5 millimeters [about 0.1 to 0.2 inch] in diameter. The juxtaposition of the flake scars restored the implement's sharp edge." According to Rutot, this type of retouching is, without a doubt, clearly distinguishable from the retouching performed for accommodation of the hand. Rutot (1907, p. 458) stated: "Retouching for accommodation of the hand involved hammering and blunting various

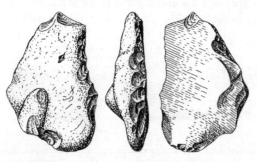

Figure 4.29. Three views of a side scraper (*racloir*) found below the Late Oligocene sands at Boncelles, Belgium (Rutot 1907, p. 458).

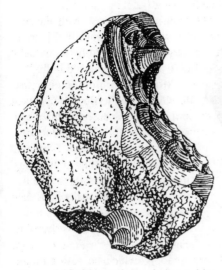

Figure 4.30. This tool was designated by Rutot as a notched side scraper (*racloir à encoche*). Scrapers of this type are commonly found in Late Pleistocene assemblages. This tool was recovered from below the Late Oligocene sands at Boncelles, Belgium (Rutot 1907, p. 458).

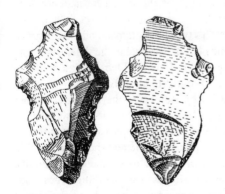

Figure 4.31. A double scraper (*racloir double*) from below the Late Oligocene sands at Boncelles, Belgium. Retouching of the two notches in the middle allowed it to be comfortably gripped. Marks of utilization are visible at the top and bottom (Rutot 1907, p. 459).

sharp edges that were either harmful or not usable. But retouching for sharpening was performed to resharpen, by repeated blows in a single direction, an edge dulled by use. One is therefore able to recognize the two types of retouching."

Rutot (1907, p. 458) pointed out that a good piece of flint can be resharpened several times. But he added "the accumulation of retouching rapidly broadens the original sharp angle of the edge, and when the angle surpasses 45 degrees, the edge offers such resistance that no retouching can be executed, and the implement, now irreparable, is discarded."

Rutot (1907, p. 459) then described another type of *racloir* discovered at the Boncelles sites: "Frequently the working edge is not straight; it is finished by means of retouching into one or more concave notches, probably for the purpose of scraping long round objects. This is the notched *racloir* [Figure 4.30]. Some are made from natural flakes, others from flakes derived from deliberate percussion."

At Boncelles, *racloirs* with two scraping edges, or double *racloirs,* were also found. About this type of implement Rutot (1907, pp. 459–460) said: "I have provided an illustration of an interesting example [Figure 4.31]. It could be held in the hand, between the thumb and forefinger, at the points nicely indicated by the two lateral notches; the other double *racloirs,* in the form of pointed flakes with two sharp edges, resemble the true 'Mousterian points.' Mostly they

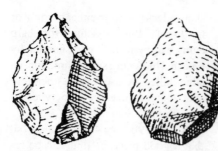

Figure 4.32. This implement was taken from below the Late Oligocene sands at Boncelles, Belgium (Rutot 1907, p. 460). Rutot said it resembled a Mousterian point from the Late Pleistocene of Europe. The implement's ventral surface (right) shows a bulb of percussion.

look, as is the case with the one shown in . . . [Figure 4.32], as if they were detached by percussion and show a pronounced bulb."

Mousterian implements are found in Late Pleistocene contexts of Europe. It is the resemblance of some of the flint implements discovered at the Boncelles, Belgium, site to Late Pleistocene implements that causes us to classify this industry among the crude paleoliths. Another specimen looking very much like a Mousterian point is shown in Figure 4.33.

Rutot also described a special category of tools, which he called mixed implements (Figure 4.34), because they looked as if they could have been employed in more than one fashion. Rutot (1907, p. 460) stated: "They tend to have on the sharp edge a point formed by the intersection of two straight edges, or more frequently, two notches, made by retouching. These implements might be said to resemble awls, but in general the point is too short or rounded. In fact, although the notches are the result of deliberate flaking and retouching, the point seems to be merely the incidental byproduct of the intersection of the two notches."

Rutot (1907, p. 460) went on to say: "This type of implement, of quite singular form, is quite abundant in the old Eolithic period, very rare in the Paleolithic, and again quite abundant in the Neolithic, particularly in the Flensian assemblages. Good examples also appear among the tools of the modern Tasmanians."

The next type of implement discussed by Rutot was the *grattoir,* another broad category of scraper. According to Rutot (1907, p. 462), the *grattoir* differed from the *racloir* in that "its working edge is employed longitudinally in relation to the direction of the force of application, whereas the *racloir* is held between the thumb and forefinger in such a manner to set the working edge transverse to the direction of the force. When being used, the working edges of the *racloir* and the *grattoir* are thus situated perpendicular to each other." Rutot observed that in order to help the user direct and push the cutting edge of the *grattoir,* these implements in many cases had special notches to accommodate the thumb and forefinger (Figure 4.35b), this in addition to the usual removal of sharp edges to facilitate gripping. At the Boncelles site in Belgium, from strata dated to the Oligocene, there were unearthed a variety of *grattoirs* (Figure 4.35), including the especially large specimen shown in Figure 4.36.

Figure 4.33. A *racloir* from below the Late Oligocene sands at Boncelles, Belgium. Rutot (1907, p. 460) observed it looked very much like a Mousterian point from the Late Pleistocene of Europe.

Figure 4.34. This pointed flint implement was discovered in a stratigraphic position below the Late Oligocene sands at Boncelles, Belgium (Rutot 1907, p. 461). The ventral surface (right) of this tool shows a well-developed bulb of percussion with an eraillure. According to Rutot, this type of implement is common in Neolithic and modern assemblages.

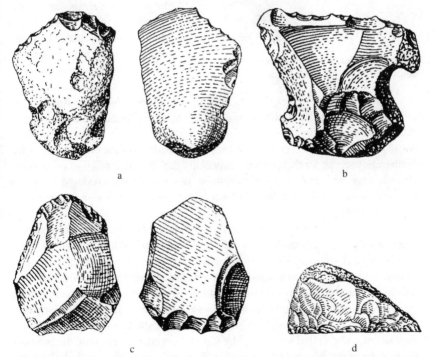

a

b

c

d

Figure 4.35. End scrapers (*grattoirs*) from below the Late Oligocene sands at Boncelles, Belgium: (a) two views of a *grattoir,* the ventral surface of which (right) shows a bulb of percussion; (b) *grattoir* with curved indentations for gripping; (c) two views of a double *grattoir,* with the chipping on each of the two working edges confined to one side of the flake; (d) *grattoir* with finely retouched working edge (Rutot 1907, pp. 462–464).

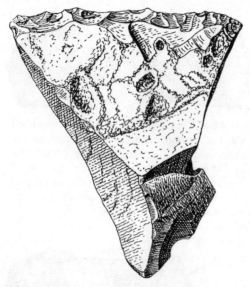

Figure 4.36. Large end scraper (*grattoir*) from below the Late Oligocene sands at Boncelles, Belgium (Rutot 1907, p. 463).

Rutot (1907, pp. 462, 464) noted: "In the case of *grattoirs* as well as *racloirs,* there are some that bear very well marked bulbs of percussion. I do not, however, consider these flakes to have been intentionally made for use as implements. I believe that the flakes with the bulb of percussion were detached involuntarily from the edges of anvils while they were being struck by hammer stones. These detached flakes were usable as tools just as were the sharp natural flakes found nearby. And they were in fact used like them, but they were not deliberately struck for this purpose."

It is difficult, however, to comprehend how Rutot could tell what was going through the minds of his ancient toolmakers as they struck flakes and worked them into implements. Specifically, we wonder how Rutot could say with such certainty that the flakes made into implements were not deliberately struck for that purpose, especially the ones with bulbs of percussion. Here it may be recalled that the bulb of percussion is considered by many authorities, such as Leland W. Patterson (1983), to be a clear sign of intentional controlled flaking.

Rutot was probably attempting to fit the evidence before him within his own framework of evolutionary ideas. He apparently wanted to characterize the makers of the Oligocene industry of Boncelles as more primitive than the makers of industries at Pleistocene sites. But leaving aside Rutot's evolutionary expectations, we can see no reason to rule out the possibility that some of the Boncelles specimens are tools intentionally made from flakes struck for specific purposes.

Rutot then described *perçoirs,* which might be called awls or borers. "These instruments, also called *poinçons,*" he stated, "are characterized by the presence of a sharp point, obtained by intentional modification of a natural flake that already has a somewhat pointed shape. This modified point is situated indifferently in regard to the axis of the instrument, sometimes in a position oblique to the axis" (Rutot 1907, p. 464). An instrument with an oblique point is shown in Figure 4.37, along with two awls with straight points.

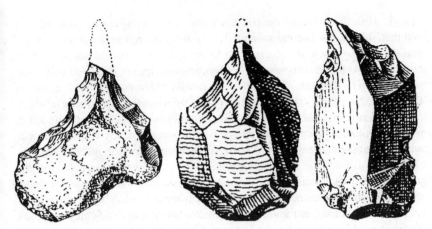

Figure 4.37. Three awls (*percoirs*) from below the Late Oligocene sands at Boncelles, Belgium (Rutot 1907, p. 465).

According to Rutot (1907, pp. 464–465), the Boncelles toolmakers had two ways of modifying a naturally pointed flake to make an awl: "Sometimes the chipping on the two edges making the point was done on just one side of the flake. But sometimes one edge was chipped on the flake's front side, and the other edge was chipped on the flake's back side. This procedure is convenient because it allows all the blows to be struck in the same position and the same direction. In effect, when the first edge is chipped, one flips the implement and chips in the same place on the other edge to make a point." Rutot showed a find with this kind of chipping, unlikely to have occurred by random natural battering (Figure 4.38). Rutot

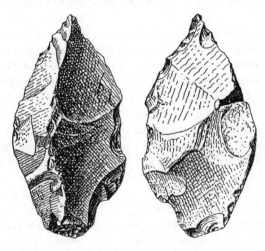

Figure 4.38. An awl discovered below the Late Oligocene sands at Boncelles, Belgium (Rutot 1907, p. 465). The chipping on one edge of the point is on the dorsal surface of the implement (left), while the chipping for the other edge of the point is on the ventral surface (right). According to Rutot, this pattern illustrates the use of a specific chipping technique that allowed the maker of the implement to chip one edge of the point, turn the implement over, and chip the other edge of the point from the same position and direction.

remarked that the point obtained by this method of chipping can easily be broken and that, in fact, most of the specimens of awls recovered at Boncelles do have broken points.

Rutot also noted the presence at Boncelles of objects that appeared to be *pierres de jet*—throwing stones or sling stones. "Throwing stones," observed Rutot (1907, p. 466), "are polyhedral pieces of stone that present an irregular combination of natural and artificial surfaces. They are somewhat rounded in shape and of small volume, appropriate for throwing violently with the hand or with a sling. Such a weapon would strike in such a manner as to produce not only shock from impact but also cutting from the rotation of the sharp edges of the projectile. The flint industry of Boncelles contains many such polyhedral stones that give every appearance of being throwing stones."

Rutot concluded that flint objects with certain characteristics may very well have been used by the ancient inhabitants of Boncelles to make fire. "Not only in Eolithic series, but in Paleolithic and Neolithic assemblages," stated Rutot (1907, p. 467), "one encounters pieces of flint which along one side bear traces of numerous and repeated violent blows, distributed in groups, each group presenting a series of blows arranged in the same direction. Furthermore, each distinct group has its traces of blows arranged in a direction different from that of the other groups." These marks could be interpreted as the result of attempts to strike sparks from the pieces of flint. In French, flints used to ignite fire are called *briquets*.

According to Rutot, these peculiarly marked stones might superficially resemble other tool types such as anvils, *racloirs,* or *grattoirs*. But he pointed out that "they are different from these in the violence and the irregularity of the blows inflicted upon them and also by the presence of the flint cortex on the surface marked by the blows, which eliminates any supposition that these are actual cutting implements" (Rutot 1907, p. 467). The working edges of implements are almost always free of cortex.

Regarding his hypothesis that the pieces of flint in question might have been used for making fire, Rutot (1907, p. 467) mentioned in a footnote: "The same idea has been nicely expressed by E. Lartet and Christy in *Reliquiae aquitanicae,* pages 85–86 and also pages 138–140. One sees that some Mousterian specimens are represented as *briquets* for making fire, and the very interesting explanation is given that the fire was obtained not only by friction of flint and pyrite but by flint against flint. A note calls attention to the fact that in England, in Norfolk and Suffolk, up until a century ago, people used the friction of two flints to obtain fire. Dried moss was used as the combustible substance while one rapidly moved two pieces of flint together."

All in all, Rutot believed the present-day implements that the objects in question most singularly resembled were *briquets,* flints used for making fire.

Rutot (1907, pp. 467–478) wrote: "One could respond that it is a bit rash to think that the primitive humans of Boncelles made fire; nevertheless, I have some reasons to think that they did have knowledge of the usage of fire, but the moment has not come to introduce them. In any case, the humans of Mesvin and Reutel did know how to make fire, and we encounter in the debris of their industries stones that look like *briquets*. At Boncelles, stones of the exact same type are found, and these also appear to have been used as *briquets*. We therefore believe it is useful to point out, with some reserve, and by means of comparison, that the stones with special signs of usage and flaking at Boncelles could in fact be either *briquets* or *pierres à feu* (fire stones)."

Rutot (1907, p. 468) then stated: "So we have now conducted our review of the variegated industry of the intelligent beings of the Oligocene, and we are justifiably astonished at their expertise, given the vast duration of time that has elapsed since they were present. On the other hand, when we examine the industry of the recent Tasmanians, which has been brought to light by the research of Dr. Noetling, then we are no less justifiably astonished to see its extraordinarily primitive and rudimentary character. So the truth, after direct comparison, is that the two industries are exactly the same [Figure 4.39] and that the Tasmanians, now annihilated, but still in existence just sixty years ago, were at the same level

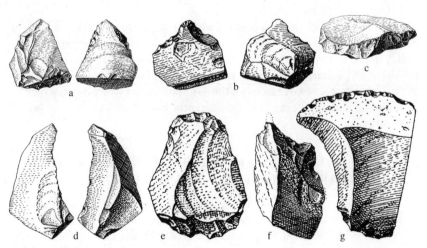

Figure 4.39. Implements manufactured by native Tasmanians in recent historical times (Rutot 1907, pp. 470–477). Rutot said they resembled almost exactly the tools from the Oligocene period at Boncelles, Belgium. (a) Side scraper (*racloir*), compare Figure 4.33. (b) Pointed implement (*perçoir*), compare Figure 4.34. (c) Anvil (*enclume*), compare Figure 4.27. (d) Stone knife (*couteau*), compare Figure 4.28. (e) Double end scraper (*grattoir double*), compare Figure 4.35c. (f) Awl (*perçoir*), compare Figure 4.37. (g) End scraper (*grattoir*), compare Figure 4.36.

of culture as the very primitive inhabitants of Boncelles and the Hautes Fagnes." Only the materials from which the Tasmanian tools were made were different—quartzite, diabase, granite, and similar types of rock rather than flint.

At some Tasmanian campsites, noted Rutot, Klaatsch found vast numbers of stone implements, attributing this accumulation to the long period of habitation. Rutot pointed out that some opponents of anomalously old early stone industries had traditionally used the very large numbers of specimens recovered at various sites as an argument against their being the product of human industry. Rutot (1907, p. 482) believed that Klaatsch's observations proved this objection invalid.

Rutot (1907, pp. 480–481) then clearly framed the essential question posed by his discoveries: "When we take into consideration the analogies, or rather the identities, between the Oligocene eoliths of Boncelles and the modern eoliths of the Tasmanians, we find ourselves confronted with a grave problem—the existence in the Oligocene of beings intelligent enough to manufacture and use definite and variegated types of implements. Who was the intelligent being? Was it merely a precursor of the human kind, or was it already human? This is a grave problem—an idea that cannot but astonish us and attract the attention and the interest of all those who make the science of humanity the object of their study and meditation" (Rutot 1907, pp. 480–481).

It might be a shock to many persons with scientific training that a statement like this could have appeared in a scientific journal in the twentieth century. Today mainstream scientists do not give any consideration at all to the possibility of a human—or even protohuman—presence in the Oligocene. We believe there are two reasons for this—unfamiliarity with evidence such as Rutot's and unquestioning faith in currently held views on human origin and antiquity.

4.5 DISCOVERIES BY FREUDENBERG NEAR ANTWERP (EARLY PLIOCENE TO LATE MIOCENE)

In addition to being the site of Rutot's finds in Oligocene strata, Belgium was also the site of another intriguing series of discoveries. In February and March of 1918, Wilhelm Freudenberg, a geologist attached to the German army, was conducting test borings for military purposes in Tertiary formations west of Antwerp. In clay pits at Hol, near St. Gillis, and at other locations, Freudenberg discovered flint objects he believed to be implements, along with cut bones and shells.

Most of the objects came from sedimentary deposits of the Scaldisian marine stage, which Freudenberg (1919, p. 2) regarded as Middle Pliocene. But according to modern authorities, the Scaldisian spans the Early Pliocene and Late

Miocene (Klein 1973, table 6; Savage and Russell 1983, p. 294). The Scaldisian is thus dated at 4–7 million years (Klein 1973, table 6). Freudenberg (1919, p. 9) suggested that the objects he discovered may have dated to the period just before the Scaldisian marine transgression, which, if true, would give them an age of 7 million years or more.

4.5.1 Flint Implements

Freudenberg believed some of the flint implements he found had been used to open shells. One such implement (Figure 4.40) came from a cavity in the top part of the Scaldisian formation at Koefering, where it was found along with broken shells (Freudenberg 1919, p. 18).

In describing a second shell-opening tool (Figure 4.41), Freudenberg (1919, p. 20) stated: "It comes from the Scaldisian sands of Mosselbank and was found

together with many Pliocene molluscs in excavations for fortifications on the outskirts of Antwerp. It is a typical hook-shaped shell opener, found among broken Pliocene shells, especially the broken shells of *Cyprina tumida.* The shell heap appears to have been a Tertiary kitchen midden. The length of the shell opener is 9 centimeters [3.5 inches], when one includes the missing end section." In addition, Freudenberg uncovered some burned flints, which he considered to be evidence that intelligent beings had used fire during the Tertiary in Belgium.

Figure 4.40. This object, characterized by W. Freudenberg (1919, p. 16) as an implement for opening shells, was discovered in a Scaldisian formation (4–7 million years B.P.) at Koefering, near Antwerp, Belgium. The left end of the specimen object appears to be the working edge.

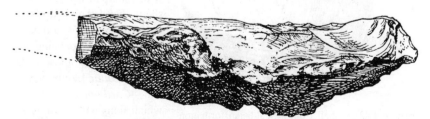

Figure 4.41. A implement for opening shells, from a Scaldisian formation at Mosselbank, near Antwerp, Belgium (Freudenberg 1919, p. 16). Along with the implement, which could be from Early Pliocene to Late Miocene in age, many broken shells were found.

4.5.2 Cut Shells

Of special interest were the numerous shells collected by Freudenberg from the Scaldisian sands at Vracene and Mosselbank, where fortifications were being constructed. About his discoveries, Freudenberg (1919, p. 39) wrote: "The shell heaps of Koefering and Mosselbank near Vracene have yielded countless examples of *Cyprina islandica* and *Cyprina tumida* broken while living and also a shell opener of shiny, patinated flint, like those found at Hol."

Freudenberg (1919, p. 39) further stated: "The examination of the shell materials from Vracene and Hol that I undertook in the beginning of 1919 at Göttingen proved the correctness of my initial judgement that the shell beds were a kitchen midden. In cleaning off the yellow quartz sand and clay, I found many intentional incisions, mostly on the rear part of the shells, quite near the hinge [Figure 4.42]. This was particularly clear on the two *Cyprina* species. On the extinct *Cyprina tumida* specimens, the forward closing muscle was cut through quite regularly by an incision. . . .The incision could only have been made with the help of a sharp flint knife or a shark tooth (we find here teeth of *Oxyrhina hastalis* Ag.). The intentional nature of this action is quite apparent. I have 7 left half-shells of *Cyprina tumida* and 9 right half-shells with the same kind of incision near the depression in the shell that marks the point of attachment of the forward closing muscle."

Describing the incisions themselves, Freudenberg (1919, pp. 39–40) wrote: "The inner surfaces of the cuts on the shells of *Cyprina tumida* are smooth and bear the same yellow-white weathered surface as the other old surfaces and breaks on any part of the shell. The length of the cut marks is a few millimeters, seldom more than half a centimeter. The incisions on the shells of *Cyprina tumida* with well-preserved cut marks are sharply V-shaped, such as could only have been made with a sharp instrument. Other shells that are almost always found broken, as would be expected if they were being used for food, include those of the extinct *Voluta* Lamberti Sow. and *Cardium decorativum*, which along with *Cardium edule* and *C. echinatum*, could have served as edible shell fish." The sharp cut marks

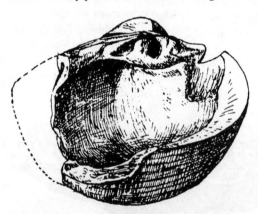

Figure 4.42. A shell from a Scaldisian formation (Early Pliocene–Late Miocene) near Antwerp, Belgium, with a cut mark to the right of the hinge (Freudenberg 1919, p. 33).

found near the hinges of the shells collected by Freudenberg would appear to be more consistent with human work than the action of shellfish-consuming creatures such as otters.

Freudenberg also found many oysters with broken and cut shells. Of *Ostrea edulis* L. var. *ungulata* Nyst, Freudenberg (1919, p. 45) wrote: "I dug up 20 flat right half-shells and about half as many arched left half-shells. Many shells show puncture marks made by sharp, pointed objects, perhaps shark teeth used as tools. From the position of these marks on the edges of the shells, it is obvious they were intended to force them open. The marks sometimes repeat themselves in the same place, giving the impression of premeditated work. The marks are always found on the flat half-shells rather than the curved half-shells, which would be harder to pierce. Splintering is found only on the inner surface of the puncture marks, from which one can conclude that the sharp body that made them entered from outside. All this rules out a posthumous injury, because a dead shellfish opens its shell, and in that case any kind of shell-opening operation would have been pointless."

Summarizing his report on shellfish, Freudenberg (1919, p. 50) said: "The number of extinct species is half the total, 27 of 54. Thus the late Tertiary date of the site is not in doubt. The existence of a shellfish-eating population on the Flemish coast in the late Tertiary is also not in doubt."

4.5.3 Incised Bones

In addition to cut shells, Freudenberg also found cut bones of marine mammals. Among them was part of the upper jaw of a member of the porpoise family, probably related to *Lagenocetus latifrons* Gray. The surface of the jaw is flat and bears upon it a series of incisions. Freudenberg believed the incisions had been purposefully made. In a taphonomic analysis of the jaw, he stated: "Were these grooves not seen as intentional work, but rather as the selective corrosion of the bone through chemical or mechanical means (such as the dissolving action of mineral salts or the friction of sand), then one would expect that the grooves would reach as far down as the nourishment channels (Haversian canals) that run through the bone and there find their end. In reality the grooves cut straight through the nourishment channels; they are also independent of the fine bone structure" (Freudenberg 1919, p. 22). Freudenberg said that this jaw may have been used as some kind of a press.

As discussed in Section 2.11, incisions such as those reported by Freudenberg on the porpoise bone might have been the result of shark bites. In his report, Freudenberg did not mention this possibility, which thus needs further investigation. Still, if the shells from the Scaldisian at Vracene and Mosselbank are taken as bearing intentional cut marks, then this strengthens the possibility that the incisions on the porpoise bones may also have been made by tools.

Marked and polished whale bones were also discovered, along with bones of other marine mammals. Freudenberg (1919, p. 28) wrote: "Artificially broken long bones of walruses and seals are found directly on top of the Septarian clay (Middle Oligocene). These bone fragments were found embedded in clayey greensand, some of which has hardened into limonite on the bones. The bones bear the deep impact marks of blows that could have been made by stone hammers. The depth of the marks varied with the strength of the blows."

4.5.4 Possible Human Footprints

Further confirmation of a human presence came in the form of partial footprints, apparently made when humanlike feet compressed pieces of clay. From a clay pit at Hol, located just south of the road leading westward from St. Gillis to Meuleken, Freudenberg (1919, p. 3) recovered one impression of the ball of a foot and four impressions of toes (Figure 4.43).

The stone bed in which the footprints were found was judged to be Scaldisian on the basis of the shell fauna. The Scaldisian sediments, as previously mentioned, were deposited in the time period from the Early Pliocene through the Late Miocene. The footprints would thus be at least 4–7 million years old. Freudenberg (1919, p. 9), however, believed they were probably made during the period immediately preceding the Scaldisian marine transgression, and were later incorporated into the Scaldisian formation in which were found. This would make the footprints somewhat more than 7 million years old. Freudenberg conducted a dermatoglyphic analysis of the prints, as carefully as modern physical anthropologists.

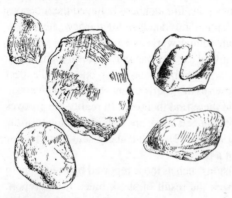

Figure 4.43. Five partial foot impressions from a Scaldisian formation (Early Pliocene to Late Miocene) at Hol, near Antwerp, Belgium (Freudenberg 1919, p. 9).

About the impression of the ball of a right foot, Freudenberg (1919, p. 11) stated: "There are on the left side signs of displaced grains of sand as well as imprints of the dermal ridges and lines of the skin of the foot, and these show a movement from left to right, or from inside to outside, as would result from the normal movement of the foot in walking."

Continuing his dermatoglyphic analysis of the print, Freudenberg (1919, p. 11) stated: "One notices that the right, or outer, side of the impression of the ball of the foot is also covered with impressions of fine lines like those of the foot of a humanlike being." According to Freudenberg (1919, p. 11), the pattern of the lines matched that of modern humans and was distinct from that of apes.

Freudenberg (1919, p. 12) added: "The number of lines found in the space of a millimeter is the same on the fossil impression as on the ball of the foot of a modern human adult. The fossil has about 2 lines in 1 mm (10 in 5 mm), in some places fewer. In adult humans, I have measured 4 lines in 2 mm, 5 in 2 mm, and 6 in 2 mm, giving an average of 2–3 lines per mm." Freudenberg (1919, p. 12) then mentioned another significant feature of the impression: "The outlets of the sweat glands are perhaps to be recognized on the fossil impression as tiny bumps, arranged in rows."

Freudenberg, having described the impression of the ball of the foot, then turned to the impressions of toes. Concerning an imprint of the fourth and fifth toes of a left foot, Freudenberg (1919, pp. 13–14) noted: "The length of an impression of the little toe, measured on the inner side, is, for a 4-year-old child, about 18 mm. The same measurement on the fossil impression is 15 mm. . . . There are also to be observed the impressions of dermal ridges on the imprints of the toes. They are arranged in the same pattern as on the foot of a human child, in that they radiate in all directions from the juncture of the fifth and fourth toes. As in the case of the human child, there are 6–7 dermal ridges per 2 mm at this place. Furthermore, there are properly oriented wrinkles of the skin."

Freudenberg (1919, p. 14) then stated: "The most important discovery on the fossil toe impression is the shortness of the fifth toe, which is reminiscent of the little toe of the human being. The anthropoid apes, including the gorilla, have long little toes. The foot structure of the genus *Homo* was in the Middle Pliocene already the same as today. The big toe was also short and broad, relative to that of apes, as shown on a somewhat fragmentary impression from Hol, which appears to be that of a left big toe."

4.5.5 The Identity of Freudenberg's Palaeanthropus

In his conclusion, Freudenberg (1919, p. 52) stated: "It stands without doubt that the sites at Hol and Koefering are part of the Scaldisian formation of the Middle Pliocene [Early Pliocene to Late Miocene according to modern authorities]. The geological age of *Palaeanthropus,* the Flemish Tertiary man, dates back this far, if not into older times. This conclusion is especially supported by the fact that bones of Pliocene marine mammals provided Tertiary man with raw materials for his implements and Pliocene shellfish served as his food. Furthermore, the fossil footprints of a humanlike being are found among the Middle Pliocene beach pebbles of Hol."

Freudenberg (1919, pp. 52–53) directed the attention of his readers to supporting evidence from England—the carved shell discovered by Henry Stopes in the Pliocene Red Crag (Section 2.15), as well as the flint implements found in the same formation by J. Reid Moir (Section 3.3) and the cut bones reported by Fisher (Section 2.16). As we have demonstrated, there is abundant evidence, of all kinds, in favor of a human presence dating to the Pliocene and earlier. In this context, the discoveries of Freudenberg are not at all surprising.

We should point out, however, that Freudenberg (1919, p. 12) was an evolutionist and believed that his Tertiary man must have been a very small hominid, about 1 meter (3.28 feet) tall, displaying, in addition to its humanlike feet, a combination of apelike and human features. Altogether, Freudenberg's description of his Flemish Tertiary man seems reminiscent of Johanson's portrayal of *Australopithecus afarensis* (Lucy). Even if Freudenberg's hypothetical picture of a primitive hominid with humanlike feet is accepted, one would not, according to current paleoanthropological doctrine, expect to find any australopithecines in Belgium during the Late Miocene, at the onset of the Scaldisian, over 7 million years ago. The oldest australopithecines date back only about 4 million years in Africa.

But a late Scaldisian (Early Pliocene) date of 4 million years for a Flemish australopithecine would be within the range of possibility. It should be kept in mind that African mammals such as the hippopotamus ranged as far north as England during the Pliocene and the interglacials of the Pleistocene. Modern paleoanthropologists might, therefore, have good reason to give some serious consideration to Freudenberg's reports, but unfortunately, the knowledge filtration process has, over the course of this century, resulted in the reports disappearing from view.

Thus far we have been going along with Freudenberg's assumption that the humanlike partial footprints from the Scaldisian of Belgium were made by a small primitive hominid. But there is another possibility. There are today, in Africa and the Phillipines, pygmy tribes, with adult males standing less than five feet tall and females even shorter. The proposal that a pygmy human being rather than an australopithecine made the footprints found by Freudenberg has the advantage of being consistent with the whole spectrum of evidence—stone tools, incised bones, isolated signs of fire, and artificially opened shells. Australopithecines are not known to have manufactured stone tools or used fire. And, as we shall see in Section 11.10, the toes of australopithecines are noticeably longer than those of modern *Homo sapiens,* while the little toe of the Belgian hominid is similar in length to that of modern humans.

Freudenberg's principal reason for concluding that the being that left the footprints was quite small had to do with certain measurements he made. He ascertained that the radius of curvature of the imprint of the ball of a foot excavated at Hol was similar to that of a human child 4 years of age (Freudenberg 1919, pp. 10–11). The radius of curvature is the radius of a circle that would fit a section of the curve of the print.

Another feature of the same imprint led Freudenberg to conclude that the creature, despite its short stature, was an adult. In the fossil imprint of the ball of a foot, he found 2 dermal ridges per millimeter. Human adults have 2–3 ridges per millimeter in this part of the foot whereas human children have about 4 ridges per millimeter. Freudenberg therefore believed that the creature must have been an adult, although the radius of curvature of the ball of the foot indicated it was only about 1 meter tall—the height of a human 4-year-old.

But other measurements reported by Freudenberg suggest the adult forms were taller. One of the toe impressions from the Scaldisian was about the same size as that of a human 4-year-old, indicating the creature stood about 1 meter high. On this impression Freudenberg (1919, p. 14) counted 3.0–3.5 ridges per millimeter. The toe impressions of human children have the same number of ridges in that location (Freudenberg 1919, p. 14). This suggests that the creature that made this print was not an adult but a child. Thus when it grew to adult size, it would have been somewhat taller than 1 meter.

To a modern reader, Freudenberg's reports are bound to seem somewhat idiosyncratic. Nevertheless, Freudenberg does provide yet another example of a professional scientist reporting in a scientific journal finds that today would not be given a moment's serious consideration.

4.6 CENTRAL ITALY (LATE PLIOCENE)

In 1871, Professor G. Ponzi (1873, p. 53) presented to the meeting in Bologna of the International Congress of Prehistoric Anthropology and Archeology the following report about evidence for Tertiary humans in central Italy: "The very ancient rocks of subappenine Italy that contain human vestiges are breccias which, reposing on the Pliocene yellow sands, can be referred to the end of the Pliocene or beginning of the Quaternary. These vestiges consist of one flint evidently worked into a triangular pointed shape, extracted from the breccia of the 'Acquatraversa sur la Voie Cassienne,' by the geologists de Verneuil and Mantovani, and several other flints, of almost the same type, collected by Rossi and Nicolucci in the breccia of Janicule." A breccia is a deposit composed of rock fragments in a fine-grained matrix of hardened sand or clay.

The Acquatraversan erosional phase, during which the breccia was laid down on the yellow sand, can still be regarded as Pliocene. Nilsson (1983, p. 95) stated that the Acquatraversan erosion "is taken to predate a volcanite with a radiometric date of 2.3 million years." This indicates that the stone tools embedded in the Acquatraversan breccias could be at least that old. The yellow sands, upon which the breccias are found, are most likely those of the Astian (Piacenzian) stage of the Late Pliocene (Nilsson 1983, p. 83). This case, although maybe not as strong as others just discussed, is nevertheless worthy of attention and study.

4.7 STONE TOOLS FROM BURMA (MIOCENE)

At the end of 1894 and beginning of 1895, scientific journals announced the discovery of worked flints in Tertiary formations in Burma, then part of the British Indian empire. The implements were reported by Fritz Noetling, a paleontologist and Fellow of the Geological Society who served as director of the Geological Survey of India in the region of Yenangyaung. Noetling (1894, p. 101) stated in the *Record of the Geological Survey of India:* "While engaged in mapping out a part of the Yenangyoung [Yenangyaung] oil-field my attention was particularly directed to the collecting of vertebrate remains, which are rather common in certain strata around Yenangyoung. One of the most conspicuous beds . . . is a ferruginous conglomerate, upwards of ten feet in thickness. This bed may be distinguished a long distance off as a dull-red band, running, in a continuous line, across ravines and hills. Besides numerous other vertebrate remains, such as *Rhinoceros perimense,* etc., one of the commonest species is *Hippotherium* [*Hipparion*] *antelopinum* Caut. and Falc. of which numerous isolated teeth can be found." Modern authorities still date the Yenangyaung fauna to the Late or Middle Miocene (Savage and Russell 1983, pp. 247, 326).

While picking up a lower molar of *Hipparion antelopinum,* Noetling noticed a rectangular flint object (Figure 4.44). He later described the object: "The two long edges run nearly parallel and are sharp and cutting. This flake affords particular interest in as much as the two faces must have been produced by an action, which is difficult to explain by natural causes" (Noetling 1894, p. 101).

Figure 4.44. Two sides of a flint implement from the Miocene Yenangyaung formation in Burma (Noetling 1894, plate 1).

Each face, one concave and the other convex, has two planes meeting in the middle to form an edge, giving the piece four plane surfaces. Noetling wrote: "Let us consider the convex face [on the left in Figure 4.44] first; it will be seen that one side is smooth, apparently produced by the chipping off of a single flake, while the other side shows that at least four smaller flakes have been chipped off at a right angle to the first one." Many authorities see chipping at right angles as a good sign of human work, as natural random battering tends to produce chipping at a variety of angles. In addition, such random battering also removes sharp edges. Noetling (1894, pp. 101–102) further stated: "The concave face which is however much damaged at one side must have been produced by the chipping off of two longitudinal flakes. The shape of this specimen reminds me very much of the chipped flint described in Volume I of the Records, Geological Survey of India, and discovered in the Pleistocene of the Nerbudda river, the artificial origin of which nobody seems to have ever doubted."

Noetling (1894, p. 101) searched further and found about a dozen more chipped pieces of flint. Some of these he categorized as "irregularly shaped." According to Noetling (1894, p. 101), the edges were "sharp and cutting." The remainder were "triangular flakes," about which he stated: "The lateral edges are straight, sharp, and cutting." Noetling (1894, p. 101) thought one of the triangular flakes shown in his illustrations (Figure 4.45) was "particularly remarkable" because "it shows that the upper face must have been produced by the repeated chipping off of thin flakes."

Figure 4.45. A flint tool from the Miocene Yenangyaung formation of Burma (Noetling 1894, plate 1).

Analyzing the stratigraphic position of his discoveries, Noetling stated that the ferruginous conglomerate containing the chipped flints was surmounted by a 4,620-foot-thick formation (Group C) composed of yellow sandstone alternating with beds of light brown clay. Noetling wrote (1894, p. 102): "A superficial examination of the vertebrate remains shows that the fauna is nearly identical with that of the Siwaliks, or in other words, that Group C . . . must be of upper miocene if not pliocene age. We must therefore claim either pliocene or at the latest upper miocene age for the ferruginous conglomerate in which the chipped flints have been found. But whatsoever their particular age be, it is certain that a considerable amount of time must have elapsed since the deposit of a series of strata of more than 4,620 foot thickness, containing numerous genera of animals which are now-a-days either entirely extinct, or at least no longer living in India, which rests upon it [the ferruginous conglomerate]."

W. T. Blanford believed that *Rhinoceros perimensis* and *Hipparion antelopinum,* fossils of which accompanied the flints, characterized in India the Pliocene rather than the Miocene (G. de Mortillet and A. de Mortillet 1900, pp. 90–91). A Pliocene date (2–5 million years) would, however, still be quite anomalous, considering the now-dominant view that toolmaking beings (*Homo erectus*) first migrated from Africa around 1 million years ago. Blanford, however appears to have been wrong. According to modern authorities *Rhinoceros (Aceratherium) perimensis* and *Hipparion antelopinum* occur in Late Miocene assemblages of Asia, including India (Savage and Russell 1983, pp. 283–284). Furthermore, as previously mentioned, the Yenangyaung fauna in general is today regarded as Miocene (Savage and Russell 1983, pp. 247, 326). This would place toolmaking creatures in Burma over 5 million years ago.

According to de Mortillet, R. D. Oldham observed flints similar to Noetling's on a plateau rising above the location of Noetling's discovery. Oldham wanted to use this fact to dispute the age of Noetling's flints, but it is not clear why the presence of flints on the plateau should invalidate Noetling's statements about the stratigraphic position of his discoveries.

How certain was the stratigraphic position of Noetling's flints? According to de Mortillet, the strata in which the flints were found appeared to Oldham to be loosely compacted conglomerates, which suggested that the flints might have been introduced in recent times. But Noetling (1894, pp. 102–103) offered this account: "Having now described the geological position of the strata in which the chipped flints were found, there still remains the question to be discussed whether they were really found *in situ,* or not. To this I can only answer that to the best of my knowledge they were really found *in situ.* . . . The exact spot where the flints were found is marked on my geological map of the Yenangyoung oil-field with No. 49 and is situated on the steep eastern slope of a ravine, high above its bottom, but below the edge in such a position that it is inconceivable how the flints should have been brought there by any foreign agency. There is no room for any dwelling place in this narrow gorge, nor was there ever any; it is further impossible from the way in which the flints were found that they could have been brought to that place by a flood. If I weigh all the evidence, quite apart from the fact that I actually dug them out of the bed, it is my strong belief that they were *in situ* when found." It should be remembered that these statements were made by a professional paleontologist who was a member of the Geological Society of London and the Geological Survey of India.

In conclusion, Noetling (1894, p. 103) said: "As to their nature whether artificial or not, I do not want to express an opinion; all I can say is, that if flints of this shape can be produced by natural causes, a good many chipped flints hitherto considered as undoubtedly artificial products are open to grave doubts as to their origin." We agree with this statement. In our review of controversial evidence, we are not so much insisting on the Tertiary date or human manufacture of particular stone objects as insisting on consistent application of standards for evaluating such evidence. We have found that such consistent application is lacking, that prejudice and preconception very often come into play. This raises serious questions about the empiric method as the primary cognitive tool for understanding human origins and antiquity.

4.8 TOOLS FROM BLACK'S FORK RIVER, WYOMING (MIDDLE PLEISTOCENE)

In 1932, Edison Lohr and Harold Dunning, two amateur archeologists, found many stone tools on the high terraces of the Black's Fork River in Wyoming, U.S.A. We may recall from our discussion of the eoliths of the Kent Plateau, England, that high river terraces are older than lower terraces. The stone implements found by Lohr and Dunning appeared to be of Middle Pleistocene age, which would be anomalous for North America.

Lohr and Dunning showed the tools they collected to E. B. Renaud, a professor of anthropology at the University of Denver. Renaud, who was also director of the Archaeological Survey of the High Western Plains, then organized an expedition to the region where the tools were found. During the summer of 1933, Renaud's party collected specimens from the ancient river terraces between the towns of Granger and Lyman.

Renaud, who had been trained in Europe under Henri Breuil, characterized the implements as similar to those of the early European Paleolithic (Minshall 1989, p. 86). Among the specimens were crude handaxes and other flaked implements representative of those frequently attributed to *Homo erectus*. In 1933, Renaud said the tools would "suggest a cultural complex in America similar to that of Europe, and also a possible great antiquity for these artifacts" (Minshall 1989, p. 86).

The reaction from anthropologists in America was negative. Renaud wrote in 1938 that his report had been "harshly criticized by one of the irreconcilable opponents of the antiquity of man in America, who had seen neither the sites nor the specimens" (Minshall 1989, p. 87).

In response, Renaud mounted three more expeditions, collecting more tools, which he studied carefully, comparing them with artifacts of similar age from France and England. Although many experts from outside America agreed with him that the tools represented a genuine industry, American scientists have continued their opposition to the present day.

The most common reaction is to explain that the crude Paleolithic specimens are in fact blanks (unworked flakes) dropped fairly recently by Indian tool-makers. Opposing this hypothesis, Herbert L. Minshall (1989, p. 87) stated that the tools "show heavy stream abrasion" even though they are "fixed in desert pavements on ancient flood plain surfaces that could not have had streams for over 150,000 years." In 1938, E. H. Stephens, a geologist at the Colorado School of Mines, visited the sites where the tools had been found. According to Stephens, the high flood plain terraces dated to the Illinoian glacial period (Minshall 1989, p. 88). This would mean they were formed from 125,000 to 190,000 years ago, and perhaps even further back in time (Minshall 1989, p. 88).

If found at a site of similar age in Africa or Europe or China, stone tools like those found by Renaud would not be a source of controversy. But their presence in Wyoming is certainly very much unexpected at 125,000 to 190,000 years ago. The view now dominant is that humans entered North America not earlier than about 30,000 years ago at most. And before that there was no migration of any other hominid.

Renaud's discoveries were therefore either ignored or explained away. Stephens and others suggested that the abrasion on the implements was the result of windblown sand rather than water. In 1957, Marie Wormington stated: "It is true that many of Renaud's artifacts were found on high terraces and showed definite signs of abrasion. If it could be proven that this was the result of water action it might provide some evidence of age, for a considerable length of time has elapsed since water last reached these terraces. However, if the smoothing was due to wind erosion it provides no evidence of real antiquity" (Minshall 1989, pp. 89–90).

In reply Minshall (1989, p. 90) observed: "The specimens were abraded on all sides, top and bottom, ventral and dorsal surfaces equally. That is extremely unlikely for windblown dust to achieve on heavy stone tools lying in heavy gravel but expectable on objects subjected to surf or heavy stream action. Having examined thousands of stone tools on desert surfaces, I can testify that all-over wind abrasion is rare under any circumstances, is only present on specimens lying in loose sand, and never appears on heavy gravel inclusions."

Minshall (1989, p. 91) also noted that the tools were covered with a thick mineral coating of "desert varnish." This varnish, which takes a long time to accumulate, was thicker than that on tools found on lower, and hence more recent, terraces in the same region.

The cumulative evidence appears to rule out the suggestion that the implements discovered by Renaud were blanks dropped fairly recently on the high desert floodplain terraces. But Minshall (1989, p. 87) noted: "The reaction of American scientists to Renaud's interpretation of the Black's Fork collections as evidences of great antiquity was, and has continued to be for over half a century, one of general skepticism and disbelief, even though probably not one in a thousand archaeologists has visited the site nor seen the artifacts."

According to Minshall, the tools found by Renaud were the work of *Homo erectus,* who may have entered North America during a time of lowered sea levels in the Middle Pleistocene. Minshall believed this was also true of stone tools found at other locations of similar age, such as Calico, and his own excavation at Buchanan Canyon, both in southern California.

Minshall was, however, skeptical of another Middle Pleistocene site. In January 1990, Minshall told one of us (Thompson) that he was not inclined to accept as genuine the technologically advanced stone tools found at Hueyatlaco in Mexico (Section 5.4.4). Hueyatlaco was determined to be about 250,000 years old—roughly contemporary with Black's Fork, Calico, and other sites with primitive stone tools that Minshall was prepared to accept. But the advanced stone tools found at Hueyatlaco were characteristic of *Homo sapiens sapiens,* and were thus not easy to attribute to *Homo erectus.* Minshall's response to Hueyatlaco was to suggest, without supporting evidence, that the stratigraphy had been misinterpreted and that the animal bones used to date the site, as well as the sophisticated stone artifacts, had been "washed onto the site from different sources" (Minshall 1989, p. 93). This shows that researchers who accept some anomalies may rule out others using the double standard method.

5

Advanced Paleoliths and Neoliths

Having reviewed the crudest of the anomalously old stone tools (the eoliths) and then the crude paleoliths, we shall now proceed to examine advanced paleoliths and neoliths. Here once more we face difficulties in classification, as many of the discoveries we shall be considering involve implements of various levels of sophistication. The deciding factor for including a group of implements in the category of advanced paleoliths is that a number of specimens represent a clear technical advance over the crude paleoliths discussed in the last chapter. For example, the stone tool industries discovered by Florentino and Carlos Ameghino in Argentina include many implements that might be classed among the eoliths or crude paleoliths; nevertheless, they also include implements of a higher order, such as presumed projectile points and bolas. In this chapter, we shall first discuss the discoveries of Florentino Ameghino, as well as the attacks upon them by A. Hrdlicka and W. H. Holmes. Next we shall consider the finds of Carlos Ameghino, which provide some of the most solid and convincing evidence for a fully human presence in the Pliocene. We shall then proceed to anomalous finds made at sites in North America, including Hueyatlaco, Mexico; Sandia Cave, New Mexico; Sheguiandah, Ontario; Lewisville, Texas; and Timlin, New York. We shall conclude with the Neolithic finds from the Tertiary gold-bearing gravels of the California gold rush country.

5.1 DISCOVERIES OF FLORENTINO AMEGHINO IN ARGENTINA

In the late nineteenth and early twentieth centuries, Florentino Ameghino thoroughly investigated and described the stratigraphy and fossil fauna of the coastal provinces of Argentina. He thereby became an internationally known and respected paleontologist. Ameghino's controversial discoveries of stone implements, carved bones, and other signs of a human presence in Argentina during the Pliocene, Miocene, and earlier periods served to increase his worldwide fame.

291

5.1.1 Monte Hermoso (Middle and Early Pliocene)

Among the most significant examples of human work reported by Florentino Ameghino are those he discovered in 1887 at Monte Hermoso, on the coast of Argentina about 60 kilometers (37 miles) northeast of Bahia Blanca. Here is how F. Ameghino (1908, p. 105) recounted the circumstances of his first discoveries at Monte Hermoso, which were made in a formation he regarded as Miocene: "During an exploratory visit, which lasted from the end of February to the beginning of March of 1887, we had the good fortune to find remains that demonstrated the existence of an intelligent being contemporary with . . . extinct fauna at this site. These vestiges consisted of fragments of *tierra cocida* (burned earth), *fogones* (hearths), *escoria* (glassy, melted earth), bones that had been split and burned, and worked stone. These discoveries caused me such surprise and appeared so important that I immediately wrote up my impressions and sent them to the journal *La Nación,* which published them on March 10, 1887."

In another description of the initial discoveries made at Monte Hermoso, written in 1889, F. Ameghino (1911, p. 74) commented: "I was occupied in extracting part of a skeleton of *Macrauchenia antiqua* [a camellike Pliocene mammal] when I was surprised to see a piece of yellow-red stone among the bones. I picked it up and immediately recognized it as an irregular fragment of quartzite, displaying positive and negative bulbs of percussion, a striking platform, and eraillure. These features indicated in an irrefutable manner that I had found a stone object worked by an intelligent being during the Miocene period. I continued my work and soon found similar objects. Doubt was not possible, and on the same day, March 4, 1887, I communicated to *La Nación* the discovery of objects evidently worked by an intelligent being in the Miocene formations of Argentina." F. Ameghino (1911, p. 74) added: "Later, at my instigation, the Museum of La Plata sent to the same place, for the purpose of collecting fossils, the preparator Santiago Pozzi, who found objects similar to mine."

Summarizing the Monte Hermoso evidence, F. Ameghino (1911, pp. 52–53) said: "The presence of man, or rather his precursor, at this ancient site, is demonstrated by the presence of crudely worked flints, like those of the Miocene of Portugal, carved bones, burned bones, and burned earth proceeding from ancient fireplaces, in which earth containing a substantial quantity of sand came in contact with fire so intense that it was partially vitrified."

Regarding the fireplaces, F. Ameghino (1911, p. 52) stated: "In this part of the formation there are no traces of volcanic activity, nor deposits of lignite, nor any vestiges of vegetation that might have sustained accidental fires with the rare property of occurring at intervals consecutive with the successive depositing of the strata at the site. Furthermore, these fireplaces, by the rarest of coincidences, are accompanied by burned bones. The temperature of the fires was so high, that in the pieces of burned earth there have formed spherical cavities, resulting from the expansion of air or the special gases produced by combustion of the substances contained in the earth."

F. Ameghino, who was, like most scientists of his time, committed to the concept of evolution, wrote: "The vestiges belong to such a distant epoch that I do not dare to consider them as proof of the existence of man, but rather as remains of 'a being more or less resembling man, directly ancestral to man of the modern type'" (1908, p. 105). After two years of research, Ameghino decided the intelligent being that made the artifacts at Monte Hermoso was of a different genus than modern humans and their immediate ancestors. Among the fossils recovered from Monte Hermoso was a hominid atlas (the first bone of the spinal column, at the base of the skull). Ameghino thought it displayed primitive features, but A. Hrdlicka judged it to be fully human (Section 6.2.4). This strongly suggests that beings of modern human type were responsible for the artifacts and signs of fire discovered in the Montehermosan formation.

Although Ameghino thought the Montehermosan formation to be Miocene, modern authorities place it in the Early Pliocene. According to E. Anderson (1984, p. 41) the stratigraphic sequence of the Argentine coastal region can be dated in the following way: the Ensenadan at .4–1.5 million years, the Uquian at 1.5–2.5 million years, and the Chapadmalalan at 2.5–3.0 million years. The Montehermosan precedes the Chapadmalalan in the general Argentine sequence, and thus it would be over 3 million years old.

Other authorities (Marshall *et al.* 1982, p. 1352) give a slightly different chronology, placing the Ensenadan formation at .4–1.0 million years, the Uquian at 1–2 million years, the Chapadmalalan at 2–3 million years, and the Montehermosan at 3–5 million years. A potassium-argon date of 3.59 million years has been obtained for materials from the Montehermosan formation (Savage and Russell 1983, p. 347).

The antiquity of the Montehermosan formation is further supported by the character of its fossil mammalian bones. Paleontologists believe that during the early part of the Tertiary, North America and South America were separated by water and developed distinct mammalian populations. For example, huge ground sloths not found in North America populated South America. When a land bridge eventually formed, North American mammals migrated south, and South American mammals moved north. Modern authorities (Marshall *et al.* 1982, p. 1351) say that the Panamanian land bridge, which allowed the exchange of mammals between North America and South America, appeared 3 million years ago, just after the period represented by the Montehermosan formation. According to Ameghino (1912, p. 64), the fauna of Monte Hermoso reveals "the complete absence of North American types." Ameghino's discoveries in the

Montehermosan formation—including stone tools, modified animal bones, signs of fire, and human skeletal remains—thus suggest a human presence in Argentina more than 3 million years ago.

5.1.2 Hrdlicka Attempts to Discredit Ameghino

Ameghino's discoveries at Monte Hermoso and elsewhere in the Tertiary formations of Argentina attracted the interest of several European scientists, especially those who were attempting to demonstrate the existence of Tertiary humans on the basis of the European discoveries discussed in preceding chapters.

Ales Hrdlicka, an anthropologist at the Smithsonian Institution, also took great, though unsympathetic, interest in Ameghino's discoveries. Hrdlicka found the degree of support they enjoyed among professional scientists, particularly in Europe, dismaying. In addition to being opposed to the existence of Tertiary humans, Hrdlicka was also extremely hostile to any reports of a human presence in the Americas earlier than a few thousand years before the present. After building an immense reputation by discrediting, with questionable arguments, all such reports from North America, Hrdlicka then turned his attention to the much-discussed South American discoveries of Florentino Ameghino. Hrdlicka was most concerned about the human skeletal remains reported by Ameghino (Sections 6.1.5, 6.2.4), but he also scrutinized Ameghino's discoveries of stone tools and other cultural remains.

In 1910, Hrdlicka visited Argentina, and Florentino Ameghino himself accompanied him to Monte Hermoso. Hrdlicka took an interesting approach to the discoveries that were made at that site. In his book *Early Man in South America* (1912), Hrdlicka barely mentioned the stone implements and other evidence of human occupation previously uncovered by Ameghino in the Montehermosan formation.

"In 1887," wrote Hrdlicka (1912, p. 346), "F. Ameghino announced the discovery, in the barranca of Monte Hermoso, a low cliff facing the sea in the central part of the coast of the Province of Buenos Aires, of vestiges of 'a being, more or less closely related to actual man, who was a direct forerunner of the existing humanity.' These vestiges consisted of fragments of 'tierra cocida, fogónes (fire places)—some of the latter vitrified and having the appearance of scoria—split and burnt bones (of animals) and worked stones.'"

Hrdlicka said nothing more about these particular discoveries of Amegh-ino—not even to dispute them. Instead, he devoted dozens of pages to casting doubt on subsequent, and less convincing, discoveries Ameghino made in the Puelchean, a more recent formation overlying the Pliocene Montehermosan at Monte Hermoso.

The Puelchean formation would, according to modern nomenclature, be included within the Uquian. Savage and Russell (1983) stated that the Uquian comprises several formations including the "Pulchense." The Puelchean would thus fall within the Uquian time range, estimated at 1.5–2.5 million years (Anderson 1984) or 1–2 million years (Marshall *et al.* 1982).

Apparently, Hrdlicka believed his lengthy refutation of the finds from the Puelchean formation was sufficient to discredit the finds in the far older Montehermosan formation at the same site. This tactic is often used to cast doubt on anomalous discoveries—criticize the weakest evidence in detail and ignore the strongest evidence as much as possible. Nevertheless, there is much evidence to suggest that the Puelchean finds, as well as the Montehermosan finds, were genuine.

In and of themselves, the Puelchean discoveries at Monto Hermoso are not of paramount interest to us. If accepted, they merely add to our already abundant stock of evidence for a human presence in the Early Pleistocene. But as a well-documented example of how scientists treat anomalous evidence, the case is significant. We shall therefore take the trouble to examine in detail the short-comings of Hrdlicka's attempts to discredit the Puelchean implements.

As mentioned above, Hrdlicka and F. Ameghino together visited the Monte Hermoso site. Hrdlicka reproduced an English translation of Ameghino's report of their excursion. Ameghino stated: "On the 11th of June, in the afternoon, we visited Monte Hermoso, where with difficulty we were able to stay a couple of hours. . . . The deposits of sands and sandy ground which rest above the Hermosean and constitute the Puelchean stratum, formerly visible over a small space of only about 40 meters [about 131 feet], now appear exposed along the barranca for several hundred meters and also to a greater extent vertically" (Hrdlicka 1912, p. 105). Ameghino collected a number of implements from "the superior part of this formation" (Hrdlicka 1912, p. 105).

The stone implements recovered by Ameghino from the upper section of the Puelchean formation at Monte Hermoso were very crude. Judging from his descriptions, they appear to resemble the pebble tools of the Oldowan industry of East Africa. Ameghino characterized the Puelchean specimens as fragments of "water-worn pebbles of quartzite" (Hrdlicka 1912, p. 105). To Ameghino, it was clear that the implements had been deliberately struck from quartzite pebbles: "The larger number of these fragments preserve still on one or two of their faces the natural surface of the rolled pebble, and on this surface are always observed scratches, bruises, abrasions, dints, etc., produced by strong and repeated blows given with other stones" (Hrdlicka 1912, p. 105). Further-more, the sharp cutting edges of the implements, according to Ameghino, showed "irregularities, denticulation, and other effects produced by use" (Hrdlicka 1912, p. 105).

Ameghino noted: "these broken quartzites, however rustic they may appear, are surely the work of man or his precursor, for there can not be opposed to them the objections which are being made to the eoliths. In this case there can be no question of pressure by the rocks, of shocks produced by stones driven by water or due to falling stones, because, I repeat, they are loose in the sand, and are all separated from one another" (Hrdlicka 1912, p. 106).

Hrdlicka interpreted the finds in another way. Significantly, he did not dispute the human manufacture of even the crudest specimens. Instead, Hrdlicka, an anthropologist with little experience of South American paleontology, offered a different analysis of the stratigraphy than did Ameghino, a professional paleontologist who had devoted decades to the study of the formations in question.

Hrdlicka (1912, p. 118) said in his book: "The writer found that the Monte Hermoso formation exposed in the now famous barranca was covered by more recent material. On the old formation rests a layer of volcanic ash, then some stratified sand, while the highest part is formed of a stratum of gravelly sand continuous with the base of the sand dune situated above and a little farther inland from the edge of the barranca." Ameghino had said the volcanic ash, stratified sand, and gravelly sand comprised the Puelchean formation, overlying the Pliocene Montehermosan. Hrdlicka disputed the inclusion of the uppermost layer of gravelly sand within the superior part of the Puelchean formation. He observed: "The last-named surface material [the gravelly sand] is unstratified and somewhat packed, but in no way consolidated, and bears every evidence of being very recent. It crumbles over the clearly marked, ancient Monte Hermosean deposit, and in falling down becomes here and there lodged on the shelves or in the depressions of the old formation" (Hrdlicka 1912, pp. 118–119).

Hrdlicka (1912, p. 119) then recalled: "In common with Professor Ameghino the writer found in such crumbled down material some large irregular and entirely fresh-looking fragments or chips of quartzite which indicate plainly the work of man. One of the heavier fragments had been employed as a hammer, portions of the periphery being distinctly worn by use. In addition, he found on one of the upper ledges a well-finished scraper of jasper [a variety of quartz]. Subsequently he extracted a number of quartzite chips or fragments from the more gravelly part of the uppermost deposit itself, within 18 inches of the surface."

It is significant that Hrdlicka reported he extracted stone artifacts not just from crumbled material on ledges, but from within the upper deposit of gravelly sand itself. The fact that they were recovered a full 18 inches from the surface, upon which the recent sand dunes rested, shows they were an integral part of a distinct stratum. Ameghino said the stratum from which he (and Hrdlicka, it seems) took implements was part of the Puelchean formation, which according to modern opinion could be from 1.0 to 2.5 million years old.

5.1.3 Willis Stacks the Geological Deck

Hrdlicka, perhaps worried that his visit to Monte Hermoso had led him into a deadly ambush, wanted to suggest that the implements found there, by himself as well as Ameghino, were recent. As we have seen, he attempted to do this by casting doubt on the age of the stratum from which the tools had been taken. Hrdlicka received support in this regard from his companion, the geologist Bailey Willis.

Willis wrote: "Monte Hermoso is a dune on the southern coast of Buenos Aires. It surmounts a short section of the Pampean terrane, which is exposed by wave erosion in a low bluff along the shore. First described by Darwin, it has since been visited by many geologists who have studied the Pampean. . . . The Pampean terrane, which forms the base of the section, contains a notable fauna and the geologic age of the formation has been much discussed. General opinion places it among the lowest or as the lowest of the divisions of the Pampean and Ameghino regards it as Miocene" (Hrdlicka 1912, pp. 361–362). This lower Pampean formation is the Montehermosan, now regarded as Middle to Early Pliocene, and above it lies the Puelchean, which could be Early Pleistocene or Late Pliocene. According to Florentino Ameghino, the layer from which both he and Hrdlicka extracted stone implements during their excursion to Monte Hermoso represented the "superior part" of the Puelchean formation (Hrdlicka 1912, p. 105).

Describing the Puelchean, Willis stated: "The Puelchean consists of the stratified, slightly indurated, gray sands or sandstone, both above and below the volcanic ash, marked by very striking cross stratification and uniformity of gray color and grain. The writer regards it as an eolian formation. Later in the season, when studying the section exposed along the Rio Colorado from the delta to Pichi-Mahuida, he observed a very similar sandstone, which might be correlated with the Puelchean on grounds of lithologic identity. It is a thick widespread formation which is regarded as a Tertiary sandstone. The Puelchean, if the same, represents only a thin edge of it" (Hrdlicka 1912, p. 363).

We may note that the Puelchean sands, which Willis, with some hesitation, accepted as Tertiary, are characterized by their gray color, quite different from the underlying Montehermosan loess, which is yellow-brown. Willis then described the so-called recent topmost layer, apparently included by Ameghino in the Puelchean formation, as "a layer of 15 to 40 cm. [6 to 16 inches] thick composed of gray sand, angular pieces of gray sandstone and pebbles, some fractured by man" (Hrdlicka 1912, p. 362).

Willis elsewhere remarked that the top layer of gray implement-bearing sand is separated from the lower layers of the Puelchean by an "unconformity by erosion" (Hrdlicka 1912, p. 363). An unconformity is a lack of continuity in deposition between strata in contact with each other, corresponding to a period of nondeposition, weathering, or, as in this case, erosion. The unconformity may represent a long break in time, or a very short one. So by itself, the presence of an unconformity by erosion should not allow Willis to so greatly separate the top layer of gray gravelly sand from the underlying formation. He appears to have used the mere presence of stone tools to perform the desired operation, which was necessary to save Hrdlicka the deadly embarrassment of having discovered stone implements in an unexpectedly old formation in South America. For judging how much time might be represented between the formations lying above and below the line of unconformity, the surest indicator is faunal remains. Willis, however, did not mention them. It is thus unclear how much time might be represented by the unconformity.

Willis then stated about the upper layer of gray gravelly sand and the underlying Puelchean: "The two are identical in constitution; they are both eolian and may exhibit similar structures; the Recent formation may be consolidated almost or quite to the firmness of the older one. The unconformity between them suffices to establish the difference in age and is unmistakable when clearly shown in section" (Hrdlicka 1912, p. 363).

Let us carefully consider just what Willis is asking us to accept. First of all he admitted that the two strata are identical in composition, which would seem to be very much in favor of Ameghino, who considered the top layer to be part of the Puelchean. And, given the evidence, why not?

But for argument's sake, let us accept Willis's version. The unconformity by erosion that he proposed would mark a gap of 1–2 million years, as the upper gray layer is supposedly recent, while the gray sand layer below it, identical in composition, is referred to the Early Pleistocene–Late Pliocene Uquian formation.

Another, and perhaps more likely scenario, would be that the two gray sand layers, identical in composition, are separated by an unconformity representing a relatively short episode of erosion in the Early Pleistocene or Late Pliocene.

As mentioned previously, Willis could have accurately determined how long a period was represented by the unconformity only by examining animal fossils above and below the line of unconformity. If the fossils in the layer above the unconformity were all recent, only then would he have been justified in concluding that this layer was recent. But Willis did not make the slightest attempt to establish this. In the absence of such an age determination (which today might be made by radiometric methods), the implement-bearing layer could very well be about the same age as the Puelchean formation below the unconformity, which it greatly resembles in content and texture.

Here is how Willis attempted to eliminate this alternative: "hand-chipped stones associated with the sands would mark them as recent, such objects being common in the belt of sand dunes which the Indians were in the habit of using as a line of march and cover in attacking Argentine settlements" (Hrdlicka 1912, p. 363). Willis simply assumed that the stone tools were recent and that the layer in which they were found also had to be recent. It would appear, however, that the implement-bearing gray gravelly sand may actually belong to the Puelchean formation, as Ameghino believed, and that the stone implements found there could be as much as 2.0 or even 2.5 million years old.

In short, the question of the age of the implement-bearing stratum below the dune sand at Monte Hermoso remains open. Ameghino's assertion that it belonged to the Puelchean was not conclusive, but neither was the attempt by Willis and Hrdlicka to assign it to the most recent historical times. Since the stratigraphic units in question contain layers of volcanic ash, their ages could be investigated by applying the potassium-argon test, which is specifically used for dating volcanic material. It may also be possible to make a determination by conducting a more thorough search for faunal evidence. In short, the question is still open and should still be a matter of active research. But the report by Willis and Hrdlicka succeeded in closing the books on this intriguing case.

5.1.4 A Demolition Job by W. H. Holmes

Samples of stone tools from Monte Hermoso and other sites on the Argentine coast were sent by Hrdlicka to Washington, where W. H. Holmes of the Smithsonian Institution examined them. Concerning the attribution of any great antiquity to the implements, Holmes was as hostile as Hrdlicka or Willis. In opening his report, included by Hrdlicka (1912, p. 125) in *Early Man in South America,* Holmes stated: "No attempt is made in these notes to consider or weigh the published data relating to the stone implements of Argentina. The collections at hand are classified and briefly described, and such conclusions are drawn as seem warranted by their character and manner of occurrence." In other words, Holmes plainly intended to completely ignore the reports of Ameghino and other professional scientists, who had given detailed evidence for the Early Pleistocene or Pliocene age of the stone artifacts.

We may recall that Hrdlicka, in the company of Ameghino, personally extracted stone tools at a depth of 1.5 feet in the upper layer of the Early Pleistocene–Late Pliocene Puelchean formation at Monte Hermoso (Hrdlicka 1912, p. 104). This fact was subsequently reported by Ameghino in a scientific publication. Hrdlicka and his associates were anxious to discredit this report. If accepted, Ameghino's report on the discoveries he and Hrdlicka made together at Monte Hermoso would have contradicted the entire substance of the book Hrdlicka was then writing. Hrdlicka's book was specifically designed to prove that the only early inhabitants of South America had been the Indians, who had arrived within the past few thousand years.

We detect a slight sense of panic in the following passages, hastily added by Holmes to the end of his report on the stone tools from Argentina. Holmes wrote: "Subsequent to the completion of the foregoing pages Doctor Hrdlicka drew attention to certain specimens collected by him along the barranca at Monte Hermoso, which had escaped particular notice on the writer's part. Attention was directed also to a brief pamphlet just received from Dr. Ameghino, describing a series of similar specimens collected by him while examining this same barranca in company with Doctor Hrdlicka. Considering the nature of the specimens and the manner of their occurrence, the observations and interpretations of Doctor Ameghino are so remarkable that the writer is constrained to refer to them in some detail" (Hrdlicka 1912, p. 149). Otherwise, Hrdlicka's whole project would be shot to pieces. A report showing that Hrdlicka had, in the company of Ameghino, himself extracted undisputed stone tools from an Early Pleistocene or Late Pliocene formation would in itself have destroyed everything Hrdlicka had tried to accomplish by publishing *Early Man in South America,* which was nothing less than a polite but thorough demolition of Ameghino's work.

Holmes wrote: "The objects in question are about 20 freshly-fractured chips and fragments of coarse, partially fire-reddened quartzite, a larger fragment of the same material used as a hammer, and a knife or scraper of jasper. All were found in a surface layer of gravelly sand capping the Monte Hermoso barranca,

or on the broken face of the barranca itself. The latter were picked up on the ledges of the bluff face, where they had cascaded from above. The jasper knife or scraper is of a type familiar in the coast region as well as in Patagonia" (Hrdlicka 1912, pp. 149–150). It should, however, be kept in mind that it is principally the objects found *in situ* that concern us. The implements found lying on ledges might very well have been recent.

Holmes suggested in every possible way that all of the objects, even those found *in situ,* were of recent origin, pointing to their discovery in a "surface formation." He also characterized most of the pieces of stone not as implements but rather as the "shop refuse" of recent tribes (Hrdlicka 1912, p. 150). This latter conclusion was apparently an attempt to contradict Ameghino's view that the crude nature of the objects was supportive of their being of extremely great antiquity.

Holmes stated: "The inclusion of such objects in superficial deposits which are subject to rearrangement by the winds and by gravity is a perfectly normal and commonplace occurrence" (Hrdlicka 1912, p. 150). As we have seen, it is not certain that the top layer of Ameghino's Puelchean formation, in which implements were found by Hrdlicka at a depth of 1.5 feet, should be classified as such a superficial deposit, especially one that could be rearranged by the wind. Even the large dune surmounting the stratum in which the implements were found was covered with grass and fixed (Hrdlicka 1912, p. 363).

Getting to the real heart of the matter, Holmes stated: "Such differences as may arise between the writer's interpretation and those of Doctor Ameghino are probably due in large measure to the fact that the points of view assumed in approaching the problem of culture and antiquity are widely at variance. Doctor Ameghino takes for granted the presence in Argentina of peoples of great antiquity and extremely primitive forms of culture and so does not hesitate to assign finds of objects displaying primitive characteristics to unidentified peoples and to great antiquity, or to assume their manufacture by methods supposed to characterize the dawn of the manual arts. To him all this is a simple and reasonable procedure" (Hrdlicka 1912, p. 150). This is not a fair characterization of Ameghino's work, for it is quite clear that in addition to the form of tools he also took into consideration their geological position, which for him served as the chief indicator of their age. If one finds stone implements in geological strata of a certain age, one is certainly justified in attributing them to a people that lived at that time. It would appear that accusations of bias and preconception are more properly directed at those who, like Holmes, Hrdlicka, and Willis, assume from the start that the human occupation of North America and South America goes back no further than a few thousand years, and who therefore dismiss, in various unfair ways, the extensive evidence that indicates a much more ancient human presence.

Holmes directly revealed his prejudice: "The writer finds it more logical to begin with the known populations of the region whose culture is familiar to us and which furnishes lithic artifacts ranging in form from the simplest fractured stone to the well-made and polished implement, and prefers to interpret the finds made, unless sufficient evidence is offered to the contrary, in the illuminating light of known conditions and of well-ascertained facts rather than to refer them to hypothetic races haled up from the distant past" (Hrdlicka 1912, p. 150). Scientists are certainly entitled to their predispositions, which play a covert but substantial role in their supposedly objective evaluation of evidence. In this case, however, Holmes's overt preferences appear to have played too exclusive and dominant a role. To be sure, Holmes offered the condition that Ameghino's stone implements must be attributed to modern Indians unless "sufficient evidence is offered to the contrary." But what is sufficient contrary evidence? For someone with a strong negative bias, no contrary evidence will prove sufficient.

Holmes stated in his report: "Nothing short of perfectly authenticated finds of objects of art in undisturbed formations of fully established geologic age will justify science in accepting the theory of Quaternary or Tertiary occupants for Argentina" (Hrdlicka 1912, p. 149). Ameghino, of course, fully believed he had satisfied these criteria. Paleontological truth, it would appear, is, like beauty, in the eye of the beholder. Furthermore, as we have documented previously, objects of human industry have elsewhere been discovered by professional scientists in undisturbed formations of great antiquity, and yet reasons were still found to reject them. For example, we have Ribeiro's testimony that he extracted flint implements from within the interior of Miocene limestone formations in Portugal (Section 4.1), and yet opponents nevertheless found ample reason to disagree with his interpretation of their age. It seems clear that Holmes was selectively requiring an impossibly stringent standard of proof for evidence that challenged his preferred views.

5.1.5 Other Finds by F. Ameghino

What do modern authorities have to say about Florentino Ameghino? Not much, because most modern authorities will not even have heard of Ameghino or his discoveries—both buried many decades ago. But if we go back to the 1950s, we can find some references to Ameghino by one of the scientists who did the burying—Marcellin Boule, author of the classic text *Fossil Men*. After pointing out that Ameghino, like his contemporaries in Europe, had discovered stone implements and other evidence for a human presence in the Pliocene and Miocene, Boule added: "Ameghino also recorded facts of the same kind from much more ancient deposits dating, according to him, from the Oligocene and even from the Eocene. He claimed that they were rudimentary implements manufactured and used by the small apes of these remote periods, the supposed ancestors of the human kind. These statements are not even worthy of discussion" (Boule and Vallois 1957, p. 491). Boule may be commended for his candor, which demonstrates the parochialism sometimes manifest in the scientific mentality. In all fairness, why should not evidence presented by a professional scientist at least be considered and discussed, even if it does completely contradict accepted views?

Of course, this does not mean that one should uncritically accept everything Ameghino said. For example, Ameghino wanted to attribute some of his older stone tools to primitive apelike precursors of modern humans. But as we have several times noted, even the simplest types of tools are made and used today by culturally primitive yet fully human peoples. Furthermore, there is from parts of the world other than Argentina abundant evidence that points to a fully human presence throughout the Tertiary. One would therefore be fully justified in leaving open the possibility that humans of the fully modern type were responsible for the manufacture of any of the tools found by Ameghino in Argentina, including the oldest.

Indeed, one of the main tactics employed by Hrdlicka against Ameghino was to show that the fossil bones of presumed Tertiary human precursors found by Ameghino were in fact identical to those of morphologically modern humans

(Sections 6.1.5, 6.2.4). For Hrdlicka, who firmly believed in the recent origin of the human species, this meant that Ameghino's fossils were also recent. But it could also mean something else, namely, that the skeletal remains were from the Tertiary, as Ameghino so ably maintained, and were, as Hrdlicka so ably demonstrated, anatomically modern.

The complete original reports of the finds Ameghino regarded as Eocene and Oligocene have proved very difficult to track down. As of this writing, we have bibliographical references giving the titles of these publications, which were small pamphlets of 8 pages each, apparently presented as papers at a scientific conference (F. Ameghino 1910a; 1910b). Ameghino did, however, refer to the discoveries described in these two papers in an article that appeared in 1912. "Recently," he wrote, "I have published a report on new materials, very well substantiated, found in the Entrerrean formation" (F. Ameghino 1912, p. 74). According to Ameghino, the Entrerrean formation could be assigned to the Late Oligocene, or perhaps the Early Miocene. He then mentioned a second report about discoveries in a formation he regarded as Late Eocene, the Santacrucian.

Today the Santacrucian formation, which Ameghino considered Late Eocene, is referred to the Early and Middle Miocene (Marshall *et al.* 1977, p. 1326). It would thus be about 15–25 million years old. We have not encountered any mention of the Entrerrean in the current literature we have examined, but since this formation comes before the Monte Hermosan, it would be at least Late Miocene, over 5 million years old.

In the two reports published in 1910, Ameghino had apparently discussed only stone tools. Afterwards, Ameghino found other signs of a human presence. F. Ameghino (1912, p. 72) therefore wrote: "I can announce that I possess from these two formations even newer materials still more demonstrative than those I have published. Regarding this new material, I am not bringing into consideration more eoliths, which we find in our formations at the close of the Eocene and which differ from those of Boncelles in Belgium in that they are of much smaller size. Instead I base my assertions on bones that have been incised, cut, scraped, and split and on the vestiges of fire, found in the same beds as the bones." The modified bones and signs of intentional use of fire found along with stone tools at these two sites support the idea that anatomically modern humans may have been present in Argentina prior to the time of the Montehermosan, which is considered to be 3–5 million years old.

5.1.6 Evidence for the Intentional Use of Fire

Let us now consider in detail an important category of evidence accompanying Ameghino's discoveries of stone tools—signs of intentional use of fire. At various locations, along with stone tools, Florentino Ameghino discovered, remnants of hearths, in the form of burned earth (*tierra cocida*), slag (*escoria*), charcoal, and burned animal bones. This combination of evidence tends to strongly confirm the view that the tools were manufactured by human beings in the distant past. In some cases, Ameghino interpreted the presence of scoria (slag) and burned earth as signs of grass fires intentionally set by primitive hunters.

Ameghino gave great importance to his discoveries of burned earth and slag. While in Argentina, Hrdlicka and Willis therefore collected many such specimens. At Miramar, for example, Willis found broken chunks of red *tierra cocida* and pieces of heavy black scoria 8–10 centimeters [3–4 inches] in diameter, all of which "occurred in the undisturbed Pampean" (Hrdlicka 1912, p. 47).

Some scientists thought the Miramar *tierra cocida* and slag were the product of volcanoes. But Whitman Cross of the U.S. Geological Survey had conducted studies of the slag and burned earth. Willis stated: "According to Mr. Cross . . . they are probably not volcanic" (Hrdlicka 1912, p. 47). Some authors suggested grass fires as the cause. Cross tested this idea by burning the most common Pampas grass (*cortadera*) on samples of earth, but this produced only a very thin layer of hardened earth, with no bricklike *tierra cocida* or melted scoria. But Willis, while visiting the Rio Colorado region of Argentina, observed another kind of grass, called *esparto,* that grows more deeply into the earth, and saw a place where it had burned. At this location, he observed one could pick up pieces of brick-colored earth up to 10 centimeters [4 inches] in diameter. Some of the pieces were penetrated with grass roots and carbonized grass, as in the case of some of the specimens described by F. Ameghino (Hrdlicka 1912, pp. 46–48).

In his reports about Monte Hermoso and other Argentine sites, Ameghino had noted the presence of similar specimens. He said that Dr. Gustave Steinmann came to Argentina during an expedition to South America, and in 1906 visited the *barrancas* of the Atlantic coast near Cabo Corrientes, accompanied by Santiago Roth and Robert Lehmann-Nitsche. F. Ameghino (1908, p. 106) stated: "These gentlemen discovered in the *barrancas* pieces of burned and partially vitrified earth, reporting specimens resembling or identical to those from the beds at Monte Hermoso, which I had attributed to the action of man and presented as proof of his existence in that distant epoch."

But Steinmann believed that humans had appeared in South America only in recent times. F. Ameghino (1908, p. 106) noted: "In a report presented during the course of the past year by Dr. Steinmann at the Geological Society of Berlin, he stated that these reputed vestiges of *Homo americanus* were in fact natural productions that appeared to be caused artificially only in the imaginations of recent immigrants of the species *Homo europaeus*. According to Dr. Steinmann, the specimens were pieces of volcanic lava which had arrived there through the air or more probably by means of water currents." The nearest volcanoes, however, were a thousand kilometers (621 miles) from the Atlantic coast, in the

Cordillera, the mountain range running the length of western Argentina. Still, Steinmann believed that small pieces of scoria were transported by rivers.

F. Ameghino (1911, pp. 68–69) responded: "Although all the strange affirmations of M. Steinmann will be refuted in detail in a monograph I am preparing, the facts have been so misrepresented by him that I cannot restrain myself from remarking that all that he has said in connection with the relative antiquity of man in South America and Europe is a natural result of his preconceived ideas. For Steinmann, the presence in true geological formations of scoria is an illusion, and the supposed formations do not actually exist. The pieces of scoria he encountered may be no bigger than nuts or somewhat bigger. But I have found masses of burned earth weighing many kilograms, the transport of which from the Cordillera to the places in which they are found, by means of movement through the air or by rivers, is impossible. Contrary to his statements, the scoria are accompanied by, that is to say, they are embedded in the same strata with, other vestiges of the activity of man (burned and broken bones, etc.)."

F. Ameghino (1908, p. 106) further stated about Steinmann's hypothesis: "Fantastic though it may be, this opinion is not completely new; I mentioned it 18 years ago, but did not consider it worth much discussion. Dr. Steinmann, in characterizing these vestiges as volcanic lava, has proceeded with excessive haste. What he has characterized as volcanic lava is a product resulting from the burning of fires intentionally set in dry grass." Ameghino noted that modern Indians sometimes burn dry Pampas grass to drive out small game for hunting, producing fused earth, which, because of the holes left by the roots, resembles lava. He held that the ancient Tertiary inhabitants of Argentina had done the same (F. Ameghino 1908, p. 107). Of course, one could also propose that the grass fires could have been started by lightning strikes.

But these light and porous specimens of burned earth were not the only kind found by Ameghino. Other specimens, from a variety of sites along the coast, were harder and more solid. Noting this distinction in his own research, Hrdlicka (1912, p. 50) stated: "Small particles and occasionally larger masses of *tierra cocida*, were found by Mr. Willis or the writer in a number of localities along the coast from northeast of Miramar to Monte Hermoso, and were relatively abundant in the deposits exposed in the *barrancas* at the former locality. They occur at different depths from the surface, to below the sea level at ordinary low tide. The pieces collected are all compact, with the exception of two or three that show on one side a transition to scoria. While there is a general resemblance, they all differ in aspect and weight from the very porous, light products of the burning of the *esparto* grass, collected by Mr. Willis on the Colorado."

As we have seen, Ameghino thought some of the compact pieces of scoria and burned earth were the remnants of *fogones*, or fireplaces, rather than grass fires, which may, it seems, have been set naturally rather than by humans. But Willis rejected human action in all cases. About some specimens from Monte Hermoso, Willis stated: "Through the courtesy of Doctor Ameghino the writer saw at Buenos Aires 10 pieces of burnt clay which would appear to have formed a layer about 10 by 15 cm. [4 by 6 inches] in area and about 5 to 10 mm. [.2 to .4 inch] thick, collected by Ameghino from the Monte Hermosean formation below high-tide level. As stated in describing certain observations on burnt earth of the Pampaean, the writer finds that clays of that formation may be burnt without the agency of man, and he does not attach any significance to the occurrence of burnt earth as an evidence of man's existence in the Miocene (?) 'Monte Hermosean'" (Hrdlicka 1912, p. 364).

Willis also stated: "In order to prove that man maintained a fire which burned a particular mass of *tierra cocida* it would be necessary to bring independent evidence of his handiwork" (Hrdlicka 1912, p. 364). In many cases Ameghino did, however, supply such independent evidence. Hrdlicka himself noted that "burnt bones, carbon, and other substances that might possibly be due to man have been found at or near fogónes" (Hrdlicka 1912, p. 50).

Willis was quick, perhaps too quick, to dismiss this evidence. He wrote: "Two classes of facts have been cited to demonstrate his [man's] agency: The presence of supposed artifacts and the arrangement of a mass of burnt clay; chief among the former are split, broken, or scratched fragments of bone, and it appears to the writer that these may be referred, with greater probability, to weathering, biting, gnawing, and accidents incident to the wanderings of bones, as strata were eroded and redeposited" (Hrdlicka 1912, p. 48). Willis's remarks about the bones are extremely suspicious, especially when considered in the light of our discussion of the treatment of such evidence in Chapter 2. Also, it should be kept in mind that Willis was a geologist, with no particular training in the study of incised bones. Any fair-minded investigator would want to have a careful look at those bones before accepting Willis's characterization of them.

Willis then stated: "Certainly the proofs of man's agency should be un-controvertible and the possibility of explanation by other than human action should be positively excluded, before the conclusion that he intentionally or incidentally burned the earth can be accepted" (Hrdlicka 1912, p. 48). Here Willis is demanding a level of certainty that empirical evidence relevant to paleoanthropology is incapable of providing. Scientists representing an estab-lishment view often dismiss anomalous evidence by requiring it to meet a higher standard of proof than the conventionally accepted evidence.

It is, however, possible that the compact burned earth and slag were not the product of campfires, as proposed by Ameghino. Hrdlicka observed some contemporary fire sites, noting that reddening and blackening of the earth was produced, but no cohesion. This suggested the improbability that the compact *tierra cocida* resulted from campfires (Hrdlicka 1912, pp. 49–50). Furthermore, specimens of *tierra cocida* were sent to Washington, D.C., where they were examined by Frederick Eugene Wright and Clarence N. Fenner of the Geophysical Laboratory of the Carnegie Institution. These researchers reported that the *tierra cocida* was composed of Pampean loess heated at 850–1050 degrees Centigrade, a temperature they said was too high to be attributed to either grass fires or small wood bonfires (Hrdlicka 1912, p. 88).

Evidence for a more intensive fire was suggested by the presence of the scoriae, or pieces of slag. According to the report of the Geophysical Laboratory, the scoriae examined there were not of volcanic origin. Wright and Fenner noted that the scoriae "do not agree with any known eruptive rock or lava in their microscopic features" (Hrdlicka 1912, p. 94).

Wright and Fenner went on to note some puzzling features of the scoriae. First they were a melted loess, but the melted loess was not composed of the same materials as the layers of loess from which the scoriae had been extracted. To Wright and Fenner, this indicated the scoriae had not been produced by fire in that locality. Second, although the glassy scoriae contained iron compounds, they were not reddish in color, as would be the case if the iron compounds had been exposed to oxygen. This indicated that the scoriae were not formed by the action of fire in the open air. The scientists of the Geophysical Laboratory, straining for an explanation, suggested that the scoriae were produced underground by an extrusion of molten lava from deep within the earth, which melted a loess different from that found in the surface layers (Hrdlicka 1912, pp. 93–97).

But there are many difficulties with such an explanation. First of all, as noted by Wright and Fenner, there was no sign of any extrusion of lava in the strata throughout which the scoriae were found scattered. The researchers of the Geophysical Laboratory nevertheless stuck to their opinion that contact of loess with molten lava was the most likely cause of the scoriae. But they had to go to great lengths to explain away the absence of any normal lava at the sites from which the scoriae had come: "it may be that the volcanic extrusion was of the explosive type, whereby the lava . . . was shattered and reduced to dust, which fell to the surface as volcanic ash and now constitutes an integral part of the loess formation. Under these conditions the cooler, viscous, melted loess fragments would remain intact and be ejected as scoriae and resist attrition and breaking down more effectively than the shattered volcanic lava" (Hrdlicka 1912, p. 96).

5.1.7 Primitive Kilns and Foundries?

The lava hypothesis of Wright and Fenner involves a quite extraordinary chain of speculative reasoning. There is, however, a possible explanation for the burned earth and slag that places considerably less strain on the limits of credibility—namely that they might be the result of intentional fire of a type other than campfires. Even today, one can observe inhabitants of many areas of the world making use of primitive foundries and kilns. Let us therefore consider the hypothesis that the burned earth and slag present on the Argentine coast are the byproducts of crude iron smelting furnaces. This idea was suggested to us by Arlington H. Mallery's book *Lost America,* which describes primitive iron furnaces discovered in Ohio and other locations in North America. Mallery thought the makers of the furnaces came from Europe. Since the type of process used in these foundries went out of use in Europe before the time of Columbus, Mallery therefore concluded that the furnaces he found in America must have been used by pre-Columbian European immigrants. And this, according to standard views of history, is unexpected. Admittedly, the kiln or foundry hypothesis is speculative, but no more so than the disappearing lava hypothesis offered by Wright and Fenner.

Mallery (1951, p. 100) stated: "The earliest iron-smelting furnaces in both the Old World and the New were merely shallow pits with rounded bottoms located on the hilltops. In order to catch the usual up-draft of air from the valley below for combustion, they were built close to the edge of the hillside facing the prevailing winds." In Argentina, the prevailing winds are the southeast trades that blow in from the ocean, so it seems the coastal slopes would be suitable for natural draft furnaces. Mallery (1951, p. 199) further stated: "The bottoms of these pit-furnaces were frequently covered with a layer of clay spread evenly to form a rounded basin from six to twelve inches deep."

Describing the smelting process, Mallery (1951, pp. 197–198) stated: "iron smelting was performed in three distinct stages utilizing, as a rule, bog ore from swamps. The ore was first piled up in heaps on layers of wood fagots and heated or calcined until it was red. It was then mixed with fuel and burned in a smelting furnace operated at a temperature below the melting point (about 2100 degrees) of cast iron. At or below this temperature, the fusible material in the ore became a fluid slag which seeped down and formed a pool in the pit of the furnace. The iron and mineral oxides in the ore were carried down with the slag and collected in a porous lump or bloom at the bottom of the pool. When the melt was completed the fire was quenched with water and the iron-workers lifted the bloom, still red hot, out of the furnace. It was then beaten with stones or heavy hammers to squeeze out some of the contained slag. In the finishing stage the bloom was usually taken to a smithy, reheated in a smaller furnace or forge, and hammered to squeeze out more of the slag, the process being repeated until forgeable wrought iron was obtained."

What exactly is bog ore? Mallery (1951, p. 199) explained: "Bog ore is a yellowish-brown, clay-like material composed mainly of clay, loam, and hydrated oxides of iron. Some pottery maker who attempted to use bog ore instead of clay for his pots may have discovered the iron-extracting process. . . . Even now, the small closed furnaces used by the Agaria in India and, until recently, by the Liberian natives, resemble pottery kilns."

As it turns out, there is an iron-rich earth at Miramar and other localities on the coast. For example, Wright and Fenner analyzed specimens from Miramar, describing them as "brown ferruginous earth" with "pronounced accumulation of limonitic material" (Hrdlicka 1912, p. 70). Limonite is an iron ore. Wright and Fenner also observed: "Brown ferruginous earths have also been considered *tierra cocida* by some investigators. A careful microscopic examination of these specimens has shown that they are simply loess in which ferruginous material abounds" (Hrdlicka 1912, p. 89). It is possible that these ferruginous earths could have served as the raw material for iron smelting.

A key indicator is the iron content of the slag left over from smelting. Mallery (1951, p. 200) pointed out: "The iron content of the slag . . . in the mounds of England, Belgium, Scandinavia, Virginia, and the Ohio Valley is very high—from 10 per cent to 60 per cent. Slag produced in modern blast furnaces, which have been in general use since the fourteenth century, seldom contains more than one per cent iron." He then gave a specific example: "On top of Ohio's Spruce Hill is an extensive deposit of slag. In this deposit are several low mounds composed mainly of typical hearth-pit slag, which tests show has an iron content of about ten per cent. Cutting a short trench into this heap, I uncovered the edge of a twelve-inch slab of clay. In the heap were large pieces of slag, lumps of red-burned bog ore, charcoal and glazed stone" (Mallery 1951, p. 204).

How does this compare with the slag found on the Argentine coast? Chemical analysis of a scoria sample from north of Necochea revealed 9.79 percent iron compounds (Hrdlicka 1912, p. 81). Another piece of scoria from San Blas, north of Rio Negro gave 9.71 percent iron compounds (Hrdlicka 1912, p. 86). Several other samples yielded at least 5 percent iron compounds.

The following description of a crude furnace uncovered in Sweden is interesting when compared with the evidence discovered in Argentina. John Nihlen stated: "The owner of the farm found some pieces of slag on a hill about two hundred meters south of the farm. In a smaller pit here was found under the grass, one-half meter [20 inches] deep, a large amount of slag pieces, such as iron slag in chunks of glazed pieces mixed with or attached to pieces of hard-burned red clay. At the bottom of the pit was dark sand and a few cinders of charcoal but no real burned material. Around the pieces of slag were some round stones but no real construction of stone" (Mallery 1951, p. 204).

Of particular interest in the above statement are the pieces of scoria "mixed with or attached to pieces of hard-burned red clay." At Miramar, reported Wright and Fenner, Hrdlicka and Willis collected some specimens of *"tierra cocida* and scoriae combined" (Hrdlicka 1912, p. 73). Wright and Fenner described a particularly interesting example: "The hard specimen shows a regular and uniform transition from a dark-gray scoria filled with small vesicles to a brick-red material, which bears a close resemblance to some of the specimens of baked earth. It is different from the latter, however, in this respect that, while the baked earths have a close, compact texture, the portion of this specimen which resembles them most . . . is filled with minute holes and is distinctly glassy in character. . . . A careful determination of the mineral fragments in the black and the red portions of the specimen proved them to be of the same general size and kind. . . . Superficially the red portion of this specimen resembles the baked earths, but closer examination has shown it to be distinctly different. Its glassy, vesicular texture throughout is indicative of melting; the red coloration may be the result of alteration or oxidation, whereby magnetite has been changed to the red oxide of iron" (Hrdlicka 1912, pp. 73–74).

At another location in Sweden, John Nihlen discovered another furnace, and described it as follows: "While the gravel was being dug, pieces of slag were found here and there, none of them collected in heaps nor visible on the surface. . . . It [the furnace] was about one meter [about 3 feet] wide in the upper part and narrowed slightly downward, being cup-formed at the bottom. The sides were made of round or flat gray stone which were laid in clay which also covered large parts of the inside. Probably the lining was not over the stones. The bottom of the furnace . . . consisted of a ten-centimeter [4 inch] layer of hard and partially burned clay. It could almost have been taken out. In the cup-formed lower part there still remained a ten-centimeter layer of slag, bog ore, and charcoal. The depth of the furnace was about one meter. . . . it had been a simple earth furnace without a blast intake, built of stone and clay and with a thick bottom of burned clay" (Mallery 1951, p. 201).

Here we take note of the furnace bottom, which consisted of a "ten-centimeter layer of hard and partially burned clay." Willis described a similar section of hardened red earth found in the Chapadmalalan beds of a seaside *barranca,* or cliff, at Miramar. The Chapadmalalan, said by Ameghino to be of Late Miocene age, is dated by modern authorities to the Late Pliocene (about 2–3 million years before the present). According to Willis, the section of burned earth was approximately 1 meter, or just over a yard, long and 30 centimeters, or about a foot, deep. The upper part was of red clay, passing into a dark brown and black mass that faded into the brown loess. Willis stated: "The principle mass of red clay is 60 cm. [about 2 feet] long and 10 cm. [about 4 inches] thick" (Hrdlicka 1912, p. 46). Willis attributed this particular specimen to a process of chemical dehydration, but admitted that its "coloring might have been occasioned by a fire burning on the surface that is now red" (Hrdlicka 1912, p. 46).

It would, however, have taken an extraordinarily hot fire to produce the observed effects. Wright and Fenner stated: "The assumption that the large specimens of *tierra cocida* were formed simply by the action of open fires is hardly possible in view of the quantity of heat involved, which must have acted through a period of time on large masses of material to have produced the effects observed" (Hrdlicka 1912, p. 85). They further stated: "Many of the specimens of *tierra cocida* are so large and compact that one is forced, in explaining their mode of formation, to assume long-continued and confined heating at a fairly high temperature, such as would be encountered near the contact of an intrusive igneous or volcanic mass, but not beneath an open fire made of grass or small timber" (Hrdlicka 1912, p. 89). But there was no evidence of intrusive volcanic masses at the sites under consideration, and "long-continued and confined heating at a fairly high temperature" is characteristic of a kiln or furnace.

The furnace hypothesis would explain the dark gray rather than red color of some of the scoria. Wright and Fenner, in conducting thermal experiments, noted that when small samples of loess were burned they turned red because all the loess particles were exposed to oxygen. But when larger masses were burned, oxygen did not reach the interior, which remained gray, like some of the Argentine scoria (Hrdlicka 1912, p. 88). As we have seen, the smelting process outlined by Mallery involved burning large masses of ore, the interior of which may have remained gray. Furthermore, the primitive furnaces operated on the principle of reduction rather than oxidation, which would also account for grayish rather than reddish slag.

In summary, we propose the following. While some types of *tierra cocida* might have been produced by grass fires, campfires, or perhaps even chemical dehydration, the thick, hard, red pieces of *tierra cocida* at the Argentine sites might well have been the burned earth that lined the bottom of primitive smelting furnaces. Samples of this burned earth were, according to Wright and Fenner, generally of low iron content (Hrdlicka 1912, pp. 88–89). Also found at the Argentine sites were pieces of brown ferruginous earths, which might represent unburned ore. Wright and Fenner did not give a chemical analysis of these earths, but they would appear to be of high iron content. The gray scoria and the gray-and-red scoria, as indicated by their iron content of about 10 percent, could represent the slag from primitive iron smelting furnaces operated on the Argentine coast several million years ago. Other specimens of scoria might have been produced in connection with pottery kilns.

5.1.8 Ameghino on the South American Origins of Hominids

Florentino Ameghino proposed that human beings evolved on the South American continent and then migrated first to North America, and thence by separate routes to Europe and Asia. At this point, many will doubtlessly conclude that Ameghino was simply an overly patriotic Argentine nationalist promoting the totally absurd view that humans originated in the country of his own birth.

But the same skeptics accept without similar reserve the claims of scientists such as Leakey, Broom, and Dart, who resided in former British possessions in Africa and proposed that the human race just happened to originate there.

In fact, paleoanthropologists the world over have a tendency to claim their homelands as the cradle (or one of the cradles) of humanity. Scientists from China, India, and the former Soviet Union (Section 3.6.4) maintain such conceptions. Underlying most such claims is the assumption of a monogenetic evolutionary origin of the human race—that human beings evolved only once from apelike ancestors within a certain region and then radiated from there to populate the rest of the world. Today, the dominant view in science is that the first apelike human ancestor (*Australopithecus*) arose in Africa in the Pliocene. In the Early Pleistocene, this creature attained protohuman status (as *Homo habilis* and then *Homo erectus*). Further evolutionary progress in Africa resulted in the emergence of *Homo sapiens sapiens* about 100,000 years ago.

As we have seen, an important confirmation of human or protohuman status is the presence of stone tools, such as those found in Bed I of Olduvai Gorge. These tools, among the oldest given unqualified recognition, are attributed to *Homo habilis* in the Early Pleistocene. According to the monogenetic evolutionary assumptions underlying modern paleoanthropology, one should expect to find tools dating from the very early Pleistocene only in eastern or southern Africa.

Florentino Ameghino, however, discovered stone implements in strata dating back to the Early Pliocene (3–5 million years B.P.) and even as far back as the Miocene—and in Argentina instead of Africa. Along with stone implements, Ameghino found abundant signs of human occupation, such as evidence of fire, burned bones, incised bones, and human skeletal remains.

So one might wonder, do we intend to give support to Ameghino's claims that humankind originated in Argentina? Hardly. But we do feel that the evidence uncovered by Ameghino lends strong support to the conclusion that the whole concept of a monogenetic evolutionary origin for humanity, be it in Kenya, Argentina, Siberia, China, or Kashmir, is incorrect. If at various points around the world one can find stone implements and other evidence for the presence of human beings dating back as far as 20 million years, there is good reason to suspect that the current picture of human origins and antiquity is completely wrong.

It just might be that the version of human ancestry promoted by the dominant Anglo-American school of anthropology, namely that humans originated in former British possessions in Africa, deserves no greater credibility than an Argentine scientist's claim that humans originated in South America. Indeed, humans may not have evolved at all. Human beings may have been present on this planet, in their current form and at essentially the same level of cultural advancement, for as far back in time as we can carry our investigations. That is what the totality of the evidence—not the carefully edited selection of evidence found in current textbooks—actually suggests. Or to put it another way: the hypothesis that human beings of the fully modern type have existed on this planet for several millions of years accounts for all the available evidence, in the form of stone implements, incised bones, and human skeletal remains, more fully than the modern evolutionary theory, which survives only by discarding, under various excuses, a vast number of discoveries made by scientists over the past 150 or so years. The discoveries of Florentino Ameghino are a case in point.

5.2 TOOLS FOUND BY CARLOS AMEGHINO AT MIRAMAR (PLIOCENE)

After Ales Hrdlicka's attack on the discoveries of Florentino Ameghino, Ameghino's brother Carlos launched a new series of investigations on the Argentine coast south of Buenos Aires. From 1912 to 1914, Carlos Ameghino and his associates, working on behalf of the natural history museums of Buenos Aires and La Plata, discovered stone tools in the Pliocene strata of a *barranca,* or cliff, extending along the seaside at Miramar.

5.2.1 A Commission of Geologists Confirms Age of Site

In order to confirm the age of the implements, Carlos Ameghino invited a commission of four geologists to give their opinion. The geologists were Santiago Roth, chief of the paleontology section of the Museum of La Plata and director of the Bureau of Geology and Mines for the province of Buenos Aires; Lutz Witte, a geologist of the Bureau of Geology and Mines for the province of Buenos Aires; Walther Schiller, chief of the mineralogy section of the Museum of La Plata and consultant to the National Bureau of Geology and Mines; and Moises Kantor, chief of the geology section of the Museum of La Plata.

In their report, the commission of geologists (Roth *et al.* 1915, p. 419) first told what they were asked to investigate: "The two questions are: (1) Were the objects in question found in primary deposits, that is to say, were they covered over at the time the deposits were being laid down, or is there reason to doubt this and to suppose instead that the objects were buried by a different cause at that site, at a time later than the formation of the respective deposits? (2) Concerning the stratigraphic position of the beds that contain the objects, can it be determined if they correspond to levels of the Eopampean horizon (the Montehermosan of F. Ameghino); or were the respective sediments more recently deposited against an ancient *barranca,* or in an eroded valley or some other depression of the earth that corresponds to the later part of the Pampean series?"

TABLE 5.1

Stratigraphy of Coastal Provinces of Argentina

Est. Age (Years B.P.)	Marshall et. al. 1982	F. Ameghino 1909	Roth et. al. 1915
		Platean	Postpampean
	Lujanian	Lujanian	Neopampean
		Bonarian	
.4 million	Ensenadan	Ensenadan	Mesopampean
1.0 million			
	Uquian	Puelchean	?
2.0 million			
	Chapadmalalan	Chapadmalalan	Eopampean
3.0 million			
	Montehermosan	Montehermosan	
5.0 million			

The report (Roth et al. 1915, p. 420) then went on to describe the stratigraphy of the *barranca* in some detail, making use of a nomenclature different from that used by either Ameghino or modern authorities (Table 5.1): "The cliff displayed four Pampean horizons: Eopampean (the Montehermosan and Chapadmalalan of F. Ameghino); Mesopampean (Ensenadan); Neopampean (Bonarian and Lujanian); and Postpampean (Platean)."

Of special interest were the Mesopampean and Eopampean. The Mesopampean was a bank of water-redeposited loess, 3–4 meters [10–13 feet] thick, extending 500 meters [1640 feet] between two transverse valleys interrupting the *barranca*. Of the Mesopampean layer, the commission said: "C. Ameghino, Schiller, and Roth agree that the bank in question corresponds to the Ensenadan level in the subdivisions of the Pampean made by F. Ameghino" (Roth et al. 1915, p. 420).

"In some parts there are layers of rounded stones," reported the commission of geologists. "Also present is freshwater limestone, very common in the Mesopampean. The loess is traversed in all directions by veins or seams of calcareous tufa, which frequently form in such beds. These stratigraphic and lithological   conditions make it impossible to suppose that cavities were formed and refilled after the Mesopampean formation was initially deposited" (Roth *et al.* 1915, p. 420).

This is an important consideration. One might say that the presence of stone implements in the Eopampean strata below the Mesopampean could be accounted for in the following way. Imagine a fairly recent Indian settlement on the top of the *barranca*. The villagers leave stone tools on the surface. Later, an arroyo forms in the *barranca*, cutting through the Mesopampean layers into the Eopampean formation. Stone tools are washed into the bottom of the arroyo. Later, the arroyo is refilled, leaving stone tools in the Eopampean layer. As we shall see, this is exactly the sort of challenge that would be made (Section 5.2.3).

According to the commission, the geological evidence ruled out such cutting and refilling (Roth *et al.* 1915, p. 420). The Mesopampean formation contained distinct layers of stones, seams of calcareous tufa (a porous limestonelike material), and deposits of *tosca* (a hard limestone deposit). All these would have been noticeably disturbed if cut by a gully that was later refilled.

The commission report then turned to the layers that contained implements: "The base of the *barranca* is formed of Eopampean deposits. Carlos Ameghino, Schiller, and Roth declare that the geological characteristics of this deposit are exactly like those of the loess found at the base of the Lobería *barranca* south of Mar del Plata, where F. Ameghino originally established the Chapadmalalan formation. In both locales one encounters, according to Carlos Ameghino and Roth, remains of mammals typical of this level, among them an abundance of *Pachyrucos* [a small rabbitlike creature]" (Roth *et al.* 1915, p. 421).

The geologists went on to say: "The banks of calcareous tufa are almost completely absent, and in general the *tosca* is much scarcer than in the upper beds; but despite this, the loess forms a very consistent mass and to break it required a pick and crowbar. The general aspect is of an eolian loess, formed of very homogeneously pulverized mineral substances" (Roth *et al.* 1915, p. 421). The identification of the Eopampean as eolian (wind-deposited) loess is important, for later an opponent (Romero 1918) would charge the layers were of marine origin (Section 5.2.3).

The geologists noted that the loess contained pieces of burned earth and scoria (slag). "Moreover," said the report, "at distinct locales in the *barranca*, the original investigators found objects, instruments, and weapons of stone, fabricated by different techniques" (Roth *et al.* 1915, p. 421).

"The first deposit of implement-bearing loess to be examined," said the geologists, "was approximately 50 meters [164 feet] from the small drainage channel that exists in the slope of this *barranca* and more or less 1 meter [about 3 feet] lower than the limit of the horizon between the Mesopampean and the Eopampean" (Roth *et al.* 1915, p. 421).

The geologists then recounted how the initial discoveries took place: "The first objects were discovered, according to Torres and Ameghino, when Lorenzo Parodi [a collector employed by the natural history museums of Buenos Aires and La Plata] attempted to extract a piece of slag. Parodi's pick struck a hard object, which when uncovered, turned out to be a bola stone. It was extracted encased in a chunk of loess, and it is preserved in the same condition in the Museum of Natural History in Buenos Aires. Later, Torres, Ameghino, and Doello-Jurado, digging at the same site, discovered other stone objects and instruments, and finally Parodi very recently discovered a round stone and a flint knife in place, and left them there, following the instructions he had received, in order that they could be extracted in the presence of this commission of geologists" (Roth *et al.* 1915, p. 421). It is apparent from this description that the excavation was carried out with some degree of professionalism—the commission of geologists was able to study implements *in situ*.

The report then conclusively answered the first of the questions the geologists were asked: "This commission . . . after examining the place where the artifacts in question were found, gave their unanimous opinion that if the sediments had shifted after the time of deposition, the members would have been able to see some alterations in the texture of the bed, but they were not able to observe any such alterations. The lithological composition of the sediments and the texture of the deposit that contained the artifacts did not demonstrate any difference in character from the loess of this horizon. All of those present declared that the stone artifacts . . . were found in intact, undisturbed terrain, in primary position. Based on this fact, the first question posed may be answered: visual inspection of the site where the artifacts were found has not given us any reason to suppose that the artifacts have been buried by any means whatsoever at a time after the formation of the bed. They are found in primary position and, for that reason, should be considered objects of human industry, contemporary with the geological level in which they were deposited" (Roth *et al.* 1915, p. 422).

The report was equally conclusive about the second matter under consideration: "In respect to the second question, whether it is possible that there was in this place a juxtaposition of strata, or if it is possible that the layer containing the artifacts was deposited up against an old cliff and therefore corresponds to one of the most recent levels of the Pampean formation, we firmly declare: the stratigraphic conditions at this place are so clear as to present no difficulty in resolving any such problem. In the first place, the above-mentioned layer of freshwater limestone, which corresponds to the Mesopampean horizon and is found directly above the Eopampean deposits, has not suffered any alteration. Nor has it been possible to find in any part of the bed in question any refilling of gullies or caverns after the bed's formation. Furthermore, the face of the cliff is quite vertical, which allows one to see clearly that the sediments found in the lower part were not deposited against an old cliff of the Mesopampean horizon. Rather the beds of the Mesopampean horizon pass in all places above the deposits in question. The undulations and irregularities presented by the horizon between the Mesopampean and Eopampean beds are filled with the calcareous tufa above mentioned, allowing one to distinguish a discordance between the two horizons. The committee of geologists is in accord that the second point in question can be defined in the following way: that the objects of human industry encountered in this place are situated in deposits of loess characteristic of the Eopampean horizon, which constitute the base of the *barranca*; and that the stratigraphic relationships allow us to establish with scientific certainty that there exists here no juxtaposition of newer layers and older ones" (Roth *et al.* 1915, pp. 422–423).

The Eopampean layers in which the implements were found correspond to the Chapadmalalan formation, sometimes called the Chapadmalean or Chapalmalean. Modern authorities assign the Chapadmalalan formation a Late Pliocene age of 2.5–3.0 million years (Anderson 1984, p. 41) or 2.0–3.0 million years (Marshall *et al.* 1982, p. 1352). In their world survey of Pliocene mammalian fauna, Savage and Russell (1983) list Miramar as a Chapadmalalan site.

Concerning the objects they examined *in situ* at Miramar, the commission of geologists reported: "The round bola stone, which was discovered in the loess and which was extracted in the presence of the commission, did not display any sign of human work; but from its form and size it would appear to have served as a weapon, like the other bola previously discovered in the same stratum. The flint knife had fallen out onto the ground, but the place where it had been situated was noticeable; it presents every indication of having been fabricated by percussion and pressure" (Roth *et al.* 1915, p. 423).

As stated before, burned earth (*tierra cocida*) and slag (*escoria*) were both found at the Miramar site. Earlier Hrdlicka and other researchers (Sections 5.1.6, 5.1.7) had rejected the possibility that the burned earth and slag discovered at various Argentine sites, including the *barranca* at Miramar, could have resulted from fires of human origin. But the members of the commission of geologists reported: "Digging with a pick at the same spot where the bola and knife were found, someone discovered in the presence of the commission other flat stones, of the type that the Indians use to make fire" (Roth *et al.* 1915, p. 423).

Those committed to the standard view of human evolution will reflexively attempt to explain away all this evidence, but to those with more open minds, the facts clearly suggest that humans, capable of manufacturing tools and using fire, lived in Argentina about 2–3 million years ago in the Late Pliocene. The report further stated: "Carlos Ameghino, who continued the excavation, encountered another stone of small size, completely round and smooth, presenting the characteristics of a stone subjected to intentional work" (Roth *et al.* 1915, p. 423).

The commission made another find confirming its views: "About 50 meters [164 feet] from this site, in a layer still lower, there were found fossil remains of a species of the suborder Gravigrada [ground sloths]. During the excavation, there were discovered in this spot, in the presence of the commission, other round stones associated with the fossil remains. . . . Considering all the circumstances surrounding this discovery, as well as the condition of the objects and their stratigraphic relation to the bed, the commission is of the opinion that they are objects manufactured by humans who lived at the time of the geological period corresponding to the Chapadmalalan" (Roth *et al.* 1915, p. 423).

5.2.2 A Stone Point Embedded in a Toxodon Femur (Pliocene)

After the commission left for Buenos Aires, Carlos Ameghino remained at Miramar conducting further excavations in the Chapadmalalan beds just northeast of the spot where the bola stone and flint knife had been found. Ameghino uncovered many fossils of animals characteristic of the Chapadmalalan, such as *Pachyrucos,* a rabbitlike creature, and *Dicoelophoros,* a ratlike rodent. These animals were absent from the overlying Mesopampean beds (C. Ameghino 1915, p. 438).

From the top of the Late Pliocene Chapadmalalan layers, Ameghino extracted the femur of a toxodon, an extinct South American hoofed mammal, resembling a furry, short-legged, hornless rhinoceros. Ameghino discovered embedded in the toxodon femur a stone arrowhead or lance point (Figure 5.1), giving evidence for culturally advanced humans 2–3 million years ago in Argentina. Those who are committed to the view that *Homo sapiens sapiens* evolved about 100,000 years ago in Africa will likely attribute Ameghino's discovery to an intrusion from upper levels. But we would request such persons to, at least for a moment, set aside their preconceptions and withhold judgement while considering the facts of this remarkable case.

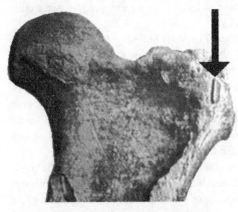

Figure 5.1. This toxodon thighbone (femur), with a stone projectile point embedded in it, was discovered in a Pliocene formation at Miramar, Argentina (C. Ameghino 1915, photograph 2).

Significantly, the toxodon femur was not discovered alone.

C. Ameghino (1915, pp. 438–439) reported: "As we proceeded with the excavation, there also appeared in the *barranca* almost all of the bones of the rear leg of the toxodon, still articulated and conserved in their relative positions. This is very evident proof that the toxodon femur was buried in the terrain contemporaneously with the formation of the bed and that it has not since been subjected to movement. In addition to the femur, the proximal extremity of which scarcely cropped out of the *barranca*, the bones, which, as mentioned, appeared articulated, included, the tibia and fibula, the calcaneum [heel bone], the scaphoid and other pieces of the tarsus [ankle], and finally some metatarsals. All the facts make it absolutely certain that these remains were found in their primary position. Their condition is identical to that of all the fossils that appear in this part of the *barranca* and in those parts that extend many leagues to the north. There have been discovered in this same *barranca*, on many occasions, perfectly articulated skeletons of animals from the same period as the toxodon. One of the most notable is a skeleton of *Pachyrucos*, which was discovered and extracted by the naturalist M. Doello-Jurado." From C. Ameghino's description, it is clear that the femur with the stone point embedded in it was the femur that was part of the articulated leg.

In December of 1914, Carlos Ameghino, with Carlos Bruch, Luis Maria Torres, and Santiago Roth, visited Miramar to mark and photograph the exact location where the toxodon femur had been found. C. Ameghino (1915, p. 439) stated: "Like the previous visits, this last visit was full of surprises. . . . When we arrived at the spot of the latest discoveries and continued the excavations, we uncovered more and more intentionally worked stones, convincing us we had come upon a veritable workshop of that distant epoch."

The many implements, including anvils and hammer stones, resembled, in form and lithic material, those of Florentino Ameghino's *piedra hendida* (broken stone) industry, discovered in the same region. Carlos Ameghino and Roth continued their investigations to the south at Mar del Sur and found stone tools in the Ensenadan level. The identification of the formation as Ensenadan had been accomplished previously, by the excavation from the same bed of a portion of the skeleton of *Typotherium cristatum*, a rodentlike mammal attaining the size of a small bear.

Taken together, the discoveries from the formations at Miramar and Mar del Sur, and other locations on the Argentine coast, are significant in that they show continuous habitation of the region by humans, from the Pliocene to recent historical times, with scarcely any change in the inhabitants' mode of living.

Returning to the toxodon bones found at Miramar, we find that Carlos Ameghino anticipated accusations that the bones had worked their way into the Chapadmalalan formation from above. In his report he stated: "The bones are of a dirty whitish color, characteristic of this stratum, and not blackish, from the magnesium oxides in the Ensenadan" (C. Ameghino 1915, p. 442). This tended to rule out any suggestion that the toxodon bones had been mixed into the Chapadmalalan from upper beds of lesser age. Ameghino further pointed out that some of the hollow parts of the bones were filled with the Chapadmalalan loess. If the bones were derived from another level, one might expect them to be filled with a different kind of material. Of course, even if the bones had worked there way in from the Ensenadan, they would still be anomalously old. Dates for the Ensenadan range from 0.4–1.0 million years (Marshall *et al.* 1982, p. 1352) to 0.4–1.5 million years (Anderson 1984, p. 41).

In describing the nature of the loess in which the bones had been discovered, C. Ameghino (1915, p. 442) said: "The terrain surrounding these remains is a loess exceedingly fine and pulverized, a true aeolian loess, fairly well decalcified and of reddish grey tint, a loess which, as we have said, corresponds to the Chapadmalalan" (C. Ameghino 1915, p. 442). Furthermore, as we have seen (Section 5.2.1), a commission of geologists had confirmed that the Chapadmalalan beds at Miramar were intact, showing no signs of disturbance.

Those who want to dispute the great age attributed to the toxodon femur will nevertheless point out that the toxodon survived until just a few thousand years ago in South America. They will say: "Of course, these early researchers were often surprised to find evidence of a human presence in connection with remains of the toxodon, an animal they thought typical of the Pliocene or Miocene, but since that time scientists have discovered the truth—that the toxodon roamed South America until quite recently." The clear implication is that if early researchers had been aware of this fact, they would certainly have hesitated to make claims for the great antiquity of humans based, for example, on the association of stone tools with toxodon bones.

But the fact that the toxodon lived until the Holocene does not rule out the discovery of toxodon bones in older strata, such as the Pliocene, for the toxodon definitely lived during that period. But the survival of the toxodon does allow critics to cast suspicion on finds such as Ameghino's, despite the fact that such finds were made in clear stratigraphic contexts.

Early researchers were often aware that mammals characteristic of ancient strata persisted until recent times. This is certainly true of Carlos Ameghino (1915, p. 442), who reported that the toxodon he found at Miramar, an adult specimen, was smaller than those in the upper, more recent levels of the Pampean stratigraphic sequence. This indicated it was a distinct, older species. Carlos Ameghino (1915, p. 442) believed his Miramar toxodon was of the Chapadmalalan species *Toxodon chapalmalensis,* first identified by F. Ameghino, and characterized by its small size.

Furthermore, Carlos Ameghino (1915, p. 443) directly compared his Chapadmalalan toxodon femur with femurs of toxodon species from more recent

Pampean formations and observed: "The femur of Miramar is on the whole smaller and more slender." Ameghino then reported more details showing how the femur he found in the Late Pliocene Chapadmalalan of Miramar differed from that of *Toxodon burmeisteri* of more recent Pampean levels.

None of this is mentioned by later critics. Boule, for example, in dismissing the toxodon femur, simply stated that the toxodon persisted in South America until fairly recent times (Boule and Vallois 1957, p. 492). But that does not invalidate Carlos Ameghino's conclusions. Toxodons did exist in the Pliocene, and according to Ameghino, the toxodon femur he recovered at Miramar was from a Pliocene species of toxodon. This information was available to Boule, yet he did not mention it. One could therefore say that Boule's presentation was dishonest. In order to have made a fair challenge to Ameghino, Boule should have demonstrated that Ameghino was incorrect in asserting that the femur he discovered was characteristic of a Pliocene species of toxodon.

In researching this book, we have learned that statements found in textbooks and scientific papers cannot always be trusted to give fair and accurate information about key discoveries. One quickly discovers that apparently objective statements reflect personal bias and prejudice and are often deliberately misleading. Nevertheless, Boule was not guilty of one of the most effective techniques for dealing with disconcerting evidence—complete omission.

Concerning the toxodon discovery at Miramar, we again emphasize that the bones of an entire leg were found articulated (in their natural relative positions) in the Late Pliocene Chapadmalalan formation. This indicates that the animal died in the Late Pliocene and that its bones were incorporated into the formation at that time. If the bones of a toxodon from a much later period had somehow been washed into the Chapadmalalan, one would not expect them to have been articulated.

Carlos Ameghino (1915, p. 445) then described the stone point found embedded in the femur: "This is a flake of quartzite obtained by percussion, a single blow, and retouched along its lateral edges, but only on one surface, and afterward pointed at its two extremities by the same process of retouch, giving it a form approximating a willow leaf, therefore resembling the double points of the Solutrean type, which have been designated *feuille de saule*. . . . by all these details we can recognize that we are confronted with a point of the Mousterian type of the European Paleolithic period." That such a point should be found in a formation dating back as much as 3 million years provokes serious questions about the version of human evolution presented by the modern scientific establishment, which holds that 3 million years ago we should find only the most primitive australopithecines at the vanguard of the hominid line.

Near the end of his discussion about the discovery at Miramar of the projectile point embedded in the toxodon femur, Carlos Ameghino (1915, p. 447) made some statements about Ales Hrdlicka, who, as we have seen, had attempted to demolish the work of Florentino Ameghino: "We cannot remain silent about the book recently published in this connection by Ales Hrdlicka and his collaborators (*Early Man in South America,* Washington, 1912). This work, apparently impartial and conscientious, serves, on the contrary, to reveal, especially in reference to the evidence for fossil man in this part of the Americas, the preconceived ideas of its authors. The authors did not spend in the terrain itself the time materially necessary to arrive at any judgement, as we had a chance to personally observe, since we accompanied them on many excursions. Without ignoring any part of the truth that this work may contain, we are convinced the conclusions of Hrdlicka are completely exaggerated. And the main proof of this is the report of the commission of geologists [Section 5.2.1]." The weight of evidence suggests Carlos Ameghino's statements about Hrdlicka's book are fully warranted.

Carlos Ameghino (1915, p. 449), in concluding his report on the projectile point found in the Miramar toxodon femur, stated that "at least since the Chapadmalalan, that is, the Late Miocene [Late Pliocene say modern authorities], there have existed in this territory humans of the type *Homo sapiens,* who, as surprising as it may seem, were possessed of a grade of culture and advancement comparable to the most recent prehistoric inhabitants of the region."

5.2.3 Romero's Critique of the Miramar Site

Carlos Ameghino's views about the antiquity of humans in Argentina were challenged by Antonio Romero. In a paper published in the *Anales de la Sociedad Científica Argentina,* Romero (1918) contradicted not only Carlos Ameghino, but his more famous late brother, Florentino Ameghino, who had for many years conducted research establishing a human presence in Argentina during the Tertiary. Quite apart from his work in paleoanthropology, Florentino Ameghino had gained an international reputation in the fields of paleontology and geology. A great deal of Argentine national pride was thus invested in Florentino Ameghino, who had almost singlehandedly focused the attention of the world's scientific community on his country. Romero was therefore very careful to frame his criticism of the Ameghinos with attention to Argentine patriotic sensibilities.

Early in the twentieth century, a dominant group within the scientific community was trying to "bury" evidence suggesting a human presence in the Tertiary. Romero was a supporter of this policy. In his paper, Romero (1918, p. 22), called special attention to the book *Fossil Man,* recently published by Hugo Obermaier, a noted European scientist who dismissed F. Ameghino's conclusions about a human presence in the Miocene and Pliocene of Argentina. Taking Obermaier's view as correct, and representative of responsible scientific opinion, Romero suggested that Carlos Ameghino and his supporters, by insisting on a human presence in the Tertiary of Argentina, were bringing ridicule and discredit upon the Argentine nation. Concerning Florentino Ameghino, Romero pleaded that Argentine science should continue to hold him in high regard for his valuable and quite extensive work in the areas of geology and paleontology, but that it was now time to set aside his unfortunate conclusions in the area of paleoanthropology, and thus preserve his reputation as a great scientist. Romero (1918, p. 15) wrote: "We now have to consolidate the monument of his work, casting out the fantastic discoveries that have so much preoccupied simpleminded spirits and cost our country so much, greatly injuring the work of the great scientist and his contribution to our culture."

As part of his investigation of Carlos Ameghino's discoveries, Romero visited the Miramar area. There he took time to view the fairly recent stone implements displayed in the small museum of Jose Maria Dupuy, a local collector. The implements had been gathered from the *paraderos* (settlements) of the coastal Indians. Noting the similarity of Dupuy's specimens to those Carlos Ameghino had forwarded to the Museum of National History in Buenos Aires from his exacavations in the Chapadmalalan formation at Miramar, Romero (1918, p. 12) stated that he was "convinced they were made by the same artificers who made those that are considered to belong to a fanciful epoch." In other words, Romero believed that Carlos Ameghino's discoveries were manufactured by Indians in relatively recent times.

Romero (1918, p. 15) went on to state that notwithstanding the similarity of C. Ameghino's Miramar implements to objects recently manufactured by Indians "there are arguments of a more fundamental order that we intend to pose in support of our thesis." Romero (1918, p. 15) said he would "demonstrate with incontestable facts, in the plain light of truth, that it was false to suppose that artifacts discovered at Las Brusquitas [Miramar], resembling classical types of the Neolithic age, can be attributed to human beings that existed in the Miocene [Late Pliocene according to modern estimates]."

After reading Romero's combative introductory remarks, one might expect to find in his report some cogent geological reasoning backed up with convincing facts. Instead one finds assertions backed up with little more than some unique and fanciful views of the geological history of the Miramar coastal region.

About the fossil remains from the *barranca*, Romero (1918, p. 24) said: "All the evidence relevant to this investigation demonstrates quite well that the bones are not from animals that died *in situ*, but are instead from skeletons of animals transported great distances, fractured, and dispersed by water." Then what about the almost complete rear leg and foot of a toxodon found by Carlos Ameghino? It hardly seems likely that flowing water brought together the several bones comprising the leg and foot and deposited them in their natural connection.

In this regard, Romero (1918, p. 24) said: "The discovery of bones more or less complete, and part of one skeleton, signifies that the bones were brought to their final resting place in that condition and not that the animal perished there." In this case, one would have to suppose that a detached rear leg of a toxodon, still covered with flesh was transported by water and somehow deposited in the lower levels at the Miramar site. But all this shows is that the animal died shortly before its leg wound up in the bed where it was found.

Romero implied that both the bones and the bed were recent. But according to Carlos Ameghino, the toxodon bones were from a Pliocene species of toxodon. Furthermore, modern authorities (Savage and Russell 1983, p. 365) still list the Chapadmalalan at Miramar as a Pliocene formation containing a distinct Pliocene fauna.

Romero (1918, p. 24), however, insisted: "If you find the fossils of distinct epochs in different levels of the *barranca,* that does not signify a succession of epochs there, because water may have elsewhere eroded very ancient fossil-bearing deposits of previous epochs, depositing the older fossils at the base of the *barranca.* I mention a case demonstrating this fact: the sea brings up fossil molluscs onto the beach, and my daughter found the foot bone of a great edentate, rolled up on the beach by the waves." The Edentata are an order of New World mammals that includes the sloths and armadillos. Romero was trying to build a case that the formations identified as Chapadmalalan and Ensenadan at Miramar were not really ancient, even though they contained fossils characteristic of the Late Pliocene and Middle Pleistocene, respectively. In making his case, Romero attributed remarkable capabilities to the action of sea waves and rivers. If Romero is to be taken at his word, he seems to have been implying that the random movements of water could selectively deposit fossils of certain periods in a definite sequence so as to mimic actual geological formations of those periods.

But Romero's speculative proposal appears incapable of accounting for the arrangement of fossils and sediments in such a way as to reproduce a series of actual geological formations, even in a relatively confined area. And here we are talking about a section of cliff extending for several hundred meters. Significantly, these same formations at Miramar had been extensively studied on several occasions by different professional geologists and paleontologists, none of whom viewed them in the manner suggested by Romero. Modern authorities also disagree with Romero.

In his attack on Carlos Ameghino, Romero sought to demonstrate that the implement-bearing beds of ancient Chapadmalalan loess at Miramar were fairly recent marine deposits. As evidence he cited the particular nature of the rounded stones that marked the boundary between the Chapadmalalan and the overlying Ensenadan. Romero (1918, p. 28) believed that their pattern of distribution, in an almost unbroken band along the entire formation, indicated they were pebbles formed by the action of waves and deposited on a beach. The large undulations now observed in the layer of stones were, according to Romero, caused by the action of later mountain-building forces in the region.

But Bailey Willis, no friend of Tertiary humans in Argentina, had earlier given a different interpretation of the undulating layer of stones in the *barranca* at Miramar. Willis, who had investigated several other *barrancas* on the same coast, wrote: "The sections were carefully studied in each locality, but since we require here only an illustration . . . it will suffice to describe a characteristic relation observed in the Barrancas del Norte, north of Mar del Plata. . . . The upper surface of the basal formation in the Barrancas del Norte [the Chapadmalalan] is eroded and the hollows are filled by later deposits, sometimes of one character, sometimes of another. It will be seen that the formation was carved by an agent that undercut the sides and rounded the bottoms of the hollows, leaving masses with sharp points or edges in relief. Wind produces these effects in this material, whereas water cuts channels having nearly vertical walls. Thus it would appear that wind erosion is favored. . . . The phenomenon recurs . . . in other exposures of the formation at Miramar" (Hrdlicka 1912, pp. 22–23). Romero thought the same hollows to be the result of marine action and mountain building rather than wind.

Then regarding the stone layer itself, found in the hollows, Willis wrote: "The pebbles . . . could have been formed only by wind action, since the loess of which they consist would readily melt down in water and lose its form. The formation thus suggests arid conditions" (Hrdlicka 1912, p. 24). Willis, a member of the U.S. Geological Survey, was an expert in the study of loess formations, having conducted extensive investigations in the course of geological expeditions in North America and China.

Romero's qualifications are unknown to us, and his view that the pebbles are a sign of marine action seems in great disharmony with the geological evidence. The same is true of his assertion that the Chapadmalalan at Miramar is actually a fairly recent marine mud deposited against the base of the cliff. This opinion was based upon visual inspection of a piece of sediment from an excavation at Miramar. Romero (1918, p. 31), stated about this chunk: "It is constituted principally of a mixture of clayey elements and sand, very fine, and is deposited uniformly in layers about 1 mm thick, which indicates successive, slow, tranquil deposition in a bay. Throughout the piece are many holes .25–1.0 mm in diameter, forming small tunnels in the direction of the plane of stratification, and on close observation you can see traces of organic remains of annelids." On the strength of this one piece of sediment, which he did not demonstrate to be typical of the entire deposit, Romero (1918, p. 31) then concluded: "It is obvious the bed was deposited on the sea bottom, and any animal bones or human artifacts found in the beds were brought there by the action of waves or were washed down from the cliff and covered up." Romero neglect to mention that annelids include not only marine worms but ordinary earthworms.

Furthermore, Willis described the formation at the base of the cliffs as follows: "At Miramar . . . the formation . . . consists of loess-like alluvium, the surface of which has been eroded and filled in by wind" (Hrdlicka 1912, p. 27). Describing the Chapadmalalan layer in the Barrancas del Norte, which he regarded as continuous with that at Miramar, Willis wrote: "The writer is inclined to regard this formation and similar deposits as due to river work on confluent flood plains" (Hrdlicka 1912, p. 23). A river flood plain is generally covered with water during only a small part of the year, and not every year. Such conditions are very favorable for fossilization of animal remains in primary position, especially when conditions are becoming more arid. At such times, animal remains may be buried during floods and remain undisturbed for long periods of time because of lower water levels. In short, Willis gave no hint that the deposit at the base of the cliff at the Miramar site was a recent marine formation.

The incorrectness of Romero's interpretation of the stratigraphy at Miramar is confirmed by modern researchers, who identify the formation at the base of the cliff as Chapadmalalan and assign it to the Late Pliocene, making it 2–3 million years old (Savage and Russell 1983, p. 365).

Considering Romero's farfetched and strained geological reasoning, one would certainly have a right to be cautious in accepting his conclusion about the stratigraphic position of the artifacts in the *barranca* at Miramar: "Visual inspection demonstrates that the artifacts discovered were interred at a time after the formation of the bed, and that they are in a secondary position in relation to the formation, because of an intrusion resulting from erosion at this place" (Romero 1918, p. 27). Here Romero mockingly reproduced the language of the report by the commission of geologists, who concluded the stone implements of Miramar were found in primary position (Section 5.2.1).

Romero, however, did not provide a great deal of evidence in favor of his point of view. In addition to his sea wave hypothesis, Romero suggested that there had been massive resorting and shifting of the beds in the *barranca,* making it possible that implements and animals bones from surface layers had become mixed into the lower levels of the cliff. But the only facts that he could bring forward to support this conclusion were two extremely minor dislocations of strata.   •

Some distance to the left of the spot where the commission of geologists extracted a bola stone from the Chapadmalalan level of the *barranca,* there is a place where a section of a layer of stones in the formation departs slightly from the horizontal (Romero 1918, p. 28). This dislocation occurs near the place where the *barranca* is interrupted by a transverse valley. As might be expected, part of the *barranca* slopes down to the left at this point, but at the place where the bola stone was extracted, the horizontal stratigraphy remained intact. At another place in the *barranca*, a small portion of a layer of stones departed only 16 degrees from the horizontal (Romero 1918, p. 29).

On the basis of these two relatively inconsequential observations, Romero suggested that all the strata exposed in the barranca had been subjected to extreme dislocations. This would have allowed the intrusion into the lower levels of stone tools from relatively recent Indian settlements that might have existed above the cliffs. Romero (1918, p. 30) asserted: "I have demonstrated . . . that the artifacts had been·intrusively buried in the strata called Chapadmalalan, and with this demonstration, which is irrefutable, I have also buried the opinion of the experts." But from photographs and descriptions of the stratigraphy by many other geologists, including Willis, it appears that the normal sequence of Pampean beds in the *barranca* at Miramar was intact in locations where discoveries were made.

Romero then continued with another barrage of unsatisfactory objections. He pointed out that some of the stone tools were found at the base of the Chapadmalalan bed and others in the middle, while the toxodon femur with the embedded projectile point was found at the top of the Chapadmalalan. Romero (1918, p. 33) thought the fact that the artifacts were distributed vertically within a limited horizontal space argued against their being in primary position. He would have preferred to see them all distributed in one horizontal plane. Why he thought this is not clear. If, as Carlos Ameghino believed, the place had been continuously inhabited from the Miocene until the recent past, by humans maintaining a constant level of cultural advancement, then one might expect to find just such a distribution of artifacts as was actually discovered.

Romero noted that stone tools resembling those extracted from the *barranca* are found on the surface, in a valley, slightly inland, that runs parallel to the barranca. He joked with his guide: "Are these Miocene implements?" But even today African tribal people use stone tools as crude as the Olduvai Gorge pebble choppers, which are attributed to human precursors living almost 2 million years ago. In other words, humans and humanlike beings in Africa have been making the same kinds of stone tools for at least 2 million years. It is, therefore, not valid to argue that the tools from the Chapadmalalan at Miramar must be recent because they resemble tools made by modern Indians in the same region.

5.2.4 Boule on the Toxodon Femur with Arrowhead

Now that we have considered Romero's objections to Carlos Ameghino's discoveries, let us turn our attention to a rare mid-twentieth-century review of the toxodon femur with the projectile point embedded in it. In the 1957 posthumous edition of *Fossil Men,* revised by H. V. Vallois, Marcellin Boule said that after the original discovery of the toxodon femur, Carlos Ameghino found in the Chapadmalalan at Miramar an intact section of a toxodon's vertebral column, in which two stone projectile points were embedded. Boule stated: "These discoveries were disputed. Reliable geologists affirmed that the objects came from the upper beds, which formed the site of a *paradero* or ancient Indian settlement, and that they were found today in the Tertiary bed only as a consequence of disturbances and resortings which that bed had suffered" (Boule and Vallois 1957, p. 492). Here Boule footnoted as a reference only the 1918 report by Romero! Boule did not mention the commission of four highly qualified geologists who reached a conclusion exactly opposite that of Romero, perhaps because they were, in his opinion, not "reliable." However, having closely studied Romero's geological conclusions, particularly in light of those of Bailey Willis and modern researchers, we are mystified that Romero should be characterized as "reliable."

Boule added: "The archaeological data support this conclusion, for the same Tertiary bed yielded dressed and polished stones, *bolas* and *boladeras,* identical with those used as missiles by the Indians" (Boule and Vallois 1957, p. 492). Boule said that Eric Boman, an "excellent enthnographer," had documented these facts.

Could human beings have lived continuously in Argentina since the Tertiary and not changed their technology? Why not, especially if, as certified by a commission of geologists (Section 5.2.1), implements were found *in situ* in beds of Pliocene antiquity? The fact that these implements were identical to those used by more recent inhabitants of the same region poses no barrier to acceptance of their Tertiary age. Modern tribal people in various parts of the world fashion stone implements indistinguishable from those recognized as having been manufactured 2 million years ago. We should also point out that in 1921 a fully human fossil jaw was found in the Chapadmalalan at Miramar (Section 6.2.5).

In his statements about the Miramar finds, Boule provides a classic case of prejudice and preconception masquerading as scientific objectivity. In Boule's book, all evidence for a human presence in the Tertiary formations of Argentina was dismissed on theoretical grounds and by ignoring crucial observations reported by competent scientists who happened to hold forbidden views. For example, Boule said nothing at all about the above-mentioned discovery of a human jaw in the Chapadmalalan at Miramar. We should thus be extremely careful in accepting the statements one finds in famous textbooks as the final word in paleoanthropology.

It is common to find scientists who disagree with certain controversial evidence taking the same approach as Boule. One mentions an exceptional discovery, one states that it was disputed for some time, and then one cites an authority (such as Romero) who supposedly conclusively settled the matter, once and for all. But we have found that when one takes the time to dig up the report that, like Romero's, supposedly delivered the coup de grace, it often fails to make a convincing case.

5.2.5 Boman, the Excellent Ethnographer

What was true of Romero's report is also true of Boman's. Boule, we have seen, advertised Boman as an "excellent ethnographer." But in examining Boman's report, the reason for Boule's favorable judgement becomes apparent. Throughout his paper, which attacked Florentino Ameghino's theories and Carlos Ameghino's discoveries at Miramar, Boman, taking the role of a dutiful disciple, regularly cited Boule as an authority. As might be expected, Boman also quoted extensively from Hrdlicka's lengthy negative critique of Florentino Ameghino's work. Nevertheless, Boman, despite his negative attitude, inadvertently managed to give some of the best possible evidence for a human presence in Argentina during the Pliocene.

Boman (1921, p. 336) wrote: "Before November 1913, at which time commenced the discoveries of vestiges of human industry in the Chapadmalalan of Miramar, the theories of F. Ameghino could be considered to have been definitely rejected. But now there was reason to question whether these additional discoveries did not constitute a new proof for the existence of Tertiary man in South America. Carlos Ameghino, brother of Florentino, announced these discoveries in various summary and preliminary reports. These notices were received with ironic skepticism in the few scientific journals that continued publication during the war in Europe."

At that time, Boman (1921, p. 337) wrote a short article reviewing Carlos Ameghino's finds, citing negative assessments by Antonio A. Romero and the Italian anthropologist and geologist Guido Bonarelli, who believed the objects were not found *in situ*. Boman later stated: "I must observe that at the time I wrote my article, I had not yet visited Miramar and was thus guided by the facts furnished to me by Carlos Ameghino and others who had personally visited the site. I also personally inspected the objects that had been gathered there."

Boman (1921, p. 337) then carefully yet deliberately raised the possibility of fraud by Lorenzo Parodi, the collector who worked for Carlos Ameghino: "Regarding the intervention of Lorenzo Parodi in the discoveries . . . I had no right to express any suspicions about him, because Carlos Ameghino had spoken highly of him, assuring me that he was as honest and trustworthy a man as could be found." Boman (1921, p. 341) added: "I do not have any personal reason to doubt the honesty of Parodi, but generally speaking, a person in his condition participates in discoveries of this kind without any scientific interest. Instead, such persons tend to be solely interested in obtaining money and keeping their employment. Therefore, it is not possible to do anything but raise suspicions about fraud. Concerning the question of where it is possible to obtain objects for fraudulent introduction into the Chapadmalalan strata, that is a problem easily resolved. A couple of miles from the discoveries exists a *paradero,* an abandoned Indian settlement, exposed on the surface and relatively modern—about four or five hundred years old—where there exist many objects identical to those found in the Chapadmalalan strata."

Boman (1921, p. 342) went on to describe his own visit to the Miramar site on November 22, 1920: "Parodi had given a report of a stone ball, uncovered by the surf and still encrusted in the *barranca.* Carlos Ameghino invited various persons to witness its extraction, and I went there along with Dr. Estanislao S. Zeballos, ex-minister of foreign affairs; Dr. H. von Ihering, ex-director of the Museum of São Paulo in Brazil; and Dr. R. Lehmann-Nitsche, the well known anthropologist." At the Miramar *barranca,* Boman (1921, p. 343) convinced himself that the geological information earlier reported by Carlos Ameghino was essentially correct. Boman's admission confirms our assessment that the contrary views of Romero are not to be given much credibility (Section 5.2.3).

"Arriving at the final point of our journey," wrote Boman (1921, p. 343), "Parodi showed us a stone object encrusted in a perpendicular section of the *barranca,* where there was a slight concavity, apparently produced by the action of waves. This object presented a visible surface only 2 centimeters [just under an inch] in diameter. Parodi proceeded to remove some of the surrounding earth so it could be photographed, and at that time it could be seen that the object was a stone ball with an equatorial groove of the kind found on bola stones. Photographs were taken of the ball *in situ,* the *barranca,* and the persons present, and then the bola stone was extracted. It was so firmly situated in the hard earth that it was necessary to use sufficient force with cutting tools in order to break it out little by little."

Boman then confirmed the position of the bola stone (Figure 5.2a), which was found in the *barranca* about a meter (about 3 feet) above the beach sand. Boman (1921, pp. 343–344) stated: "The *barranca* consists of Ensenadan above and Chapadmalalan below. The boundary between the two levels is undoubtedly a little confused. . . . Be that as it may, it appears to me that there is no doubt that the bola stone was found in the Chapadmalalan layers, which were compact and homogeneous." It bears repeating that this description invalidates the views of Romero (1918), who had sought to demonstrate that the Chapadmalalan formations at Miramar were recent marine deposits. Boman's account therefore also discredits Boule, who relied solely upon Romero in his own attempt to dismiss the discovery at Miramar of the toxodon femur and vertebral column, both with stone arrowheads embedded in them (Section 5.2.4).

Boman (1921, p. 344) then told of another discovery: "Later, at my direction, Parodi continued to attack the *barranca* with a pick at the same point where the bola stone was discovered, when suddenly and unexpectedly, there appeared a second ball ten centimeters lower than the first. . . . It is more like a grinding stone than a bola." This tool (Figure 5.2b) was found at a depth of 10 centimeters (4 inches) in the face of the cliff. Boman (1921, p. 345) said it was "artificially worn." Still later Boman and Parodi discovered another stone ball (Figure 5.2c), 200 meters from the first ones, and about half a meter lower in the *barranca* (Boman 1921, p. 344). Of this last discovery at Miramar, Boman (1921, p. 346) said "there is no doubt that the ball has been rounded by the hand of man."

Boman then discussed the materials from which the implements had been made. The first bola was of quartzite, which can be found at Mar del Plata, about 15 miles northeast of Miramar. It was more difficult to account for the presence at Miramar of the diabase, from which the other two bolas were made. The nearest place where diabase could be found was near Rio Negro, about 300 miles to the southwest, from where it could have been carried up along the coast. The other possibility was the mountains of the Cordillera, deemed by Boman to be too far away—over 600 miles.

Altogether, the circumstances of discovery greatly favored a Pliocene date for the Miramar bolas. Boman (1921, p. 347) reported: "Dr. Lehmann-Nitsche has said that according to his opinion the stone balls we extracted were found *in situ*, are contemporary with the Chapadmalalan terrain, and were not introduced at any later time. Dr. von Ihering is less categorical in this regard. Concerning myself, I can declare that I did not observe any sign that indicated a later introduction. The bolas were firmly in place in the very hard terrain that enclosed them, and there was no sign of there having been any disturbance of the earth that covered them."

Boman (1921, p. 347) then artfully raised, as previously, the suspicion of cheating: "I have exchanged opinions with various colleagues about the possibility there could have been any kind of fraud involved in the circumstances under consideration, and we came to the conclusion that this possibility cannot be completely excluded. One could drill in the *barranca* a hole of the required size, introduce the object, and then carefully cover it with some dampened earth, the same removed in making the hole.

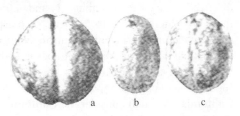

a   b   c

Figure 5.2. These stone bolas were extracted from the Late Pliocene Chapadmalalan formation at Miramar, Argentina, in the presence of ethnographer Eric Boman (1924, p. 345).

It could then be left to the waves, which periodically strike the *barranca,* to smooth and harden the earth, in such a manner that after a few months or a year it would appear as if nothing had touched the *barranca.* It would be interesting to verify this experimentally."

Boman (1921, pp. 347–348) then went on to cast doubt on another discovery made at Miramar: "In the Museo de La Plata, I have made an experiment of a similar nature, relative to a specimen discovered in the Chapadmalalan of Miramar—the femur of toxodon which has embedded in its trochanter the point of an arrow made of quartzite. I searched the museum collection for a toxodon femur of the same size and state of fossilization, and drove a similar quartzite point into the corresponding region of the bone. C. Heredia, then secretary of the museum, who studied this piece for a long time on his desk, said he could not distinguish it from the original." However, Boman (1921, p. 348) himself admitted: "But this experiment does not demonstrate anything other than the possibility of an exact imitation, and is not a conclusive proof that the point of the arrowhead was introduced into the femur of Miramar when it was already in a fossil state."

Boman (1921, p. 348) added: "Concerning the question of the authenticity of the finds from the Chapadmalalan strata at Miramar, in the final analysis there undoubtedly exists no conclusive proof of fraud. On the contrary many of the circumstances speak strongly in favor of their authenticity."

Despite this remarkable admission, Boman (1921, p. 348) could not resist once more raising the possibility of fraud or incompetence: "Nevertheless, the manner in which the discoveries were made, and above all, the continuous involvement of a person such as Parodi, necessarily give rise to suspicions. I do not believe that there is anyone in the world of science who could accept without the most careful consideration the above-mentioned discoveries as authentic proofs of nothing less than the existence of humans in South America during the Tertiary epoch."

Boman (1921, pp. 348–349) then wrote: "In North America many analogous discoveries have been unanimously and definitively rejected because they were made by illiterate workers, miners, or prospectors of various kinds. Modern science requires stringent scientific control of the facts that serve as the foundations of its conclusions. It does not admit the affirmations and stories of ordinary persons, and stories in newspapers convince no one." Here Boman footnoted a negative report by Holmes on the auriferous gravel finds in California. We shall consider the California discoveries in some detail later in this chapter (Section 5.5), but for now we shall simply forewarn the reader that Holmes's dismissals are themselves open to question.

It is difficult to see why Boman should have been so skeptical of Parodi. One could argue that Parodi would not have wanted to jeopardize his secure and longstanding employment as a museum collector by manufacturing fake discoveries. In any case, the museum professionals insisted that Parodi leave any objects of human industry in place so they could be photographed, examined, and removed by experts. This procedure is superior to that employed by scientists involved in many famous discoveries that are used to uphold the currently accepted scenario of human evolution. For example, most of the *Homo erectus* discoveries reported by von Koenigswald in Java were made by native diggers, who, unlike Parodi, did not leave the fossils *in situ* but sent them in crates to von Koenigswald, who often stayed in places far from the sites. Also, many fossil hominid discoveries in Africa were made in a manner similar to that employed at Miramar—native diggers uncovered fossils or stone tools and left them in place to be examined by professional scientists. It may further be noted that the famous Venus of Willendorf, a Neolithic statuette from Europe, was discovered by a road workman. It is obvious that if one were to apply Boman's extreme skepticism across the board one could raise suspicions of fraud about almost every paleoanthropological discovery ever made. Boman himself recognized this.

Boman admitted that his principal reasons for not accepting the Miramar discoveries were theoretical. Boman (1921, pp. 349–350) wrote: "If one were able to prove in an evident manner the authenticity of the discoveries in the Chapadmalalan of Miramar and the Tertiary age of these strata, this would provide proof not only of the existence of Tertiary humans in South America but also of a thing very strange—the identity of their artifacts with those of the modern Indians. Can anyone imagine that Miocene humans [Pliocene according to modern estimation] made polished bola stones with grooves around the middle? In response to this question, I can do nothing but repeat the point I made at the end of my last publication on Miramar, which has also been reproduced by Boule in his book on fossil man: 'The principal difficulty in accepting a Tertiary age for the objects we have finished enumerating consists in that without exception all the objects unearthed from the Chapadmalalan at Miramar are absolutely similar to like objects found in all parts of the surface and uppermost strata of the Pampas and Patagonia. Is it possible that man could have lived in the Pampas from the Miocene to the time of the Spanish conquest, without changing his customs and without perfecting his primitive industry in some fashion?'" But why not? As previously mentioned, scientists in Africa have found that modern tribal people make crude stone tools almost identical to those recovered from geological contexts 2 million or more years old, in the same localities.

Ironically, Boman's testimony provides, even for skeptics, very strong evidence for the presence of toolmaking human beings in Argentina as much as 3 million years ago. Even if, for the sake of argument, one admits that the first bola stone recovered during Boman's visit to Miramar was planted by the collector Parodi, how can one explain the second and third finds? These were instigated not by the collector Parodi but by Boman himself, on the spot and without any warning. Significantly, they were completely hidden from view, and Parodi did not even hint at their existence.

Altogether, it appears that Boule (Section 5.2.4), Romero (Section 5.2.3), and Boman did very little to discredit the discoveries of Carlos Ameghino and others at the Miramar site. In fact, Boman gave first-class evidence for the existence of bola makers there in the Pliocene period.

5.3 OTHER BOLAS AND BOLALIKE IMPLEMENTS

The bolas of Miramar are significant in that they point to the existence of human beings of a high level of culture during the Pliocene, and perhaps even earlier, in South America. Similar implements have been found in Africa and Europe in formations of similar Pliocene age. This refutes the suggestion that the bolas discovered in the Pliocene Chapadmalalan of Miramar must be recent because of their resemblance to modern Argentine Indian bolas.

Bolas have also been found in Middle Pleistocene formations. In North America, bolas have been recovered from the Calico site, dated at about 200,000 years (Minshall 1989, p. 110). Bolas have also been found in China at the Gehe site, dated at about 600,000 years (Minshall 1989, p. 38), and at the similarly ancient Lantien site (Minshall 1989, p. 40).

Taken together, these round projectile stones, found in widely distant parts of the world in Pleistocene, Pliocene, and perhaps earlier geological contexts powerfully challenge the currently accepted notions of human origins and antiquity. In particular, the Pliocene discoveries of bolas strongly contradict the idea that 2–3 million years ago only very primitive protohuman hominids were living, and these only in Africa. The use of bola stones requires complex behavior generally associated with *Homo sapiens sapiens*. Let us now give some detailed attention to two significant cases of bolas and bolalike implements—the Bramford sling stone and the bolas of Olduvai Gorge.

5.3.1 The Sling Stone from Bramford, England (Pliocene to Eocene)

In 1926, one of J. Reid Moir's assistants uncovered a particularly interesting object from below the Pliocene Red Crag. Moir had been conducting excavations in the Red Crag and detritus bed below the Red Crag, which were exposed in a brick-earth pit on the north bank of the River Gipping at Bramford, near Ipswich. Moir (1929, p. 63) wrote: "The beds surmounting the loamy sand at Pit No. 2, Bramford, do not exhibit signs of glacial disturbance such as might have ploughed into the detritus-bed, and rearranged it with later material. The conclusion, therefore, must be that the object now to be described which was removed from the detritus-bed by my trained excavator, John Baxter, formed an integral part of that deposit."

Moir recalled that Baxter once gave him a small oval object that did not seem to warrant close inspection. Three years later, however, the round stone object (Figure 5.3) was noticed by Henri Breuil: "While I was staying in Ipswich with my friend J. Reid Moir, we were examining together a drawer of objects from the base of the Red Crag at Bramford, when J. Reid Moir showed me a singular egg-shaped object, which had been picked up on account of its unusual shape. Even at first sight it appeared to me to present artificial striations and facets, and I therefore examined it more closely with a mineralogist's lens [Figure 5.4]. This

Figure 5.3. A sling stone from the detritus bed beneath the Red Crag at Bramford, England (Moir 1929, p. 64). At least Pliocene in age, the sling stone could be as old as the Eocene.

examination showed me that my first impression was fully justified, and that the object had been shaped by the hand of man. . . . The whole surface . . . has been scraped with a flint, in such a way that it is covered with a series of facets running fairly regularly from end to end. . . . The scraping described above covers the whole surface of the object, and penetrates into its irregularities. As it stands,

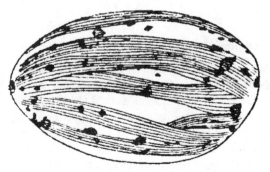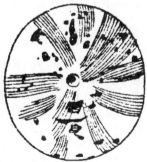

Figure 5.4. A drawing showing marks of intentional shaping on the sling stone from the detritus bed beneath the Red Crag at Bramford, England (Moir 1929, p. 65).

the object is entirely artificial, and, although somewhat smaller, it recalls the steatite sling stones of New Caledonia" (Moir 1929, p. 63). According to Moir (1929, p. 64), several other archeologists had confirmed Breuil's hypotheses. Moir, who believed the object had been shaped when soft, performed experiments with clay and flint, and he obtained results that were very much the same.

Moir (1929, p. 65) wrote: "it becomes clear that the presence of this object at such an horizon . . . points to the fact that man of the Pliocene period had already progressed some distance upon the evolutionary path, as it seems impossible to imagine any ape-like creature producing artifacts such as have now been found in the detritus bed." Sling stones or bola stones represent a level of technological sophistication universally associated with modern *Homo sapiens*. It may be recalled that the detritus bed below the Red Crag contains fossils and sediments from habitable land surfaces ranging from Pliocene to Eocene in age. Therefore the Bramford sling stone could be anywhere from 2 to 55 million years old.

It is altogether remarkable that almost without exception scientists have ignored the Bramford sling stone. It was found by a trained excavator, reported by a reputable archeologist, and examined by many experts including the famous Professor Breuil of the Institute of Human Paleontology in France. Some might object that it was found by a hired digger and not immediately noticed. But many of the Java *Homo erectus* fossils reported by von Koenigswald, which now figure prominently in every textbook of general paleoanthropology, were uncovered by native collectors. This is also true of the Petralona skull, found by Greek peasants in a cave. These cases, and many others like them, will be discussed in coming chapters. If these finds are accepted, despite the questionable circumstances of their discovery, then the Bramford sling stone deserves equal treatment. Otherwise we have another good example of scientists applying a double standard in the treatment of anomalous evidence.

5.3.2 Bolas from Olduvai Gorge (Early Pleistocene)

In 1956, G. H. R. von Koenigswald described some human artifacts that were discovered in the lower levels of the Olduvai Gorge site in Tanzania, Africa. "Apart from these archaic handaxes," wrote von Koenigswald (1956, p. 170), "the same levels have yielded numbers of stones that have been chipped until they were roughly spherical. . . . These stones are enormously widespread in Africa, occurring both in the north and south; indeed, at Ain Hanech, east of Algiers, they are the only signs of human culture that have been found there in association with fossil remains of elephant and giraffe. They are believed to be an extremely primitive form of throwing ball. Stone balls of this type, known to them as *bolas*, are still used by native hunters in South America. They are tied in little leather bags and two or three of them are attached to a long cord. Holding one ball in his hand, the hunter whirls the other one or two around his head and then lets fly."

An early hominid might have had enough intelligence to use bolas, but only *Homo sapiens sapiens* is actually known to have used them. Bolas are not unequivocally associated with the fossil remains of any other hominid. The objects reported by von Koenigswald, if used in the same manner as South American bolas, imply that their makers were adept not only at stoneworking but leatherworking as well.

All this becomes problematic, however, when one considers that Bed I at Olduvai, where stone balls were found, is 1.7–2.0 million years old. According to standard views on human evolution, only *Australopithecus* and *Homo habilis* should have been around at that time. At present, there is not any definite evidence that *Australopithecus* used tools, and *Homo habilis* is not generally thought to have been capable of employing a technology as sophisticated as that represented by bola stones, if that is what the objects really are. Some scientists doubt that *Homo habilis* was a toolmaker at all, and want to attribute tools found in the same level as *Homo habilis* to early representatives of *Homo erectus*.

Once more we find ourselves confronted with a situation that calls for an obvious, but forbidden, suggestion—perhaps there were creatures of modern human capability at Olduvai during the earliest Pleistocene. After all, the present inhabitants of the same region, as well as people in other parts of the world, make and use tools like the pebble choppers found in Bed I of Olduvai Gorge. Any crude stone tool now attributed to *Homo habilis* or *Homo erectus* could, therefore, also be attributed to *Homo sapiens*.

Those who find this suggestion incredible will doubtlessly respond that there is no fossil evidence to support such a conclusion. In terms of evidence currently accepted, that is certainly true. But if we widen our horizons somewhat, we encounter Reck's skeleton, fully human, recovered from upper Bed II, right at Olduvai Gorge (Section 11.1). And not far away, at Kanam, Louis Leakey, according to a commission of scientists, discovered a fully human jaw in Early Pleistocene sediments, equivalent in age to Bed I (Section 11.2.3). In more recent times, humanlike femurs have been discovered in East Africa, in Early Pleistocene contexts (Section 11.6.3). These isolated femurs were originally attributed to *Homo habilis,* but the subsequent discovery of a relatively complete skeleton of a *Homo habilis* individual has shown the *Homo habilis* anatomy, including the femur, to be somewhat apelike. This opens the possibility that the humanlike femurs once attributed to *Homo habilis* might have belonged to anatomically modern human beings living in East Africa during the Early Pleistocene (Section 11.7.1). If we expand the range of our search to other parts of the world, we can multiply the number of examples of fully human fossil remains from the Early Pleistocene and earlier. In this context, the bola stones of Olduvai do not seem out of place.

But perhaps the objects are not bolas. To this possibility Mary Leakey (1971, p. 262) replied: "Although there is no direct evidence that spheroids were used as bolas, no alternative explanation has yet been put forward to account for the numbers of these tools and for the fact that many have been carefully and accurately shaped. If they were intended to be used merely as missiles, with little chance of recovery, it seems unlikely that so much time and care would have been spent on their manufacture." Mary Leakey (1971, p. 266) added: "Their use as bola stones has been strongly supported by L. S. B. Leakey and may well be correct."

It should also be noted that Louis Leakey (1960a, p. 1051) claimed to have found "a genuine bone tool" in the same level as the bola stones. Leakey (1960a, p. 1051) said: "This would appear to be some sort of a 'lissoir' for working leather. It postulates a more evolved way of life for the makers of the Oldowan culture than most of us would have expected."

The complex behavior required for making and using bolas seems clearly out of character for either *Australopithecus* or *Homo habilis,* both of which were quite apelike. As far as *Homo erectus* is concerned, this creature is not generally portrayed using bolas. If use of bolas were to be attributed to *Homo erectus,* this would require a substantial redefinition of his technological capabilities. It thus appears that the bola stones may point to the existence in the African Early Pleistocene of a being with the intellectual and physical abilities of *Homo sapiens.* This, of course, would do severe damage to the whole picture of human evolution. The sling stone discovered below the Red Crag, with its possible age of 2.5 million years, could be just as damaging to the evolutionary hypothesis.

5.4 RELATIVELY ADVANCED NORTH AMERICAN PALEOLITHIC FINDS

We shall now examine some relatively advanced anomalous Paleolithic implements from North America, beginning with those found at Sheguiandah, Canada, on Manitoulin Island in northern Lake Huron. Many of these North American discoveries are not particularly old, but they are nonetheless significant because they give insight into the inner workings of archeology and paleoanthropology. We have already seen how the scientific community suppresses data with uncomfortable implications for the currently dominant picture of human evolution. And now we shall encounter revelations of another aspect of this—the personal distress and bitterness experienced by scientists unfortunate enough to make anomalous discoveries.

5.4.1 Sheguiandah: Archeology as a Vendetta

The excavations at Sheguiandah were carried out between 1951 and 1955 by Thomas E. Lee, an anthropologist at the National Museum of Canada. The upper layers of the site contained, at a depth of approximately 6 inches (Level III), a variety of projectile points (Figure 5.5).

According to Lee, excavation exposed an implement-bearing layer of unsorted sediments, apparently a glacial till. Ordinary sediments deposited by water tend to be sorted into distinct layers of sand and gravel. Deposits laid down by receding glaciers are generally not sorted in this fashion. Since at Sheguiandah stone tools were found in an unsorted till, the implication was that human beings had lived in the area during or before the time of the last glaciation. Further study showed that there was a second layer of till, which also contained artifacts.

Figure 5.5. Projectile point from Level III of the Sheguiandah site, Manitoulin Island, Ontario, Canada (T. E. Lee 1983, p. 61).

Figure 5.6. Bifacially chipped implement from upper glacial till (Level IV) at the Sheguiandah site (T. E. Lee 1983, p. 64).

Among the stone implements found in the upper section of glacial till, Level IV, were several large, thin, bifacial implements (Figure 5.6). T. E. Lee (1983, pp. 64–65) said about the bifaces: "Many retain some portion of a large bulb of percussion at one end. . . . Secondary chipping is prominent. . . . An interesting feature of several bifaces is the curious shoulder produced at one end. . . . Some of the double-shouldered tools show unmistakable evidence of use as scrapers, presumably hafted." In addition, Lee (1983, p. 65) stated: "A few cutting and scraping tools have been found in Level IV. Two examples show fine cutting edges resulting from removal of small flakes from both sides of one edge."

The lower section of till, Level V, produced small thick bifaces and man-made flakes (Figure 5.7). The artifacts found in Level V were fewer in number than those in Level IV (T. E. Lee 1983, p. 66).

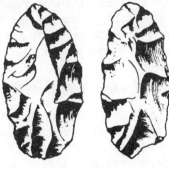

Figure 5.7. Quartzite bifaces from the lower glacial till (Level V) at Sheguiandah (T. E. Lee 1983, p. 66). Geologist John Sanford (1971) argued these tools and the one in Figure 5.6, were at least 65,000 years old.

Stone implements were also discovered in the layers beneath the tills. The layer immediately below the lower till, a meltwater deposit, covered a pavement of boulders. In and just beneath the boulder pavement were discovered one notched biface and several scrapers. Below the boulder pavement were silty stratified clays, with some cobblestones and boulders. From the upper part of the stratified clays, apparently deposited in a lake, came a broken bifacial implement and several stone flakes apparently struck by human beings (T. E. Lee 1983, p. 49).

How old were the tools? In his first reports, Lee was indefinite. Yet it seemed to him that some Sheguiandah artifacts were older than standard views about the peopling of the New World would allow. Lee (1972, p. 30) stated: "It is impossible to set a maximum age with certainty. . . . Of the four geologists most closely concerned—Dr. John Sanford of Wayne [State] University, Dr. Bruce Liberty and Dr. Jean Terasmae, both formerly of the G.S.C. [Geological Survey of Canada], and Dr. Ernst Antevs of Arizona—all but Dr. Antevs suggested that the site might extend back to interglacial times. Opinions differed as to whether that was 30,000 or 100,000 years ago. Dr. Antevs favored an interstadial for the appearance of man . . . estimated by him at 30,000 years ago. On his advice the group, in close communication, made public their conclusion: 'a minimum of 30,000 years.'" In another paper, Lee (1981) said some of the geologists had suggested that the implements were perhaps 150,000 years old.

From this point on the story becomes murky. Lee's discovery was obviously controversial, pointing to a human presence in North America far earlier than most scientists thought possible. John Sanford nevertheless continued to support Lee's position. He provided geological evidence and arguments suggesting the Sheguiandah site was quite old. But the view advocated by Lee and Sanford did not receive serious consideration from other scientists. Instead, political maneuvers and ridicule were employed to discredit Lee.

5.4.1.1 Sanford Presents Evidence in Favor of Lee

Sanford (1971) gave strong arguments for an early Wisconsin or Sangamon interglacial date for the tools in and below the tills at Sheguiandah. The reasoning he used was somewhat complex, reflecting the intricate series of Wisconsin glacial events at the site.

The Wisconsin, the final North American glacial age, is divided into three periods—early, middle, and late (or main). The entire Wisconsin glaciation was preceded by the Sangamon interglacial.

The geology of the Pleistocene glacial episodes is undergoing constant revision. In fact, some experts would favor scrapping the traditional system of four principal glaciations (the Günz, Mindel, Riss, and Würm of Europe and their

North American equivalents) for a system of alternating warm and cold periods of shorter duration. This system is said to more accurately reflect the evidence obtained from oxygen isotope studies of ocean core samples (Evans 1971). Even so, most authorities continue to make use of the traditional nomenclature, and we have chosen to do the same.

The early Wisconsin was dominated by glacial advances from centers north of the Great Lakes down into Ohio, Indiana, and other states. In the eastern Great Lakes region, the ice front was divided into three principal lobes (the Huron, Erie, and Ontario Lobes), which tended to advance and retreat together. The Huron Lobe is the one that covered Sheguiandah, located on Manitoulin Island in northern Lake Huron.

The middle Wisconsin was a period of significant glacial retreats, which took place during interstadials, or warm periods. The interstadial retreats were interrupted by some partial readvances.

During the late Wisconsin, the glaciers again advanced, this time to their maximum extent, after which they finally retreated, leaving the present Great Lakes.

According to Sanford, the presence of tools in the tills indicated that Manitoulin Island must have been habitable (not covered by ice or water) at certain periods. During these times, people quarried stone and made tools. After these periods of habitation, glacial advances mixed the tools lying on the ground with stone and earth. When the glaciers retreated this material was deposited as till. So the most important problem facing Sanford was to identify the times when it was possible for toolmakers to have lived in the vicinity of Sheguiandah and the times that glaciers subsequently advanced over the habitation sites.

Supporters of the dominant view about the peopling of the New World would want the habitation dates to be as recent as possible. This is because they believe that a human presence in the New World does not go much further back than 12,000 years. It should be kept in mind, however, that such a recent period of habitation at Sheguiandah must have been followed by a glacial advance and retreat; otherwise, one would not find tools in glacial till.

Was there in fact such a situation within the past 12,000 or so years at Sheguiandah—a period when Manitoulin Island was habitable followed by a period of glacial advance and retreat? In the 1950s, when the site was discovered, it was thought there were two relatively recent glacial advances and retreats that might have reached Sheguiandah—the Cochrane advance, at maybe 8,000 years ago (Nilsson 1983, p. 390), and the Valders advance, at around 11,000 thousand years ago (Dreimanis and Goldthwait 1973, p. 81). These advances were thought to have taken place after the main Wisconsin ice sheet retreated north of Manitoulin Island during the final part of the late Wisconsin. One might therefore propose that the tools found in the glacial till were manufactured in a warm period before the Cochrane advance or before the Valders advance. The Two Creeks interstadial has been mentioned.

But current geological opinion argues against this. First of all, during the Two Creeks interstadial, Sheguiandah appears to have been under ice (Hough 1958, p. 288). And when the ice finally retreated, it apparently did not come back (and deposit till). Also, recent authorities do not find evidence for either the Cochrane or Valders advances in the Lake Huron region (Dreimanis and Goldthwait 1973, pp. 71–72, 95–96; Nilsson 1983, p. 390). According to this view, around 11,000 or 12,000 years ago, the retreating Wisconsin ice sheet passed north of the region now occupied by Lake Huron, apparently without advancing again (Dreimanis and Goldthwait 1973, pp. 95–96). Furthermore, as the ice passed north of the present Lake Huron basin, it appears that Manitoulin Island remained under a body of water called Lake Algonquin. Lake Algonquin was a proglacial lake, one that forms at the front of an advancing or retreating glacier.

"The position of the ice front is speculative," said Sanford (1971, p. 12). "However, if the map presented by Hough (1958, fig. 62, p. 288) can be considered as summarizing the opinion of geologists, and I believe that it can, Sheguiandah would have been covered by ice during the Two Creeks interval. On the other hand, let us suppose the ice had melted sufficiently so that the front was farther north; and supposing that the area would have been habitable so far as climate is concerned, even though the ice front was not very far away. What are the chances people could have lived on the island? They are extremely slight, because of the probability that the island would have been well covered by water."

Thus far in our review we have found no situation within the past 12,000 years that would account for stone tools in glacial till at Sheguiandah on Manitoulin Island in northern Lake Huron. What about in earlier late Wisconsin times? Sanford (1971, p. 3) stated: "We do not know whether there was an earlier late-Wisconsin interstadial during which the site was uncovered and suitable for occupancy. The literature is commonly indefinite on this point, but the series of charts by Hough (1958, figs. 53–75) appear to summarize general opinion quite well. The literature indicates that the site was covered by either ice or water throughout late Wisconsin time until the lowering of Lake Algonquin."

Sanford's judgement is confirmed in a later study by two experts on the Wisconsin glaciation—A. Dreimanis of the University of Western Ontario and R. P. Goldthwait of Ohio State University. In a report titled "Wisconsin Glaciation in the Huron, Erie, and Ontario Lobes," published by the Geological Society of America, Dreimanis and Goldthwait (1973) provided a chart showing the changing position of the ice front during the entire Wisconsin glaciation (our Table 5.2, Figure 5.8). According to Dreimanis and Goldthwait (1973, p. 81), the ice front depicted on the chart "represents the advances and retreats of the

TABLE 5.2

Position of the Ice Front During the Wisconsin Glaciation

Glacial Period	Years B.P.	Central Ohio, Indiana	Lake Erie Basin	Lake Ontario Basin	St. Lawrence Lowlands
Late Wisconsin	10,000				
	20,000				
Mid Wisconsin	30,000				
	40,000				
	50,000			St. Pierre Interstadial	
Early Wisconsin	60,000				
	70,000				
Sangamon Interglacial	80,000				

Manitoulin Island, in northern Lake Huron, is at the same latitude as the middle part of the St. Lawrence Lowlands. As can be seen from this table (after Dreimanis and Goldthwait 1973, p. 81), the ice front was well south of this region during the entire Wisconsin glaciation, except for the St. Pierre interstadial at 65,000–70,000 years B.P. Tools found in glacial till at Sheguiandah, on Manitoulin Island, were probably made during this period or during the preceding Sangamon interglacial (Sanford 1971).

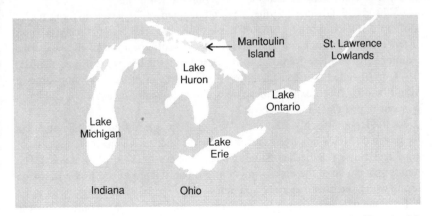

Figure 5.8. The Great Lakes region, showing Manitoulin Island, where the Sheguiandah site is located.

Ontario-Erie Lobe, including also participation of the Georgian Bay and Huron Lobes." The latter two lobes are the ones that covered Manitoulin Island and the Sheguiandah site. From the chart and discussion supplied by Dreimanis and Goldthwait, it can be concluded that Manitoulin Island was under ice during the entire period from about 10,000 years ago back to the time of the St. Pierre interstadial, which ended about 65,000 years ago in the early Wisconsin.

Prior to the St. Pierre interstadial came the first Wisconsin ice advance, but Dreimanis and Goldthwait (1973, p. 81) said "there is no evidence that this first advance of the ice sheet reached the Great Lakes." Thus the Sheguiandah site was habitable during and before the St. Pierre interstadial.

This brings us to the Sangamon interglacial period, which preceded the Wisconsin glaciation. According to some, the Sangamon interglacial extended from 75,000 to 100,000 years ago, while others say it extended from 75,000 to 125,000 years ago. The latter conclusion is based on the fact that the Gulf of Mexico provides fossil and other evidence for a warm climate during that period (Nilsson 1983, p. 455). So the most likely period for the manufacture of the stone tools found at Sheguiandah extends from the St. Pierre interstadial back through the Sangamon interglacial, perhaps as far back as 125,000 years ago.

Therefore, according to the view outlined by Sanford, we can envision the following series of events to account for the artifact-bearing geological formations observed at Sheguiandah. During the Sangamon interglacial or the earliest part of Wisconsin time, humans manufactured tools on and north of Manitoulin Island. As the ice sheet advanced after the St. Pierre interstadial, a proglacial lake formed in front of the glaciers. This proglacial lake, which covered the Sheguiandah site, deposited sediments. Perhaps a few tools discarded by humans traveling over the lake were incorporated into the sediments.

As the glacier advanced further toward Sheguiandah, it picked up tools, rocks, and earth. By ice rafting, some of these tools and rocks were floated a short distance into the proglacial lake and dropped, settling on top of the lacustrine sediments, which Lee found at the lowest levels of his excavations. Sanford (1971, p. 6) stated: "In consideration of the character of the overlying strata it would be difficult to explain these beds chronologically other than as an early Wisconsin, pro-glacial lake deposit."

As the ice approached, the early Wisconsin proglacial lake at Sheguiandah disappeared. Meltwater from the ice front created the glacio-fluvial layers of stone and clay over the lacrustine deposits. Sanford (1971, p. 6) stated: "The boulder pavement at the base of this unit probably represents a time of erosion during which the finer materials were removed."

Implements were found by Lee among and below the boulders in the pavement. Sanford (1971, p. 6) stated: "The presence of a few artifacts under boulders in the pavement is difficult to explain except by the same mechanism that provided for them in the still higher tills. They were picked up by advancing ice from a cultural site that existed prior to this glacial advance, and therefore were available for incorporation in both the outwash materials and the till deposited by the melting ice. The stratigraphic relations indicate an early Wisconsin age for this horizon. Erosion responsible for the boulder bed may have taken place during an early Wisconsin interstadial or even during a very minor recession of the ice front during a time of increasing glaciation."

Meltwater from the glacier deposited over the boulder pavement more glacio-fluvial materials, which Sanford (1971, p. 6) characterized as "outwash materials from the front of the advancing ice." Sanford stated (1971, p. 12): "The glacio-fluvial materials underlying tills at the habitation area certainly must be as old as early Wisconsin in age."

Finally, the glaciers themselves advanced over the site and then retreated, leaving the lower layer of till, which contains stone tools. Then the region was again briefly inhabited by a group of humans who made a different kind of stone tool. The early Wisconsin glaciers advanced once more, and when they finally receded tens of thousands of years later, in late Wisconsin times, they left the upper till—along with the artifacts they picked up in the early Wisconsin.

Sanford (1971, p. 14) stated: "the artifacts in the lower till layer, and probably those in the uppermost, certainly date from early rather than late Wisconsin time. . . . A Sangamon age for the earliest artifacts at Sheguiandah would appear more logical than an early Wisconsin date, at this latitude."

After the final glacial retreat, the site was covered by Lake Algonquin, which receded about 9,000 years ago, leaving Manitoulin Island as we now know it. Indians inhabited the island and left stone tools, including projectile points, now found in the surface layers above the glacial tills.

It is interesting to note that Sanford, unlike Lee, believed that the projectile point horizon (Level III) was also glacial in origin. According to Sanford (1983, p. 83): "It seems reasonably certain that the approximately six inches of surface material overlying the projectile point horizon was originally deposited as till. . . . [its] general character shows a similarity to the underlying tills and indicates a common genesis. The artifacts show it to be a definite stratigraphic unit." If Sanford's view is accepted, then the very advanced projectile points (Figures 5.5, 5.9) should, like the tools in the layers below them, also be early Wisconsin or Sangamon in age.

So it seems there is very good evidence for the presence of tool-making humans at Sheguiandah at

Figure 5.9. Quartzite projectile point recovered from Level III of the Sheguiandah site, Manitoulin Island, Ontario, Canada (T. E. Lee 1983, p. 62).

least as far back as the St. Pierre interstadial, which ended 65,000 years ago. The implements could possibly have been manufactured during the Sangamon interglacial, which means they could be as much as 125,000 years old. The presence of relatively advanced stone tools in the St. Pierre interstadial or Sangamon interglacial of North America would, according to currently dominant views, be quite unexpected.

5.4.1.2 How Lee Was Treated

We shall now present Thomas E. Lee's account of how his discoveries were received. Although this history will not be found in standard archeological publications, it is worth careful study. Lee's experiences shed light on how the scientific process works in practice. We shall leave it to the reader to decide whether or not his complaints are justified.

Lee (1966a, pp. 18–19) recalled: "Several prominent geologists who examined the numerous excavations in progress during four years at Sheguiandah privately expressed the belief that the lower levels of the Sheguiandah site are interglacial. Such was the climate in professional circles—one of jealousy, hostility, skepticism, antagonism, obstructionism, and persecution—that, on the advice of the famed authority, Dr. Ernst Antevs of Arizona, a lesser date of '30,000 years minimum' was advanced in print by some of the geologists to avoid ridicule and to gain partial acceptance from the more serious scholars. But even that minimum was too much for the protagonists of the 'fluted-point-first-Americans' myth. The site's discoverer [Lee] was hounded from his Civil Service position into prolonged unemployment; publication outlets were cut off; the evidence was misrepresented by several prominent authors among the Brahmins; the tons of artifacts vanished into storage bins of the National Museum of Canada; for refusing to fire the discoverer, the Director of the National Museum [Dr. Jacques Rousseau], who had proposed having a monograph on the site published, was himself fired and driven into exile; official positions of prestige and power were exercised in an effort to gain control over just six Sheguiandah specimens that had not gone under cover; and the site has been turned into a tourist resort. All of this, without the profession, in four long years, bothering to take a look, when there was still time to look. Sheguiandah would have forced embarrassing admissions that the Brahmins did not know everything. It would have forced the re-writing of almost every book in the business. It had to be killed. It was killed."

Lee's account was supported by Dr. Carl B. Compton, who wrote in *The Interamerican* (January 1966, p. 8): "When Thomas E. Lee found artifactual material in glacial till at Sheguiandah some years ago and when the age was estimated by several well-known and respected geologists at more than 30,000, the Brahmins presented their well-known 'Berlin Wall' to 'contain' this heresy" (T. E. Lee 1966b). Compton obviously thought that Lee was the victim of a power play in a scholarly community divided into hostile factions.

Here are additional comments by T. E. Lee (1964, p. 24) on his fate: "I, even as a professional archaeologist and officer of the National Museum of Canada over most of a nine-year period, found my work subjected to discrimination, whisper campaigns, and behind-the-scenes throat-cutting of the most contemptible and despicable order." Eventually, Lee had no choice except to resign from his position at the museum. He recalled: "My own resignation was in protest against the activities of R. S. MacNeish and was forced upon me by an impossible ultimatum delivered to me by that same Director of Natural History" (T. E. Lee 1964, p. 28).

Lee experienced great difficulty in getting his reports on his discoveries at Sheguiandah published through the National Museum of Canada. He wrote: "By depriving me of all essential services, burdening me with routine cataloguing, and closing publication outlets to me, every effort was made in the National Museum of Canada and in its string of satellites to block such publication. . . . I was hounded from my Canadian government position by certain American citizens on both sides of the border and driven into eight long years of blacklisting and enforced unemployment" (T. E. Lee 1974, p. 23). He also said that papers written by him were "filed away and lost" in the museum (T. E. Lee 1964, p. 24).

Having failed to get his reports into print in government publications, Lee, as a private citizen, experienced similar difficulties with standard scientific journals. Expressing his frustration, he wrote: "A nervous or timid editor, his senses acutely attuned to the smell of danger to position, security, reputation, or censure, submits copies of a suspect paper to one or two advisors whom he considers well placed to pass safe judgement. They read it, or perhaps only skim through it looking for a few choice phrases that can be challenged or used against the author (their opinions were formed long in advance, on the basis of what came over the grapevine or was picked up in the smoke-filled back rooms at conferences—little bits of gossip that would tell them that the writer was far-out, a maverick, or an untouchable). Then, with a few cutting, unchallenged, and entirely unsupported statements, they 'kill' the paper. The beauty—and the viciousness—of the system lies in the fact that they remain forever anonymous. The author may damn and fume and even correctly guess at their identity—but he is helpless. When in the course of time he is dead and safely buried—and proven to have been right—he will either be ignored or said to have been right, but for the wrong reasons" (T. E. Lee 1977, p. 2).

Most of the key reports about Sheguiandah were published in the *Anthropological Journal of Canada,* which Lee himself founded and edited. Lee died in 1982, and the journal was then edited for a short time by his son, Robert E. Lee.

After Lee left the National Museum of Canada, he eventually obtained a teaching position at Laval University. In 1980, he received for review a book (*Initiation a l'archeologie*) by René Lévesque, a former student. Lévesque had written on the title page: "I hope not to lose your friendship with my book. But the eternal enemies are still on the war path. I am honored to be with you in that fight" (T. E. Lee 1981, p. 18). Somewhat puzzled, Lee paged through the book. It contained a list of important archeological sites in North America, but the list did not include Sheguiandah. Nor was there any discussion of Sheguiandah in the text. Lee (1981, p. 19) found this strange, because Lévesque, his student, knew well "the inescapable proofs that put Sheguiandah back in the 150,000-year range, as determined by a number of geologists, both Canadian and American." Lee noticed, however, a very complete list of his works on Sheguiandah in the bibliography in the back of the book. "It should be clear by now," wrote Lee (1981, p. 19), "to everyone who reads this in the far corners of the earth, that a hatchet job was done . . . to eliminate Sheguiandah from the text, with the Bibliographie being overlooked, either in haste or arrogance."

Of course, it has not been possible for establishment scientists to completely avoid mentioning Sheguiandah, but when they do, they tend to downplay, ignore, or misrepresent any evidence for an unusually great age for the site.

Lee's son Robert wrote: "Sheguiandah is erroneously explained to students as an example of postglacial mudflow rather than Wisconsin glacial till; the reports, they are told, are too badly written to be worth reading, if indeed their existence is acknowledged" (R. E. Lee 1983, p. 11).

The original reports are, however, not so badly written, and give cogent arguments against the mudflow hypothesis. The elder Lee (1983, p. 58) wrote that many geologists "have stated that the deposits would definitely be called glacial till were it not for the presence of artifacts within them. This has been the reaction of almost all visiting geologists."

To Thomas E. Lee (1983, p. 58), the signs of the glacial origin of the deposits in question were unmistakable: "Among the indications which point to till are the lenses of fine gravels and sands observed in the lower half of the deposits. Such lenses are typical of till."

Any kind of mudflow or soil creep (solifluxion) would have required a slope of an appropriate inclination in the immediate area of the site, but no such slope was evident. The paths of flows from more distant high areas were blocked by transverse ridges of quartzite bedrock. Furthermore, according to T. E. Lee (1983, p. 58), the deposits in question were not of the same type as those resulting from solifluxion.

Lee (1983, p. 58) added: "An adequate explanation must also include the evidence in the Middle Quarry area near the high point of the hill, where unsorted artifact-bearing deposits are perched on the top of a ridge. There is no place from which soil could have crept, unless we can conceive of it crossing a swamp and climbing a ridge." Lee (1983, p. 59) concluded the deposits had been left on the ridge, and elsewhere, by glaciers.

Lee also considered other possible explanations for the presence of stone implements deep in unsorted deposits. One was that the deposits had been churned by the action of frost. But Lee (1983, pp. 59–60) pointed out: "If frost action has been churning the unsorted deposits, it is difficult to see why the artifact assemblages are not thoroughly mixed. . . . Frost churning cannot account for superimposed assemblages, typologically and quantitatively different, within the unsorted deposits. The occurrence of undisturbed horizontal lenses of fine gravels and sands in the lower half of the till deposits, sometimes with artifacts directly under them, is conclusive evidence that frost action did not severely affect the lower beds." The introduction of tools into the deposits by tree roots was considered and rejected for the same reasons.

Summarizing his findings, T. E. Lee (1983, p. 71) wrote: "Various explanations for the unsorted artifacts-bearing deposits have been considered, including tree plowing, root slumping, beach action, ice rafting, frost action, viscous flow, and soil creep. Although these factors may have been operative in a minor way, they do not explain the main body of the observed evidence. The suggestion of glacial till, on the other hand, is favored and supported by the nature of the deposits and their peculiar position on the site; the faceted stones, many of which are striated; the distribution and condition of encased artifacts; the occurrence of sand 'lumps'; and the presence of certain horizontal lenses of sorted sands, which are typical of till."

Sanford visited the Sheguiandah site several times during the period 1952–1957. In agreement with Lee, Sanford (1983, p. 82), a professional geologist, found the unsorted artifact-bearing layers to be glacial till: "There is no doubt in the writer's mind that this is till, although its origin has been questioned. It is made of a heterogeneous mixture of material ranging from clay to boulders." Sanford, for the same reasons as Lee, believed that neither frost, nor root action, nor mudflows could explain the formation of the layers in question and the presence in them of stone tools. The layers of till were thin, but Sanford said this was to be expected because previous glaciations had stripped the region of materials that could have been incorporated into the tills.

Sanford (1971, p. 7) wrote: "Perhaps the best corroboration of these unsorted deposits as ice-laid till was the visit of some 40 or 50 geologists to the site in 1954 during the annual field trip of the Michigan Basin Geological Society. At that time the excavation was open and the till could be seen. The sediments were presented to this group in the field as till deposits, and there was no expressed dissension from the explanation. Certainly had there been any room for doubt as to the nature of these deposits it would have been expressed at this time."

The belief that the Sheguiandah deposits are something other than glacial till is not confined to scientists holding the view that humans entered North America no earlier than 12,000 years ago. Even maverick researchers who accept a far more ancient date of entry hesitate about Sheguiandah because of the advanced nature of the stone tools found there. According to these researchers, early Americans, living in early Wisconsin and pre-Wisconsin times at sites such as Buchanan Canyon in San Diego and Calico in southern California, had a very primitive level of culture and used only the crudest kind of stone tools. From one advocate of this point of view, we have heard that the Sheguiandah tills are actually storm-driven beach deposits, perhaps 10,000 years old at most. This despite the fact that T. E. Lee (1983, p. 67) said "beach action has been considered and rejected." The reports of Lee and Sanford, full of references to old beaches on Manitoulin Island, show that they were fully acquainted with beach deposits and could make evaluations concerning their presence at particular locations. Arguing against the storm-driven beach deposit hypothesis is the fact that the stone tools found in the till are "not severely ground, battered or smashed" and that their "edges are often sharp" (T. E. Lee 1983, p. 58).

In cases of controversial claims, such as those made by Lee in connection with Sheguiandah, it is, of course, to be expected that counterarguments will be presented. But when such counterarguments and critiques are repeated blindly as conclusive verdicts, this tends to prevent any genuine discussion of the real issues. We might imagine the following classroom scene. "Tools found in till at Sheguiandah?" says the professor. "That's nonsense. Everyone knows the so-called till is just a beach deposit." Even a student with some genuine interest in the matter might hesitate to raise further questions because of fear of ridicule.

If one approach is to deny that the unsorted tool-bearing deposits are till, another is to demand excessively high levels of proof for a human presence at the site at the designated time. James B. Griffin, an anthropologist at the University of Michigan, believed that the most certain date for the entry of humans into the New World was 12,000 years ago, although he admitted growing evidence favored a 20,000-year date. Griffin (1979, pp. 43–44) added: "There are, however, some evaluations of the antiquity of man in the New World measured in the high tens of thousands and to hundreds of thousands of years or more on what I regard as either provocative or very slim evidence. These are simply not regarded as demonstrated by a large number of competent authorities."

This is how the social process of science works. Someone might present a good case, but if a consensus of established authorities does not support it, it goes by the wayside.

Griffin (1979, p. 46) further stated: "There are a large number of locations in North America for which considerable antiquity has been claimed as places inhabited by early Indians. Even whole books have been published on nonsites.

The reasons it is now difficult or impossible to include such 'sites' varies from location to location; a detailed dissent is not within the scope of this chapter." Griffin included Sheguiandah in the category of a nonsite. Of course, it is understandable that Griffin may not have had space in a chapter authored by him in a collective work to give a detailed discussion of why Sheguiandah should not be considered quite old. But he should have at least given some reference to where such a detailed dissent might be found. This he failed to do. As yet, we have found no detailed refutation in print, by Griffin or anyone else, of Sanford's analysis of the site's geology.

If, according to Griffin, Sheguiandah is a nonsite, then what is a real site? Griffin (1979, p. 44) stated that a proper site must possess "a clearly identifiable geologic context. . . . with no possibility of intrusion or secondary deposition." He also insisted that a proper site must be studied by several geologists expert in the particular formations present there, and that there must be substantial agreement among these experts. Furthermore, there must be "a range of tool forms and debris . . . well preserved animal remains . . . pollen studies . . . macrobotanical materials . . . human skeletal remains. Griffin also required dating by radiocarbon and other methods. Although Griffin (1979, p. 44) himself admitted "this is an ideal model for an 'Early Man' site," he nevertheless insisted that "insofar as finds that have been proposed fail to satisfy such criteria they are inevitably open to question, rejection, or suspended judgement." The problem here is that practically none of the locations where major paleoanthropological discoveries have been made would qualify as genuine sites. This includes many sites crucial to the picture of human evolution so carefully built up over the past century. For example, most of the African discoveries of *Australopithecus, Homo habilis,* and *Homo erectus* have occurred not in "clearly identifiable" geological contexts, but on the surface or in cave deposits, which are notoriously difficult to interpret geologically. Most of the Java *Homo erectus* finds also occurred on the surface, in poorly specified locations. At none of the places of these discoveries can one find the combination of factors Griffin deemed necessary for a proper site.

In this regard, it is interesting to note that the Sheguiandah site appears to satisfy most of Griffin's stringent requirements. Implements were found in a geological context clearer than that of many accepted sites. Several geologists expert in North American glacial deposits did apparently agree on an age in excess of 30,000 years. Evidence suggested there was no secondary deposition or intrusion. A variety of tool types were found, pollen studies and radiocarbon tests were performed, and macrobotanical materials (peat) were present. The only things absent were human and animal bones.

A few years after dismissing Sheguiandah as a nonsite, Griffin grudgingly accepted it as a recent site. In reading the report containing this admission, one gets the impression that the only tools found were those lying on or near the surface, and that the Sheguiandah site can best be dated with reference to peat bogs that formed after Manitoulin Island emerged from Lake Algonquin about 9,000 years ago. There is not the slightest hint that tools were found in glacial till and meltwater deposits. Griffin (1983, p. 247) stated, referring to Sheguiandah and two neighboring locales: "A reasonable estimate of the age of these sites would be from about 7000 to 6000 B.C., a time when there is a high pollen count from a nearby bog at Sheguiandah. These sites are almost certainly not earlier than the period of the lowest level of Lake Algonquin."

In 1974, a similar approach was taken by P. L. Storck of the Royal Ontario Museum in Toronto. He listed Sheguiandah as a Shield Archaic site. The Shield Archaic is a recent, and broadly defined, Indian stone tool culture that spread across much of central Canada. Lee protested, pointing out the absurdity of treating the tools from Sheguiandah as if they all belonged to a single unit of recent historical time. To do so would mean ignoring the obviously stratified nature of the Sheguiandah site, with tools of distinct types found on the surface and at different levels below the surface—within glacial tills, meltwater deposits, and lacrustine sediments (T. E. Lee 1974).

The Shield Archaic culture was preceded by Paleo-Indian cultures in Canada, and it may thus be called a post-Paleo-Indian culture. According to Lee, the Paleo-Indian culture is represented at Sheguiandah by the upper projectile point level, lying above the glacial tills. The Shield Archaic came later, and might be represented by the tools found lying on the surface at Sheguiandah. In any case, both the Shield Archaic and Paleo-Indian cultures came after the glacial period. Disputing the Shield Archaic labeling of Sheguiandah, T. E. Lee (1974, p. 24) asked the following questions: "Where do we go to find post-Paleo-Indian in and beneath primary glacial tills, in meltwater deposits, and beneath boulder pavings? Shall we consult the geologists who extensively and intensively studied the site during four years while the trenches were open? Will their opinions carry weight in the face of a young archaeologist's statement? Four of them—the most closely involved out of a hundred geologists who saw the trenches—Dr. Sanford, Dr. Antevs, Dr. Terasmae, and Dr. Liberty put the age of the site at 'a minimum of 30,000 years.' They did so on the advice of old Dr. Antevs, who warned them that the profession was not prepared to accept an older date, and for the most part would balk at 30,000 (as they did)."

In recent years, a minority of archeologists have begun to accept sites showing a human presence in North America over 30,000 years ago. It is noteworthy that few of these archeologists mention the Sheguiandah site, testifying to the effectiveness of the suppression of reports regarding it. An exception is W. N. Irving of the University of Toronto. As early as 1971, he was drawing attention to sites at Old Crow River and Edmonton that yielded deliberately fractured bone from the middle and early Wisconsin respectively (Irving 1971, pp. 69, 71). The Edmonton site may have been late Sangamon interglacial.

Irving (1971, p. 71) then wrote: "I think our recent findings require that Sheguiandah be reexamined, for the investigations there were not completed. No one has yet suggested an age of 30,000 years or more for Sheguiandah, and I do not do so now, but I would like very much to know how old it really is, and what is there." Irving was apparently unaware of Sanford's work, or deliberately avoided saying anything about it.

The most favorable review of Sheguiandah we have yet been able to locate comes from José Luis Lorenzo, of the National Institute for Anthropology and History, in Mexico City. He wrote (Lorenzo 1978, p. 4): "The site is a complex one with several levels of likely occupancy due to the fact that there is a type of quartzite on the island that is an excellent material for artifacts. Various series of artifacts were also found mixed with glacial debris at the bottom of the stratigraphy. All the studies on the glacial ecology of the area indicate that the remains not mixed with till are later than 12,500 years ago, whereas those that were mixed go back over 30,000 years according to available data (Prest 1969; Flint 1971; Dreimanis and Goldthwait 1973)."

It would seem that the Sheguiandah site deserves more attention than it has thus far received. The discoverer, Thomas E. Lee, certainly felt frustrated because of this. Looking back to the time when it first became apparent to him that stone implements were being found in glacial till, T. E. Lee (1968, p. 22) wrote: "At this point, a wiser man would have filled the trenches and crept away in the night, saying nothing. Books had been written, lectures had been given, pronouncements made, and armchairs comfortably filled. . . . Indeed, while visiting the site, one prominent anthropologist, after exclaiming in disbelief, 'You aren't finding anything down *there?*' and being told by the foreman, 'The hell we aren't! Get down in here and look for yourself!,' urged me to forget all about what was in the glacial deposits and to concentrate upon the more recent materials overlying them. Today, 13 years after vigorous professional efforts succeeded in halting the investigation of that great site, the same arguments and distortions are spreading through the literature. . . . The sacred cow must be defended, and to hell with the facts."

5.4.2 Lewisville: The Vendetta Goes On
(Late Pleistocene)

In 1958, at a site near Lewisville, Texas, stone tools and burned animal bones were found in association with hearths. Later, as the excavation progressed, radiocarbon dates of at least 38,000 years were announced for charcoal from the hearths. Still later, a Clovis point was found. Herbert Alexander, who was a graduate student in archeology at the time, recalled how this sequence of finds was received. "On a number of occasions," stated Alexander (1978, p. 20), "I had the opportunity to listen to faculty and visitors discuss their visits to this site. The opinions voiced at that time were that the hearths were man-made, and the faunal associations valid. Once the dates were announced, however, some opinions were changed and after the Clovis point was found, the process of picking and ignoring began in earnest. Those who had previously accepted the hearths and/ or faunal associations began to question their memories."

Finding a Clovis point in a layer 38,000 years old was disturbing, because orthodox anthropologists date the first Clovis points at 12,000 years, marking the entry of humans into North America. Some critics responded to the Lewisville find by alleging that the Clovis point had been planted as a hoax.

After mentioning a number of similar cases of ignored or derided discoveries, Alexander (1978, p. 22) recalled a suggestion that "in order to decide issues of early man, we may soon require attorneys for advocacy." This may not be a bad idea in a field of science like archeology, where opinions determine the status of facts, and facts resolve into networks of interpretation. Attorneys and courts may aid archeologists in arriving more smoothly at the consensus among scholars that passes for the scientific truth in this field. But Alexander noted that a court system requires a jury, and the first question asked of a prospective juror is, "Have you made up your mind on the case?" Very few archeologists have not made up their minds on the date humans first entered North America.

5.4.3 Timlin, New York (Late Pleistocene)

The idea that Clovis-type projectile points represent the earliest tools in the New World is challenged by an excavation at the Timlin site in the Catskill mountains of New York State. In the mid-1970s, tools closely resembling the Upper Acheulean tools of Europe were found there. In the Old World, Acheulean tools are routinely attributed to *Homo erectus*. But such attribution is uncertain because skeletal remains are usually absent at tool sites. The Catskill tools have been dated to some 70,000 years B.P. on the basis of glacial geology. An interesting feature of the Timlin site is that investigators have been able to trace sequences of stone tool cultures from the "Upper Acheulean" level up to the recent Archaic period (Raemsch and Vernon 1977).

5.4.4 Hueyatlaco, Mexico (Middle Pleistocene)

In the 1960s, highly sophisticated stone tools (Figure 5.10) rivaling the best work of Cro-magnon man in Europe were unearthed by Juan Armenta Camacho and Cynthia Irwin-Williams at Hueyatlaco, near Valsequillo, 75 miles southeast of Mexico City. Stone tools of a somewhat cruder nature were found at the nearby site of El Horno. At both the Hueyatlaco and El Horno sites, the stratigraphic location of the implements does not seem to be in doubt. However, these artifacts do have a very controversial feature: a team of geologists, some working for the U.S. Geological Survey, gave them dates of about 250,000 years B.P. This team, working under a grant from the National Science Foundation,

Figure 5.10. Stone tools found at Hueyatlaco, Mexico, a site dated at about 250,000 years by a team from the United States Geological Survey.

consisted of Harold Malde and Virginia Steen-McIntyre, both of the U.S. Geological Survey, and the late Roald Fryxell of Washington State University.

These geologists said four different dating methods independently yielded an anomalously great age for the artifacts found near Valsequillo (Steen-McIntyre *et al.* 1981). The dating methods used were (1) uranium series dating, (2) fission track dating, (3) tephra hydration dating, and (4) study of mineral weathering. The carbon 14 and potassium-argon methods were not applicable at the Hueyatlaco and El Horno sites, and paleomagnetic measurements did not provide any useful information.

As might be imagined, the date of about 250,000 years obtained for Hueyatlaco by the U.S. Geological Survey team provoked a great deal of controversy. If accepted, it would have revolutionized not only New World anthropology but the whole picture of human origins. Human beings capable of making the sophisticated tools found at Hueyatlaco are not thought to have come into existence until about 100,000 years ago in Africa.

Of course, it is possible to dispute the dates reported by the U.S. Geological Survey team. But something more than a legitimate scientific disagreement over dating techniques appears to have been involved in the treatment of Hueyatlaco, as we shall see from the testimony of Virginia Steen-McIntyre. First, however, we shall examine how the anomalously old dates for the site were obtained.

5.4.4.1 The Uranium Series Dating of the Hueyatlaco Site

The principal technique used for dating materials from Hueyatlaco and El Horno was the uranium series method. The tests were performed by Barney J. Szabo of the U.S Geological Survey (Szabo *et al.* 1969). In this section, we will discuss Szabo's results in some detail to show they support his dates. In particular, Szabo's data suggest that leaching of uranium from the sample could not have produced erroneously old dates, as some have hypothesized. Readers uninterested in the technicalities may proceed to the next section.

The uranium series technique is based on the fact that each of several isotopes of uranium spontaneously breaks down into a distinct series of byproducts. At the Hueyatlaco and El Horno sites, Szabo was concerned with uranium 238 and uranium 235.

Uranium 238 decays to uranium 234, with a half life of 4.51 billion years, and uranium 234 decays to thorium 230, with a half life of 248,000 years. Thorium 230 in turn decays to radium 226, with a half life of 75,000 years (Considine 1976, p. 1866).

Uranium 235 decays to protactinium 231, with a half life of 707 million years, and protactinium 231 decays to actinium 227, with a half life of 32,500 years (Considine 1976, p. 1868).

The concept of a half life can be explained as follows. Say you start out with one pound of uranium 234, with a half life of 248,000 years. After 248,000 years, you would have a half pound of uranium 234, along with some thorium and radium. After another 248,000 years, you would have a quarter pound of uranium 234, with more thorium and radium, after another 248,000 years, an eighth of a pound of uranium and still more thorium and radium, and so on.

Small amounts of the uranium isotopes that form the starting points of our two series (uranium 238 and uranium 235) occur naturally in water, yet their decay products, thorium and protactinium respectively, are not found in water (Gowlett 1984, p. 86). Certain types of rocks (such as travertines, tufas, and concretions) form when inorganic carbonates precipitate out of water. During this precipitation, small quantities of uranium are included within the rock, but no thorium or protactinium. Hence, under ideal conditions, all of the thorium and protactinium found within such rocks comes from the decay of uranium isotopes. Also, bones that are soaked in uranium-bearing water tend to absorb uranium, which decays and produces byproducts.

Since the half lives of uranium, thorium, and protactinium are known, scientists say that by measuring the amounts of these elements present within a sample they can calculate the age of the sample. The more decay products present in the sample, the older it is. Determining the exact age of the sample is complicated by the fact that uranium and its byproducts may migrate in or out of the sample. An open system is one in which such migration occurs; a closed system is one in which migration does not occur.

Uranium series tests were applied to samples from Hueyatlaco and the nearby site of El Horno (Szabo *et al.* 1969). In obtaining these dates, both the uranium 234/thorium and the uranium 235/protactinium series were used, and they yielded results that were in substantial agreement with each other.

Calculations yielded dates of about 245,000 years B.P. for sample MB3 (a camel pelvis) from Unit C of the Hueyatlaco site. Unit C is the uppermost layer at Hueyatlaco, and was found to contain highly sophisticated stone tools. This layer is underlain by Unit E, which contained similar tools, and Unit I, which contained tools of a simpler mode of manufacture. Unit E and Unit I are separated by a stratigraphic discontinuity, which suggests that Unit I is considerably older than Unit E. In other words, 245,000 years is a minimum age for the site, the lower levels of which could be substantially older.

The uranium series method gave open and closed system estimates of over 280,000 years for sample MB8, a mastodon tooth, from El Horno. The El Horno site is at a lower stratigraphic level than any of the Hueyatlaco layers, and contained tools similar to those of Unit I, the lowest tool-bearing layer at Hueyatlaco. One wonders what remains of human culture might be found at levels lower, and hence older, than El Horno.

Szabo reported that he used calculations based on both open and closed systems in obtaining the above uranium series test results. Nevertheless, some scientists have suggested that these dates are in error because uranium and its decay products may have migrated into or out of the samples over the course of their interment to a greater extent than Szabo supposed. Cynthia Irwin-Williams, who originally discovered the tools, suggested that the real age of the samples should be around 25,000 years. But careful study of the data supplied by Szabo, who performed the uranium series tests, appears to rule out the hypothesis that migration caused falsely old dates.

There are two ways that a falsely old age can be obtained by uranium series dating—outward migration of uranium or inward migration of byproducts. If uranium has migrated out, this will result in a higher ratio of byproducts (thorium or protactinium) to uranium in a sample, and hence a greater than normal age for the sample. If byproducts (thorium or protactinium) have migrated in, that will, of course, also result in a higher than normal ratio of byproducts, and hence a greater age for the sample. This latter alternative is, however, highly unlikely since both thorium and protactinium are virtually insoluble in water.

Furthermore, thorium 230, the isotope produced by the decay of uranium 234, is always accompanied in nature by the far more common isotope thorium 232. So let us suppose that the Hueyatlaco bone samples are in fact very young. Let us also suppose, although it is quite unlikely, that thorium 230 and thorium 232 have migrated into the bone, giving a falsely old age. In this case, one would expect to find a low ratio of thorium 230 to thorium 232, because thorium 232 is more common than thorium 230. But it was reported (Szabo *et al.* 1969, p. 243) that the ratio of thorium 230 to thorium 232 in the samples under consideration was "unusually high," which indicates that virtually all the thorium 230 measured in the samples was produced by the decay of uranium 234.

We have thus established that the uranium byproducts thorium and protactinium most probably did not migrate into the samples. That means that the hypothesis of a falsely old age depends on uranium migrating, or leaching, out of the samples.

In order to investigate the possibility of uranium leaching out of a sample, one of us (Thompson) analyzed two of several possible models—one in which leaching takes place at the end of the period of burial and one in which leaching is continuous throughout the period of burial. We shall now briefly discuss the results of these calculations.

Let us first consider the model in which leaching of uranium took place at the end of the period of burial. Taking bone sample MB3, we assumed, as claimed by Cynthia Irwin-Williams, that its real age is only 25,000 years instead of roughly 245,000 years. Then we computed the amount of leaching that must have taken place in order to give a date of 245,000 years for sample MB3, using the ratio of protactinium to uranium 235.

Sample MB3 was also originally dated at roughly 245,000 years using the ratio of thorium to uranium 234. So when we plugged the leaching factor for uranium 235 into the uranium 234 series equations we expected that the ratio of thorium to uranium 234 would yield a date of 25,000 years. Here we assumed that uranium 234 and uranium 235 are chemically identical (as atomic theory says they are) and that any leaching process would affect them equally. But the ratio of thorium to uranium 234, when calculated using the standard equations for radioactive decay, yielded an age of 52,451 years instead of 25,000 years. This result calls into question the leaching hypothesis.

A similar result was obtained when we reversed the order of the calculations. First we computed a uranium leaching factor using the ratio of thorium to uranium 234 and a sample age of 25,000 years. Using this uranium leaching factor, we then computed an age for sample MB3 based on the ratio of protactinium to uranium 235. This procedure yielded a date of 11,675 years rather than the expected 25,000 years.

Either way, these results are not consistent with the idea that the sample was deposited only 25,000 years ago, and that the leaching of uranium occurred fairly recently, at the end of the period of burial. According to our model, we should expect both sets of uranium series computations, done using the standard equations for radioactive decay, to yield results near 25,000 years. But they did not.

When we performed the computations assuming that leaching of uranium took place continuously rather than at the end of the period the bone sample was buried, similar results were obtained. In summary, the hypothesis that uranium leached out of the samples (either all at the end or continuously throughout the period of burial), and that the samples are therefore only 25,000 years old, is not consistent with the activity ratios reported for these samples.

At this point, one might raise the following objection. Admittedly, the date of 25,000 years suggested by Cynthia Irwin-Williams does not give good results in the above analysis. If we assume leaching of uranium 234 and uranium 235 took place, we would expect the computations for the thorium/uranium 234 and protactinium/uranium 235 ratios to yield the same results—25,000 years. But they did not. Then what about some other relatively young date? Could it be possible that using this alternative young date, good agreement might result?

We varied the assumed age for sample MB3, using the continuous leaching model, from 25,000 years through 250,000 years to see at what age the protactinium/uranium 235 age agreed best with the thorium/uranium 234 age. For assumed protactinium/uranium 235 ages from 25,000 up to 140,000 years the protactinium/uranium 235 ages disagreed with the thorium/uranium 234 ages by more than 30 percent. For a protactinium age of 180,000 years the thorium age disagreed by 20 percent, and the difference dropped as the assumed protactinium age increased. At 235,000 years the two differed by only .2 percent and at 245,000 years they differed by 3.1 percent. Thus the data reported by Szabo strongly support a date of around 235,000 years B.P. for the upper artifact-bearing layer (Unit C) at Hueyatlaco.

The same calculation was performed for sample MB8 using the continuous leaching model. The protactinium age was varied from 25,000 years through 370,000 years. We found that the thorium dates disagreed with the protactinium dates by more than 30 percent from 25,000 up to 260,000 years. At 300,000 years the two disagreed by 16 percent and this difference decreased to .32 percent at 355,000 years. Thus the activity ratios reported by Szabo strongly support an age of about 355,000 years for this sample from the site of El Horno, even if we assume, for the sake of argument, there was continuous leaching of uranium. Szabo pointed out that "sample MB8 was a tooth fragment from a butchered mastodon at El Horno, the oldest known site, and was therefore itself an artifact" (Szabo *et al.* 1969, p. 240).

Uranium series dating methods were also applied to bone samples from the nearby Caulapan site, yielding dates of about 20,000 years B.P. These agreed nicely with carbon 14 dates of 21,850 and 30,600 years from this site. We should note that the Caulapan carbon 14 date of 21,850 years applies to mollusk shells associated with the single artifact found at this site. Cynthia Irwin-Williams (1978, p. 22; 1981, p. 258) maintained that this was the only valid date for any Valsequillo artifact, and it should therefore be used for Hueyatlaco and El Horno as well. But the U.S. Geological Survey team could find no "geological basis by which the relation of the Caulapan deposits to Hueyatlaco can be determined" (Malde and Steen-McIntyre1981, p. 420). Therefore Irwin-Williams's dating of the Hueyatlaco and El Horno sites at about 25,000 years is not justified.

5.4.4.2 Other Methods Used for Dating Hueyatlaco and El Horno

In addition to the uranium series method, the team of geologists used fission track counting, tephra hydration dating, and mineral weathering analysis to assign dates to the implement-bearing layers at Hueyatlaco and El Horno.

Fission track dating is based on the accumulation of radioactive decay tracks in volcanic mineral crystals as a function of time. The more tracks, the older the crystals. When the age of crystals in a volcanic deposit has thus been determined, it is possible to assign appropriate dates to implements or fossils found beneath the volcanic layer in question. The fission track method was applied to two volcanic layers (the Tetela mud and the Hueyatlaco ash), situated above the most recent Hueyatlaco artifacts. The fission track dates for these layers should give a minimum age for all of the Hueyatlaco tools. The fission track dates were 260,000 to 940,000 years for the Tetela mud and 170,000 to 570,000 years for the Hueyatlaco ash. The considerable ranges in the dates were attributed to statistical effects due to the small number of fission tracks that were counted (Malde and Steen-McIntyre 1981, p. 419). The date ranges for the two volcanic layers overlap in the interval from 260,000 years to 570,000 years B.P.

Tephra hydration dating is a relatively new technique. It relies on the fact that volcanic glass, or tephra, slowly absorbs water. For this method to be feasible, it is necessary to have independently dated control samples of volcanic glass with the same chemical properties and geological situation as the samples to be dated. In this case, control samples were taken from the nearby La Malinche volcano. The method gave a date of about 250,000 years B.P. for tephra deposits associated with the Hueyatlaco artifacts (Steen-McIntyre et al. 1981, p. 13).

The final method of dating, the study of the weathering of a volcanic mineral, hypersthene, gives only a relative measure of age. As time passes, exposed crystals of this mineral are slowly etched, leaving a "picket fence" profile when viewed under a microscope. At the nearby human occupation site of Tlapacoya, this etching was rare and incipient in volcanic deposits dated by carbon 14 to about 23,000 years B.P. In contrast, the etching was pronounced in volcanic deposits associated with the Hueyatlaco artifacts. This suggested that the Hueyatlaco artifacts must have an age considerably greater than 23,000 years (Steen-McIntyre et al. 1981, p. 11).

A final consideration in the dating of the Hueyatlaco artifacts is that they were found buried beneath at least 10 meters (33 feet) of sediment. Geological study showed that these strata had to accumulate before being cut by the nearby Atoyac River, which has carved a valley 50 meters (164 feet) in depth (Steen-McIntyre et al. 1981, p. 10).

In other words, the geological history of the site would go something like this. The artifacts were left on an ancient land surface. Layers of sediment were deposited over them. Then the river began to cut through the layers of sediment.

Given this sequence, it is possible to estimate the age of the tools. Two elements are required. The first is the time required to deposit at least 10 meters of sediments over the tools. The second is the time that the river took to cut its valley, which is now 50 meters deep. If one could estimate the time it took the river to cut its valley and then add the time it took to deposit at least 10 meters of sediments over the tools, then one would have a rough date for the tools.

Since the valley and its side channels have gentle slopes, it is not likely that the river has exhibited an unusually high rate of erosion. But even if we assume a rather high rate of erosion, as in the Colorado River valley, the river Atoyac would have required around 150,000 years to carve out its present channel (Steen-McIntyre *et al.* 1981, p. 10). Add to this the time originally required to deposit 10 or more meters of sediment over the tools, and it can thus be seen that the geology of Hueyatlaco and the Rio Atoyac valley corroborates the ancient date obtained by the four dating methods discussed previously.

We have examined in some detail the cases of Hueyatlaco and El Horno in order to show that the dates for stone tools from these sites were solidly based on serious scientific analysis, more rigorous than in many accepted dating studies. However, due to the anomalous character of the 250,000-year figure, this dating has proven to be extremely controversial. The daters declared themselves to be "painfully aware" of the dilemma they had caused and "perplexed" about how to resolve it. Roald Fryxell said: "We have no reason to suppose that over decades, actually hundreds of years, of research in archaeology in the Old and New World our understanding of human prehistory is so inaccurate that we suddenly discover that our past understanding is all wrong. . . . On the other hand, the more geological information we've accumulated, the more difficult it is to explain how multiple methods of dating which are independent of each other might be in error by the same magnitude" (*Denver Post,* November 13, 1973).

According to Cynthia Irwin-Williams, the date of 250,000 years was impossible: "These tools surely were not in use at Valsequillo more than 200,000 years before the date generally accepted for development of analogous stone tools in the Old World, nor indeed more than 150,000 years before the appearance of *Homo sapiens*" (Szabo *et al.* 1969, p. 241).

Negative responses to the dating of the Valsequillo sites of Hueyatlaco and El Horno arise from acceptance of a theory of human evolution that was established by unwarranted elimination of extensive evidence for the extreme antiquity of humans in both the Old and New Worlds. In light of the total evidence, a date of 250,000 years B.P. for sophisticated stone tools is not greatly surprising. Ironically, in the treatment of the Valsequillo findings by the scientific community, we see the same tendency to suppress unwanted evidence that eliminated the earlier material and thereby rendered the Valsequillo dates unbelievable.

5.4.4.3 Negative Reception of the Hueyatlaco Evidence

Virginia Steen-McIntyre has sent us some of her correspondence, which documents the difficulties she had in publishing her findings on Hueyatlaco. We shall now introduce excerpts from this correspondence. Our purpose in doing so is to clarify how anomalous evidence is treated by the scientific community.

We have already shown that much evidence for the presence of anatomically and culturally modern humans in the Tertiary epoch was suppressed in the late nineteenth and early twentieth centuries, mainly because it conflicted with emerging theories of human evolution. Some might object that we have misinterpreted what went on in that period, taking the normal scientific procedures scientists use in differentiating good evidence from bad as some kind of diabolical plot to distort the truth. Others will maintain that even if good evidence was in fact rejected for reasons that appear unscientific in hindsight, this just does not happen any more. But the case of Hueyatlaco (along with Texas Street, Sheguiandah, Calico, and Lewisville) demonstrates otherwise.

Among the social processes that discourage acceptance and reporting of anomalous evidence are ridicule and gossip, including attacks on character and accusations of incompetence. Furthermore, discoveries have almost no impact in the world of science unless they are published in standard journals. The editorial process, especially the practice of anonymous peer review, often presents an insurmountable obstacle. Some submissions are met with a wall of silence. Others are shunted around for months, from editor to editor. Sometimes manuscripts are mysteriously lost in the shuffle. And while positive reports of anomalous evidence are subjected to protracted review and/or rejection, negative critiques are sometimes rushed into print. Occasionally, a maverick report eventually does appear in a journal, but only after it has gone through such extensive modification that the original message has become totally obscured—by editorial deletions and, in some cases, rewriting of data.

Thomas E. Lee's attempts to get articles about Sheguiandah published (Section 5.4.1.2) exemplify what can happen. Also, we heard from a paleontologist at the San Diego Museum of Natural History that a forthcoming paper by other researchers on an incised elephant bone found in the Anza-Borrego Desert of southern California and dated at over 250,000 years would never make it past peer review (Section 2.3). The word was out that the article was coming, and competent authorities had already decided what would happen to it.

Virginia Steen-McIntyre experienced many of the above-mentioned social pressures and obstacles. In a note to a colleague (July 10, 1976), she stated: "I had found out through backfence gossip that Hal [Malde], Roald [Fryxell], and I are considered opportunists and publicity seekers in some circles, because of Hueyatlaco, and I am still smarting from the blow."

The publication of a paper by Steen-McIntyre and her colleagues on Hueyat-laco was inexplicably held up for years. The paper was first presented in 1975 at a joint meeting of the Southwestern Anthropological Association and the Societe Mexicana de Antropologia and was to appear in a symposium volume. Four years later, Steen-McIntyre wrote (March 29, 1979) to H. J. Fullbright of the Los Alamos Scientific Laboratory, one of the editors of the forever forthcoming book: "We received your name and address from Dave Snow, who said you were the one to contact about the publication date for the SWAA-SMA symposium volume. We hope that it is soon! I personally have been put in an awkward position by the publication delay. Our joint article on the Hueyatlaco site is a real bombshell. It would place man in the New World 10x earlier than many archaeologists would like to believe. Worse, the bifacial tools that were found *in situ* are thought by most to be a sign of *H. sapiens.* According to present theory, *H.s.* had not even evolved at that time, and certainly not in the New World."

Steen-McIntyre continued, explaining: "Archaeologists are in a considerable uproar over Hueyatlaco—they refuse even to consider it. I've learned from second-hand sources that I'm considered by various members of the profession to be 1) incompetent; 2) a news monger; 3) an opportunist; 4) dishonest; 5) a fool. Obviously, none of these opinions is helping my professional reputation! My only hope to clear my name is to get the Hueyatlaco article into print so that folks can judge the evidence for themselves. (Geologists have no trouble with it.) The longer the delay, the more archaeologists will be convinced that the whole thing is just a crass attempt of another egomaniac for publicity. I'm quite certain the archaeologist who was in charge of the excavations and who no longer corre-sponds with me feels this way."

Steen-McIntyre, upon receiving no answer to this and other requests for information, withdrew the article. Later she got a letter from Roger A. Morris of Los Alamos, who explained that he had taken the liberty of opening a letter addressed to Fullbright, who had been transferred to another group of researchers. Morris said he would return her manuscript, but it never came.

A year later, Steen-McIntyre wrote (February 8, 1980) to Steve Porter, editor of *Quaternary Research,* about having her article printed. She first explained its status. "It's been languishing down in Los Alamos for almost five years, awaiting publication as part of a symposium volume. During that time I have written or called a dozen times to learn the status of the volume only to receive no response. (The original editor was always 'in conference' or 'out of the office' or would 'return my call,' which he never did.) In the meantime, there's been a lot of false information circulated about the site and the work we did there in 1973. Especially damaging is an article by Cynthia Irwin-Williams published in 1978 (Summary of archaeological evidence from the Valsequillo Region, Puebla, Mexico, in *Cultural Continuity in Mesoamerica,* Brownman, D.L. ed., Mouton).

In it she discounts Szabo's uranium-series dates (concordant) on butchered bone supplied by herself because she doesn't believe in the method. She does the same with Naeser's 2 sigma zircon fission track dates for two tephra layers that we proved by a cross-trench and direct tracing of the stratigraphy to overlie beds exposed in the archaeological trenches. Needless to say, she never showed us a draft of this ms or even told us she planned to publish anything on Hueyatlaco!"

Steen-McIntyre added: "The ms I'd like to submit gives the geologic evidence. It's pretty clear-cut, and if it weren't for the fact a lot of anthropology textbooks will have to be rewritten, I don't think we would have had any problems getting the archaeologists to accept it. As it is, no anthro journal will touch it with a ten foot pole. Right now I don't even have a copy to send you. The editor's copy is still in Santa Fe and my working copy disappeared into the office of *Science 80* (AAAS) months ago and, despite howls and threats, has yet to be returned."

Steve Porter wrote to Steen-McIntyre (February 25, 1980), replying that he would consider the controversial article for publication. But he said he could "well imagine that objective reviews may be a bit difficult to obtain from certain archaeologists." The usual procedure in scientific publishing is for an article to be submitted to several other scientists for peer review. It is not hard to imagine how an entrenched scientific orthodoxy could manipulate this process to keep unwanted information out of scientific journals. The manner in which reports by Thomas E. Lee about the Sheguiandah site were kept out of standard publications provides a good example of this (Section 5.4.1.2).

Steen-McIntyre wrote to Porter (March 4, 1980): "Often it is next to impossible to get a controversial paper published that even indirectly challenges current archaeological dogma; George Carter is a case in point!" In a letter to Steen-McIntyre, Carter had called the dominant clique of New World archeologists "priests of the High Doctrine" and complained that they bragged among themselves about having blocked him from publishing in the major journals. He compared his treatment to a modern Inquisition. Steen-McIntyre then stated: "I had thought to circumvent these 'true believers' by publishing in an obscure symposium volume, but no such luck."

The competence of Steen-McIntyre's associates was also called into question. Steen-McIntyre informed Porter: "there's the old saw that Fryx wasn't in his right mind when he did the work. Those folks forget that I saw the stratigraphy too, and once you get into a cross-trench, it was relatively simple, thanks to a magnesium-stained bed that traced on the excavation wall like a pencil mark!"

On March 30, 1981, Steen-McIntyre wrote to Estella Leopold, the associate editor of *Quaternary Research:* "The problem as I see it is much bigger than Hueyatlaco. It concerns the manipulation of scientific thought through the suppression of 'Enigmatic Data,' data that challenges the prevailing mode of thinking. Hueyatlaco certainly does that! Not being an anthropologist, I didn't realize the full significance of our dates back in 1973, nor how deeply woven into our thought the current theory of human evolution had become. Our work at Hueyatlaco has been rejected by most archaeologists because it contradicts that theory, period. Their reasoning is circular. *H. sapiens sapiens* evolved *ca.* 30,000–50,000 years ago in Eurasia. Therefore any *H.s.s.* tools 250,000 years old found in Mexico are impossible because *H.s.s.* evolved *ca* 30,000– . . . etc. Such thinking makes for self-satisfied archaeologists but *lousy* science!"

As demonstrated in this book, the stone tools of Hueyatlaco are not an isolated example of "impossible" evidence that challenges the recent origin of *Homo sapiens* by a Darwinian evolutionary process. We have already discussed numerous examples of such impossible evidence from the Pliocene, Miocene, and earlier periods. And there is much more to come in the remainder of this volume. We have simply paused briefly in order to demonstrate that the suppression of such evidence did not end with the nineteenth century—it has continued to the present day. We also take the current examples of suppression of anomalous evidence as confirmation that our interpretation of what went on in the nineteenth century (and early twentieth century) is in fact correct.

On May 18, 1981, Steen-McIntyre wrote to Estella Leopold and Steve Porter about "suppression of data on Hueyatlaco and other possible Pre-Wisconsinian Early Man sites in the New World by unethical means." She told how she had submitted a general paper on her dating techniques to be included in a volume in a scientific series. Steen-McIntyre then learned from the editor that "he had decided to 'drastically edit' this manuscript, essentially by deleting most of the section on Hueyatlaco and by treating the remainder in a negative way." In her letter to Leopold and Porter, Steen-McIntyre stated: "I protested strongly, and he agreed to reinsert some of the deleted material, but only in a way that will hold both me and my research up for laughter and ridicule." In a note to our researcher, Steve Bernath, dated January 29, 1989, Steen-McIntyre explained that the editor had, in the course of his drastic editing, altered one of her data tables. According to Steen-McIntyre: "when I threatened him he replaced the missing material [in the text] but forgot to retype the table."

Steen-McIntyre's case is not unique. Some American scientists reporting anomalous evidence for a human presence in North America have found it necessary to publish overseas. Steen-McIntyre said in her letter to Leopold and Porter that "Roy Schlemon, a pedologist who has helped date Calico and who is working at other sites in Southern California . . . had been publishing outside the country." A pedologist is a scientist who studies soils.

Eventually, *Quaternary Research* (1981) published an article by Virginia Steen-McIntyre, Roald Fryxell, and Harold E. Malde. It upheld an age of

250,000 years for the Hueyatlaco site. Of course, it is always possible to raise objections to archeological dates, and Cynthia Irwin-Williams (1981) did so in a letter responding to Steen-McIntyre, Fryxell, and Malde. Her objections were answered point for point in a counter-letter (Malde and Steen-McIntyre 1981). But Irwin-Williams did not relent. She, and the American archeological community in general, have continued to reject the dating of Hueyatlaco carried out by Steen-McIntyre and her colleagues on the U.S. Geological Survey team.

As in the case of Sheguiandah, the anomalous findings at Hueyatlaco resulted in personal abuse and professional penalties for those who dared to present and defend them in the scientific literature. This involved withholding of funds and loss of job, facilities, and reputation for at least one of the geologists involved in the dating project (Steen-McIntyre, personal communication).

The case of Virginia Steen-McIntyre opens a rare window into the actual social processes of data suppression in paleoanthropology, processes that involve a great deal of hurt and conflict. In general, however, this goes on behind the scenes, and the public sees only the end result—the carefully edited journals and books that have passed the censors.

A final note—we ourselves once tried to secure permission to reproduce photographs of the Hueyatlaco artifacts in a publication. We were informed that permission would be granted only if we gave a date of no more than 30,000 years for the artifacts. But permission would be denied if we intended to cite a "lunatic fringe date" of 250,000 years. We grant that the 250,000-year date may be wrong. But is it really appropriate to apply the term "lunatic fringe" to studies such as the one carried out by Steen-McIntyre and her colleagues?

5.4.5 Sandia Cave, New Mexico (Middle Pleistocene)

In 1975, quite by accident, Virginia-Steen McIntyre learned of the existence of another site with an impossibly early date for stone tools in North America—Sandia Cave, New Mexico, U.S.A., where the implements, of advanced type (Folsom points), were discovered beneath a layer of stalagmite considered to be 250,000 years old. One such tool is shown in Figure 5.11.

In a letter to Henry P. Schwartz, the Canadian geologist who had dated the stalagmite, Virginia Steen-McIntyre wrote (July 10, 1976): "For the life of me, I can't remember if it was you or one of your colleagues I talked to at the 1975 Penrose Conference (Mammoth Lakes, California). The fellow I spoke to as we waited in line for lunch mentioned a uranium series date on the stalagmite layer above artifacts at Sandia Cave that was very upsetting to him—it disagreed violently with the commonly held hypothesis for the date of entry of man into the New World. When he mentioned a date of a quarter million years or thereabouts, I nearly dropped my tray. Not so much in shock at the age, but that this date agreed so well with dates we have on a controversial Early Man site in Central Mexico.... Needless to say, I'd be interested to learn more about your date and your feelings about it!" According to Steen-McIntyre, she did not receive an answer to this letter.

After writing to the chief archeological investigator at the Sandia site for information about the dating, Steen-McIntyre received this reply (July 2, 1976): "I hope you don't use this 'can of worms' to prove anything until after we have had a chance to evaluate it."

Figure 5.11. A Folsom blade embedded in the lower surface of a travertine crust from Sandia Cave, New Mexico (*Smithsonian Miscellaneous Collections,* vol. 99, no. 23, plate 7). The layer of travertine is said to be 250,000 years old.

Four years later, Virginia Steen-McIntyre wrote (February 8, 1980) in a letter to Steve Porter, editor of *Quaternary Research:* "Did you know they now have a 250,000 year date on the stalagmite layer in Sandia Cave, N.M., the one that sealed off sediments that contained leaf-shaped points and fire hearths? I've been trying to get more information out of Vance Haynes, who collected the samples, and Dr. Schwarz at McMaster [University], who ran the date, but so far no luck."

Steen-McIntyre sent us some reports and photos of the Sandia artifacts and said in an accompanying note: "Talk about a study in frustration! Read the enclosed, then look at that picture of the 'folsom' blade imbedded in the travertine crust (stalagmite layer, 250,000 years). The geochemists are sure of their date (oral communication, GSA meeting, 1978), but archaeologists have convinced them the artifacts and charcoal lenses beneath the travertine are the result of rodent activity. The archaeologists who have seen the evidence are sure of the presence of artifacts beneath the crust, but believe the date is wrong! But what about the artifacts *cemented in* the crust?"

The Sandia Cave discoveries, along with the finds made at Hueyatlaco (Section 5.4.4), Calico (Section 3.8.3), and Toca da Esperança (Section 3.8.4), strongly suggest a human presence over 200,000 years ago in the Americas. This challenges not only the orthodox time estimate for the entry of *Homo sapiens* into North America (12,000 years ago) but also the whole picture of human evolution, which has *Homo sapiens* arising from *Homo erectus* in Africa about 100,000 years ago.

5.5 NEOLITHIC TOOLS FROM THE TERTIARY AURIFEROUS GRAVELS OF CALIFORNIA

In 1849, gold was discovered in the gravels of ancient riverbeds on the slopes of the Sierra Nevada Mountains in central California, drawing hordes of rowdy adventurers to places like Brandy City, Last Chance, Lost Camp, You Bet, and Poker Flat. At first, solitary miners panned for flakes and nuggets in the gravels that had found their way into the present stream beds. But soon gold-mining companies brought more extensive resources into play, some sinking shafts into mountainsides, following the gravel deposits wherever they led, while others washed the auriferous (gold-bearing) gravels from hillsides with high pressure jets of water.

Occasionally, the miners would find stone artifacts, and, more rarely, human fossils (Section 6.2.6). Altogether, miners found hundreds of stone implements—mortars, pestles, platters, grinders, and so forth. Many of the specimens found their way into the collection of Mr. C. D. Voy, a part-time employee of the California Geological Survey. Voy's collection eventually came into the possession of the University of California, and the most significant artifacts were reported to the scientific community by J. D. Whitney, then the state geologist of California.

The finds occurred in three situations: (1) in surface deposits of gravel; (2) in gravels washed from hillsides by hydraulic mining; and (3) in underground deposits of gravel reached by mine shafts and tunnels. The artifacts from surface deposits and hydraulic mining were of doubtful age, but the artifacts from deep mine shafts and tunnels could be more securely dated because the gold-bearing gravels lay underneath thick layers of volcanic material.

5.5.1 The Age of the Auriferous Gravels

J. D. Whitney thought the geological evidence indicated the auriferous gravels, and the sophisticated stone tools found in them, were at least Pliocene in age. But modern geologists think some of the gravel deposits, which lie beneath volcanic formations, are much older.

According to Paul C. Bateman and Clyde Wahrhaftig (1966), R. N. Norris (1976), and William B. Clark (1979), the majority of the gold-bearing gravels were laid down in stream channels during the Eocene and Early Oligocene. These are called the prevolcanic auriferous gravels. During the Oligocene, Miocene, and Pliocene, volcanic activity in the same region covered some of the auriferous gravels with deposits of rhyolite, andesite, and latite.

In particular, widespread andesitic mudflows and conglomerates were deposited during the Miocene. These attained a considerable thickness, varying from more than 3,000 feet along the crest of the Sierras to 500 feet in the foothills. The volcanic flows were so extensive that they almost completely buried the bedrock landscape of the northern Sierra Nevada mountain region.

Although intense at times, the volcanic activity in the Sierra Nevada Mountains was not continuous, allowing rivers to carve new channels and canyons. These rivers often redistributed old gravels laid down in the Eocene and Early Oligocene periods. So below the volcanic formations, the most recent of which are Early Pleistocene (Jenkins 1970, p. 25), there can now be found auriferous gravel deposits that were laid down in stream beds during the Eocene, Oligocene, Miocene, and Pliocene periods. Over the course of time, rivers carved deep channels up to a couple of thousand feet below the level of the prevolcanic gravels. This allowed Gold Rush miners to reach the auriferous gravels by digging horizontal tunnels into the sides of the channels. The advanced stone tools found in these tunnels could be from Eocene to Pliocene in age.

5.5.2 Discoveries of Doubtful Age

Before discussing the most significant discoveries, the ones made in mines extending into gold-bearing gravels beneath ancient lava flows, let us briefly examine why it is not possible to attribute any great age to the artifacts found elsewhere in the gold mining region.

Some stone implements were found in the sluices of the hydraulic mines, where powerful jets of water were directed at entire hillsides. William H. Holmes, of the Smithsonian Institution, pointed out that recently abandoned Indian villages were often found on the slopes above the open mines and that it was quite possible that modern stone implements were washed into the Tertiary gravels below (1899, p. 445).

Other artifacts were found deep within surface deposits of Tertiary gravel. For example, at Gold Springs, a little west of Columbia, Mr. Lot Cannell stated that he discovered stone mortars and platters along with the bones and teeth of mastodons in gold-bearing gravels, approximately 90 feet below the surface (Whitney 1880, p. 262). At first glance, this discovery, so deep in the gravel, promises extreme antiquity. After all, mastodons existed in North America as far back as the Miocene.

But investigators such as William J. Sinclair determined that implements found deep in such deposits might be recent. Sinclair (1908, p. 112) wrote: "The underlying Carboniferous limestone has been eroded into fantastic shapes by percolating waters during or after the deposition of the auriferous wash. . . . In a limestone region with underground drainage, it is quite apparent that implements of human manufacture which happened to be scattered on the surface would stand an excellent chance of reaching deeper levels through the many sink holes affording drainage ways to surface waters."

So although the Gold Springs gravel itself might have been Tertiary in age it is possible, in light of Sinclair's observations about sinkholes, that the implements found deep in the gravels might have worked their way down from the surface in relatively recent times. Therefore, all we can safely conclude is that the stone implements found at Gold Springs might be anywhere from several million to several thousand years old. The same is true of discoveries made at Kincaid Flat, Oregon Bar, and several other localities where the gold-bearing gravels were not capped by volcanic deposits of known age.

5.5.3 Tuolumne Table Mountain

Finds from mine shafts can be dated more securely than those from hydraulic mines and surface deposits of gravel. Many shafts were sunk at Table Mountain in Tuolumne County. Whitney and others reported that miners found stone tools and human bones (Section 6.2.6) there, in the gold-bearing gravels sealed beneath thick layers of a volcanic material called latite. In many cases, the mine shafts extended horizontally for hundreds of feet beneath the latite cap at a depth of over 100 feet below the latite (Figure 5.12).

Tuolumne Table Mountain was created by a massive latite flow which moved down the Cataract Channel, a Miocene course of the Stanislaus River, forcing the river into a new channel. According to R. M. Norris (1976, p. 43), the latite lava cap is 9 million years old and is 300 feet thick in the vicinity of the town of Sonora. Slemmons (1966, p. 200) gave dates for the latite cap and underlying strata at Tuolumne Table Mountain (Table 5.3).

Discoveries from the auriferous gravels just above the bedrock are probably 33.2 to 55 million years old, but discoveries from auriferous gravels whose positions are not specified may be anywhere from 9 to 55 million years old.

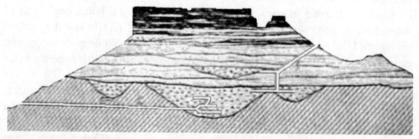

Figure 5.12. Side view of Table Mountain, Tuolumne County, California, showing mines penetrating into Tertiary gravel deposits beneath the lava cap, shown in black (Holmes 1899, p. 450).

TABLE 5.3

Age of Strata at Tuolumne Table Mountain

Age (millions of years)	Description of Formation
9.0	Table Mountain latite member
9.0–21.1	Andesitic tuffs, breccias, and sediments
21.1–33.2	Rhyolite tuffs
33.2–55.0	Prevolcanic auriferous gravels
>55.0	Bedrock

5.5.4 Dr. Snell's Collection

The more important discoveries from Tuolumne Table Mountain add up to a considerable weight of evidence. Whitney personally examined a collection of Tuolumne Table Mountain artifacts belonging to Dr. Perez Snell, of Sonora, California. About this collection of artifacts, Whitney (1880, p. 264) stated: "In Dr. Snell's collection . . . there were several objects which were marked as having come *'from under Table Mountain.'*" C. D. Voy said: "Among them was a piece of stone apparently designed as a handle for a bow. It was made of silicious slate and had little notches at the end, which appear to have been formed for tying the stone to the bow. There were also one or two spear heads, from six to eight inches long, and several scoops or ladles, with well shaped handles" (Whitney 1880, p. 264).

As can be seen from Whitney's statements about Dr. Snell's collection, there is not much in the way of direct testimony about the discoverers and original stratigraphic positions of the implements. There was, however, one exception. "This was," wrote Whitney (1880, p. 264), "a stone muller, or some kind of utensil which had apparently been used for grinding. It was carefully examined by the writer, and recognized as unquestionably of artificial origin. In regard to this implement Dr. Snell informed the writer that he took it with his own hands from a car-load of 'dirt' coming out from under Table Mountain." A human jaw, inspected by Whitney, was also present in the collection of Dr. Snell. The jaw was given to Dr. Snell by miners, who claimed that the jaw had came from the gravels beneath the basalt cap at Table Mountain in Tuolumne County (Becker 1891, p. 193).

5.5.5 The Walton Mortar

A better-documented discovery from Tuolumne Table Mountain was made by Mr. Albert G. Walton, one of the owners of the Valentine claim. Walton found a stone mortar, 15 inches in diameter, in gold-bearing gravels 180 feet below the surface and also beneath the latite cap. Significantly, the find of the mortar occurred in a "drift," a mine passageway leading horizontally from the bottom of the main vertical shaft of the Valentine mine. This tends to rule out the possibility that the mortar might have fallen in from above. Furthermore, the vertical shaft "was boarded up to the top, so that nothing could have fallen in from the surface during the working under ground" (Whitney 1880, p. 265). In fact, Walton, who found the mortar, was the carpenter responsible for timbering the shaft. A piece of a fossil human skull was also recovered from the Valentine mine (Section 6.2.6.3).

William J. Sinclair (1908, p. 115) later claimed that many of the drift tunnels from other mines near the Valentine shaft were connected. Sinclair granted that the Valentine vertical shaft may have been, as Whitney stated, securely boarded up to the top, so that nothing could fall in from the surface. But he proposed that objects still could have found their way into the Valentine underground tunnel from some other tunnels. Sinclair did not, however, offer any specific evidence that any tunnels were connected with the Valentine drift tunnels at the time the discoveries were made. In fact, Sinclair admitted that when he visited the area in 1902 he was not even able to find the Valentine shaft. It appears that Sinclair simply used his vague retrospective conjectures about possible invalidating circumstances to dismiss Walton's report of his discovery. Operating in this manner, one could find good reason to dismiss any paleoanthropological discovery ever made.

Another author suggested that prehistoric miners, perhaps from the known culture centers in Mexico or Central America, left the stone artifacts in the course of gold-mining operations conducted in California (Southall 1882, p. 197). We are, however, aware of only a single report of a mine existing before the California Gold Rush of the 1850s. This one mine (Southall 1882, p. 198) is insufficient to explain mortars and other implements found in many separated locations. The proposal that there may have been numerous mines that were collapsed and therefore escaped detection is highly improbable. The Gold Rush miners, being expert in such matters, would most likely have detected them, especially since collapsed mine shafts would have posed a threat to their lives in the form of cave-ins.

Some critics have contended that the mortars were carried by Indians into mines dug during the Gold Rush days. But the Indians of those times did not possess portable mortars (Section 5.5.13). And even if they did possess portable mortars, it is unlikely they would have carried them into the mines. Mortars were generally used for the grinding of raw acorns, a laborious, time-consuming task not likely to have been performed in the cramped, dark, dangerous confines of a working mine shaft.

The mortars found in the mines do resemble those used by some California Indians in the recent past. But the mortars from the mines also resemble those made by primitive people in other parts of the world, at various times in the past. In fact, it is likely that any human beings, living anywhere and at any time, would, when faced with the task of making grinding tools, come up with tools very much like the mortars and pestles from the mines. Therefore, the resemblance of the mortars found in the mines to those used by California Indians in recent times is not proof the mortars from the mines are also recent. They could have been made, as the evidence suggests, millions of years ago.

5.5.6 The Carvin Hatchet

Another find at Tuolumne Table Mountain was reported by James Carvin in 1871: "This is to certify that I, the undersigned, did about the year 1858, dig out of some mining claims known as the Stanislaus Company, situated in Table Mountain, Tuolumne County, opposite O'Byrn's Ferry, on the Stanislaus River, a stone hatchet . . . with a hole through it for a handle, near the middle. Its size was four inches across the edge, and length about six inches. It had evidently been made by human hands. The above relic was found from sixty to seventy-five feet from the surface in gravel, under the basalt, and about 300 feet from the mouth of the tunnel. There were also some mortars found, at about the same time and place" (Whitney 1880, pp. 274–275).

5.5.7 The Stevens Stone Bead

In 1870, Oliver W. Stevens submitted the following notarized affidavit: "This is to certify that I, the undersigned, did about the year 1853, visit the Sonora Tunnel, situated at and in Table Mountain, about one half a mile north and west of Shaw's Flat, and at that time there was a car-load of auriferous gravel coming out of said Sonora Tunnel. And I, the undersigned, did pick out of said gravel (which came from under the basalt and out of the tunnel about two hundred feet in, at the depth of about one hundred and twenty-five feet) a mastodon tooth in a good state of preservation, which afterwards was partly broken, in the hollow of which was sulphuret of iron [iron sulfide, or pyrite]. And at the same time I found with it some relic that resembled a large stone bead, made perhaps of alabaster, about one and a half inches long, and about one and one fourth inches in diameter, with a hole through it one fourth of an inch in size, which no doubt had been used, some time, to put a string through. I also certify that I gave the specimens to C. D. Voy, about the year 1864, to put in his collection" (Whitney 1880, p. 266). Voy visited the site and confirmed the geological details.

Whitney (1880, p. 266) later wrote: "The bead was carefully examined by the writer. It is correctly described above, except that the material of which it is made is white marble, not alabaster. It had evidently been much handled, and unfortunately cleaned of the incrusting material; but quite distinct traces of a former filling of the hole with sulphuret of iron were still visible. The mastodon tooth bore, also, as stated by Mr. Stevens, evident marks of an incrustation of the same mineral; and it may be added that several of the bones, which are said to have come from under Table Mountain, have been found to have more or less abundant crystallizations of pyrites in the cellular portions. There can be no question of the artificial character of the so-called bead. It is regularly and symmetrically shaped, and looks as if intended for an ornament."

William J. Sinclair, of the University of California, objected (1908, p. 115): "Little dependence, as an evidence of antiquity, can be placed on the presence of pyrite in the hollow of the marble bead reported by Whitney from the gravels of this mine. The rapidity with which secondary pyrite forms is well known."

But the real significance of Whitney's remark about the presence of pyrite in the hollow of the bead is not that it proves, in and of itself, great age. Instead, it confirms that the bead examined by Whitney was the same one described by Stevens. And Stevens testified in his affidavit that he personally found the bead in a carload of rock and gravel from deep within the mine, below the latite cap of Table Mountain. In the absence of more exact information, this means that the bead would be at least 9 million years old and perhaps as much as 55 million years old.

But Sinclair (1908, pp. 115–116) objected: "If this degree of association with the gravel is to be accepted as proof of antiquity, we would be justified in supposing that any object of recent manufacture acquired under similar circumstances was as old as the gravels."

Of course, if one were convinced that an object was "of recent manufacture," then no circumstance of acquisition whatsoever—even the most perfect—would compel one to suppose it was as old as the gravels. But if the object provided no clear evidence for its date of manufacture, then circumstances of acquisition like those encountered in the case of the marble bead would argue strongly in favor of an age equivalent to that of the Tertiary gravels.

So here we have a typical example of the unfair treatment of anomalous evidence. Sinclair attempted to raise unreasonable doubt and suspicion about the origin of the white marble bead, even though the initial report that it came from Tertiary gold-bearing deposits was credible. But in the cases of many accepted discoveries, the circumstances of discovery are similar to that of the marble bead.

For example, at Border Cave in South Africa, *Homo sapiens sapiens* fossils were taken from piles of rock excavated from mines years earlier. The fossils were then assigned dates of about 100,000 years, principally because of their association with certain kinds of rock. The scientists who assigned the dates wrote: "Border Cave 1 and 2 comprise an adult male cranial vault and a partial adult female mandible respectively. These fragments were all displaced from their original contexts in 1940 during the removal of 'fertilizer' from Horton's Pit. . . . Cooke *et al.* claimed that the character of the soil adhesions in small interstices of the skull was only matched by a distinctive 'chocolate coloured layer' corresponding to the base of our [layer] IGBS.LR" (Beaumont *et al.* 1978, p. 414).

The Heidelberg jaw was discovered by workmen in a gravel pit, with no scientist present, and was assigned a Middle Pleistocene date. Furthermore, most African hominid fossils, including those of Lucy (*Australopithecus afarensis*), were discovered on the surface and were assigned specific dates because of their loose association with certain exposed strata. In Java, also, most of the *Homo erectus* discoveries occurred on the surface, and, in addition, they were found by paid native collectors, who shipped the fossils in crates to distant scientists for study.

If Sinclair's strict standards were to be applied to these finds, they also should have to be rejected as evidence for hominids of any particular antiquity. In other words, most of the evidence upon which the current picture of human evolution is based would have to be thrown out.

And this takes us back to the central theme of this book. We are not promoting any particular discovery or set of discoveries. Rather, we are looking at the entire body of evidence relating to human origins and antiquity and asking for consistent application of standards for acceptance and rejection of evidence. Our historical survey has led us to conclude that up to now scientists have not impartially applied such standards. This raises some legitimate doubts about the trustworthiness of the evolutionary lineages that have been erected upon such a shaky evidential foundation.

Indeed, we find that when all the available evidence is considered impartially, an evolutionary picture of human origins fails to emerge. On the one hand, if we apply the tactic of extreme skepticism equally to all available evidence, we wind up with such an insufficiency of facts that it becomes next to impossible to say anything at all about human origins. On the other hand, if we take a more liberal, yet evenhanded, approach to the totality of evidence, we are confronted with facts demonstrative of a human presence in remote geological ages, as far back as the Eocene, and even further.

5.5.8 The Pierce Mortar

In 1870, Llewellyn Pierce gave the following written testimony (Whitney 1880, p. 266): "This is to certify that I, the undersigned, have this day given to Mr. C. D. Voy, to be preserved in his collection of ancient stone relics, a certain stone mortar, which has evidently been made by human hands, which was dug up by me, about the year 1862, under Table Mountain, in gravel, at a depth of about 200 feet from the surface, under the basalt, which was over sixty feet deep, and about 1,800 feet in from the mouth of the tunnel. Found in the claim known as the Boston Tunnel Company." Whitney (1880, p. 266) said the mortar was 31.5 inches in circumference. Voy visited the site and saw the approximate place where the object was found (Whitney 1880, p. 267).

William J. Sinclair interviewed Llewellyn Pierce in 1902, a good 40 years after the original discovery was made. Sinclair (1908, p. 116) wrote: "The mortar from the Boston claim was [according to Pierce] as large as a sixteen-gallon milk bucket and would weigh about seventy-five pounds. It was found in hard gravel under the cement, and was taken out by Mr. Pierce while he was sitting on a candle box, breasting [sic] out gravel. . . . The mortar preserved in Voy's collection is an oval boulder of hornblende andesite into which a hole has been worked, about four and three-quarters inches in greatest width, and three and three-quarters inches deep, dimensions to which those of a sixteen-gallon bucket must be regarded as a rather liberal approximation." This last sarcastic remark appears calculated to cast doubt on Pierce's testimony. But it should be recalled that the entire mortar was 31.5 inches in circumference, which is close to the size of the mouth of a tall sixteen gallon milk bucket commonly used in dairies.

Sinclair (1908, p. 117) added: "The deep gravels in the bottom of the Table Mountain channels, tapped by the Boston Tunnel and other workings, are largely inaccessible, but so far as known are not volcanic. The incongruity of associating an andesitic mortar . . . with the old prevolcanic gravels is at once apparent. The andesitic sands and gravels of Table Mountain lie above the auriferous gravel channels in which these relics were supposed to occur." If Sinclair is correct that the mortar was found in the prevolcanic gravel, then it would be 33–55 million years old (Table 5.3, p. 371).

But what was the source of the andesite from which the Pierce mortar was made? The prevolcanic auriferous gravels contained boulders of different kinds of rock formed in previous ages, so who can say that there were no isolated andesite boulders in the ancient river channels? Furthermore, there may have been deposits of andesite as old as the prevolcanic gravels in other nearby areas of the Sierra Nevada mountains, and therefore andesite boulders or finished andesite mortars could have been transported by human agency to the region of Tuolumne Table Mountain.

In fact, Durrell (1966, pp. 187–189) reported four nearby sites, all north of Tuolumne Table Mountain, which are just as old as the prevolcanic auriferous gravels and contain deposits of hornblende andesite. These are the Wheatland Formation, at 100 miles; the Reeds Creek Formation, at 100 miles; Oroville Table Mountain, at 140 miles; and the Lovejoy Formation, at 200 miles.

Good portable andesite mortars might have been a valuable trade item, and might have been transported good distances by rafts or boats, or even by foot. In a study of California Indians, R. F. Heizer and M. A. Whipple reported the presence of basalt mortars in Marin County, north of San Francisco. The mortars ranged in weight from 20 to 125 pounds. Heizer and Whipple (1951, p. 298) stated: "Each of these pieces must have been carried to the spot from not less than 25 miles away, no mean task for the slightly built, barefoot Indians. Stone is completely lacking in the alluvial deposits of the valley floodplain of the Sacramento and San Joaquin delta region." According to Heizer and Whipple, such manufactured objects were frequently traded by Indians for unfinished raw materials.

Hence none of Sinclair's arguments are strong enough to invalidate the testimony indicating that the Pierce mortar was deposited in the Table Mountain gravels during Tertiary times. The general tone of Sinclair's paper indicates that he was strongly biased against the possibility of toolmaking humans living in the Tertiary, and that he was searching for any excuse to discredit these discoveries.

According to Sinclair (1908, pp. 116–117), Pierce found another artifact along with the mortar: "The writer was shown a small oval tablet of dark colored slate with a melon and leaf carved in bas-relief. Mr. Pierce claimed to have found this in the same gravels as the mortar, and, he thought, probably at the same time. This tablet shows no signs of wear by gravel. The scratches are all recent defacements. The carving shows very evident traces of a steel knife blade and was conceived and executed by an artist of considerable ability."

Sinclair stated that this carving could not really have been as old as the Tertiary gravels in which it was discovered. It appears that Sinclair brought the carved tablet into his discussion simply for the purpose of distracting attention from the mortar reported by Pierce. Such tactics are often encountered in critiques of anomalous evidence.

Sinclair provided no account of the exact features of the slate tablet that led him to conclude it had been carved with a steel blade. Therefore, he may have been wrong about the type of implement that was used. Furthermore, the level of human technological achievements in the Tertiary was then, and still is, very much an open question. If the slate tablet was in fact discovered, with the mortar, in prevolanic gravels deep under the latite cap of Tuolumne Table Mountain, beneath a hard layer of "cement," and if the tablet does in fact display definite signs of carving by a steel blade, then one would be justified in concluding that

human beings of a relatively high level of cultural achievement were present
between 33 million and 55 million years ago. In fact, the carved tablet could be
taken as proof that the artisan used steel tools. Sinclair also said that the tablet
showed no signs of wear by gravel. But perhaps it was not moved very far by the
action of the Tertiary river and therefore remained unabraded. Or perhaps the
tablet could have been dropped into a gravel deposit of a dry channel of a shifting
stream. This would also explain why it showed no signs of excessive wear.

5.5.9 The Neale Discoveries

On August 2, 1890, J. H. Neale signed the following statement about
discoveries made by him: "In 1877 Mr. J. H. Neale was superintendent of the
Montezuma Tunnel Company, and ran the Montezuma tunnel into the gravel
underlying the lava of Table Mountain, Tuolumne County. . . . At a distance of
between 1400 and 1500 feet from the mouth of the tunnel, or of between 200 and
300 feet beyond the edge of the solid lava, Mr. Neale saw several spear-heads,
of some dark rock and nearly one foot in length. On exploring further, he himself
found a small mortar three or four inches in diameter and of irregular shape. This
was discovered within a foot or two of the spear-heads. He then found a large well-
formed pestle, now the property of Dr. R. I. Bromley, and near by a large and very
regular mortar, also at present the property of Dr. Bromley." This last mortar and
pestle are shown in Figure 5.13.

Neale's affidavit continued: "All of these
relics were found the same afternoon, and were all
within a few feet of one another and close to the
bed-rock, perhaps within a foot of it. Mr. Neale
declares that it is utterly impossible that these
relics can have reached the position in which they
were found excepting at the time the gravel was
deposited, and before the lava cap formed. There
was not the slightest trace of any disturbance of
the mass or of any natural fissure into it by which
access could have been obtained either there or in
the neighborhood" (Sinclair 1908, pp. 117–118).

Figure 5.13. This mortar and pestle (Holmes 1899, plate XIII) were found by J. H. Neale, who removed them from a mine tunnel penetrating Tertiary deposits (33–55 million years old) under Table Mountain, Tuolumne County, California.

The position of the artifacts in gravel "close to the
bed-rock" at Tuolumne Table Mountain indicates
they were 33–55 million years old.

In 1898, William H. Holmes decided to interview Neale and in 1899 published the following
summary of Neale's testimony: "One of the miners coming out to lunch at noon brought with him

to the superintendent's office a stone mortar and a broken pestle which he said had been dug up in the deepest part of the tunnel, some 1500 feet from the mouth of the mine. Mr. Neale advised him on returning to work to look out for other utensils in the same place, and agreeable to his expectations two others were secured, a small ovoid mortar, 5 or 6 inches in diameter, and a flattish mortar or dish, 7 or 8 inches in diameter. These have since been lost to sight. On another occasion a lot of obsidian blades, or spear-heads, eleven in number and averaging 10 inches in length, were brought to him by workmen from the mine. They had been found in what Mr. Neale called a 'side channel,' that is, the bed of a branch of the main Tertiary stream about a thousand feet in from the mouth of the tunnel, and 200 or 300 feet vertically from the surface of the mountain slope. . . . Four or five of the specimens were given to Mr. C. D. Voy, the collector. . . . Some had one notch, some had two notches, and others were plain leaf-shaped blades" (Sinclair 1908, pp. 118–119; Holmes 1899, pp. 452–453).

As can be seen, there are significant differences between the account given by Holmes and the earlier affidavit of Neale. In particular, Holmes (1899, p. 453) said: "In his conversation with me he did not claim to have been in the mine when the finds were made." This might be interpreted to mean that Neale had lied in his original statement. Here, however, the following points need to be carefully considered. The just-quoted passages from Holmes are not the words of Neale but of Holmes (1899, p. 452), who said: "His [Neale's] statements, written down in my notebook during and immediately following the interview, were to the following effect." It is not clear what liberties Holmes may have taken in his representation of Neale's conversations with him. It is interesting that Holmes did not say that Neale denied that he entered the mine; Holmes merely said he did not positively state that he did enter, which leaves open the possibility that perhaps he did. It is thus debatable whether one should place more confidence in Holmes's indirect summary of Neale's words than in Neale's own notarized affidavit, signed by him. Significantly, we have no confirmation from Neale himself that Holmes's version of their conversation was correct.

That Holmes may have been mistaken is certainly indicated by a subsequent interview with Neale conducted by William J. Sinclair in 1902. Summarizing Neale's remarks, Sinclair (1908, p. 119) wrote: "A certain miner (Joe), working on the day shift in the Montezuma Tunnel, brought out a stone dish or platter about two inches thick. Joe was advised to look for more in the same place. At the time, they were working in caving ground. Mr. Neale went on the night shift and in excavating to set a timber, 'hooked up' one of the obsidian spear points. With the exception of the one brought out by Joe, all the implements were found personally by Mr. Neale, at one time, in a space about six feet in diameter on the shore of the channel. The implements were in gravel close to the bed-rock and were mixed with a substance like charcoal." When all the testimony is duly weighed, it appears that Neale himself did enter the mine and find stone implements in place in the gravel.

As in the case of the Pierce discoveries, Sinclair (1908, p. 119) observed "there is involved the anomaly of two late volcanic rock types, andesite and obsidian, occurring in the prevolcanic gravels." Holmes raised the same objection. He asserted that according to geologists andesite "is not found in the formations of the particular region" until a period long after the gold-bearing gravels were deposited (Holmes 1899, p. 426). He added: "The objects being generally large, it is not to be supposed for a moment that they could have been brought from a distance" (Holmes 1899, p. 426). About obsidian, he stated it "is known only as a late product, having its origin in the most recent flows of the Sierra" (Holmes 1899, p. 426).

Concerning the andesite, we have already noted that there occur in the same region andesite deposits of the same age as the prevolcanic gravels at Tuolumne Table Mountain. Furthermore, the fact that andesite artifacts were found in more than one mine shaft under Tuolumne Table Mountain strengthens the supposition that boulders of andesite may have been present in the rivers that deposited the prevolcanic gravels. Furthermore, andesite mortars, although heavy, may have been transported by boat or raft, or even by foot.

As far as the obsidian spearheads are concerned, it is well established that Neolithic cultures all over the world have traded such objects over extended areas. Thus even if no raw obsidian was locally available, that would pose no obstacle to the presence of finished obsidian blades in the lowermost prevolcanic gravels at Tuolumne Table Mountain, which are 33–55 million years old.

In countering Neale's direct testimony that he found stone tools in prevolcanic gravels at Tuolumne Table Mountain, Holmes and Sinclair could, in the end, raise only the vague suspicion that the objects had somehow been recently introduced into the Montezuma mine. Sinclair (1908, p. 120) stated: "There was every indication of a former Indian camp site in this vicinity. Half an hour's search resulted in the discovery of a pestle and a flat stone muller, a few yards north of the mine buildings. Similar discoveries were reported by Holmes. South of the tunnel, a large permanent mortar was found. The material of this mortar is latite from the cliff above. It is quite possible that the implements mentioned by Mr. Neale came from this Indian camp."

In similar fashion, Holmes (1899, pp. 451–452) questioned: "Is it not more reasonable to suppose that some of the typical implements of the Indians living at the mouth of Montezuma mine should have been carried in for one purpose or another, embedded in the gravels, and afterwards dug up and carried out to the superintendent than that the implements of a Tertiary race should have been left in the bed of a Tertiary torrent to be brought out as good as new, after the lapse of vast periods of time, into the camp of a modern community using identical forms?" But the reasonableness of Holmes's supposition is questionable. There is, in fact, ample reason to believe that the implements found by Neale were not carried into the Montezuma shaft but were deposited in Tertiary times. First of all, in the passage quoted above, Sinclair referred to a large, immovable, permanent mortar found near the mine entrance, but the mortars found by Neale in the mine were portable mortars. Also, mortars much like those found in the California mines have been discovered at various sites around the world, including Jarmo and Beidha in the Middle East. This shows that such stone mortars are likely to have been made by any people, living at any time or place.

Furthermore, it has been shown in Africa that modern tribes use the same kind of cobble implements found in the lower levels of Olduvai Gorge. The similarity of the Olduvai implements to modern ones in use in the same region did not prevent acceptance of their Early Pleistocene antiquity. Therefore, the similarity of implements found in the prevolcanic gravels at Tuolumne Table Mountain to those found on the surface in the same region should not be taken as sufficient cause to deny their great age.

About the obsidian spearheads found by Neale, Holmes (1899, p. 453) reported: "Desiring to find out more concerning these objects . . . he [Neale] showed them to the Indians who chanced to be present, but, strangely enough, they expressed great fear of them, refusing to touch them or even speak about them; but finally, when asked whether they had any idea whence they came, said they had seen such implements far away in the mountains, but declined to speak of the place further or to undertake to procure others."

Holmes (1899, p. 453) then stated: "I was not surprised when a few days later it was learned that obsidian blades of identical pattern were now and then found with Digger Indian remains in the burial pits of the region. The inference to be drawn from these facts is that the implements brought to Mr. Neale had been obtained from one of the burial places in the vicinity by the miners." Here we must discount Holmes's inference that the implements were brought to Neale by miners. We have established that Neale's statements in his original affidavit, as confirmed by his later statements recorded by Sinclair, are deserving of credence, and these statements show quite clearly that Neale himself found the implements in the gravels. Holmes (1899, p. 453) then stated: "How the eleven large spearheads got into the mine, or whether they came from the mine at all, are queries that I shall not assume to answer."

Using Holmes's methods, it is clear that one could discredit any paleoanthropological discovery ever made: one could simply refuse to believe the evidence as reported, and put forward all kinds of vague alternative explanations, without answering legitimate questions about them. In the case under consideration, there is credible testimony by a reliable observer, Neale, that the implements were in fact found in the mine; therefore Holmes should not have failed to assume the burden of answering the queries he raised. Indeed, his failure to do so raises justifiable doubt about the value of his queries.

Holmes (1899, p. 453) further wrote about the obsidian implements: "that they came from the bed of a Tertiary torrent seems highly improbable; for how could a cache of eleven, slender, leaf-like implements remain unscattered under these conditions; how could fragile glass blades stand the crushing and grinding of a torrent bed; or how could so large a number of brittle blades remain unbroken under the pick of the miner working in a dark tunnel?" As often as such objections are raised, we can answer, first of all, that one can imagine many circumstances in which a cache of implements might have remained undamaged in the bed of a Tertiary stream. Just for example, let us suppose that in Tertiary times a trading party, while crossing or navigating a stream, lost a number of obsidian blades securely wrapped in hide or cloth. The package of obsidian blades may have been rather quickly covered by gravel in a deep hole in the stream bed and remained there relatively undamaged until recovered tens of millions of years later. As to how the implements could have remained unbroken as they were being uncovered, answering that question poses no insuperable difficulties. As soon as Neale became aware of the presence of the blades, he could have, and apparently did, exercise sufficient caution to preserve the obsidian implements intact. Maybe he even broke some of them.

In a paper read before the American Geological Society and published in its journal, geologist George F. Becker (1891, pp. 192–193) said: "It would have been more satisfactory to me individually if I had myself dug out these implements, but I am unable to discover any reason why Mr. Neale's statement is not exactly as good evidence to the rest of the world as my own would be. He was as competent as I to detect any fissure from the surface or any ancient workings, which the miner recognizes instantly and dreads profoundly. Some one may possibly suggest that Mr. Neale's workmen 'planted' the implements, but no one familiar with mining will entertain such a suggestion for a moment. . . . The auriferous gravel is hard picking, in large part it requires blasting, and even a very incompetent supervisor could not possibly be deceived in this way. . . . In short, there is, in my opinion, no escape from the conclusion that the implements mentioned in Mr. Neale's statement actually occurred near the bottom of the gravels, and that they were deposited where they were found at the same time with the adjoining pebbles and matrix."

5.5.10 The King Pestle

Although the tools discussed so far were found by miners, there is one case of a stone tool being found in place by a scientist. In 1891, George F. Becker told the American Geological Society that in the spring of 1869, Clarence King, director of the Survey of the Fortieth Parallel, and a respected geologist, was conducting research at Tuolumne Table Mountain. Becker (1891, pp. 193–194) stated: "At one point, close to the high bluff of basalt capping, a recent wash had swept away all talus and exposed the underlying compact, hard, auriferous gravel beds, which were beyond all question in place. In examining the exposure for fossils he [King] observed the fractured end of what appeared to be a cylindrical mass of stone. The mass he forced out of its place with considerable difficulty on account of the hardness of the gravel in which it was tightly wedged. It left behind a perfect cast of itself in the matrix and proved to be part of a polished stone implement, no doubt a pestle [Figure 5.14]." The facts recorded by Becker tend to rule out the phenomenon of secondary deposition—i.e., that the pestle had fallen from a higher, more recent layer and become recemented in the lower, older layer. Becker (1891, p. 194) added: "Mr. King is perfectly sure this implement was in place and that it formed an original part of the gravels in which he found it. It is difficult to imagine a more satisfactory evidence than this of the occurrence of implements in the auriferous, pre-glacial, sub-basaltic gravels." From this description and the modern geological dating of the Table Mountain strata, it is apparent that the object was over 9 million years old.

Even Holmes (1899, p. 453) had to admit that the King pestle, which was placed in the collection of the Smithsonian Institution, "may not be challenged with impunity." Holmes searched the site very carefully and noted the presence of some modern Indian mealing stones, but nothing else. He stated: "I tried to learn whether it was possible that one of these objects could have become embedded in the exposed tufa deposits in recent or comparatively recent times, for such embedding sometimes results from resetting or recementing of loose materials, but no definite result was reached" (Holmes 1899, p. 454). One may rest assured that if Holmes had found the slightest evidence of recementing, he would have seized the opportunity to cast suspicion upon the pestle discovered by King.

Figure 5.14. Left: Broken stone pestle found by Clarence King of the U.S. Geological Survey (Holmes 1899, p. 455). King personally extracted it from Tertiary deposits at Table Mountain, Tuolumne County, California. Right: A modern Indian pestle.

Unable, however, to find anything to discredit the report, Holmes (1899, p. 454) was reduced to wondering "that Mr. King failed to publish it—that he failed to give to the world what could well claim to be the most important observation ever made by a geologist bearing upon the history of the human race, leaving it to come out through the agency of Dr. Becker, twenty-five years later." But Becker (1891, p. 194) noted in his report: "I have submitted this statement of his discovery to Mr. King, who pronounces it correct."

Sinclair (1908, pp. 113–114) nevertheless attempted to raise doubts about the King pestle. He stated: "As a geologist, Mr. King was a reliable observer and able to determine whether or not the implement was in place and formed an integral part of the mass of gravel in which it was imbedded. Secondary cementation does not seem to have been taken into consideration. On many of the outcrops of andesitic sandstone in the vicinity of this locality, secondary cementation is in progress, indurating the soft sands into a hard rock to the depth of at least an inch. It is unfortunate that the matrix containing the impression of this relic was not preserved. As it is, there is no way of confirming the discovery. We have nothing but the specimen and the published account to work from."

In response to Sinclair's insinuations, the following points may be made. First of all, if as Sinclair himself stated, King was "a reliable observer," it is extremely unlikely, next to impossible, that he did not consider the obvious likelihood of recementing. Second, the original description of the find, recorded by Becker, and attested to by King, stated that at first only the cylindrical broken end of the pestle was visible and that the rest of the pestle was embedded in the hardened gravel. A photograph of the pestle (Holmes 1899, plate XIV), shown three-quarters size, indicates that the pestle was at least four inches long. This means that the pestle was probably embedded at least a couple of inches in the hardened gravel. Sinclair, searching for evidence of secondary cementation, noted that this occurred to a depth of "at least" one inch on some of the rock surfaces. The recemented material was of sand. But the pestle found by King was embedded in a hard deposit of auriferous gravel that had only recently been exposed. Sinclair said it was unfortunate that the gravel matrix containing the cast of the pestle was not available for inspection, thereby implying that perhaps the pestle was embedded in something other than the auriferous gravel. But King's statements, as recorded by Becker, make it clear that the pestle was found in the hard gravel deposits. Even Holmes, it should be remembered, hesitated to affirm that the pestle had been recemented onto the gravel in recent times.

Sinclair asserted that there was no way to confirm the authenticity of the discovery of the King pestle because all we now have is the specimen and the published report! The absurdity of this statement becomes apparent when we consider that specimens and reports of the circumstances of their discovery are all that we have to work from in almost all paleoanthropological discoveries ever made. Using Sinclair's logic, we could assert that there is no way to confirm any of them. For example, *Pithecanthropus erectus* was discovered by Dubois in Java during the 1890s. By 1908, when Sinclair wrote about the King pestle, all that was left of Java man were the specimens, stored in a Dutch museum, and the published reports. Therefore the Java man discovery might also have been judged unconfirmable. But this Sinclair did not do. Why? Favored evidence, it appears, can pass where unfavored evidence cannot. That is one of the main messages of this book. Paleoanthropologists frequently apply very subjective standards in the process of accepting and rejecting evidence.

5.5.11 Finds at San Andreas and Spanish Creek

The next set of reports describes discoveries that were made under intact volcanic layers at places other than under the latite cap of Tuolumne Table Mountain. Whitney (1880, pp. 273–274) described some of these discoveries and their geological setting as follows: "The fact that human implements had been found in some of the mining claims near San Andreas, in gravel under the volcanic strata, was repeatedly mentioned to the writer by persons living in that vicinity, and Mr. Voy was successful in finding some of the parties personally concerned in these finds, and getting their written testimony in regard to them. . . . Through all the higher southeastern portion of this county [Calaveras] the streams run in deep parallel cañons, quite close to each other, and having the ridges between them capped with volcanic overflows, all seeming to form part of the grand lava system which has spread far down the Sierra slope from the vicinity of Silver Mountain. In the vicinity of San Andreas the volcanic accumulations consist of alternating layers of sand, gravel, and volcanic ashes and conglomerates, overlying, as usual in the Sierra, gravel deposits more or less auriferous, the pay gravel being usually quite thin, and the whole series of detrital and volcanic materials reaching a thickness, in places, of from 150 to 200 feet."

Some evidence for the San Andreas discoveries came from R. D. Hubbard and John Showalter, who on January 3, 1871 provided C. D. Voy with the following statement: "This is to certify that we, the undersigned, proprietors of the Gravel claims known as Marshall & Company's, situated near the town of San Andreas, do know of stone mortars and other stone relics, which had evidently been made by human hands, being found in these claims, about the years 1860 and 1869, under about these different formations:

		Feet.
1.	Coarse gravel	5
2.	Sand and gravel	100
3.	Brown gravel	20
4.	'Cement' sand	4
5.	Bluish volcanic sand	15
6.	Pay gravel	6
	Total	150

The above [mentioned relics] were found in bed No. 6" (Whitney 1880, p. 274).

According to George Saucedo, a geologist of the California State Division of Mines and Geology, the pay gravels at the Marshall mine are probably Miocene or older (personal communication, 1989). The artifacts found in Bed 6 would, therefore, be in excess of 5 million years old.

W. O. Swenson, the justice of the peace who notarized the statement of Hubbard and Showalter, added: "I certify that I have seen one of the above described mortars, taken from said claims, and know the above to be true."

Another set of discoveries was made nearby. Whitney (1880, p. 274) stated: "In Smilow & Company's claim, on Gold Hill, about one mile west of Marshall & Company's, stone mortars were found at a depth of about one hundred feet in pay gravel, under the volcanic, the formation being closely similar to that of the last-mentioned locality [San Andreas]. This find is vouched for by Mr. Smilow himself."

From El Dorado County came a report, by a Mr. Ford, that "near the head of Spanish Creek a perfect mortar and pestle were once found in the gravel beneath the volcanic matter" (Whitney 1880, p. 277).

5.5.12 Discoveries at Cherokee

In 1875, Amos Bowman, a part-time assistant to the Geological Survey of California, told of finds made at Cherokee, a few miles north of Oroville, in Butte County: "One of the mortars, found by Mr. R. C. Pulham, of the Spring Valley Mining Company, was taken out of a shaft he dug himself in 1853, and was found, according to his testimony, twelve feet underneath undisturbed strata. . . . About 300 feet east of this shaft Mr. Frederic Eaholtz took out in 1853 a similar mortar at a greater depth. I visited both places with Mr. Pulham, and found several mortars still lying around on the top of the blue-gravel bench which is not yet mined away." The blue gravels, in which Pulham and Eaholtz discovered mortars, were "immediately underlying the auriferous gravel formation and the volcanic outflows" near Cherokee (Whitney 1880, p. 278).

Eaholtz gave information of further discoveries at another site near Cherokee. Bowman stated: "he told me further that, in 1858, while engaged with Wilson and Abbott in mining in the southwesterly part of the Sugar Loaf, he found in place, forty feet under the surface, a mortar of the same sort in unbroken blue gravel. This blue gravel nowhere comes to the surface, and it extends with the before-mentioned white and yellow gravel, under the Sugar Loaf, and under the Oroville volcanic *mesa*. It appeared only on the bottom of this claim. He was picking the blue gravel to pieces with a pick, when he found the mortar, which was a portion of the mass of cemented boulders and sand. He picked it out with his own hands" (Whitney 1880, p. 278). There were similar cases from Trinity and Siskiyou counties (Whitney 1880, p. 278).

George Saucedo of the California Division of Mines and Geology (personal communication, 1989) reported that the blue gravel is older than 23.8 million years. According to a study by R. S. Creely (1965), published in the bulletin of the California Division of Mines and Geology, the blue gravel is Eocene, or over 38 million years old. The implements found within the blue gravel would thus appear to be at least 23 million years old.

5.5.13 Evolutionary Preconceptions of Holmes and Sinclair

In light of the evidence we have presented, it is hard to justify the sustained opposition to the California finds by Holmes and Sinclair, who, as we have seen, were very reluctant to accept them as evidence of humans living in Tertiary times. Let us now review their five principal arguments.

(1) Holmes and Sinclair proposed that the discoveries of stone artifacts may have been the result of trickery by miners. But it is hard to see how or why such practical jokers could have slipped unseen into dozens of different mines over a distance of 100 miles, depositing numerous stone artifacts over a period of many years, or that so many miners would have assisted persons engaged in such trickery by not reporting them. Presumably, the motive would have been to deceive anthropologists, but this would have been a lot of work (some of the artifacts weighed 30 or more pounds) for what would seem to be an insignificant reward.

(2) Holmes (1899, p. 471) stated that none of the stone mortars showed evidence of unusual age or evidence of "wear and tear that would come from transportation in Tertiary torrents." But one would not expect such simple, durable mortars to show much evidence of age; once buried they could remain undamaged for millions of years. As far as "Tertiary torrents" are concerned, why should we assume that rivers were always torrents during Tertiary times? Perhaps, as in the case of most other rivers within our experience, these rivers sometimes flowed swiftly and fiercely but at other times slowly and calmly. Furthermore, it is possible that stone implements were dropped into the streams at a point very close to the place where they became lodged in the gravels, or perhaps they were dropped on the banks of the streams. In either case, one would not expect to find many signs of "wear and tear."

(3) Were the stone mortars perhaps carried into the mines by Indians living nearby? Holmes (1899, pp. 449–450) suggested this was the case: "the mountain Indians were in those days very numerous about the mining camps. The men were employed to a considerable extent in the mines, and it is entirely reasonable to suppose that their implements and utensils would at times be carried into the mines, perhaps to prepare or contain food, or perhaps as a natural proceeding with half-nomadic peoples habitually carrying their property about."

But Whitney (1880, p. 279) said of the portable mortars found in the mine shafts: "They are not in use at present among the Indians of that part of California where the implement in question is so abundantly found. The Digger Indians seem now, for some unknown reason, to prefer cavities worn in the rock in place, and in these the writer has often seen them crushing their acorns; but never once has he found them using the portable mortar."

Holmes (1899, p. 447) countered that Whitney "made the mistake of supposing they used only fixed mortars, that is, those worked in the surface of large masses or outcrops of rock. The fact is that portable mortars and grinding stones of diversified form are and have been used by Indians in all parts of California."

But modern authorities agree with Whitney. Glenn J. Farris, a California state archeologist, wrote to our researcher Steve Bernath: "Generally speaking the Indians of the gold rush period used bedrock mortars rather than the portable cobble mortars. The only instance I know of use of portable mortars in this area was to grind pine seeds into a pine nut butter, but I would see no reason for them to be carried into the mines" (personal communication, April 11, 1985).

In another letter, W. Turrentine Jackson, professor of history at the University of California at Davis, stated: "the Indians rarely transported a mortar for the grinding of acorns because of the heavy weight." Jackson also contradicted the proposition, voiced by Holmes thirty or forty years after the fact, that Indians remained in the mining region: "During the gold rush era the Indians were driven from the mining region, and they seldom came into contact with the forty-niners. I seriously doubt that any Indians had mortars of a portable nature in the mining areas. Certainly they would not have taken them onto a property while the miners were still operating" (personal communication, March 19, 1985). All in all, Holmes's arguments against the Tertiary age of the stone artifacts from the auriferous gravels are not very convincing.

(4) Holmes and Sinclair were unable to believe that humans of the modern type could have existed millions of years ago. And even if they did exist, their implements could not, they believed, have remained the same from then until now. The implements from the ancient gold-bearing gravels closely resembled those used by Indians in relatively recent times. According to evolutionary principles, the implements should have been much different. This suggested to Holmes, Sinclair, and Hrdlicka that the implements from the gold-bearing gravels were in fact of recent manufacture. But on examining these implements, both the ones from the gold-bearing gravels and those known to have been made by Indians in recent historical times, we see that they are simple artifacts of a kind that would naturally have been manufactured by any Neolithic-type culture anywhere in the world and at any time down through history. For example, stone artifacts from Neolithic sites at Beidha in the Middle East and the Nakura site in East Africa (Figure 5.15) are very much like those known to have been made by

California Indians in recent historical times. According to standard views of human prehistory, the cultures of the ancient Middle East and East Africa have no direct relation to those of the American Indians. This means that the mortars of California and the Middle East, although very similar in appearance, were developed independently. So if different peoples separated by thousands of miles independently developed similar implements, this suggests that different peoples separated by millions of years could have done the same.

(5) A final objection was that the artifacts were generally found by inexpert persons who could have been deceived by fraud, but George F. Becker, a professional geologist, disagreed. "Now, so far as the detection of a fraud is concerned," said Becker (1891, pp. 192–193), "a good miner, regularly employed in superintending the workings would be much more competent than the average geological visitor. The superintendent sees day by day every foot of new ground exposed, and it is his business to become thoroughly acquainted with its character, while he is familiar with every device for 'salting' a claim. . . . It is therefore an argument in favor of the authenticity of the implements that they have been found by miners." Deception by miners was unlikely, especially in the

Figure 5.15. Left: Stone bowl from Nakura, Kenya (L. Leakey 1931, p. 219). Center: Pestle from the Beidha site, in the Middle East (Singh 1974, p. 29). Right: A mortar and pestle from the Beidha site (Singh 1974, p. 29). Neolithic implements like these, manufactured at various locations around the world during the Late Pleistocene and Holocene, resemble implements found in Tertiary gravels in California gold mines (compare Figures 5.13 and 5.14). Holmes and Sinclair thought the implements found in the mines were recent because they looked like implements used by the California Indians of recent historical times. But the fact that such implements have been manufactured by African and Middle Eastern peoples, with no connection to the California Indians, shows that the mortars and pestles from the mines are of a type that might be made by people living at any time and any place. Thus the resemblance of tools from the California gold mines to recent tools found in the same region does not rule out their great antiquity.

many cases of tools found firmly embedded in the compact gravel.

S. Laing agreed with Becker about the unlikelihood of deception. Laing (1894, p. 387) wrote: "A conspiracy has been imagined of many hundreds of ignorant miners, living hundreds of miles apart, to hoax scientists, or make a trade of forging implements, which is about as probable as the theory that the paleolithic remains of the Old World were all forged by the devil, and buried in Quaternary strata in order to discredit the Mosaic account of creation." Regarding forgery, it is significant that no money was ever asked for any one of the artifacts.

Having closely examined the arguments put forward by Sinclair and Holmes, we find it apparent that their positions were based more on prejudice than on sound scientific reasoning. One might ask why Holmes and Sinclair were so determined to discredit Whitney's evidence for the existence of Tertiary humans. The following statement by Holmes (1899, p. 424) provides an essential clue: "If these forms are really of Tertiary origin, we have here one of the greatest marvels yet encountered by science; and perhaps if Professor Whitney had fully appreciated the story of human evolution as it is understood to-day, he would have hesitated to announce the conclusions formulated, notwithstanding the imposing array of testimony with which he was confronted." In other words, if the facts do not fit the favored theory, the facts, even an imposing array of them, must go. A more reasonable attitude was taken by Becker (1891, p. 190), who wrote of the implements found in the prevolcanic auriferous gravels of California: "If such an association of remains actually occurs, theories must be modified to fit the fact."

In his reports about the Tertiary discoveries in California, Whitney mentioned evidence from other parts of the world indicating the existence of culturally advanced humans in the Pliocene and Miocene periods. In 1871, according to Whitney (1880, p. 282), Portuguese geologist Carlos Ribeiro published a report (Section 4.1) that "cut flints, evidently the work of human hands, have been found in abundance in the Pliocene and Miocene even, of Portugal." Whitney faulted Charles Lyell for not mentioning Ribeiro's report in his authoritative survey, *The Antiquity of Man.* This is a valid criticism, demonstrating that right from the beginning of the scientific study of human evolution, uncomfortable evidence was simply ignored.

Alfred Russell Wallace, who shares with Darwin the credit for formulating the theory of evolution by natural selection, expressed dismay that evidence for anatomically modern humans existing in the Tertiary tended to be "attacked with all the weapons of doubt, accusation, and ridicule" (1887, p. 667).

In a detailed survey of the evidence for the great antiquity of humans in North America, Wallace gave considerable weight to Whitney's record of the discoveries in California of human fossils and stone artifacts from the Tertiary. In light of the incredulity with which the auriferous gravel finds and others like them were received in certain quarters, Wallace (1887, p. 679) advised that "the proper way to treat evidence as to man's antiquity is to place it on record, and admit it provisionally wherever it would be held adequate in the case of other animals; not, as is too often now the case, to ignore it as unworthy of acceptance or subject its discoverers to indiscriminate accusations of being impostors or the victims of impostors."

Wallace, an evolutionist, charged scientists who automatically rejected evidence for the extreme antiquity of anatomically modern humans with playing "into the hands of those who can adduce his recent origin and unchangeability as an argument against the descent of man from the lower animals" (Wallace 1887, p. 679). But, humanity's extreme antiquity, as demonstrated by the evidence cited by Wallace, also challenges the idea of human evolution from animals. This was certainly recognized by critics such as W. H. Holmes.

It is not hard to see why a supporter of the idea of human evolution, such as Holmes, would want to do everything possible to discredit information pushing the existence of humans in their present form too far into the past. Why did Holmes feel so confident about doing so? One reason was the discovery in 1891, by Eugene Dubois, of Java man (*Pithecanthropus erectus*), hailed as the much sought after missing link connecting modern humans with supposedly ancestral apelike creatures. Holmes (1899, p. 470) stated that Whitney's evidence "stands absolutely alone" and that "it implies a human race older by at least one-half than *Pithecanthropus erectus* of Dubois, which may be regarded as an incipient form of human creature only." For those who accepted the controversial Java man (Chapter 7), any evidence suggesting the modern human type existed before him had to be cut down, and Holmes (1899, p. 448) was one of the principal hatchet men. Holmes stated about the California finds: "It is probable that without positive reinforcement the evidence would gradually lose its hold and disappear; but science cannot afford to await this tedious process of selection, and some attempt to hasten a decision is demanded."

Holmes and his partner Hrdlicka warred long and hard to discredit all evidence for a human presence in the Americas any further back than four or five thousand years ago. During the nineteenth century, an extensive amount of evidence demonstrating a human presence far into the Tertiary had been amassed. But by the beginning of the twentieth century, it had become apparent to many American scientists that the decks had to be cleared.

Sinclair assisted in this task. In his introductory remarks to his paper on the California finds, Sinclair (1908, p. 108) wrote: "In working on the general problem of the time of man's appearance in the California region, the Department of Anthropology of the University of California has taken up, as a necessary part of the investigation, a review of the evidence relating to the so-called auriferous gravel relics. The writer was commissioned to visit the localities where the discoveries of human remains reported by Whitney and others were made."

Translation: Responsible scientists had concluded that modern human beings evolved from *Pithecanthropus erectus,* discovered in Java in 1891, in Middle Pleistocene formations. It was therefore an embarrassment to the University of California that it had in its collections stone implements said to date back well into the Tertiary. Further complicating the matter was the fact that the Tertiary age of these implements (and their human manufacturers) was vigorously advocated by the state geologist of California and other scientists. This contradicted the emerging picture of human evolution in general, as well as the increasingly accepted view that humans entered the Americas only recently. Sinclair was thus "commissioned" to do the required demolition job, and he did it well.

As might be guessed, Sinclair shared the evolutionary bias of Holmes, and it was this bias, more than anything else, that determined his negative attitude toward the California evidence. Sinclair (1908, pp. 129–130) wrote: "The occurrence in the older auriferous gravels of human remains indicative of a state of culture and a degree of physical development equal to that of the existing Indians of the Sierra Nevada would necessitate placing the origin of the human race in an exceedingly remote geological period. This is contrary to all precedent in the history of organisms, which teaches that mammalian species are short-lived. In North America, there are abundant remains of the lower animals preserved in deposits ranging from the Eocene to the Pleistocene. In all these deposits, excepting those of late Pleistocene age, the remains of man or any creature directly ancestral to man are conspicuously absent. No remains of Anthropoidea (from which man is doubtless derived), are known on this continent."

There are a number of comments that can be made about Sinclair's statements. As far as the history of organisms is concerned, we propose that our investigation of human antiquity shows that the history of other organisms might be different than Sinclair supposed. We began this present study for the purpose of evaluating the claim that all available evidence supports an evolutionary view of human origins, with modern humans descending quite recently from more apelike predecessors. We have determined, after thorough investigation, that such is not the case, that there is in fact abundant evidence that human beings of modern type have coexisted with more apelike creatures as far back in time as we care to extend our research. This clearly contradicts the usual claims made by evolutionists. We are therefore not certain what an objective evaluation of the fossil evidence for the history of other mammalian species might reveal.

Sinclair further maintained that human remains are absent from all North American deposits "excepting those of late Pleistocene age." He also stated: "It has been reported on the preceding pages that a large proportion of the implements reported from the gravels are from those of the rhyolitic and intervolcanic epochs. This would mean that man of a type as high as the existing race was a contemporary of the three-toed horse and other primitive forms of the late Miocene and early Pliocene, a thesis to which all geological and biological evidence is opposed" (Sinclair 1908, p. 130).

But if one were to agree that the California auriferous gravel discoveries should be rejected just because nothing like them had been discovered before, one would then be obliged to reject any fundamentally new paleontological discoveries whatsoever. Of course, it should be pointed out that at the time Sinclair was writing there was in fact abundant evidence, from North America, South America, and Europe, attesting to a human presence in the Early Pleistocene, Pliocene, Miocene, and earlier geological periods. If Sinclair was aware of this evidence, which he should have been, he simply chose to ignore it. Concerning Sinclair's assertion that no evidence of any anthropoid creatures had been found in North America, even this is contrary to the facts (Chapter 10).

In short, there was good reason to accept the California finds, as well as the many other discoveries we have reviewed in our discussion of anomalous stone tool industries. Nevertheless, in the early part of the twentieth century, the intellectual climate favored the views of Holmes and Sinclair. Tertiary stone implements just like those of modern humans? Soon it became uncomfortable to report, unfashionable to defend, and convenient to forget such things. Such views remain in force today, so much so that discoveries that even slightly challenge dominant views about human prehistory are effectively suppressed.

Concluding our study of anomalous stone tool industries, let us review some of the main points of interest. (1) Anomalous stone tool industries are not rare, isolated occurrences. The cases we have discussed form a massive body of evidence. Although many discoveries occurred in the nineteenth and early twentieth centuries, they have continued to occur up to the present. (2) Anomalous stone tool industries from very early times are not limited to eoliths, the human manufacture of which is considered by some as doubtful. Artifacts of undoubted human manufacture, similar to those of the finest Neolithic craftsmanship, are known to occur in geological contexts of extreme antiquity, as demonstrated by the California discoveries. (3) The much-debated eoliths are comparable to many unquestioningly accepted crude stone tool industries. Furthermore, eoliths appear to bear signs of intentional work not encountered in rocks broken by purely natural forces. (4) The scientific reporting of anomalous stone tool industries, even in the nineteenth century, was rigorous and of high quality. (5) It is apparent that preconceptions about human evolution have played an important role in the suppression of reports of anomalous stone tool industries. Such suppression continues to the present day.

6

Anomalous Human Skeletal Remains

In the nineteenth and early twentieth centuries quite a number of scientists found stone implements and other artifacts in Tertiary and early Quaternary formations. Scientists also discovered anatomically modern human skeletal remains in similarly ancient geological contexts.

Although these human bones originally attracted considerable attention, they are now practically unknown. Most current literature gives one the impression that after the discovery of the first Neanderthal in the 1850s no significant skeletal finds were made until the discovery of Java man in the 1890s.

For example, in describing the aftermath of the first Neanderthal find in 1856, anthropologist Jeffrey Goodman (1982, p. 56) wrote: "In the decades that followed, only discoveries of very old and crudely fashioned stone tools were made." As we shall see, this is simply not true. Why, then, do we rarely, if ever, encounter discussions of the skeletal finds of this period in modern paleoanthropological literature? One reason may be that these finds contradict the current scenario of human evolution.

We shall now consider these skeletal remains, some more challenging to the accepted views of human evolution than others. These anomalous discoveries are not numerous, but then the accepted hominid skeletal remains enshrined in museums around the world are also limited in number. More than one author has declared that the essential skeletal evidence supporting the idea that human beings evolved from apelike creatures would fit on a billiard table or two.

R. N. Vasishat (1985, p. 1) stated that the "fossil primate record is poor and within it, the record of fossil man still poorer." He added, however, that the few remains that have been recovered "when considered in the context of evolutionary evidences, known for other vertebrates with a better fossil record, allow us justifiable efforts at primate phylogenetic restorations" (Vasishat 1985, p. 1).

At first glance, the hominid fossils mentioned by Vasishat seem to support the phylogenetic restorations one usually encounters in textbooks and museums, but these restorations of evolutionary lineages fall apart when we include the human skeletal remains presented in this chapter.

395

In discussing these human bones, we shall focus first on their circumstances of discovery and the resulting stratigraphic age determinations, which are beyond the range modern evolutionary theory would permit. We shall begin with the least anomalous discoveries and then discuss those that are more so. In Appendix 1, we will review negative critiques by modern scientists, who have used chemical and radiometric methods to discredit the anomalously old stratigraphic dates assigned to some of the skeletal remains.

6.1 MIDDLE AND EARLY PLEISTOCENE DISCOVERIES

The first finds we shall consider are from the Middle and Early Pleistocene. The Trenton femur, if correctly dated, would be about 100,000 years old, which is anomalous for North America. The Galley Hill skeleton, from England, and the Moulin Quignon jaw, the Clichy skeleton, and the La Denise skull fragment, from France, are of ambiguous age, but are nevertheless relevant to our study of how scientists treat paleoanthropological evidence. The Ipswich skeleton appears to place anatomically modern humans in England during the Hoxnian interglacial, over 300,000 years ago. Many other Middle Pleistocene sites in Europe are linked with *Homo erectus,* even though no skeletal remains have been found. We argue that the tools and other artifacts found at these sites could just as well be attributed to anatomically modern *Homo sapiens.* A fully modern skull found by workmen excavating a dry dock in the harbor of Buenos Aires, Argentina takes us back to the Early Pleistocene. We shall also discuss a very primitive skullcap from Brazil, indicating the presence of creatures resembling *Homo erectus* in South America.

In our discussion of twentieth-century African discoveries (Chapter 11), we shall review three additional finds. These are the human skeleton recovered by H. Reck from an Early Pleistocene formation at Olduvai Gorge, Tanzania (Section 11.2), the human jaw discovered by Louis Leakey in an Early Pleistocene formation at Kanam, Kenya (Section 11.3), and the human skull fragments discovered by Louis Leakey in a Middle Pleistocene formation at Kanjera, Kenya (Section 11.3). We have chosen to discuss these three anomalous cases in Chapter 11 rather than here because they are closely connected to the accounts of conventionally accepted African finds.

6.1.1 The Trenton Human Bones (Middle Pleistocene)

On December 1, 1899, Ernest Volk, a collector working for the Peabody Museum of American Archaeology and Ethnology at Harvard University, discovered a human femur in a railroad cut south of Hancock Avenue within the city limits of Trenton, New Jersey. The femur was found lying on a small ledge,

91 inches beneath the surface. Volk (1911, p. 115) stated: "About four inches over or above the bone . . . was a place about the length of the bone where it evidently had fallen out of." This impression was overlain by the following strata: 7 inches of surface black soil, 16–20 inches of yellow loam with wa.er-worn pebbles, 44 inches of coarse gravel cemented together with red clay, and 21 inches of clean sand with red bands lying close together (Volk 1911, p. 116).

The human femur was found towards the bottom of the clean sand stratum, and it was photographed in that spot by Volk, who declared that the overlying strata immediately above and for some distance on either side of the find were undisturbed.

The fossil femur from Trenton was examined by two famous anthropologists, F. W. Putnam of the Peabody Natural History Museum at Harvard University and A. Hrdlicka of the Smithsonian Institution. Both of them declared the bone to be human. According to Hrdlicka (1907, p. 46), Putnam reported on the femur to the American Association for the Advancement of Science. Volk (1911, p. 117) wrote: "It was found to be part of the left femur of a human being, that had been cut off square at one end; the cellular structure had been gouged out to enlarge the opening and it had been perforated in two places; it had apparently been the handle of some implement." Volk said that the femur was thoroughly fossilized.

On December 7, 1899, Volk returned to the railway cut. About 24 feet west of the spot where he found the fossilized femur, and in the same layer, Volk recovered two fragments of a human skull (part of the parietal). The strata immediately overhead and for some distance on either side were said to be undisturbed.

Volk (1911, p. 118) stated: "That these human bones did not come from the upper deposits is made more probable by the fact that wherever . . . human skeletons have been found they have invariably been stained by the deposit in which they had been lying, but these fragments were nearly white and chalky." The upper deposits were reddish and yellowish.

Hrdlicka (1907, p. 46) stated that the stratum in which the Trenton femur was found lay underneath a deposit of glacial gravel. This would put the Trenton femur well back into the Pleistocene period. We have already discussed the views of Hrdlicka (Section 5.1.2), who labored hard to prove that human beings entered North America and South America only quite recently, during the Holocene. Since the Trenton femur was like that of modern humans, Hrdlicka suspected it was of recent age. He expected that a genuinely ancient human femur should display primitive features. Hrdlicka (1907, p. 46) therefore said about the Trenton femur: "The antiquity of this specimen must rest on the geological evidence alone."

Hrdlicka, however, was apparently unable to point out anything strikingly wrong with the geological evidence. The femur had been found in undisturbed

Pleistocene interglacial deposits by a reputable collector for a prestigious university. Consequently, Hrdlicka did not directly dispute the femur's Pleistocene interglacial age, but one gets the impression he felt further research would prove it recent. Hrdlicka (1907) did not mention the skull fragments Volk found.

In a letter dated July 30, 1987, Ron Witte of the New Jersey Geological Survey told us that the stratum containing the Trenton femur and skull fragments is from the Sangamon interglacial and is about 107,000 years old. According to standard ideas, human beings of modern type arose in southern Africa about 100,000 years ago and migrated to America at most 30,000 years ago.

6.1.2 Some Middle Pleistocene Skeletal Remains from Europe

During the nineteenth century, several discoveries of human skeletal remains were made in Middle Pleistocene formations in Europe. The reports we have studied raise doubts about the true age of these bones. We have nevertheless included them in our discussion for the sake of completeness. The presence of these skeletons in Middle Pleistocene strata could be attributed to recent intrusive burial, mistakes in reporting, or fraud. Nonetheless, there are reasons for thinking that the skeletons might in fact be of Middle Pleistocene age. We shall now briefly review some of the more noteworthy cases.

6.1.2.1 Galley Hill

In 1888, workmen removing deposits at Galley Hill, near London, England, exposed a bed of chalk. The overlying layers of sand, loam, and gravel were about 10 or 11 feet thick. One workman, Jack Allsop, informed Robert Elliott, a collector of prehistoric items, that he had discovered a human skeleton firmly embedded in these deposits about 8 feet below the surface and about 2 feet above the chalk bed (Keith 1928, pp. 250–266).

According to Elliott, Allsop had removed the skull but left the rest of the skeleton in place. Elliott stated that he saw the skeleton firmly embedded in the stratum: "I carefully examined the section on either side of the remains, for some distance, drawing the attention of my son, Richard, who was with me, and of Jack Allsop to it. It presented an unbroken face of gravel, stratified horizontally in bands of sand, small shingle, gravel, and, lower down, beds of clay and clayey loam, with occasional stones in it—and it was in and below this that the remains were found. We carefully looked for any signs of the section being disturbed, but failed: the stratification being unbroken, and much the same as the section in the angle of the pit remaining to this day" (Keith 1928, p. 253).

Elliott then removed the skeleton and later gave it to E. T. Newton (1895), who published a report granting it great age. An independent observer, a schoolmaster named M. H. Heys, reported that he had also seen the bones embedded in undisturbed deposits before Elliott removed the skeleton. Heys did not know Elliott at the time he examined the bones, and in fact he saw the bones before Elliott saw them. Heys reported that he saw the skull *in situ* just after it was exposed by a workman excavating the deposits and before it was removed from them.

Heys said about the bones: "No doubt could possibly arise to the observation of an ordinary intelligent person of their deposition contemporaneously with that of the gravel, for there was a bed of loam, in the base of which these human relics were embedded. The underneath part of the skull, as far as I could see, was resting on a sandy gravel. The stratum of loam was undisturbed. This undisturbed state of the stratum was so palpable to the workman that he said, 'The man or animal was not buried by anybody.' The gravel underneath the skull, of which I took particular notice, was stratified and undisturbed" (Keith 1928, p. 255).

Numerous stone tools were recovered from the Galley Hill site. Newton (1895, p. 521) reported: "Mr. Elliott has obtained several types of implements from this pit; namely, tongue- or spear-shaped forms, ovoid implements, hand hatchets, chipping tools, drills or borers, and flakes of various kinds." Newton (1895, p. 521) added: "There are also many rude flakes and roughly-chipped flints in this gravel, the human origin of which might be doubted if they were found alone; and occasionally deeply-stained primitive forms are met with, similar to those found by Mr. B. Harrison on the high plateau near Ightham."

According to Stuart Fleming (1976, p. 189), the stratum in which the Galley Hill skeleton was discovered is more than 100,000 years old. K. P. Oakley and M. F. A. Montagu (1949, p. 34) commented that this stratum is Middle Pleistocene and is "broadly contemporary with the Swanscombe skull." Oakley (1980, p. 26) and Gowlett (1984, p. 87) considered the Swanscombe skull, found only a short distance from Galley Hill, to be from the Holstein interglacial, which occurred about 330,000 years ago. The Galley Hill skeleton, if roughly contemporary with Swanscombe, would be of the same age.

In terms of anatomy, the Galley Hill skeleton was judged to be of the modern human type (Newton 1895, Keith 1928, Oakley and Montagu 1949). Most scientists now think that anatomically modern humans (*Homo sapiens sapiens*) originated in Africa around 100,000 years ago. They say that *Homo sapiens sapiens* eventually entered Europe in the form of Cro-Magnon man approximately 30,000 years ago, replacing the Neanderthals (Gowlett 1984, p. 118). Fully modern and found in strata contemporaneous with the Swanscombe site (about 330,000 years B.P.), the Galley Hill skeleton thus presents an anomaly.

Just what do modern paleoanthropologists say about the Galley Hill skeleton? Despite the stratigraphic evidence reported by Heys and Elliott, Oakley and Montagu (1949) concluded that the skeleton must have been recently buried in the Middle Pleistocene deposits. They considered the bones, which were not fossilized, to be only a few thousand years old. This is also the opinion of almost all anthropologists today.

The Galley Hill bones had a nitrogen content similar to that of fairly recent bones from other sites in England. Nitrogen is one of the constituent elements of protein, which normally decays with the passage of time. But there are many recorded cases of proteins being preserved in fossils for millions of years. Because the degree of nitrogen preservation may vary from site to site, one cannot say for certain that the relatively high nitrogen content of the Galley Hill bones means they are recent. The Galley Hill bones were found in clayey sediments known to preserve protein.

Oakley and Montagu (1949) found the Galley Hill human bones had a fluorine content similar to that of Late Pleistocene and Holocene (recent) bones from other sites. It is known that bones absorb fluorine from groundwater. But the fluorine content of groundwater may vary widely from place to place and this makes comparison of fluorine contents of bones from different sites an unreliable indicator of their relative ages.

Later, the British Museum Research Laboratory (Barker and Mackey 1961) obtained a carbon 14 date of 3,310 years for the Galley Hill skeleton. But this test was performed using methods now considered unreliable. Also, it is highly probable that the Galley Hill bones, kept in a museum for 80 years, were contaminated with recent carbon, causing the test to give a falsely young date.

For a more detailed discussion of the above-mentioned tests, see Appendix 1. Although modern paleoanthropologists have great confidence in these tests, there are good reasons for thinking that they are at least as imperfect and subject to error as older methods of dating, such as stratigraphic observation. Thus chemical and radiometric test results do not automatically invalidate stratigraphic observations with which they may be in disagreement.

In attempting to discredit the testimony of Elliott and Heys, who said no signs of burial were evident at Galley Hill, Oakley and Montagu (1949) and Oakley (1980) offered several arguments in addition to their chemical and radiometric tests.

Oakley and Montagu suggested (1949, p. 37) that by the time Elliott and Heys saw the skeleton "it is probable that the bulk of any evidence of burial had already been destroyed by the gravel digger."

Oakley and Montagu (1949, p. 36) also stated that many fragments of animal bones had been found in the sands and gravels at the Barnfield and Rickson's pits (both are a half mile from Galley Hill), whereas no animal bones had been found at the Galley Hill pit. From these facts, they concluded that originally there may have been a substantial number of animal bones in the Galley Hill deposits. They hypothesized that later these animal bones were all decalcified, or dissolved away, by the groundwaters. Hence the Galley Hill skeleton must have been recently introduced into the Middle Pleistocene gravels, after all the genuine Middle Pleistocene bones had been dissolved away. If the skeleton were really of Middle Pleistocene antiquity, it should have been dissolved away like the rest of the bones.

According to Oakley and Montagu (1949, p. 36), "This point does not seem to have been considered by previous investigators." That is not, however, accurate. E. T. Newton, the scientist who published the original report about the Galley Hill skeleton, was well aware of the significance of the absence of animal bones at Galley Hill and other places in the immediate vicinity. Newton wrote (1895, p. 524): "The rarity of bones in these high-level gravels suggests the possibility of their having been removed by the continued percolation of water during the long period which has elapsed since they were deposited. It still further suggests that, if any human bones had been deposited with the gravel in Paleolithic times, they would long since have disappeared. However bones of certain extinct mammals, *Elephas, Rhinoceros, Hippopotamus,* and *Felis leo,* have occasionally been found, although generally in a much decayed condition, and the circumstances sufficiently favourable for their preservation may have obtained in other places also."

Elliott reported that when the Galley Hill skeleton was uncovered "the bones were so friable and fragile that many went to pieces as soon as touched" (Newton 1895, p. 518; Keith 1928, p. 253). The decayed condition of the human bones thus matches that of the other rare occurrences of genuinely old mammalian bones in the immediate vicinity of the Galley Hill site.

Oakley and Montagu considered the possibility that the Galley Hill skeleton had been protected from percolating groundwaters by the loam layer in which it was embedded, but they concluded that this loam layer was "permeable." Yet Newton (1895, p. 524) stated: "it is clear from Mr. Elliott's letter, and from my own observation in the pit, that patches of more clayey deposit do here and there occur, one such having been noticed very near where the skeleton was found." Clay is less permeable to water than loam, and, as noted in Appendix 1, is responsible for many remarkable cases of organic preservation.

Oakley and Montagu argued that the relatively complete nature of the Galley Hill skeleton was a sure sign that it was deliberately buried. The postcranial bones found were two partial humeri, two partial femurs, two partial tibiae, and some small fragments of the ribs and hip. Completely missing were almost all of the ribs, the backbone, the forearms, hands, and feet. In the case of Lucy, the most famous specimen of *Australopithecus afarensis,* more of the skeleton was preserved (Section 11.9.3). And no one has yet suggested that Australopithecines buried their dead. Scientists have also discovered fairly complete skeletal remains of *Homo erectus* (Section 7.1.8) and *Homo habilis* (Section 11.7) individuals.

These cases, as all paleoanthropologists would agree, definitely do not involve deliberate burial. It is thus possible for relatively complete hominid skeletons to be preserved apart from burial.

Throughout their report, Oakley and Montagu returned to the suggestion that the Galley Hill skeleton must have been a burial, and this may in fact be true. But the burial may not have been recent. Sir Arthur Keith (1928, p. 259) suggested: "Weighing all the evidence, we are forced to the conclusion that the Galley Hill skeleton represents a man . . . buried when the lower gravel formed a land surface."

To sum it up, the arguments presented by Oakley and Montagu suggest the Galley Hill skeleton may have been a recent burial. But these arguments are not conclusive enough to invalidate the stratigraphic observations of Elliott and Heys, who, like Keith, were convinced the Galley Hill skeleton was genuinely ancient.

As can be seen, old bones point beyond themselves, quite obliquely, to events in the remote and inaccessible past. Controversy about their age is almost certain to arise, and in many cases the available evidence is insufficient to allow disputes to be definitely settled. This would appear to be true of Galley Hill. Since 1949, most scientists have, however, followed the lead of Oakley and Montagu in assigning the Galley Hill skeleton a recent date.

6.1.2.2 The Moulin Quignon Jaw: A Possible Case of Forgery

In 1863, Boucher de Perthes discovered an anatomically modern human jaw in the Moulin Quignon pit at Abbeville, France. He removed it from a layer of black sand and gravel that also contained stone implements of the Acheulean type (Keith 1928, p. 270). This black layer was 16.5 feet below the surface of the pit. According to Gowlett (1984, p. 88), the Acheulean sites at Abbeville are of the same age as the Holstein interglacial, and would thus be about 330,000 years old.

Upon hearing of the discovery of the Abbeville jaw and tools, a group of distinguished British geologists visited Abbeville and were at first favorably impressed (Keith 1928, p. 271). Later, however, it was alleged that some of the stone implements in Boucher de Perthes's collection were forgeries "foisted on him by the workmen" (Keith 1928, p. 271). The British scientists began to doubt the authenticity of the jaw. Taking a tooth found with the jaw back to England, they cut it open and were surprised at how well preserved and fresh it appeared. Also they determined that it contained 8 percent "animal matter" (organic matter in today's terms). Sir Arthur Keith pointed out, however, that in the same museum where the scientists met there were animal bones of Pleistocene age, prepared for display by John Hunter in 1792, containing up to 30 percent animal matter. There are also reports of bones from the Late Pliocene Red Crag formation with up to 8 percent animal matter (Osborn 1921, p. 568).

It is not clear exactly how old the bones prepared by Hunter actually were—they might have been Late Pleistocene, perhaps as little as 10,000 years old. Even so, the general point that Keith was making is relevant. As we show in Appendix 1, there is much evidence that the amount of organic matter remaining in a bone (as measured by nitrogen content) is not always a reliable indicator of a bone's age. Neither is the degree of fossilization. The rate at which a bone's organic matter decays, or the rate at which minerals accumulate in a bone, varies greatly from one location to another.

According to Ronald Millar (1972, p. 72), the Moulin Quignon jaw had a coloring "which was found to be superficial" and "was easily scrubbed from one of the portions of bone, revealing a surface which bore little of the erosion common in old bones." Some took this to be an indication of forgery. But Keith (1928, p. 272) interpreted this differently: "The mandible was originally covered by the black specks of the stratum in which it lay. Mr. Busk found he could brush these specks off; that does not invalidate its authenticity."

Prestwich was also said to have discovered that the flint tools from Moulin Quignon had a superficial coloring that could easily be washed off. But other pieces of flint (not artifacts) from the same site had a coloring that could not be scrubbed off. This was also taken by British scientists as an indication of forgery.

In May 1863, British geologists met with their French counterparts in Paris to jointly decide the status of the jaw. According to Keith, the French maintained the jaw was authentic despite arguments by the British that it was a forgery.

Keith (1928, p. 271) stated: "French anthropologists continued to believe in the authenticity of the jaw until between 1880 and 1890, when they ceased to include it in the list of discoveries of ancient man. At the present time opinion is almost unanimous in regarding the Moulin Quignon jaw as a worthless relic. We see that its relegation to oblivion begins when the belief became fixed that Neanderthal man represented a Pleistocene phase in the evolution of modern races. That opinion, we have seen, is no longer tenable."

In other words, scientists who believed the Neanderthals were the immediate ancestors of *Homo sapiens* could not accommodate the Moulin Quignon jaw because it would have meant that anatomically modern human beings were in existence before the Neanderthals. Today, the idea that the Neanderthals were the direct ancestors of the modern human type is out of vogue, but this in itself does not clear the way for acceptance of the Abbeville jaw, which if genuine, would be over 300,000 years old.

From the information we now have at our disposal, it is difficult to form a definite opinion about the authenticity of the Moulin Quignon jaw. Even if we accept that the jaw and the many flint implements found along with it were fakes, what does this tell us about the nature of paleoanthropological evidence? As we shall see, the Moulin Quignon jaw and tools, if they were forgeries, are not alone. Piltdown man (Chapter 8) was accepted for 40 years before being dismissed as an elaborate hoax.

6.1.2.3 *The Clichy Skeleton*

In 1868, Eugene Bertrand reported to the Anthropological Society of Paris that on April 18 of that year he found parts of a human skull, along with a femur, tibia, and some foot bones, in a quarry on the Avenue de Clichy. According to Keith (1928, pp. 276–277), the bones were found 5.25 meters (17.3 feet) beneath the surface, in a grey loam. Bertrand (1868, pp. 329–330) reported a similar depth but said that the bones were found in a reddish clayey sand layer within the grey loam. MacCurdy (1924a, p. 413) said that the bones were found in "a band of reddish sand at the base of the gray diluvium." A workman at the site reported that this reddish band was 10 or 20 centimeters (about 4–8 inches) in thickness (Bertrand 1868, p. 332). Keith believed that the age of the stratum in which the human bones were found was roughly the same age as the layer in which the Galley Hill skeleton was discovered. We recall from Section 6.1.2.1 that this layer is, according to current estimates, approximately 330,000 years old. The depth at which the Clichy human fossils were found (over 17 feet) argues against the recent intrusive burial hypothesis, and furthermore there was no mention of any disturbance in the overlying strata.

But Gabriel de Mortillet (Bertrand 1868, p. 332) said that a workman at the quarry on the Avenue de Clichy told him that he had stashed a skeleton in the pit. Skeptical about the reliability of testimony by workmen, de Mortillet asked him for proof. According to de Mortillet, the workman responded by telling him that he had taken the skeleton from a layer of reddish sediments in the upper part of the quarry. This explained why the bones found by Bertrand were reddish in color. The workman said that the layer of reddish material in the lower levels of the quarry was too thin to contain the bones.

According to de Mortillet, the workman further said that in the upper levels of the quarry, bones from the same animal were sometimes found together, whereas in the lower levels the fossil remains of mammals were always mixed and scattered. Thus the fact that the human bones were found "piled up in a little space" indicated they were not originally part of the layer in which they were found. On the basis of the workman's statements, de Mortillet concluded that the skeleton said to have been stashed by the workman was the same as the one found by Bertrand.

There is, of course, no guarantee that the workman was speaking the truth—de Mortillet himself frankly admitted the unreliability of testimony from workmen.

On the other hand, it is possible that the workman was being honest, and if this is the case then the Clichy skeleton could be only a few thousand or a few hundred years old.

De Mortillet was convinced by the workman's report, but the facts the workman reported can be interpreted differently. The workman suggested that the layer in which Bertrand said he found the bones was too thin to have colored them. But according to the workman, the maximum depth of the layer in question was 8 inches, which seems enough to have accommodated the fragmentary skeletal remains reported by Bertrand.

The workman's objection that it was unprecedented for several bones of the same creature to be found together in the lower layers of the quarry is of questionable significance. Eugene Bertrand (1868) said that he had evidence, which he planned to show to the Anthropological Society, that it was common for bones of the same animal to be found together in the lower layers as well as the upper layers.

Even after hearing de Mortillet relate the workman's story about stashing the bones of the Clichy skeleton, a number of scientists remained convinced Bertrand's discovery was genuine. For example, Professor Hamy (Bertrand 1868, p. 335) said: "Mr. Bertrand's discovery seems to me to be so much less debatable in that it is not the first of this kind at Avenue de Clichy. Indeed, our esteemed colleague, Mr. Reboux, found in that same locality, and almost at the same depth (4.20 meters), human bones that he has given me to study."

Hamy was not alone in accepting the Clichy find. Keith (1928, p. 276) reported that almost all authorities in France believed that the Clichy skeleton was as old as the layer in which Bertrand said it was found. Keith mentioned, however, that later on, after accepting the Neanderthals as the Pleistocene ancestors of modern humans, French anthropologists dropped the Clichy skeleton, which predated the Neanderthals, from the list of bona fide discoveries. A representative of the modern human type should not have been existing before his supposed ancestors. The Neanderthals are thought to have existed from 30,000 to 150,000 years ago. If the Clichy skeleton is about the same age as the Swanscombe skull, as suggested by Keith, it would be over 300,000 years old.

In his remarks to the Anthropological Society, Bertrand provided additional evidence for the great antiquity of the Clichy skeleton. He stated that he found a human cubitus, or ulna, in the stratum containing the other bones of the Clichy human skeleton. The ulna, the larger of the two long bones of the forearm, is located on the side opposite the thumb. When Bertrand tried to extract the ulna it crumbled into dust. He offered this as proof that the Clichy human skeleton must have been native to the layer in which it was found. Apparently, Bertrand reasoned that a bone as fragile as the decayed ulna could not possibly have been removed from an upper layer of the quarry and inserted into the lower layer in which he found it—it would certainly have been destroyed in the process. This indicated that the ulna belonged to the stratum in which Bertrand found it, as did the other human bones.

So in the case of the Clichy site, we have testimony indicating a recent age for the human skeleton found there, but at the same time there are good arguments that it was Middle Pleistocene.

6.1.2.4 *La Denise, France*

In the 1840s, pieces of human bone were discovered in the midst of volcanic strata at La Denise, France. Of particular interest was the frontal of a human skull. Keith (1928, p. 279) stated that the frontal "differs in no essential particular from the frontal bone of a modern skull."

The frontal was reported to have been taken from a limonite bed of considerable age. De Mortillet (1883, p. 241) wrote: "That the human frontal bone, now in the collection of M. Pichot, was in fact from the bed of argillaceous [clayey] limonite is perfectly established by deep incrustations of limonite on the interior of the bone."

In 1926, the French researcher C. Deperet reported to the French Academy of Science on the stratigraphy at La Denise. Deperet (1926, pp. 358–361) said the source of the human fossils was a layer of sediment deposited in a lake that formed after a Pliocene volcanic eruption and before the resumption of volcanic activity in the Pleistocene. According to Deperet, river deposits over the basalt from the last eruptions contained an Aurignacian fauna—horses, rhinoceroses, mammoths, hyenas, etc. This meant the last eruptions at La Denise occurred not later than the Late Pleistocene. Deperet's report thus suggests the presence of humans of the modern type at La Denise at some time during the Pleistocene, between 30,000 years ago (the latest time for the last eruptions) and 2 million years ago (the earliest time for the first eruptions). We would like to obtain more detailed stratigraphic evidence for the age of the argillaceous limonite layer from which the La Denise jaw was taken. Over a century after the La Denise fossils were discovered, they were tested chemically by K. P. Oakley. We shall discuss these test results, said to confirm a recent age for the fossils, in Appendix 1.

6.1.3 The Ipswich Skeleton (Middle Middle Pleistocene)

In 1911, J. Reid Moir discovered an anatomically modern human skeleton beneath a layer of glacial boulder clay near the town of Ipswich, East Anglia, in England. Reading through various secondary accounts, we learned that J. Reid Moir later changed his mind about the skeleton, declaring it recent. We thus did not consider the Ipswich skeleton for inclusion in this book. But after further investigation, we determined that the Ipswich skeleton could be genuinely old.

The key fact reported by Moir was that the Ipswich skeleton was found below a layer of boulder clay. The Boulder Clay of East Anglia overlies the Middle Pleistocene Cromer Forest Bed formation, which in turn overlies the Late Pliocene Red Crag. According to modern opinion, the Boulder Clay (a glacial deposit) could be as much as .4 million years old (Table 2.1, p. 78).

The Ipswich skeleton was discovered in a pit located at a brick field overlooking the valley of the river Gipping. Sir Arthur Keith (1928, p. 293) stated: "Passing northwards through Ipswich the traveller soon leaves the town and valley and finds himself on a plateau, about 150 feet above the level of the sea, and covered everywhere by a thick stratum of chalky boulder clay, varying in depth from 15 to 25 feet. . . . At the brick-field the chalky boulder clay has become reduced to a stratum of about 4 feet in thickness. . . . That the stratum at the brick-field represents a direct extension of the great sheet of boulder clay, Mr. Moir proved by sinking a series of pits from the brick-field to the crown of the plateau. In the map prepared by the officers of the Geological Survey the chalky boulder clay is shown to extend to the pit."

The skeleton was found at a depth of 1.38 meters (about 4.5 feet), between the boulder clay and some underlying glacial sands. Moir was aware of the possibility that the skeleton might represent a recent burial. Therefore, according to Keith (1928, pp. 294–295), Moir "took every means of verifying the unbroken and undisturbed nature of the stratum in and under which the skeleton lay."

Keith supervised the removal of the skeleton from its matrix at the Royal College of Surgeons. Keith (1928, p. 295) stated that "a whole skeleton was represented, and that it was placed on the right side in the ultra-contracted posture." To Keith the evidence suggested a burial from an ancient land surface. "At least it was not made from the present land surface," he said, "for the overlying stratum was intact" (Keith 1928, p. 295).

As for the condition of the bones, Keith said it was similar to that of Pleistocene animal fossils found elsewhere in the glacial sands. He noted: "The substance of the bones is grey and chalky in appearance, crumbling to white dry dust on pressure. The bones, when dissolved in hydrochloric acid, leave no animal matrix behind" (Keith 1928, p. 296).

The Ipswich skeleton was that of a man about 5 feet 10 inches tall. The brain capacity was 1430 cc, about average for modern humans, and according to Keith (1928, p. 297) "all the characters of the skull are those we are familiar with in modern man."

The discovery, however, inspired intense opposition. Keith (1928, p. 299) questioned: "if . . . the Ipswich skeleton had shown characters as distinctive as those of Neanderthal man . . . would anyone have doubted its age was older than the deposition of the boulder clay?" Keith (1928, p. 299) answered: "I do not think the age would then have been called into question. But under the presumption that the modern type of man is also modern in origin, a degree of high antiquity is denied to such specimens."

Keith (1928, p. 299) suggested: "It is, therefore, all the more important that every discovery of human remains, made in circumstances which make their high antiquity a reasonable presumption, should be placed on record, with no fact kept back." We fully agree with Keith on this point, and indeed, his suggestion has been one of the operating principles governing the compilation of the material in this book. The nature of paleoanthropological evidence is that it is never absolutely conclusive. There is always a chance that new evidence or new methods of analysis might result in a reevaluation of previous discoveries. It is therefore valuable to keep the details of controversial finds readily at hand for future generations of researchers.

Despite opposition, Moir initially stuck to his guns, holding that the Ipswich skeleton was genuinely old. What then happened to change his mind? His own report, published in *Nature* in 1916, tells the story. Moir (1916, p. 109) conducted further excavations in the area and reported: "These investigations have shown that at about the level at which the skeleton rested the scanty remains of a 'floor' are present, and that the few associated flint implements appear to be the same as others found on an old occupation-level in the adjacent valley. This occupation-level is in all probability referable to the early Aurignac period, and it appears that the person whose remains were discovered was buried in this old land surface. The material which has since covered the ancient 'floor' may be regarded as a sludge, formed largely of re-made boulder clay, and its deposition was probably associated with a period of low temperatures occurring in post-chalky boulder clay times." The Aurignacian stage in Europe occurred about 30,000 years ago (Gowlett 1984, p. 122) and is identified with *Homo sapiens sapiens.*

In Moir's statements we find nothing that compels us to accept a recent age for the skeleton. We know from our discussion of stone tools (Section 3.3.1) that Moir believed in a temporal succession of tool types, the older being more primitive than the recent. In other words, Moir was operating under the influence of an evolutionary preconception. But our own review of stone tools led us to the conclusion that it is not possible to place such implements in a temporal sequence simply on the basis of their degree of sophistication. Modern human beings are known to make very crude stone tool implements, and we have evidence that very sophisticated tools, comparable to those of Aurignacian Europe, turn up all over the world, in very distant times. In the 1960s, such implements were discovered at Hueyatlaco, Mexico, in strata yielding a uranium series age of over 200,000 years (Section 5.4.4). During the nineteenth century, very advanced stone objects turned up in the California gold mines, in gravels that might be as old as the

Eocene (Section 5.5). Therefore, we cannot agree with Moir that the discovery of tools of advanced type at the same level as the Ipswich skeleton was sufficient reason to reinterpret the site stratigraphy to bring the age of the skeleton into harmony with the supposed age of the tools.

Moir's reinterpretation of the boulder clay over the skeleton was not, it appears, based on any compelling geological evidence. In fact, he gave no geological reasons whatsoever in support of his conclusion that the boulder clay was a recently deposited sludge. Therefore, the simplest hypothesis is that it really was a layer of intact glacial boulder clay, as originally reported by Moir and recorded by the British Geological Survey on its detailed map of the region.

The age of the skeleton thus depends on the age of the boulder clay. Over the years, the age of the boulder clay in East Anglia has, however, been a matter of controversy. During the 1920s, Moir proposed that there were two glacial boulder clays in East Anglia, one laid down during the Mindel glaciation, and the other during the subsequent Riss glaciation (Keith 1928, pp. 302–303). If this scheme is accepted, the boulder clay that covered the Ipswich skeleton would belong to the Riss glaciation. We note, however, that the Riss glacial period extended from about 125,000 to over 300,000 years ago, which would still give a considerable antiquity to the Ipswich skeleton.

But it appears that Moir was wrong about there being evidence for a Riss glaciation at Ipswich. The English glaciation equivalent to the European Mindel glaciation is the Anglian. The Anglian was followed by the Hoxnian interglacial. The next English glacial period, corresponding to the European Riss glaciation, used to be called the Gippingian, following Moir's interpretation of the glacial deposits by the Gipping River near Ipswich. But according a modern English authority, D. Q. Bowen (1980, p. 420), geologists have "shown that the Gipping Till of Essex was Anglian in age."

The "necessary replacement" (Bowen 1980, p. 420) for the Gippingian glaciation was the Wolstonian. The clearest evidence for the Riss-equivalent Wolstonian glaciation is in the region of Birmingham, quite far from Ipswich, but there are some rather unclear signs of the Wolstonian in Mildenhall, about 40 miles to the northwest of Ipswich (Bowen 1980, p. 420). There is no sign of the Wolstonian at Ipswich itself. The final English glaciation, corresponding to the Würm glaciation of continental Europe, was the Devensian, which did not come down as far as Ipswich (Bowen 1980, p. 421; Nilsson 1983, p. 113). A modern authority on the Pleistocene geology of England stated: "nowhere in East Anglia is it possible to demonstrate post-Hoxnian and pre-Devensian glaciation on stratigraphic grounds" (Bowen 1980, p. 420).

In other words, there was, in the opinion of modern authorities, no Riss-equivalent glaciation (the Wolstonian) at Ipswich. Neither did the subsequent Devensian ice sheet reach Ipswich. What this means is that the Boulder Clay at

Ipswich can only be referred to the Anglian glaciation. Therefore the glacial sands in which the Ipswich skeleton was found must have been laid down between the onset of the Anglian glaciation, about 400,000 years ago, and onset of the Hoxnian interglacial, about 330,000 years ago. It would thus appear that the Ipswich skeleton is between 330,000 and 400,000 years old. Some authorities (Gowlett 1984, p. 87) put the onset of the Mindel glaciation (equivalent to the Anglian) at about 600,000 years, which would give the Ipswich skeleton an age potentially that great. Yet human beings of modern type are not thought to have appeared in Western Europe before 30,000 years ago (Gowlett 1984, p. 118).

From the story of the Ipswich skeleton, we learn that discoverers of anomalies, such as Moir, can be the victims of prejudices as strong as those of their opponents. For mavericks and establishment figures alike, evolutionary preconceptions block proper evaluation of paleoanthropological evidence. On the one hand, we find ideas about the recent evolution of the modern human form prevented certain scientists from believing that the Ipswich skeleton might be truly ancient. And on the other hand, ideas about the progressive evolution of stone tools influenced Moir to revise downward the age of his own discovery.

6.1.4 Possible Early Man Sites with No Skeletal Remains

If anatomically modern humans were present during the Middle and Early Pleistocene in Europe and elsewhere, why, one might ask, are scientists no longer finding any evidence of this? Instead, today's scientists are continually finding *Homo erectus* sites of Middle and Early Pleistocene age.

But here we run into a problem. There are many Middle and Early Pleistocene sites at which scientists have found stone tools but no hominid bones. The artifacts at these sites, mostly of a type called Acheulian, are nevertheless attributed to *Homo erectus*. But from a strictly objective point of view, in the absence of hominid fossils, the Acheulian artifacts could just as well be attributed to *Homo sapiens sapiens*. This is true regardless of the level of sophistication of the tools, since anatomically modern humans are known to make and use tools of the crudest sort.

Let us now review some cases demonstrating the difficulties one encounters in determining who made artifacts found at a site. In Section 5.4.3, we considered Acheulian tools from Timlin, New York. Mainstream authorities would surely attribute these to anatomically modern humans. The same is true the Acheulian-type tool from at Black's Fork River, Wyoming, U.S.A. (Section 4.8). Likewise it should be possible to attribute European Acheulian tools to *Homo sapiens sapiens* at sites with no hominid skeletal remains.

In Chapter 3, we discussed the Alabama pebble tools from the U.S.A., which are similar to the crude Oldowan tools of Africa (Section 3.8.5). The pebble tools at Olduvai Gorge are said to be the work of *Homo habilis* and are considered to be close to 2 million years old. Mainstream authorities, if they recognized the Alabama tools at all, would say they were not more than 12,000 years old and would attribute them to *Homo sapiens sapiens.*

We have also discussed the Late Pleistocene site at Monte Verde, Chile, where scientists found Oldowan-type tools hafted to wooden handles, along with other cultural remains typical of modern humans (Section 3.8.6). The preservation of wood in such circumstances is rare. It is therefore possible that the tools could have been found without the wooden handles and other perishable artifacts indicating a high level of culture. By analogy, we should be open to the possibility that at other sites where Oldowan tools are found, perishable artifacts typical of humans with a high degree of culture have been lost. This may be true even for Early Pleistocene sites. Normally, Oldowan tools, if found alone in an Early Pleistocene context, would not be attributed to anatomically modern humans. But the example of Monte Verde demonstrates that for sites where Oldowan tools alone are discovered it is possible that a variegated *Homo sapiens sapiens* culture was originally present but was not entirely preserved.

But even when signs of higher culture are present, scientists with strongly held preconceived ideas often fail to imagine that beings on the level of *Homo sapiens sapiens* might have been responsible for them. This is true for the Middle Pleistocene site of Terra Amata in southern France. Here, according to the discoverer, Henry de Lumley (1969, p. 42), campsites were established by bands of hominids 300,000 years ago. An account of the Terra Amata finds in a Time-Life book (*The Emergence of Man*) gave a date of 400,000 years.

At the ancient seashore site, de Lumley found oval patterns of post holes and stone circles indicating that the hominids erected temporary shelters and built fires. Also found were bone tools. Among them was one apparently used as an awl, perhaps to sew skins. Impressions found in the old land surface at the site were said to demonstrate that the hominids slept or sat on hides. Stone implements were also found, including an object described as a projectile point, made from volcanic rock obtained from the Esterel region, 30 miles away.

Significantly, no hominid fossils were found at the Terra Amata site in France. De Lumley (1969, p. 45) did, however, report "the imprint of a right foot, 9.5 inches long, preserved in the sand of a dune." The print was described as "human" in the Time-Life book. It was also said that the foot of the Terra Amata hominid was "arched to support his whole weight" and "had lost all trace of the ability to grasp that ape feet possess" (Time-Life 1973, p. 12). In a welcome display of scientific reserve, de Lumley, in his 1969 article about the Terra Amata discoveries published in *Scientific American,* did not identify the type of hominid that occupied the temporary habitation site on the shores of the Mediterranean.

However, the authors of the Time-Life study were not so inhibited. They wrote: "What sort of man visited this cove on the coast of Europe 400,000 years ago? Who was he? Although he came each spring for many years, the fossils he left behind him included no human bones—only a single human footprint in the hard sand. . . . He was the first man. He is known, in the scheme of evolution, as *Homo erectus,* or upright man. He was the direct descendant of *Australopithecus,* a creature considered the missing link between the apes and man" (Time-Life 1973, pp. 11–12). Judging from the available reports, the footprint is not different from that of a modern human being. The Time-Life book's assertion that *Homo erectus* was the inhabitant of Terra Amata is therefore unjustified.

At the Torralba, Spain, site, estimated to be about 300,000 years old, stone tools have been found in connection with fossil bones of elephants. Some scientists have interpreted Torralba as a *Homo erectus* kill-site. But as in the case of Terra Amata, no hominid fossils were found there. Only preconceived ideas about human evolution allowed scientists to attribute the Torralba tools and elephant bones to *Homo erectus.*

One skeptical researcher, Lewis Binford, even disagreed that Torralba was a kill-site. During the Middle Pleistocene, the area was a boggy marsh. Binford (1981, p. 16) pointed out that elephant fossils are generally found by water margins, because that is where they tend to die. Furthermore, the sediments at Torralba, according to F. Clark Howell, were deposited over many tens of thousands of years (Binford 1981, p. 16). Over this time, 115 elephants died. Assuming the sediments were deposited in just one 10,000-year period, Binford calculated that one elephant died every 87 years. Natural deaths, by disease, old age, or predators, he said, could very well account for the accumulation of elephant bodies at that rate. During the same 10,000 years, 611 stone tools accumulated at the site. This again, is not very many, considering the time involved—about 6 tools per century. So the association of stone tools and elephant bones could be purely accidental.

Reacting to the standard view of what happened at Torralba, Binford (1981, pp. 17–18) stated: "Man killed the animals while executing game drives—possibly aided by fire—butchered them, and carried the meat away—truly extraordinary. . . . This is a truly remarkable set of conclusions to draw from the Torralba data. . . . Given an aggregation of stone tools—evidence of hominid behavior—it is assumed that all other remains associated with the stone tools are also a by-product of human behavior. The researchers of Torralba have certainly made this assumption. Pleistocene archaeologists need to abandon such an approach."

Examples such as Torralba and Terra Amata could be multiplied, for at most paleoanthropological sites, no hominid bones are found. The artifacts at these sites are attributed to *Homo habilis, Homo erectus,* the Neanderthals, or *Homo sapiens* on the basis of their presumed age or their level of workmanship. But this practice, strictly speaking, is not justifiable. Therefore, many Early and Middle Pleistocene sites currently identified with *Homo erectus,* for example, could just as well be identified with anatomically modern *Homo sapiens.*

6.1.5 A Human Skull from the Early Pleistocene at Buenos Aires

In 1896, workers excavating a dry dock in Buenos Aires found a human skull (Figure 6.1). They took it from the rudder pit at the bottom of the excavation, after breaking through a layer of a hard, limestonelike substance called *tosca.* The level at which the skull was found was 11 meters (36 feet) below the bed of the river La Plata (Hrdlicka 1912, p. 318).

The workers who found the skull gave it to Mr. Junor, their supervisor, a senior member of the public works division of the Port of Buenos Aires (Hrdlicka 1912, p. 318). Information about the skull was furnished to the Argentine paleontologist Florentino Ameghino by Mr. Edward Marsh Simpson, an engineer for Charles H. Walker & Co. of London, the company contracted to excavate the port of Buenos Aires (Ameghino 1909, p. 108; Hrdlicka 1912, p. 319). In the opinion of Ameghino, the skull removed from the rudder pit belonged to a Pliocene precursor of *Homo sapiens.* He called this precursor *Diprothomo platensis.*

A. Hrdlicka (1912, p. 319) wrote: "Professor Ameghino [1909, p. 121] concludes from the information obtained from Mr. Simpson alone that the fragments of the skull came from the lower portion of the rudder-pit in Dry Dock No. 1 and from beneath the *tosca.* He states further, however, that beneath the *tosca* was found a layer of quartzy sand followed by a stratum of grey clay, and that it was in this layer of grey clay, 50 cm. [about 20 inches] below the floor of the dry dock, that the skull-cap of the *Diprothomo* was discovered."

Hrdlicka (1912, p. 321) said about Ameghino's opinion on the age of the deposits below the *tosca:* "The gray clay he identifies as belonging to the upper-most portion of the Pre-Ensenadean stratum, which is

Figure 6.1. Human skull taken from an Early Pleistocene formation in Buenos Aires, Argentina (Hrdlicka 1912, plate 49).

the most inferior part of the Pampean formation, and belongs to the base of the Pliocene." The base of the Pliocene is now dated at approximately 5 million years before the present. But modern authorities say that the Ensenadan began 1.5 million years ago (Anderson 1984, p. 41) or 1 million years ago (Marshall *et al.* 1982, p. 1352). The Pre-Ensenadan stratum in which the Buenos Aires skull was found would thus be at least 1.0–1.5 million years old. Even at 1 million years, the presence of a fully modern human skull anywhere in the world—what to speak of South America—is highly anomalous.

In the course of his investigation, Hrdlicka (1912, p. 319) found Mr. Simpson and Mr. Junor, and through them located Mr. J. E. Clark of Bahia Blanca, the foreman of the laborers who found the skull. Simpson revealed that he had not been present at the dry dock at the time the discovery had been made. He had simply received a report, but noted that he had been told there was more than one skull (Hrdlicka 1912, p. 319).

Hrdlicka (1912, p. 320) then reported the substance of the discussion he had with Mr. Junor: "Mr. Junor states that he did not see the find, but was told of it the next day, or perhaps the second day after, by the foreman, Mr. Clark. . . . The foreman brought Mr. Junor two pieces of the skull, and the latter saved them because they were said to have come from beneath the *tosca,* giving them later to the Museo Nacional. . . . As to the place from which the bones came, he remembers having been informed that the workmen had gotten through the floor of the dock into a sort of quicksand when the bones were encountered. . . . The bones must have been just beneath the *tosca,* for a small quantity of *tosca* was adhering to them. . . . He did not examine the site from which the skull fragments given him were supposed to have come. No inquiries were made of the laborers."

Relating the testimony of Clark, Hrdlicka (1912, p. 320) stated: "Mr. Clark states in his letter that the skull 'was found at the commencement of the Rudder Pit at dock bottom'; he 'is quite sure the skull was found at the Rudder Pit and under *tosca*'; and 'it was the only one found in that locality, but there was another skull found in the sand at the entrance to Dock No. 4.'"

Bailey Willis, the geologist who accompanied Hrdlicka on his expedition to Argentina, related this account of the interviews they had made: "Mr. Junor was found at his home in Flores, a suburb of Buenos Aires, on the evening of May 7, 1910, and we were most courteously received. He appeared to be about 70 years of age, of sanguine temperament, still enthusiastic as in youth, and an ardent believer of the antiquity of man in Argentina. He recited freely his recollection of the finding of the skull, stating in substance: The piece of skull was brought to him by the foreman of a gang of workmen who were digging out the rudder-pit. He (Mr. Junor) was very much occupied at the time by duties of supervision of construction and did not see the skull taken out, nor did he examine the place afterward to see where it came from; but he had no doubt that it came out of the well, 'probably' from between a layer of *tosca* and the underlying sand. The skull was said to have been found by a workman, who passed it to the foreman, who in turn gave it to Mr. Junor. The workman cannot now be identified. It does not appear that he was ever questioned as to how the bone was found. . . . On one point Mr. Junor was positive: The fragment of skull was taken out of the well. And although this statement rests on the say-so of the foreman who was told so by a workman, it appears to be the one item in the early history of the find that is not open to serious doubt" (Hrdlicka 1912, pp. 343–344).

Some will be critical of the fact that the skull was not found in place by a scientist. One should note, however, that the Heidelberg jaw, which is accepted by paleoanthropologists as genuine, was uncovered by a workman in a sand pit in Germany and turned over to a foreman, who in turn brought it to the attention of a local professor (Section 7.2). Many of the *Homo erectus* specimens from Java, reports of which have found their way into all authoritative textbooks, were collected by Javanese laborers while scientists were absent from the sites (Section 7.3). A more recent example is the Petralona skull, which Greek villagers found in a spot not clearly designated. Despite this, scientists, on stratigraphic grounds, assign the Petralona skull an age of 200,000 to 300,000 years (Gowlett 1984, p. 87), and use it as evidence for an evolutionary transition from *Homo erectus* to *Homo sapiens*.

Willis speculated that the Buenos Aires skull had somehow arrived quite recently in the position in which it was found. In discussing this possibility, Willis gave the known details of the construction of the dry dock. First an embankment was built to keep out the river, and to keep the excavation dry there was a pump operating from the sump or well in the lowest place. Then a concrete floor was laid, and concrete walls were built. Finally the rudder pit was dug. It was during the digging of the rudder pit that the skull was found. Willis suggested: "Any objects contained in the material excavated [in the course of building the dry dock] or in the standing earth exposed at the side might have found their way into the close vicinity of the rudder-pit, if not into the pit itself" (Hrdlicka 1912, p. 344). But by the time the pit was dug, the concrete walls and floor of the dry dock were already completed, which would mean there should not have been much dirt piled around on the floor or exposed on the walls. Plus there is abundant testimony that the skull was in fact found under a hard layer of *tosca* and was not lying loose at the top of the pit before digging took place.

Willis added: "We were told by Dr. Francesco Moreno that he, when a boy, used to go swimming where the dry dock now is, in deep pools" (Hrdlicka 1912, p. 345). Willis noted that a few kilometers away from the dry dock, there is a place where the river "has worked out deep irregular holes into which anything like the skull-cap called *Diprothomo* would readily sink and where it would become buried lower than the surface of the Pampaean, but beneath recent river mud" (Hrdlicka 1912, p. 345). There is, however, no basis for saying this occurred at the dry dock, where the skull was found beneath a layer of *tosca* in an excavation fully 11 meters (36 feet) below the present bed of the river La Plata.

As previously mentioned, Ameghino thought his *Diprothomo* represented an ancestral form of human. According to Hrdlicka (1912, p. 323), he believed the skull's capacity was only 1100 cc, compared to 1400 cc for an average *Homo sapiens,* and that it had a low vault. Hrdlicka (1912, p. 325) stated: "The writer reached Buenos Aires with the foregoing data before him and in consequence thereof with very eager expectations. But when the specimen itself was placed before him by Professor Ameghino there followed a rapid disenchantment."

Hrdlicka (1912, p. 326) noted: "In a detailed study of the specimen it soon became plain that almost the entire original description by Ameghino had miscarried by reason of the fragment having been placed and considered in a wrong position. . . . The accidental and faulty position of the fragment . . . had caused the forehead to appear much lower than it is. . . . these results of faulty orientation combined have helped to make the specimen look extraordinary and primitive, even unhuman." Hrdlicka's views on the positioning of the skull fragment were supported in an independent report by G. Schwalbe of Germany (Hrdlicka 1912, p. 343).

Describing the skull in its new orientation, Hrdlicka (1912, p. 332) wrote: "It was fairly but not very high; its capacity was surely not below 1,350, more probably between 1,400 and 1,500 cc." Hrdlicka (1912, p. 332) further stated: "Every feature shows it to be a portion of the skull of man himself; it bears no evidence of having belonged to an early or physically primitive man, but to a well-developed and physically modern-like human individual."

A firm believer in evolution and the recent origin of the human species, Hrdlicka (1912, p. 2) stated: "no conclusion can be more firmly founded than that man is the product of an extraordinary progressive differentiation from some anthropogenic stock, which developed somewhere in the later Tertiary, among the primates."

Hrdlicka therefore believed that fossils of human ancestors from the Tertiary and Quaternary should be to a greater or lesser degree apelike, as confirmed by the discovery of *Pithecanthropus erectus* in Java in 1891. Any fossils of anatomically modern human appearance from the Tertiary and Quaternary had to be explained away as intrusive burials or hoaxes. Hrdlicka's prejudice is evident in the following statement (1912, p. 2): "to establish beyond doubt the geological antiquity of human remains, it should be shown conclusively that the specimen or specimens were found in geologically ancient deposits, whose age is confirmed by the presence of paleontologic remains; and the bones should present evidence of organic as well as inorganic alterations, and show also morphologic characteristics referable to an earlier type. In addition, it is necessary to prove in every case by unexceptional evidence that the human remains were not introduced, either purposely or accidentally, in later times into the formation in which discovered."

Hrdlicka (1912, p. 2) amplified this view in another statement: "On the basis of what is positively known to-day in regard to early man, and with the present scientific views regarding man's evolution, the anthropologist has a right to expect that human bones, particularly crania, exceeding a few thousand years in age, and more especially those of geologic antiquity, shall present marked morphological differences, and that these differences shall point in the direction of more primitive forms."

Hrdlicka (1912, p. 3) further stated: "The antiquity, therefore, of any human skeletal remains which do not present marked differences from those of modern man may be regarded, on morphologic grounds, as only insignificant geologically, not reaching in time, in all probability, beyond the modern, still unfinished, geologic formations. Should other claims be made in any case, the burden of proof would rest heavily on those advancing them." Here we have a very clear formulation of the dubious principle of dating by morphology. We also see the application of a double standard in the treatment of evidence, with finds contrary to evolutionary expectations being subjected to much more rigorous scrutiny than finds conforming to evolutionary expectations.

Hrdlicka's views, which, in modified form, remain in force today among the vast majority of scientists concerned with human origins and antiquity, represent a perversion of the scientific method. Hrdlicka and those who shared his methodology were not prepared to impartially consider the facts and construct a theory upon the foundation of all the available evidence. Rather they allowed their theoretical biases to determine what evidence should be considered valid. Modern students of paleoanthropology may justifiably question whether their scientific predecessors have bequeathed to them a body of evidence that accurately reflects the truth, as far as it can be known by empiric methods, about human origins and antiquity. They should carefully consider the fact that the evidence that has come down to them has been selected from a larger body of evidence according to the criteria established by persons such as Hrdlicka. One purpose of this book is to acquaint modern students of paleoanthropology with that larger body of evidence and allow them to make their own decisions about the worth of the portion of it that was rejected.

It is abundantly clear that Hrdlicka harbored a strong prejudice that any reputedly ancient human remains must display primitive features. The *Diprothomo* skull from the dry dock excavation in the harbor of Buenos Aires did not display such features. Therefore, according to Hrdlicka, it could not possibly be as ancient as the Early Pleistocene stratum in which it was discovered. Willis, in his role as Hrdlicka's geological assistant, offered some purely speculative alternative explanations about how the skull may have found its way into the formation.

Of course, Ameghino had his own prejudices. Like Hrdlicka, he was committed to evolutionary ideas, but whereas Hrdlicka believed that *Homo sapiens* had evolved in the Old World and only recently emigrated to the Americas, Ameghino believed man had evolved in South America. Therefore, Ameghino had wanted his *Diprothomo* to be appropriately primitive for its Early Pliocene age (Early Pleistocene by modern reckoning). Hrdlicka, however, showed the skull was actually not different from that of *Homo sapiens sapiens*.

Putting aside prejudice and preconception, it seems that the bare facts, as far as we can ascertain them, support the view that human beings physiologically indistinguishable from *Homo sapiens sapiens* were present in Argentina during the Early Pleistocene. This supposition, although in clear contradiction to presently accepted accounts of human evolution, fits in quite well with the overwhelming mass of evidence detailed in the preceding chapters.

Before moving on, let us consider another South American find with unsettling implications for current thinking about human evolution in general and the populating of the New World in particular.

6.1.6 The Lagoa Santa Calotte

In 1970, Canadian archeologist Alan Lyle Bryan found a highly mineralized calotte (skullcap) with "very thick walls and exceptionally heavy brow ridges" in a paleontological collection from caves in the Lagoa Santa region of Brazil. This skullcap could not be given a date, since the cave excavations had not been stratigraphically controlled, but the fossil's morphology was reminiscent of *Homo erectus*. Bryan stated that he left the skullcap in a local museum, but unfortunately it was later lost. When Bryan (1978) showed photographs of the skullcap to several American physical anthropologists, they were unable to believe it could have come from the Americas, and proposed that it was either a fake, a cast, or possibly a skullcap from Europe that had somehow been introduced into the Brazilian collection examined by Bryan.

But Bryan countered that both he and his wife, who also saw the skullcap, had abundant experience with human fossil bones. And they were both quite sure that the skullcap could not have been a fake or a cast—it was a genuine, highly fossilized human skullcap.

That the Lagoa Santa calotte was not a European fossil, accidentally introduced into the Brazilian collection, was supported, said Bryan, by the fact that it differed in several important measurements from known European skulls.

Also, it was similar to other skulls found in the Sumidouro cave in the Lagoa Santa region during the 1930s. Bryan (1978) reported that pieces of similar skulls were found more recently in the same cave and were being studied by Marilia Carvalho de Mello e Alvim at the National Museum in Rio de Janeiro.

What is the significance of the Lagoa Santa calotte? The presence of hominids with *Homo erectus* features in Brazil at any time in the past is highly anomalous. Paleoanthropologists holding standard views say that only anatomically modern humans ever came to the Americas. The methodology of science allows for views to change, but the kind of change inherent in accepting the presence of *Homo erectus* in the New World would be revolutionary.

Of course, there are now a few paleoanthropologists who propose that *Homo erectus* was responsible for the crude stone tools at sites such as Toca da Esperança in Brazil (Section 3.8.4) and Calico in California (Section 3.8.3). If the view that *Homo erectus* was responsible for tools at certain very ancient sites in the Americas were to be become more widely accepted, this might have the beneficial effect of encouraging the far more radical changes in view that would be required to accommodate the evidence for the presence of anatomically modern humans in the early Quaternary and Tertiary.

Finally, we wonder how such an important fossil as the Lagoa Santa skullcap could have been lost in the museum where it was being kept. A similar thing happened to the postcranial portion of the skeleton discovered by H. Reck at Olduvai Gorge (Section 11.1.5). In the case of Bryan's and Reck's discoveries, we at least had a chance to hear about them before they disappeared. But we suspect that other fossils have escaped our attention because they were misplaced in museums or were perhaps intentionally discarded—without report.

6.2 FOSSIL HUMAN REMAINS FROM TERTIARY FORMATIONS

Having reviewed human skeletal remains from the Middle and Early Pleistocene, we shall now consider discoveries from the Tertiary. Of course, modern authorities, almost without exception, are convinced there were no humans in the Tertiary. In *Fossil Men,* Boule and Vallois (1957, p. 108), in predicting what sort of fossils might turn up to fill the gaps in the record of human ancestry, said: "Even in the Pliocene, what we shall meet will no longer be—or rather, will not yet be—true Hominids. They will be the ancestors of the Prehominians, the ancestors of the Australopithecines— or even these Australopithecines themselves—all of them forms so apelike that to call them human would be to give this term an extension that would deprive it of all logical meaning." Here we have yet another example of evolutionary preconceptions dictating what kind of evidence is safely discoverable.

6.2.1 The Foxhall Jaw (Late Pliocene)

We have already discussed J. Reid Moir's reports about stone implements and
hearths discovered at Foxhall, England, in the Late Pliocene Red Crag formation
(Section 3.3.4). Earlier, in 1855, a human jaw was discovered at Foxhall by
workers digging for coprolites (phosphate-rich nodules) in a quarry on Mr. Law's
farm.

John Taylor, the town druggist, purchased the Foxhall jaw (Figure 6.2) from
a workman who wanted a glass of beer, and Taylor called it to the attention of
Robert H. Collyer, an American physician then residing in London. Collyer,
having acquired the fossil, visited the quarry on Mr. Law's farm and noted that
the coprolite bed, from which the jaw was said to have been taken, was 16 feet
below the surface. The condition of the jaw, thoroughly infiltrated with iron
oxide, was consistent with incorporation in the coprolite bed. Collyer said that
the Foxhall jaw was "the oldest relic of the human animal in existence" (Osborn
1921, p. 567). The 16-foot level at Foxhall is the same from which Moir (1924,
p. 647) later recovered stone tools and signs of fire. Anything found at this level,
considered an old land surface, would be at least 2.5 million years old.

Aware that he was in the possession of a fossil of great significance, Collyer
brought the jaw to the curator of the Royal College of Surgeons, who suggested
that he show it to Richard Owen. Collyer delivered the jaw to Owen, who kept
it for two years without giving a report. In 1859, Collyer retrieved the jaw, and
then took it, in turn, to Sir John Prestwich and Thomas Henry Huxley.

In April of 1863, Collyer displayed the fossil jaw at a meeting of the
Ethnological Society of London, at which the prominent geologists Charles Lyell
and Roderick Murchison were present. George Busk, a paleontologist, said at the
meeting that the fossil bone from Foxhall was "the jaw of some old woman,
perhaps from some Roman burial ground," but he later withdrew this skeptical
statement (Osborn 1921,
p.567).

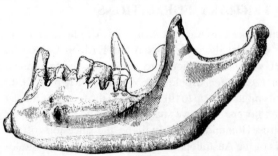

Huxley, who was also
present at the meeting,
visited Collyer the next
morning to further ex-
amine the jaw. At that
time, Huxley said that it
was "most extraordi-
nary," but in May of 1863

Figure 6.2. Human jaw discovered in 1855 in the Late
Pliocene Red Crag formation at Foxhall, England (Osborn
1921, p. 568).

he wrote that the mor-
phology of the bone did
not indicate it belonged

"to an extinct or aberrant race of mankind," adding that "the condition of the bone is not such as I should expect a crag fossil to be" (Osborn 1921, p. 568). The jaw then passed into the hands of Hugh Falconer and eventually wound up in the possession of George Busk, who showed it to de Quatrefages and other French scientists. In July of 1863, Busk stated the jaw was of "very great antiquity" (Osborn 1921, p. 568) but not necessarily from the coprolite bed at Foxhall.

American paleontologist Henry Fairfield Osborn, writing in the 1920s about Moir's finds of flint tools in the same area where the Foxhall jaw was uncovered, wondered why the above-mentioned scientists did not take the trouble to visit the site. They disbelieved, said Osborn (1921, p. 568), "probably *because the shape of the jaw was not primitive* and the degree of mineralization was not such as positively to prove it a fossil. He [Collyer] had a chemical analysis made that showed that the jaw was largely mineralized, but retained 8 per cent of animal matter." But Moir reported that chemical analysis of bones from the Red Crag demonstrated that many of them had up to 6.5 percent animal matter (Osborn 1921, p. 568).

After some time, the jaw mysteriously disappeared, as did Collyer himself. All that is now known of Collyer is that he was a graduate of the Berkshire School of Medicine, once located at Pittsfield, Massachusetts, and that he was a friend of Dr. Morton, who was a craniologist and member of the Academy of Natural Sciences of Philadelphia. All that remains of the jaw is a detailed drawing made in 1867 by Collyer and the scant published record of the controversy surrounding it (Osborn 1921, p. 569).

The fossil jaw from the Red Crag at Foxhall is almost never mentioned by modern authorities, and those who do mention it are invariably scornful. For example, we find in *Fossil Men,* by Boule and Vallois (1957, p. 107), this statement: "It requires a total lack of critical sense to pay any heed to such a piece of evidence as this."

But, as we have often pointed out, many conventionally accepted bones and artifacts have been found by uneducated workers or in other dubious ways. For example, most of the *Homo erectus* finds from Java were made by unsupervised, paid native collectors (Section 7.3). And the Heidelberg *Homo erectus* jaw was found by German workmen, whose foreman later turned it over to scientists (Section 7.2).

If scientists can seriously consider these discoveries, then why can they not seriously consider the Foxhall jaw as well? One might object that the Java *Homo erectus* fossils and the Heidelberg *Homo erectus* jaw are still available for inspection, while the Foxhall jaw has vanished. But the original Peking *Homo erectus* fossils disappeared from China during World War II (Section 9.1.12); yet they are still accepted as evidence for human evolution.

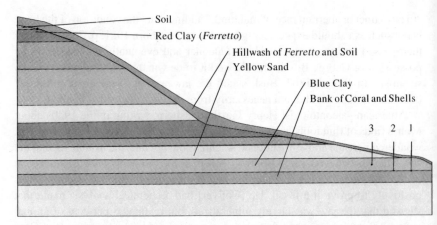

Figure 6.3. This section of the Colle de Vento, near Castenedolo, Italy (after Sergi 1884, p. 313), shows the general stratigraphic position of human skeletal remains found there. (1) The human fossils found by geologist G. Ragazzoni in 1860 lay on the bank of coral and shells, at a place where it was surmounted by Middle Pliocene blue clay, which was itself covered by red clay (*ferretto*) washed from the top of the hill. (2) On January 2 and January 25, 1880, more human fossils, representing three individuals (a man and two children), were found about 15 meters (49 feet) from the 1860 site. The bones lay on the bank of coral, and were covered by about 2 meters (7 feet) of Pliocene blue clay, surmounted by a red layer of *ferretto*. (3) On February 16, 1880, the bones of a woman were found at a depth of 1 meter (3 feet) in the blue clay, which was overlain by a layer of yellow sand and a layer of bright red *ferretto*. In all three cases, Ragazzoni looked for signs of burial and found none.

6.2.2 Human Skeletons from Castenedolo, Italy (Middle Pliocene)

One of the more significant Tertiary finds turned up in Italy. Millions of years ago, during the Pliocene period, a warm sea washed the southern slopes of the Alps, depositing layers of coral and molluscs. Late in the summer of 1860, Professor Giuseppe Ragazzoni, a geologist and teacher at the Technical Institute of Brescia, traveled to the nearby locale of Castenedolo, about 10 kilometers (roughly 6 miles) southeast of Brescia, to gather fossil shells in the Pliocene strata exposed in a pit at the base of a low hill, the Colle de Vento (Figure 6.3).

Ragazzoni (1880, p. 120) reported: "Searching along a bank of coral for shells, there came into my hand the top portion of a cranium, completely filled with pieces of coral cemented with the blue-green clay characteristic of that formation. Astonished, I continued the search, and in addition to the top portion of the cranium I found other bones of the thorax and limbs, which quite apparently belonged to an individual of the human species."

Ragazzoni took the bones to the geologists A. Stoppani and G. Curioni. According to Ragazzoni (1880, p. 121), their reaction was negative: "Not giving much credence to the circumstances of discovery, they expressed the opinion that the bones, instead of being those of a very ancient individual, were from a very recent burial in that terrain."

"I then threw the bones away," stated Ragazzoni (1880, p. 121), "not without regret, because I found them lying among the coral and marine shells, appearing, despite the views of the two able scientists, as if transported by the ocean waves and covered with coral, shells, and clay."

But that was not the end of the story. Ragazzoni could not get out of his mind the idea that the bones he had found belonged to a human being who lived during the Pliocene. "Therefore," he wrote, "I returned a little later to the same site, and was able to find some more fragments of bone in the same condition as those first discovered" (Ragazzoni 1880, p. 121).

In 1875, Carlo Germani, on the advice of Ragazzoni, purchased land at Castenedolo for the purpose of selling the phosphate-rich shelly clay to local farmers for use as fertilizer. Ragazzoni stated (1880, p. 121): "I explained to Germani about the bones I had found, and strongly advised him to be vigilant while making his excavations and to show me any new human remains."

A few years later, Germani noticed some bones. Ragazzoni recalled (1880, p. 121): "In December of 1879, Germani made an excavation, about 15 meters [49 feet] from the first place, to the northwest, and on January 2, 1880 announced to me the discovery of human bones between the bank of coral and the overlying shelly clay. The next day, I went there with my assistant Vincenzo Fracassi, in order to remove the bones with my own hands. These were: pieces of the left parietal, fragments of the occipital, the left temporal, the front part of the lower jaw with a canine, two loose molars, a cervical vertebra, fragments of vertebrae and ribs, part of the ilium, pieces of humerus, ulna, radius, femur, tibia, fibula; and a tarsal and two phalanges." More discoveries were to follow: "On the 25th of the same month, Carlo Germani brought me two fragments of lower jaw, and some teeth of smaller size and different shape than the first, found at a distance of 2 meters [7 feet] from them, but at the same depth. Uncertain whether they belonged to a young individual or to an anthropomorphic ape, I returned once more to Castenedolo with Signor Germani, and was able to collect: a great quantity of upper cranial fragments (which I suspected belonged to two individuals), the left orbital of the frontal, two parietals, a fragment of the upper jaw with two molars, other free teeth, and fragments of ribs and limbs. All of them were completely covered with and penetrated by the clay and small fragments of coral and shells, which removed any suspicion that the bones were those of persons buried in graves, and on the contrary confirmed the fact of their transport by the waves of the sea" (Ragazzoni 1880, p. 122).

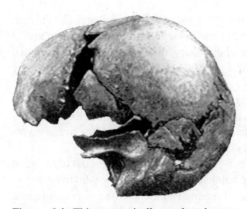

Figure 6.4. This anatomically modern human skull (Sergi 1884, plate 1) was found in 1880, at Castenedolo, Italy. The stratum from which it was taken is assigned to the Astian stage of the Pliocene (Oakley 1980, p. 46). According to modern authorities (Harland *et al.* 1982, p. 110), the Astian belongs to the Middle Pliocene, which would give the skull an age of 3–4 million years.

On February 16, Germani advised Ragazzoni that a complete skeleton was discovered. Ragazzoni (1880, p. 122) journeyed to the site and supervised the excavation, instructing the workmen to "use the greatest diligence so as to be able to ascertain as clearly and exactly as possible the reality of the facts." According to Ragazzoni (1880, p. 122), they "removed the strata successively from higher to lower, with the intent of exposing the entire skeleton." This was accomplished. About the remains, Ragazzoni (1880, pp. 122–123) wrote: "The skeleton, slightly inclined to the southeast, appeared to have been subjected to a kind of pressure in an oblique direction from south to north by movement of the strata in which it was found; consequently, it was from the region of the pelvis that we recovered the majority of the ribs, which appeared to have been crushed from above. The cranium was bent somewhat to the right. The lower jaw was detached and the separated facial bones were encased in a mass of blue-green clay penetrating the cavity of the cranium, which presented a variety of fractures." The cranium, as restored by G. Sergi (Figure 6.4), was indistinguishable from that of a modern woman.

Ragazzoni (1880, p. 121) then stated: "I desired to make photographs, but the perverse winter weather prevented it. In spite of the bad weather, the next morning I returned to the site with my son Pietro, and resolved to remove the entire skeleton despite the icy rain, which did, however, by penetrating the clay, make it easier to take out the bones." Ragazzoni (1880, p. 123) wrote: "Differing from the set of remains found in 1860, and the other two found earlier this year [1880], the complete skeleton was found in the middle of the layer of blue clay . . . over which passes a layer of medium yellow sand." The other skeletons were found lower in the blue clay, at the point where it meets the underlying bank of coral and shells (Figure 6.3). Ragazzoni (1880, p. 123) added: "The stratum of blue clay, which is over 1 meter [3 feet] thick, has preserved its uniform stratification, and does not show any sign of disturbance. In accordance with the judgement of the excavator himself, who is not preoccupied with any preconceived ideas, the skeleton was very likely deposited in a kind of marine mud and not buried at a later time, for in this case one would have been able to detect traces of the overlying yellow sand and the iron-red clay called *ferretto,* which forms the top part of the hill, and which by successive flooding has washed down and covered the lower formations of conglomerate and sand that cover the shelly Subappenine blue clays."

From the above statements by Ragazzoni, it appears that the finds of 1860 and 1880 (except the complete female skeleton) were made in places on the slope of the hill where the layer of yellow sand covering the blue clay had been stripped away by erosion (Sergi 1884, p. 314). The hill at Castenedolo (Figure 6.3) was approximately 25 meters (82 feet) high. The top layer was recent soil. Below that, on the summit of the hill, was the red *ferretto.* Next in sequence came layers of glacial deposits and conglomerate. Below these came several layers of sand and clay. Then came the above-mentioned yellow sand, followed by the blue Pliocene clay in which the skeletal remains were discovered. Ragazzoni (1880, p. 126) indicated that even in the places where the blue clay had been exposed, the rain had washed down a surface layer of red *ferretto* deposits. Thus, for all the fossil discoveries, a layer of bright red clay was apparently lying above the blue clay. Any burial would have certainly produced a noticeable mixing of different colored materials in the otherwise undisturbed blue clay layer, and Ragazzoni, a geologist, testified that there was no sign of such mixing.

Of course, one could always propose that the skeletons (other than the adult female skeleton) were buried in the blue clay at a time when the red hillwash was not present. But this is unlikely. The red *ferretto* at the top of the hill would, it seems, have been continually carried down the hillside by rain or melting snow. Only under some unusual circumstance would it not have been present over the blue clay. Also, the blue clay had its own stratification, any disturbance of which would have been noticed.

Ragazzoni (1880, p. 126) then dealt with another possible objection to his conclusion that the human bones from Castenedolo were as old as the Pliocene layer in which they were found. Perhaps streams had stripped away the layers covering the blue clay and penetrated part way into the blue clay itself. The human bones could then have been washed into hollows, and new material could have been deposited over them. This could explain why there were no signs of burial. But Ragazzoni (1880, p. 126) said that it was highly unlikely that the human fossils had been washed recently into the positions in which they were found: "The fossil remains discovered on January 2 and January 25 lay at a depth of approximately 2 meters. The bones were situated at the boundary between the bank of shells and coral and the overlying blue clay. They were dispersed, as if scattered by the waves of the sea among the shells. The way they were situated allows one to entirely exclude any later mixing or disturbance of the strata."

Ragazzoni (1880, p. 126) further stated: "The skeleton found on the 16th of February occurred at a depth of over 1 meter in the blue clay, which appeared to have covered it in a state of slow deposition." Slow deposition of the clay, which Ragazzoni (1880, p. 123) said was stratified, ruled out the hypothesis that the skeleton had recently been washed into the blue clay by a torrential stream. Ragazzoni (1880, p. 126) added that the blue clay "was also in such a condition as to exclude any rearrangement by human agency." At the place where this complete human skeleton was discovered, the blue clay was still covered by a layer of yellow sand and a layer of red *ferretto*. The absence of any mixture of yellow and red materials in the blue clay eliminated the idea of recent intrusive burial.

Ragazzoni (1880, p. 126) concluded: "These facts demonstrate the existence of man in Lombardy during the Early Pliocene." He stated elsewhere in his report: "To render it perfectly clear to anyone that the terrain in which the bones and skeleton were found belongs to the Early Pliocene, I thought it convenient to offer a sample of the fossils that exist there in abundance" (Ragazzoni 1880, p. 124). He then referred his readers to an accompanying illustration of fossil Pliocene shells. Geologists who examined the blue clay layer of the Colle de Vento at Castenedolo, including Professor G. B. Cacciamali, agreed that it belonged to the Astian stage of the Pliocene (Oakley 1980, p. 46). Modern opinion places the Astian in the Middle Pliocene (Harland *et al.* 1982, p. 110), which would give the discoveries from Castenedolo an age of about 3–4 million years.

In 1883, Professor Giuseppe Sergi, an anatomist from the University of Rome, visited Ragazzoni and personally examined the human remains at the Technical Institute of Brescia. After studying the bones, he determined they represented four individuals—an adult male, an adult female, and two children.

Sergi also visited the site at Castenedolo. He wrote (1884, p. 315): "I went there accompanied by Ragazzoni, on the 14th of April. The trench that had been excavated in 1880 was still there, and the strata were clearly visible in their geological succession. In order to see still better, we cut a fresh vertical section down to the bank of coral. . . . The terrain was undisturbed, and Professor Ragazzoni said to me that I was seeing the undisturbed clay just as he had found it when he extracted the skeletons. And what was true of the clay, was also true of the underlying strata, which were also found intact, with no sign of resorting." In his report, Sergi (1884, p. 315) also wrote: "What is, one might demand, the guarantee of the authenticity of a discovery of this type? I believe that any doubt can be removed if the person who made the discovery adopted the necessary methods and noted all the circumstances with due care and conscientiousness. Professor Ragazzoni is a geologist and was quite familiar with the stratigraphic conditions of that region, and all of Lombardy, and would have been able to immediately recognize any movement of the terrain or signs that the blue clay had been mixed with materials from the overlying strata."

Sergi (1884, pp. 315–316) added: "It is especially noteworthy that the color and structure of the strata in question are quite different. If a hole had been excavated for a burial, then it would not have been refilled exactly as before. The clay from the upper surface layers, recognizable by its intense red color, would have been mixed in. Such discoloration and disturbance of the strata would not have escaped the notice of even an ordinary person what to speak of a trained geologist. One may also note that we are not talking about just one small area from which the bones of a single individual were taken, but also of a larger area, many square meters in size, from which the remains of three other individuals, lying close together, were taken. If these latter three were burials why was there not observed any artificial displacement of the overlying strata? Were such signs of burial overlooked? That might have happened in the case of a single unexpected discovery, made too suddenly to properly observe the circumstances; but as we have seen, the excavations were planned in advance and carried out with all due caution at different times and in various conditions, allowing sufficient time for observation and examination. Signs of mixing of the strata may have been missed the first time, but certainly not the second, third, and fourth times."

Gabriel de Mortillet (1883, pp. 71–72) did not believe that the Castenedolo skeletons were truly of Pliocene antiquity. Responding to de Mortillet's negative opinion, Sergi (1884, p. 316) wrote: "De Mortillet, in connection with this discovery, did not attempt to dispute the fact that there was no sign of disturbance in the strata. He said, however, that this observation was not sufficient to rule out burial, because 'the action of the sea would have dispersed the bones of the skeletons.' We note, however, the presence of four individuals—two adults (a male and female) and two children, indicating a family shipwrecked on a Pliocene coastline. The bones of all of the skeletons—except for one—were in fact dispersed, which accounts for the fact that they were discovered at successive times, with the fragments found in diverse parts of an extensive area. These bones were found as if disseminated across a single flat surface. Not all of the bones were found for each individual, but some only, which doubtlessly means that the others were dispersed elsewhere. By some circumstance, the last skeleton happened to come to rest at one spot, where, as could be seen when it was excavated, it was covered by layers of sedimentary deposits. Professor Ragazzoni told me that he found the bones in a uniform, compact mass of clay, like a fly that happened to fall into soft soap, where it remained caught."

Sergi (1884, pp. 314–315) noted: "the almost entirely preserved female skeleton was not found in a posture indicating ordinary burial, but overturned; I saw the front part of the skull with the face inside the posterior cavity, and all the head entangled in the greenish clay, from which I patiently separated it. The other parts of the skeleton were also like this, and I still have the vertebral column with the ribs in a mass of clay, and the bones of one hand in the same clay."

Sergi (1884, p. 316) concluded: "From all that I heard and saw, I came to the conclusion that the skeletons of Castenedolo are from the geological era to which the strata of blue clay and the marine shell bed may be referred, and they are an irrefutable document for the existence of man in the Tertiary epoch, man of a character fully human, and not a precursor." To Sergi, the Castenedolo skeletons suggested that the creatures responsible for the Tertiary flints and cut bones found by other researchers (Chapters 2–5) might have been fully human rather than apelike ancestors, as scientists such as de Mortillet had proposed.

Sergi pointed out that the scientific community had recognized the existence of human beings in the Pleistocene only after great controversy. "But no sooner than that fact was accepted," wrote Sergi (1884, p. 303), "human artifacts from the Tertiary began to appear. This development was, however, confronted with obstacles and opposition arising not only from the prejudice expected of common men, but also from prejudice within the scientific community. Science has no problem finding existing species of shells in strata millions of years old, and some living mammals are also represented in the Tertiary, but man himself, it is thought, must be quite recent."

Sergi (1884, pp. 303–304) stated: "There were presented at various academies and congresses the artifacts of Tertiary man, consisting of impressions, incisions, cuts, and scraping on bone and stone, including flints chipped by the hand of man, and there followed only negativity. And when there was no other reason to negate, it was simply said: 'I don't believe it.' The reporting of actual human remains—crania and other bones—was received with irony and rejected with dogmatic incredulity."

Sergi (1884, p. 304) then recounted how the artifacts of Tertiary man gradually won a degree of positive recognition: "One set of facts was not able to be rejected, although it took much time to be accepted, and that was the chipped flints discovered in Tertiary deposits at Thenay (Loire-et-Cher) by Bourgeois. At the congress at Paris in 1867, Bourgeois was not believed; but Worsae soon declared his support, and shortly thereafter de Mortillet and others did the same. At the congress in Brussels in 1872, the question was discussed, and the adherents increased. This prepared the way for Rames, who discovered worked flint and quartzite in the conglomerates of Cantal at Aurillac."

Continuing his review, Sergi (1884, p. 304) said: "The Tertiary flints of Portugal also encountered great resistance. But C. Ribeiro did research of unequaled value and effect. Yet only at the congress of 1880 at Lisbon did his discoveries achieve complete recognition, especially after a commission of scientists, in the course of a visit to Monte Redondo at Otta, found a flint implement in place, still embedded in the conglomerate. Professor Bellucci had the fortune to make this discovery and report it." It is quite remarkable that most modern students of paleoanthropology are unaware of the sequence of discoveries discussed by Sergi.

Sergi (1884, p. 304) then stated: "Taking all this into consideration, it is possible to conclude and affirm without hesitation that man appeared not only in the Quaternary epoch, but that the signs of his existence certainly extend back into the Tertiary epoch."

After pointing out that some scientists wanted to attribute the Tertiary flint implements and other artifacts to a hypothetical apelike human precursor, Sergi (1884, p. 305) wrote: "Therefore it became important to consider human skeletal remains, but no acceptable ones had been found. This is the reason for the negative opinions of de Mortillet and Hovelacque. But neither was there much fossil evidence to back up the proposed precursor of man." Java man, the first scientifically accepted ape-man, was not uncovered until 1891, seven years after Sergi presented the report we have been reviewing.

De Mortillet, it may be recalled, believed the fossil record showed that mammals displayed extensive and progressive evolutionary development from primitive forms in the Tertiary up to more advanced forms in the present. Accepting this sequence as a paleontological law, de Mortillet anticipated that any fossils of Tertiary human ancestors would be very primitive and apelike. It could not be otherwise.

But Sergi pointed out that some Tertiary mammals (such as the mastodon) had survived without much change well into the Quaternary (Pleistocene) in Italy and Spain. Also, Sergi (1884, p. 306) reported that in the United States geologists had discovered in Late Miocene formations some fossil wolf jaws that were indistinguishable from those of living wolves.

Sergi (1884, p. 309) therefore stated: "the tendency to reject, by reason of theoretical preconceptions, any discoveries that can demonstrate a human presence in the Tertiary is, I believe, a kind of scientific prejudice. Natural science should be stripped of this prejudice." This prejudice was, however, not overcome, and it persists today. Sergi (1884, p. 310) wrote: "By means of a despotic scientific prejudice, call it what you will, every discovery of human remains in the Pliocene has been discredited."

But Sergi was not alone in his acceptance of Ragazzoni's discoveries at Castenedolo. De Quatrefages, familiar to us from our review of stone implements, also accepted them. Concerning the female skeleton uncovered at Castenedolo, he said in his book *Racès Humaines:* "The deposit was removed in successive horizontal layers, and not the least trace was found of the beds having been mixed or disturbed" (Laing 1894, p. 371). De Quatrefages further stated: "there exists no serious reason for doubting the discovery of M. Ragazzoni, and . . . if made in a Quaternary deposit no one would have thought of contesting its accuracy. Nothing, therefore, can be opposed to it but theoretical *a priori* objections, similar to those which so long repelled the existence of Quaternary man" (Laing 1893, p. 119).

In 1889, an additional human skeleton was discovered at Castenedolo. This find introduced an element of confusion about the discoveries of 1880.

Ragazzoni invited G. Sergi and A. Issel to examine the new skeleton, which had been found in an ancient oyster bed. Sergi (1912) reported that both he and Issel believed this new 1889 skeleton to be a recent intrusion into the Pliocene layers "because the almost intact skeleton lay on its back in a fissure of the oyster bed and showed signs of having been buried" (Cousins 1971, p. 53).

Issel (1889) therefore reported that this new skeleton was not of Pliocene age, but was much younger. Issel (1889, p. 109) concluded that the 1880 discoveries were also recent burials. Concerning the dispersal of the bones of some of the skeletons found in 1880, he suggested this might have been caused by agricultural work (Issel 1889, p. 109). In a footnote, Issel (1889, p. 109) claimed that Sergi agreed with him that none of the skeletons found at Castenedolo were of Pliocene age. For the scientific community, this apparently resolved the ongoing controversy.

But Sergi (1912) later wrote that Issel was mistaken. Despite his views on the 1889 skeleton, Sergi said he had never given up his conviction that the 1880 bones were Pliocene. "Today I declare that the one thing does not invalidate the other" (Cousins 1971, p. 54). Sergi (1912) then added: "In any case this new pseudo-discovery [of 1889] gave a decisive blow to the first, and from this a deeper silence, like that of a grave, fell on the Castenedolo man; I had neither heart nor reason to exhume him. . . . Since then no one has spoken any more of the Castenedolo man [except to cast doubt upon him]" (Cousins 1971, p. 54).

A good example of the unfair treatment given to the Castenedolo finds may be found in Professor R. A. S. Macalister's *Textbook of European Archaeology,* written in 1921. Macalister (1921, p. 183) admitted that the Castenedolo finds "whatever we may think of them, have to be treated seriously." He noted that they were "unearthed by a competent geologist, Ragazzoni . . . and examined by a competent anatomist, Sergi." Still he could not accept their Pliocene age. Faced with the uncomfortable facts, Macalister (1921, p. 183) claimed "there must be something wrong somewhere." First of all the bones were anatomically modern. "Now, if they really belonged to the stratum in which they were found," wrote Macalister (1921, p. 184), "this would imply an extraordinarily long standstill for evolution. It is much more likely that there is something amiss with the observations." Macalister (1921, p. 185) also said: "the acceptance of a Pliocene date for the Castenedolo skeletons would create so many insoluble problems that we can hardly hesitate in choosing between the alternatives of adopting or rejecting their authenticity." Here once more we find a scientist's preconceived ideas about evolution influencing him to reject skeletal evidence that would otherwise be considered of good quality.

Equally unfriendly to Tertiary stone tools, Macalister (1921, p. 185) protested: "On the one hand, we are asked to believe in eoliths; on the other hand we are introduced to highly advanced and intellectual people like those of Castenedolo. The two are incompatible. The quest for Tertiary Man is a game at which the player must be fair; he cannot win both ways. Let him become an Eolithist if he see fit, but let him then give up all expectation of finding a Tertiary man with a fully-developed mental equipment. Or let him seek a Tertiary *Man,* but he must then throw his eoliths and all the rest of his ballast overboard."

There is, however, no fundamental incompatibility between advanced intellectual capabilities and the manufacture of crude stone tools—even today tribal people in various parts of the world, with the same brain capacity as modern city dwellers, make such implements. Also, there is no reason why anatomically modern humans could not have coexisted with more apelike creatures in the Tertiary, just as humans today coexist with gorillas, chimpanzees, and gibbons.

Macalister cited Issel (1889) in support of his attempt to discredit the Castenedolo finds, apparently not aware that according to Sergi (1912) Issel's 1889 report discredited only the 1889 skeleton. For example, Macalister (1921, p. 184), referring to all of the Castenedolo finds, wrote: "examination of the bones and their setting, by Issel of Geneva, revealed the fact that the strata were full of marine deposits, and that everything solid within them, *except the human bones,* shewed marine incrustations." While it is true that Issel (1889, p. 108) reported that the bones of the skeleton uncovered in 1889 were smooth and free of incrustations, the same cannot be said of the earlier discoveries, which both Ragazzoni (1880, pp. 120, 122) and Sergi (1884, pp. 311, 312) said were incrusted with blue Pliocene clay and pieces of shells and coral.

Another example of the unfair treatment given the Castenedolo discoveries is found in *Fossil Men.* In this book, Boule and Vallois (1957, p. 107) stated that "it seems certain that at Castenedolo, as at Savona [Section 6.2.3], we are dealing with more or less recent burials." But in *Fossil Men,* Boule and Vallois devoted only one paragraph to Castenedolo, and did not mention the undisturbed layers lying over the skeletons or the scattered and incomplete state of some of the skeletons—information that tends to rule out intrusive burial.

Boule and Vallois (1957, p. 107) noted: "In 1889, the discovery of a new skeleton was the subject of an official report by Professor Issel, who then observed that the various fossils from this deposit were all impregnated with salt, with the sole exception of the human bones." Here Boule and Vallois implied that what was true of the bones found in 1889 was also true of the bones found previously. But in his 1889 report, Issel described in this connection only the bones found in 1889. In fact, Issel did not even mention the word *salt,* referring instead to marine incrustations—which were, as above mentioned, present on the bones found in 1860 and 1880.

Scientists have employed chemical and radiometric tests to deny a Pliocene age to the Castenedolo bones. K. P. Oakley (1980, p. 40) found the Castenedolo bones had a nitrogen content similar to that of bones from Late Pleistocene and Holocene Italian sites and thus concluded the Castenedolo bones were recent. But as previously mentioned, in connection with Galley Hill, the degree of nitrogen preservation in bone can vary widely from site to site, making such comparisons unreliable as age indicators. The Castenedolo bones were found in clay, a substance known to preserve nitrogen-containing bone proteins.

The Castenedolo bones had a fluorine content that Oakley (1980, p. 42) considered relatively high for bones he thought were recent. Oakley explained this discrepancy by positing higher past levels of fluorine in the Castenedolo groundwater. But this was simply guesswork. The Castenedolo bones also had an unexpected high concentration of uranium, consistent with great age.

A carbon 14 test yielded an age of 958 years for some of the Castenedolo bones. But, as in the case of Galley Hill, the methods employed are now considered unreliable and the bones themselves were very likely contaminated with recent carbon, causing the test to yield a falsely young age. For a more detailed discussion of the chemical and radiometric testing of the Castenedolo bones, see Appendix 1.

The case of Castenedolo demonstrates the shortcomings of the methodology employed by paleoanthropologists. The initial attribution of a Pliocene age to the discoveries of 1860 and 1880 appears to have been amply justified. The finds were made by a trained geologist, G. Ragazzoni, who carefully observed the stratigraphy at the site. He especially searched for signs of intrusive burial, and observed none. Ragazzoni duly reported his findings to his fellow scientists in scientific journals. But because the remains were modern in morphology they came under intense negative scrutiny. As Macalister put it, there had to be something wrong.

The account of human origins now dominant in the scientific community is the product of attitudes such as Macalister's. For the last century, the idea of progressive evolution of the human type from more apelike ancestors has guided the acceptance and rejection of evidence. Evidence that contradicts the idea of human evolution is carefully screened out. Therefore, when one reads textbooks about human evolution, one may think, "Well, the idea of human evolution must be true because all the evidence supports it." But such textbook presentations are misleading, for it is the unquestioned belief that humans did in fact evolve from apelike ancestors that has determined what evidence should be included and how it should be arranged and interpreted.

6.2.3 A Skeleton from Savona, Italy (Middle Pliocene)

We now turn our attention to another Pliocene find, made at Savona, a town on the Italian Riviera, about 30 miles west of Genoa. In the 1850s, a church was being built on a hill bearing the same name as the one at Castenedolo (Colle de Vento). During the construction, workmen discovered a human skeleton at the bottom of a trench 3 meters (10 feet) deep. De Mortillet (1883, p. 70) reported: "Its bones were found in their natural connection, encased in a very compact and characteristic Pliocene marl, which also contained many other fossils typical of the Astian stage of the Pliocene." This would make the Savona skeleton the same age as the Castenedolo skeletons—Middle Pliocene.

Arthur Issel communicated details of the Savona find to the members of the International Congress of Prehistoric Anthropology and Archeology at Paris in 1867. In favor of the authenticity of the discovery, Issel (1868) declared that "the man of Colle de Vento was contemporary with the strata in which he was found" (de Mortillet 1883, p. 70). Issel said it was unfortunate that a trained geologist was not present to confirm that the strata were undisturbed and that the human bones had been buried at the same time as the animals bones found at the same level.

De Mortillet (1883, pp. 70–71), after mentioning that most of the skeleton was later lost, stated: "The bones that remain show that the individual was of small stature, much smaller than the present Ligurians. At first glance, the lower jaw appears to have a special primitive character, but the more one examines it the more one sees that the effects are those of breaking or wear rather than actual features. One sees that its features are analogous to those of modern jaws. The Pliocene strata are characterized by many marine shells, but they also contain terrestrial fossils including bones of rhinoceros as well as remains of plants. But the mammalian bones are scattered and separated, while the human bones preserve their natural connections. Does this not prove that instead of the remains of a human cadaver tossing in the waves of a Pliocene sea, we are simply in the presence of a later burial of undetermined date? Much desiring to clarify this question, Issel in 1874 began an excavation 1 meter [3 feet] distant from the foundation of the church, almost at the spot where the human bones were uncovered. He observed that at this point the Tertiary beds were completely intact, without a trace of disturbance. Unfortunately, those members of the religious order owning the land stopped him from continuing the excavation, which had then reached the 2-meter [7-foot] level."

In his report, Issel catalogued the remaining bones of the Savona skeleton: a fragment of the right parietal, some pieces of the jaw, a fragment of humerus, a clavicle, the head of a femur, and some finger bones. Issel (1868, p. 77) pointed out that "the material embedded in the fractures of the bones is the same as that in the Pliocene strata." To Issel (1868, p. 78) the bones seemed "slightly different and smaller than those of modern man."

At the International Congress of Prehistoric Anthropology and Archeology at Bologna in 1871, Father Deo Gratias (D. Perrando), a priest who had been present at the time of the discovery of the human skeleton at Savona, gave a report indicating that it was not an intrusive burial. Deo Gratias, a student of paleontology, explained that in 1851 the sisters of the Misericorde of Savona had wanted to build a church next to their convent. G. B. Mogliolo undertook the work, under the direction of Giuseppe Cortesé. Antonio Brilla, a sculptor and artist, assisted in the excavations. So, in addition to the workers, these three educated gentlemen regularly monitored the excavations. Brilla, in particular, was specifically looking for fossils.

At a depth of 3 meters (10 feet), the excavators discovered an object that Brilla thought might be a piece of ancient pottery, but it turned out to be part of a skull. Despite their astonishment, the gentlemen present did not report the find to professional scientists and allowed the workers to continue. They later uncovered a skeleton. The fragmentary remains were taken to Brilla's studio. Deo Gratias wanted them, but Brilla kept them to use as models for some of his works. Brilla did, however, eventually give Deo Gratias some pieces of bone. These would appear to be the bones listed by Issel. The rest of the bones were then lost.

Deo Gratias (1873, pp. 419–420) stated: "It is unfortunate an experienced naturalist was not there, but on the basis of testimony by Brilla and the workers who excavated the skeleton here is what is known. The body was discovered in an outstretched position, with the arms extending forward, the head slightly bent forward and down, the body very much elevated relative to the legs, like a man in the water. Can we suppose a body was buried in such a position? Is it not, on the contrary, the position of a body abandoned to the mercy of the water? The fact that the skeleton was found on the side of a rock in the bed of clay makes it probable that it was washed against this obstacle."

Deo Gratias (1873, p. 419) further stated: "Had it been a burial we would expect to find the upper layers mixed with the lower. The upper layers contain white quartzite sands. The result of mixing would have been the definite lightening of a closely circumscribed region of the Pliocene clay sufficient to cause some doubts in the spectators that it was genuinely ancient, as they affirmed. The biggest and smallest cavities of the human bones are filled with compacted Pliocene clay. This could only have happened when the clay was in a muddy consistency, during Pliocene times." Deo Gratias pointed out that the layers of Pliocene clay, now hard and dry, were situated on a hill, which meant they would be well drained.

De Mortillet, and later Boule and Vallois (1957, p. 106), argued that since the mammal bones in the stratum were scattered, whereas the human bones were found in natural connection, this indicated that the latter must be a recent intrusive burial. But the following points all argue strongly against the intrusive burial hypothesis: (1) the lack of material from the higher stratum mixed in with that of the lower stratum; (2) the depth of 3 meters (10 feet)—rather deep for a burial, at least from the present land surface; (3) the position of the skeleton, face down when discovered.

How then do we explain the scattered mammal bones? The site was once covered by the shallow shoreline waters of a Pliocene sea, as shown by the presence of characteristic shells. Animals could have died on the land, and their isolated bones could have been washed into the sea and incorporated into the formation. The human bones, found in natural connection, could have come to rest in the same marine formation as a result of someone drowning there during the Pliocene. This combination of events accounts for the presence of a relatively complete human skeleton amid scattered animal bones, without recourse to the hypothesis of recent intrusive burial. Keep in mind that the posture of the skeleton, face down and with limbs outstretched, was like that of a drowned corpse rather than one deliberately buried.

The very infrequent references to the Savona skeleton in current textbooks are predictably negative, and just as predictably flawed in their presentation of the facts. For example, Boule and Vallois (1957, p. 106) claimed: "No stratigraphic study of the formation was made." This statement of theirs is, however, inaccurate, as can be seen from the above-mentioned reports, which established the Pliocene age and undisturbed condition of the layer containing the skeleton.

6.2.4 A Human Vertebra from Monte Hermoso
(Early Pliocene)

Having discussed the discovery of flint tools and signs of intentional use of fire at Monte Hermoso in Argentina (Section 5.1.1), we will now consider a human bone found there. Dr. F. Ameghino (1908, pp. 106–107) reported: "The precursor of man who burned the pampas grass, who made fire in hearths, chipped flint implements, and burned and split the bones of animals he hunted, has also left some of his own fossil bones." He was speaking of a human atlas (the first, or topmost, vertebra of the spinal column) collected by Santiago Pozzi, an employee of the Museo de La Plata (F. Ameghino 1908, p. 174). According to Ameghino (1908, p. 107), the atlas was from the Pliocene Montehermosan formation at Monte Hermoso.

A. Hrdlicka wrote (1912, p. 346): "some time in the [eighteen] eighties (the exact date is not known), an employee of the Museo de La Plata made for that institution at Monte Hermoso a collection of fossils. Among these bones was found at the museum a humanlike atlas of subaverage size."

"When this atlas was seen by Señor Moreno, at that time the director of the La Plata Museum," wrote Hrdlicka (1912, p. 346), "it was still partially enveloped in yellowish or yellowish-brown earth." The Montehermosan is a yellow-brown loess. There are no other beds of that color at Monte Hermoso, according to a detailed description of the site stratigraphy compiled by geologist Bailey Willis (Hrdlicka 1912, p. 362). As previously mentioned (Section 5.1.1), the Montehermosan dates back about 3–5 million years before the present, and belongs to the Early Pliocene (Marshall *et al.* 1982).

In a footnote, Hrdlicka (1912, p. 346) added: "Ameghino (*Tetraprothomo*, etc., p. 174) says that the specimen was 'still in a portion of the rock' but Señor Moreno expressly stated to the writer that it was in 'earth' which held together but was not solidified. Whether or not this earth was sandy can not now be definitely determined. The fact that later the bone was cleanly disengaged from the mass shows further that it could not have been in 'rock.' Roth speaks of the bone as having been enveloped in 'loess' (in Lehmann-Nitsche, *Nouvelles recherches*, etc., p. 386)." The Montehermosan is the only loess formation at Monte Hermoso (Hrdlicka 1912, p. 362).

Hrdlicka (1912, pp. 346–347) then traced the further history of the atlas: "Soon after its discovery the specimen was forgotten and lay unnoticed in the collections of the museum for many years, until finally it was observed accidentally by Santiago Roth, who freed it from the 'loess,' and seeing that the specimen appeared to be a human atlas of small size transferred it to the anthropologic collections of the institution. There again it lay for several years longer without receiving any special consideration, until a new discovery at Monte Hermoso attracted to it the attention of Ameghino. Through Lehmann-Nitsche Ameghino borrowed the specimen, studied it in detail, and published a description of it in his memoir on the *Tetraprothomo,* identifying the bone with that particular hypothetic genus of man's precursors. At the same time a study of the atlas was undertaken and published by Lehmann-Nitsche, who in turn attributed it to 'a Tertiary primate of Monte Hermoso, the *Homo neogaeus.*'" It may be noted that the Gibraltar skull lay for many years in the garrison museum before it was recognized as a Neanderthal specimen. Also, several *Homo erectus* femurs from Java were shipped to Holland in boxes of bones. They went unrecognized and uncataloged for several decades after they were unearthed, but are now listed in textbooks with other accepted finds. The number of similar cases could be expanded, the point being that scientists have become aware of many fully accepted fossil finds in the same way as the Monte Hermoso atlas.

At a later date, another bone turned up. Hrdlicka (1912, p. 347) wrote: "Sometime during the early years of the present century Carlos Ameghino discovered in the same barranca of Monte Hermoso a peculiar bone, which eventually was referred to a supposed ancient parental form of man. It was a portion of the fossil femur of a being which F. Ameghino identified as a very ancient forerunner of man, the *Tetraprothomo argentinus*."

Hrdlicka felt the femur belonged to something other than a human being. Hrdlicka (1912, p. 376) wrote: "The femur of the *Tetraprothomo* bears only a slight resemblance to that of man or the anthropoid apes, and but little greater to that of the lower monkeys. It presents no feature which would make obligatory or even possible its inclusion in the Primate class, but on the other hand it shows many features which approximate it to a distant family of mammals. The class of mammals with which the characteristics of the femur connect it most closely are the carnivores, and among these especially the cats."

As far as the atlas was concerned, Ameghino and others thought it displayed some primitive features, but extensive analysis by Hrdlicka (1912, p. 364) led him to conclude: "The bone is submedium in size and rather massive, but is in every respect human. An extensive comparison with human and other mammalian atlases settles its human provenience beyond question. It is more or less distant morphologically from the atlases of all the anthropoid apes and still more so from those of the monkeys, while the atlases of the Carnivora and other mammals present such differences that a comparison becomes entirely superfluous." It is fairly obvious what Hrdlicka was trying to do. Ameghino had pointed to primitive features in the atlas, with a view toward attributing it to a precursor of the modern human race, a species that lived in Argentina during the Early Miocene (the Early Pliocene according to modern estimates). For Hrdlicka, it was sufficient to show that the bone was completely modern in character. Hrdlicka was an evolutionist and believed the laws of biological development required that the human form should, as we proceed back in time, become more and more primitive. If the bone was of the fully modern human type, then no matter what layer it was found in, it had to be of recent origin. There was no doubt about it. Such a bone's presence in an ancient stratum always could be, indeed had to be, explained as some kind of intrusion.

Along these lines, Hrdlicka (1912, p. 384) wrote about the atlas: "Its extraction is problematical, but even if found in quite intimate relation with the real Monte Hermosean loess, it is not necessarily old. It may well have been derived from the dune above the Monte Hermoso barrancas, which, as shown before, contain numerous traces of the modern native of the coast, and which fall from the crumbling edge above the ledges into pockets of the lower ancient formation." But there is another possible explanation: human beings of the modern physiological type were living over 3 million years ago in Argentina. This is supported by the fact that the atlas showed signs of having been thoroughly embedded in sediments from the Montehermosan formation.

All in all, Hrdlicka (1912, p. 384) felt that the Monte Hermoso atlas was worthy of being "dropped of necessity into obscurity." That is exactly what happened. The atlas was dropped into obscurity. It had to be done. Otherwise, Hrdlicka's claim that humans only recently entered the Americas would have been placed on very shaky ground. Certainly there are many who will insist that the Monte Hermoso atlas remain in the obscurity into which it was of necessity dropped. Evidence for a fully human presence 3 million or more years ago, in Argentina of all places, is still not welcome in mainstream paleoanthropology.

6.2.5 A Jaw Fragment from Miramar, Argentina (Late Pliocene)

Early in the twentieth century, fossil human skeletal remains were found in the Late Pliocene Chapadmalalan formation at Miramar, Argentina. Previously, stone tools and a mammalian bone with an arrow head embedded in it had been discovered at this site (Section 5.2). Hugo Obermaier (1924, p. 306) wrote: "in 1921 M. A. Vignati discovered further human remains at Miramar, not far from Buenos Aires, consisting of a fragment of lower jaw with two molars still in it. According to Vignati it came from the geologic formation of Chapalmalal." We have not been able to locate Vignati's original report on this find, potentially an important paleoanthropological discovery. But we have found a report about the jaw fragment by another South American scientist, E. Boman.

Boman (1921, pp. 341-342) stated: "From the publication of my article in the *Journal de la Société des Americanistes de Paris* up to the time of my visit to Miramar last year, some other objects have been discovered there. Those that have attracted the most attention are two human lower molars (2nd and 3rd right), which were adhering to a small fragment of mandible. Parodi found them, according to the report, embedded in the *barranca,* at great depth in the Chapadmalalan strata, at about the level of the sea. Parodi extracted the object from its position and took it to town, where he showed it to Dr. F. Kühn, who at the time he saw it concluded it was of some importance. Kühn advised him to inform Carlos Ameghino, who came to Miramar to take possession of the teeth."

The discovery appears quite significant—human fossil remains found in the Chapadmalalan—a formation which Anderson (1984, p. 41) gave an age of 2.5–3.0 million years and which Marshall *et al.* (1982, p. 1352) gave an age of 2.0–3.0 million years.

Boman, however, treated this evidence in a manner typical of those sharing his views. He stated: "The newspapers published bombastic articles about 'the most ancient human remains in the world.' But all who examined the molars found them to be identical to the corresponding molars of modern human beings. Human beings existing at that time would have been contemporary with their 'precursor,' the mysterious *Tetraprothomo*" (Boman 1921, pp. 341–342). In the opinion of Florentino Ameghino, *Tetraprothomo* was a primitive apelike ancestor of anatomically modern humans, which he thought evolved in South America.

Boman took it for granted that the fully human nature of the Miramar jaw fragment unequivocally insured its recent date. But nothing Boman said excludes the possibility that the Miramar fossil demonstrates a fully human presence in the Pliocene of Argentina.

Boman mockingly suggested that the Miramar jaw fragment, if one could imagine it was genuinely old, would contradict Florentino Ameghino's theory that human beings evolved from apelike ancestors in Argentina, a theory Boman regarded as fanciful. But Boman neglected the possibility that the discovery of a fully human jaw in the Chapadmalalan formation might contradict his own views, and those of others, who believed that *Homo sapiens* evolved quite recently. The presence of *Homo sapiens* 2–3 million years ago in Argentina would have invalidated the entire story of human evolution then, and now, accepted as fact.

6.2.6 Human Skeletal Remains from the California Gold Country (Pliocene to Eocene)

In the preceding chapter (Section 5.5), we discussed the numerous stone implements discovered in the auriferous gravels of the Sierra Nevada Mountains of California. Some of these implements were found beneath the latite cap of Table Mountain in Tuolumne County. We noted that this latite cap has yielded radiometric dates of 9 million years, while the prevolcanic auriferous gravels lying just above the bedrock have yielded dates of 33–55 million years. Now we will describe human skeletal remains that have been discovered beneath the latite cap of Tuolumne Table Mountain, and elsewhere in California. We will begin our review with the Calaveras skull, the history of which is colorful but inconclusive. The accounts of the remaining discoveries, although less entertaining, provide better evidence for a human presence in the Tertiary.

6.2.6.1 The Calaveras Skull

The most notorious fossil discovered in the Gold Rush mines of California was the Calaveras skull. The State Geologist of California, J. D. Whitney (1880, pp. 267–273), described the circumstances surrounding this find.

In February 1866, Mr. Mattison, the principal owner of the mine on Bald Hill, near Angels Creek, removed this skull from a layer of gravel 130 feet below the surface. The gravel was near the bedrock, underneath several distinct layers of volcanic material. Volcanic eruptions began in this region during the Oligocene, continued through the Miocene, and ended in the Pliocene (Clark 1979, p. 147).

Since the skull occurred near the bottom of the sequence of interspersed gravel and lava layers at Bald Hill, it would seem likely that the gravel in which the skull was found was older than the Pliocene, perhaps much older.

After finding the skull, Mattison later carried it to Mr. Scribner, an agent of Wells, Fargo and Co.'s Express at Angels. Mr. Scribner's clerk, Mr. Matthews, cleaned off part of the incrustations covering most of the fossil. Upon recognizing that it was part of a human skull, he sent it to Dr. Jones, who lived in the nearby village of Murphy's and was an enthusiastic collector of such items. Then Dr. Jones wrote to the office of the Geological Survey in San Francisco, and after receiving a reply, he forwarded the skull to this office, where it was examined by Whitney. Whitney at once made the journey to Murphy's and Angels, where he personally questioned Mr. Mattison, who confirmed the report that was given by Dr. Jones. Both Scribner and Jones were personally known to Whitney and were regarded by him as trustworthy.

On July 16, 1866, Whitney presented to the California Academy of Sciences a report on the Calaveras skull, affirming that it was found in Pliocene strata. The skull caused a great sensation in America.

According to Whitney (1880, p. 270), "The religious press in this country took the matter up . . . and were quite unanimous in declaring the Calaveras skull to be a 'hoax.'" One paper reported: "We believe the whole story worthy of no scientific credence, and are also more fully established in this belief by the declaration of an able Congregationalist minister, who has preached some time in the region, and who told us that the miners freely told him that they purposely got up the whole affair as a joke on Professor Whitney." Another religious paper (the *Congregationalist,* Sept. 27, 1867) reported that the skull "had been placed [in the mine] by some mischievous miners as a hoax upon one of their own number, who was of an anti-Scriptural and geologic turn of mind. He swallowed the hoax and carried the news to Professor Whitney, who thereupon secured the skull for the State Museum" (Whitney 1880, p. 270).

The image of the rough and ready humorists of the rip roaring Gold Rush mining camps having a good joke at the expense of a stuffy geologist is reflected in the following verses excerpted from Bret Harte's poem "The Pliocene Skull" (Harte 1912, pp. 280–281):

> "Speak, O man, less recent! Fragmentary fossil!
> Primal pioneer of pliocene formation,
> Hid in lowest drifts below the earliest stratum
> Of volcanic tufa!

> "Older than the beasts, the oldest Paleotherium;
> Older than the trees, the oldest Cryptogami;
> Older than the hills, those infantile eruptions
> Of earth's epidermis!

"Eo—Mio—Plio—whatsoe'er the 'cene' was
That those vacant sockets filled with awe and wonder,—
Whether shores Devonian or Silurian beaches,—
Tell us thy strange story!

"Speak, thou awful vestige of the earth's creation,
Solitary fragment of remains organic!
Tell the wondrous secret of thy past existence,—
Speak! thou oldest primate!"

Even as I gazed, a thrill of the maxilla,
And a lateral movement of the condyloid process,
With post-pliocene sounds of healthy mastication,
Ground the teeth together.

And from that imperfect dental exhibition,
Stained with express juices of the weed nicotine,
Came these hollow accents, blent with softer murmurs
Of expectoration:

"Which my name is Bowers, and my crust was busted
Falling down a shaft in Calaveras County;
But I'd take it kindly if you'd send the pieces
Home to old Missouri!"

Whitney noted that the hoax stories did not arise until after his discovery was publicized widely in newspapers. Some of the hoax stories were propagated not by Western poets and preachers but by scientists such as William H. Holmes.

Holmes, an anthropologist, worked for the Smithsonian Institution, founded in 1846 with a half-million dollar bequest from James Smithson, an English scientist and inventor. As late as the 1890s, the Calaveras skull was still a matter of great interest and debate within the scientific community. Holmes, who tended to doubt the skull's Tertiary age, wanted to put the matter to rest, once and for all. During a visit to Calaveras County, he gathered testimony from some people who were acquainted with Mr. Scribner and Dr. Jones, and this testimony raised the possibility that the skull examined by Whitney was not a genuine Tertiary fossil (Holmes 1899, pp. 459–464).

Mr. J. L. Sperry, the keeper of the lone hotel in Murphy's, recalled that one day he had seen Dr. Jones, whose office faced the hotel, come out shouting and hurl a broken skull into the street. Sperry asked Jones what the fuss was about. Jones explained that he felt he had been the victim of a practical joke by Scribner, who had sent him a supposedly ancient skull that now appeared to be a fake. But then Jones reconsidered the matter, picked up the skull, and later sent it to Whitney (Holmes 1899, p. 459).

Furthermore, in 1908, William J. Sinclair, a California archeologist, reported receiving an article by Rev. W. H. Dyer from the *Tuolumne Independent* of

September 14, 1901. In this article (Sinclair 1908, p. 128), Dyer stated that he had been present when Mr. Scribner and two friends retold "the story of the skull, which they had planted deep in the bottom of the shaft where it astonished the miner, the curious public and the wondering scientists." Dyer later told Sinclair he had learned from Scribner's sister that his relatives "have long known as a joke of his, the planting of a skull in a mine" (Sinclair 1908, p. 129).

But there are many different sides to the story. Holmes reported the efforts of Dr. A. S. Hudson to solve the Calaveras mystery. In 1883, Dr. Hudson received a letter from Dr. John Walker of Sonora. In this letter, Walker related how he had tried to convince J. D. Whitney that the Calaveras skull had originally been found in an Indian grave at Salt Spring Valley and not in Mattison's mine on Bald Mountain. Walker subscribed to the view that the whole incident was "a fabrication and a joke" (Holmes 1899, p. 460). Hudson visited Walker, but found he had little evidence to back up his claims.

Hudson then went to Angels to talk to Scribner, the alleged prankster, who, according to Holmes, "assured him that Dr. Walker was wrong, and that no deception whatever had been practiced" (Holmes 1899, p. 460). Dr. Hudson then interviewed Mattison and his wife, and they confirmed that he had brought the incrusted skull home from his mine, where he had found it at a depth of 128 feet. It had remained in the Mattison household for a year. When shown a picture of the skull from Whitney's book, Mrs. Mattison recognized the skull as the same one she had kept for a year (Holmes 1899, p. 461). Feeling "perplexed and discouraged" by the seemingly "incomplete and incoherent" stories, Dr. Hudson returned to his office (Holmes 1899, p. 461).

Two weeks later, Scribner appeared and gave more information. Hudson wrote: "It seems, as time went on, Mrs. Mattison, an orderly housekeeper, began to take a dislike to the untidy thing—an unwashed dead head in her house—and made a complaint. It was more in the way than of use or ornament, and she decided to get rid of it. Thereupon her husband, like a proper acquiescing partner in life, carried it to Mr. Scribner's store" (Holmes 1899, p. 461).

Scribner related to Hudson that his partner, Mr. Henry Matthews, was angry at Dr. Jones for giving him some unpleasant medicine. Therefore, as a kind of a joke, Matthews sent the skull in a sack with some lumps of rock and petrified wood to the office of Dr. Jones, who was known to be a collector of geological curiosities. Dr. Jones, apparently thinking the skull to be recent and of little value, is then said to have tossed the skull out into his back yard, where it remained for several months. Then, while visiting Dr. Jones, Mr. Mattison saw the object, and upon recognizing it stated it was the same skull he had removed from his mine. Appreciating the relic in a new light, Dr. Jones then forwarded it to Whitney. So according to Dr. Hudson, there was some joking involved, but the motive was "not to play upon the spirit of scientific inquiry" but rather an attempt by

Mr. Matthews to get even with Dr. Jones (Holmes 1899, p. 463).

Additional stories open up the possibility that the skull was exchanged with another one while it was at Mr. Scribner's store. Holmes spoke with George Stickle, the postmaster at Angels Camp, who told him that the Calaveras skull had actually been brought to him by Mr. J. L. Boone, from an Indian burial place in Salt Spring Valley, 12 miles from Angels Camp. The likelihood of such a thing happening cannot be easily dismissed. As Holmes (1899, p. 463) noted: "There were ancient skulls in plenty in this region in early times, and the valley and the county received their name Calaveras—which in Spanish signifies skulls—from this circumstance. The Indians of the high sierra do not bury their dead, but cast them into pits, caverns, holes in the rock, and deep gorges. . . . Skulls were plentiful at Angels in those days."

After remaining in his store for a few weeks, said Stickle, the skull fell into the hands of Scribner and his fun-loving friends, who were always pulling practical jokes on each other (Holmes 1899, p. 463). Stickle also testified that the skull taken from Mattison's mine was whole and white in color, and did not at all resemble the skull sent by Dr. Jones to Whitney. Yet Dr. Hudson reported that when Mrs. Mattison was shown a photograph of the Calaveras skull she identified it as the same one she had kept in her home. These stories are rather sketchy and incomplete, but at any rate there appears to be some doubt about the real age of the skull examined by Whitney.

After visiting Calaveras county, Holmes (1899, p. 469) examined the actual Calaveras skull at the Peabody Museum in Cambridge, Massachusetts, and concluded that "the skull was never carried and broken in a Tertiary torrent, that it never came from the old gravels in the Mattison mine, and that it does not in any way represent a Tertiary race of men." Some testimony supporting this conclusion comes from persons who examined the matrix of pebbles and earth in which the Calaveras skull had been discovered. Dr. F. W. Putnam of Harvard University's Peabody Museum of Natural History testified: "Had it been taken from the shaft there probably would have been some trace of gravel, such as is found in the beds through which the shaft was sunk, mixed with the materials taken from the skull by Professors Whitney and Wyman, but no such gravel has been found in the several examinations which have been made of the matrix" (Sinclair 1908, p. 129). Professor William J. Sinclair of the University of California also personally examined the matrix and concluded that it "is not strictly a gravel" and that "the material is dissimilar in every respect to either of the gravels exposed on Bald Hill. In every respect it is comparable to a cave breccia" (1908, p. 126). A breccia is a deposit of various kinds of stone fragments mixed in a matrix of sand or clay. Sinclair believed that tiny fragments of bone, belonging to humans and small mammals, found adhering to the skull, along with a decorative bead found inside it, all reported by Whitney, were evidence of a recent cave origin.

On the other hand, Holmes (1899, p. 467) reported: "Dr. D. H. Dall states that while in San Francisco in 1866, he compared the material attached to the skull with portions of the gravel from the mine and that they were alike in all essentials." And W. O. Ayres (1882, p. 853), writing in the *American Naturalist*, stated: "I saw it and examined it carefully at the time when it first reached Professor Whitney's hands. It was not only incrusted with sand and gravel, but its cavities were crowded with the same material; and that material was of a peculiar sort, a sort which I had occasion to know thoroughly. It was the common 'cement' or 'dirt' of the miners; that known in books as the auriferous gravel." Ayres, a competent observer, intimately familiar with the region, should have been able to distinguish a recent cave breccia from Pliocene or Eocene auriferous gravels.

But even if it were true that some auriferous gravel was adhering to the skull, that would not have satisfied Holmes (1899, p. 467), who stated that "the peculiar agglomeration of earth, pebbles, and bones is readily explained by referring to the conditions existing in the limestone caverns and crevices of the region where the calcareous accretions bind together bones, gravel (very generally present), cave earth, and whatever happens to be properly associated, in just such a manner as that illustrated in the specimen under discussion." Yet if we prefer to listen to Ayres (1882, p. 853), we learn: "It has been said that it is a modern skull which has been incrusted after a few years of interment. This assertion, however, is never made by anyone knowing the region. The gravel has not the slightest tendency toward an action of that sort. . . . the hollows of the skull were crowded with the solidified and cemented sand, in such a way as they could have been only by its being driven into them in a semi-fluid mass, a condition the gravels have never had since they were first laid down."

Whitney (1880, p. 271), in his original description of the fossil, observed that the Calaveras skull was highly fossilized. This is certainly consistent with great age, however, as Holmes pointed out, it is also true that bones can become fossilized over the course of a few hundred or thousand years. Yet geologist George Becker (1891, p. 195) reported: "I find that many good judges are fully persuaded of the authenticity of the Calaveras skull, and Messrs. Clarence King, O. C. Marsh, F. W. Putnam, and W.H. Dall have each assured me that this bone was found in place in the gravel beneath the lava." Becker added that this statement was made with the permission of the authorities named. Clarence King, as mentioned previously, was a famous geologist attached to the U.S. Geological Survey. O. C. Marsh, a paleontologist, was one of the pioneer dinosaur fossil hunters, and served as president of the National Academy of Sciences from 1883 to 1895. But F. W. Putnam of Harvard's Peabody Museum, as we have seen, later changed his mind, saying that the matrix of the skull appeared to be a cave deposit.

Can it really be said with certainty that the Calaveras skull was either genuine or a hoax? The evidence is so contradictory and confusing that although the skull could have come from an Indian burial cave we might regard with suspicion anyone who comes forward with any kind of definite conclusion. The reader may pause to contemplate what steps one would take to make one's own determination of the true age of the Calaveras skull.

It should, however, be kept in mind that the Calaveras skull was not an isolated discovery. Great numbers of stone implements were found in nearby deposits of similar age. And, as we shall see in the next sections of this chapter, additional human skeletal remains were also uncovered in the same region. The reports of these discoveries, although brief, are more satisfactory than the reports concerning the Calaveras skull. The reports are simpler, providing no basis for charges of fraud—unless one wants to argue that California gold miners suffered from a massive paleoanthropological hoax obsession.

Similar discoveries, although not quite as old as those from California, were made elsewhere in the world, as at Castenedolo. In light of this, the Calaveras skull cannot be dismissed without the most careful consideration. As Sir Arthur Keith (1928, p. 471) put it: "The story of the Calaveras skull . . . cannot be passed over. It is the 'bogey' which haunts the student of early man . . . taxing the powers of belief of every expert almost to the breaking point."

Furthermore, it seems the evolutionary preconceptions of Holmes, Hrdlicka, and others were partly responsible for the scientific community's rejection of the Calaveras skull, as well as other anomalously old human fossils. We have documented the opinions of Holmes and Hrdlicka in our discussion of the stone implements discovered in the California auriferous gravels (Section 5.5.13) and in our discussion of the Buenos Aires skull (6.1.5). Concerning the Calaveras skull, James Southall (1882, p. 199) said, in a paper delivered at the Victoria Institute in London, England: "If the human skull was exactly the same at the beginning of the Pliocene, or the close of the Miocene, that it is now; on the theory of evolution, how shall we explain the absence of all progress or change? and what margin of time is there for man's development from the generalised lemurs of the Eocene? There is no doubt whatever that the confirmation of Professor Whitney's opinion as to the age of this skull would be fatal to the evolution theory."

In this regard, Laing (1894, p. 389) wrote: "if we accept . . . the skulls of Castelnedolo [sic] and Calaveras, which are supported by such extremely strong evidence, it would seem that as we recede in time, instead of getting nearer to the 'missing link,' we get further from it. This, and this alone, throws doubt on evidence which would otherwise seem to be irresistible." In other words, the fact that the discoveries violated evolutionary expectations was sufficient to overrule all other testimony.

It is indeed true that J. D. Whitney's reports of skeletal remains and artifacts, which imply that anatomically modern human beings existed in California over 9 million years ago, do call into question the theory of human evolution, as presently understood. How can humans not have changed over that vast period of time? Whitney was certainly aware of the implications of his findings. Writing 11 years before the discovery of the Java ape-man, *Pithecanthropus erectus,* he stated: "All the investigations of geologists and ethnologists thus far have failed to obtain satisfactory evidence of the existence at a previous epoch of any type of being connecting man with the inferior animals, or decidedly lower in grade than races now inhabiting portions of the earth, or anything that we fail to recognize instantly as man" (Whitney 1880, p. 286). More explicitly, Whitney concluded: "Man, thus far, is nothing but man, whether found in Pliocene, Post-pliocene, or recent formations" (1880, p. 288). He did admit the chance that some precursor of modern humanity might someday be found in strata older than Pliocene, but his tone in presenting this possibility suggested a challenge to his opponents rather than a fervent and soon-to-be-fulfilled hope of his own.

In the decades following Whitney's statements, fossils displaying varying degrees of apelike and humanlike features did in fact come to light in Pliocene and post-Pliocene formations. But their discovery does not, however, automatically eliminate the many remains of anatomically modern humans discovered in the same, and earlier, formations. Nevertheless, the anomalously old human discoveries were eliminated by advocates of the recent evolution of the modern human type. If this elimination had not occurred, it would not have been possible to speak of the newly discovered ape-man-like creatures as precursors of *Homo sapiens sapiens,* human beings of the modern type.

6.2.6.2 Captain Akey's Report

On January 1, 1873, the president of the Boston Society of Natural History read extracts from a letter by Dr. C. F. Winslow about a discovery of human bones at Table Mountain in Tuolumne County. The find was made in 1855 or 1856, and the details of it were communicated to Winslow by Capt. Akey, who had witnessed it. The discovery took place about 10 years before J. D. Whitney first reported on the famous Calaveras skull. Regarding the finds Whitney described, Winslow (1873, p. 257) wrote: "some distrust as to their identity has been entertained in certain scientific circles. The verification of such discoveries is all important to the interests of science, and I take great pleasure in communicating another fact to the Society of the same character; and in order that the record may in this instance be placed beyond dispute, I have requested my informant to substantiate his statement made to me in due legal form before a notary public."

Winslow (1873, pp. 257–258) then went on to relate: "During my visit to this mining camp I have become acquainted with Capt. David B. Akey, formerly commanding officer of a California volunteer company, and well known to many persons of note in that State, and in the course of my conversation with him I learned that in 1855 and 1856 he was engaged with other miners in running drifts into Table Mountain in Tuolumne County at the depth of about two hundred feet from its brow, in search of placer gold. He states that in a tunnel run into the mountain at the distance of about fifty feet from that upon which he was employed, and at the same level, a complete human skeleton was found and taken out by miners personally known to him, but whose names he does not now recollect. He did not see the bones in place, but he saw them after they were brought down from the tunnel to a neighboring cabin. All the bones of the skeleton apparently were brought down in the arms of miners and placed in a box, and it was the opinion of those present that the skeleton must have been perfect as it laid in the drift. He does not know what became of the bones, but can affirm to the truth of this discovery, and that the bones were those of a human skeleton, in an excellent state of preservation. The skull was broken in on the right temple, where there was a small hole, as if a part of the skull was gone, but he cannot tell whether this fracture occurred before the excavation or was made by the miners. . . . He thinks that the depth from the surface at which this skeleton was found was two hundred feet, and from one hundred and eighty to two hundred feet from the opening cut or face of the tunnel. The bones were in a moist condition, found among the gravel and very near the bed rock, and water was running out of the tunnel. There was a petrified pine tree, from sixty to eighty feet in length and between two and three feet in diameter at the butt, lying near this skeleton. Mr. Akey went into the tunnel with the miners, and they pointed out to him the place where the skeleton was found. He saw the tree in place and broke specimens from it. He cannot remember the name of this tunnel, but it was about a quarter of a mile east of the Rough and Ready tunnel and opposite Turner's Flat, another well known point. He cannot tell the sex of the skeleton, but it was of medium size. The bones were altogether, and not separated, when found."

Winslow (1873, p. 258) added: "On the same level at which this skeleton was found, but from other tunnels, Mr. Akey saw many bones of animals taken, but no other human remains. Among those remains were mastodon's teeth and bones of animals smaller than mastodons, the names of which he does not know. . . . Overlying these placer deposits and organic remains was volcanic matter consisting of lava or of 'honey-combed' material." Akey swore to the truth of these statements before a notary at Bear Gulch (Winslow 1873, p. 259).

The gravel just above the bedrock at Tuolumne Table Mountain, where the skeleton was found, is said to be between 33 and 55 million years old (Slemmons 1966, p. 200). This must be the age of the skeleton unless it was introduced into the gravels at a later time, and we are not aware of any evidence indicating such an intrusion. The reported presence of mastodon teeth "on the same level . . . but from other tunnels" is interesting. Mastodons are generally thought to have appeared in North America during the Miocene, but if mastodon teeth were in fact found near the bedrock at Tuolumne Table Mountain, they would be considerably older— Early Oligocene or Eocene.

6.2.6.3 The Hubbs Skull Fragment

In 1868, J. D. Whitney reported on the Calaveras skull to the American Association for the Advancement of Science. Soon thereafter, Dr. J. Wyman informed him that in the collection of the Museum of the Natural History Society of Boston there was a skull fragment that Wyman, one of America's leading craniologists (Holmes 1899, p. 456), said was human. The fossil was labeled as follows: "From a shaft in Table Mountain, 180 feet below the surface, in gold drift, among rolled stones and near mastodon debris. Overlying strata of basaltic compactness and hardness. Found July, 1857. Given to Rev. C. F. Winslow by Hon. Paul K. Hubbs, August, 1857." Another fragment, from the same skull, and similarly labeled, was to be found at the Museum of the Philadelphia Academy of Natural Sciences.

The proceedings of the Boston Natural History Society (Volume VI, p. 278, October 7, 1857) contain a message that Winslow sent to Boston along with the first skull fragment. Winslow stated: "I sent by a friend, who was going to Boston this morning, a precious relic of the human race of earlier times, found recently in California, 180 feet below the surface of Table Mountain. . . . My friend Colonel Hubbs, whose gold claims in the mountains seem to have given him much knowledge of this singular locality, writes that the fragment was brought up in the pay dirt (the miner's name for the placer gold-drift) of the Columbia Claim, and that the various strata passed through in sinking the shaft consisted of volcanic formations exclusively" (Whitney 1880, p. 264).

Whitney, in California, then began his own investigation. He learned that Hubbs was "a well-known citizen of Vallejo, California, and a former State Superintendent of Education" (Whitney 1880, p. 264). Whitney got from Hubbs a detailed written account of the discovery, which occurred in the Valentine Shaft, south of Shaw's Flat. Whitney (1880, p. 265) stated: "The essential facts are, that the Valentine Shaft was vertical, that it was boarded up to the top, so that nothing could have fallen in from the surface during the working under ground, which was carried on in the gravel channel exclusively, after the shaft had been sunk. There can be no doubt that the specimen came from the drift in the channel under Table Mountain, as affirmed by Mr. Hubbs." The skull fragment was found in a horizontal mine shaft (or drift) leading from the main vertical shaft, at a depth of 180 feet from the surface (Whitney 1880, p. 265). Hubbs stated that he "saw the portion of skull immediately after its being taken out of the sluice into which it had been shoveled" (Whitney 1880, p. 265). Adhering to the bone was the characteristic gold-bearing gravel. Whitney (1880, p. 265) commented: "It is clear from Mr. Hubbs's statements that the fragment was raised from the stratum of pay gravel, and that it was noticed when the contents of the bucket were dumped into the head of the sluice, and either picked up by Mr. Hubbs, or by some one else, who happened to be standing by, and who handed it to him on the spot."

Independent corroborating evidence came from Mr. Albert G. Walton, one of the owners of the Valentine claim, at which the skull fragment was discovered. Mr. Walton reported that a stone mortar 15 inches in diameter was found in the Valentine mine, in gold-bearing gravels 180 feet below the surface and also below the latite cap of Tuolumne Table Mountain (Whitney 1880, p. 265).

When Sinclair (1908, p. 115) visited Table Mountain in 1902, he found that many of the drift mines south of Shaw's Flat were connected. Thus, according to Sinclair, Whitney's statement that the Valentine shaft was securely boarded up to the top so that nothing could fall in from the surface did not rule out the possibility that objects could have found their way into the Valentine underground tunnel from some other tunnels.

But Sinclair did not prove that there were in fact such interconnections between the tunnels at the time the discoveries were made in 1857. Perhaps the interconnections between the tunnels he observed in 1902 were made after the discoveries. Furthermore, Sinclair (1908, p. 115) admitted that during his 1902 visit he was not even able to find the old Valentine shaft. This means he had no direct evidence that the Valentine mine shafts were connected to any others. Finally, even if there were tunnels that connected with the drift tunnel running from the main Valentine shaft, this does not invalidate Hubbs's report. Whitney (1880, p. 265) observed that all the mines near the Valentine mine were "working through vertical shafts." One would have to imagine that somehow or other a fragment of skull was dropped into one of these vertical shafts and that it was transported some distance along a horizontal tunnel. It is hard to see how this could happen, because material from the horizontal tunnel, as it was being excavated, would have been taken back toward the vertical shaft.

So Sinclair was not able to confirm, by direct inspection or testimony, his claim that the horizontal drift tunnels running from the Valentine vertical shaft were in fact connected to other tunnels. His objection thus appears to be simply a weak and highly speculative attempt to discredit a discovery he opposed on theoretical grounds. The gravels in which the skull fragment was embedded lay 180 feet below the surface and beneath the latite cap of Table Mountain. The skull fragment could thus be from 9 million to 55 million years old.

Whitney (1880, p. 265), in his discussion of this find, noted: "It is clear that, had it not been for the accidental presence of Mr. Hubbs on the spot, at the time the piece of skull was found, we should never have heard anything of it. And if Mr. Hubbs had not given it to an enthusiastic observer, like Dr. Winslow, it would probably never have come to the notice of scientific men. One should bear in mind how few of the discoveries of human relics or remains which are made are likely ever to be heard of beyond a very limited area, even under the most favorable circumstances, as is well illustrated by the facts in this case."

6.2.6.4 *A Human Jaw from Below Table Mountain*

J. D. Whitney (1880, p. 264) personally examined a collection belonging to Dr. Snell, consisting of stone spoons, handles, spearheads, and a human jaw—all found in the auriferous gravels beneath the latite cap of Tuolumne Table Mountain. The jaw measured 5.5 inches across from condyle to condyle, which is within the normal human range. Whitney (1880, p. 288) remarked that all the human fossils uncovered in the gold-mining region, including this one, were of the anatomically modern type. The gravels from which the jaw came could be anywhere from 9 to 55 million years old.

6.2.6.5 *Human Bones from the Missouri Tunnel*

Whitney reported several discoveries from Placer County. In particular, he gave this account of human bones that were found in the Missouri tunnel: "The Missouri Tunnel runs from the Devil's Cañon southerly into the ridge between it and the Middle Fork of the American River, a little above Yankee Jim's. This region has been described in the preceding pages as deeply covered with volcanic materials. In this tunnel, under the lava, two bones had been found . . . which were pronounced by Dr. Fagan to be human. One was said to be a leg bone; of the character of the other nothing was remembered. The above information was obtained by Mr. Goodyear from Mr. Samuel Bowman, of whose intelligence and truthfulness the writer has received good accounts from a personal friend well acquainted with him. Dr. Fagan was at that time one of the best known physicians of the region" (Whitney 1880, p. 277).

In October 1989, our researcher (Stephen Bernath) contacted the California Division of Mines and Geology regarding the age of the deposits at the place where the Missouri tunnel was located. George Saucedo informed him that the andesitic deposits in that vicinity are probably part of a larger formation that has yielded a potassium-argon date of 8.7 million years ago. Thus the human skeletal remains found under the andesitic deposits in that location would have an age of over 8.7 million years.

6.2.6.6 Dr. Boyce's Discovery

Professor Whitney (1880, p. 276) reported a discovery made in 1853 by a physician named Dr. H. H. Boyce at Clay Hill in El Dorado County, California. In 1870, Dr. Boyce wrote to Whitney (1880, p. 276), who had requested information: "While engaged in the business of mining in the spring of 1853, I purchased an interest in a claim on this hill, on condition that it prospected sufficiently well to warrant working it. The owner and myself accordingly proceeded to sink a shaft for the purpose of working it. It was while doing so that we discovered the bones to which you refer. Clay Hill is one of a series of elevations which constitute the water-shed between Placerville Creek and Big Cañon, and is capped with a stratum of basaltic lava, some eight feet thick. Beneath this there are some thirty feet of sand, gravel, and clay. The country-rock is slightly capped on this, as on most of the elevations, the slope being towards the centre of the hill. Resting on the rock and extending about two feet above it, was a dense stratum of clay. It was in this clay that we came across the bones. While emptying the tub, I saw some pieces of material which on examination I discovered were pieces of bones; and, on further search, I found the scapula, clavicle, and parts of the first, second, and third ribs of the right side of a human skeleton. They were quite firmly cemented together; but on exposure to the air began to crumble. We made no further discoveries." According to Whitney (1880, p. 276), Boyce "stated there could be no mistake about the character of the bones, and that he had made a special study of human anatomy."

Sinclair (1908, p. 123) reported that he examined Clay Hill in 1902 and found "no basalt capping appeared either on the hill or anywhere in the vicinity." He did, however, note the presence of "a small area of andesitic breccia on the top of the hill" (Sinclair 1908, p. 123). Both andesite and basalt are dark greyish volcanic rocks; thus it is possible that Boyce, not a trained geologist, may have mistaken the andesite for basalt. Whitney (1880, p. 276) said that Dr. Boyce's "description of the geology of Clay Hill agrees, in the main, with that given by Mr. Goodyear, who states that the deposit on the bed-rock was from twenty-five to thirty feet thick, all but the lower five feet consisting of 'mountain gravel', a local name for the volcanic material capping the hills in that vicinity."

According to the United States Geological Survey Map made by W. Lindgren and H. Turner in July 1893, the andesitic deposits on the top of Clay Hill are Pliocene or Miocene in age—therefore the stratum in which the human bones were found must be at least as old.

But Sinclair persistently attempted to cast whatever doubt he could on the discovery. He said he could not locate the clay stratum said to have contained the bones "owing to the heavy talus slopes" (Sinclair 1908, p. 123). He further stated: "The impression conveyed . . . is that the skeleton found by Dr. Boyce was at a depth of thirty-eight feet, in undisturbed strata under eight feet of so-called basalt. There is nothing, however, in the letter to show that this was the section passed through in sinking the Boyce shaft" (Sinclair 1908, p. 123). Because of the ambiguity about the exact location of the shaft, Sinclair thus concluded (1908, p. 123): "The skeleton may have been found in such a place and at such a depth in the clay that the possibility of recent interment would have to be considered. As the evidence is presented, we are not justified in regarding the skeleton from Clay Hill as of great antiquity."

The points raised by Sinclair are valid, and we agree that there are reasons to doubt the antiquity of the skeletal remains found at Clay Hill. Yet the presence of heavy talus slopes, with so much rock that Sinclair was not able to gain access to the stratum of clay at the base of the hill, seems to argue against, rather than for, the possibility of a recent burial into the clay from the slope of the hill. Also, if there were a recent burial, it is peculiar that so few bones were recovered.

This brings us to the end of our review of fossil human skeletal remains from the auriferous gravels of California. Despite the imperfections of the evidence, one thing is certain—human bones were found in the Tertiary gravels, dating as far back as the Eocene. How the bones got there is open to question. The reports of the discoveries are sometimes vague and inconclusive, yet they are suggestive of something other than pranks by miners or recent intrusive burials by Indians. The presence of numerous stone tools, incontestably of human manufacture, in the same formations, lends additional credibility to the finds.

In an address to the American Association for the Advancement of Science, delivered in August, 1879, O. C. Marsh, president of the Association and one of America's foremost paleontologists, said about Tertiary man: "The proof offered on this point by Professor J. D. Whitney in his recent work (*Aurif. Gravels of Sierra Nevada*) is so strong, and his careful, conscientious method of investigation so well known, that his conclusions seem irresistible. . . . At present, the known facts indicate that the American beds containing human remains and works of man, are as old as the Pliocene of Europe. The existence of man in the Tertiary period seems now fairly established" (Southall 1882, p. 196).

6.2.7 More European Discoveries (Miocene and Eocene)

More evidence for human beings in the early and middle Tertiary comes from Europe. According to de Mortillet, M. Quiquerez reported the discovery of a skeleton at Delémont in Switzerland in ferruginous clays said to be Late Eocene. About this find, de Mortillet (1883, p. 72) simply said one should be suspicious of human skeletons found with the bones in natural connection. De Mortillet (1883, p. 72) further stated that one should be cautious about a similarly complete skeleton found by Garrigou in Miocene strata at Midi de France.

It is possible, however, that these skeletons were from individuals buried during the Eocene or Miocene periods. A burial does not necessarily have to be recent. The truly frustrating thing about finds such as these is that we are not able to get more information about them. We find only a brief mention by an author bent on discrediting them. Because such finds seemed doubtful to scientists like de Mortillet, they went undocumented and uninvestigated, and were quickly forgotten. How many such finds have been made? We may never know. In contrast, finds which conform to accepted theories are thoroughly investigated, safely enshrined in museums, and widely taught to millions around the world.

We are now nearing the end of our survey of evidence for Tertiary man uncovered by scientists in the nineteenth and early twentieth centuries. Much of this evidence is comparable to (or better than) the evidence used by paleoanthropologists in support of the standard scenario of human evolution. For example, in the case of Castenedolo, human skeletal remains were taken from undisturbed Pliocene formations by a professional geologist. By way of contrast, most of the Java man discoveries reported by von Koenigswald and others (Sections 7.3, 7.4) were made at poorly specified locations by paid native collectors, with no scientist present. Yet the Castenedolo find is rarely mentioned in standard textbooks, while the Java *Homo erectus* finds are routinely reported.

Over time, the scientific community eliminated Castenedolo and other discoveries discussed in this chapter from the realm of serious consideration. In 1924, in one of the final published discussions of this material, Hugo Obermaier offered a decidedly negative opinion about human beings in the Tertiary. "A fact of such transcendent importance would be demonstrated beyond question by the discovery of human skeletons of Tertiary age, but up to the present time none of the supposed discoveries of this nature is sufficiently well proved to withstand any serious scientific investigation. Neither the 'Eocene' skeleton of Delemont in Switzerland, nor the 'Pliocene' remains of Colle del Vento near Savona, Liguria, nor those of Matera, all in Italy, have supplied any data for the solving of this interesting problem—being therefore relegated to oblivion, even as the Indian skull of Calaveras, California. Neither has it been possible to prove that the discoveries of F. Ameghino in South America during the last fifteen years . . . are of Tertiary age as claimed" (Obermaier 1924, p. 2).

It is questionable whether the evidence mentioned by Obermaier, and additional evidence presented in this chapter (such as the Castenedolo finds), should have been "relegated to oblivion." Is it really the case that there were no valid scientific grounds for considering this evidence? It would appear that it was simply the great age of the discoveries, an age that conflicted with accepted ideas about human evolution, that was the real problem. In any case, science has quite effectively buried this disconcerting evidence. For example, we have so far been unable to find any other data on the Matera skeleton referred to by Obermaier.

6.3 PRE-TERTIARY DISCOVERIES

We shall now consider rare cases of anatomical evidence for the presence of human beings in pre-Tertiary geological contexts. As we have seen in earlier chapters, some scientists believed ape-men existed as far back as the Miocene and Eocene. A few bold thinkers even proposed that fully human beings were alive during those periods. But now we are going to proceed into times still more remote. Since most scientists had trouble with Tertiary humans, we can just imagine how difficult it would have been for them to give any serious consideration to the cases we are about to discuss. One is tempted not to mention such finds as these because they seem unbelievable. But the result of such a policy would be that we discuss evidence only for things we already believe. And unless our current beliefs represent reality in total, this would not be a wise thing to do.

6.3.1 Macoupin, Illinois (Carboniferous)

In December of 1862, the following brief but intriguing report appeared in a journal called *The Geologist*: "In Macoupin county, Illinois, the bones of a man were recently found on a coal-bed capped with two feet of slate rock, ninety feet below the surface of the earth. . . . The bones, when found, were covered with a crust or coating of hard glossy matter, as black as coal itself, but when scraped away left the bones white and natural."

We wrote to the State Geological Survey Division of the Illinois Department of Energy and Natural Resources for information about the age of the coal in which the bones were found. We received the following response from C. Brian Trask of the Geological Survey, who wrote in a letter dated July 9, 1985: "In response to your inquiry concerning age of coal, the youngest bituminous coal beds in Illinois are found in the upper Pennsylvanian system. . . . The coal mined in the 1860's in Macoupin County is probably the Herrin (No. 6) Coal, although the Colchester (No. 2) Coal occurs at this depth locally in the western part of the county. The Herrin Coal is late Desmoinesian (middle to late Westphalian D) in age." In North America, the Pennsylvanian makes up the latter half of the Carboniferous, which extends from 286 million to 360 million years ago. From the information provided by Trask, it would thus appear that the coal in which the Macoupin County skeleton was found is at least 286 million years old and might be as much as 320 million years old.

6.3.2 Human Footprints from the Carboniferous

Our final examples of anomalous pre-Tertiary evidence are not in the category of fossil human bones, but rather in the category of fossil humanlike footprints. Professor W. G. Burroughs, head of the department of geology at

Berea College in Berea, Kentucky, reported (1938, p. 46): "during the beginning of the Upper Carboniferous (Coal Age) Period, creatures that walked on their two hind legs and had human-like feet, left tracks on a sand beach in Rockcastle County, Kentucky. This was the period known as the Age of Amphibians when animals moved about on four legs or more rarely hopped, and their feet did not have a human appearance. But in Rockcastle, Jackson and several other counties in Kentucky, as well as in places from Pennsylvania to Missouri inclusive, creatures that had feet strangely human in appearance and that walked on two hind legs did exist. The writer has proved the existence of these creatures in Kentucky. With the cooperation of Dr. C. W. Gilmore, Curator of Vertebrate Paleontology, Smithsonian Institution, it has been shown that similar creatures lived in Pennsylvania and Missouri."

The Upper Carboniferous (the Pennsylvanian) began about 320 million years ago (Harland *et al.* 1982, p. 94). It is thought that the first animals capable of walking erect, the pseudosuchian thecodonts, appeared around 210 million years ago (Desmond 1976, p. 86). These lizardlike creatures, capable of running on their hind legs, would not have left any tail marks since they carried their tails aloft. But their feet did not look at all like those of human beings; rather they resembled those of birds. Scientists say the first appearance of apelike beings was not until around 37 million years ago, and it was not until around 4 million years ago that most scientists would expect to find footprints anything like those reported by Burroughs from the Carboniferous of Kentucky.

Burroughs (1938, p. 46) stated: "The footprints are sunk into the horizontal surface of an outcrop of hard, massive grey sandstone on the O. Finnell farm. There are three pairs of tracks showing left and right footprints. . . . Each footprint has five toes and a distinct arch. The toes are spread apart like those of a human being who has never worn shoes." Kent Previette (1953) wrote: "Scientists and travelers who have seen the tracks which he [Burroughs] proved to be genuine, or studied photographs of them, state that they resemble those of the most primitive people of the Andes, the aboriginal Chinese, and the South Sea islanders—all being people who have never worn shoes."

Giving more details about the prints, Burroughs (1938, p. 46) stated: "The length of the foot from the heel to the end of the longest toe is nine and one-half inches though this length varies slightly in different tracks. The width across the ball of the foot is 4.1 inches while the width including the spread of the toes is about six inches. The foot curves back like a human foot to a human appearing heel." These humanlike tracks are thus quite distinct, unlike the more famous but indistinct Paluxy "man tracks" reported in Biblical creationist literature.

David L. Bushnell, an ethnologist with the Smithsonian Institution suggested the prints were carved by Indians (*Science News Letter* 1938a, p. 372). In ruling out this hypothesis, Dr. Burroughs (1938, pp. 46–47) used a microscope to study the prints and noted: "The sand grains within the tracks are closer together than the sand grains of the rock just outside the tracks due to the pressure of the creatures' feet. Even the sand grains in the arch of one of the best preserved tracks are not as close together as in the heel of the same track, though closer together than the sand outside the track. This is because there was more pressure upon the heel than beneath the arch of the foot. In comparing the texture of sandstone only the same kind of grains and combinations of grains within and outside of the tracks are considered. The sandstone adjacent to many of the tracks is uprolled due to the damp, loose sand having been pushed up around the foot as the foot sank into the sand. The forward part of one track is covered by solid Pottsville sandstone only a few days or weeks younger than the sandstone in which is the track. Another track nearby is also partially covered by solid Pottsville sandstone of the Coal Age." These facts led Burroughs to conclude that the humanlike footprints were formed by compression in the soft, wet sand before it consolidated into rock some 300 million years ago.

Two doctors from the town of Berea, Alson Baker and A. F. Cornelius, also counted the sand grains per unit area under magnification and arrived at the same result as had Dr. Burroughs. They reported: "We examined the arrangement of the sand grains in the deepest portions of the prints, with special attention to the heels. The sand grains in the bottoms of the prints were much more closely packed than those in the slopes, and those in the slopes were more closely packed than those in the rock an inch from the margins of the prints, or at any other point. Each member of the party certified and checked these findings and we all agree that the imprints were made by pressure when the sand was soft and wet. The fact that the sand grains in the bottoms and slopes of the imprints are of exactly the same kind as those in all other parts of the rock surface examined, seems to prove conclusively that the closer arrangement observed was not due to any possible drifting in of extraneous material" (*Science News Letter* 1938a, p. 372).

Burroughs also consulted a sculptor. Kent Previette (1953) wrote: "The sculptor said that carving in that kind of sandstone could not have been done without leaving artificial marks. Enlarged photomicrographs and enlarged infrared photographs failed to reveal any 'indications of carving or cutting of any kind.'"

If the prints were not carvings, were they left by a nonhuman Carboniferous species? The most advanced land animals then existing were amphibians that resembled crocodiles and moved about on four legs. But Burroughs (1938, p. 47) wrote: "There are no indications of front feet although the rock is large enough to have recorded front feet if front feet had been used to move about. In the pair of footprints that show the left and right feet about parallel to each other, the distance between the feet is about the same as that of a normal human being. Nowhere on this rock nor on another rock outcrop that also has numerous similar tracks upon its surface, is there any sign that these creatures had tails." Nor were there any belly marks (Previette 1953).

Burroughs (1938, p. 47) added: "The creatures that made the tracks have not as yet been identified, but a name for these creatures has been chosen by the writer with the co-operation of Dr. Frank Thone, Editor in Biology, Science Service, Washington, D.C., Dr. C. W. Gilmore, Curator of Vertebrate Paleontology, Smithsonian Institution, and Miss Charlotte Ludlum, Professor of Latin, Berea College. The name chosen is *Phenanthropus Mirabilis.*" The word *phenanthropus* means "looks human," and *mirabilis* means "remarkable."

Burroughs himself stopped short of claiming that the prints were made by humans, but his presentation leaves one with the strong impression that they were human. When asked about them, Burroughs said, "They look human. That is what makes them especially interesting, as man according to some textbooks has been here only a million and a half years" (Previette 1953). But mainstream science reacted predictably to any suggestion, that the prints were made by humans. *Science News Letter* (1938b) published an article titled "Human-Like Tracks in Stone Are Riddle to Scientists." A subtitle stated: "They Can't Be Human Because They Are Much Too Old—But What Strange Biped Amphibian Can Have Made Them?" Despite the doubts of scientists, the Burroughs footprints continued to attract public attention, which might explain why geologist Albert G. Ingalls felt compelled to set matters straight in *Scientific American.*

Ingalls (1940, p. 14) stated that a scientist, confronted with the suggestion that the tracks were human, would have little choice but to reply: "What? You want *man* in the *Carboniferous?* Entirely and absolutely—totally and completely—impossible. We admit we don't know exactly what made the prints, but we do know one agency that didn't, and that is man in the Carboniferous."

But what about scientific detachment—the willingness to give up established ideas or tentative hypotheses when confronted with contrary evidence? Ingalls (1940, p. 14) wrote: "Science is like the streets of New York: it is never finished, and is always being torn up, often in a major way. . . . Nevertheless, asking the scientist for man in the Carboniferous is like asking the historian for Diesel engines in ancient Sumeria. The comparison is no exaggeration but an under-statement. If man, or even his ape ancestor, or even that ape ancestor's early mammalian ancestor, existed as far back as in the Carboniferous Period in any shape, then the whole science of geology is so completely wrong that all the geologists will resign their jobs and take up truck driving. Hence, for the present at least, science rejects the attractive explanation that man made these mysterious prints in the mud of the Carboniferous with his feet."

Ingalls thought the prints were made by some as yet unknown kind of amphibian. "Science has no proof that these tracks were not made by one or more of these animals—one with which it is not yet familiar—for it does not know everything. Professor W. G. Burroughs, Berea College, Kentucky, geologist, champions this theory, supported by the paleontologist Charles W. Gilmore at the United States Museum" (Ingalls 1940, p. 14). Here Ingalls appears to have put his own interpretation on Burroughs's ambiguous testimony, bringing the wayward researcher firmly back within the bounds of scientific sanity.

We should note that scientists do not really take the amphibian theory seriously. Human-sized Carboniferous bipedal amphibians do not fit into the accepted scheme of evolution much better than Carboniferous human beings— they *wreak havoc* with our ideas of early amphibians, requiring a host of evolutionary developments we now know nothing about.

Ingalls (1940, p. 14) wrote: "What science does know is that, anyway, unless 2 and 2 are 7, and unless the Sumerians had airplanes and radios and listened to Amos and Andy, these prints were not made by any Carboniferous Period man."

6.3.3 A Central Asian Footprint (Jurassic)

The Moscow News (1983, no. 24, p. 10) gave a brief but intriguing report on what appeared to be a human footprint in 150-million-year-old Jurassic rock next to a giant three-toed dinosaur footprint. The discovery occurred in the Turkmen Republic in what was then the southeastern USSR. Professor Amanniyazov, corresponding member of the Turkmen SSR Academy of Sciences, said that although the print resembled a human footprint, there was no conclusive proof that it was made by a human being. This discovery has not received much attention, but then, given the current mindset of the scientific community, such neglect is to be expected. We only know of a few cases of such extremely anomalous discoveries, but considering that many such discoveries probably go unreported we wonder how many there actually might be.

6.4 CONCLUSION

The evidence reviewed in Chapters 2–6 suggests the existence of anatomically modern humans as far back as the early Tertiary. None of this evidence tends to be reported in modern textbooks on anthropology. Should it be reported? We leave it to the reader to decide. If taken seriously, this evidence would certainly challenge the currently dominant understanding of human origins and antiquity, but perhaps this topic is not as thoroughly understood as some believe. The cultural evidence we have considered, in the form of stone tools and incised bones, suggests a relatively primitive level of advancement. There is, however, evidence that suggests a higher level of cultural achievement. But unlike the evidence considered in Chapters 2–6, much of this evidence was never reported by scientists. For a review of this controversial evidence see Appendix 2.

Part II
ACCEPTED EVIDENCE

7

Java Man

In the preceding chapters, we have reviewed three categories of anomalous evidence relating to human origins and antiquity—human skeletal remains, incised bones, and stone implements of various kinds. At the end of the nineteenth century, on the basis of such evidence, a consensus was building within an influential portion of the scientific community that human beings of the modern type had existed as far back as the Pliocene and Miocene periods—and perhaps even earlier.

Anthropologist Frank Spencer (1984, pp. 13–14) stated: "From accumulating skeletal evidence it appeared as if the modern human skeleton extended far back in time, an apparent fact which led many workers to either abandon or modify their views on human evolution. One such apostate was Alfred Russell Wallace (1823–1913). In 1887, Wallace examined the evidence for early man in the New World, and . . . found not only considerable evidence of antiquity for the available specimens, but also a continuity of type through time. In an effort to explain this, Wallace suggested that . . . man, through culture, had been essentially partitioned from the vagaries of natural selection and was, thereby, a unique creation of the biotic realm."

To Darwin, this was heresy of the worst sort. But Spencer (1984, p. 14) noted that Wallace's challenge to evolutionary doctrine "lost some of its potency as well as a few of its supporters when news began circulating of the discovery of a remarkable hominid fossil in Java." Considering the striking way in which the Java man fossils were employed in discrediting and suppressing evidence for the great antiquity of the modern human form, we shall now review their history.

We will discuss the initial discoveries made by Eugene Dubois in the 1890s, the discoveries made by G. H. R. von Koenigswald in the 1930s and 1940s, and the discoveries made by other researchers since 1950. We will then discuss the chemical and radiometric dating of these discoveries, and conclude with a critique of standard scientific presentations of the Java *Homo erectus* evidence. In this chapter, we shall also discuss the Heidelberg jaw, discovered not long after the original Java man finds and also classified as *Homo erectus*.

461

In succeeding chapters, we will examine other paleoanthropological evidence currently employed by scientists to support their hypothesis that the modern human form evolved within the past 100,000 years from more primitive hominid ancestors. We will focus on discoveries made in China (Chapter 9) and Africa (Chapter 11). In addition to this conventionally accepted evidence, we will also examine the controversial Piltdown case (Chapter 8) and evidence for living ape-men (Chapter 10).

7.1 DUBOIS AND PITHECANTHROPUS ERECTUS

The city of Bandung lies in the high cool uplands of western Java. From there a road leads eastward, down to the steaming plain of Leles, continuing on to the district town of Madiun. The green, forested peaks of the volcanos Mt. Lawu and Mt. Willis rise against the brilliantly blue tropical sky. Pushing onward one arrives at the *kampong,* or village, of Trinil, surrounded by fields of rice and sugar cane, as well as groves of coconut trees. Past the village, the road ends on a high bank overlooking the Solo River. Here one encounters a small stone monument, marked with an arrow pointing toward a sand pit on the opposite bank. The monument also carries a cryptic German inscription, "P.e. 175 m ONO 1891/93," indicating that *Pithecanthropus erectus* was found 175 meters east northeast from this spot, during the years 1891–1893.

The discoverer of *Pithecanthropus erectus* was Eugene Dubois, born in Eijsden, Holland, in 1858, the year before Darwin published *The Origin of Species.* As a boy, Dubois explored the nearby limestone quarries, filling his pockets with fossils. Although the son of devout Dutch Catholics, the idea of evolution, especially as it applied to the question of human origins, fascinated him. His imagination was quickened by this passage in A. R. Wallace's *Malay Archipelago* (1869): "With what interest must every naturalist look forward to the time when the caves and tertiary deposits of the tropics may be thoroughly examined and the past history and earliest appearance of the great man-like apes be at length made known."

After studying medicine and natural history at the University of Amsterdam, Dubois became a lecturer in anatomy at the Royal Normal School in 1886. But his real love remained evolution. Dubois knew that Darwin's opponents were constantly pointing out the almost complete lack of fossil evidence for human evolution. He carefully studied the principal evidence then available—the bones of Neanderthal specimens. These were regarded by most authorities (among them Thomas Huxley) as too close to the modern human type to be considered truly intermediate between fossil apes and modern humans. The German scientist Ernst Haeckel had, however, predicted that the bones of a real missing link would eventually be found. Haeckel even commissioned a painting of the creature, whom he called *Pithecanthropus* (in Greek, *pitheko* means "ape," and *anthropus* means "man"). Influenced by Haeckel's vision of *Pithecanthropus,* Dubois resolved to someday find the ape-man's bones.

Mindful of Darwin's suggestion that humanity's forbearers lived in "some warm, forest-clad land," Dubois became convinced *Pithecanthropus* would be found in Africa or the East Indies. Because he could more easily reach the East Indies, then under Dutch rule, he decided to journey there and begin his quest. He applied first to private philanthropists and the government, requesting financing for a scientific expedition, but was turned down. He then accepted an appointment as an army surgeon in Sumatra. With his friends doubting his sanity, he gave up his comfortable post as a college lecturer and with his young wife set sail for the East Indies in December 1887 on the *S. S. Princess Amalie.*

7.1.1 Initial Discoveries

In 1888, Dubois found himself stationed at a small military hospital in the interior of Sumatra. His exact movements during this period remain somewhat unclear, but from a variety of accounts the following general sequence emerges. The year he arrived in Sumatra, Dubois published a scientific paper titled "On the need for an investigation of an Ice Age Fauna in the Dutch East Indies, and especially in Sumatra." He was, of course, primarily interested in finding the remains of human ancestors. Dubois wrote: "Since all apes—and notably the anthropoid apes—are inhabitants of the tropics, and since man's forerunners, as they have gradually lost their coat of hair, must certainly have continued to live in warm regions, we are inescapably led toward the tropics as the area in which we may expect to find the fossilized precursors of man" (von Koenigswald 1956, p. 28).

Dubois's writings attracted the attention of officials of the colonial Mining Authority. In its first quarterly report for the year 1889, the Mining Authority informed its readers that Dubois had been authorized to undertake paleontological research in Sumatra. In his spare time, and using his own funds, Dubois investigated Sumatran caves, finding fossils of rhino and elephant, and the teeth of an orangutan, but no hominid remains.

In 1890, after suffering an attack of malaria, Dubois was placed on inactive duty and transferred from Sumatra to Java, where the climate was somewhat drier and healthier. He and his wife set up housekeeping in Tulungagung, on eastern Java's southern coast. The Mining Authority gave him permission to carry out his paleontological explorations in Java, supplying him with two sergeants from the corps of military engineers and a crew of fifty convict laborers. At the nearby marble quarry at Wadjak, Dubois turned up two fossil human skulls, both modern in type (related to the Australian aborigines) and therefore not worthy of consideration as ancestral ape-men. Interestingly enough, Dubois did not report these skulls to the scientific world until 1922.

In November 1890, at Kedungbrubus, Dubois made another find—a fossil jaw with part of a tooth root embedded in it. In a preliminary report, he judged it to be human (von Koenigswald 1956, p. 31). This specimen was not fully described until 1924, at which time Dubois designated it *Pithecanthropus*.

7.1.2 The Discoveries at Trinil

During the dry season of 1891, Dubois conducted excavations on the bank of the Solo River in central Java, near the village of Trinil. His laborers took out many fossil animal bones. In September, they turned up a particularly interesting item—a primate tooth, apparently a third upper right molar, or wisdom tooth. Dubois, believing he had come upon the remains of an extinct giant chimpanzee, ordered his laborers to concentrate their work around the place where the tooth had turned up. In October, they found what appeared to be a turtle shell. But when Dubois inspected it, he saw it was actually the top part of a cranium (Figure 7.1), heavily fossilized and having the same color as the volcanic soil. The fragment's most distinctive feature was the large, protruding ridge over the eye sockets, leading Dubois to suspect the cranium had belonged to an ape. The onset of the rainy season then brought an end to the year's digging. In a report published in the government mining bulletin, Dubois said about the tooth and skullcap, "That both specimens come from a great manlike ape was at once clear" (Time-Life 1973, p. 40). There was no suggestion that the fossils belonged to a creature transitional to humans. The term "manlike" (or "anthropoid") is widely used with reference to modern apes such as chimpanzees, gorillas, and orangutans, although these are not considered ancestral to human beings.

In August 1892, Dubois returned to Trinil and found there—among bones of deer, rhinoceroses, hyenas, crocodiles, pigs, tigers, and extinct elephants—a fossilized humanlike femur (thighbone). This femur (Figure 7.2) was found about 45 feet from where the skullcap and molar were dug up. Later another molar was found about 10 feet from the skullcap. Dubois believed the molars, skull, and femur all came from the same animal, which he still considered to be an extinct giant chimpanzee (von Koenigswald 1956, p. 31).

Figure 7.1. *Pithecanthropus* skullcap discovered by Eugene Dubois in 1891 in Java (Wendt 1972, p. 167).

The British researcher Richard Carrington (1963, p. 84) stated in his book *A Million Years of Man*: "Dubois was at first inclined to regard his skull cap and teeth as belonging to a chimpanzee, in spite of the fact that there is no known evidence that this ape or any of its ancestors ever lived in Asia.

Figure 7.2. Thighbone found by Eugene Dubois at Trinil, Java (Boule 1923, p. 100). Dubois attributed it to *Pithecanthropus erectus.*

But on reflection, and after corresponding with the great Ernst Haeckel, Professor of Zoology at the University of Jena, he declared them to belong to a creature which seemed admirably suited to the role of the 'missing link.'" We have not found any correspondence Dubois may have exchanged with Haeckel, but if further research were to turn it up, it would add considerably to our knowledge of the circumstances surrounding the birth of *Pithecanthropus erectus*. Obviously, both men had a substantial emotional and intellectual stake in finding an ape-man specimen. Haeckel, on hearing from Dubois of his discovery, telegraphed this message: "From the inventor of *Pithecanthropus* to his happy discoverer!" (Wendt 1972, p. 167).

It was only in 1894 that Dubois finally published a complete report of his discovery, titled "*Pithecanthropus erectus,* a Man-like Species of Transitional Anthropoid from Java." Therein he wrote: "*Pithecanthropus* is the transitional form which, in accordance with the doctrine of evolution, must have existed between man and the anthropoids" (von Koenigswald 1956, p. 31). *Pithecanthropus erectus,* we should carefully note, had itself undergone an evolutionary transition within the mind of Dubois, from fossil chimpanzee to transitional anthropoid.

What factors, other than Haeckel's influence, led Dubois to consider his specimen transitional between fossil apes and modern humans? Dubois found that the volume of the *Pithecanthropus* skull was in the range of 800–1000 cubic centimeters. Modern apes average 500 cubic centimeters, while modern human skulls average 1400 cubic centimeters, thus placing the Trinil skull midway between them. To Dubois, this indicated an evolutionary relationship. But logically speaking, one could have creatures with different sizes of brains without having to posit an evolutionary progression from small to large. Furthermore, in the Pleistocene many mammalian species were represented by forms much larger than today's. Thus the *Pithecanthropus* skull might belong not to a transitional anthropoid but to an exceptionally large Middle Pleistocene gibbon, with a skull bigger than that of modern gibbons.

Today, anthropologists still routinely describe an evolutionary progression of hominid skulls, increasing in size with the passage of time—from Early Pleistocene *Australopithecus* (first discovered in 1924), to Middle Pleistocene Java man (now known as *Homo erectus*), to Late Pleistocene *Homo sapiens sapiens*. But the sequence is preserved only at the cost of eliminating skulls that disrupt it. For example, the Castenedolo skull, discussed in Chapter 6, is older than that of Java man but is larger in cranial capacity. In fact, it is fully human in size and morphology. Even one such exception is sufficient to invalidate the whole proposed evolutionary sequence.

Dubois observed that although the Trinil skull was very apelike in some of its features, such as the prominent brow ridges, the thighbone was almost human. This indicated that *Pithecanthropus* had walked upright, hence the species designation *erectus*. It is important, however, to keep in mind that the femur of *Pithecanthropus erectus* was found fully 45 feet from the place where the skull was unearthed, in a stratum containing hundreds of other animal bones. This circumstance makes doubtful the claim that both the thighbone and the skull actually belonged to the same creature or even the same species.

7.1.3 Reports Reach Europe

When Dubois's reports began reaching Europe, they received much attention. In *Meeting Prehistoric Man,* von Koenigswald (1956, p. 26) commented on Java man's significance: "Dubois's find came at just the right moment: at a time when the conflict around Darwinism was at its height. For the scientific world it constituted the first concrete proof that man is subject not only to biological but also to paleontological laws." The discoverer of Lucy, Donald C. Johanson, in describing the expectant mood of scientists in the late nineteenth century, wrote: "If the theory of evolution had any validity whatsoever, then human fossils would have to reveal an increasing retreat toward primitiveness as one tracked them deeper into time" (Johanson and Edey 1981, p. 30). *Pithecanthropus erectus* appeared to amply satisfy this requirement, and even today, it is advertised (under the name *Homo erectus*) as a critical piece of evidence confirming the theory of evolution.

Haeckel, of course, was among those celebrating *Pithecanthropus* as the strongest proof to date of human evolution. "Now the state of affairs in this great battle for truth has been radically altered by Eugene Dubois's discovery of the fossil *Pithecanthropus erectus,*" proclaimed the triumphant Haeckel. "He has actually provided us with the bones of the ape-man I had postulated. This find is more important to anthropology than the much-lauded discovery of the X-ray was to physics" (Wendt 1972, p. 167). Haeckel would also state that Java man "was truly a Pliocene remainder of that famous group of the higher Catarrhines

[Old World apes], which were the pithecoid ancestors of man. He is indeed the long-searched-for Missing Link" (Bowden 1977, p. 128). There is an almost religious tone of prophecy and fulfillment in Haeckel's remarks. But Haeckel had a history of overstating physiological evidence to support the doctrine of evolution; an academic court at the University of Jena once found him guilty of falsifying drawings of embryos of various animals in order to demonstrate his particular view of the origin of species (Section 1.3).

7.1.4 Dubois Journeys to Europe with Java Man

In 1895, Dubois decided to return to Europe to display his *Pithecanthropus* to what he was certain would be an admiring and supportive audience of scientists. Taking 215 cases of other fossils, he boarded a ship along with his family. During a storm at sea, Dubois was especially concerned about his prized *Pithecanthropus erectus* specimens. Standing with the box containing *Pithecanthropus,* Dubois said to his wife, "If something happens, you're to take care of the children. I've got to look after this" (Time-Life 1973, p. 44).

Soon after arriving in Europe, Dubois exhibited his specimens and presented reports at the Third International Congress of Zoology at Leyden, Holland. Although some of the scientists present at the Congress were, like Haeckel, anxious to support the discovery as a fossil ape-man, others thought it merely an ape, while still others challenged the idea that the bones belonged to the same individual.

Dubois exhibited his treasured bones at Paris, London, and Berlin. In December of 1895, experts from around the world gathered at the Berlin Society for Anthropology, Ethnology, and Prehistory to pass judgement on Dubois's *Pithecanthropus* specimens. The president of the Society, Dr. Virchow, refused to chair the meeting. In the controversy-ridden discussion that followed, the Swiss anatomist Kollman said the creature was an ape. Virchow himself said that the femur was human, and further stated: "The skull has a deep suture between the low vault and the upper edge of the orbits. Such a suture is found only in apes, not in man. Thus the skull must belong to an ape. In my opinion this creature was an animal, a giant gibbon, in fact. The thigh-bone has not the slightest connection with the skull" (Wendt 1972, pp. 167–168). This opinion contrasted strikingly with that of Haeckel and others, who remained convinced that Dubois's Java man was a genuine human ancestor.

As Dubois traveled from city to city, carrying his *Pithecanthropus* fossils with him, controversy continued. Some were suspicious because the discoveries had been made in Java without any opportunity for confirmation by other scientists on the scene. Nevertheless, Dubois repeatedly defended his ape-man interpretation of the fossils. For example, when Sir Arthur Keith of Britain maintained that

Pithecanthropus erectus was actually just a somewhat primitive human, Dubois personally brought his fossils for Keith to examine, but even after seeing them Keith maintained his dissent (Goodman 1982, p. 60).

Dubois carried his bones around with him on his paleontological pilgrimage in a battered suitcase. Once, in Paris, he went to show the bones to Leonce Pierre Manouvrier, a noted French anthropologist. In Manouvrier's laboratory, the two talked till midnight and then went to a restaurant. Upon leaving, Dubois realized that he had forgotten his suitcase. Rushing back to the restaurant, Dubois asked desperately, "Where is *Pithecanthropus!*" It turned out that a waiter had the suitcase. Dubois hurriedly opened it, assuring himself that the fossils were still inside. Manouvrier suggested Dubois sleep that night with the bones under his pillow (Time-Life 1973, p. 45).

Dubois and Manouvrier attempted a reconstruction of the whole Java man skull, including the facial region, for which no bones were actually discovered. It is apparent that the entire jaw and facial structure were simply imagined (Boule 1923, p. 105). Some authorities thought Dubois and Manouvrier had reconstructed the skull improperly, making the cranial curve appear too low. A reconstruction that featured a higher skull profile made *Pithecanthropus* appear much more human (von Koenigswald 1956). Another imaginative reconstruction of the Java man skull was attempted by J. H. W. McGregor (Osborn 1916, p. 79). About a full-scale statue of Java man, Boule (1923, p. 105) stated: "Dubois ventured still further in the realm of imagination when he exhibited at the International Exhibition of 1900, in the Dutch Indies pavilion, a painted model of *Pithecanthropus* as he appeared in life."

In light of the incompleteness of the Java man skeletal remains and the doubtful circumstances of their discovery, it is amazing numerous scientists accepted the hypothetical *Pithecanthropus erectus* and wrote so many books about him.

7.1.5 The Selenka Expedition

To resolve some of the questions surrounding the *Pithecanthropus* fossils and their discovery, Emil Selenka, professor of zoology at Munich University in Germany, prepared a full-fledged expedition to Java, but he died before it departed. His wife, Professor Lenore Selenka, took over the effort and conducted excavations at Trinil in the years 1907–1908, employing 75 laborers to hunt for more *Pithecanthropus erectus* fossils. Altogether, Selenka's team of geologists and paleontologists sent back to Europe 43 boxes of fossils, but they included not a single new fragment of *Pithecanthropus*. Sir Arthur Keith (1911) reviewed the results of the Selenka expedition in the journal *Nature* and reported that the geological stratification at Trinil was unclear. Three of the geologists with the expedition thought the deposits were Pleistocene, perhaps recent, but two other experts agreed with Dubois that they might be Pliocene. As far as the bones themselves were concerned, their age was uncertain. Some of the geologists believed volcanic activity could have caused their fossilization, which therefore was not a sure sign of great age. The bones might have been recent, and might have been mixed in with older fossils by floods. The report mentioned a flood that occurred in Java in 1909, sending mud slides down the volcanic mountainsides, killing 500 persons and sweeping away entire villages. Perhaps even more troubling was the discovery in the Trinil strata of signs of a human presence—splintered animal bones, charcoal, and foundations of hearths. Signs like this led Lenore Selenka to conclude that humans and *Pithecanthropus erectus* were contemporary (Bowden 1977, pp. 134–135). The implications of all this for an evolutionary interpretation of Dubois's *Pithecanthropus* specimens were, and still are, unsettling.

Furthermore George Grant MacCurdy, a Yale professor of anthropology, wrote in his book *Human Origins* (1924a, p. 316): "The Selenka expedition of 1907–1908 . . . secured a tooth which is said by Walkoff to be definitely human. It is a third molar from a neighboring stream bed and from deposits older (Pliocene) than those in which *Pithecanthropus erectus* was found. Should this tooth prove to be human, *Pithecanthropus* could no longer be regarded as a precursor of man. Instead it would simply give us the cross section of a different limb of the primate tree whose branches now represent the various types of Hominidae." The beds referred to by MacCurdy as being older than the *Pithecanthropus erectus* deposits might be the Djetis Beds of the Putjangan formation, now placed in the Early Pleistocene or in the early Middle Pleistocene (Section 7.5.1).

In the aftermath of the Selenka expedition, tourists began coming to Java to look at the place where Java man had been discovered. They found the site littered with hundreds of beer bottles left by the thirsty scientists. As might be expected, many of the pilgrims were hoping they might stumble upon a *Pithecanthropus* bone. The local residents, who would find all kinds of bones washed out of the ground after floods, obliged them by selling them assorted pieces of skeletons. On December 27, 1926, a newspaper in Batavia announced that Dr. C. E. J. Heberlein had found at Trinil a new skull of *Pithecanthropus.* But it turned out to be a large ball-like joint from the leg bone of a fossil elephant.

7.1.6 Dubois Withdraws from the Battle

Meanwhile, the status of Dubois's ape-man remained somewhat controversial. Surveying the range of opinion about *Pithecanthropus,* Berlin zoologist Wilhelm Dames gathered statements from 25 scientists: three said *Pithecanthropus* was an ape, five said it was human, six said it was an ape-man, six said it was a missing link, and two said it was a link between the missing link and man. Virchow had said: "All I can do is warn against drawing decisive conclusions from these few pieces of bone about the greatest question facing us in the study of our creation. *Pithecanthropus* will remain doubtful as a transitional form until someone can demonstrate how this transition, which to me is conceivable only in my dreams, actually came true" (Wendt 1972, p. 169).

But although Virchow and others maintained their doubts, many scientists followed Haeckel in hailing Java man as stunning proof of Darwin's theory. Some used Java man to discredit evidence for a fully human presence in the Tertiary. As we learned in Section 5.5.13, W. H. Holmes (1899, p. 470) dismissed discoveries of stone tools in the Tertiary auriferous gravels of California because they "implied a human race older by at least one-half than *Pithecanthropus erectus* of Dubois, which may be regarded as an incipient form of human creature only."

At a certain point, Dubois became completely disappointed with the mixed reception the scientific community gave to his *Pithecanthropus*. He stopped showing his specimens. Some say that he kept them for some time beneath the floorboards in his home. In any case, they remained hidden from view for some 25 years, until 1932.

During and after the period of withdrawal, the controversies concerning *Pithecanthropus* continued. Marcellin Boule, director of the Institute of Human Paleontology in Paris, reported (1923, p. 96), as had other scientists, that the layer in which the *Pithecanthropus* skullcap and femur were said to have been found contained numerous fossil bones of fish, reptiles, and mammals. Why, therefore, should anyone believe the skullcap and femur came from the same individual or even the same species? Boule, like Virchow (Section 7.1.4), stated that the femur was identical to that of a modern human whereas the skullcap resembled that of an ape, possibly a large gibbon. Dr. F. Weidenreich, honorary director of the Cenozoic Research Laboratory at Peiping Union Medical College, also stated (1941, p. 70) that there was no justification for attributing the femur and the skullcap to the same individual. The femur, Weidenreich said, was very similar to that of a modern human, and its original position in the strata was not securely established. Modern researchers have employed chemical dating techniques in order to determine whether or not the original *Pithecanthropus* skull and femur were both contemporary with the Middle Pleistocene Trinil fauna, but the results were inconclusive (Section 7.5.2).

7.1.7 More Femurs

The belated revelation that more femurs had been discovered in Java further complicated the issue. In 1932, Dr. Bernsen and Eugene Dubois recovered three femurs from a box of fossil mammalian bones in the Leiden Museum in the Netherlands. The box contained specimens said to have been excavated in 1900 by Dubois's assistant, Mr. Kriele, from the same Trinil deposits on the left bank of the Solo river that had yielded Dubois's first Java man finds. Dr. Bernsen died very shortly thereafter, without providing further information about the details of this museum discovery.

Dubois (1932, p. 719) stated that he was not present when the femurs were taken out by Kriele. Therefore the exact location of the femurs in the excavation, which was 75 meters (246 feet) long by 6–14 meters (20–46 feet) wide, was unknown to him. According to standard paleontological procedures, this uncertainty greatly reduces the value of the bones as evidence of any sort. Nevertheless, as we shall see, authorities later assigned these femurs to a particular stratum without mentioning the dubious circumstances of their discovery in boxes of fossils over 30 years after they were originally excavated (Section 7.6). Moreover, G. H. R. von Koenigswald (1956, p. 36) reported that Dubois's collection "comprised finds from various sites and various ages, which are very inadequately distinguished, because some of the labels got lost."

Eugene Dubois (1934, p. 139) reported that in December of 1932 he discovered a fragment of a fourth new femur in the same collection in which the others had been found. Once again, Dubois pointed out that the original place of excavation was unknown. In August 1935, a museum employee named Van der Steen handed Dubois yet another femur fragment from the collection, but Dubois said that this bone was "certainly not from Trinil but from another part of the Kendeng region." Dubois (1935, p. 850) speculated that it may have been found at Kedungbrubus, but he admitted that he was not really sure.

The existence of these additional femurs has important implications for the original *Pithecanthropus* skull and femur found by Dubois in the 1890s. As we have seen, the fact that the apelike skull and humanlike femur were found at a great distance from each other is sufficient to suggest that they belonged not to one ape-man creature but to two different creatures, one apelike and the other fully human. In response, one might argue, as Dubois's supporters did, that the odds of an apelike creature leaving a skull and no leg bones and a man leaving a leg bone and no skull so close to each other were remote. If it were not possible to prove there were two creatures, it would be best to assign the bones to one creature. Dubois suggested that the bones were found separated because the *Pithecanthropus* had been dismembered by a crocodile (Bowden 1977, p. 127). But if you throw in more humanlike femurs, that argument loses a great deal of its force. Where were the other skulls? Were they apelike skulls, like the one found? And what about the skull that was found? Does it really go with the femur that was found 45 feet away? Or does it belong with one of the other femurs that later turned up? Or maybe with a femur of an entirely different sort?

7.1.8 Are the Trinil Femurs Human?

M. H. Day and T. I. Molleson (1973, p. 151) concluded that "the gross anatomy, radiological [X-ray] anatomy, and microscopical anatomy of the Trinil femora does not distinguish them significantly from modern human femora." They also said that *Homo erectus* femurs from China (Zhoukoudian) and Africa (Olduvai Hominid 28) "are anatomically similar, and distinct from those of Trinil" (Day and Molleson 1973, p. 152).

In 1984, Richard Leakey and three American scientists discovered an almost complete skeleton of *Homo erectus* in Kenya. Examining the leg bones, these scientists found that the femurs differed substantially from those of modern human beings: "The biomechanical neck length of 85 mm is well over 3 standard deviations from the mean of a sample of *H. sapiens.* As well as having a long femoral neck, the neck-shaft angle is very small at 110 degrees, being 5 standard deviations from the mean of the same *H. sapiens* population" (Brown *et al.* 1985, p. 791). About the Java discoveries, the authors stated: "From Trinil, Indonesia, there are several fragmentary and one complete (but pathological) femora. Despite the fact that it was these specimens that led to the species name [*Pithecanthropus erectus*], there are doubts as to whether they are *H. erectus* with the most recent consensus being that they probably are not" (Brown *et al.* 1985, p. 789).

In summary, Brown *et al.* (1985) and Day and Molleson (1973) agreed that the Trinil femurs were not like those of *Homo erectus,* while Day and Molleson (1973) said the Trinil femurs were like those of modern *Homo sapiens.*

What is to be made of these revelations? The Java thighbones have traditionally been taken as evidence of an ape-man (*Pithecanthropus erectus,* now called *Homo erectus*) existing around 800,000 years ago in the Middle Pleistocene. Should we now accept them as evidence for anatomically modern humans existing 800,000 years ago? Perhaps wisely, Brown and his associates offered in their report no suggestions about the real age of the human femurs found at Trinil. There is safety in silence when confronting disconcerting paleontological anomalies.

Some have said that the femurs were mixed in from higher levels. Of course, if one insists that the humanlike Trinil femurs were mixed in from higher levels, then why not the *Pithecanthropus* skull as well? That would eliminate entirely the original Java man find, long advertised as solid proof of human evolution.

7.1.9 Dubois Backs Away from His Original Claims

Late in his life, Dubois concluded that the skullcap of his beloved *Pithecan-thropus* belonged to a large gibbon, an ape not thought by evolutionists to be closely related to humans (Gowlett 1984, p. 17). But the heretofore skeptical scientific community was not about to say good-bye to Java man, for by this time *Pithecanthropus* was firmly entrenched in the ancestry of modern *Homo sapiens.* Dubois's denials were dismissed as the whims of a cantankerous old man. If anything, the scientific community wanted to remove any remaining doubts about the nature and authenticity of Java man. This, it was hoped, would fortify the whole concept of Darwinian evolution, of which human evolution was the most highly publicized and controversial aspect.

Despite the doubts about the Trinil find expressed by Dubois himself in his later years, and by other scientists from the 1890s to the present, public presentations remain unchanged. Visitors to museums around the world still find models of the Trinil skullcap and femur portrayed as belonging to the same Middle Pleistocene *Homo erectus* individual. In 1984, the much-advertised *Ancestors* exhibit, at the Museum of Natural History in New York brought together from around the world the major fossil evidence for human evolution, including prominently displayed casts of the Trinil skullcap and femur.

7.2 THE HEIDELBERG JAW

In addition to Dubois's Java man discoveries, further evidence relating to human evolution turned up in the form of the Heidelberg jaw. On October 21, 1907, Daniel Hartmann, a workman at a sand pit at Mauer, near Heidelberg, Germany, discovered a large jawbone at the bottom of the excavation, at a depth of 82 feet. The workmen were on the lookout for bones, and many other nonhuman fossils had already been found there and turned over to the geology department at the nearby University of Heidelberg. The workman then brought the jaw (Figure 7.3) over to J. Rösch, the owner of the pit, who sent a message to Dr. Otto Schoetensack: "For twenty long years you have sought some trace of early man in my pit . . . yesterday we found it. A lower jaw belonging to early man has been found on the floor of the pit, in a very good state of preservation" (Wendt 1972, p. 161).

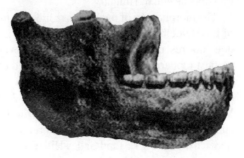

Professor Schoetensack designated the creature *Homo heidelbergensis,* dating it using the accompanying fossils to the Günz-Mindel interglacial period. David Pilbeam (1972, p. 169) said: "It appears to date from the Mindel

Figure 7.3. The Heidelberg mandible, discovered in 1907 at Mauer, near Heidelberg, Germany (Osborn 1916, p. 98).

glaciation, and its age is somewhere between 250,000 and 450,000 years."

The German anthropologist Johannes Ranke, an opponent of evolution, wrote in the 1920s that the Heidelberg jaw belonged to a representative of *Homo sapiens* rather than an apelike predecessor (Wendt 1972, p. 162).

Even today, the Heidelberg jaw remains somewhat of a morphological mystery. The thickness of the mandible and the apparent lack of a chin are features common in *Homo erectus*. But mandibles of some modern Australian aboriginals are also massive compared to jaws of modern Europeans and have chins that are less well developed (Le Gros Clark and Campbell 1978, p. 96, figure 11).

According to Frank E. Poirier (1977, p. 213), the teeth in the Heidelberg jaw are closer in size to those of modern *Homo sapiens* than those of Asian *Homo erectus* (Java man and Peking man). T. W. Phenice of Michigan State University wrote (1972, p. 64): "the teeth are remarkably like those of modern man in almost every respect, including size and cusp patterns."

Modern opinion thus confirms Ranke, who wrote in 1922: "The teeth are typically human; the canines do not project above the level of the other teeth, and the third molar, which in primitive races of men—for instance often in the aboriginal Australians—is similar in size to or even larger than the second, is smaller in the Heidelberg jaw, just as in our more advanced races today" (Wendt 1972, p. 162).

Many *Homo erectus* jaws are characterized by projecting canines and a diastema, a gap in the teeth that accommodates the tip of a projecting canine. The fact that these features were absent in the Heidelberg jaw, and other considerations, led Poirier (1977, p. 213) to question: "Is Heidelberg a representative of *Homo erectus* or a primitive member of the species *H. sapiens*?"

The Heidelberg jaw is one of the few European fossils generally attributed to *Homo erectus*. Another is the Vértesszöllös occipital fragment, from a Middle Pleistocene site in Hungary.

The morphology of the Vértesszöllös occipital is even more puzzling than that of the Heidelberg jaw. David Pilbeam (1972, p. 169) wrote: "the occipital bone does not resemble that of *H. erectus*, or even archaic man, but instead that of earliest modern man. Such forms are dated elsewhere as no older than 100,000 years." Pilbeam believed the Vértesszöllös occipital to be approximately the same age as the Heidelberg jaw, between 250,000 and 450,000 years old. If the Vértesszöllös occipital is modern in form, it helps confirm the genuineness of anatomically modern human skeletal remains of similar age found in England at Ipswich (Section 6.1.3) and Galley Hill (6.1.2.1).

Returning to the Heidelberg jaw, we note that the circumstances of discovery were less than perfect. If an anatomically modern human jaw had been found by a workman in the same sand pit, it would have been subjected to merciless criticism and judged recent. After all, no scientists were present at the moment of discovery. But the Heidelberg jaw, because it fits, however imperfectly, within the bounds of evolutionary expectations, has been granted a dispensation.

7.3 FURTHER JAVA MAN DISCOVERIES BY VON KOENIGSWALD

In 1929, another ancient human ancestor was discovered, this time in China. Eventually, scientists would group Java man, Heidelberg man, and Peking man together as examples of *Homo erectus,* the direct ancestor of *Homo sapiens.* But initially, the common features and evolutionary status of the Indonesian, Chinese, and German fossils were not obvious, and paleoanthropologists felt it particularly necessary to clarify the status of Java man.

In 1930, Gustav Heinrich Ralph von Koenigswald of the Geological Survey of the Netherlands East Indies was dispatched to Java. In his book *Meeting Prehistoric Man,* von Koenigswald (1956, p. 55) wrote: "despite the discovery of Pekin man, it remained necessary to find a further *Pithecanthropus* sufficiently complete to prove the human character of this disputed fossil."

Upon personally examining Dubois's *Pithecanthropus* skull, von Koenigswald (1956, p. 33) had noted: "It is no more than a calvarium from which the most important parts are missing—the temporal region, which is essential to an accurate assessment of its nature." Dubois had attributed three teeth (two large molars and one premolar) to his Java man specimen, but von Koenigswald believed that only one of them belonged to *Pithecanthropus.* The others were apparently from the jaw of an orangutan. Von Koenigswald (1956, p. 34) concluded: "It therefore becomes manifest on what shaky ground Dubois erected his hypothetical building, and we can only wonder at the boldness and tenacity with which he defended his *Pithecanthropus.*"

Von Koenigswald, like Dubois, was fascinated by fossils as a youth, having also gathered a collection of ancient bones, teeth, and shells. He managed to put himself through a university education in Germany during the troubled years following the First World War, and upon graduation obtained a position as a museum assistant in Munich. In 1930, he took an opportunity to join the Geological Survey in the Dutch East Indies, where his finds would eventually gain him a reputation as one of the twentieth century's greatest fossil hunters.

7.3.1 The Ngandong Fossils

Von Koenigswald arrived in Java in January 1931. In August of that same year, one of von Koenigswald's colleagues, the Dutch archeologist ter Haar, was surveying the Kendeng Hills region near Trinil. He set up camp at the *kampong* of Ngandong on the River Solo. One evening at sundown, while going to the river for a bath, he happened upon a terrace of old river gravels, from which he pulled out a buffalo skull and some other bones. A trained native collector, or *mantri*, named Samsi, who was employed by the Geological Survey, was given the job of excavating the site. Samsi dutifully sent boxes of fossils back to the city of Bandung, where they were examined by Dr. W. F. F. Oppenoorth, the head of the Geological Survey. On September 15, 1931, Oppenoorth examined a specimen labeled by Samsi as a tiger skull and determined it was actually the major portion of a humanlike braincase. More fragments turned up in the boxes of bones arriving in Bandung, and others were turned over to Oppenoorth at the Ngandong site. Von Koenigswald (1956, pp. 65–77) classified the Solo specimens discovered in the fall of 1931 as a Javanese variety of Neanderthal, appearing later in time than *Pithecanthropus erectus*.

7.3.2 First Find at Sangiran

Gradually, the history of human ancestors in Java seemed to be clearing up, but more work was needed. In 1934, von Koenigswald journeyed to Sangiran, a site west of Trinil on the Solo River. He took with him several Javanese workers, including his trained collector, Atma, who also served as von Koenigswald's cook and laundryman in the field.

Von Koenigswald (1956, p. 88) wrote: "There was great rejoicing in the *kampong* over our arrival. The men gathered all the jaws and teeth they could lay hands on and offered to sell them to us. Even the women and girls, who are generally so retiring, took part." When one considers that most of the finds attributed to von Koenigswald were actually made by local villagers or native collectors, who were paid by the piece in most cases, the scene described cannot but cause some degree of uneasiness.

At the end of 1935, because of funding cutbacks in the midst of the worldwide economic depression, von Koenigswald's position with Java's Geological Survey was terminated. Undeterred, von Koenigswald kept his servant Atma and others working at Sangiran, financing their activities with contributions from his wife and colleagues in Java.

Uncovered during this period was what appeared to be the fossilized right half of the upper jaw of an adult *Pithecanthropus erectus*. This fossil jaw from Sangiran is designated S1a in Table 7.2 on p. 498. An examination of many reports by von Koenigswald has failed to turn up any description by him of exactly how this specimen was found. But the British researcher K. P. Oakley and his associates stated (Oakley *et al.* 1975, p. 108) that the fossil was found in 1936 on the surface of exposed lake deposits east of Kalijoso in central Java by collectors employed by von Koenigswald.

Considering that Sla was a surface find, it is surprising that modern authorities (Oakley *et al.* 1975, p. 109) have concluded that this fossil is of the same Middle to Early Pleistocene age as the exposed Djetis beds where it was found (see Section 7.5.1 for more on the age of the Djetis beds). The fact that the Sla upper jaw fragment is fossilized is not a guarantee of any great age, because there is evidence that bone can be fossilized in periods as short as a few hundred years.

It might be argued that in the Sangiran region there are no strata younger than those of the Middle Pleistocene Kabuh formation, which lie over the Djetis beds of the Putjangan formation. And therefore the jaw should be at least Middle Pleistocene in age. But the Sla jaw was said to have been found by paid collectors, who may have imported it from almost anywhere.

As we have seen in our discussions of anomalous discoveries made in Argentina and elsewhere, professional scientists sometimes question the credibility and honesty of paid collectors. Boman (Section 5.2.5), for example, said such persons are always suspect. If so, that judgement should also automatically apply to the collectors who found the Sla jaw and other Java *Homo erectus* fossils, all of which are completely accepted by the scientific community.

A more reasonable approach would be to separately evaluate the qualification of the collectors involved in particular discoveries. Lorenzo Parodi, the collector who worked for C. Ameghino at Miramar, Argentina, it may be recalled, left his discoveries in place for scientists to photograph *in situ* and excavate, and was not known to have engaged in any kind of deception over the course of a long career.

As we shall see in connection with the subsequent hominid finds reported by von Koenigswald, his Javanese collectors were often implicated in questionable behavior. Therefore, we do not really know the exact place of discovery of the Sla jaw reportedly found at Kalijoso in 1936.

In addition, we must keep in mind that scientists themselves are not always honest. We shall explore this subject in some detail in Chapter 8, which deals with the infamous Piltdown incident.

At this point, an anthropologist might observe that the Sla jaw fragment exhibits the features of *Homo erectus*, as *Pithecanthropus erectus* is now known. Hence it must have been deposited at least several hundred thousand years ago, despite the fact that it was found on the surface. But this is reasoning from theory to fact, not from fact to theory. What if there were existing in geologically recent times, or even today, a rare species of hominid having physical features similar to those of *Homo erectus*? In that case one could not automatically assign a date to a given bone based on the physical features of that bone. In Chapter 10 can be found evidence suggesting that a creature like *Homo erectus* has lived in recent times and in fact may be alive today.

As far as chemical dating of finds such as the Sla jaw is concerned, we discuss this in Sections 7.5.2 and 7.5.3.

We do not insist that the S1a jaw is recent. It might very well be several hundred thousand years old, and we would have no problem with that. What we do object to is the uneven application of standards for evaluating paleoanthropological evidence. In affirming the contemporaneity of the S1a jaw with the Djetis beds, scientists have applied such standards more leniently than they have in denying great antiquity to the anomalous finds discussed in previous chapters.

7.3.3 The Role of the Carnegie Institution

During the difficult year of 1936, in the course of which the fossil jaw discussed above was uncovered, the unemployed von Koenigswald received a remarkable visitor—Pierre Teilhard de Chardin, whom von Koenigswald himself had invited to come and inspect his discoveries in Java. Teilhard de Chardin, a world-famous archeologist and Jesuit priest, had been working in Peking (now Beijing), where he had participated in the Peking man excavations at Choukoutien (now Zhoukoudian). One reason for his coming to Java was that he desired to establish a link between Peking man and Java man.

During his visit to Java, Teilhard de Chardin advised von Koenigswald to write to John C. Merriam, the president of the Carnegie Institution (Cuenot 1958). Von Koenigswald did so, informing Merriam that he was on the verge of making important new *Pithecanthropus* finds. Teilhard de Chardin, who was personally acquainted with most of the leading paleoanthropologists of the day, also wrote in support. Teilhard de Chardin's biographer Cuenot (1958, p. 163) stated: "One has the impression of a vast web, of which Teilhard held in parts the threads, where he served as a liaison agent, or better still, as chief of staff, able, like a magician, to make American money flow, or at least to channel it for the greatest good of paleontology."

Merriam responded positively to von Koenigswald's letter, inviting him to come to Philadelphia in March 1937 to attend the Symposium on Early Man, sponsored by the Carnegie Institution. There von Koenigswald joined many of the world's leading scientists working in the field of human prehistory.

One of the central purposes of the meeting was to form an executive committee for the Carnegie Institution's financing of paleoanthropological research. Suddenly, the impoverished von Koenigswald found himself appointed a research associate of the Carnegie Institution and in possession of a large budget.

Considering the critical role played by private foundations in the financing of research in human evolution, it might be valuable at this point to further consider the motives of the foundations and their executives. The Carnegie Institution and John C. Merriam provide an excellent case study. In Chapter 9, we will examine the Rockefeller Foundation's role in financing the excavation of Peking man.

The Carnegie Institution was founded in January 1902 in Washington, D.C., and a revised charter approved by Congress became effective in 1904. The Institution was governed by a board of 24 trustees, with an executive committee meeting throughout the year, and was organized into 12 departments of scientific investigation, including experimental evolution. The Institution also funded the Mt. Wilson Observatory, where the first systematic research leading to the idea that we live in an expanding universe was conducted. Thus the Carnegie Institution was actively involved in two areas, namely evolution and the big bang universe, that lie at the heart of the scientific cosmological vision that has replaced earlier religiously inspired cosmologies.

It is significant that for Andrew Carnegie and others like him, the impulse to charity, traditionally directed toward social welfare, religion, hospitals, and general education, was now being channeled into scientific research, laboratories, and observatories. This reflected the dominant position that science and its world view, including evolution, were coming to occupy in society, particularly within the minds of its wealthiest and most influential members, many of whom saw science as the best hope for human progress.

John C. Merriam, president of the Carnegie Institution, believed that science had "contributed very largely to the building of basic philosophies and beliefs" (1938, p. 2531), and his support for von Koenigswald's fossil-hunting expeditions in Java should be seen in this context. A foundation like the Carnegie Institution had the means to use science to influence philosophy and belief by selectively funding certain areas of research and publicizing the results. "The number of matters which might be investigated is infinite," wrote Merriam (1938, p. 2507). "But it is expedient in each period to consider what questions may have largest use in furtherance of knowledge for the benefit to mankind at that particular time."

The question of human evolution satisfied this requirement. "Having spent a considerable part of my life in advancing studies on the history of life," said Merriam (1938, p. 2529), "I have been thoroughly saturated with the idea that evolution, or the principle of continuing growth and development, constitutes one of the most important truths obtained from all knowledge."

By training a paleontologist, Merriam was also by faith a Christian. But his Christianity definitely took a back seat to his science. "My first contact with science," Merriam (1938, pp. 2041–2042) recalled in a 1931 speech, "was when I came home from grammar school to report to my mother that the teacher had talked to us for fifteen minutes about the idea that the days of creation described in Genesis were long periods of creation and not the days of twenty-four hours. My mother and I held a consultation—she being a Scotch Presbyterian—and agreed that this was rank heresy. But a seed had been sown. I have been backing away from that position through subsequent decades. I realize now that the elements of science, so far as creation is concerned, represent the uncontaminated and unmodified record of what the Creator did."

Having dispensed with scriptural accounts of creation, Merriam managed to turn Darwinian evolution into a kind of religion. At a convocation address at the George Washington University in 1924, Merriam (1938, p. 1956) said of evolution, "There is nothing contributing to the support of our lives in a spiritual sense that seems so clearly indispensable as that which makes us look forward to continuing growth or improvement."

He held that science would give man the opportunity to take on a godlike role in guiding that future development. "Research is the means by which man will assist in his own further evolution," said Merriam (1938, pp. 2541–2542) in a 1925 address to the Carnegie Institution's Board of Trustees. He went on to say: "I believe that if he [man] had open to him a choice between further evolution directed by some Being distant from us, which would merely carry him along with the current; or as an alternative could choose a situation in which that outside power would fix the laws and permit him to use them, man would say, 'I prefer to assume some responsibility in this scheme.'"

"According to the ancient story," Merriam continued, "man was driven from the Garden of Eden lest he might learn too much; he was banished so that he might become master of himself. A flaming sword was placed at the east gate, and he was ordered to work, to till the ground, until he could come to know the value of his strength. He is now learning to plough the fields about him, shaping his life in accordance with the laws of nature. In some distant age a book may be written in which it will be stated that man came at last to a stage where he returned to the Garden, and at the east gate seized the flaming sword, the sword that symbolized control, to carry it as a torch guiding him to the tree of life." Seizing the flaming sword and marching to take control of the tree of life? One wonders if there would be enough room in Eden for both God and a hard-charging scientific superachiever like Merriam.

7.3.4 Back to Java

Armed with Carnegie grant money, von Koenigswald returned to Java in June of 1937. Immediately upon his arrival, he hired hundreds of natives and sent them out in force to find more fossils.

Meanwhile, in the course of looking through baskets of fossils gathered at Sangiran during his absence, von Koenigswald came upon a large, fossilized, lower right jaw fragment (S1b in Table 7.2, p. 498). Von Koenigswald stated (1940a, p. 142) that the fossil had been lying on the surface at the time it was discovered. Von Koenigswald then asked his native collectors to specify the exact location at which it was discovered, and they informed him that it was found at a place where the Djetis beds of the Putjangan formation are exposed. Von Koenigswald searched this area, but he stated that he was unable to locate the exact spot at which this fossil was said to have been found.

Adhering to the S1b jaw fragment was a fine-grained conglomerate, the presence of which caused von Koenigswald (1937, p. 884) to conclude that the fossil had originally been embedded in the Kabuh formation, which lies above the Putjangan formation. It is in the early Middle Pleistocene Trinil beds of the Kabuh formation that Dubois reported he found the original *Pithecanthropus* specimens.

Two years later, in 1939, after the conglomerate surrounding the S1b jaw fragment had been removed in the Cenozoic Research Laboratory in Peking, it was observed that the fossil had fine cracks in it. Such cracks are typically reported on specimens that have been embedded in a clayey layer, such as one would find in the Putjangan formation. Also, this fossil was more heavily fossilized than most bones found in the Kabuh formation. In light of these new facts, von Koenigswald (1940a, p. 142) reversed himself and declared that the S1b jaw must have come from the Putjangan formation, considered early Middle Pleistocene or late Early Pleistocene (Section 7.5.1).

A radically different opinion was expressed by Dubois, who thought that the jaw belonged to *Homo soloensis* (the Javanese Neanderthal) and therefore was only about 100,000 years old (von Koenigswald 1956, p. 93). Von Koenigswald countered that such recent layers were not present at Sangiran.

But the fact remains that the jaw was said to have been discovered by native collectors on the surface at a location the collectors themselves could not clearly remember, and thus we do not know for sure where the jaw was originally situated. It is apparent from the above discussion that the actual age of the jaw is unknown.

At Modjokerto, in 1936, one of von Koenigswald's native collectors, Andojo, discovered the skullcap of a young hominid and labeled it an orangutan. On unpacking the specimen, von Koenigswald reported that the skull exhibited features that are typically human, and not those of an ape; yet its brain capacity was smaller than that of an anatomically modern human of corresponding age. H. DeTerra (1943, p. 443) stated that "since the facial part and the base of the skull are missing, its true phylogenetic rank is unknown." But today most paleoanthropologists believe several features of the skull indicate it belonged to a *Homo erectus* child.

Von Koenigswald, relying on Andojo's statement that the skull was dug up from a depth of 3 feet at Modjokerto, concluded that the skull was found embedded in the Putjangan layers. The real location of this skull was, however, known only by Andojo, because von Koenigswald was not present at the time of the discovery. During the course of most of the Sangiran finds, von Koenigswald remained at Bandung, about 200 miles away, although he would sometimes travel to the fossil beds after being notified of a discovery. Andojo's credibility is suspect, because as we shall see in the description of the next discovery, the Javanese collectors employed by von Koenigswald were sometimes driven, by a desire for extra profit, to engage in deception and trickery.

In the fall of 1937, one of von Koenigswald's collectors, Atma, mailed him a temporal bone that apparently belonged to a thick, fossilized, hominid cranium. This specimen (S2 in Table 7.2, p. 498) was said to have been discovered near the bank of a river named the Kali Tjemoro, at the point where it breaks through the sandstone of the Kabuh formation at Sangiran.

Von Koenigswald took the night train to central Java and arrived at the site the next morning. "We mobilized the maximum number of collectors," stated von Koenigswald (1956, pp. 95–96). "I had brought the fragment back with me, showed it round, and promised 10 cents for every additional piece belonging to the skull. That was a lot of money, for an ordinary tooth brought in only ½ cent or 1 cent. We had to keep the price so low because we were compelled to pay cash for every find; for when a Javanese has found three teeth he just won't collect any more until these three teeth have been sold. Consequently we were forced to buy an enormous mass of broken and worthless dental remains and throw them away in Bandung—if we had left them at Sangiran they would have been offered to us for sale again and again."

The highly motivated crew quickly turned up the desired skull fragments. Von Koenigswald (1947, p. 15) would later recall: "There, on the banks of a small river, nearly dry at that season, lay the fragments of a skull, washed out of the sandstones and conglomerates that contained the Trinil fauna. With a whole bunch of excited natives, we crept up the hillside, collecting every bone fragment we could discover. I had promised the sum of ten cents for every fragment belonging to that human skull. But I had underestimated the 'big-business' ability of my brown collectors. The result was terrible! Behind my back they broke the larger fragments into pieces in order to increase the number of sales! . . . We collected about 40 fragments, of which 30 belonged to the skull. . . . They formed a fine, nearly complete Pithecanthropus skullcap. Now, at last, we had him!"

How did von Koenigswald know that the fragments found on the surface of a hill really belonged, as he claimed, to the Middle Pleistocene Kabuh formation? Perhaps the native collectors found a skull elsewhere and broke it apart, sending one piece to von Koenigswald and scattering the rest by the banks of the Kali Tjemoro.

Von Koenigswald constructed a skull from the 30 fragments he had collected, calling it *Pithecanthropus II*, and sent a preliminary report to Dubois. The skull (S2 in Table 7.2, p. 498) was much more complete than the original skullcap found by Dubois at Trinil. Von Koenigswald (1956, pp. 97–99) had always thought that Dubois had reconstructed his *Pithecanthropus* skull with too low a profile, and believed the *Pithecanthropus* skull fragments he had just found allowed a more humanlike interpretation. Dubois, who by this time had concluded his original *Pithecanthropus* was merely a fossil ape (von Koenigswald 1956, p. 55), disagreed with von Koenigswald's reconstruction and published an accusation that he had indulged in fakery. He later retracted this indictment and said that the mistakes he saw in von Koenigswald's reconstruction were probably not deliberate.

But von Koenigswald's position was gaining support. Franz Weidenreich, supervisor of the Peking man excavations at Choukoutien, stated (1938, p. 378) in the prestigious journal *Nature* that von Koenigswald's new finds had definitely established *Pithecanthropus* as a human precursor and not a gibbon as claimed by Dubois.

Weidenreich journeyed to Java and participated in another discovery, known as *Pithecanthropus III* (S3 in Table 7.2, p. 498). This find consisted of many skull fragments, adding up to the right parietal bone, part of the left parietal bone, and a small piece of the occipital bone of a juvenile individual. Von Koenigswald stated (1940a, p. 102) that most of these fragments were found by his collectors in July 1938 in the southern sector of the Sangiran dome. A few were discovered by himself and Weidenreich in the course of their visit to the site in September 1938. Von Koenigswald (1940a) went on to describe the condition of these skull fragments. They were found on the surface, and they were extensively corroded. In fact, von Koenigswald wrote that these pieces of bone must have been lying around on the surface for a long time, because roots of grasses were penetrating a piece of one of the fragments. Despite this, von Koenigswald (1940a, p. 103; 1956, p. 101) and Le Gros Clark and Campbell (1978, p. 94) stated that *Pithecanthropus III* is from the Kabuh formation. But considering that this specimen was discovered on the surface, there is room to doubt this. In his 1956 book *Meeting Prehistoric Man,* von Koenigswald failed to mention that the *Pithecanthropus III* skull was found on the surface, thus misleading the reader into accepting it as strong evidence for a missing link in the Middle Pleistocene.

7.3.5 A Meeting in Peking

In January 1939, von Koenigswald and Weidenreich met at the Cenozoic Research Laboratory in Peking to directly compare fossils of *Pithecanthropus* and *Sinanthropus,* as Peking man was known in scientific circles. Peking man was represented by some fairly complete skulls with features thought to be markedly human. The humanlike nature of Peking man was further supported by the presence of crude stone implements and evidence of fire in the excavation at

Choukoutien. All of this indicated *Sinanthropus* was much more than an ape. At their meeting, von Koenigswald and Weidenreich agreed that *Pithecanthropus* and *Sinanthropus* were anatomically very closely related. So if Peking man was a distant ancestor of the human type, then so was Java man. Von Koenigswald (1956, pp. 47–48) wrote: "the cranial curve of Peking man was exactly similar to that of the disputed Javanese *Pithecanthropus*. Since there could be no doubt that the Peking man, despite all his primitive characteristics, was genuinely human, practically all Dubois's opponents were convinced by this new find that *Pithecanthropus,* too, must have been human."

Von Koenigswald's characterization of *Pithecanthropus* and *Sinanthropus* as "genuinely human" refers only to certain departures from apelike morphology in the direction of humanlike morphology. But such departures do not require one to conclude that modern humans descended from *Pithecanthropus* or *Sinanthropus.* Furthermore, one cannot rule out the possibility that humans of modern type existed contemporaneously with or previous to *Pithecanthropus* and *Sinanthropus.* As we have seen, there is much evidence demonstrating this latter possibility.

Concerning the alleged identity of *Pithecanthropus* and *Sinanthropus,* Dubois himself was not convinced. "The discovery of Peking man might have been expected to represent a great triumph for Dubois, who had up till then been exerting every ounce of his authority to convince the world that the disputed [Java man] fossil was human," stated von Koenigswald (1956, p. 55). "Curiously enough, this was not how Dubois saw it. Till the end of his life [in 1940] he refused to recognize any affinity between *Sinanthropus* and his *Pithecanthropus.* He described *Sinanthropus* as a degenerate Neanderthaler, and suddenly decided that his own find must be ascribed to a gibbon-like ape."

While in Peking, von Koenigswald received from his collector Rusman a new *Pithecanthropus* fossil, a thickly encrusted upper jaw. Later von Koenigswald's Javanese servants sent from the same site another piece of rock that appeared to match the broken piece of rock in which the jaw had been embedded. In this new piece of rock were found the fragments of the rear part of a cranium. Although the cranium had no direct connection with the upper jaw, both were attributed to the same individual, *Pithecanthropus IV* (von Koenigswald 1956, pp. 105–106; Oakley *et al.* 1975, p. 109).

Von Koenigswald said that the jaw was found in the upper layer of the Black Clay stratum of the Putjangan formation at Sangiran (Weidenreich 1945, p. 14). Although von Koenigswald reported on this find in five different publications (1939 pp. 926–929, with Weidenreich as coauthor; 1940a, p. 52; 1947, p. 48; 1949a, p. 92; 1956, pp. 105–111), in none of these did he state the exact location in the stratum of the jaw or the rear braincase. He did not state if the stratum was undisturbed or at what depth the fossils (S4 in Table 7.2, p. 498) were found.

Considering the importance of these discoveries, the reader is entitled to the detailed stratigraphic information that should have been provided.

One can simply imagine what might have happened if the discovery had been a fossil skull of the modern human type. Authorities such as Hrdlicka would have pointed out that it had been discovered by a native collector and not a trained scientist, that the exact location of its discovery was unknown, and that there was, therefore, sufficient reason to reject the find. But sloppiness that would be fatal to an anomalous find is easily tolerated in the case of a find that fits in nicely with accepted ideas about human evolution.

7.3.6 Weidenreich's Reconstruction

In 1945, Weidenreich used the S1b lower jaw, found in 1936 near Kalijoso, and the upper jaw and rear braincase of the so-called *Pithecanthropus IV* individual (S4) to put together his famous *Pithecanthropus robustus* reconstruction. This was surely a strange mix. Von Koenigswald and more recent authors such as Le Gros Clark and Campbell (1978) uncritically accepted that the rear braincase and upper jaw belonged to the same individual although no empirical data was ever brought forward to prove this. The S1b lower jaw was found at a different site. Furthermore, in Weidenreich's *Pithecanthropus robustus* reconstruction, there were no fossils for the front half of the cranium and the upper part of the face. Von Koenigswald (1949a, p. 92) concluded that Weidenreich modeled the facial part of the *Pithecanthropus robustus* skull after the Peking man fossils and the front half of the cranium after Dubois's original *Pithecanthropus* skullcap. Despite the considerable guesswork involved with this reconstruction (Figure 7.4) many paleoanthropologists have accepted it as valid.

Thirty years later, however, Grover S. Krantz presented a very convincing case that the upper jaw used in the reconstruction

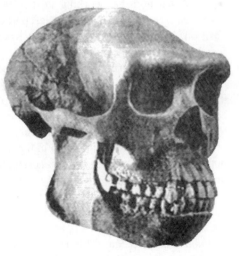

Figure 7.4. Reconstruction of the Java man (*Pithecanthropus robustus*) skull by Franz Weidenreich (1945, plate 4). The only bone fragments used in the reconstruction comprised the rear part of the cranium and the upper and lower jaws. These were from different sites.

did not belong to the same individual as the rear part of the skull.

After making detailed measurements, Krantz (1975, pp. 363–365) concluded that the upper jaw (palate) was much too wide to fit with the rear braincase. Weidenreich appears to have realized this back in 1945 when he did the reconstruction. Krantz noted that Weidenreich artificially spread apart two key bones of the rear braincase (called the mandibular fossae) "without any explanation." Krantz (1975, p. 366) stated that the mandibular fossae, even after Weidenreich spread them apart, were nevertheless "much too close together to accommodate the palate breadth."

It is interesting that in the same volume in which Krantz's report appeared, von Koenigswald stated that the width of the palate as Krantz had measured it—94 millimeters—was incorrect (Tuttle 1975, p. 377). Von Koenigswald arrived at a width narrower than 94 millimeters, which he said allowed the palate to perfectly fit the braincase. But following von Koenigswald's statement came a statement by Krantz saying that his own measurement of the palate width was correct and was identical to Weidenreich's 1945 measurement of 94 millimeters. According to Krantz, before the upper jaw was discovered it was broken through the socket of the first left incisor; then the greater part of its left half was shifted toward the midline and fossilized in that position. Krantz reported that both he (working in the 1970s) and Weidenreich (working in 1945) realized this and moved the left part of this upper jaw back to its original position. Krantz pointed out that von Koenigswald apparently measured the fossil in its distorted condition (without moving the left half back) and that von Koenigswald's measurement was therefore wrong. Following Krantz's statement there was no published reply by von Koenigswald. It seems, therefore, that von Koenigswald was wrong in his measurement and that Krantz was correct. This supports Krantz's contention that the upper jaw does not belong to the same individual as the rear braincase.

According to Krantz, the teeth in the upper jaw were positioned like those in the upper jaw of an ape such as an orangutan. Krantz (1975, p. 369) therefore proposed "that the palate previously related to the Javan *Homo erectus* skull IV should be removed from this association" and "assigned to the genus *Pongo,* large Asiatic apes." If Krantz's assessment is correct, then both Weidenreich and von Koenigswald were apparently unable to distinguish an ape palate from a *Homo erectus* palate. This is especially damaging to Weidenreich's skull reconstruction, which included the questionable palate.

7.3.7 More Discoveries by Von Koenigswald

West of Trinil there is an area where the Kabuh formation comes to the surface. At this location primitive stone tools are also present. Von Koenigswald

(1940b) stated that in this area a fragment of a heavy mandible (S5 in Table 7.2, p. 498) came to light in 1939.

By now the reader may be hoping that there might be a detailed report of the strata in which this fossil was discovered so that a proper date can be assigned to it. Such hopes must remain unfulfilled in this case. Von Koenigswald (1949b, p. 110) stated explicitly that this fossil was called *"Pithecanthropus dubius"* because its original position was unknown. In a later report, von Koenigswald (1968a, p. 102) flatly admitted that this fossil was a surface find. He thought that it must originally have come from the Black Clay stratum of the Putjangan formation, but he was not sure. Therefore, this fossil cannot be assigned to a particular point in geological time, which makes it next to useless as paleoanthropological evidence.

In 1941, one of von Koenigswald's native collectors, at Sangiran, sent to him, at Bandung, a fragment of a gigantic lower jaw (S6 in Table 7.2, p. 498). According to von Koenigswald (1956, p. 111), it displayed the unmistakable features of a human ancestor's jaw. He named the jaw's owner *Meganthropus palaeojavanicus* ("giant man of ancient Java") because the jaw was twice the size of a typical modern human jaw.

Von Koenigswald believed that the S6 jaw was discovered in the Putjangan formation near the site where the Modjokerto child's skull was found. A careful search of original reports has not revealed a description of the exact location at which the S6 jaw was found, or who discovered it. If von Koenigswald did report the exact circumstances of this find then it is a well-kept secret. He discussed *Meganthropus* in at least three reports (von Koenigswald 1956, pp. 111–113; 1949a, p. 92; 1949b, p. 107); however, in none of these did he inform the reader of the details of the fossil's original location. All he said was that it came from the Putjangan formation, but no further information was supplied. Hence all we really know for certain is that some unnamed collector mailed a jaw fragment to von Koenigswald. Its age, from a strictly scientific standpoint, remains a mystery.

Meganthropus, in the opinion of von Koenigswald, was a giant offshoot from the main line of human evolution. Von Koenigswald had also found some large humanlike fossil teeth, which he attributed to an even larger creature called *Gigantopithecus.* According to von Koenigswald, *Gigantopithecus* was a large and relatively recent ape. But Weidenreich, after examining the *Meganthropus* jaws and the *Gigantopithecus* teeth, came up with another theory. He proposed that both creatures were direct human ancestors. According to Weidenreich, *Homo sapiens* evolved from *Gigantopithecus* by way of *Meganthropus* and *Pithecanthropus* (Simons and Ettel 1970, p. 77). Each species was smaller than the next. Most modern authorities, however, consider *Gigantopithecus* to be a variety of ape, living in the Middle to Early Pleistocene, and not directly related to humans. The *Meganthropus* jaws are now thought to be much more like those of Java man (*Homo erectus*) than von Koenigswald originally believed. Some researchers, however, have suggested that *Meganthropus* fossils might be classified as *Australopithecus* (Jacob 1973, p. 475). This is intriguing, because according to standard opinion, *Australopithecus* never left its African home.

7.4 LATER DISCOVERIES IN JAVA

Meganthropus was the last major discovery reported by von Koenigswald, but the search for more bones of Java man has continued up to the present. We shall now discuss the most important of the later finds, which are uniformly accepted as evidence for *Homo erectus* in the Javanese Middle and Early Pleistocene (Le Gros Clark and Campbell 1978, p. 94). The discoveries were all made in the Sangiran region.

In September 1952, P. Marks, a member of the science faculty at the University of Indonesia at Bandung, happened to pass by the fossil beds at Sangiran, at which time a local villager handed him a large fragment of a heavily fossilized mandible (S8 in Table 7.2, p. 498). Later, Marks analyzed this mandible and concluded it belonged to von Koenigswald's *Meganthropus*. Marks (1953, p. 26) stated that the jaw fragment was found lying loose on the surface north of the village of Glagahombo, on a slope of strongly cemented conglomerate, consisting mainly of small pebbles of volcanic origin. Numerous fragments of vertebrate bones were present within this conglomerate. Marks noted that the material clinging to the S8 jaw fragment was of the same type as the conglomerate of the slope. He added, however, that it was not possible "to collect associated vertebrate remains of stratigraphical value." From the information provided by Marks in his report, one cannot assign a specific age to this fossil. In light of this, it is surprising that Marks advocated that this fossil was from the Middle Pleistocene Kabuh formation and that this judgement is accepted without question by other modern authorities.

In 1960, near the village of Mlandingan in the vicinity of Sangiran, a villager discovered a highly fossilized right mandible (S9 in Table 7.2, p. 498) on the surface of a hill slope. The bones of other vertebrates had also been discovered in this area. The S9 fossil consisted of the right half of the jaw and contained five teeth. T. Jacob, of the department of physical anthropology at Gadjah Madah University in Jogjakarta, Indonesia, proposed that the S9 mandible belonged to one of the species of *Pithecanthropus*.

This right mandible was incrusted by a matrix containing foraminifera (small marine organisms) that S. Sartono (1974) reported were exactly the same as the foraminifera of the Putjangan formation, which is considered older than the Middle Pleistocene Kabuh formation (Section 7.5.1).

Jacob (1964) reported that in August 1963 an Indonesian farmer discovered fragments of a fossilized skull "in the Sangiran dome area while working in the field." When assembled, these skull fragments formed what appeared to be a skull (S10 in Table 7.2, p. 498) similar to the type that is designated as *Homo erectus.*

Although Jacob stated that this skullcap was deposited in the Kabuh formation during the Middle Pleistocene, he gave no more information than "a farmer discovered fragments of a fossil skull in the Sangiran dome area while working in the field." Jacob (1964) did not state the exact position of the fragments when found. All we really know is that a farmer discovered some fossil skull fragments that were most likely on or close to the surface.

There were two other reports on the S10 find: Sartono (1964) and Jacob (1966). The report by Sartono (1964, p. 3) provided a diagram of the beds of the Kabuh formation and the overlying Notopuro beds. The S10 skull was assigned to layer 8 in the Kabuh formation. In the general area of the discovery, layer 8 was at the surface. Sartono (1964) did not tell whether the skull was found lying loose on the ground or firmly embedded in layer 8.

Jacob (1966, p. 244) reported that the fragments making up the S10 skull were discovered in two successive months. The first group of fragments was found in July 1963 and was reported by Sartono (1964). The second group was found in August 1963 and was reported in Jacob (1964). Unfortunately, Jacob (1966) did not provide any more definite information about the location and situation of either group of fragments at the time of their discovery than can be found in Sartono (1964) or in his own earlier report (Jacob 1964).

Jacob (1973, p. 476) made this interesting remark about Sangiran, where the S10 find and all the other finds discussed in this section were made: "The site seems to be still promising, but presents special problems. . . . This is mainly due to the site being inhabited by people, many of whom are collectors who had been trained in identifying important fossils. Chief collectors always try to get the most out of the Primate fossils found accidentally by primary discoverers (Jacob 1964). In addition, they may not report the exact site of the find, lest they lose one potential source of income. Occasionally, they may not sell all the fragments found on the first purchase, but try to keep a few pieces to sell at a higher price at a later opportunity."

Concerning another find in the Sangiran region, S. Sartono (1967) stated that on January 30, 1965 pieces of a skullcap were obtained by one of his collectors (unnamed) from a local villager, who was also not named. The fossils came into Sartono's possession towards the end of February 1965. This discovery consisted of 44 fragments of a skullcap, which were "collected all around the site of the skull in the field." After restoration, the skullcap (S12 in Table 7.2, p. 498) consisted of both parietals, the left temporal, part of the left mastoid bone, a large part of the occipital bone, and a small part of the frontal bone. The S12 skull exhibited the features typical of *Homo erectus*.

The pieces of this skullcap were found on the surface (Sartono 1967, p. 85) of a slope southwest of the village of Putjung. At this point, the Kabuh formation was exposed. Sartono stated that the skull was deposited at the same time as the early Middle Pleistocene Kabuh beds. Modern authorities have uncritically accepted Sartono's age estimate for the S12 skull. But if the skull had been of modern morphology, one suspects that the same authorities would have used the fact that it was found on the surface to rule out a Middle Pleistocene age for it.

On July 20, 1969, one of Sartono's collectors gave him a small fragment of an upper jaw that contained two upper left premolars (S15 in Table 7.2, p. 498). From the jaw's dimensions and shape, Sartono concluded that it belonged to a member of the species *Pithecanthropus modjokertensis*. But modern authorities tend to classify all the *Pithecanthropus* species of earlier researchers (*P. erectus, P. modjokertensis, P. soloensis,* and *P. robustus*) as *Homo erectus*.

The S15 jaw was said to have been found lying loose on the surface near Ngrejeng village on the northern part of the Sangiran dome. The rock matrix that incrusted it consisted of a grey clay stone. Because the Putjangan beds at this point also have a layer of grey clay stone, Sartono (1974) assumed that this specimen was originally embedded in that layer. Sartono and most other scientists believed the Putjangan formation to be older than the Middle Pleistocene Kabuh formation.

There are several reasons why one should be hesitant to accept that the S15 jaw was contemporaneous with the Putjangan formation. First, the credibility of the collector is unknown. He might have obtained this bone from any place where there is a grey clay stone. Second, even if S15 was discovered at the site reported above, because it was found on the surface it may not have been originally a part of the Putjangan beds at this point. Holmes or Hrdlicka, confronted with evidence for anatomically modern human beings in a situation like this, would probably have proclaimed that the fossil had been recently introduced. As always, our point is that a double standard should not be employed in the evaluation of paleoanthropological evidence—an impossibly strict standard for anomalous evidence and an exceedingly lenient standard for acceptable evidence.

Sartono (1972) reported that on September 13, 1969, Mr. Towikromo, a resident of the village of Putjung, accidentally discovered a fossil hominid skull when his plow broke through its crest. According to Sartono (1972, p. 124), the skull (S17 in Table 7.2, p. 498) had a low vault.

There are several aspects of Sartono's 1972 report that seem unclear. Sartono stated that Mr. Towikromo was using an iron tool for cultivation of his land when his tool contacted the skull. Sartono also stated explicitly that the skull was embedded in the Kabuh beds, which are here made of sandstone. Sartono (1972, p. 124) said: "This sandstone forms the base of the blind valley and contains the skull."

It does not seem probable that the farmer was plowing sandstone. Perhaps the lower part of the skull was embedded in the very topmost layer of the Kabuh sandstone, and the upper part of the skull was projecting up into a thin, recently deposited layer of soil that was being cultivated. This may be true, but Sartono (1972) did not state this in his report. It is also not clear just how firmly the skull was embedded in the sandstone layer. Perhaps it belonged entirely to a recent soil layer. We must bear in mind that Sartono did not find this skull himself. It is therefore quite possible that it was not firmly embedded in the Kabuh sandstone.

In order to clear up these uncertainties in Sartono's 1972 report, letters were written in 1985 to both him and to T. Jacob for further information about this and several other important discoveries reported by them from Java. No answers were received. One can easily find dozens of popular books describing *Homo erectus* and how this hominid lived between 0.5 and 2 million years ago in Java, but finding a report describing how a particular fossil was situated when discovered is often quite difficult. Many of the popular books do not describe the original situation of the fossil. Nor do they give references to reports by the original discoverer. Sometimes these books do give references to reports. But upon reading these reports, one finds no information about the original position of the fossil. Sometimes the references are to reports in journals that are not easily found even at the libraries of major universities. It is therefore often difficult to obtain information describing the original stratigraphic position of a fossil. This means that the fossil cannot be properly assigned a geological age, and if it cannot be assigned a geological age it cannot be inserted into a proposed evolutionary sequence. Nevertheless, this is exactly what has been done.

7.5 CHEMICAL AND RADIOMETRIC DATING OF THE JAVA HOMO ERECTUS FINDS

We shall now discuss issues related to the potassium-argon dating of the formations yielding hominid fossils in Java, as well as attempts to date the fossils themselves by various chemical and radiometric methods. See Appendix 1 for general information about these methods.

7.5.1 The Ages of the Kabuh and Putjangan Formations

In the foregoing discussion, we have several times referred to the Kabuh formation and the Putjangan formation. The original *Pithecanthropus* finds of Dubois were from the Trinil beds of the Kabuh formation. Some of the subsequent finds by von Koenigswald and later researchers were also assigned to the Kabuh formation. Others were assigned to the Djetis beds of the Putjangan formation. Many researchers have attempted to establish an age difference between the two formations, placing the Kabuh formation in the Middle Pleistocene and the Putjangan formation in the Early Pleistocene.

As we have seen (Section 7.1.5), Eugene Dubois originally attempted, on faunal grounds, to classify the Trinil beds of the Kabuh formation as Pliocene (Boule 1923, p. 98), but later researchers have characterized the Trinil fauna as post-Villafranchian (Le Gros Clark and Campbell 1978, p. 91) or Middle Pleistocene (Hooijer 1951, p. 273; 1956, p. 5).

The Trinil beds of the Kabuh formation have also been dated by the potassium-argon method. Potassium-argon dating relies on the fact that volcanic materials contain potassium 40, which decays into the gas argon 40. The argon gas remains trapped in crystals of volcanic material. By comparing the ratios of potassium 40 and argon 40 gas in a sample, one can date volcanic materials (for a fuller discussion of potassium-argon dating, see Section 11.6.5.1). Basalt at Mount Muria, from a layer above the *Pithecanthropus erectus* level of the Trinil beds, yielded an age of 500,000 years, while tektites (pieces of glass produced by meteors) from the Trinil beds yielded potassium-argon dates of 710,000 years (von Koenigswald 1968b, p. 201; Jacob 1973, p. 477). Further potassium-argon tests by G. H. Curtis (Jacob 1973, p. 477) on pumice from Trinil beds at Tanjung and Putjung, where the S10 and S12 fossils were found, gave similar ages. Jacob (1973, p. 477) said the average for the above four dates for the Trinil beds is 830,000 years. This would put the Trinil beds in the early Middle Pleistocene.

As far as the Putjangan formation is concerned, it was originally classified as Early Pleistocene on faunal grounds by von Koenigswald. But D. A. Hooijer (1956) objected to this. He pointed out that both the Trinil beds of the Kabuh formation and the Djetis beds of the Putjangan formation share fossil species characteristic of the *Stegodon-Ailuropoda* mammalian fauna of southern China, generally recognized as Middle Pleistocene (Hooijer 1956, p. 7).

According to Hooijer (1956, p. 8), von Koenigswald principally employed molluscan stratigraphy in assigning an Early Pleistocene age to the Djetis beds. But Hooijer (1956, p. 9), after citing authorities on Javan molluscs, stated: "there is no reason to attach greater importance to the mollusks than to certain mammalian genera of long standing in establishing Pleistocene correlations." Hooijer (1956, p. 9) concluded that the Trinil and Djetis beds were of roughly the same Middle Pleistocene age.

Attempts were later made to establish the geological age of the Djetis beds using the potassium-argon dating method. The Djetis beds of the Putjangan formation near Modjokerto yielded an Early Pleistocene potassium-argon date of about 1.9 million years (Jacob 1973, p. 477; Jacob and Curtis 1971; Jacob 1972).

The date of 1.9 million years is significant for the following reasons. As we have seen, many *Homo erectus* fossils (previously designated *Pithecanthropus* and *Meganthropus*) have been assigned to the Djetis beds. If these fossils are given an age of 1.9 million years, this makes them older than the oldest African *Homo erectus* finds, which are about 1.6 million years old (Brown *et al.* 1985, p. 788). According to standard views, *Homo erectus* evolved in Africa and did not migrate out of Africa until about 1 million years ago.

Also, some researchers have suggested that von Koenigswald's *Meganthropus* might be classified as *Australopithecus* (Jacob 1973, p. 475; Jacob and Curtis 1971). If one accepts this opinion, this means that Javan representatives of *Australopithecus* arrived from Africa before 1.9 million years ago or that *Australopithecus* evolved separately in Java. Both hypotheses are in conflict with standard views on human evolution.

It should be kept in mind, however, that the potassium-argon technique that gave the 1.9-million-year date is not any more reliable than the other dating techniques we discuss in Appendix 1. Jacob and Curtis (1971), who attempted to date most of the hominid sites in Java, found "it has been difficult to obtain meaningful dates from most samples." In other words, dates were obtained, but they deviated so greatly from what was expected that Jacob and Curtis (1971) had to attribute the unsatisfactory results to "contaminants."

Moveover, according to Nilsson (1983, p. 329): "A much lower [potassium-argon] date for the Djetis Beds, less than 1 million years, is indicated by later studies (Bartstra 1978)." This agrees with Hooijer's conclusion that the Djetis beds are, like the Trinil beds, early Middle Pleistocene. Finally, M. H. Day and T. I. Molleson (1973, p. 147) reported that fluorine content "analyses of bones from both the Djetis and the Trinil faunas at Sangiran showed that it was not possible to distinguish, analytically, the two assemblages at this site."

7.5.2 Chemical Dating of the Trinil Femurs

In Section 7.1.8, we learned that the Trinil femurs are indistinguishable from those of modern humans and distinct from those of *Homo erectus*. This has led some to suggest that the Trinil femurs do not belong with the *Pithecanthropus* skull and were perhaps mixed into the early Middle Pleistocene Trinil bone bed from higher levels (Day and Molleson 1973, p. 152). Another possibility is that anatomically modern humans were living alongside ape-man-like creatures during the early Middle Pleistocene in Java. In light of the evidence presented in this book, this would not be out of the question.

The fluorine content test has often been used to determine if bones from the same site are of the same age. Bones absorb fluorine from groundwaters, and thus if bones contain similar percentages of fluorine (relative to the bones' phosphate content) this suggests such bones have been buried for the same amount of time.

M. H. Day and T. I. Molleson (1973) analyzed the Trinil skullcap and femurs and found they contained roughly the same ratio of fluorine to phosphate. Middle Pleistocene mammalian fossils at Trinil contained a fluorine-to-phosphate ratio similar to that of the skullcap and femurs. Day and Molleson (1973, p. 146) stated that their results (Table 7.1) "apparently indicated the contemporaneity of the calotte and femora with the Trinil fauna."

If the Trinil femurs are distinct from those of *Homo erectus* and identical to those of *Homo sapiens sapiens,* as Day and Molleson (1973, p. 128) reported, then the fluorine content of the femurs is consistent with the view that anatomically modern humans existed in Java during the early Middle Pleistocene, about 800,000 years ago.

Day and Molleson (1973, p. 147) suggested that Holocene bones from the Trinil site might, like the Java man fossils, also have fluorine-to-phosphate ratios similar to those of the Middle Pleistocene animal bones, making the fluorine test useless here. In discussing the La Denise human bones (Appendix 2), Oakley pointed out that the rate of fluorine absorption in volcanic areas, such as Java, tends to be quite erratic, allowing bones of widely differing ages to have similar fluorine contents. This could not be directly demonstrated at the Trinil site, because there only the Middle Pleistocene beds contain fossils.

Day and Molleson (1973, p. 148) showed that Holocene and Late Pleistocene beds at other sites in Java contained bones with fluorine-to-phosphate ratios similar to those of the Trinil bones (Table 7.1). But they admitted (Day and Molleson 1973, p. 144) that the fluorine-to-phosphate ratios of bones from other sites "would not be directly comparable" with those of bones from the Trinil site. This is because the fluorine absorption rate of bone depends upon factors that can vary from site to site. Such factors include the groundwater's fluorine content, the groundwater's rate of flow, the nature of the sediments, and the type of bone.

Therefore, the fluorine content test results reported by Day and Molleson remain consistent with (but are not proof of) an early Middle Pleistocene age of about 800,000 years for the anatomically modern human Trinil femurs.

A nitrogen content test was also performed on the Trinil bones. Dubois had boiled the skullcap and the first femur in glue. Day and Molleson (1973, p. 147) attempted to correct for this by "pre-treating the samples in order to remove soluble nitrogen before analysis." Results showed that all of the Trinil bones listed in Table 7.1 had very little nitrogen left in them. This is consistent with all of the bones being of the same early Middle Pleistocene age, although Day and Molleson (1973, p. 148) did report that nitrogen in bone is lost so rapidly in Java that even Holocene bones often have no nitrogen. The uranium contents of the Trinil hominid bones and fauna were all almost zero, again consistent with (although not proof of) their being of the same early Middle Pleistocene age.

TABLE 7.1

Fluorine Analysis of Bone from Java Sites

Site	$(F\%/P_2O_5\%)' 100$
Trinil (early Middle Pleistocene)	
Pithecanthropus skullcap	4.1
Femur I shaft	5.8
Femur II shaft	6.2
Femur III shaft	5.7
Femur IV shaft	5.8
Femur V shaft	5.8
Bibos mandible	6.9
Axis antler	6.2
Wadjak (Holocene)	4.0, 8.2
Ngandong (Late Pleistocene)	5.6, 9.3

This data is from Day and Molleson (1973, pp. 147–148). The figures represent the ratio of fluorine (F) to phosphate (P_2O_5) in the bones.

7.5.3 Uranium Content Testing of the Sangiran Fossils

In our discussion of the Sangiran hominid finds reported by von Koenigswald (Section 7.3) and later researchers (Section 7.4), we learned that almost all occurred on the surface. We suggested that this made their real age uncertain.

Some might infer a Middle or Early Pleistocene date for the Sangiran fossils on the basis of their equivalent U_3O_8 (uranium oxide) content. For example, the S1a upper jaw discovered in 1936 has an equivalent uranium content of 25 parts per million (ppm), somewhat less than that of 63 ppm for a *Cervus* (deer) antler from the same general region (Oakley *et al.* 1975, p. 109).

Sangiran 1a (S1a) was found on the surface, and von Koenigswald simply assumed it weathered out of the Putjangan formation. Therefore, we are not able to measure the concentration of uranium, either in the groundwater or in other fossils, at the precise location where Sangiran 1a, if in fact from the Putjangan formation, lay buried. If we were able to compare the uranium contents of the S1a jaw and other bones from the same spot and found them similar, that would be consistent with, although not proof of, the view that they were of the same age.

But lacking such evidence, the reported uranium content for the Sla jaw itself gives little reason to suggest that the Sla jaw is as old as the Putjangan formation. We cannot exclude the possibility that Sangiran 1a is a very young bone that was originally situated in a stratum through which uranium-rich waters percolated.

Day and Molleson (1973, p. 148) reported that two Late Pleistocene bones from Ngandong had uranium contents of 25 and 30 ppm. These figures are not directly comparable with the uranium content of 25 ppm for the supposedly Early Pleistocene Sangiran 1a jaw, but they do demonstrate the difficulty in inter-preting the kind of uranium content data reported by Oakley *et al.* (1975).

The above line of reasoning is also applicable to the other Sangiran hominid fossils, since they were also found on the surface or in other dubious ways.

7.6 MISLEADING PRESENTATIONS OF THE JAVA MAN EVIDENCE

Most books dealing with the subject of human evolution present what appears at first glance to be an impressive weight of evidence for *Homo erectus* in Java between 0.5 and 2.0 million years ago. One such book is *The Fossil Evidence for Human Evolution* (1978), by W. E. Le Gros Clark, professor of anatomy at Oxford University, and Bernard G. Campbell, adjunct professor of anthropology at the University of California at Los Angeles. An impressive table showing discoveries of *Homo erectus* is presented in their book (Le Gros Clark and Campbell 1978, p. 94). These discoveries (Table 7.2, p. 498) have been used widely to support the belief that man has evolved from an apelike being.

T3 is the femur found by Dubois at a distance of 45 feet from the original cranium, T2. We have already discussed how unjustified it is to assign these two bones to the same individual (Sections 7.1.7, 7.1.8). Yet ignoring many important facts, Le Gros Clark and Campbell stated (1978, p. 91) that "the accumulation of evidence speaks so strongly for their natural association that this has become generally accepted."

T6, T7, T8, and T9 are the femurs found in boxes of fossils in Holland over 30 years after they were originally excavated in Java. Le Gros Clark and Campbell apparently ignored Dubois's statement that he himself did not excavate them, and that the original location of the femurs was unknown. We may also recall von Koenigswald's statement that the femurs were from Dubois's general collection, which contained fossils from "various sites and various ages which are very inadequately distinguished because some of the labels got lost." Nevertheless, Le Gros Clark and Campbell (1978, p. 94) assumed that these femurs came from the Trinil beds of the Kabuh formation. But Day and Molleson (1973, p. 130) observed: "if the rigorous criteria that are demanded in modern excavations were applied to all of the Trinil material subsequent to the calotte and

Femur I, it would all be rejected as of doubtful provenance and unknown stratigraphy."

Fossil M1 and fossils Sla through S6 are those discovered by Javanese native collectors employed by von Koenigswald. Only one of them (M1) was reported to have been discovered buried in the stratum to which it is assigned, and even this report is subject to question. The remaining fossils of the S series are the ones reported by Marks, Sartono, and Jacob, and the majority of these were surface finds by villagers and farmers, who sold the fossils, perhaps by way of middlemen, to the scientists. One familiar with the way these specimens were found can only wonder at the intellectual dishonesty manifest in Table 7.2 (p. 498), which gives the impression that the fossils were all found in strata of definite age.

In our discussion of the Sheguiandah site in Canada, where anomalously old stone tools were found by T. E. Lee, we found that an establishment scientist, James B. Griffin, dismissed the discovery because the site did not conform to certain very strict standards (Section 5.4.1.2, pp. 350–352). Griffin and others like him demand to see an intact habitation site, in a clearly defined geological context, complete with stone tools, skeletal remains, signs of deliberate use of fire, remains of animals and plants used as food, and more (1979, p. 44). Otherwise, there is always the chance that an isolated artifact or bone might be intrusive in the layer in which it was found. Griffin felt no hesitation whatsoever in using his criteria to reject as nonsites dozens of places in North American where anomalously old traces of humans had been discovered.

Should not the same strict standards apply in Java? One might argue that Griffin's requirements were intended for Indian sites in North America and not for *Homo erectus* sites in other parts of the world. But according to standard opinion, *Homo erectus* was, like *Homo sapiens,* a toolmaker and user of fire, as shown by *Homo erectus* sites in China, Africa, and Europe. One might therefore expect to find the same kinds of artifacts and signs of habitation at a *Homo erectus* site as at a *Homo sapiens* site. As we have seen, none of the *Homo erectus* sites in Java (over twenty) conform to Griffin's criteria and should therefore be classed as nonsites. No cultural remains whatsoever have been found along with the Java *Homo erectus* fossils, most of which were surface finds.

We regard Griffin's approach as extreme. However, our main objection is not to the stringency of his requirements but to the fairness of their application. If one decides to employ Griffin's criteria, one should do so in all similar cases or none at all. Obviously, if one were to universally apply Griffin's criteria, much of the paleoanthropological evidence currently accepted by scientists, such as the Java *Homo erectus* evidence, would have to be thrown out. Since that has not been done, we believe Griffin's strict standards should not be selectively applied to eliminate anomalous discoveries, such as Sheguiandah. Applying the more lenient criteria by which the Java finds have been accepted by the scientific

TABLE 7.2

Fossil Hominids from Java

Stratigraphic Unit	Sites	Age Bracket
Trinil (Kabuh Formation)	Sangiran S2 Adult female calotte (1937) S3 Juvenile calotte (1938) S8 Right mandible (1952) S10 Adult male calotte (1963) S12 Old male calotte (1965) S15 Maxilla (1969) S17 Cranium (1969) S21 Mandible (1973) Trinil T2 Calotte (1892) = *Pithecanthropus* T3,T6,T7,T8,T9 Femora Kedung Brubus KB1 Right juvenile mandible (1890)	0.7–1.3 million years, (potassium-argon date of about .83 million years)
Djetis (Putjangan Formation)	Sangiran S1a Right maxilla (1936) S1b Right mandible (1936) S4 Adult male calvaria & maxilla (1938–39) = *P. robustus* S5 Right mandible (1939) = *P. dubius* S6 Right mandible (1941) = *Meganthropus* S9 Right mandible (1960) S22 Maxilla, mandible (1974) Modjokerto M1 Child, 7 years, calvaria (1936)	1.3–2.0 million years, (potassium-argon date of about 1.9 million years)

This table is reproduced from Le Gros Clark and Campbell (1978, p. 94). Calotte, cranium, and calvaria mean skull, mandible means lower jaw, maxilla means upper jaw, and femora means thighbones.

community as evidence for *Homo erectus* in the Middle Pleistocene, the Sheguiandah evidence, and other anomalous evidence, should also be accepted.

Although Le Gros Clark and Campbell (1978, p. 93) noted that Hooijer (1951) had said the real location of many of von Koenigswald's finds was unknown, they nevertheless accepted that the fossils must have come from Middle or Early Pleistocene formations, which they designated 0.7–1.3 and 1.3–2.0 million years of age.

The ages given by Le Gros Clark and Campbell, derived from the potassium-argon dates discussed in Section 7.5.1, refer only to the age of the volcanic soils, and not to the bones themselves. Potassium-argon dates have meaning only if the bones were found securely in place within or beneath the layers of dated volcanic material. But the vast majority of fossils listed in Table 7.2 were surface finds, rendering their assigned potassium-argon dates meaningless.

Concerning the age of 1.3–2.0 million years given by Le Gros Clark and Campbell for the Djetis beds of the Putjangan formation, we note that this is based on the potassium-argon date of 1.9 million years reported by Jacob and Curtis (1971). But Bartstra (1978) obtained a potassium-argon age of less than 1 million years (Section 7.5.1). As we have seen (Section 7.5.1), other researchers have reported that the fauna of the Djetis and Trinil beds are quite similar and that the bones have similar fluorine-to-phosphate ratios.

Le Gros Clark and Campbell (1978, p. 92) concluded that "at this early time there existed in Java hominids with a type of femur indistinguishable from that of *Homo sapiens,* though all the cranial remains so far found emphasize the extra-ordinarily primitive characters of the skull and dentition." All in all, the pre-sentation by Le Gros Clark and Campbell was quite misleading. They left the reader with the impression that cranial remains found in Java can be definitely associated with the femurs when such is not the case. Furthermore, discoveries in China and Africa, as previously noted (Section 7.1.8), have shown that *Homo erectus* femurs are different from those collected by Dubois in Java.

Judging strictly by the hominid fossil evidence from Java, all we can say is the following. As far as the surface finds are concerned, these are all cranial and dental remains, the morphology of which is primarily apelike with some humanlike features. Because their original stratigraphic position is unknown, these fossils simply indicate the presence in Java, at some unknown time in the past, of a creature with a head displaying some apelike and humanlike features.

The original *Pithecanthropus* skull (T2) and femur (T3) reported by Dubois were found *in situ,* and thus there is at least some basis for saying they are perhaps as old as the early Middle Pleistocene Trinil beds of the Kabuh formation. The original position of the other femurs is poorly documented, but they are said to have been excavated from the same Trinil beds as T2 and T3 (Section 7.1.7). In any case, the original femur (T3), described as fully human, was not found in close connection with the primitive skull and displays anatomical features that distinguish it from the femur of *Homo erectus*. There is, therefore, no good reason to connect the skull with the T3 femur or any of the other femurs, all of which are described as identical to those of anatomically modern humans. Consequently, the T2 skull and T3 femur can be said to indicate the presence of two kinds of hominids in Java during the early Middle Pleistocene—one with an apelike head and the other with legs like those of anatomically modern humans. Following the typical practice of giving a species identification on the basis of partial skeletal remains, we can say that the T3 femur provides evidence for the presence of *Homo sapiens sapiens* in Java around 800,000 years ago. Up to now, no creature except *Homo sapiens sapiens* is known to have possessed the kind of femur found in the early Middle Pleistocene Trinil beds of Java.

8

The Piltdown Showdown

After Eugene Dubois's discovery of Java man in the 1890s, the hunt for fossils to fill the evolutionary gaps between ancient apelike hominids and modern *Homo sapiens* intensified. It was in this era of strong anticipation that a sensational find was made in England—Piltdown man, a creature with a humanlike skull and apelike jaw.

The outlines of the Piltdown story are familiar to both the proponents and opponents of the Darwinian theory of human evolution. The fossils, the first of which were discovered by Charles Dawson in the years 1908–1911, were declared forgeries in the 1950s by scientists of the British Museum. This allowed the critics of Darwinian evolution to challenge the credibility of the scientists who for several decades had placed the Piltdown fossils in evolutionary family trees.

Scientists, on the other hand, were quick to point out that they themselves exposed the fraud. Some sought to identify the forger as Dawson, an eccentric amateur, or Pierre Teilhard de Chardin, a Catholic priest-paleontologist with mystical ideas about evolution, thus absolving the "real" scientists involved in the discovery.

In one sense, it would be possible to leave the story of Piltdown at this and go on with our survey of paleoanthropological evidence. But a deeper look at Piltdown man and the controversies surrounding him will prove worthwhile, giving us greater insight into how facts relating to human evolution are established and disestablished.

Contrary to the general impression that fossils speak with utmost certainty and conviction, the intricate network of circumstances connected with a paleoanthropological discovery can preclude any simple understanding. Such ambiguity is especially to be expected in the case of a carefully planned forgery, if that is what the Piltdown episode represents. But as a general rule, even "ordinary" paleoanthropological finds are enveloped in multiple layers of uncertainty. As we trace the detailed history of the Piltdown controversy it becomes clear that the line between fact and forgery is often indistinct.

8.1 DAWSON GETS A SKULL

Sometime around the year 1908, Charles Dawson, a lawyer and amateur anthropologist, noticed that a country road near Piltdown, in Sussex, was being mended with flint gravel. Always on the lookout for flint tools, Dawson inquired from the workmen and learned that the flint came from a pit on a nearby estate, Barkham Manor, owned by Mr. R. Kenward, with whom Dawson was acquainted. Dawson visited the pit and asked two workers there to be on the lookout for any implements or fossils that might turn up. In 1913, Dawson wrote: "Upon one of my subsequent visits to the pit, one of the men handed to me a small portion of an unusually thick human parietal bone. I immediately made a search but could find nothing more. . . . It was not until some years later, in the autumn of 1911, on a visit to the spot, that I picked up, among the rain-washed spoil-heaps of the gravel pit, another and larger piece belonging to the frontal region of the same skull" (Dawson and Woodward 1913, p. 117). Dawson noted that the pit contained pieces of flint much the same in color as the skull fragments.

Dawson was not a simple amateur. He had been elected a Fellow of the Geological Society and had for 30 years contributed specimens to the British Museum as honorary collector (Weiner 1955, p. 83). Furthermore, he had cultivated a close friendship with Sir Arthur Smith Woodward, keeper of the Geological Department at the British Museum and a fellow of the Royal Society. In February 1912, Dawson wrote a letter to Woodward at the British Museum, telling how he had "come across a very old Pleistocene bed . . . which I think is going to be very interesting . . . with part of a thick human skull in it . . . part of a human skull which will rival *Homo heidelbergensis*" (Bowden 1977, p. 40). Altogether, Dawson had found five pieces of the skull. In order to harden them, he soaked them in a solution of potassium dichromate.

On Saturday, June 2, 1912, Woodward and Dawson, accompanied by Pierre Teilhard de Chardin, a student at a local Jesuit seminary, began excavations at Piltdown and were rewarded with some new discoveries. On the very first day, they found another piece of skull. More followed. Dawson later wrote: "Apparently the whole or greater portion of the human skull had been shattered by the workmen, who had thrown away the pieces unnoticed. Of these we recovered, from the spoil-heaps, as many fragments as possible. In a somewhat deeper depression of the undisturbed gravel I found the right half of a human mandible. So far as I could judge, guiding myself by the position of a tree 3 or 4 yards away, the spot was identical with that upon which the men were at work when the first portion of the cranium was found several years ago. Dr. Woodward also dug up a small portion of the occipital bone of the skull from within a yard of the point where the jaw was discovered, and at precisely the same level. The jaw appeared to have been broken at the symphysis and abraded, perhaps when it lay fixed in the gravel, and before its complete deposition. The fragments of the cranium show little or no sign of rolling or other abrasion, save an incision at the back of the parietal, probably caused by a workman's pick" (Dawson and Woodward 1913, p. 121). A total of nine fossil skull pieces were found, five by Dawson alone and an additional four after Woodward joined the excavation.

Dawson and Woodward decided to keep their discovery quiet until such time as they would officially announce it, but news of the fossils circulated privately among scientists with interest in human prehistory. Sir Ray Lankester wrote to J. Reid Moir in 1912: "It seems possible that it is our Pliocene Man—the maker of rostro-carinate flints! At any rate if they say to us 'you say we call in vague, unknown agencies such as torrents and pressure to produce these flints by natural force, but you are in the same position of calling in a hypothetical man. You have no other evidence that such a man was there!' Now we can say, 'Here he is.' It is wonderful that, after so many years, man's bones should turn up in a gravel. I do not despair now of you finding a sub-Crag human cranium and lower jaw. You must keep this dark for a month or so yet as the discoverers will not be ready to publish before that lapse of time and more will be found some day in the same place" (Millar 1972, p. 125).

Others also received previews of the coming attraction, among them, Lewis Abbott, an amateur geologist associated with Benjamin Harrison of Ightham. Harrison's eoliths (Section 3.2) had, like Moir's Red Crag tools (Section 3.3), convinced many researchers that human fossils would be found in southern England's Pliocene and Early Pleistocene formations. After consulting Abbott about the Piltdown fossils, Dawson wrote to Woodward, "Abbott is in no doubt. They are man and man all over" (Weiner 1955, p. 100). Sir Arthur Keith also appears to have heard whispers of the Piltdown discoveries, for in a paper presented at the British Association meeting of 1912, he spoke of new fossil evidence for human beings of the modern type in Britain in the Middle Pleistocene, predating the Neanderthals of Europe (Millar 1972, p. 108).

In December 1912, Dawson and Woodward presented their formal report on the fossils they had discovered at the Piltdown site to the Geological Society of London. The report was published in the journal of the Society in 1913. Concerning the geological context of the discovery, Dawson and Woodward (1913, p. 119) stated: "At Piltdown the gravel-bed occurs beneath a few inches of the surface-soil, and varies in thickness from 3 to 5 feet; it is deposited upon an uneven bottom, consisting of hard yellow sandstone of the Tunbridge Wells Sands (Hastings Beds). . . . Portions of the bed are rather finely stratified, and the materials are usually cemented together by iron oxide, so that a pick is often needed to dislodge portions—more especially at one particular horizon near the base. It is in this last mentioned stratum that all the fossil bones and teeth discovered *in situ* by us have occurred." They added: "The gravel is situated on a well-defined plateau of large area, lying above the 100-foot contour line, averaging about 120 feet at Piltdown, and lies about 80 feet above the level of the main stream of the Ouse" (Dawson and Woodward 1913, p. 119).

In addition to the human fossils, the 1912 excavations at Piltdown yielded a variety of mammalian fossils. Dawson listed them as: "two small broken pieces of a molar tooth of a rather early Pliocene type of elephant, also a much-rolled cusp of a molar of *Mastodon,* portions of two teeth of *Hippopotamus,* and two molar teeth of a Pleistocene beaver." He added: "In the adjacent field to the west, on the surface close to the hedge dividing it from the gravel bed, we found portions of a red deer's antler and the tooth of a Pleistocene horse. These may have been thrown away by the workmen, or may have been turned up by a plough. . . . in the spoil heaps occurred part of a deer's metatarsal. . . . All the specimens are highly mineralized with iron oxide" (Dawson and Woodward 1913, p. 121).

Stone tools were also found: "Among the flints we found several undoubted flint implements, besides numerous 'Eoliths.' The workmanship of the former is similar to that of the Chellean or pre-Chellean stage" (Dawson and Woodward 1913, p. 122). In a footnote, Dawson stated: "Father P. Teilhard, S.J., who accompanied us on one occasion, discovered one of the implements *in situ* in the middle stratum of the gravel-bed, also a portion of the tooth of a Pliocene elephant from the lowest bed" (Dawson and Woodward 1913, p. 122).

The report of Dawson and Woodward (1913, p. 123) concluded: "It is clear that this stratified gravel at Piltdown is of Pleistocene age, but that it contains, in its lowest stratum, animal remains derived from some destroyed Pliocene deposit probably situated not far away, and consisting of worn and broken fragments. These were mixed with fragments of early Pleistocene mammalia in a better state of preservation, and both forms were associated with the human skull and mandible, which show no more wear and tear than they might have received *in situ.* Associated with these animal remains are 'Eoliths,' both in a rolled and an unrolled condition; the former are doubtless derived from an older drift and the latter in their present form are of the age of the existing deposit. In the same bed, in only a very slightly higher stratum, occurred a flint implement, the workmanship of which resembles that of implements found at Chelles; and among the spoil-heaps were found others of a similar, though perhaps earlier, stage. From these facts it appears probable that the skull and mandible cannot safely be described as being earlier than the first half of the Pleistocene Epoch. The individual probably lived during a warm cycle in that age."

In the decades that followed, many scientists agreed with Dawson and Woodward that the Piltdown man fossils belonged to the Early Pleistocene fauna, contemporary with the Piltdown gravels. Others, such as Sir Arthur Keith and A. T. Hopwood thought the Piltdown man fossils belonged with the older Late Pliocene (or Villafranchian) fauna that had apparently been washed into the

Piltdown gravels from an older horizon (Oakley and Hoskins 1950, p. 379).

From the beginning, the Piltdown skull was deemed morphologically human-like, although there was some disagreement about the cranial capacity. In 1913, Woodward estimated the brain capacity at 1,070 cc, perhaps more (Dawson and Woodward 1913, p. 126). This falls well below the average adult male human capacity of about 1,500 cc. But Sir Arthur Keith later proposed a reconstruction of the skull that yielded a brain capacity of 1,500 cc, matching the average adult male human capacity (Dawson and Woodward 1914, p. 98). Interestingly enough, von Koenigswald (1956, p. 179) said that Keith's reconstruction actually yielded a brain capacity of 1,370 cc, and Keith said that the original reconstruction by Dawson and Woodward was around 1,200 cc (Dawson and Woodward 1914, p. 98).

On the human appearance of the skull, Woodward stated in 1913: "A detailed examination of the several bones of the skull is interesting, as proving the typically human character of nearly all the features that they exhibit. . . . there cannot have been any prominent or thickened supraorbital ridge, and the missing region above the glabella may be restored on the plan of an ordinary modern human skull" (Dawson and Woodward 1913, p. 127).

Woodward then compared the apelike Piltdown jaw with the Heidelberg jaw, which is larger and heavier than the Piltdown specimen. "When it is remembered that *Eoanthropus dawsoni* and *H. heidelbergensis* are almost (if not absolutely) of the same geological age," he wrote, "we are thus led to the interesting conclusion that at the end of the Pliocene Epoch the representatives of man in Western Europe were already differentiated into widely divergent groups" (Dawson and Woodward 1913, pp. 137–138).

In addition, Woodward observed that the humanlike skull of Piltdown man was quite different from the more recent skulls of Java man and Neanderthal man, with their low foreheads and prominent brow ridges.

Woodward believed that in general the evolution of a species mirrored the growth of an individual of that species from birth to adulthood. For example, infant apes have rounded skulls, with high foreheads and almost no brow ridges, whereas adult apes have low foreheads with prominent brow ridges. Woodward therefore predicted that the skulls of adult apes from the early Tertiary, when discovered, would be much like those of modern infant apes.

"Hence," stated Woodward, "it seems reasonable to interpret the Piltdown skull as exhibiting a closer resemblance to the skulls of the truly ancestral mid-Tertiary apes than any fossil human skull hitherto found. If this view be accepted, the Piltdown type has gradually become modified into the later Mousterian type [the Neanderthals] by a series of changes similar to those passed through by the early apes as they evolved into the typical modern apes, and corresponding with the stages in the development of the skull in an existing ape-individual. It tends to support the theory that Mousterian man was a degenerate offshoot of early man, and probably became extinct; while surviving man may have arisen directly from the primitive source of which the Piltdown skull provides the first discovered evidence" (Dawson and Woodward 1913, pp. 138–139).

Woodward had come up with his own theory about human evolution, which he thus wanted to support by fossil evidence, however limited and fragmentary. Today, a version of Woodward's proposed lineage survives in the widely accepted idea that *Homo sapiens sapiens* and *Homo sapiens neanderthalensis* are both descendants of a species called archaic or early *Homo sapiens*. Not at all widely accepted, but quite close to Woodward's idea, is Louis Leakey's proposal that both *Homo erectus* and the Neanderthals are side branches from the main line of human evolution (Section 11.4.3). But all of these proposed evolutionary lineages ignore the evidence, catalogued in this book, for the presence of anatomically modern humans in periods earlier than the Pleistocene.

8.2 REACTIONS TO PILTDOWN MAN

The notes of the discussion following the presentation made by Dawson and Woodward at the meeting of the Geological Society in December of 1912 stated: "Prof. A. Keith regarded the discovery of fossil human remains just announced as by far the most important ever made in England, and of equal, if not greater consequence than any other discovery yet made, either at home or abroad" (Dawson and Woodward 1913, p. 148).

Sir Ray Lankester, who had earlier written a favorable note to J. Reid Moir about the newly discovered Piltdown man, now expressed an opinion that the jaw and skull might not be from the same individual (Dawson and Woodward 1913, p. 148). David Waterston, professor of anatomy at King's College, also thought the jaw did not belong to the skull. He believed it had probably washed down from some older Pliocene stratum along with other mammalian fossils (Weiner 1955, p. 7; Dawson and Woodward 1913). Waterston felt that connecting the jaw with the skull was akin to linking a chimpanzee's foot with a human leg (Millar 1972, p. 140). If Waterston was correct, he was confronted with a skull that appeared to be very much like that of a human and was quite possibly from the Early Pleistocene.

So right from the start, some experts were uncomfortable with the seeming incompatibility between the humanlike skull and apelike jaw of the Piltdown man (Figure 8.1). Sir Grafton Eliot Smith, an expert in brain physiology, tried to defuse this doubt. Smith wrote in an appendix to the report by Dawson and Woodward (1913, p. 146) that the cranial cast of Piltdown man "presents more primitive features than any known human brain or cranial cast." This was quite a remarkable judgement considering the otherwise almost unanimous view that the skull itself was very much like that of a human being. Smith added: "we must consider this as being the most primitive and most simian human brain so far recorded; one, moreover, such as might reasonably have been expected to be associated in one and the same individual with the [apelike] mandible and which so definitely indicates the zoological rank of its original possessor" (Dawson and Woodward 1913, p. 147). But according to modern scientists, the Piltdown skull is a fairly recent *Homo sapiens sapiens*

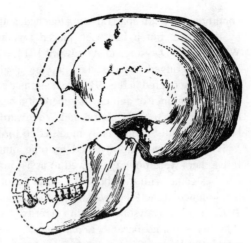

Figure 8.1. Restoration of the Piltdown skull and jaw by Dawson and Woodward (1914, p. 89).

skull that was planted by a hoaxer. If we accept this, that means Smith, a renowned expert, was seeing simian features where none factually existed.

8.3 A CANINE TOOTH AND NOSE BONES

It was hoped that future discoveries would clarify the exact status of Piltdown man. The canine teeth, which are more pointed in the apes than in human beings, were missing from the Piltdown jaw. Woodward thought a canine would eventually turn up, and even made a model of how a Piltdown man canine should look (Bowden 1977, p. 5).

On August 29, 1913, Teilhard de Chardin did in fact find a canine tooth in a heap of gravel from the Piltdown excavation site, near the place where the mandible had been uncovered (Dawson and Woodward 1914, p. 85). The point of the tooth was worn and flattened like that of a human canine. Woodward (Dawson and Woodward 1914, p. 87) stated: "In the upper half of the outer face the thin layer of enamel is . . . marked by the usual faint transverse striations (or imbrications)." Such markings are characteristic of human canines. According to von Koenigswald (1956, p. 159), it was not clear whether the tooth was an upper or lower canine, but the British scientists placed it in the lower jaw discovered at Piltdown.

Some nose bones were also found. Dawson stated: "While our laborer was digging the disturbed gravel within 2 or 3 feet from the spot where the mandible was found, I saw two human nasal bones lying together with the remains of a turbinated bone beneath them *in situ.* The turbinal, however, was in such bad condition that it fell apart on being touched, and had to be recovered in fragments by the sieve; but it has been pieced together satisfactorily by Mrs. Smith Woodward" (Dawson and Woodward 1914, p. 85). Turbinals are thin, platelike bones with ridged surfaces; they line the nasal chambers.

Also discovered in the 1913 excavations were a tooth of *Stegodon* (an extinct elephant), an incisor and jaw fragment of a beaver, a fragment of a rhino tooth, and more flint tools (Dawson and Woodward 1914, pp. 84–85). A mastodon bone, apparently intentionally modified to form a pointed tool, was also found

By this time, Piltdown had become quite a tourist attraction. Visiting researchers were politely allowed to assist in the ongoing excavations. Motor coaches came with members of natural history societies. Dawson even had a picnic lunch at the Piltdown site for the Geological Society of London (Millar 1972, p. 132). Dawson achieved celebrity. Indeed, the scientific name for the Piltdown hominid became *Eoanthropus dawsoni*, meaning "Dawson's dawn man." But Dawson's enjoyment of his fame was short-lived; he died in 1916.

8.4 A SECOND DAWN MAN DISCOVERY

Doubts persisted that the jaw and skull of *Eoanthropus* belonged to the same creature, but these doubts weakened when Woodward (1917) reported the 1915 discovery of a second set of fossils about 2 miles from the original Piltdown site.

Woodward (1917, p. 3) stated: "One large field, about 2 miles from the Piltdown pit, had especially attracted Mr. Dawson's attention, and he and I examined it several times without success during the spring and autumn of 1914. When, however, in the course of farming, the stones had been raked off the ground and brought together into heaps, Mr. Dawson was able to search the material more satisfactorily; and early in 1915 he was so fortunate as to find here two well-fossilized pieces of human skull and a molar tooth, which he immediately recognized as belonging to at least one more individual of *Eoanthropus dawsoni*. Shortly afterwards, in the same gravel, a friend met with part of the lower molar of an indeterminable species of rhinoceros, as highly mineralized as the specimens previously found at Piltdown itself."

Woodward (1917, p. 3) added: "The most important fragment of human skull is part of . . . a right frontal bone. . . . It is in exactly the same mineralized condition as the original skull of *Eoanthropus*, and deeply stained with iron-oxide." The second fragment was from the occipital, the bone of the lower rear portion of the skull.

The tooth found at what came to be called the Piltdown II site was a left lower first molar, which according to Woodward (1917, p. 5) was "stained brown with iron oxide in the usual manner."

The report on the fossils found at the Piltdown II site included these remarks by W. P. Pycraft about the molar found there: "If the new tooth be compared with the corresponding molars of a Melanesian, a Tasmanian, and a Chimpanzee, of approximately the same size, it will readily be recognized as essentially human. In the considerable depth of the crown and its gradual passage into the root, it agrees with the human tooth and differs from that of the Chimpanzee, in which the crown is very brachyodont [broad] and overhangs the root. . . . These comparisons are made because it has been stated that the molar teeth in the Piltdown mandible are those of a Chimpanzee" (Woodward 1917, p. 6).

Gerritt T. Miller, of the Smithsonian Institution, had sent Pycraft a chimpanzee jaw with molars flattened by wear, like those in the original Piltdown jaw. The molars of human beings are generally worn flat, while the surfaces of ape and chimp molars are usually more pointed (Figure 8.2). The flat molars of the otherwise chimpanzeelike Piltdown mandible were taken as a sign that the mandible was not that of a chimpanzee or other member of the ape family. So by presenting a chimp jaw with flattened molars, Miller was implying that the Piltdown mandible might still be that of an ape rather than an early human. This would mean that the Piltdown cranium and jaw belonged to two different creatures, the former to a human and the latter to an ape.

Pycraft replied that the flat molar surfaces on the chimpanzee jaw Miller had sent him were due "not to normal wear, but to some interference in the normal 'bite.'" Pycraft added: "In no other chimpanzee that the speaker had examined had he ever found anything in the matter of wear comparable with the molars of Mr. Miller's specimen. These are quite abnormal in this regard, and therefore of no value as evidence that the Piltdown teeth might, even in the wear of their crowns, agree with the teeth of chimpanzees" (Woodward 1917, p. 6).

In the published summary of the discussion among scientists that took place following Arthur Smith Woodward's report on the Piltdown II discoveries, it is recorded: "Prof. A. Keith said that these further Piltdown 'finds' established beyond any doubt that *Eoanthropus* was a very clearly differentiated type of being—in his opinion a truly human type" (Woodward 1917, p. 6). In the discussion, Sir Ray Lankester stated: "The present 'find' therefore makes it impossible to regard the Piltdown man as an isolated abnormal individual" (Woodward 1917, p. 6).

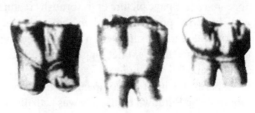

Figure 8.2. The crown of a human molar (middle) is generally worn flat, while the crown of a chimpanzee molar (right) generally remains pointed. In this respect, a Piltdown II molar (left) resembles a human molar (Woodward 1917, plate 1).

8.5 ONE CREATURE OR TWO?

Keith and Lankester, like Dawson and Woodward, accepted the idea that the humanlike skull and apelike jaw belonged to the same creature, which represented an Early Pleistocene ancestor of *Homo sapiens*. After all, what were the odds of finding a human skull and an ape's jaw in such close proximity, with no sign of the matching human jaw and ape's skull? But two German scientists interpreted the Piltdown finds somewhat differently. Franz Weidenreich said the molar of the Piltdown II specimen was human, indicating Piltdown II was a fully human find. As for the original Piltdown jaw, von Koenigswald (1956, p. 179) informs us: "A pupil of Weidenreich's wanted to assign the mandible to a new ape, *Boreopithecus,* the 'Northern Ape.' Weidenreich compared it in the first place to the orang-utan, because, like the latter, it lacked certain muscle-attachments on the under margin of the jaw." If Weidenreich's view were to have been accepted, this would have left scientists with a fully human skull and an ape jaw, from an ape living at the same time as Piltdown man. Still, H. Weinert thought the original Piltdown jaw could easily be reconstructed as human (Weiner *et al.* 1955, p. 231). It is interesting to note the widely varying interpretations by professional anthropologists.

So after years of study, debate continued about whether the jaw and skull belonged to the same creature. Ales Hrdlicka, among other American scientists, was convinced they were from different creatures. Eager to resolve the controversy in his own mind, the famous American anthropologist Henry Fairfield Osborn, accompanied by two other scientists, journeyed to England to view the Piltdown fossils. Osborn was no stranger to controversy. Around this same time he proposed a new hominid genus on the basis of a single molar found by a geologist in western Nebraska. The ape-man was named *Hesperopithecus* (Cousins 1971, pp. 40–41). In England, Sir Grafton Elliott Smith wrote a three-page article for the *Illustrated London News* (June 24, 1922), in which there appeared a full-page picture of the brutish creature walking along carrying a club, with his wife preparing food in the background. Later, Dr. W. K. Gregory demonstrated to the satisfaction of the scientific community that the Nebraska tooth belonged to an extinct pig (Cousins 1971, p. 40). Thereafter everyone was silent about *Hesperopithecus.*

Osborn and his two companions had tended to favor Gerritt S. Miller's proposal that Piltdown man's jaw was actually that of a separate chimpanzeelike creature. In his book *Ancient Hunters,* W. J. Sollas (1924, pp. 189–190) wrote: "As a consequence, Profs. Osborn, Matthews and McGregor, who had previously been much impressed by Mr. Miller's observations, took the opportunity when they last visited Europe to make a special pilgrimage to the British Museum in order that they might see and handle the actual bones themselves of the Piltdown man, previously known to them only as represented by plaster casts. The result was eminently satisfactory, the doubts these observers had previously entertained were dissipated and they fully recognised that the jaw and skull had rightly been assigned to a single individual" (Sollas 1924, pp. 189–190).

In his book *Man Rises to Parnassus,* Osborn (1927, pp. 45–74) recalled how he had approached the British Museum feeling greatly thankful this treasure house had been spared destruction from German bombardment during the Zeppelin raids of World War I. After spending several hours with Woodward examining the Piltdown fossils, and finally concurring that the jaw and skull belonged to the same creature, Osborn recalled the opening words of a prayer sung at Yale: "Paradoxical as it may appear O Lord, it is nevertheless true."

8.6 THE EFFECT OF NEW DISCOVERIES ON PILTDOWN MAN

But as more hominid fossils were found, the Piltdown fossil, with its *Homo sapiens* type of cranium, introduced a great deal of uncertainty into the construction of the line of human evolution. At Choukoutien (now Zhoukoudian), near Peking (now Beijing), researchers initially uncovered a primitive-looking jaw resembling that of Piltdown man. But when the first Peking man skull was uncovered in 1929, it had the low forehead and pronounced brow ridge of *Pithecanthropus erectus* of Java, now classified with Peking man as *Homo erectus* (Millar 1972, p. 173). In the same decade, Raymond Dart uncovered the first *Australopithecus* specimens in Africa. Other *Australopithecus* finds followed, and like Java man and Peking man they also had low foreheads and prominent brow ridges. Most British anthropologists, however, decided that *Australopithecus* was an apelike creature that was not a human ancestor. That lessened the threat to Piltdown man, who was, nevertheless, beginning to seem out of place.

In spite of the new evidence, Sir Arthur Smith Woodward remained a champion of Piltdown man until his death. Von Koenigswald (1956, p. 182) wrote: "Sir Arthur Smith Woodward was so convinced of the significance of Piltdown man that he had a small house built at Haywards Heath, not far from the site of the find, so that he could always keep an eye on it. He was a man with a strong sense of fair play, and when he felt he had been passed over on the occasion of a promotion, he left his beloved British Museum, never to set foot in it again. From now on he dedicated his whole life to Piltdown man. When we visited him at Haywards Heath in 1937 he talked of nothing else. . . . In spite of the bad weather we had to go out in a taxi to the site of the discovery. Standing under a big umbrella, Sir Arthur showed us the spot at which he had unearthed the celebrated find."

But after World War II, new finds by Robert Broom led the British to change their minds about *Australopithecus*. Sir Arthur Keith telegrammed Broom: "All my landmarks have gone, you have found what I never thought could be found: a man-like jaw associated with an ape-like skull—the exact reverse of the Piltdown evidence" (Goodman 1982, p. 94). So now what was to be done with Piltdown man, who was thought to be as old as the *Australopithecus* finds that had by then been made?

In addition to *Australopithecus*, some of the anomalous human fossils discussed previously in this book also appeared to contradict the evidence provided by Piltdown man. In his book *Meeting Prehistoric Man*, von Koenigswald (1956, pp. 179–180) addressed this problem stating: "Apart from Piltdown man there was a whole series of allegedly very ancient *sapiens* forms, none of which, however, possessed such a simian jaw. The Foxhall mandible, which already has a chin, is said to have come from deep in the Red Crag on the East Anglian coast. Then there is the complete skeleton from Galley Hill near Northfleet in Kent, the finds at Denise in southern France, and various others. In the past, there was no conclusive method of determining the age of skeletal remains. Since man is in the habit of burying his dead, human bones occur in strata of differing ages. In most cases, of course, it is not difficult to ascertain whether remains are those of a modern interment or not. But there are finds that remain doubtful, and these have misled certain anthropologists into ascribing a very great geological age to *Homo sapiens* as such. This is naturally very important for the interpretation of our evolutionary history; for if we [i.e., *Homo sapiens*] really go back to the Tertiary, then all the forms of primordial man discovered up to the present are merely parallel forms, interesting in themselves but without any bearing on the history of our own stock." This conclusion becomes even stronger when we consider discoveries von Koenigswald neglected to mention, such as those made at the Castenedolo site (Section 6.2.2). Putting all the evidence on the table, we appear to be dealing with the coexistence of various fully human and ape-man-like forms rather than evolutionary relationships in which primitive ape-men clearly give rise to the modern human type..

Von Koenigswald (1956, p. 180) added: "Hence Piltdown man, more than any other find, introduced an element of uncertainty into our ideas on the course of human evolution; and anyone who takes the trouble to read several books on fossil men will see to his horror that practically every author holds a different view as to the connexions between the known human forms. It is tempting to take refuge in the theory of 'parallel evolution,' according to which *Homo sapiens* is derived direct from *Eoanthropus* ("Dawn Man") and all other early hominids are simply 'collateral forms.'" Parallel evolution, of the kind von Koenigswald described above, is exactly the position that some British anthropologists maintained for several decades.

8.7 MARSTON'S CRUSADE

Meanwhile, an English dentist named Alvan Marston kept badgering British scientists about Piltdown man, contending that something was not quite right about the fossils. In 1935, Marston discovered a human skull at Swanscombe, accompanied by fossil bones of 26 kinds of Middle Pleistocene animals. Desiring that his discovery be hailed as "the oldest Englishman," Marston challenged the age of the Piltdown fossils (Johanson and Edey 1981, 79–80).

In 1949, Marston convinced Kenneth P. Oakley of the British Museum to test both the Swanscombe and Piltdown fossils with the newly developed fluorine content method. The Swanscombe skull had the same fluorine content as the fossil animal bones found at the same site, thus confirming its Middle Pleistocene antiquity. The test results for the Piltdown specimens were more confusing.

Oakley, it should be mentioned, apparently had his own suspicions about Piltdown man. Oakley and Hoskins, coauthors of the fluorine content test report, wrote (1950, p. 379) that "the anatomical features of *Eoanthropus* (assuming the material to represent one creature) are wholly contrary to what discoveries in the Far East and in Africa have led us to expect in an early Pleistocene hominid."

Oakley tested the Piltdown fossils in order to determine whether the cranium and jaw of Piltdown man really belonged together. The fluorine content of four of the original Piltdown cranial bones ranged from 0.1 to 0.4 percent. The jaw yielded a fluorine content of 0.2 percent, suggesting it belonged with the skull. The bones from the second Piltdown locality gave similar results.

The fluorine content of some of the Piltdown animal bones was for the most part substantially higher, with one group (Early Pleistocene forms) ranging from 1.9 to 3.1 percent. But another group (Middle to Late Pleistocene forms) ranged from 0.1 to 1.5 percent (Oakley and Hoskins 1950, p. 381). Oakley concluded: "Comparison of the fluorine values of the specimens attributed to *Eoanthropus* and of the bones and teeth the geological ages of which are certain leaves little doubt that: (1) all the specimens of *Eoanthropus* including the remains of the second skull found two miles away, are contemporaneous; (2) *Eoanthropus* is, at the earliest Middle Pleistocene" (Oakley and Hoskins 1950, p. 381).

Giving a more exact estimate of the date, based in part upon his interpretation of the geological age of the Piltdown gravels, Oakley wrote: "*Eoanthropus* may be provisionally referred to the last warm interglacial period (Riss-Würm interglacial); that is, early Upper Pleistocene, although here it should be noted that some authorities count Riss-Würm as Middle Pleistocene" (Oakley and Hoskins 1950, p. 382).

Oakley's opinion that the Piltdown man fossils belonged to the last interglacial period is consistent with the view of R. G. West (1968, p. 343), an expert on the Pleistocene geology of England, who assigned the Piltdown gravels to the Late or Middle Pleistocene. Modern authorities place the last interglacial at 75,000 to 125,000 years ago, spanning the boundary of the Late Pleistocene and Middle Pleistocene. This is quite a bit more recent than the Early Pleistocene date originally ascribed to the Piltdown fossils, but it is still anomalously old for a skull of the fully human type in England. According to current theory, *Homo sapiens sapiens* arose in Africa about 100,000 years ago and only much later migrated to Europe, at around 30,000 years ago (Gowlett 1984, p. 118).

Oakley apparently still accepted that the skull and jaw belonged to the same creature. He suggested that "Piltdown man, far from being a primitive type, may have been a late specialized hominid which evolved in comparative isolation" (Goodman 1983, p. 100).

This did not entirely satisfy Marston, who was convinced the Piltdown jaw and skull were from completely different creatures. From his knowledge of medicine and dentistry, Marston concluded that the skull, with its closed sutures, was that of a mature human, while the jaw, with its incompletely developed molars, was from an immature ape (Goodman 1983, p. 101). He also felt that the dark staining of the bones, taken as a sign of great antiquity, was caused by Dawson soaking them in a solution of potassium dichromate to harden them.

8.8 EVIDENCE OF FORGERY

Marston's ongoing campaign about the Piltdown fossils eventually drew the attention of J. S. Weiner, an Oxford anthropologist. Weiner himself soon became convinced that something was wrong with the Piltdown fossils. He noticed on teeth abrasion marks that to him indicated artificial filing. As early as 1916, C. W. Lyne, a dentist, had noted that the Piltdown molars, which apparently belonged to a fairly young individual, should not be as worn down as they appeared (Goodman 1983, p. 102).

J. S. Weiner reported his suspicions to W. E. Le Gros Clark, head of the anthropology department at Oxford University, but Le Gros Clark was at first skeptical. On August 5, 1953, Weiner and Oakley met with Le Gros Clark at the British Museum, where Oakley removed the actual Piltdown specimens from a safe so they could examine the controversial relics. At this point, Weiner (1955, pp. 44–45) presented to Le Gros Clark a chimpanzee tooth he had taken from a museum collection and then filed and stained. The resemblance to the Piltdown molar was so striking that Le Gros Clark authorized a full investigation of all the Piltdown fossils (Weiner *et al.* 1953, pp. 141–142). One wonders if this step would have been taken if the Piltdown man fossils had fit more comfortably within the emerging hominid evolutionary progression.

In any case, a second fluorine content test, using new techniques, was applied to the Piltdown human fossils. Three pieces of the Piltdown skull now yielded a fluorine content of .1 percent. By this time, all the other fossil material from Piltdown was also suspect. Consequently, the Piltdown skull was compared with Late Pleistocene fossils from other sites in the same region, which showed a minimum fluorine content of .1 percent. But the Piltdown jaw and teeth yielded a much lower fluorine content of .01–.04 percent. A modern chimpanzee tooth had .06 percent fluorine. Because fluorine content increases with the passing of time, the results indicated a younger age for the jaw and teeth than the skull. The test results were reported in a paper authored by Weiner, Le Gros Clark, and Oakley, who stated: "the results leave no doubt that, whereas the Piltdown cranium may well be Upper Pleistocene as claimed in 1950 [by Oakley], the mandible, canine tooth and isolated molar are quite modern" (Weiner *et al.* 1953, p. 143). The conclusion that the jaw and cranial bones were of different ages is a correct application of the fluorine content test, which, as we have seen, is best used as an indicator of the relative ages of bones found in close proximity in the same deposit. However, the attribution of a Late Pleistocene date to the skull by comparison with fossils from other sites is not quite as sound. As we have seen, the fluorine content in groundwater and its rate of absorption are subject to quite a bit of variation at different sites and over long periods of time, making relative dating estimates by this method alone subject to doubt. Thus the fluorine content test results do not completely rule out an earlier—perhaps Middle Pleistocene— date for the cranial bones discovered at Piltdown.

Regarding the two fluorine content tests by Oakley, we see that the first indicated both the skull and jaw were of the same age whereas the second indicated they were of different ages. It was stated that the second set of tests made use of new techniques—that happened to give a desired result. This sort of thing occurs quite often in paleoanthropology—researchers run and rerun tests, or refine their methods, until an acceptable result is achieved. Then they stop. In such cases, it seems the test is calibrated against a theoretical expectation.

Nitrogen content tests were also run on the Piltdown fossils. Examining the results, Weiner found that the skull bones contained 0.6–1.4 percent nitrogen while the jaw contained 3.9 percent and the dentine portion of some of the Piltdown teeth contained 4.2–5.1 percent. The test results therefore showed that the cranial fragments were of a different age than the jaw and teeth, demonstrating they were from different creatures. Modern bone contains about 4–5 percent nitrogen, and the content decreases with age. So it appeared the jaw and teeth were quite recent, while the skull was older (von Koenigswald 1956, p. 181; Weiner *et al.* 1953, p. 144).

By including for comparison in their report a nitrogen content of 0.7 percent for a Late Pleistocene bone from London, Weiner and his coauthors indicated the Piltdown skull fragments, with a similar nitrogen content, were probably also Late Pleistocene in age. However, as discussed in connection with the Galley Hill fossils in Chapter 6 and Appendix 1, the rate of nitrogen decay is subject to many variables. This greatly reduces the usefulness of comparing nitrogen contents of bones from different sites. In any case, the results of the nitrogen content test still allowed one to believe that the skull, at least, was native to the Piltdown gravels.

But finally even the skull came under suspicion. Weiner and his associates wrote in a lengthy report published by the British Museum: "As the fluorine and nitrogen content of the cranial bones were consistent with their being fairly ancient [1950 report], it seemed at first that the hoax had been based on a genuine discovery of portions of a skull in the gravel, and that animal remains and implements had been subsequently 'planted' to suggest that it was Pliocene or Early Pleistocene in age. As the investigations proceeded the skull too became suspect. Dr. G. F. Claringbull carried out an X-ray crystallographic analysis of these bones and found that their main mineral constituent, hydroxy-apatite, had been partly replaced by gypsum. Studies of the chemical conditions in the Piltdown sub-soil and ground-water showed that such an unusual alteration could not have taken place naturally in the Piltdown gravel. Dr. M. H. Hey then demonstrated that when sub-fossil bones are artificially iron-stained by soaking them in strong iron sulphate solutions this alteration does occur. Thus it is now clear that the cranial bones had been artificially stained to match the gravel, and 'planted' at the site with all the other finds" (Weiner *et al.* 1955, p. 257).

8.9 WAS THE PILTDOWN SKULL GENUINE?

Despite the evidence presented in the British Museum report, it can still be argued that the skull was originally from the Piltdown gravels. All of the skull pieces were darkly iron-stained throughout, while the jaw bone, also said to be a forgery, had only a surface stain (Bowden 1977, p. 13). Furthermore, a chemical analysis of the first skull fragments discovered by Dawson showed that they had a very high iron content of 8 percent, compared to only 2–3 percent for the jaw (Weiner *et al.* 1953, p. 145). This evidence suggests that the skull fragments acquired their iron-staining (penetrating the entire bone and contributing 8 percent iron to the bones' total mineral content) from a long stay in the iron-rich gravels at Piltdown. The jaw, with simply a surface stain and much smaller iron content, appears to be of a different origin.

If the skull fragments were native to the Piltdown gravels and were not artificially stained as suggested by Weiner and his associates, then how is one to explain the gypsum (calcium sulfate) in the skull fragments? One possibility is that Dawson used sulfate compounds (along with or in addition to potassium dichromate) while chemically treating the bones to harden them after their excavation, thus converting part of the bones' hydroxy-apatite into gypsum.

Another option is that the gypsum accumulated while the skull was still in the

Piltdown gravels. The British Museum scientists claimed that the concentration of sulfates at Piltdown was too low for this to have happened. But Bowden (1977, p. 15) observed that sulfates were present in the area's groundwater at 63 parts per million and that the Piltdown gravel had a sulfate content of 3.9 milligrams per 100 grams. Admitting these concentrations were not high, Bowden said they could have been considerably higher in the past. We note that Oakley appealed to higher past concentrations of fluorine in groundwater to explain an abnormally high fluorine content for the Castenedolo bones (Appendix 1).

Significantly, the Piltdown jaw contained no gypsum. The fact that gypsum is present in all of the skull fragments but not in the jaw is consistent with the hypothesis that the skull fragments were originally from the Piltdown gravel while the jaw was not.

Chromium was present in the five skull fragments found by Dawson alone, before he was joined by Woodward. This can be explained by the known fact that Dawson dipped the fragments in potassium dichromate to harden them after they were excavated. The additional skull fragments found by Dawson and Woodward together did not contain any chromium.

The jaw did have chromium, apparently resulting from an iron-staining technique involving the use of an iron compound and potassium dichromate.

To summarize, it may be that the skull was native to the Piltdown gravels and became thoroughly impregnated with iron over the course of a long period of time. During this same period of time, some of the calcium phosphate in the bone was transformed into calcium sulfate (gypsum) by the action of sulfates in the gravel and groundwater. Some of the skull fragments were later soaked by Dawson in potassium dichromate. This would account for the presence in them of chromium. The fragments found later by Dawson and Woodward together were not soaked in potassium dichromate and hence had no chromium in them. The jaw, on the other hand, was artificially iron-stained, resulting in only a superficial coloration. The staining technique involved the use of a chromium compound, which accounts for the presence of chromium in the jaw, but the staining technique did not produce any gypsum.

Alternatively, if one accepts that the iron-staining of the skull fragments (as well as the jaw) was accomplished by forgery, then one has to assume that the forger used three different staining techniques: (1) According to the British Museum scientists, the primary staining technique involved the use of an iron sulfate solution with potassium dichromate as an oxidizer, yielding gypsum (calcium sulfate) as a byproduct. This would account for the presence of gypsum and chromium in the five iron-stained skull fragments first found by Dawson. (2) The four skull fragments found by Dawson and Woodward together contained gypsum but no chromium (Weiner *et al.* 1955, p. 269; Woodward 1948, p. 10). These must have been stained by another method. In connection with some

Piltdown beavers' teeth, which also contained gypsum but no chromium, the British Museum report said: "These were presumably stained by another technique, which dispensed with the use of a dichromate solution as an oxidizer" (Weiner *et al.* 1955, p. 252). (3) The jaw, which contained chromium but no gypsum, must have been stained by a method that involved use of iron and chromium compounds, but which did not produce gypsum. It is hard to see why a forger would have used so many methods when one would have sufficed. We must also wonder why the forger carelessly stained the jaw to a far lesser extent than the skull, thus risking detection.

Additional evidence, in the form of eyewitness testimony, suggests that the skull was in fact originally from the Piltdown gravels. The eyewitness was Mabel Kenward, daughter of Robert Kenward, the owner of Barkham Manor. On February 23, 1955, the *Telegraph* published a letter from Miss Kenward that contained this statement: "One day when they were digging in the unmoved gravel, one of the workmen saw what he called a coconut. He broke it with his pick, kept one piece and threw the rest away" (Vere 1959, p. 4). Particularly significant was the testimony that the gravel was unmoved.

Francis Vere amplified this information in a book published the same year that Miss Kenward wrote her letter: "the first discovery of the skull was most graphically described to me by Miss Mabel Kenward herself. She remembers seeing from her window her father, Mr. Robert Kenward, standing by the pit looking at the workmen, while they were digging in the gravel. One of them said there was something just like a coconut in the pit, and her father said that they should take care how they got it out, but before he could stop them, a blow from the pick shattered the skull and pieces flew in all directions. He picked up as many pieces as he could find and came into the house, whereupon Miss Kenward exclaimed, 'What on earth have you loaded up your pockets with all those old stones for?' He laid them out on the table and looked at them, but later returned them to the workmen, telling them to give them to Mr. Dawson next time he came. She could not say, of course, whether all the pieces were given to Dawson by the workmen. Presumably, as recounted by Miss Kenward and Woodward, the workman kept one piece which he later handed to Dawson, and threw away the rest" (Bowden 1977, p. 12).

Even Weiner himself (1955, p. 193) wrote: "we cannot easily dismiss the story of the gravel diggers and their 'coconut' as pure invention, a plausible tale put about to furnish an acceptable history for the pieces. . . . Dawson told frequently of the labourers' part (even if he did not clearly record the coconut episode) in the next few years and could hardly have had reason to fear anyone's seeking confirmation of the men. Granting, then the probability that the workmen did find a portion of skull, it is still conceivable that what they found was not the semi-fossil *Eoanthropus* but some very recent and quite ordinary burial." Weiner suggested that the culprit, whoever he may have been, could have then substituted treated skull pieces for the ones actually found. But if the workmen were dealing with "a very recent and quite ordinary burial" then where were the rest of the bones of the corpse? At least some of the harder bones, like the femurs, should have remained.

In the end, Weiner suggested that an entire fake skull was planted, and the workmen broke it. But Mabel Kenward testified that the surface where the workman started digging was unbroken. The spot was also on the manor grounds, quite near the house, it appears from the testimony, and the gravel was so compact and cemented that it took a pick to break into it. It would not have been easy, it seems, for some unknown person to enter onto private property in England and excavate a deep hole with a pick by the drive near a manor house, any time of day or night, without being questioned. In any case, if we accept the plant hypothesis in this instance, then practically any discovery of human fossil remains anywhere could also be said to have been a plant and forgery.

What about the altered animal fossils allegedly planted at Piltdown? Those could have been introduced without arousing suspicion, because after the skull's discovery the investigators would often be searching through the gravel, which was already broken up.

Robert Essex, a scientist personally acquainted with Dawson in the years 1912 to 1915, provided interesting testimony about the Piltdown jaw, or jaws, as it turns out. Essex wrote in 1955: "Another jaw not mentioned by Dr. Weiner came from Piltdown much more human than the ape's jaw, and therefore much more likely to belong to the Piltdown skull parts which are admittedly human. I saw and handled that jaw and know in whose bag it came to Dawson's office. The jaw was also seen by Mr. H. H. Wakefield, then an articled clerk of Dawson's, and he has given written evidence of seeing it. Dawson never saw it, and the owner himself probably never knew until 1953 that anybody but himself had seen it" (Bowden 1977, p. 37).

Essex then gave more details. At the time, he had been science master at a local grammar school, located near Dawson's office. Essex stated: "One day when I was passing I was beckoned in by one of the clerks whom I knew well. He had called me in to show me a fossil half-jaw much more human than an ape's and with three molars firmly fixed in it. When I asked where this object came from, the answer was 'Piltdown.' According to the clerk, it had been brought down by one of the 'diggers' who, when he called and asked for Mr. Dawson, was carrying a bag such as might be used for carrying tools. When he was told that Mr. Dawson was busy in court he said he would leave the bag and come back. When he had gone, the clerk opened the bag and saw this jaw. Seeing me passing he had called me in. I told him he had better put it back and that Mr. Dawson would be cross if he knew. I found afterwards that when the 'digger' returned, Mr. Dawson was still busy in court, so he picked up his bag and left" (Bowden 1977, p. 37). Essex later saw photographs of the Piltdown jaw. Noting the jaw was not the same one he had seen in Dawson's office, he communicated this information to the British Museum.

These reports are significant for the following reason. It is unlikely that a forger would have planted a human jaw at Piltdown, along with everything else. So the story about the discovery of a human jaw tends to confirm the view that the human skull found at Piltdown was native to the gravels. Even if we grant that every other bone connected with Piltdown is a forgery, if the skull was found *in situ*, we are confronted with what could be one more case of *Homo sapiens sapiens* remains from the late Middle Pleistocene or early Late Pleistocene.

8.10 THE IDENTITY OF THE FORGER

Most recent writing, totally accepting that all the Piltdown fossils and implements were fraudulent, has focused on identifying the culprit. Weiner and Oakley insinuated that Dawson, the amateur paleontologist, was to blame. Woodward, the professional scientist, was absolved.

But Dawson's honor was defended by, of all people, Alvan Marston. At a meeting of the Geological Society of London, held on November 25, 1953, Weiner and Oakley made a slide presentation detailing the evidence that led them to conclude the Piltdown jaw had been forged from a modern orangutan jaw. Marston, who strongly objected to any suggestion that Dawson was guilty of forgery, showed his own slides proving that the jaw, though that of an ape, was a genuine fossil with no sign of deliberate staining or filing. Marston apparently believed that although Dawson may have been mistaken about the age of the jaw, and in connecting it with the human cranium found with it, he was not a forger. Marston accused the British Museum of making Dawson a scapegoat. Dawson could not fight back, but he, Marston, could. Marston shouted, "Let them try and tackle me!" A British author wrote: "The hubbub at this meeting was wrongly reported in the United States to have developed into a series of fist fights" (Millar 1972, pp. 218–219).

Von Koenigswald, like Weiner and Oakley, blamed Dawson: "It is certainly not nice to accuse a dead man who cannot defend himself; but everything quite clearly points to his responsibility for the forgery. Indeed, it has now turned out that neither the fossils nor the tools belong to this locality at all, and that the whole find was carefully planted" (Wendt 1972, p. 154). But as we have seen, the evidence is not so clear cut in this regard. Indeed, it is difficult to see how anyone could say that anything "quite clearly points" to anything in regard to the Piltdown controversy.

For example, if Dawson were involved in chemical forgery, why did he immediately send the five pieces of skull originally found at Piltdown to a public chemist for analysis? Furthermore, Dawson openly admitted he had treated the fragments with potassium dichromate to harden them, and it was well known that he was performing some experiments on staining techniques in his offices (Bowden 1977, p. 26). If Dawson had really been involved in some deliberate fraudulent staining and planting of the original skull pieces and all the other fossils found at Piltdown, it seems he would have been more careful.

Furthermore, it would appear that the Piltdown forgery (even excluding the skull) demanded extensive technical knowledge and capability—beyond that seemingly possessed by Dawson, an amateur anthropologist. Gavin De Beer, a director of the British Museum of Natural History, wrote in a foreword to a report by Weiner, Le Gros Clark, and Oakley: "We are now in a position to give an account of the full extent of the Piltdown hoax. The mandible has been shown by . . . anatomical and X-ray evidence to be almost certainly that of an immature orang-utan; that it is entirely Recent has been confirmed by a number of microchemical tests, as well as by the electron-microscope demonstration of organic (collagen) fibers; the black coating on the canine tooth, originally assumed to be an iron encrustation, is a paint (probably Vandyke brown); the so-called turbinal bone is shown by its texture not to be a turbinal bone at all, but thin fragments of probably non-human limb-bone; all the associated flint implements have been artificially iron-stained; the bone implement was shaped by a steel knife; the whole of the associated fauna must have been 'planted,' and it is concluded from radioactivity tests and fluorine analysis that some of the specimens are of foreign origin" (Weiner *et al.* 1955, p. 228). It appears that a professional scientist, who had access to rare fossils and knew how to select them and modify them to give the impression of a genuine faunal assemblage of the proper age, had to be involved in the Piltdown episode.

Some have tried to make a case against Teilhard de Chardin, who had studied at a Jesuit college near Piltdown and who had become acquainted with Dawson as early as 1909. A *Stegodon* tooth found at Piltdown was believed by Weiner and his associates to have come from a North African site that might have been visited by Teilhard de Chardin in the period from 1906 to 1908, during which time he was a lecturer at Cairo University (Millar 1972, p. 232).

Woodward is another suspect. Over the course of several decades, he tightly controlled access to the original Piltdown fossils, which were stored under his care in the British Museum. This could be interpreted as an attempt to prevent evidence of forgery from being noticed by other scientists. It is interesting, for example, that Woodward originally reported the following facts about the Piltdown site: "Portions of the bed are rather finely stratified, and the materials are usually cemented together by iron oxide, so that a pick is often needed to dislodge portions—more especially at one particular horizon near the base. It is in this last mentioned stratum that all the fossil bones and teeth discovered *in situ* by us have occurred" (Dawson and Woodward 1913, p. 119). If all the Piltdown fossils are fakes, then why did Woodward not notice something was wrong? There are three possible reasons: (1) the planting was done in an incredibly clever fashion, so as to exactly reproduce the fine stratification and the hard, unbroken consistency of the bed in question; (2) Woodward, supposedly an expert, missed obvious signs of planting; (3) Woodward was involved in the planting. It is also suspicious that Woodward so tightly controlled access to the original Piltdown specimens, compelling all but a select few researchers to examine only casts.

Ronald Millar, author of *The Piltdown Men*, suspected Grafton Eliot Smith. Having a dislike for Woodward, Smith may have decided to entrap him with an elegant deception. Smith, like Teilhard de Chardin, had spent time in Egypt, and so had access to fossils that could have been planted at Piltdown (Johanson and Edey 1981, pp. 81–82).

Frank Spencer, a professor of anthropology at Queens College of the City University of New York, has written a book that blames Sir Arthur Keith, conservator of the Hunterian Museum of the Royal College of Surgeons, for the Piltdown forgery (Wilford 1990). Keith believed that modern humans evolved earlier than other scientists could accept, and this, according to Spencer, impelled him to conspire with Dawson to plant evidence favoring his hypothesis.

Another suspect was William Sollas, a professor of geology at Cambridge. He was named in a tape-recorded message left by English geologist James Douglas, who died in 1979 at age 93. Sollas disliked Woodward, who had criticized a method developed by Sollas for making plaster casts of fossils. Douglas recalled he had sent mastodon teeth like those found at Piltdown to Sollas from Bolivia and that Sollas had also received some potassium dichromate, the chemical apparently used in staining many of the Piltdown specimens. According to Douglas, Sollas had also "borrowed" some ape teeth from the Oxford museum collection (Johanson and Edey 1981, p. 83). According to this view, Sollas secretly enjoyed seeing Woodward duped by the Piltdown forgeries.

But if Piltdown does represent a forgery, it is likely that something more than personal revenge was involved. Spencer said that the evidence "had been tailored to withstand scientific scrutiny and thereby promote a particular interpretation of the human fossil record" (Wilford 1990).

Possible motivations for forgery by a professional scientist may be sensed when we consider the inadequacies of the evidence for human evolution that had accumulated by the beginning of the twentieth century. Darwin had published *The Origin of Species* in 1859, setting off almost immediately a search for fossil evidence connecting *Homo sapiens* with the ancient Miocene apes. Leaving aside the discoveries suggesting the presence of fully modern humans in the Pliocene and Miocene, Java man and the Heidelberg jaw were the only fossil discoveries that science had come up with. And as we have seen in Chapter 7, Java man in particular did not enjoy unanimous support within the scientific community.

Right from the start there were ominous suggestions that the apelike skull did not really belong with the humanlike thighbone found 45 feet away from it. As we have seen in this chapter, a number of scientists in England and America, such as Arthur Smith Woodward, Grafton Eliot Smith, and Sir Arthur Keith were developing alternative views of human evolution in which the formation of a high-browed humanlike cranium preceded the formation of a humanlike jaw. Java man, however, showed a low-browed cranium like that of an ape.

Since so many modern scientists have indulged in speculation about the identity and motives of the presumed Piltdown forger, we would also like to introduce a tentative hypothesis. Consider the following scenario. Workmen at Barkham Manor actually discovered a genuine Middle Pleistocene skull, in the manner described by Mabel Kenward. Pieces of it were given to Dawson. Dawson, who had regularly been communicating with Woodward, notified him. Woodward, who had been developing his own theory of human evolution and who was very worried about science's lack of evidence for human evolution after 50 years of research, planned and implemented the forgery. He did not act alone, but in concert with a select number of scientists at the British Museum, who assisted in acquiring the specimens and preparing them so as to withstand the investigations of scientists not in on the secret.

Oakley, who played a big role in the Piltdown exposé, himself wrote (Boule and Vallois 1957, p. 3): "The Trinil [Java man] material was tantalizingly incomplete, and for many scientists it was inadequate as confirmation of Darwin's view of human evolution. I have sometimes wondered whether it was a misguided impatience for the discovery of a more acceptable 'missing link' that formed one of the tangled skein of motives behind the Piltdown Forgery (1912)."

Weiner also admitted the possibility: "Behind it all we sense, therefore, a strong and impelling motive. . . . The planning . . . must betoken a motive more driving than a mere hoax or prank. . . . There could have been a mad desire to assist the doctrine of human evolution by furnishing the 'requisite' 'missing link.' . . . Piltdown might have offered irresistible attraction to some fanatical biologist to make good what Nature had created but omitted to preserve" (Weiner 1955, pp. 117–118).

Unfortunately for the hypothetical conspirators, the discoveries that turned up over the next few decades did not support the evolutionary theory represented by the Piltdown forgery. The discoveries of new specimens of Java man and Peking man, as well as the *Australopithecus* finds in Africa, were accepted by many scientists as proving the low-browed ape-man ancestor hypothesis, the very idea the high-browed Piltdown man was meant to discredit and replace.

Time passed, and the difficulties in constructing a viable evolutionary lineage for the fossil hominids increased. At a critical moment, the remaining insiders in the British Museum chose to act. Perhaps enlisting unwitting colleagues, they organized a systematic exposé of the forgery the Museum had perpetrated earlier in the century. In the course of this exposé, perhaps some of the specimens were further modified by chemical and physical means to lend credence to the idea of forgery.

The idea of a group of conspirators operating out of the British Museum, perpetrating a forgery and then later exposing the same, is bound to strike many as farfetched. But it is founded upon as much, or as little, evidence as the indictments made by others. Doubt has been cast on so many British scientists individually, including some from the British Museum, that this conspiracy theory does not really enlarge the circle of possible wrongdoers.

Perhaps there were no conspirators at the British Museum. But according to many scientists, someone with scientific training, acting alone or with others, did carry out a very successful forgery.

Gavin De Beer, a director of the British Museum of Natural History, believed the methods employed in uncovering of the Piltdown hoax would "make a successful repetition of a similar type of forgery virtually impossible in the future" (Weiner *et al.* 1955, p. 228). But a forger with knowledge of modern chemical and radiometric dating methods could manufacture a fake that would not be easily detectable. Indeed, we can hardly be certain that there is not another Piltdown-like forgery in one of the world's great museums, just waiting to be uncovered.

The impact of Piltdown remains, therefore, damaging. But incidents of this sort appear to be rare, given our present knowledge. There is, however, another more insidious and pervasive kind of cheating—the routine editing and reclassifying of data according to rigid theoretical preconceptions.

Vayson de Pradenne, of the Ecole d'Anthropologie in Paris, wrote in his book *Fraudes Archéologiques* (1925): "one often finds men of science possessed by a pre-conceived idea, who, without committing real frauds, do not hesitate to give observed facts a twist in the direction which agrees with their theories. A man may imagine, for example, that the law of progress in pre-historic industries must show itself everywhere and always in the smallest details. Seeing the simultaneous presence in a deposit of carefully finished artefacts and others of a coarser type, he decides that there must be two levels: the lower one yielding the coarser specimens. He will class his finds according to their *type*, not according to the *stratum* in which he found them. If at the base he finds a finely worked implement he will declare there has been accidental penetration and that the specimen must be re-integrated with the site of its origin by placing it with the items from the higher levels. He will end with real trickery in the stratigraphic presentation of his specimens; trickery in aid of a pre-conceived idea, but more or less unconsciously done by a man of good faith whom no one would call fraudulent. The case is often seen, and if I mention no names it is not because I do not know any" (Vere 1959, pp. 1–2).

This sort of thing goes on not just in the British Museum, but in all museums, universities, and other centers of paleoanthropological research the world over. Although each separate incident of knowledge filtration seems minor, the cumulative effect is overwhelming, serving to radically distort and obscure our picture of human origins and antiquity.

An abundance of facts suggests that beings quite like ourselves have been around as far back as we care to look—in the Pliocene, Miocene, Oligocene, Eocene, and beyond. Remains of apes and apelike men are also found throughout the same expanse of time. So perhaps all kinds of hominids have coexisted throughout history. If one considers all the available evidence, that is the clearest picture that emerges. It is only by eliminating a great quantity of evidence— keeping only the fossils and artifacts that conform to preconceived notions—that one can construct an evolutionary sequence. Such unwarranted elimination of evidence, evidence as solidly researched as anything now accepted, represents a kind of deception carried out by scientists desiring to maintain a certain theoretical point of view. This deception is apparently not the result of an deliberately organized plot, as with the Piltdown man forgery (if that is what Piltdown man was). It is instead the inevitable outcome of social processes of knowledge filtration operating within the scientific community.

But although there may be a lot of unconscious fraud in paleoanthropology, the case of Piltdown demonstrates that the field also has instances of deception of the most deliberate and calculating sort.

9

Peking Man and Other Finds in China

After the discoveries of Java man and Piltdown man, ideas about human evolution remained unsettled. Dubois's *Pithecanthropus erectus* fossils did not win complete acceptance among the scientific community, and Piltdown simply complicated the matter. Scientists waited eagerly for the next important discoveries—which they hoped would clarify the evolutionary development of the Hominidae. Many thought the desired hominid fossils would be found in China.

Eventually, such fossils did turn up, at Choukoutien, near Peking. The creature to which the bones originally belonged was designated Peking man or *Sinanthropus*. The Peking man fossils were lost to science during World War II, but more fossil discoveries were made in the postwar era. In this chapter, we will discuss the controversial nature of the Peking man fossils and the questionable practice of dating later Chinese hominid fossils by their morphology, in the absence of more secure means of determining their actual age.

In the course of this discussion, the reader will be confronted with various spellings of names of Chinese geographical locations and scientists. Over the years, scholars have adopted different conventions for rendering Chinese names into English. For example, Peking is now spelled Beijing. And Choukoutien is now spelled Zhoukoudian. In the first part of the chapter, we use Peking and Choukoutien, and in the later part of the chapter we use Beijing and Zhoukoudian. As far as names of scientists are concerned, what Westerners regard as the last name comes first in the Chinese name. For example, Wu Rukang will be listed in our bibliography as Wu, R. Complicating the matter are variant spellings, such as Woo Jukang. In our text and citations, we will use the modern spellings of most authors. In the bibliography, we will also give the modern spelling with the older variant in brackets: Wu, R. [Woo, J.].

9.1 DISCOVERIES AT CHOUKOUTIEN

The ancient Chinese called fossils dragon bones. Believing dragon bones to possess curative powers, Chinese druggists have for centuries powdered them for

527

use in remedies and potions. For early Western paleontologists, Chinese drug shops therefore provided an unexpected hunting ground.

In 1900, Dr. K. A. Haberer collected mammalian fossils from Chinese druggists and sent them to the University of Munich, where they were studied and catalogued by Max Schlosser in *The Fossil Mammals of China* (1903). Among the specimens, Schlosser found a tooth from the Peking area that appeared to be a "left upper third molar, either of a man or hitherto unknown anthropoid ape" (Goodman 1983, p. 63). Schlosser suggested China would be a good place to search for primitive man.

Among those who agreed with Schlosser was Gunnar Andersson, a Swedish geologist employed by the Geological Survey of China. Andersson, a keen hunter of dragon bones, traced out their sources from druggists and other informants. He then carried out excavations for fossils. Andersson was particularly interested in discoveries that might increase the evidence for the theory of human evolution. In his paleoanthropological research, Andersson enjoyed the support of the Swedish government, members of the Swedish royal family, and wealthy patrons such as Ivar Kreuger, who monopolized match-manufacturing in several countries.

In 1918, Andersson visited a place called Chikushan, or Chicken Bone Hill, near the village of Choukoutien, 25 miles southwest of Peking. There, on the working face of an old limestone quarry, he saw a fissure of red clay filled with fossil bones.

In 1921, Andersson again visited the Chikushan site. He was accompanied by Otto Zdansky, an Austrian paleontologist who had been sent to assist him, and Walter M. Granger, of the American Museum of Natural History. Their first excavations were not very productive, resulting only in the discovery of some fairly recent fossils.

Then some of the local villagers told Zdansky about a nearby place with bigger dragon bones, near the small Choukoutien railway station. Here Zdansky found another limestone quarry, the walls of which, like the first, had fissures filled with red clay and broken bones. Andersson visited the site and discovered some broken pieces of quartz, which he thought might be very primitive tools. The mineral quartz did not occur naturally at the site, so Andersson reasoned that the quartz pieces must have been brought there by a hominid. Zdansky, who did not get along very well with Andersson, disagreed with this interpretation.

Andersson, however, remained convinced. Looking at the limestone wall, he said, "I have a feeling that there lies here the remains of one of our ancestors and it's only a question of finding him" (Hood 1964, p. 65). He asked Zdansky to keep searching a filled-in cave, saying, "Take your time and stick to it until the cave is emptied if need be" (Goodman 1983, p. 65).

9.1.1 The First Teeth

In 1921 and 1923, Zdansky, somewhat reluctantly, conducted brief excavations. He uncovered signs of an early human precursor—first one tooth and then a second, tentatively dated to the Early Pleistocene. Of the first tooth Zdansky said: "I recognized it at once, but said nothing" (Goodman 1983, p. 65).

Even after finding the second tooth, Zdansky kept both secret. The teeth, a lower premolar and an upper molar, were crated up with other fossils and shipped to Sweden for further study (Hood 1964, p. 66). Back in Sweden, Zdansky published a paper in 1923 on his work in China, with no mention of the teeth.

There the matter rested until 1926. In that year, the Crown Prince of Sweden, who was chairman of the Swedish China Research Committee and a patron of paleontological research, planned to visit Peking. Professor Wiman of the University of Uppsala, asked Zdansky, his former student, if he had come across anything interesting that could be presented to the Prince. Zdansky sent Wiman a report, with photographs, on the teeth he had found at Choukoutien. The report, later published in the *Bulletin of the Geological Survey of China,* was duly presented by J. Gunnar Andersson to a meeting in Peking, attended by the Crown Prince. Andersson declared in regard to the teeth: "The man I predicted had been found" (von Koenigswald 1956, p. 41).

9.1.2 Davidson Black

Another person who thought Zdansky's teeth represented clear evidence of fossil man was Davidson Black, a young Canadian physician residing in Peking.

Davidson Black graduated from the University of Toronto medical school in 1906. To satisfy his strong interest in anatomy, he took a post at Western Reserve University in Ohio, where he worked with T. Wingate Todd, a noted English anatomist.

Todd was an associate of Grafton Eliot Smith, familiar to us from our discussion of Piltdown man (Chapter 8). A forceful advocate of human evolutionary theory, Todd organized at Western Reserve University an extensive skeletal museum, including casts of bones from all known forms of fossil man. Under Todd, Davidson Black therefore had an opportunity to become acquainted with the latest developments in the field of paleoanthropology.

In 1914, Black went to Manchester, England, to work under Grafton Elliot Smith, who was then occupied with Piltdown man. Black also developed a friendly relationship with Sir Arthur Keith, accompanying him to the Piltdown site.

In a letter of recommendation, Smith wrote of Black: "during his stay in my department he has seized every opportunity of familiarizing himself with the problem of human phylogeny [evolution]" (Hood 1964, p. 27).

After returning to Western Reserve, Black read *Climate and Evolution* by William Diller Matthew. In 1911, Matthew had said in an address to the National Academy of Sciences of the United States: "All authorities are today agreed in placing the center of dispersal of the human race in Asia. Its more exact location may be differently interpreted, but the consensus of modern opinion would place it probably in or about the great plateau of central Asia" (Osborn 1928, p. 192).

Today the center of dispersal is viewed as Africa rather than central Asia, and all fossil evidence must therefore be interpreted in light of an African origin. For example, most paleoanthropologists now believe that *Homo sapiens sapiens* evolved in southern Africa about 100,000 years ago, and then spread throughout the world, diversifying into the present races. But other scientists concerned with human origins, such as Carleton S. Coon (1969), have said the fossil evidence shows that the several modern human races evolved separately from *Homo erectus* in Africa, Europe, and Asia. However, as we have several times noted, it is only by excluding or reinterpreting vast quantities of reported evidence that any evolutionary hypothesis whatsoever can be maintained.

From the time he first became acquainted with Matthew's ideas in 1915, Black intended to go to northern China to search for the center of human origins (Hood 1964, p. 35). But the First World War delayed his plans.

9.1.3 The Rockefeller Foundation Sends Black to China

In 1917, Black joined the Canadian military medical corps. Meanwhile, a friend of Black, Dr. E. V. Cowdry, was named head of the anatomy department at the Rockefeller Foundation's Peking Union Medical College. Cowdry asked Dr. Simon Flexner, director of the Rockefeller Foundation, to appoint Black as his assistant. After meeting Flexner in New York, Black was accepted and wrote to a colleague: "In addition to my work at the school I shall have the privilege of accompanying such scientific expeditions as may be organized to explore and collect material in central China, Tibet, etc." (Hood 1964, pp. 41–42).

After Rockefeller Foundation officials petitioned the Surgeon General of Canada, Black won his release from the Canadian military and proceeded to Peking, arriving in 1919.

At the Peking Union Medical College, Black did everything possible to minimize his medical duties so he could concentrate on his real interest—paleoanthropology. In November 1921, he went on a brief expedition to a site in northern China, and other expeditions followed. Black's superiors were not pleased.

In 1921, Dr. R. M. Pearce, the Rockefeller Foundation's advisor on medical education, visited Peking on an inspection tour. Afterward, Pearce wrote to

Black: "If you think of anatomy for nine months out of the year, it is no one's business what you do with the other three months in the summer in connection with anthropology, but for the next two years at least give your entire attention to anatomy" (Hood 1964, p. 55).

But gradually the Rockefeller Foundation would be won over to Black's point of view. The series of events that caused this change to take place is worth looking into.

Late in 1922, Black submitted a plan for a Siam (now Thailand) expedition to Dr. Henry S. Houghton, director of the medical school. Black expertly related his passion for paleoanthropology to the mission of the medical school. Houghton wrote to Roger Greene, the school's business director: "While I cannot be certain that the project which Black has in mind is severely practical in its nature, I must confess that I have been deeply impressed by . . . the valuable relationship he has been able to establish between our department of anatomy and the various institutions and expeditions which are doing important work in China in the fields which touch closely upon anthropology research. With these points in mind I recommend the granting of his request" (Hood 1964, p. 56). Here can be seen the importance of the intellectual prestige factor—ordinary medicine seems quite pedestrian in comparison with the quasi-religious quest for the secret of human origins, a quest that had, since Darwin's time, fired the imaginations of scientists all over the world. Houghton was clearly influenced. The expedition took place during Black's summer vacation in 1923, but unfortunately produced no results.

In 1924, Black took a year's paid leave to travel around the world, visiting early man sites, museums, and scholars in the field of human evolution. Black returned to Peking determined to give more time to his pet research projects.

9.1.4 Black and the Birth of Sinanthropus

In 1926, Black attended the scientific meeting at which J. Gunnar Andersson presented to the Crown Prince of Sweden the report on the molars found by Zdansky at Choukoutien in 1923. Excited on learning of the teeth, Black accepted a proposal by Andersson for further excavations at Choukoutien, to be carried out jointly by the Geological Survey of China and Black's department at the Peking Union Medical School. Dr. Amadeus Grabau of the Geological Survey of China called the hominid for which they would search "Peking man."

On October 27, 1926, Black wrote to Sir Arthur Keith about Zdansky's teeth: "There is great news to tell you—actual fossil remains of a man-like being have at last been found in Eastern Asia, in fact quite close to Peking. This discovery fits in exactly with the hypothesis as to the Central Asiatic origin of the Hominidae which I reviewed in my paper 'Asia and the Dispersal of Primates'"

(Hood 1964, p. 84). Black in China, like Dubois in Java, had found what he was looking for.

Hood (1964, p. 85) stated in her biography: "Black's next task was to approach the Rockefeller Foundation through Roger Greene to ask for funds with which to make a large-scale excavation at the caves of Chou-K'ou-tien. To his delight and relief a generous sum was forthcoming. This response showed a marked change in the attitude of the authorities in New York towards Black's efforts to promote research into China's prehistory from his experience in 1921."

By spring 1927, work was underway at Choukoutien, in the midst of the Chinese civil war. During several months of painstaking excavation, there were no discoveries of any hominid remains. Finally, with the cold autumn rains beginning to fall, marking the end of the first season's digging, a single hominid tooth was uncovered. On the basis of this tooth, and the two previously reported by Zdansky (now in Black's possession), Black decided to announce the discovery of a new kind of fossil hominid. He wrote in *Nature:* "The newly discovered specimen displays in the details of its morphology a number of interesting and unique characters, sufficient, it is believed, to justify the proposal of a new hominid genus *Sinanthropus,* to be represented by this material" (Black 1927, p. 954).

Black was eager to show the world his discovery. Dr. Heinrich Neckles, a friend of Black, later recalled: "One night he came to my office very excited, to show the precious tooth of *homo pekinensis.* He wanted me to advise him about the safest method to take the invaluable find to England (where he was going shortly) safe against loss or theft. I suggested a brass capsule with a screw closure and a ring at the top, with a strong ribbon through it, so he could wear it around his neck. We had a good Chinese mechanic in the Physiology Department who made a very nice capsule for him and he was as happy as a little boy" (Hood 1964, p. 90).

In the course of his travels with his newly found tooth, Black discovered that not everyone shared his enthusiasm for *Sinanthropus.* At the annual meeting of the American Association of Anatomists in 1928, some of the members heavily criticized Black for proposing a new genus on so little evidence.

In addition, Zdansky was not at all very happy regarding the purposes for which his teeth were being used: "I am indeed convinced that the existing material provides a wholly inadequate foundation for many of the various theories based upon it. . . . I decline absolutely to venture any far-reaching conclusions regarding the extremely meager material described here, and which, I think, cannot be more closely identified than as *Homo* sp. [species undetermined] . . . my purpose here is only to make it clear that my discovery of these teeth should be regarded as decidedly interesting but not of epoch-making importance" (Bowden 1977, pp. 80–81).

Regarding such criticism of Black's activities, Grafton Elliot Smith wrote: "It had no other effect upon him, beyond awaking his sympathies for anthropologists who are unfairly criticized and to make him redouble his efforts to establish the proof of his claim" (Hood 1964, p. 93).

Black kept making the rounds, showing the tooth to Ales Hrdlicka in the United States and then journeying to England, where he met Sir Arthur Keith and Sir Arthur Smith Woodward. At the British Museum, Black had casts made of the Peking man molars, for distribution to other workers. This is the kind of propaganda work necessary to bring a discovery to the attention of the scientific community. This serves to illustrate that even for a scientist political skills are not unimportant.

On returning to China, Black kept in close touch with the excavations at Choukoutien. Dynamite was used to blast out sections of rock. Crews of workers then searched through the debris, sending the larger chunks back to Peking, where any fossils were carefully extracted. The sole aim of the whole project was, of course, to find more Peking man remains. For months nothing turned up.

But Black wrote to Keith on December 5, 1928: "It would seem that there is a certain magic about the last few days of the season's work for again two days before it ended Böhlin found the right half of the lower jaw of Sinanthropus with the three permanent molars *in situ*" (Hood 1964, p. 97).

Now a financial problem loomed. The Rockefeller Foundation grant that supported the digging would run out in April of 1929. So in January, Black wrote the directors, asking them to support the Choukoutien excavations by creating a Cenozoic Research Laboratory (the Cenozoic includes the periods from the Paleocene to the Holocene). In April, Black received the funds he desired.

9.1.5 The Transformation of the Rockefeller Foundation

Just a few years before, Rockefeller Foundation officials had actively discouraged Black from becoming too involved in paleoanthropological research. Now they were backing him to the hilt, setting up an institute specifically devoted to searching for remains of fossil human ancestors. Why had the Rockefeller Foundation so changed its attitude toward Black and his work? This question bears looking into, because the financial contribution of foundations would turn out to be vital to human evolution research carried out by scientists like Black. Foundation support would also prove important in broadcasting the news of the finds and their significance to the waiting world.

As Warren Weaver, a scientist and Rockefeller Foundation official, said (1967, p. 82): "In a perfect world an idea could be born, nourished, developed and made known to everyone, criticized and perfected, and put to good use without the crude fact of financial support ever entering into the process.

Seldom, if ever, in the practical world in which we live, does this occur. The influence of money on ideas can be powerful; it can be good, or it can be downright vicious. . . . Money can be used to lure the gullible to devote their time to spiritualism, to fanatical religions, to pseudo-science, and so on."

For Weaver, biological questions were of the highest importance. Writing in 1967, Weaver stated that he regarded the highly publicized particle accelerators and space exploration programs as something akin to scientific fads. He added: "The opportunities not yet rigorously explored lie in the understanding of the nature of living things. It seemed clear in 1932, when the Rockefeller Foundation launched its quarter-century program in that area, that the biological and medical sciences were ready for a friendly invasion by the physical sciences. . . . the tools are now available for discovering, on the most disciplined and precise level of molecular actions, how man's central nervous system really operates, how he thinks, learns, remembers, and forgets. . . . Apart from the fascination of gaining some knowledge of the nature of the mind-brain-body relationship, the practical values in such studies are potentially enormous. Only thus may we gain information about our behavior of the sort that can lead to wise and beneficial control" (W. Weaver 1967, p. 203).

It thus becomes clear that at the same time the Rockefeller Foundation was channeling funds into human evolution research in China, it was in the process of developing an elaborate plan to fund biological research with a view to developing methods to effectively control human behavior. Black's research into Peking man must be seen within this context in order to be properly understood.

Over the past few decades, science has developed a comprehensive cosmology that explains the origin of human beings as the culmination of a 4-billion-year process of chemical and biological evolution on this planet, which formed in the aftermath of the Big Bang, the event that marked the beginning of the universe some 16 billion years ago. The Big Bang theory of the origin of the universe, founded upon particle physics and astronomical observations suggesting we live in an expanding cosmos, is thus inextricably connected with the theory of the biochemical evolution of all life forms, including human beings. The major foundations, especially the Rockefeller Foundation, provided key funding for the initial research supporting this materialistic cosmology, which has for all practical purposes pushed God and the soul into the realm of mythology—at least in the intellectual centers of modern civilization.

The extent of the Rockefeller Foundation's support of biological research is remarkable. The Foundation funded the fruit fly genetics work of Thomas Hunt Morgan and Theodosius Dobzhansky. Dr. Max Perutz said the Cambridge Medical Research Council Laboratory of Molecular Biology in England owed its existence to the Rockefeller Foundation. The Foundation furnished funds for the

Laboratory's X-ray diffraction equipment, which provided critical research results used by Watson and Crick in their pioneering work on DNA's helical structure (W. Weaver 1967, p. 235).

The Foundation was equally supportive of selected projects in the realm of the physical sciences. Lee A. Dubridge, President of the California Institute of Technology, wrote: "The sciences of physics and astronomy could hardly have emerged from the primitive state in which they found themselves in America in the first two decades of the twentieth century had it not been for the generosity of the great private foundations" (W. Weaver 1967, p. 252). As we have seen, the Carnegie Foundation built the Mt. Wilson Observatory. The Rockefeller Foundation built the Mt. Palomar Observatory, where much of the work on the Big Bang theory of the origin of the universe took place. The Foundation also gave funds to Ernest O. Lawrence for building the world's first particle accelerators.

If the Big Bang and biochemical evolution represent the Godless and soulless cosmology of the scientific world view, psychiatry and psychology represent its secular moral code and guidelines for practical behavior. In the early 1930s, around the time the Choukoutien excavations were in full swing, the medical division of the Rockefeller Foundation chose psychiatry as its principal focus, establishing schools of psychiatry at major medical colleges. Later the Foundation would fund the famous Kinsey reports on sexual behavior.

During the 1930s, psychiatry was fairly well dominated by the figure of Sigmund Freud, who had encountered ideas about human evolution as a youth and later wrote: "The theories of Darwin, which were then of topical interest, strongly attracted me, for they held out hopes of an extraordinary advance in our understanding of the world" (Jones 1953, pp. 27–28).

In *Totem and Tabu,* Freud explained Christianity and all organized religion in terms of his Oedipus complex. According to one of his biographers, Freud "took into account, too, the work of Charles Darwin. He recalled Darwin's conjecture that originally men had lived in hordes, each horde dominated by a single, powerful, violent, suspicious man" (Puner 1947, p. 167). In his autobiography, Freud wrote: "The father of the primal horde, since he was an unlimited despot, had seized all the women for himself; his sons, being dangerous to him as rivals, had been killed or driven away. One day, however, the sons came together and united to overwhelm, kill and devour their father, who had been their enemy but also their ideal. . . . the primal father, at once feared and hated, honored and envied, became the prototype of God himself. . . . This view of religion throws a particularly clear light upon the psychological basis of Christianity" (Puner 1947, pp. 167–168). If one takes seriously the theory of evolution, one must explain the origin of God and religion as an historical occurrence within the mind of evolving man, though perhaps not in the exact manner suggested by Freud.

The Rockefeller Foundation saw in psychiatry a way to influence human social behavior. Dr. Alan Gregg, head of the Medical Sciences Division of the Foundation, wrote: "I should not be satisfied with the definition of psychiatry as that specialty in medicine which deals with mental disorders." He believed its "province is the conduct of man, his reactions, his behavior as an indivisible sentient being with other such beings" (Fosdick 1952, p. 130). During the Second World War, Gregg served as an Army consultant and wrote of "the possibility that through psychiatric understanding our successors may be able to govern human politics and relationships more sagely" (Fosdick 1952, p. 133). The desire to bring about better human relations is certainly laudable. But our main point is that the Rockefeller Foundation scientists believed this goal could best be achieved by having science establish beneficial control over human society.

All this is quite remarkable, when one considers that John D. Rockefeller's charity was initially directed toward Baptist churches and missions. Raymond D. Fosdick, an early president of the Rockefeller Foundation, said (1952, p. 2) that both Rockefeller and his chief financial adviser, Frederick T. Gates, were "inspired by deep religious conviction." Rockefeller believed "a man should make all he can and give all he can" (Fosdick 1952, p. 6).

According to Fosdick (1952, p. 6), Rockefeller was at first "giving to a multiplicity of small causes mostly related to his church interests—schools, hospitals, and missions." As a result, he was continually being approached by Baptist ministers. To relieve Rockefeller from personally having to handle individual requests, Gates organized a system whereby Rockefeller would give a lump sum to a mission board that would distribute the funds in an appropriate fashion.

Moving on to bigger things, Rockefeller and Gates gave 35 million dollars for building the University of Chicago, which, according to Fosdick (1952, p. 7), started out as "as an idea for a Baptist institution of higher learning, under Baptist auspices and control." It is hard to imagine such a school promoting the idea that humans evolved from extinct apelike creatures. Gates, it may be noted, was formerly head of the American Baptist Education Society.

In 1913, the present Rockefeller Foundation was organized. The trustees included Frederick T. Gates; John D. Rockefeller, Jr.; Dr. Simon Flexner, head of the Rockefeller Institute for Medical Research; Henry Pratt Judson, president of the University of Chicago; Charles William Eliot, former president of Harvard; and A. Barton Hepburn, president of the Chase National Bank.

At first, the Foundation concentrated its attention on public health, medicine, agriculture, and education, avoiding anything controversial. Thus the Rockefeller Foundation began to distance itself from religion, particularly the Baptist Church. Exactly why this happened is difficult to say. Perhaps it had something to do with the fact that Rockefeller was coming to realize that his fortune was founded on exploiting the advances of modern science and technology. Perhaps it was the increasing role that science was beginning to play in the objects of traditional charitable giving—such as medicine. But whatever the reason, Rockefeller began to staff his foundation with scientists, and the giving policies reflected this change.

Even Gates, the former Baptist educator, seemed to be changing his tune. He wanted to create a nonsectarian university in China. But he noted that the "missionary bodies at home and abroad were distinctly and openly, even threateningly hostile to it as tending to infidelity" (Fosdick 1952, p. 81). Furthermore, the Chinese government wanted control, an idea that the Foundation could not support.

President Eliot, who had overseen the Harvard Medical School in Shanghai, proposed a solution: a medical college, which would serve as an opening to the rest of Western science. Fosdick (1952, p. 81) wrote: "To President Eliot there was no better subject than medicine to introduce to China the inductive method of reasoning which lies at the basis of all modern science. He thought it would be the most significant contribution that the West could make to the East." Here mechanistic science shows itself a quiet but militant ideology, skillfully, yet somewhat ruthlessly, promoted by the combined effort of scientists, educators, and wealthy industrialists, with a view towards establishing worldwide intellectual dominance.

The medical hospital strategy outlined by Eliot worked. The Chinese government approved establishment of the Peking Union Medical College under Foundation auspices. Meanwhile, Dr. Wallace Buttrick, director of Rockefeller's newly created China Medical Board, negotiated with the Protestant mission hospitals already in China. He agreed to provide financial support for these hospitals, in effect bribing them (Fosdick 1952, pp. 83–84).

In 1928, the Rockefeller Foundation and other Rockefeller charities underwent changes to reflect the growing importance of scientific research. In 1923, Wycliffe Rose, head of the General Education Board, had said: "All important fields of activity, from the breeding of bees to the administration of an empire, call for an understanding of the spirit and technique of modern science. . . . Science is the method of knowledge. It is the key to such dominion as man may ever exercise over his physical environment. Appreciation of its spirit and technique, moreover, determines the mental attitude of a people, affects the entire system of education, and carries with it the shaping of a civilization" (Fosdick 1952, p. 141).

All programs in various Rockefeller charities "relating to the advance of human knowledge" were shifted to the Rockefeller Foundation, which was organized into five divisions: international health, medical sciences, natural sciences, social sciences, and the humanities (Fosdick 1952, pp. 137–138). Each division was run by a highly competent academic and technical staff, who advised the trustees of the Foundation where to give their money. Raymond D. Fosdick, president of the Foundation at the time, said (1952, p. 140) that the year of 1928 marked "the end of an era in philanthropy." And the beginning of a new one.

The change reached right to the top, with Dr. Max Mason, a scientist himself, taking over as president. Mason, a mathematical physicist, was formerly president of the University of Chicago. According to Fosdick (1952, p. 142), Mason "emphasized the structural unity involved in the new orientation of program. It was not to be five programs, each represented by a division of the Foundation; it was to be essentially one program, directed to the general problem of human behavior, with the aim of control through understanding."

The Foundation also saw itself engaged in a kind of thought control. Fosdick (1952, p. 143) said: "The possession of funds carries with it power to establish trends and styles of intellectual endeavor."

The theme of control was echoed in 1933 by Warren Weaver, who headed the Rockefeller Foundation's natural sciences division, which funded the Cenozoic Research Laboratory in Peking. In a report to the trustees, Weaver, a mathematician from the University of Wisconsin, said: "The welfare of mankind depends in a vital way on man's understanding of himself and his physical environment. Science has made magnificent progress in the analysis and control of inanimate forces, but it has not made equal advances in the more delicate, more difficult, and more important problem of the analysis and control of animate forces" (Fosdick 1952, p. 157). The Rockefeller Foundation's annual report for the year 1933 (p. 199) asked: "Can we develop so sound and extensive a genetics that we can hope to breed in the future superior men? . . . In short, can we rationalize human behavior and create a new science of man?"

The Foundation scientists outlined a coordinated program, approved by the Foundation trustees, to attain this goal. Fosdick (1952, p. 158) stated: "the trustees, in the spring of 1933, voted to make experimental biology the field of primary interest. . . . It was conceived, moreover, as being closely linked with other aspects of the Foundation's program, notably the program in psychiatry of the Medical Sciences division and the social-science program in human relations. Biology is important because it has the potentiality of contributing to the problem of understanding ourselves, and the three programs—in widely separated fields—could be thought of as a unified endeavor to stimulate research in the sciences underlying the behavior of man."

Some commentators make light of research into the reproductive habits of earthworms and other apparently obscure research projects. But these have their purpose. According to Weaver: "Before we can be wise about so complex a subject as the behavior of a man, we obviously have to gain a tremendous amount of information and insight about living organisms in general, necessarily starting with the simpler forms of life. Experimental biology is the means for such exploration. It furnishes the basis necessary for progress in solving the sequence of problems which begins with the strictly biological and moves through the mental to the social" (Fosdick 1952, p. 158). Here once more, the intent to use science for perfecting methods of social control (and who would the controllers be but the scientists?) is stated explicitly.

And what about something as apparently innocent as stargazing through the 200-inch telescope at Mt. Palomar? Fosdick (1952, p. 179) stated: "Superficially the 200-inch and the lesser projects in astronomy which have received Foundation aid would seem to be far removed from the main interest of the Natural Sciences program. What possible relationship can there be between the stars and experimental biology?" Fosdick (1952, p. 180) answered that astronomy gives the first glimmers of regularity in nature, the understanding of which will lead to control of humanity and the universe.

It bears repeating that one should see Black's Peking man research within the larger framework of the explicitly stated goal of the Rockefeller Foundation, which reflected the implicit goal of big science—control, by scientists, of human behavior. In particular, Peking man strengthened the concept of human evolution, by which scientists attempt to determine the way we think about ourselves. Essentially, evolution defines human nature in a totally materialistic way. This materialistic definition of human nature tends to justify making the primary goal of human life the attainment of control, by science, over the visible universe.

9.1.6 An Historic Find and a Cold-Blooded Campaign

With the financial backing of the Rockefeller Foundation for the Cenozoic Research Laboratory secure, Black resumed his travels for the purpose of promoting Peking man. In May of 1929, he arrived in Java, for the Fourth Pacific Science Congress. There he was able to give a report on *Sinanthropus* before an audience that included Grafton Eliot Smith. Black stated: "Elliot Smith's cordial backing after my presentation of the material at the conference made all the difference in the world to its reception there" (Hood 1964, pp. 100–101). Nevertheless, Peking man still had not achieved the worldwide celebrity he would later enjoy. While in Java, Smith and Black visited the Trinil site, where Dubois had originally discovered *Pithecanthropus,* the southern relative of *Sinanthropus.*

Black then returned to China, where work was proceeding slowly at Choukoutien, with no new major *Sinanthropus* finds reported. Enthusiasm seemed to be waning among the workers. But then on the first of December, at the very end of the season, W. C. Pei (Pei Wenzhong) made an historic find. Pei later wrote: "At about four o'clock next afternoon I encountered the almost complete skull of *Sinanthropus*. The specimen was imbedded partly in loose sands and partly in a hard matrix so that it was possible to extricate it with relative ease" (Hood 1964, p. 104).

In order to protect the skull, Pei immediately wrapped it in paper and cloth soaked with flour paste. He then rode 25 miles on a bicycle to the Cenozoic Research Laboratory, where he presented the skull to Black, who gave him full credit for the discovery.

By early 1930, Black had published two preliminary papers on the skull and set about publicizing the find around the world. His secretary, Miss Hempel, recalled: "For weeks and months we did nothing but write letters" (Hood 1964, p. 109).

Black wrote to Dr. Pearce at the Rockefeller Foundation: "Yes, *Sinanthropus* is growing like a bally weed. I never realized how great an advertising medium primitive man (or woman) was till this skull turned up. Now everybody is crowding around to gaze that can get the least excuse to do so and it gets embarrassing at times. Being front page stuff is a new sensation and encourages a guarded manner of speech" (Hood 1964, pp. 110–111).

Black worked busily, carefully freeing the skull from its stone matrix and later making a cast of it. Copies of the cast were sent to museums all over the world. The site itself was purchased by the Geological Survey of China.

In September of 1930, Sir Grafton Elliot Smith arrived in Peking to inspect the site of the discovery and examine the fossils. During Smith's stay, Black primed him for a propaganda blitz on behalf of Peking man. Smith then departed, and apparently did his job well. In December, Black wrote an extremely candid letter to Dr. Henry Houghton, director of the Peking medical school, who was vacationing in America: "I am thrilled beyond words to know how much you enjoyed Grafton Elliot Smith. . . . he is Irish to the extent that a friend is always spoken of in lurid hyperbole and, though I love him for it, I get the collywobbles when I reflect the brazen way I have plotted to have him exercise his talent in this respect on my behalf. . . . I warned him to hold off'n me . . . but your letter makes it clear that that balloon is busted and I'm the chappee who must spend the rest of his days trying to live up to and live down the reputation acquired by his own rash act." This rash act appears to have been the *Sinanthropus* discovery.

Black went on to say: "But you, too, are dripping with the gore of the same hegoat and I love you, for your soul is white if your hood be scarlet and your aid, comfort and participation in the plot from its inception made success possible and doubly enjoyable. . . . You must admit that we have not been any blushing roses when it came to turning our wolf loose (if you don't mind mixed metaphors)—if I blushed every time I thought of the cold-blooded advertising campaign I thought of and G. E. S. has carried through, I'd be permanently purple" (Hood 1964, p. 115).

Cold-blooded advertising campaign? That is not the way most people think scientific discoveries normally make their way into academic acceptance and public notice. Black is to be commended for his forthright statements. In any case, having turned the wolf of *Sinanthropus* loose on the world, he received many honors, including appointments as honorary fellow of the Royal Anthropological Institute and honorary member of America's National Academy of Sciences (Hood 1964, p. 116). Black was later elected a fellow of the Royal Society, Britain's foremost assembly of scientists.

His newly won fame also insured continued access to Rockefeller Foundation funds. Black wrote to Sir Arthur Keith: "We had a cable from Elliot Smith yesterday so he is evidently safe home after his strenuous trip. He characteristically has not spared himself in serving the interests of the Survey and the Cenozoic Laboratory and after his popularizing *Sinanthropus* for us in America I should have a relatively easy task before me a year from now when I will have to ask for more money from the powers that be" (Hood 1964, p. 116).

Peking man had come at just the right moment for advocates of human evolution. A few years previously, in one of the most famous trials in the world's history, a Tennessee court had found John T. Scopes guilty of teaching evolution in violation of state law. Scientists wanted to fight back hard. Thus any new evidence bearing on the question of human evolution was highly welcome.

Then there had been the matter of *Hesperopithecus,* a highly publicized prehistoric ape-man constructed in the minds of paleoanthropologists from a single humanlike tooth found in Nebraska. To the embarrassment of the scientists who had promoted this human ancestor, the humanlike tooth had turned out to be that of a fossil pig.

Meanwhile, the lingering doubts and continuing controversy about Dubois's *Pithecanthropus erectus* also needed to be resolved. In short, scientists in favor of evolutionary ideas, reacting to external threat and internal disarray, were in need of a good discovery to rally their cause.

Concerning the Java fossils, Jia Lanpo wrote: "The problem of what species did the owner of the remains belong to had not been settled. Sceptics asserted that they might belong to a deformed ape, or an abnormally developed animal which had no relation to man whatsoever. The most vociferous critics were from the religious community, who held that man's ancestor was Adam, and that man's history dates back only 4,004 years before Christ. Anyone who held that those specimens were related to man was accused of being a heretic. In the end, because of the pressure or some other reason, Dubois himself gave in and stated that what he had discovered was the remains of a 'giant gibbon.' It was not until 1929 after Professor Pei Wenzhong discovered the Peking Man skullcap and later, stone artifacts and traces of the use of fire in association with it that the absurd clamor gradually died down" (Jia 1980, p. 27).

Peking man caught on like wildfire, not only in the world of science, which needed him, but among the general public as well. Therefore scientists went overboard to confirm his status as a genuine human ancestor. And as suggested by Jia, "stone artifacts and traces of the use of fire" were to be an important element in this confirmation.

9.1.7 Evidence for Fire and Stone Tools at Choukoutien

It was in 1931 that reports showing extensive use of fire and the presence of well-developed stone and bone tools at Choukoutien were first published. What is quite unusual about these announcements is that systematic excavations had been conducted at Choukoutien by competent investigators since 1927, with no mention of either fire or stone tools. For example, Black wrote in 1929 (p. 208): "though thousands of cubic meters of material from this deposit have been examined, no artifacts of any nature have yet been encountered nor has any trace of the usage of fire been observed."

On the question of tools, P. Teilhard de Chardin and C. C. Young (Yang Zhongjian) wrote: "Embedded in the fine grained material of the Lower Cave, Mr. Pei picked up an angular piece of quartz—a type of stone which is not found within one mile at least of the locality 1. Similar quartz fragments have been found from time to time in the course of the excavations, the first ones being noticed by Dr. J. G. Andersson, but none of them has ever shown any recognizable trace of artificial breaking" (1929, p. 182). In addition, Teilhard de Chardin (1965, pp. 62–63) wrote in an article published in 1930: "since the beginning of the excavation no trace has yet been found on the site suggesting the use of fire or any industry of any kind."

Grafton Eliot Smith, who had personally visited the Choukoutien site, wrote (1931, p. 36): "It is a very significant phenomenon that at Chou Kou Tien, in spite of the most careful search in the caves during the last three years, no trace whatever of implements of any sort has been found. . . . It must not be forgotten, however, that Dr. Andersson in 1921 found pieces of quartz in association with the fossil bones, and that in the later stages of the excavation Mr. Pei found further examples of this alien material. Those who have been searching in vain for evidence of human craftsmanship on this site are being forced to the conclusion that Peking Man was in such an early phase of development as not yet to have begun to shape implements of stone for the ordinary needs of his daily life."

Then Teilhard de Chardin, while visiting Paris in 1930, showed a piece of stag horn from Choukoutien to Henri Breuil, without telling Breuil its source. Breuil studied the specimen and noted that it showed signs of having been deliberately burned by fire. He also concluded it had been modified by hammering for use as a tool, and noted cut marks that appeared to have been made by a stone implement (Breuil 1932, pp. 1–2). At that point, Teilhard de Chardin revealed the source of the bone and suggested Breuil visit Choukoutien. There are mysterious undercurrents here. At this point in time, Teilhard de Chardin was on record as saying there were no signs of human industry at Choukoutien. But then why was he carrying around a rather ordinary piece of deer horn? And why did he show it to Breuil? It does not make much sense, unless we assume that Teilhard de Chardin himself had suspicions that the bone showed signs of intentional work.

While still in Paris, Teilhard de Chardin presented, at the Institute of Human Paleontology, a paper that was published the following year in *L'Anthropologie*. In this paper, Teilhard de Chardin (1931) cautiously suggested that the use of fire by *Sinanthropus* might be established after further study of blackened bones and antlers recovered from the site. Nevertheless, he still made no mention of any beds of ashes or hearths at the Choukoutien cave.

As requested by Teilhard de Chardin, Breuil visited Choukoutien in the fall of 1931 and encountered extensive signs of fire as well as stone and bone tools, many of advanced type. He reported his findings on November 3 at a meeting of the Geological Society of China in Peking, and published essentially the same material in an article for *L'Anthropologie* the following year.

Teilhard de Chardin himself had also begun to cautiously mention fire and implements in his writings, but Breuil's exceedingly direct and thorough report was explosive in its impact. Concerning the Quartz 2 level in the Kotzetang cave section, Breuil (1932, p. 3) said: "I observed the black layer indicated by Pei was a veritable hearth, or rather a hearth covered by very light-colored ash, doubtlessly mixed with clay. Soot-covered stones and burned bones were both brought out in my presence." Breuil here used the French word *foyer*, which can be variously translated as hearth, fireplace, or furnace.

Describing level 4 of the main cave, Breuil (1932, p 5) stated: "I first observed, in scraping away the surface from top to bottom, a succession of a number of ribbonlike layers of bright colors—-grey, yellow, and occasionally violet, which constitute level 4. The appearance of this uncompacted deposit is exactly that of a mass of ash derived from vegetal matter, comparable, for example, to the ash deposit, called 'ribbons' by E. Piette, in the Azilian levels of Mas d'Azil. In these masses one can observe numbers of particles of carbon, stones, occasionally in heaps, covered in soot, and fragments of burned bone. I did not encounter anything else, but I did observe between layers of ash numerous nodules and slabs of a bubbly concretion that appeared to be composed of phosphates derived from the alteration of bone. At the base of the great mass of ash, almost 7 meters [about 23 feet] in depth, one finds an ink-black layer, which, according to analysis by the Geological Survey and its color, shows itself to be composed of wood carbon reduced to particles. Worked quartz and other stones belong to this layer."

Breuil (1932, p. 5) further stated: "The layer of breccia covered by the mass of cinders is actually inaccessible, but I have examined a great heap of blocks that have been removed and broken apart for examination. These blocks are literally pastes of chipped quartz (horizon 1) and bone burned to various degrees."

In some final remarks on the evidence for fire, Breuil (1932, pp. 6–7) said: "As a result of these facts, confirmed by chemical analysis of the burned bones in Paris and Peking, it can be concluded that fire was used on a large scale at Chou Kou Tien. Perhaps the fact that such a mass of ash corresponds with a single black, carbonaceous basal level could enable one to deduce that the fire, ignited just once, was constantly maintained for a considerable period, enough to have produced the enormous accumulation of almost 7 meters that I have mentioned. This amount would actually correspond to a much greater accumulation at the time."

About the presence of stone implements, Breuil (1932, p. 7) wrote: "I might add that I collected some chipped quartz in the great mass of ash at that place. . . . at the base of that mass, lying on the stalagmitic floor . . . I extracted, along with M. Pei and Pierre Licent, a series of decomposed pieces of very compact volcanic rock. . . . It was the residue, unfortunately very much decomposed, of a great collection of tools made from large flakes of volcanic rock. I do not know if they are found in other levels."

Breuil (1932) also recorded the presence of many other stone tools, including some rounded bola stones. He reported that in some features the stone tool industry was similar to that of the Mousterian period in Europe, although he mentioned that it would be pointless to attempt to fit the Choukoutien stone tool industry exactly into the European classifications. The Mousterian period is identified with the Neanderthals.

Black, along with Teilhard de Chardin, Pei, and Yang (Young), stated in similar fashion: "In a very broad sense, the Choukoutien culture could be defined as an industry of old palaeolithic type, showing some external Mousterian analogies. But no close comparison with any Asiatic or European industry can well be made at present" (Black *et al.* 1933, p. 133).

Later investigators added considerably to the collection of stone tools from Choukoutien. To date, over 100,000 have been found, including a variety of choppers, scrapers, and small pointed flakes. According to Jia Lanpo, the most common material is quartz, followed by sandstone and opal. Jia (1980, p. 28) described a "mastery of rather complex methods of manufacture." He further stated: "The assemblage consists mainly of small tools but there are also larger ones, such as bifacial handaxes. . . . Scrapers of various types made on flakes are the most numerous. The blade after secondary working of the edge may be linear, convex, concave, multi-edged or disk-like. . . . The finest of the lot are the 'points.' About a hundred of them have been collected . . . their process of manufacture clearly indicates a higher level of skill. To make one, a flake is first struck from a core, then the edges are shaped until a slender point is achieved at one end. Up to now, nowhere in the world has yielded such finds of comparable quantity and workmanship" (Jia 1980, pp. 28–29).

Jia's description suggests a relatively advanced industry at Choukoutien, but other researchers have expressed differing opinions about the quality of the stone tools found there. David Pilbeam (1972, p. 166) quoted Kenneth Oakley as saying that the stone tools were similar to the crude Oldowan tools from Africa. Paleoanthropologists have highlighted different features of the stone industry at Choukoutien— hence one may get a completely different impression depending upon whose account one reads.

As far as bone tools were concerned, Breuil noted that the ancient population at Choukoutien had systematically employed a sizable industry. Large antlers, too big to be effectively used in one piece, had been cut down into manageable tool shapes. Since deer antlers are extremely difficult to cut, the place where an incision was to be made was first burned with fire, then a V-shaped groove was gouged out, and finally the bone was broken by a blow.

In 1931, Black, apparently embarrassed by the new revelations about fire and tools from Choukoutien, sought to explain how such important evidence had for several years escaped his attention and that of the other researchers at the site. In a report delivered at the same time as that of Breuil, Black (1931, p. 107) tried to cover himself on the critical question of fire: "From time to time since 1929 occasional specimens of apparently charred or partly calcined animal bones have been recovered from among the material excavated from the Main Deposit at Choukoutien. The physical appearance of these specimens left little room for doubt that they had been subjected at some time to the action of fire. But until the present season it has remained a question whether or not such specimens had been burned within the Choukoutien caves while the latter were occupied by *Sinanthropus* or were altered simply as the result of a surface fire from natural causes and had subsequently been washed within the deposit. In view of this uncertainty no report on these specimens has hitherto been published."

This seems unusual, especially when considered in the light of the following statement, published in 1933, by Black: "Traces of artificial fire in the Locality 1 deposit are so clear and abundant that they require only to be mentioned without any further demonstration" (Black *et al.* 1933, p. 113). If this was true in 1933, why not in 1931, or 1929, or even earlier? And even if, as Black said in his 1931 report, signs of fire had been noted but not reported because of doubts about the origin of the fire, this does not absolve him of responsibility.

The burned bones could at least have been mentioned, and the alternative explanations discussed.

Teilhard de Chardin also thought it wise to explain why he had not reported the presence of stone or bone implements at the time of his discussions with Breuil in Paris. In 1934, he stated in the journal *Revue des Questions Scientifiques* (vol. 25): "In writing my first article here on Choukoutien three years ago, I was still able to say that 'up to now', despite certain indications, no trace of industry had yet been certainly recognised in association with the bone remains of *Sinanthropus*. Two months later, returning to the site with Mr. W. C. Pei [Pei Wenzhong], the young scholar in charge of the excavation, I gathered with him *in situ* incontestable fragments of flaked stone and burnt bones. These traces had hitherto escaped attention because the works have been carried on for some years in a part of the site where they would have been extremely hard to recognise. . . . But once we recognised the first flakes of quartz, all became clear. . . . From that moment, archaeological discoveries multiplied—the most important being the discovery (Summer 1931) of a red, yellow, and black clay bed about two metres [about six and a half feet] thick, extremely rich in stone and bone debris" (Teilhard de Chardin 1965, pp. 70–71). Again, it does seem quite unusual that such experienced researchers as Teilhard de Chardin and Pei could have completely overlooked the presence of literally thousands of implements at Choukoutien.

In reference to the question of fire, Teilhard de Chardin and C. C. Young (Yang Zhongjian) wrote in 1929 about Layer 4 in the Choukoutien cave deposits: "Very conspicuous fine grained, sedimentary zone, formed by red loam and sandy clay of various colors (yellow, reddish, brown, gray, etc.) thinly bedded and interbedded. At several levels some black layers occur which are full of Rodent remains and other micro fauna. . . . Thickness 6.7 meters [22 feet]" (Teilhard de Chardin and Yang 1929, p. 181). A few years later, in 1932, this same layer would be described by Teilhard de Chardin and Pei in the *Bulletin of the Geological Society of China* (vol. 11) as "an ashy deposit" almost 7 meters deep (Bowden 1977, p. 92). The main ash piles were 300 feet long by 100 feet wide (Fix 1984, p. 118). It is quite remarkable that Teilhard and Young (Yang) could have examined this same formation in 1929 and reported on it with no suggestion at all of fire.

Concerning the failure of Teilhard de Chardin, Black, Pei, and others to report abundant tools and signs of fire at Choukoutien, there are two possible explanations. The first is the one they themselves gave—they simply overlooked the evidence or had so many doubts about it that they did not feel justified in reporting it. The second possibility is that they were very much aware of the signs of fire and stone tools, before Breuil reported them, but deliberately withheld this information.

But why? At the time the discoveries were made at Choukoutien, fire and stone tools at a site were generally taken as the work of *Homo sapiens* or Neanderthals. According to Dubois and von Koenigswald, no stone tools or signs of usage of fire were found in connection with *Pithecanthropus erectus* in Java. The Selenka expedition did report remnants of hearths at Trinil, but this information did not attain wide circulation.

So perhaps the original investigators of Choukoutien purposefully held back from reporting stone tools and fire because they were aware such things might have confused the status of *Sinanthropus*. Doubters might have very well attributed the fire and tools to a being contemporary with, yet physically and culturally more advanced than *Sinanthropus,* thus removing *Sinanthropus* from his position as a new and important human ancestor.

As we shall see, that is what did happen once the tools and signs of fire became widely known. For example, Breuil (1932, p. 14) said about the relationship of *Sinanthropus* to the tools and signs of fire: "Several distinguished scientists have independently expressed to me the thought that a being so physically removed from Man. . . . was not capable of the works I have just described. In this case, the skeletal remains of *Sinanthropus* could be considered as simple hunting trophies, attributable, as were the traces of fire and industry, to a true Man, whose remains have not yet been found." But Breuil himself thought that *Sinanthropus* was the manufacturer of tools and maker of fire at Choukoutien.

9.1.8 Recent Views

Modern investigators have tended to confirm Breuil's views. Like Breuil, they hold that certain deposits in the Choukoutien caves are deep layers of ash, indicating the massive use of fire. For example, Wu Rukang and Lin Shenglong (1983, p. 93) reported that there are four "large thick layers of ashes" and that the thickest layer is six meters [about 19.7 feet] thick in certain places.

Paleontologist Jia Lanpo stated (1975, pp. 33–36) that there are four thick layers of ash and that the layer in the upper-middle part of the cave is six meters deep and consists of beds of ash of different colors—purple, red, yellow, white, and black. He also reported burned bones, colored black, blue, white, grey, green, or dull brown. Jia believed that Peking man knew how to use but not make fire, that the fires once lit were kept burning continuously for a long time—even passed down from generation to generation. We are not, however, aware of the discovery anywhere else in the world of a cave as old as Choukoutien having such huge beds of ash, providing, of course, that the above reports, identifying the deposits as ash layers, are correct.

Father P. O'Connell, a Roman Catholic priest who lived in China during the period of the Peking man discoveries, offered an intriguing explanation for the massive ash deposits at Choukoutien. He suggested that the site had been used for producing lime for the construction of the ancient city of Cambalac, situated on land now occupied by present-day Beijing (O'Connell 1969). Lime, a caustic substance produced by heating limestone to a high temperature, is used in making mortar and plaster.

But almost all modern investigators agree with Breuil that *Sinanthropus* was responsible for the signs of fire. Breuil wrote in the 1930s: "*Sinanthropus* kindled fire and did so frequently, he used bone implements and he worked stone, just as much as the Paleolithics of the West. In spite of his skull, which so closely resembles that of *Pithecanthropus,* he was not merely a Hominian, but possessed an ingenious mind capable of inventing, and hands that were sufficiently adroit and sufficiently master of their fingers to fashion tools and weapons" (Boule and Vallois 1957, p. 144). One gets an impression of a fairly humanlike being, a hunter who brought game felled with his stone weapons back to his cave home, where he cooked the flesh on fires he kindled for that purpose.

A somewhat different view of *Sinanthropus* at Choukoutien is provided by Lewis R. Binford and Chuan Kun Ho, anthropologists at the University of New Mexico. Concerning the signs of fire, they stated: "The so-called ash layers are not hearths and may not all be ash layers. . . . There seems to be little doubt that much of the content of the so-called ash layers is largely owl or other raptor droppings. They are systematically described as dominated by rodent bone. . . . It would appear that at least some of them were originally huge guano accumulations inside the cave. In some cases, these massive organic deposits could have burned. . . . The assumption that man introduced and distributed the fire is unwarranted, as is the assumption that burned bones and other materials are there by virtue of man's cooking his meals" (Binford and Ho 1985, p. 429).

Binford and Ho's theory that the ash deposits are composed mostly of bird droppings has not received unanimous support. But their assertions about the unreliability of the common picture of Peking man drawn from the presence of bones, ashes, and hominid remains at the site are worthy of serious consideration.

For Binford and Ho, the presence of hominid bones in the caves was not a demonstration that Peking man ever permanently lived there. They gave the following information about the Peking man fossils: "It is not uncommon to find hominid remains in direct association with hyena coprolites and adjacent to cave walls, where larger bones tend to end up in animal dens. Smaller hominid bones, such as isolated teeth, already broken cranial fragments, and mandible parts, are more common in contexts that appear to represent areas near the entrances of the cave. The picture one obtains is one in which hominid carcasses or parts thereof were introduced to the active, entrance area of the cave. It is unclear whether hominids died there or parts of hominid carcasses were brought there by scavenging animals. The extreme bias in body parts [mostly skulls and lower limb bones] would favor the latter interpretation. These parts were then further dispersed within the cave, most likely by bone-carrying animals such as hyena or wolf" (Binford and Ho 1985, p. 428).

The presence of stone tools at Choukoutien is generally taken as confirmation of a picture of *Sinanthropus* as a hunter sitting around his hearth cutting up deer carcasses. But Binford and Ho felt that the kind of tools found at Choukoutien, mostly primitive scrapers and choppers, were not very well adapted to hunting. Furthermore, they pointed out: "Layers that yield hominid remains only rarely produce stone tools and almost never are they designated as ash layers. In addition, excavations conducted in areas that would have been deep in the interior of the cave, beyond the limits of natural light, may yield hominid remains but only rarely yield tools in any concentration" (Binford and Ho 1985, p. 428). In other words, there is no clear connection between the stone tools and the hominid remains.

The most that can be said of Peking man, if we confine ourselves to the actual evidence at the site, is that he was perhaps a scavenger who may or may not have used primitive stone tools to cut meat from carcasses left by carnivores in a large cave where organic materials sometimes burned for long periods. Or perhaps Peking man was himself prey to the cave's carnivores, for it seems unlikely he would have voluntarily entered such a cave, even to scavenge.

Binford and Ho did not believe there existed a bone tool industry at Choukoutien. They said Breuil's recognition of a bone tool industry was founded on "modifications that today we routinely recognize as the by-product of animal gnawing" (Binford and Ho 1985, p. 428).

In making this judgement, Binford and Ho were in agreement with the original assessment made by the members of the Cenozoic Research Laboratory. Black, Teilhard de Chardin, Young (Yang), and Pei (1933, p. 130) believed that the recurring types of broken bones, which Breuil said had been shaped intentionally, may have been shaped by purely accidental forces. They believed, however, that further research would be required before this issue could be definitely resolved.

But a modern authority, Jia Lanpo of the Republic of China's Institute for Vertebrate Paleontology and Paleoanthropology, reported, like Breuil, numerous tools shaped from deer bones (Jia 1975, p. 31). Jia believed deer antler roots may have served as hammers. Antler tines showing criss-crossed scratches may have been used for digging, and deer skullcaps may have been used as drinking bowls. "Antlers are hard to hack off," stated Jia (1980, p. 29), "but if a spot is first scorched, the cutting is much easier, and this was what Peking man did, for some of the ends bear signs of scorching."

A bone industry at Choukoutien is also recognized by a Western authority on Chinese prehistory, J. S. Aigner, who wrote (1981, p. 144): "While Breuil may have been advocating tools overly, there is no question that human alteration of bones through processing activities and to a lesser extent through use (as tools) is clearly indicated."

Unfortunately, it is not possible to completely verify these claims, because many of the older pre-World War II specimens were lost. As related by W. C. Pei (Pei Wenzhong) in a introduction to a series of photos of the bones published by Breuil in 1939: "The specimens described herein by Professor H. Breuil have been placed at the disposal of the Museum of Geological Survey of China in Nanking for exhibition purposes, but owing to the hostilities in 1937 have become lost. It is indeed most unfortunate that these valuable objects, once so carefully studied by such a leading authority in pre-history as Professor Breuil, should no longer be available to science" (Bowden 1977, p. 99). Pei, it may be noted, was not very much in favor of the bone tool industry.

9.1.9 The Fossil Bones of Sinanthropus and Signs of Cannibalism

On March 15, 1934, Davidson Black was found at his work desk, dead of a heart attack. He was clutching his reconstruction of the skull of *Sinanthropus* in his hand. Shortly after Black's death, Franz Weidenreich assumed leadership of the Cenozoic Research Laboratory and wrote a comprehensive series of reports on the Peking man fossils. According to Weidenreich, the fossil remains of *Sinanthropus* individuals, particularly the skulls, suggested they had been the victims of cannibalism.

Most of the bones discovered at Choukoutien were cranial fragments. Weidenreich (1943, p. 7) stated: "none of the 14 skulls recognised as belonging to *Sinanthropus* is complete."

Weidenreich said that the skulls had been broken after fossilization by huge masses of stone falling from the roof of the cave. He pointed out that the other animal bones found at Choukoutien were similarly crushed and fragmented. But certain aspects of the cranial fragments of *Sinanthropus* led Weidenreich to conclude that they had also been broken before being covered by material from collapsed portions of the cave.

Weidenreich particularly noted that the relatively complete skulls all lacked portions of the central part of the base. He observed that in modern Melanesian skulls "the same injuries occur as the effects of ceremonial cannibalism" (Weidenreich 1943, p. 186).

Besides the missing basal sections, Weidenreich also noted other signs that might possibly be attributed to the deliberate application of force. For example, some of the skulls showed impact marks of a type that "can only occur if the bone is still in a state of plasticity," indicating that "the injuries described must have been inflicted during life or soon after death" (Weidenreich 1943, pp. 186–187).

Weidenreich (1943, p. 188) admitted that some of these injuries might have been caused by "stones falling from the roof of the cave on the individuals living in it" or "the bites of big carnivores having their dens near-by," but pointed out that others "look like incisions made by cutting implements."

Weidenreich (1943, p. 188) observed: "Blows inflicted on living individuals or corpses by stones falling accidentally cannot be held responsible for the destruction of the base of the skull. This fact, together with the cut-like lesions, rather points to injuries incidentally practiced by man."

Some of the skull fragments showed depressions that possibly could be interpreted as animal bites. But Weidenreich (1943, p. 189) stated: "considering the size, form and thickness of the vault it is difficult to imagine how the animal could find adequate points at which to drive its teeth and crack the vault by seizing it between its upper and lower jaws."

Also, according to Weidenreich (1943, p. 189), there were no cases on record that could be cited as "as examples of bites of carnivores inflicted on completely intact human skulls."

Some of the few long bones of *Sinanthropus* found at Choukoutien also displayed signs that to Weidenreich suggested human breakage. "It seems to be certain," he said (1943, p. 189), "that the lengthwise splitting involving the greater part of the shaft bones, cannot have been executed by carnivores but must have been done by man."

Von Koenigswald (1956, p. 49) agreed with this analysis, stating: "The thigh bones of Peking man found at Chou K'ou Tien are all severely damaged and often smashed into small pieces to extract the marrow. The damage was not the work of beasts of prey, but undoubtedly of humans."

Weidenreich (1943, p. 190) then offered this summary of his observations: "My verdict is that the destruction of the base and the blows on the top of the skull are the incidental work of man, although the possibility cannot be entirely excluded that at least those lesions which indicate they were produced by pointed or blunt agents may have been caused by stones falling from the roof of the cave on a living individual. Later on the skulls were broken as carrion by carnivores, probably hyaenas, which lived in the cave and cracked the bones as long as they were fresh." It is not very likely that *Sinanthropus* and the hyenas inhabited the cave at the same time. Accepting Weidenreich's version of events, *Sinanthropus* would have been an infrequent visitor, or perhaps the cave was inhabited alternately by *Sinanthropus* and other creatures.

As to why mostly cranial fragments were found, Weidenreich believed that with the exceptions of some long bones, only heads were carried into the caves. He stated: "the strange selection of human bones we are facing at Choukoutien has been made by *Sinanthropus* himself. He hunted his own kin as he hunted other animals and treated all his victims in the same way. Whether he opened the human skulls for ritual or culinary reasons cannot be decided on the basis of the present evidence of his cultural life; but the breaking of the long-bones of animals and man alike, apparently for the purposes of removing the marrow, indicates that the latter alternative is the more likely. The remains of his meals became the prey of his predatory neighbors at the foothills of Choukoutien" (Weidenreich 1943, p. 190).

Some modern authorities have suggested that Weidenreich was mistaken in his interpretation of the fossil remains of *Sinanthropus*. Binford and Ho (1985, p. 414) pointed out that hominid skulls subjected to transport over river gravel are found with the basal section worn away. But the skulls recovered from Choukoutien were apparently not transported in this fashion.

Binford and Ho also believed that damage to one skull, which Weidenreich thought could have been caused by cutting, was typical of a kind of animal breakage. But Weidenreich (1943, p. 189) had considered this possibility and still proposed cutting by an implement as the most likely cause. Even Binford and Ho (1985, p. 415) admitted that the kind of animal breakage they were proposing had "the appearance of cut or hack marks."

Binford and Ho also disagreed with Weidenreich's view that the *Sinanthropus* long bones were deliberately broken. They stated: "Binford has examined the photographs and casts of the bones in question, and the breakage appears to be unequivocally attributable to weathering. There is no evidence that these bones were broken fresh or by percussion" (Binford and Ho 1985, p. 414).

This statement is contradicted by Weidenreich (1941, p. 5), who said about one of the femurs: "The appearance of the bony surface exposed by . . . fractures indicates that the breakage occurred prior to mineralisation. . . . The remaining surface of the bone is practically intact and scarcely weathered." It should be kept in mind that Weidenreich was working from the original fossils, while Binford could only study photographs and casts because the original specimens were lost during the Second World War.

Another modern authority objecting to Weidenreich's cannibalism interpretation was Jean S. Aigner. She suggested: "The fact that the base of several brain cases is missing is not a particularly strong point when we recall the numerous examples of Recent burials with this portion missing. The absence is due to natural processes, not artificial ones, with the part of the skull resting in contact with the ground being eroded and dissolved away" (Aigner 1981, p. 128). But the Choukoutien finds were not burials, and there is no indication that the skulls were oriented in such a way (basal section down) as to produce the effects suggested by Aigner.

Aigner (1981, p. 128) offered this conclusion: "Weidenreich's reconstruction of the practices associated—murder, severing the head, and removing the brain, dissecting the long-bones and depositing just those parts in the cave—is difficult indeed to understand in light of the strong evidence (hearths) that the cave was a habitation site and not simply a dump." However, we have already seen, according to one modern opinion (Binford and Ho 1985), that there is no real evidence for hearths at Choukoutien. It is hard to attribute the huge masses of ash found there to campfires. Furthermore, the distribution of *Sinanthropus* remains is rather indicative of a dump. As noted by Weidenreich (1943, p. 186): "The distribution of *Sinanthropus* bones both horizontally and vertically throughout the deposit is an accidental one as is that of the animal bones."

Binford and Ho (1985, p. 428), as mentioned previously, proposed that carnivores had brought the hominid bones into the caves. But Weidenreich (1935, p. 453) said: "transportation by . . . beasts of prey is impossible. . . . traces of biting and gnawing ought to have been visible on the human bones, which is not the case. Therefore, the only possibility is that man himself brought the bones into the cave, by preference brain cases and jaws."

As evidence of cannibalism, Weidenreich pointed out that the *Sinanthropus* remains were predominantly those of children and females, the easiest to kill. Weidenreich (1935, p. 456) then cited several examples of cannibalism from Europe, stating: "Matiegka . . . has reviewed rather completely human skeletal material of prehistoric times of all Europe giving testimony about cannibalism. In many of these cultural places bones have been found which were broken to pieces and mingled with those of animals, with charcoal, ashes, and stone tools or splinters of them. Very frequently there were among the human bones skulls, isolated jaws, or fragments of them. . . . In some finding places the remains of children, adolescents and sometimes also of women were prevailing. The resemblance to the conditions existing in Choukoutien is obvious."

But Marcellin Boule, director of the Institute de Paleontologie Humaine in France, suggested another possibility—namely, that *Sinanthropus* had been hunted by a more intelligent type of hominid. Boule believed that the small cranial capacity of *Sinanthropus* implied that this hominid was not sufficiently intelligent to have created the stone and bone implements that were discovered in the cave.

In his description of the stone tool industry, Boule said: "It is important to note that this industry is not primitive, since M. Breuil himself acknowledges that 'many of (its) features are not found in France until the Upper Palaeolithic'. . . . Accompanying a being like *Sinanthropus* one would have expected to find an eolithic industry, and not true gravers and scrapers and other tools 'sometimes of fine workmanship'" (Boule and Vallois 1957, p. 145). Hence Boule concluded that the Choukoutien implements and fires were created by a "true man," *Homo sapiens,* who preyed upon *Sinanthropus*. Professor Boule did not believe that *Sinanthropus* necessarily represented an intermediate link in the chain of evolution from ape to *Homo sapiens*. He believed instead that *Sinanthropus* was simply an apelike being who was hunted for food by *Homo sapiens*.

Boule stated his position quite clearly: "We may therefore ask ourselves whether it is not over-bold to consider *Sinanthropus* the monarch of Choukoutien, when he appears in its deposit only in the guise of a mere hunter's prey, on a par with the animals by which he is accompanied" (Boule and Vallois 1957, p. 145).

If the remains of *Sinanthropus* were the trophies of a more intelligent hunter, who was that hunter and where were his remains? Boule pointed out that there are many caves in Europe that have abundant products of Paleolithic human industry, but the "proportion of deposits that have yielded the skulls or skeletons of the manufacturers of this industry is infinitesimal" (Boule and Vallois 1957, p. 145).

Boule further observed: "We may say that the absence of human bone remains is the rule and their presence the exception" (Boule and Vallois 1957, p. 145). For example, Boule stated that 4,000 cubic yards of deposits were systematically and carefully removed from the Prince's Cave at Grimaldi in the hope of finding human bones; however, not a single fragment of a human bone was discovered, despite the discovery of numerous animal bones and stones shaped by humans (Boule and Vallois 1957, p. 145).

Therefore, the hypothesis that a more intelligent species of hominid hunted *Sinanthropus* at Choukoutien is not ruled out simply because its fossil bones have not yet been found at Choukoutien. From our previous chapters, it may be recalled that there is evidence, from other parts of the world, of fully human skeletal remains from periods of equal and greater antiquity than that represented by Choukoutien. For example, the fully human skeletal remains found at Castenedolo in Italy are from the Pliocene period, over 2 million years ago.

9.1.10 Discoveries in the Upper Cave

In the early 1930s, some fully human remains were found at Choukoutien, in the Upper Cave, which lies above the main deposits. Modern researchers, using a combination of carbon 14 tests and studies of faunal remains, have said these human fossils are only about 20,000 years old.

A *Sinanthropus* upper jaw was found along with the human fossils. The usual explanation is that the *Sinanthropus* jaw was derived from the Lower Cave deposits. Weidenreich (1943, p. 16) said about the jaw found in the Upper Cave: "it distinctly differs from the other bones found in this cave by the high degree of mineralization, the special color, the primitiveness of the form, the considerable size of the teeth, and the way in which the bone is broken. In all these particularities the maxilla shows a greater resemblance to the *Sinanthropus* jaws recovered from Locality 1 than to the maxillae of the Upper Cave Man which have been found in connection with their pertaining skulls."

Stone implements were also discovered in the Upper Cave. According to Pei (1939, p. 16) some of these quartz implements look "surprisingly similar to some pieces found in the much older *Sinanthropus* deposits." Pei (1939, p. 16) then added that "it is quite possible that the here described quartz implements were collected by the Upper Cave Man or introduced by natural agencies into the Upper Cave from the *Sinanthropus* deposits."

9.1.11 Our Knowledge of Peking Man

All in all, the picture we get of *Sinanthropus* is not very much like the almost human ancestral hominid seen in textbook paintings and museum exhibits—the expert hunter sitting by his hearth in his cave home. Instead, making use of all the available evidence and points of view, we see through the haze of several hundred thousand years the outlines of a somewhat apelike creature, who was most likely a scavenger who sometimes got scavenged himself—perhaps by his own kind, perhaps by a more advanced hominid.

In Boule's opinion, Weidenreich and others tended to overemphasize the humanlike features of *Sinanthropus*. Von Koenigswald (1956, p. 51) wrote: "Our real knowledge of Peking man does not amount to very much. The skull is the best-known factor, and Weidenreich used it to have a rather excessively idealized reconstruction made by the American sculptress Lucille Swan, which came to be known in Peking as 'Nelly.'" Of course, even the very term Peking "man" carries with it strong, and undeserved, overtones of human attributes and ancestorship.

This idealization was perhaps to be expected, because according to modern evolutionary theory *Sinanthropus* (or *Homo erectus*) is supposed to be the immediate ancestor of *Homo sapiens*. But Boule (1937), considering reports coming from Choukoutien, said: "To this fantastic hypothesis, that the owners of the monkey-like skulls were the authors of the large-scale industry, I take the liberty of preferring an opinion more in conformity with the conclusions from my studies, which is that the hunter was a real man and that the cut stones, etc., were his handiwork" (Fix 1984, pp. 130–131).

Why did Boule say the skull of *Sinanthropus* was monkeylike? There are several reasons. One, of course, is the large brow ridges. When the skull of *Sinanthropus* is seen from directly above, the ridges stick out like handlebars on a bicycle (Figure 9.1, p. 556). Another apelike feature is the "postorbital constriction," or narrowing of the skull in back of the eye sockets. Humans do not have this. Place your fingers at the corners of your eyes and then run them back to your temples, just above the ear. You will notice that the surface is flat.

Figure 9.1. The first *Sinanthropus* skull, discovered in 1929 at Choukoutien, viewed from above (Jia 1975, p. 17) and from the rear (Boule 1937, p. 7). Like the apes, *Sinanthropus* has enormous brow ridges and a pronounced postorbital constriction (top). Also, the *Sinanthropus* skull, seen from the rear (bottom), is narrower at the top than at the bottom, another apelike feature.

But in *Sinanthropus,* immediately in back of the eyes, there is a very pronounced indentation on either side (Figure 9.1). Another apelike feature is the general shape of the skull when seen from behind. The skull of *Sinanthropus* is somewhat narrower at the top than at the bottom (Figure 9.1). In contrast, human skulls are normally wider at the top than at the bottom (Boule and Vallois 1957, p. 135). Also, as previously noted, the walls of the *Sinanthropus* skull are twice as thick as those of the average human skull.

The capacity of the *Sinanthropus* cranium is said to average around 1000 cubic centimeters, more than the anthropoid ape average of 600 cubic centimeters but less than the human average of about 1400 cubic centimeters. There have been suggestions that Black and Weidenreich reconstructed shattered *Sinanthropus* skulls in such a way as to increase their cranial capacity above the range for apes. Concerning the initial skull recovered in 1929, Bowden, after careful study of Black's three reports and accompanying photographs, noted that the bottom edge of the reconstructed skull pictured in the last report was lower than the bottom edge shown in the first photograph, taken shortly after the skull was excavated. Bowden believed this discrepancy could be accounted for if one assumed the lower part of the skull had been carelessly cropped out of the original photograph. More likely, according to Bowden (1977, p. 118), was the possibility that the "reconstruction of the base of the skull was carried out in such a way that it was made deeper" thus yielding a larger and less apelike volume.

Boule pointed out that there was strong sexual dimorphism in *Sinanthropus.* That is to say, there was great variation between the size of males and females, much more than in human beings. This is an apelike feature, which is reflected in the *Sinanthropus* jaws found at Choukoutien (Figure 9.2). Weidenreich noticed the dimorphism when studying reconstructed jaws of a male, female, and child, which showed unusual variation in size. Boule stated: "The total result was so polymorphous that Weidenreich wondered whether it was really a single species, or at least a single race. It is a fact, however, that this polymorphism is a simian characteristic in singular contrast to the slight sexual dimorphism in human jaws" (Boule and Vallois 1957, p. 138). Another apelike feature of the *Sinanthropus* jaw is that it has several openings for the dental nerve that reaches out to the skin of the chin, while the human jaw has only one (Keith 1931, pp. 262–265). Boule also believed the dentition of *Sinanthropus* was less human than some originally thought.

Boule therefore came to the following conclusion: "By the sum total of their characters, the mandibles and teeth of *Sinanthropus* denote a large Primate that was . . . certainly less human than the Mauer [Heidelberg] jaw, which is probably older than the Peking fossils" (Boule and Vallois 1957, p. 140).

In his physiological demotion of Peking man, Boule is in harmony with some modern researchers, such as Binford and Ho, who hesitate to attribute typically human behavior to the Choukoutien hominids.

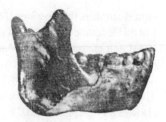

Figure 9.2. Restored jaws of an adult *Sinanthropus* male (above) and female (below). They display substantial sexual dimorphism, an apelike feature (Boule 1937, p. 13).

Binford and Ho characterized *Sinanthropus* behaviorally as a simple scavenger who was not clearly responsible for either the animal bones or beds of ashes at the cave of Choukoutien. They concluded: "What, then, was life like in the 'cave home of Beijing man?' We think we must conclude that we do not know" (Binford and Ho 1985, p. 429). We agree with this honest statement, which respects the limitations of the empiric method when applied to such questions.

9.1.12 The Fossils Disappear

As we have previously mentioned, one reason that it may be difficult to resolve many of the questions surrounding Peking man is that the original fossils are no longer available for study. By 1938, excavations at Choukoutien, under the direction of Weidenreich, were halted by guerilla warfare in the surrounding Western Hills. Later, with the Second World War well underway, Weidenreich left for the United States in April of 1941, carrying a set of casts of the Peking man fossils.

In the summer of 1941, it is said, the original bones were packed in two footlockers and delivered to Colonel Ashurst of the U.S. Marine Embassy Guard in Peking. In early December of 1941, the footlockers were reportedly placed on a train bound for the port of Chinwangtao, where they were to be loaded onto an

American ship, the *President Harrison,* as part of the U.S. evacuation from China. But on December 7, the train was intercepted, and the fossils were never seen again. In a statement published on March 22, 1951 in the *New York Times,* Pei Wenzhong (W. C. Pei) said the Americans found the fossils at the University of Tokyo after the war and secretly transported them to the American Museum of Natural History. The chairman of the department of anthropology at the Museum denied the charge (Bowden 1977, pp. 106–107).

After World War II, the Chinese Communist government continued the excavations at Choukoutien, adding a few fossils to the prewar discoveries. The present total of *Homo erectus (Sinanthropus)* discoveries since 1927 is 6 fairly complete skullcaps along with 12 other skull fragments, 15 pieces of lower jaws, 157 teeth, 3 fragments of upper arm bones, 1 clavicle, 7 fragments of thighbones, 1 fragment of a shinbone, and 1 wrist bone. These are said to represent the remains of 40 individuals (Wu and Lin 1983, p. 89). Recent opinion is that the clavicle is not from a hominid. In any case, most of the Peking man fossil bones, over 90 percent, were lost during the Second World War.

9.1.13 An Example of Intellectual Dishonesty

In an article about Zhoukoudian (Choukoutien) that appeared in the June 1983 issue of *Scientific American,* two Chinese scientists, Wu Rukang and Lin Shenglong, presented misleading evidence for human evolution.

Wu and Lin made two claims: (1) The cranial capacity of *Sinanthropus* increased from the lowest level of the Zhoukoudian excavation (460,000 years old) to the highest level (230,000 years old), indicating that *Sinanthropus* evolved towards *Homo sapiens.* (2) The type and distribution of stone tools also implied that *Sinanthropus* evolved.

In support of their first claim, Wu and Lin analyzed the cranial capacities of the 6 relatively complete *Sinanthropus* skulls found at Zhoukoudian. Wu and Lin (1983, p. 94) stated: "The measured cranial capacities are 915 cubic centimeters for the earliest skull, an average of 1075 cubic centimeters for four later skulls and 1140 cubic centimeters for the most recent one." From this set of relationships (Table 9.1, column A), Wu and Lin (1983, p. 94) concluded: "It seems the brain size increased by more than 100 cubic centimeters during the occupation of the cave."

A chart in the *Scientific American* article showed the positions and sizes of the skulls found at Zhoukoudian Locality 1. But in their explanation of this chart, Wu and Lin neglected to state that the earliest skull, found at layer 10, belonged to a child, who according to Franz Weidenreich (1935, p. 448) died at age 8 or 9, and according to Davidson Black died between ages 11 and 13. In the text of their article, Wu and Lin (1983, p. 90) did mention that one of the 6 skulls they

TABLE 9.1

**Evidence for Supposed Evolutionary Increase
in Sinanthropus Cranial Capacity at Zhoukoudian, China**

Years B.P.	Layer	A: Data Reported by Wu and Lin, 1983	B: Complete Data
230,000	1–2		
	3	1140 cc (V)	1140 cc (V)
290,000	4		
	5		
350,000	6		
	7		
420,000	8	1075 cc = average of 4 skulls	1225 cc (X), 1015 cc (XI), 1030 cc (XII), 1025 cc (II)
	9		
460,000	10	915 cc (III)	915 cc (III) child
700,000	11–13		

In *Scientific American* (June 1983), Wu Rukang and Lin Shenglong used the data in column A to suggest that *Sinanthropus* individuals evolved a larger cranial capacity during the 230,000 years they occupied the Zhoukoudian cave. But in their table Wu and Lin did not mention that the oldest skull (III) was that of a child, making it useless for comparison with the other skulls, which were those of adults. Furthermore, Wu and Lin gave an average for 4 skulls from layers 8 and 9 (II, X, XI, and XII), without mentioning that one of these skulls (X) had a cranial capacity of 1225 cc, larger than the most recent skull from layer 3. The complete data, shown in column B, reveals no evolutionary increase in cranial capacity. All of the data in the table was originally reported by Weidenreich (1935, 1943), except for the cranial capacity of the skull found at layer 3. In 1934, Weidenreich reported the discovery of some pieces of this skull, which he later designated skull V (Weidenreich 1943, p. 5). Then in 1966, Chinese paleontologists found other pieces of this same skull (Jia 1980, p. 26). The reconstruction of this skull and the cranial capacity measurement were carried out in 1966.

considered was from a child who died at age 8 or 9; yet they did not specify the level at which this skull was found. Wu and Lin acknowledged that a child's skull is smaller than an adult's. But to establish an evolutionary trend in their chart, they still compared the child's skull from layer 10 with the other skulls, which are from adults.

Wu and Lin also neglected to mention that one of the skulls discovered in layers 8 and 9 (skull X) had a cranial capacity of 1,225 cc, which is 85 cc larger than the most recent skull (V), found in layer 3. When all the data is presented, (Table 9.1, column B) it is clear that there is no steady increase in cranial capacity from 460,000 to 230,000 years ago.

Except the skull pieces from layer 3 found in 1966, Weidenreich examined all the skulls in Table 9.1. He saw no evolutionarily significant change in their general shape or cranial capacity from the bottom to the top of the excavation: "The morphological character of the *Sinanthropus* population of Locality 1, therefore, remained unchanged during the long periods of time necessary for the filling-up of the cave. Viewed from the morphological standpoint the population represents a uniform type" (Weidenreich 1935, p. 450).

Another attempt has been made, using cranial data, to establish that evolution took place during the Zhoukoudian occupation. According to W. W. Howells (1977, p. 70), Chinese paleoanthropologist Ku Yu-min believed that the skull from layer 3 should be "viewed as more progressive than the other known individuals, from the cranial capacity and various other features." Howells (1977, p. 70) described these features: "The bone of the frontal is thinner. The occipital torus is less developed than in most (skull XI excepted). . . . The frontal sinuses are larger than in all others, except skull III. The vault as reconstructed is higher. From this assessment of status . . . , the Chinese believe that the morphology of later Peking man is thus established, as showing evolutionary development within the Peking man phylum."

The above reasoning is not totally convincing. Concerning cranial capacity, the smallest adult skull (skull XI, 1015 cc) and the largest (skull X, 1225 cc) are both from layers 8 and 9. The skull from layer 3 (skull V, 1140 cc) falls within this range. This hardly demonstrates evolutionary development.

As far as other features of the skulls are concerned, one cannot make a statistically meaningful statement about them. Because of individual variations in a population, many skulls at each time horizon are needed to show that the population has undergone significant change. But at Zhoukoudian only one skull is substantially more recent than the others, and this single skull is by no means sufficient to establish a trend in any particular direction. Also, from Ku Yu-min's description above it appears that some of the supposedly progressive characteristics of the most recent skull (V) are also found in the older skulls—such as the large frontal sinuses in skull III and the less-developed occipital in skull XI.

In addition to discussing an evolutionary increase in cranial capacity, Wu and Lin noted a trend toward smaller tools in the Zhoukoudian cave deposits. They also reported that the materials used to make the tools in the recent levels were superior to those used in the older levels. The recent levels featured more high-quality quartz, more flint, and less sandstone than the earlier levels (Wu and Lin 1983, p. 92).

But a change in the technological skill of a population does not imply that this population has evolved physiologically. For example, consider residents of Germany in A.D. 1400 and residents of Germany in A.D. 1990. The technological differences are awesome—jet planes and cars instead of horses; television and telephone instead of unaided vision and voice; tanks and missiles instead of swords and bows. Yet one would be in error if one concluded that the Germans of 1990 were physiologically more evolved than the Germans of 1400. Hence, contrary to the claim of Wu and Lin, the distribution of various kinds of stone tools does not imply that *Sinanthropus* evolved.

The report of Wu and Lin, especially their claim of increased cranial capacity in *Sinanthropus* during the Zhoukoudian cave occupation, shows that one should not uncritically accept all one reads about human evolution in scientific journals. It appears the scientific community is so committed to its evolutionary doctrine that any article purporting to demonstrate it can pass without much scrutiny.

9.2 OTHER DISCOVERIES IN CHINA

Although Zhoukoudian is the most famous paleoanthropological site in China, there are many others. Discoveries at these sites have included fossils representative of early *Homo erectus, Homo erectus,* Neanderthals, and early *Homo sapiens,* thus providing an apparent evolutionary sequence. But the way in which this progression has been constructed is open to question.

9.2.1 Dating by Morphology

As we have seen in our discussion of human fossil remains discovered in China and elsewhere, it is in most cases not possible to date them with a very high degree of precision. Finds tend to occur within what we choose to call a "possible date range," and this range may be quite broad, depending upon the dating methods that are used. Such methods include chemical, radiometric, and geomagnetic dating techniques, as well as analysis of site stratigraphy, faunal remains, tool types, and the morphology of the hominid remains. Furthermore, different scientists using the same methods often come up with different age ranges for particular hominid specimens. Unless one wants to uniformly consider the age judgement given most recently by a scientist as the correct one, one is compelled to take into consideration the entire range of proposed dates.

But here one can find oneself in difficulty. Imagine that a scientist reads several reports about two hominid specimens of different morphology. On the basis of stratigraphy and faunal comparisons, they are from roughly the same period. But this period stretches over several hundred thousand years. Repeated testing by different scientists using different paleomagnetic, chemical, and radiometric methods has given a wide spread of conflicting dates within this period. Some test results indicate one specimen is the older, some that the other is the older. Analyzing all the published dates for the two specimens, our investigator finds that the possible date ranges broadly overlap. In other words, by these methods it proves impossible to determine which of the two came first.

What is to be done? In some cases, as we shall show, scientists will decide, solely on the basis of their commitment to evolution, that the morphologically more apelike specimen should be moved to the early part of its possible date range, in order to remove it from the part of its possible date range that overlaps with that of the morphologically more humanlike specimen. As part of the same procedure, the more humanlike specimen can be moved to the later, or more recent, part of its own possible date range. Thus the two specimens are temporally separated. But keep in mind the following: this sequencing operation is performed primarily on the basis of morphology, in order to preserve an evolutionary progression. It would look bad to have two forms, one generally considered ancestral to the other, existing contemporaneously.

Here is an example. Chang Kwang-chih, an anthropologist from Yale University, stated: "The faunal lists for Ma-pa, Ch'ang-yang, and Liu-chiang [hominid] finds offer no positive evidence for any precise dating. The former two fossils can be anywhere from the Middle to the Upper Pleistocene, as far as their associated fauna is concerned. . . . For a more precise placement of these three human fossils, one can only rely upon, at the present time, their own morphological features in comparison with other better-dated finds elsewhere in China" (Chang 1962, p. 757). This may be called dating by morphology.

Jean S. Aigner (1981, p. 25) stated: "In south China the faunas are apparently stable, making subdivision of the Middle Pleistocene difficult. Ordinarily the presence of an advanced hominid or relict [mammal] form is the basis for determining later and earlier periods." This is a very clear exposition of the rationale for morphological dating. The presence of an advanced hominid is taken as an unmistakable sign of a later period.

In other words, if we find an apelike hominid in connection with a certain Middle Pleistocene fauna at one site and a more humanlike hominid in connection with the same Middle Pleistocene fauna at another site, then we must, according to this system, conclude that the site with the more humanlike hominid is of a later Middle Pleistocene date than the other. The Middle Pleistocene, it may be recalled, extends from 100,000 to 1 million years ago. It is taken for granted that the two sites in question could not possibly be contemporaneous.

With this maneuver completed, the two fossil hominids, now set apart from each other temporally, are then cited in textbooks as evidence of an evolutionary progression in the Middle Pleistocene! This is an intellectually dishonest procedure. The honest thing to do would be to admit that the evidence does not allow one to say with certainty that one hominid preceded the other and that it is possible they were contemporary. This would rule out using these particular hominids to construct a temporal evolutionary sequence. All one could honestly say is that both were found in the Middle Pleistocene. For all we know, the "more advanced" humanlike hominid may have preceded the "less advanced" apelike one. But by assuming that evolution is a fact, one can then "date" the hominids by their morphology and arrange the fossil evidence in a consistent manner.

9.2.2 Tongzi, Guizhou Province

Let us now consider a specific example of the date range problem. In 1985, Qiu Zhonglang reported that in 1971 and 1972 fossil teeth of *Homo sapiens* were found in the Yanhui cave near Tongzi, in Guizhou province, southern China. The Tongzi site contained a *Stegodon-Ailuropoda* fauna. *Stegodon* is a type of extinct elephant, and *Ailuropoda* is the giant panda. This *Stegodon-Ailuropoda* fauna is typical of southern China during the Middle Pleistocene.

The complete faunal list for the Tongzi site given by Han Defen and Xu Chunhua (Han and Xu 1985, pp. 285–286) contains 24 kinds of mammals, all of which are also found in Middle (and Early) Pleistocene lists given by the same authors (Han and Xu 1985, pp. 277–283). But a great many of the genera and species listed are also known to have survived to the Late Pleistocene and the present.

The author of the report on the Tongzi discoveries stated: "the Yanhui Cave was the first site containing fossils of *Homo sapiens* discovered anywhere in the province. . . . The fauna suggests a Middle-Upper Pleistocene range, but the archaeological [human] evidence is consistent with an Upper Pleistocene age" (Qiu 1985, pp. 205–206).

In other words, the presence of *Homo sapiens* fossils was the determining factor in assigning a Late Pleistocene age to the site. This is a clear example of dating by morphology. But according to the faunal evidence reported by Qiu (1985), all that can really be said is that the age of the *Homo sapiens* fossils could be anywhere from Middle Pleistocene to Late Pleistocene.

But there is stratigraphic evidence suggesting a strictly Middle Pleistocene range. Qiu (1985, p. 206) gave the following information: "The deposits in the cave contain seven layers. The human fossils, stone artifacts, burned bones, and mammalian fossils were all unearthed in the fourth layer, a stratum of greyish-yellow sand and gravel." This concentration in a single layer suggests that the human remains and the animal fossils, all of mammals found at Middle Pleistocene sites, are roughly contemporaneous. And yellow cave deposits in South China are generally thought to be Middle Pleistocene (Han and Xu 1985, p. 273; Simons and Ettel 1970, p. 84).

Our own analysis of the faunal list also suggests it is reasonable to narrow the age range to the Middle Pleistocene. *Stegodon,* present at Tongzi, is generally said to have existed from the Pliocene to the Middle Pleistocene (Belyaeva *et al.* 1962, p. 365). In a list of animals considered important for dating sites in South China, Aigner (1978) indicated that *Stegodon orientalis* survived only to the late Middle Pleistocene, although she did place a question mark after this entry.

A strictly Middle Pleistocene age for the Tongzi cave fauna is supported by the presence of a species whose extinction by the end of the Middle Pleistocene is thought to be more definite. In her list of mammals considered important for dating sites in South China, Aigner included, in addition to *Stegodon orientalis,* other species found at Tongzi. Among them is *Megatapirus* (giant tapir), which Aigner (1981, p. 289) said is confined to the Middle Pleistocene. The species found at Tongzi is listed as *Megatapirus augustus* Matthew et Granger (Han and Xu 1985, p. 25). Aigner (1981, p. 325) characterized *Megatapirus augustus* as a "large fossil form of the mid-Middle Pleistocene south China collections." We suggest that *Megatapirus augustus* limits the most recent age of the Tongzi faunal collection to the end of the Middle Pleistocene (Figure 9.3).

Another marker fossil listed by Aigner (1981, p. 289) is *Crocuta crocuta* (the living hyena), which first appeared in China during the middle Middle Pleistocene. Since *Crocuta crocuta* is present at Tongzi, this limits the oldest age of the Tongzi fauna to the beginning of the middle Middle Pleistocene (Figure 9.3).

In summary, using *Megatapirus augustus* and *Crocuta crocuta* as marker fossils, we can conclude that the probable date range for the *Homo sapiens* fossils found at Tongzi extends from the beginning of the middle Middle Pleistocene to the end of the late Middle Pleistocene (Figure 9.3).

So Qiu (1985), in effect, extended the date ranges of some mammalian species in the *Stegodon-Ailuropoda* fauna (such as *Megatapirus augustus*) from the Middle Pleistocene into the early Late Pleistocene in order to preserve an acceptable date for the *Homo sapiens* fossils. Qiu's evolutionary preconceptions apparently demanded this operation. Once it was carried out, the Tongzi *Homo sapiens,* placed safely in the Late Pleistocene, could then be introduced into a temporal evolutionary sequence and cited as proof of human evolution. If we place Tongzi *Homo sapiens* in the older part of its true faunal date range, in the middle Middle Pleistocene, he would be contemporary with Zhoukoudian *Homo erectus.* And that would not look very good in a textbook on fossil man in China.

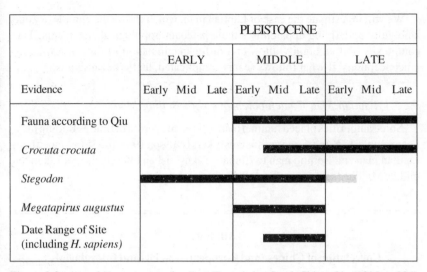

	PLEISTOCENE		
	EARLY	MIDDLE	LATE
Evidence	Early Mid Late	Early Mid Late	Early Mid Late
Fauna according to Qiu			
Crocuta crocuta			
Stegodon			
Megatapirus augustus			
Date Range of Site (including *H. sapiens*)			

Figure 9.3. Age of *Homo sapiens* fossils at Tongzi site, South China. Qiu (1985, p. 206) said the Tongzi mammalian fauna was Middle to Late Pleistocene, but used *Homo sapiens* fossils to date the site to the Late Pleistocene. But if we instead use the mammalian fauna to date the *Homo sapiens* fossils, we arrive at a different age for the site. *Stegodon* became extinct at the end of the Middle Pleistocene, possibly surviving into the early Late Pleistocene (grey part of bar) in some South China locales (Aigner 1981, p. 289). *Megatapirus augustus* (giant tapir) definitely did not survive the Middle Pleistocene (Aigner 1981, p. 289). The presence of *Stegodon* and especially *Megatapirus augustus* limit the most recent age for the Tongzi site to the end of the Middle Pleistocene. The presence of *Crocuta crocuta* (the living hyena), which first appears in the middle Middle Pleistocene (Aigner 1981, p. 289), limits the oldest age for the Tongzi site to the beginning of the middle Middle Pleistocene. Therefore, the allowed range for the *Homo sapiens* fossils at Tongzi extends from the beginning of the middle Middle Pleistocene to the end of the late Middle Pleistocene.

9.2.3 Lantian Man

Let us now consider another element in the confusing picture of the Chinese Middle Pleistocene—Lantian man. In 1963, Zhang Yuping and Huang Wanpo, of the Institute of Vertebrate Paleontology and Paleoanthropology (IVPP), discovered a *Homo erectus* mandible (lower jaw) at Chenjiawo village in Lantian county, Shaanxi province. In 1964, another team discovered the tooth of a human being at Gongwangling, also in Lantian county. Chunks of fossil-bearing rock from Gongwangling were transported to Beijing. There a hominid skullcap was discovered, along with an upper jaw bone and 3 molars, one unattached. These specimens were classified as *Homo erectus* just as the Chenjiawo jaw had been (Jia 1980, pp. 13–14).

We will investigate the case of Lantian man, which shows the complexity and ambiguity underlying apparently simple paleoanthropological statements. This complexity and ambiguity allows room for manipulation of data according to preconceptions. If what follows seems complicated, that is because it is.

9.2.3.1 Lantian Man Contemporaneous with Beijing Man?

Some authorities placed Lantian man in the same period of time as Beijing man. For example, L. Yung-Chao and his coworkers (Nilsson 1983, p. 335) assigned both Lantian man and Beijing man to China's Taku-Lushan interglacial period, in the middle Middle Pleistocene (Table 9.2).

TABLE 9.2

Correlation of Chinese and European Glacials and Interglacials

Period	European	Chinese
Holocene	Present Warm Period	Present Warm Period
Late Pleistocene	Würm Glacial	Tali Glacial
	Eemian Interglacial	Tali-Lushan Interglacial
Late Middle Pleistocene	Riss II Glacial	Lushan Glacial
	Ilford Interglacial	
	Riss I Glacial	
Middle Middle Pleistocene	Holstein Interglacial	Taku-Lushan Interglacial
	Mindel Glacial	Taku Glacial
	Cromerian Interglacial	Poyang-Taku Interglacial
Early Middle Pleistocene	Günz Glacial	Poyang Glacial

Sources for this table are Delson 1977, p. 45 and Aigner 1981, p. 32.

And Yale professor Kwang-chih Chang wrote in the 1977 edition of his book *The Archaeology of Ancient China:* "Geologically the Lan-t'ien fossils occurred in strata broadly comparable with the Chou-k'ou-tien sedimentation" (Chang 1977, p. 53). He said the Chenjiawo mandible occurred "in association with fossil remains . . . that recall the Chou-k'ou-tien fauna" (Chang 1977, p. 53). The Gongwangling skull, said Chang (1977, p. 54), was found "in association with mixed Chou-k'ou-tien-Wan Hsien fauna." Wan Hsien is a site in South China, the fauna of which includes a component comparable in age to the fauna discovered at Zhoukoudian (Aigner 1981, p. 288). Chang (1977, pp. 53–54) concluded: "both Ch'en-chia-wo and Kung-wang-ling were probably datable to the Taku-Lushan interglacial, contemporaneous with Peking Man of Chou-k'ou-tien." The Taku-Lushan interglacial is said to be equivalent to the Holstein interglacial of the European middle Middle Pleistocene (Chang 1977, p. 46). Pollen studies of the Lantian man sites indicated "prevalence of grassy species and broadleaf trees of an interglacial environment" (Chang 1977, pp. 53–54).

But Jia Lanpo (1980, p. 16) pointed out that Lantian man had thicker cranial walls than Beijing man and a much smaller cranial capacity—about 778 cubic centimeters compared with an average of over 1000 cubic centimeters for the Zhoukoudian *Homo erectus* population (Jia 1980, p. 15). Jia therefore concluded that while the jaw from Chenjiawo might be contemporaneous with Beijing man in the middle Middle Pleistocene, the skullcap from Gongwangling was older.

9.2.3.2 Morphological Dating of Lantian Man

So who was correct, Chang or Jia? J. S. Aigner discussed the Lantian man controversies in her book-length survey of Chinese discoveries. It is generally an author of broad surveys, such as Aigner, who tries to sort out conflicting reports and adjust possible date ranges so that a coherent evolutionary progression of fossil hominids in China, or elsewhere, emerges. And it is on this level that the problems of personal bias and data manipulation are most clearly evident. Adjustments are sometimes made, without adequate supporting evidence, simply to make the fossils fit some preordained scheme.

Here is how Aigner dealt with Chang, who, as we have just seen, did not accept a pre-Zhoukoudian date for the Gongwangling skull. Aigner (1981, p. 82) stated: "Chang (1977) also appears reluctant to accept the [skull's] early dating although he does accept its earliness morphologically speaking, on which point Wu (1973) and I (Aigner and Laughlin 1973) concur." This concept of "morphological earliness" assumes what must be demonstrated, providing a good example of how evolutionary prejudices distort paleoanthropological research.

With this in mind, consider Aigner's concluding statement on the skull of Lantian man (1981, p. 244): "The massive supraorbital ridges, pronounced postorbital constriction, low frontal squama, and cranial height, extraordinary thickness of the cranial wall, and small cranial capacity [778 cc] indicate this form is more primitive than both *Sinanthropus* and *Homo erectus* from Trinil. It is morphologically closer to the earlier form from the Djetis beds of Java; in my opinion it must be considered temporally earlier than any of the *Sinanthropus* remains for these same reasons." In other words, Lantian man must be dated morphologically so that he can be integrated into the existing evolutionary sequence. Otherwise, the sequence would be disrupted. Wu Rukang and Dong Xingren provided another example of this prejudice in their statement: "Among China's *H. erectus* fossils, those from Lantian exhibit more primitive morphological characteristics than do those from Zhoukoudian. . . . We believe these differences are a reflection of both spatial and temporal evolutionary diversification" (Wu, R. and Dong 1985, p. 88).

9.2.3.3 Comparison of Faunal Evidence from Gongwangling and Chenjiawo

After arguing that the primitiveness of the Gongwangling skullcap meant that it was older than the Beijing man fossils from middle Middle Pleistocene Locality 1 at Zhoukoudian, Aigner (1981, pp. 81–82) said about Gongwangling: "The other faunal remains certainly demand an early Middle Pleistocene age for the locality, not contemporaneous with Choukoutien 13 or 1 as has been suggested by some."

Aigner believed that the Chenjiawo site, where the Lantian man jaw was found, was in fact roughly contemporaneous with Zhoukoudian in the middle Middle Pleistocene. She therefore sought to prove that the Gongwangling site, with the primitive *Homo erectus* skullcap, was older than Chenjiawo, and hence older than Zhoukoudian (Aigner and Laughlin 1973, p. 102).

But W. W. Howells (1977, p. 69) noted: "According to the Chinese this is not the case, since the faunas of the two sites [Gongwangling and Chenjiawo] coincide except for a few species and contain a species of elephant unique to both; also the formation is the same at both localities and has been mapped across the interval between them." As we have seen (Section 9.2.3.1), some of the authorities who considered the jaw and skullcap to be of the same age believed they were both contemporaneous with Beijing man.

One reason why Aigner (1981, p. 329) thought the Chenjiawo jaw was younger than the Gongwangling skull concerned fossils of *Ochotonoides*, an extinct mouse hare that supposedly belongs to the early Middle Pleistocene and earlier. Fossil remains of *Ochotonoides* were found at the same level as the Gongwangling man skullcap. This would appear to place the skullcap in the early Middle Pleistocene, before the middle Middle Pleistocene Zhoukoudian occupation. But, according to Aigner, *Ochotonoides* fossils were found only below the Chenjiawo jaw, suggesting the jaw belonged to a period later than the early Middle Pleistocene—i.e., the middle Middle Pleistocene.

Aigner derived her information about the position of the *Ochotonoides* fossils at Chenjiawo from a 1975 report by the Chinese scientists Zhou Mingzhen and Li Chuankuei. But according to Wu Xinzhi and Wang Linghong, Zhou and Li "reported the discovery of an additional maxilla and mandible of the same animal from the same layer in which the *H. erectus* mandible was found. In fact, no discontinuity is discernible and this suggests the various levels in question do not represent different interglacial events as Aigner and Laughlin (1973) suggested" (Wu, X. and Wang, L. 1985, p. 37). The interglacial event to which both Gongwangling and Chenjiawo belong remains open to question, but, as we have noted, some authorities assign both Gongwangling and Chenjiawo to the same interglacial period represented at Zhoukoudian (the Holstein, or Taku-Lushan interglacial).

Given this latter view, the presence of *Ochotonoides* at both Gongwangling and Chenjiawo might be taken to indicate that this genus survived to the time of the Zhoukoudian *Homo erectus* occupation in the middle Middle Pleistocene. Interestingly enough, the initial excavators of Zhoukoudian Locality 1 listed *Ochotonoides* among the forms found there, which would appear to confirm the idea that Beijing man is contemporaneous with Lantian man. But Aigner (1981, pp. 301, 329) informs us that "the Choukoutien 1 form is revised [by her] as *Ochotona,*" a related but more recent genus that survives today in Mongolia, Tibet, and elsewhere. One cannot help but feel a little suspicious about her motives for this reclassification.

Aigner also tried to use pollen studies to support her view that Gongwangling was earlier than Chenjiawo and Zhoukoudian. Aigner (1978, p. 26) stated: "The Choukoutien assemblage is interpreted as a Holstein interglacial temperate flora. . . . Both the limited assemblages from Lantian represent temperate flora dating to the Holstein and to an earlier warm interval." The earlier warm interval would seem to be the Cromerian or Poyang-Taku interglacial (Table 9.2, p. 566). But Chang used the same pollen to studies to place both Lantian man sites in the Holstein (Taku-Lushan) interglacial (1977, pp. 53–54).

In her capacity as synthesizer of the hominid discoveries in China, Aigner very obviously wanted to adjust things so that the Lantian skullcap would appear in the geological column before Beijing man of Zhoukoudian. But her reasons for insisting on this are not immune to criticism. Altogether, there appears ample reason to suppose that Gongwangling and Chenjiawo might be of the same age, and contemporary with Zhoukoudian.

In the fourth edition of his book *The Archaeology of Ancient China,* Chang (1986, p. 38) stated: "Geologically the Lan-t'ien fossils occurred in strata broadly contemporary with the Chou-k'outien sedimentation." According to

Chang (1986, p. 38), both Zhoukoudian and the Lantian sites (Gongwangling and Chenjiawo) are part of the same reddish Li-shih loess deposits. Chang (1986, p. 38) said the Chenjiawo fauna had "elements that recall the Chou-k'ou-tien fauna." These statements appear to support his original position that Gongwangling and Chenjiawo (and hence Zhoukoudian) were nearly contemporary.

Nevertheless, Aigner's reports apparently induced Chang to change his mind about the relative ages of Gongwangling and Zhoukoudian. Chang (1986, p. 38) went on to say that the Gongwangling fauna "is apparently more archaic than that of Ch'en-chia-wo, which led some scholars to place the Kung-wang-ling cranium into a much earlier period than the Ch'en-chia-wo mandible." Chang here cited the 1973 paper by Aigner and Laughlin, which he had earlier said was mistaken. Chang added that the placement of Gongwangling before Chenjiawo and Zhoukoudian Locality 1 was "later confirmed by paleomagnetic dating."

But our own reading of the geological evidence, paleomagnetic dates, and faunal analysis presented by Aigner and Chang leads us to conclude that there is not sufficient reason to rule out the possibility that Gongwangling and Zhoukoudian Locality 1 are nearly contemporary.

9.2.3.4 Paleomagnetic Dates

The paleomagnetic evidence reported by Chang is shown in Table 9.3, along with selected dates derived from other methods. Although it is possible that Gongwangling is older than Chenjiawo and Zhoukoudian Locality 1, it is also possible to conclude from the evidence reported by Chang that all three sites are nearly contemporary.

9.2.3.5 Comparison of Faunal Evidence from Gongwangling and Zhoukoudian

We shall now give a detailed analysis of Aigner's comparison of the Gongwangling fauna with that of Zhoukoudian. Seeking to demonstrate a pre-Zhoukoudian date for the very primitive Gongwangling *Homo erectus* skullcap, Aigner (1981, p. 81) said that at Gongwangling "only 37 percent of the species are modern forms, compared to 50 percent at the 'type' locality of the Middle Pleistocene, Choukoutien 1." Clearly, such determinations depend heavily upon the faunal lists one uses for comparison. Aigner used a short list (Aigner and Laughlin 1973, p. 101; Aigner 1981, pp. 300–302). From a variety of reports (Zhou, M. *et al.* 1965, Aigner and Laughlin 1973, Aigner 1981, Chang 1977, Han and Xu 1985), we have compiled a composite master faunal list for Gongwangling. Using this longer master list (Figure 9.4, pp. 572–573), we find that at Gongwangling 23 out of 46 taxa, or 50 percent, are modern forms (marked with dots), about the same as at Zhoukoudian according to Aigner.

TABLE 9.3

**Selected Dates Obtained for Zhoukoudian Locality 1,
Gongwangling, and Chenjiawo**

Years B.P. (thousands)	Technique	Site	Source
417–507	Fission track	Zhoukoudian 1 (10)	Guo *et al.* 1980
520–610	Thermoluminescence	Zhoukoudian 1 (10)	Pei, J. 1980
530	Paleomagnetic	Chenjiawo	Cheng *et al.* 1978
650	Paleomagnetic	Chenjiawo	Ma *et al.* 1978
>510	Amino Acid	Gongwangling	Li and Lin 1979
>500–800	Paleomagnetic	Gongwangling	Ma *et al.* 1978
1,000	Paleomagnetic	Gongwangling	Cheng *et al.* 1978

The data in this table is from a book by Chang (1986, pp. 32–33). Level 10 is the oldest level at which hominid fossils have been found at Zhoukoudian Locality 1. Chang (1986 p. 34) said that the paleomagnetic studies supported an older date for Gongwangling than for Chenjiawo and Zhoukoudian Locality 1. This is true if one accepts the older of the two paleomagnetic dates for Gongwangling and rejects the younger one. But if one accepts the younger of the conflicting paleomagnetic dates for Gongwangling, the data allows the possible near contemporaneity of all three sites at about 500,000 years to 600,000 years ago, in the middle Middle Pleistocene.

Let us now further analyze the Gongwangling fauna in comparison with the Zhoukoudian fauna, and see whether or not it is possible to conclude in any other way that Gongwangling must be older. In our comparison, we have not limited ourselves to the fauna at Zhoukoudian Locality 1, where Beijing man fossils were found, but have also included the faunal lists from the nearby Zhoukoudian Localities 13 and 15. Locality 13 is about the same age as the basal part of Locality 1 (Aigner 1981, p. 32). Locality 15 is slightly more recent than Locality 1 or perhaps equivalent to its upper phase (Aigner 1981, p. 32; Pei 1939, p. 184). Zhoukoudian Localities 1, 13, and 15 all fall within the middle Middle Pleistocene (Aigner 1981, p. 32). Using their combined faunal lists, we discovered that 26 of the 46 Gongwangling taxa (about 57 percent) are found at Zhoukoudian (Figure 9.4, I).

Of the remaining taxa, 7 represent distinctly southern forms (Figure 9.4, II). One way to interpret this evidence is that the Gongwangling site represents an

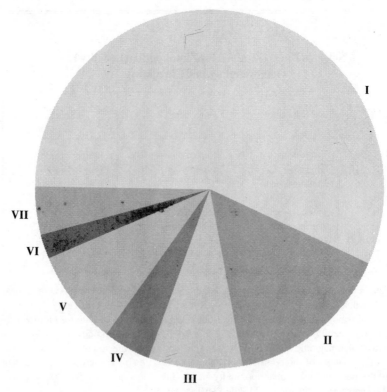

Figure 9.4. Analysis of Gongwangling fauna. In 1964, a *Homo erectus*-type cranium was found at Gongwangling. Because it was much more primitive than the *Homo erectus* crania from Zhoukoudian, some researchers thought it demanded an earlier date. These researchers sought support for their conclusion in the Gongwangling fauna (• = living forms), which they claimed was quite distinct from that of Zhoukoudian. But analysis of the Gongwangling fauna shows the following. (I) Twenty-six of the 46 taxa (57 percent) are found at Zhoukoudian Localities 1, 13, and 15. (II) Seven are typical southern forms, which would not be expected at Zhoukoudian, 500 miles to the northeast of Gongwangling. They could represent a geographical rather than a temporal variation. (III) Four of the Gongwangling taxa not found at Zhoukoudian are still living today, so they cannot be used to establish an earlier date for Gongwangling. (IV) Two species are unique to Gongwangling and thus cannot be used for relative dating. (V) Four Gongwangling species not found at Zhoukoudian are found at other Chinese sites with ages similar to Zhoukoudian. Thus far, comparison of the Gongwangling and Zhoukoudian fauna (I–V) do not establish that the Gongwangling site must be older. (VI) The only species that tends to confirm an early date is *Leptobos*, but this species differs significantly from the European type and was said by the original discoverer to resemble *Bison priscus*, which survived to the Late Pleistocene. (VII) Three species reported by original excavators at

Gongwangling and Zhoukoudian were inexplicably reclassified by later researchers. The changes tend to support a desired older date for Gongwangling, and are thus suspect. It thus appears that the Zhoukoudian *Homo erectus* and the more primitive Gongwangling *Homo erectus* were contemporary in the middle Middle Pleistocene.

I. Taxa also discovered at Zhoukoudian Localities 1,13, or 15:
- *Macaca* (monkey)
- *Neomys* (water shrew)
  Hyaena brevirostris sinensis (Chinese short-faced hyena)
  Megantheron (saber-tooth cat)
- *Felix (Panthera) pardus* (panther)
- *Felix (Panthera) tigris* (tiger)
- *Mustela* (polecat)
  Meles leucurus (hog badger)
  Canis variabilis (Chinese gray wolf)
- *Nyctereuteus sinensis* (raccoon dog)
  Equus sanmeniensis (horse)
  Sus lydekkeri (pig)
  Megaloceros (giant deer)
  Pseudaxis grayi (sika deer)
- *Gazella* (gazelle)
- *Petaurista* (flying squirrel)
- *Hystrix subcristata* (porcupine)
- *Myospalax* (mole rat)
- *Myospalax fontanieri* (mole rat)
  Myospalax tingi (mole rat)
- *Cricetulus griseus* (little hamster)
  Cricetulus varians (little hamster)
  Microtus epiraticeps (common vole)
- *Apodemus* (field mouse)
- *Gerbillus* (gerbil)
  Ochotonoides complicidens (pika)

II. Taxa typical of southern China:
  Stegodon orientalis (elephant)
- *Tapirus* (tapir)
- *Tapirus sinensis* (Chinese tapir)
  Megatapirus augustus (giant tapir)
  Nestoritherium sinensis (clawed, horselike mammal)
- *Elaphodus cephalophus* (tufted deer)
- *Capricornus sumatraensis qinlingensis* (goat)

III. Taxa not at Zhoukoudian, but still existing:
- *Scaptochirus moschatus* (musk mole)
- *Ailuropoda melanolueca fovealis* (giant panda)
- *Cervus rusa* (axis deer)
- *Ochotona* (steppe pika)

IV. Taxa unique to Gongwangling in all of China:
  Dicerorhinus lantienensis (Lantian ancient rhino)
  Leptobos brevicornus (early bison)

V. Taxa typical of middle Middle and later Pleistocene (i.e., same age as Zhoukoudian Locality 1 or younger):
  Dicerorhinus merki (early rhinoceros)
  Rhinoceros sinensis (Chinese rhinoceros)
  Arvicola terra-rubrae (water vole)
  Bahomys hypsodonta (a rodent first discovered at Lantian)

VI. *Leptobos* sp. (early bison)

VII. Taxa which had genus or species changed in some lists:
  Acinonyx pleistocaenicus -> Siva-panthera pleistocaenicus (cheetah) (Gongwangling form changed to an earlier one)
- *Ursus thibetanus* (bear) or *Ursus thibetanus kokeni* (present at Zhoukoudian) -> *Ursus etruscus* (Early Pleistocene)
  Ochotonoides -> Ochotona (Zhoukoudian form changed to a modern one)

older and warmer interglacial period than that represented by Zhoukoudian. But there is another possible interpretation—that Gongwangling, about 500 miles southwest of Zhoukoudian, is the same age as Zhoukoudian, but has southern forms because of its warmer weather. This possibility is admitted by Aigner, who said of the southern forms: "Their presence at Kungwangling . . . may be due to the more southerly location of the site" (Aigner and Laughlin 1973, p. 102).

Zhou Mingzhen, the Chinese scientist who did the initial faunal studies at Gongwangling, compared the site with others in northern China, such as Zhoukoudian: "The presence of these forms at Konwanling . . . may be interpreted as due to the more southern geographical location of the Lantian district, or to the difference in geological age of this fauna with the others, or to the insufficiency of our knowledge on the distribution of Pleistocene mammals in China in general. Probably all three of these factors are involved in this particular case" (Zhou, M. *et al.* 1965, p. 1044). As can be seen, there is wide latitude for manipulation of the Gongwangling faunal evidence in accordance with the leanings of a particular researcher and the requirements of evolutionary doctrine.

Is there any strong justification for attributing the presence of the southern forms at Gongwangling to temporal rather than geographical differences from Zhoukoudian? Specifically, are the Gongwangling southern species characteristic of pre-Zhoukoudian times? This does not appear to be the case. Of the 7 southern taxa, 4 are either recent or living forms. These comprise (1) *Capricornis sumatrensis,* the goat-antelope, or serow; (2) *Tapirus* sp.; (3) *Tapirus sinensis,* which some authorities consider a subspecies of the living *Tapirus indicus* (Zhou, M. *et al.* 1965, p. 1042); and (4) *Elaphodus cephalophus,* the tufted deer. Because they are recent or living, they cannot be used to establish an earlier dating for Gongwangling. The giant tapir *Megatapirus augustus,* according to Aigner (1981, p. 325), occurs in middle Middle Pleistocene assemblages and would thus be contemporary with the Zhoukoudian site. Aigner (1981, p. 289) also states that *Stegodon orientalis* survives through the late Middle Pleistocene.

The only southern form suggesting a pre-Zhoukoudian date for Gongwangling is *Nestoritherium sinensis,* an extinct three-clawed mammal that appears in Pliocene faunal assemblages. Aigner (1981, p. 289) suggested that *Nestoritherium* survived only to the Early Pleistocene in China, although she admitted this dating was open to question. Elwyn L. Simons and Peter C. Ettel (1970, p. 84) reported *Nestoritherium* at *Gigantopithecus* sites in South China, which they placed in the Middle Pleistocene. *Nestoritherium* also turned up in the Yenchingkuo fissures in Szechuan province, where, according to Aigner (1981, p. 288) "Kahlke distinguishes an early Cromerian-equivalent and a later Holstein-and-later component." The Cromerian interglacial is in the early Middle Pleistocene, the Holstein in the middle Middle Pleistocene, during the Zhoukoudian occupation (Table 9.2, p. 566).

All of this suggests that *Nestoritherium* could very well have appeared as a late survival in the Middle Pleistocene at Gongwangling.

At Gongwangling, *Nestoritherium,* which Zhou Mingzhen considered "unexpected . . . in the Konwanling fauna," is represented by a single "shattered and decayed" jaw fragment (Zhou, M. *et al.* 1965, p. 1041). Furthermore, Wu Rukang stated that in general the fossils found at Gongwangling "consist of odd, scattered bits and pieces which seem to have been thrown together after being washed down from the wooded areas of the southern slope of Kungwangling Hill" (Wu, R. 1966, p. 85). This opens up the possibility that fossils of different ages may have been incorporated into the deposits at the site.

We conclude, therefore, that the presence of certain southern Chinese animals at Gongwangling, absent at Zhoukoudian, may be a reflection of the more southerly location of Gongwangling rather than a difference in the sites' ages.

Moving on to the remaining Gongwangling taxa, we find that some represent species not present at Zhoukoudian but still living in China (Figure 9.4, III). That they are still living means they cannot justify an early date for Gongwangling.

Two other newly designated species (Figure 9.4, IV), *Leptobos brevicornis* Hu et Qi and *Dicerorhinus lantianensis,* are unique to Gongwangling and thus cannot be used for dating comparisons with other sites, such as Zhoukoudian.

Still other Gongwangling species (Figure 9.4, V), although not present at Zhoukoudian, are found at other middle Middle Pleistocene sites in China. Thus they establish contemporaneity with Zhoukoudian.

Some authors have tried to use the *Leptobos* sp. fossils from Gongwangling (Figure 9.4, VI) to establish a pre-Zhoukoudian date for the site. Such proposals are, however, are open to question. *Leptobos* is an extinct ox that dates back to the Pliocene and Early Pleistocene, but according to Zhou Mingzhen *et al.* (1965, p. 1043) the fossil skulls with horns discovered at Gongwangling differ substantially from those normally attributed to *Leptobos. Zhou Mingzhen et al.* (1965, p. 1043) and Aigner (1981, p. 81) stated that the Gongwangling variety is comparable to a related species, *Bison priscus = Bison palaeosinensis,* which existed throughout the Middle and Late Pleistocene (Nilsson 1983, p. 483). Thus the so-called *Leptobos* fossils cannot securely be used to establish a pre-Zhoukoudian date for Gongwangling.

Studying various Gongwangling faunal lists, we found that some researchers have changed some of the original species designations reported at Gongwangling (Zhou M. *et al.* 1965), apparently to reflect a pre-Zhoukoudian date for the site (Figure 9.4, VII). We have already discussed the implications of Aigner's reclassification of *Ochotonoides* to *Ochotona* at Zhoukoudian (Section 9.2.3.3).

Ursus thibetanus kokeni, reported by Zhou at Gongwangling (Zhou M. *et al.* 1965, p. 1040), was reclassified *Ursus* cf. *etruscus* in a later faunal list (Han and

Xu 1985, p. 281). *Ursus thibetanus kokeni* is found at Zhoukoudian Locality 1 (Han and Xu 1985, p. 282), in the middle Middle Pleistocene, while *Ursus etruscus* is an Early Pleistocene form. Zhou was quite definite that the teeth from the Gongwangling specimen were "indistinguishable from the specimens from the Kwangsi caves, which are identical with *U. thibetanus kokeni* described by Matthew and Granger from the Yenchingkou fissure deposits" (Zhou, M. *et al.* 1965, p. 1040). Aigner and Laughlin (1973, p. 101) list the Gongwangling species as *Ursus thibetanus,* the living Asiatic black bear.

Similarly, *Acinonyx pleistocaenicus* (cheetah) reported by Zhou at Gongwangling is reclassified by some as *Sivapanthera pleistocaenicus,* an Early Pleistocene form (Han and Xu 1985, p. 271). But Zhou stated that the Gongwangling *Acinonyx* fossil was "indistinguishable" from the type species, which came from a Late Pleistocene loess formation in China (Zhou, M. *et al.* 1965, p. 1041). *Acinonyx* is reported at Zhoukoudian (Zhou, M. *et. al.* 1965, p. 1045).

Having concluded our review of the Gongwangling faunal list in relation to that of Zhoukoudian, we find that the differences between the two faunas do not point in any clear fashion to a difference in their ages.

9.2.3.6 Analysis of Conflicting Opinions

We do not, however, insist that the Gongwangling *Homo erectus* skull is contemporaneous with *Homo erectus* of Zhoukoudian Locality 1. Following our standard procedure, we simply extend the probable date range of the primitive *Homo erectus* skullcap found at the Gongwangling site to include the time period represented by the Zhoukoudian occupation.

Other scientists have published reports on Lantian *Homo erectus*. For example, Wu Rukang (1965) said, on faunal and morphological grounds, that the Gongwangling skull and Chenjiawo jaw are contemporaneous but are both earlier than Zhoukoudian. As we have seen, others say Gongwangling and Chenjiawo are contemporaneous not only with each other but with Zhoukoudian. Still others say that Gongwangling predates Chenjiawo and Zhoukoudian. Opinions about the relative ages of the Lantian man skull and jaw, and their temporal relation to Beijing man of Zhoukoudian, are as diverse as they are numerous.

We have analyzed 25 reports, published between 1964 and 1986, and have graphically displayed their age estimates for the Lantian man skull and jaw (Figure 9.5). If paleoanthropology were an exact science, we should expect to find only one point for the jaw and skull marked on the chart, with a small area around it representing errors in measurement. But as one can see, this is not the case. Age estimates for the jaw and skullcap are widely distributed, with several strong convergences of opinion placing both before the time of Beijing man.

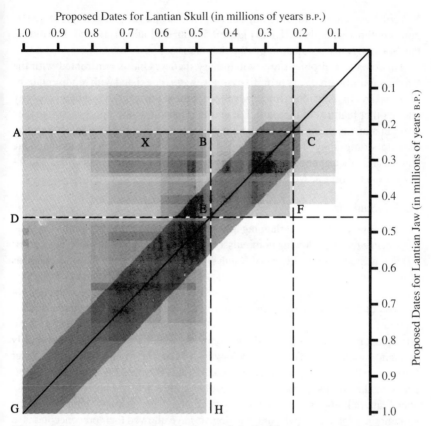

Figure 9.5. Proposed date ranges for jaw and skull of Lantian man, from 25 reports. Darker regions represent higher degrees of agreement, lighter regions lesser degrees of agreement. Dotted lines indicate the time of the *Homo erectus* occupation of Locality 1 at Zhoukoudian (.23–.46 millions years B.P.). Any point on the shaded region represents an allowed date for the skull and jaw within the range of expressed opinion. For example the point marked **X** represents an age for the skull of 0.65 million years and for the jaw of 0.25 million years. Points along the solid diagonal line show the same age for the jaw as for the skull. Aigner (1981) and others said the Lantian *Homo erectus* jaw contemporary with Zhoukoudian Locality 1, whereas the Lantian *Homo erectus* skull, more primitive than the Zhoukoudian *Homo erectus* skulls, was older (points in rectangle **ABDE** represent this opinion). Others said both the Lantian jaw and skull were earlier than Zhoukoudian Locality 1 (points in **DEGH**). But, as can be seen, there is a concentrated area of positive opinion placing both the skull and jaw from Lantian within the Zhoukoudian occupation period (points in **BCEF**). This means that two different grades of *Homo erectus* may have existed contemporaneously during the period of the Zhoukoudian occupation.

Nevertheless, another fairly strong convergence of opinion places (represented by points within rectangle **BCEF** in Figure 9.5) both the Lantian jaw and skull during the *Homo erectus* occupation of Zhoukoudian.

This creates a problem for evolutionary theory. One is confronted with the strong possibility that very primitive *Homo erectus* coexisted with representatives of *Homo erectus* considered more advanced in the significant area of brain capacity and in other features of the skull. For evolutionists, it would thus be good to date Lantian man much earlier than Zhoukoudian *Homo erectus*. But as we have seen, the site stratigraphy, faunal evidence, and dates arrived at by paleomagnetic studies and other methods also allow Lantian man and Peking man to be contemporaries in the middle Middle Pleistocene.

9.2.3.7 Summary

So now we have overlapping possible date ranges in the middle Middle Pleistocene for the following hominids: (1) Lantian man, a primitive *Homo erectus*; (2) Peking man, a more advanced *Homo erectus*; and (3) Tongzi man, described as *Homo sapiens*. We are not insisting that these beings actually coexisted. Perhaps they did, perhaps they did not. What we are insisting on is this—scientists should not propose that the hominids definitely did not coexist simply on the basis of their morphological diversity. Yet this is exactly what has happened. Scientists have arranged Chinese fossil hominids in a temporal evolutionary sequence primarily by their physical type. This methodology insures that no fossil evidence shall ever fall outside the realm of evolutionary expectations. By using morphological differences in the fossils of hominids to resolve contradictory faunal, stratigraphic, chemical, radiometric, and geomagnetic datings in harmony with a favored evolutionary sequence, paleoanthropologists have allowed their preconceptions to obscure other possibilities.

9.2.4 Maba

In 1956, peasants digging for fertilizer in a cave near Maba, in Guangdong province, southern China, found a skull that was apparently from a primitive human being. Wu Rukang thought this hominid skull displayed Neanderthaloid features: "The supra-orbital tori of this skull are heavy and project markedly both forward and sidewise" (Jia 1980, p. 41). According to Chang (1962, p. 754), the data "seem to place the Ma'pa skull within the Neanderthaloid range."

Aigner (1981, pp. 65–66) stated: "On the basis of their measurements and observations, Wu and Peng conclude the remains belong to a grade of organization similar to that of the European Neanderthals. . . . Coon (1969) agrees with the relative position of the hominid remains but emphasizes that it is not a Neanderthal in the classic sense of the word. He believes Mapa is on the threshold of modern *Homo sapiens* and is 'mostly if not entirely Mongoloid.'"

There seems to be general agreement that the Maba skull is *Homo sapiens* (Han and Xu 1985, p. 285) with some Neanderthaloid features. Coon, it may be noted, believed that *Homo erectus* evolved directly into separate races of *Homo sapiens* in different parts of the world. Thus, according to Coon, the classic Neanderthals would have been restricted to Europe.

It is easy to see that scientists, in accordance with their evolutionary expectations, would want to place the Maba specimen in the very latest Middle Pleistocene or early Late Pleistocene, after *Homo erectus*. And in fact Wu Rukang stated: "Judging from the mammalian fauna associated with the Maba skull, its geological age is probably of late Middle Pleistocene or early Late Pleistocene" (Jia 1980, p. 41). This would give it a chronometric age of about 100,000 years.

Jia Lanpo placed the Maba skull at no earlier than the Riss-Würm interglacial, in the late Middle Pleistocene (Jia 1980, p. 41). Aigner (1981, p. 65) also agreed: "The primitive hominid and fauna including *Stegodon* suggest a late Middle Pleistocene dating though Kahlke (1961) suggests a Würm-equivalent age." The Würm glaciation occurred in the early part of the Late Pleistocene.

Now let us take a close look at the associated fauna (Figure 11.6, p. 580), which Chang (1962, p. 754) said was "apparently a typical South China Middle Pleistocene assemblage." The assemblage included mostly fossils that could only be classified according to their genus (Han and Xu 1985, p. 285). All of these genera existed throughout the Pleistocene, from Early to Late.

A probable minimum age for the Maba site is provided by one of the identifiable species, *Elephas (Palaeoloxodon) namadicus* Falconer et Cautley. This elephant apparently became extinct in the Late Pleistocene (Belyaeva *et al.* 1962, p. 370). Nilsson (1983, p. 487) stated that *Palaeoloxodon namadicus* was typical of the Pleistocene interglacials, the last of which (the European Eem or Chinese Tali-Lushan) occurred about 90,000–110,000 years ago (Aigner 1981, p. 33). This a minimum age for Maba in the early Late Pleistocene.

Palaeoloxodon namadicus also occurs at Hoshantung cave near Kunming in Yunnan province (Aigner 1981, p. 293). This cave is thought to belong to the Holstein interglacial, which would make it the equivalent of Zhoukoudian Locality 1 (Aigner 1981, p. 286).

According to V. J. Maglio, an authority on elephants, *Palaeoloxodon namadicus* appears at the onset of the Middle Pleistocene, about 1 million years ago (Nilsson 1983, p. 488). In some lists, *Palaeoloxodon namadicus* also occurs in Early Pleistocene contexts (Han and Xu 1985, p. 279). The Early Pleistocene could thus be taken as a maximum age for the Maba site.

Stegodon, another extinct elephant discovered at the Maba site in China, provides an age range similar to that of *Palaeoloxodon namadicus* (Aigner 1981, p. 289).

So although Maba might be as recent as the early Late Pleistocene, the faunal evidence is also consistent with an age anywhere in the Middle Pleistocene, or even the Early Pleistocene. The principal justification for fixing the date of the Maba cave in the very latest part of the late Middle Pleistocene or in the early Late Pleistocene seems to be the morphology of the hominid remains.

W. W. Howells (1977, p. 72) stated: "The phylogenetic position of Ma-pa suggested by Woo [Neanderthal] would accord best with the date presently assigned, i.e. early late Pleistocene at latest. Viewed as a really Neanderthal-like fossil (far removed in space from any other known), an early date would seem anomalous." Maba provides another instance of morphological dating in order to preserve an evolutionary sequence. An early Late Pleistocene date was

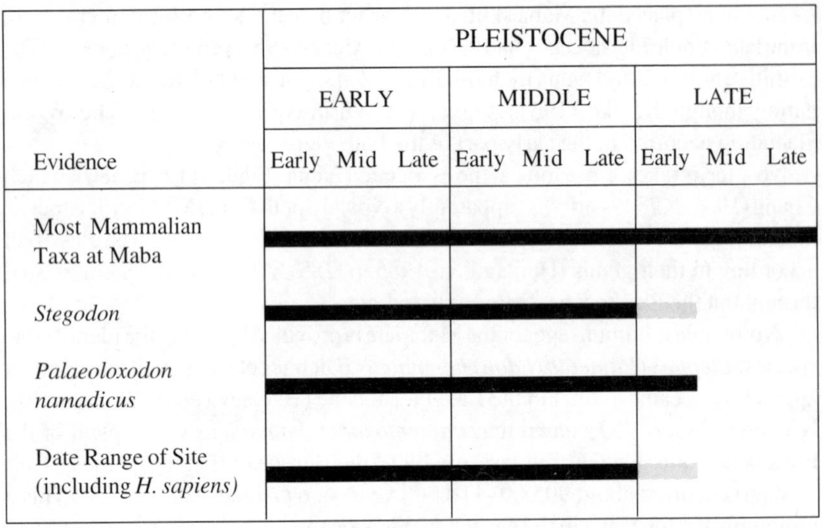

Evidence	PLEISTOCENE								
	EARLY			MIDDLE			LATE		
	Early	Mid	Late	Early	Mid	Late	Early	Mid	Late
Most Mammalian Taxa at Maba									
Stegodon									
Palaeoloxodon namadicus									
Date Range of Site (including *H. sapiens*)									

Figure 9.6. Age of *Homo sapiens* cf. *neanderthalensis* at Maba, South China. Most of the mammalian fossils from Maba were identifiable only in terms of their genus, and these genera are present throughout the Pleistocene, from Early to Late. The fauna includes *Hyaena, Felis tigris*, Mustelidae, *Ailuropoda, Ursus, Rhinoceros, Tapirus, Sus, Cervus, Bos, Hystrix*, and *Lepus*. The extinct elephants *Stegodon* and *Palaeoloxodon namadicus* provide boundaries for the age range. Both *Stegodon* and *Palaeoloxodon namadicus* are known from Early Pleistocene sites in China (Han and Xu 1985, p. 279). But according to Aigner (1981, p. 289), *Stegodon* probably became extinct in the late Middle Pleistocene, possibly surviving into the Late Pleistocene (gray part of bar). *Palaeoloxodon namadicus* apparently became extinct during the last interglacial, in the late Middle Pleistocene or early Late Pleistocene (Nilsson 1983, p. 487). The probable age range for the *Homo sapiens* skull from Maba, which is said to have Neanderthaloid features, thus extends from the early Late Pleistocene to the early Early Pleistocene.

favored. This is certainly within the realm of possibility, but a middle Middle Pleistocene date (equivalent to Zhoukoudian Locality 1) or even an Early Pleistocene date of 2 million years are also within the realm of possibility.

What conclusion may here be drawn? It would appear that at Maba we have *Homo sapiens,* with some Neanderthaloid features, existing within a possible date range that completely overlaps the *Homo erectus* presence at Zhoukoudian Locality 1. Updating our list, we now find overlapping date ranges in the middle Middle Pleistocene for: (1) primitive *Homo erectus* (Lantian); (2) *Homo erectus* (Zhoukoudian); (3) *Homo sapiens* (Tongzi); and (4) *Homo sapiens* with Neanderthaloid features (Maba).

The possibility that *Homo erectus* and more advanced hominids may have coexisted in China adds new fuel to the controversy about who was really responsible for the broken brain cases of Beijing man and the presence of advanced stone tools at Zhoukoudian Locality 1. Did several hominids, of various grades of advancement, really coexist in the middle Middle Pleistocene? We do not assert this categorically, but it is definitely within the range of possibilities suggested by the available data. In our study of the scientific literature, we have come upon no clear reason for ruling out coexistence other than the fact that the individuals are morphologically dissimilar.

Some will certainly claim that the fact of human evolution has been so conclusively established, beyond any reasonable doubt, that it is perfectly justifiable to engage in dating hominids by their morphology. But we believe this claim does not hold up under close scrutiny. As we have demonstrated in Chapters 2–6, abundant evidence contradicting current ideas about human evolution has been suppressed or forgotten. Furthermore, scientists have systematically overlooked shortcomings in the evidence that supposedly supports current evolutionary hypotheses.

If peasants digging for fertilizer in a Chinese cave had uncovered a fully human skull along with a distinctly Pliocene fauna, scientists would certainly have protested that no competent observers were present to conduct adequate stratigraphic studies. But since the Maba skull could be fitted into the standard evolutionary sequence, no one objected to its similar mode of discovery.

9.2.5 Changyang County

Even after one learns to recognize the highly questionable practice of morphological dating, one may be astonished to note how frequently it is used. In the field of human evolution research in China, it appears to be not the exception but the rule. The *Homo sapiens* upper jaw (Han and Xu 1985, p. 286) found by workers in 1956 at Longdong (Dragon Cave) in Changyang county, Hubei Province, South China, has provided many authorities with a welcome opportunity for unabashed morphological dating.

The upper jaw, judged *Homo sapiens* with some primitive features, was found in association with the typical South China Middle Pleistocene fauna including *Ailuropoda* (panda) and *Stegodon* (extinct elephant). Jia Lanpo stated: "No human fossil had been found in association with such fauna, however, until the discovery of Changyang Man. . . . The *Ailuropoda-Stegodon* fauna had been dated as Middle Pleistocene, contemporaneous with Peking Man, but new evidence puts it to be Upper Pleistocene" (Jia 1980, p. 42). Study of the faunal list (Han and Xu 1985, p. 286) shows that the only "new evidence" is the human fossil, as all the other species are representative of the Early and Middle Pleistocene. Jia concluded from the new evidence that the age of the *Stegodon-Ailuropoda* fauna should be extended to the Late Pleistocene, but an equally valid conclusion is that Changyang *Homo sapiens* was contemporary with Beijing man. Evolutionary preconceptions do not, however, easily allow this.

Chang (1962, p. 749) wrote: "This fauna is generally believed to be of Middle Pleistocene age, and the scientists working on the cave suggest a late Middle Pleistocene dating, for the morphology of the maxilla shows less 'primitive' features than does that of *Sinanthropus*." He went on to say that the upper jaw "resembles modern man in most of its features" (Chang 1962, p. 749). It is clear that Chang's primary justification for assigning Changyang *Homo sapiens* a date later than Beijing *Homo erectus* was morphological.

Aigner (1981, p. 70) joined in with her statement: "A Middle Pleistocene age is suggested by some of the fauna with the presence of the hominid which is considered near *H. sapiens* indicating a dating late in that period."

That so many scientists could confront the straightforward faunal evidence at Changyang without even considering the possibility that *Homo sapiens* coexisted in China with *Homo erectus* is amazing. In this regard, Sir Arthur Keith (1931, p. 256) wrote: "It has so often happened in the past that the discovery of human remains in a deposit has influenced expert opinion as to its age; the tendency has been to interpret geological evidence so that it would not clash flagrantly with the theory of [anatomically modern] man's recent origin."

Aigner (1981, p. 75) went on to state: "Associated with the Ch'angyang hominid are typical members of the *Stegodon-Ailuropoda* fauna. Thus, it appears that the materials date to the Holstein-equivalent or to a later interglacial, rather than to a glacial phase of the Pleistocene" (Aigner 1981, p. 75). Zhoukoudian Locality 1 is referred to the Holstein-equivalent interglacial in the Chinese middle Middle Pleistocene. But according to Aigner (1981, p. 75), "The advanced nature of the hominid remains may exclude the early dating." Is any comment required at this point?

Interestingly, Han Defen and Xu Chunhua reported *Hyaena brevirostris sinensis* fossils at Changyang (Han and Xu 1985, p. 286). Aigner (1981, pp. 289, 322) said this species is not found more recently than the Holstein interglacial, which is equivalent to the Taku-Lushan interglacial of the Chinese middle Middle Pleistocene (Table 9.2, p. 566). This should have given Aigner reason to refer Changyang *Homo sapiens* to the Taku-Lushan (Holstein) interglacial.

9.2.6 Liujiang

In 1958, workers found human fossils in the Liujiang cave in the Guangxi Zhuang Autonomous Region of South China. These included a skull, vertebrae, ribs, pelvic bones, and a right femur. Han and Xu (1985, p. 286) said the Liujiang fossils are from *Homo sapiens sapiens*. Aigner (1981, p. 63) stated: "Many measurements taken fall within the range for living Mongoloids but several are clearly in the range for 'Australoids.' . . . Wu concludes that the remains belong to an early form of (modern) *H. sapiens* and to a primitive Mongoloid."

But the anatomically modern remains were found along with a typical *Stegodon-Ailuropoda* fauna, giving a date range for the site of the entire Middle Pleistocene. The assemblage included *Hystrix* (porcupine), *Ursus* (bear), and *Sus* (hog). Since the species were not identifiable, these forms are not useful for precise dating (Han and Xu 1985, p. 286). *Ailuropoda melanoleuca fovealis* (panda) survives today and *Rhinoceros sinensis* lived until recently in China. But *Stegodon orientalis* (elephant) and *Megatapirus augustus* (giant tapir) probably did not survive past the Middle Pleistocene (Aigner 1981, pp. 289, 325), suggesting that the fossil-bearing cave deposits are at least that old. This completes the main faunal list at Liujiang, although Jia (1980, p. 46) mentions other unnamed species of cattle and deer.

"The [human] skull was found near the cave mouth no more than four meters [about 13 feet] from the find spot of the panda remains," said Jia (1980, p. 46). "The deposits there consisted of limestone, sand, and earth. The grey-brown deposits were very loose and moist in marked contrast to mammalian-fossil-bearing deposits found elsewhere in Guangxi, which are hard and yellowish. The types were obviously of different dating."

It seems Jia (1980, p. 47) assigned a recent date of about 40,000 years to the Liujiang *Homo sapiens* fossils simply because the stratum where they were found was of a different color and consistency than that found in other caves in the same province. This is a weak argument, considering that the faunal remains are typical of the Middle Pleistocene. One can easily imagine circumstances that might account for different sorts of cave deposits in different locations.

A more frank explanation of the recent dating is suggested by Chang (1962, p. 753): "Woo Ju-kang, who reported the finding of Liu-chiang Man, assumes that the fossil human skull together with that of *Ailuropoda* is later than Middle Pleistocene. As the human skull is definitely fossilized and of *Homo sapiens* type, it can be assumed that it is of late Pleistocene age." Wu Xinzhi and Zhang Zhenbiao stated: "although in 1958 most of the representatives of the fauna were thought to have been deposited during the Middle Pleistocene, Wu Rukang believes the hominid remains postdate this epoch" (Wu and Zhang 1985, p. 109). These statements imply that that the remains of the human and panda were deposited in the cave after the other mammalian fossils. One suspects, however, that if the hominid remains had been of the *Homo erectus* type, scientists would not have felt compelled to interpret the evidence in such a fashion.

Aigner (1981, p. 64) provides an example of some very finely tuned morphological dating: "Based on the descriptions of relative primitiveness of the remains noted by Wu, Coon, and Thoma, a suggested dating of 15,000 or even 25,000 and 40,000 years ago is possible." Yet the faunal evidence clearly indicates that the possible date range for Liujiang *Homo sapiens* extends far back into the Middle Pleistocene, contemporary with Lantian man and Beijing man.

9.2.7 Gigantopithecus

Also found in the Pleistocene caves of South China was *Gigantopithecus,* a very large apelike creature. Weidenreich believed *Gigantopithecus* was an ancestor of Beijing man, but modern scientists do not accept this. The time range of *Gigantopithecus* sites in China extends from the Early Pleistocene through the Middle Pleistocene (Han and Xu 1985, pp. 279–284).

9.2.8 Dali

The Dali site in Shaanxi province has yielded a skull classified as *Homo sapiens* (Han and Xu 1985, p. 284) with primitive features.

The Dali fauna (Han and Xu 1985, p. 284) includes unidentifiable species of *Palaeoloxodon* (extinct elephant), *Equus* (horse), *Rhinoceros* (rhino), *Megaloceros* (large extinct deer), and *Bubalus* (water buffalo) as well as unidentifiable genera and species of the Castoridae (beavers). All of the genera found at Dali are represented throughout the Middle Pleistocene and earlier.

Megaloceros pachyosteus (Young), one of the two identifiable species from Dali, occurs at Zhoukoudian Locality 1, and the other species, *Pseudaxis grayi* (axis deer) occurs at the Lantian man sites, said to be roughly contemporaneous with Zhoukoudian Locality 1 in the middle Middle Pleistocene, if not earlier.

Some Chinese paleoanthropologists suggest a late Middle Pleistocene age for Dali (Wu, X. and Wu, M. 1985, p. 92). While this may account for the human skull, the associated fauna does not dictate such a date. Rather it suggests for Dali *Homo sapiens* a possible date range extending further back into the Middle Pleistocene, overlapping, once more, Beijing man at Zhoukoudian Locality 1.

9.2.9 Summary of Overlapping Date Ranges

In discussing overlapping possible date ranges, we found that Beijing man *Homo erectus* at Zhoukoudian Locality 1 may very well have lived at the same time as a variety of hominids—early *Homo sapiens* (some with Neanderthaloid features), *Homo sapiens sapiens*, and primitive *Homo erectus* (Figure 9.7).

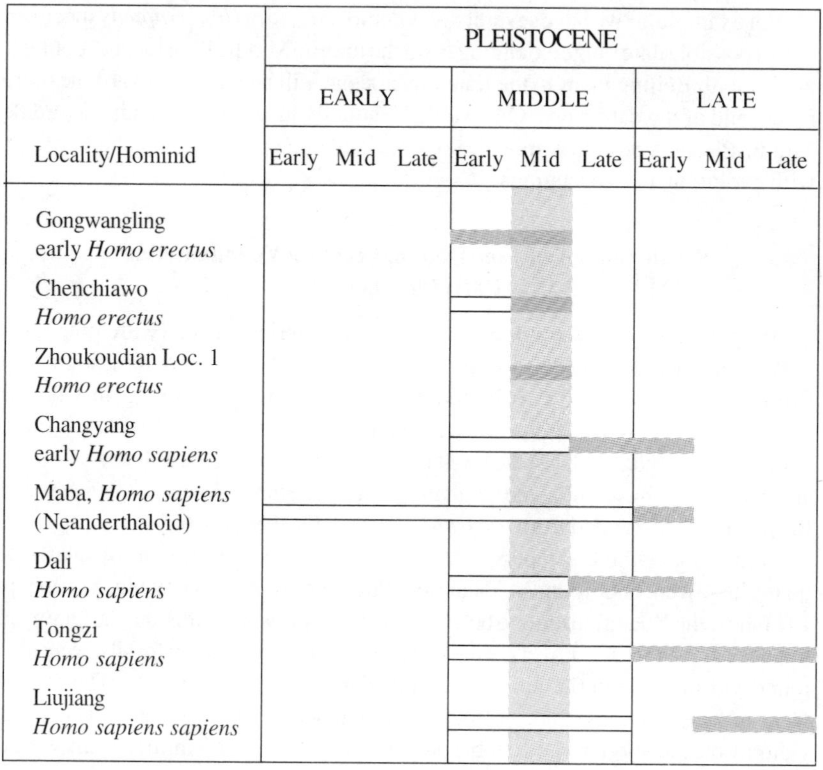

Locality/Hominid	PLEISTOCENE								
	EARLY			MIDDLE			LATE		
	Early	Mid	Late	Early	Mid	Late	Early	Mid	Late
Gongwangling early *Homo erectus*									
Chenchiawo *Homo erectus*									
Zhoukoudian Loc. 1 *Homo erectus*									
Changyang early *Homo sapiens*									
Maba, *Homo sapiens* (Neanderthaloid)									
Dali *Homo sapiens*									
Tongzi *Homo sapiens*									
Liujiang *Homo sapiens sapiens*									

Figure 9.7. The probable date ranges of Chinese hominids, as determined by their accompanying mammalian faunas, are shown. Scientists have assigned dates to the hominids, within their probable date ranges, that conform to evolutionary expectations. These dates are represented by the darker portion of each bar. For example, although the faunal date range for the Maba site extends from the Early Pleistocene to the early Late Pleistocene, scientists have used the presence of a Neanderthaloid skull to fix the date for the site in the most recent part of its date range. At Liujiang, the human fossils were given a date completely outside the faunal date range. We call this phenomenon morphological dating. But putting aside evolutionary expectations, the faunal evidence indicates that it is possible that all of the hominids were contemporary with *Homo erectus* at Zhoukoudian Locality 1 in the middle Middle Pleistocene (shaded vertical bar).

In attempting to sort out this Middle Pleistocene hominid logjam, scientists have repeatedly used the morphology of the hominid fossils to select desirable dates within the total possible faunal date ranges of the sites. In this way, they have been able to preserve an evolutionary progression of hominids. Remarkably, this artificially constructed sequence, designed to fit evolutionary expectations, is then cited as proof of the evolutionary hypothesis.

For example, as we have several times demonstrated, a *Homo sapiens* specimen with a possible date range extending from the middle Middle Pleistocene (contemporary with Beijing man) to the Late Pleistocene will be pushed toward the more recent end of the date range. One would be equally justified in selecting a middle Middle Pleistocene date within the possible date range, even though this conflicts with evolutionary expectations.

9.2.10 Stone Tools and Hominid Teeth at Yuanmou (Early Early Pleistocene)

We conclude our review of fossil hominid discoveries in China with some cases of sites regarded as Early Pleistocene. At Yuanmou, in Yunnan province, southwest China, geologists found two hominid teeth (incisors). According to Chinese scientists, these were more primitive than those of Beijing man, having a more complicated lingual surface (the lingual surface is that facing the tongue). The teeth are believed to have belonged to a very primitive *Homo erectus,* a precursor of Beijing man, descended from *Australopithecus* (Jia 1980, pp. 6–7).

Stone tools—three scrapers, a stone core, a flake, and a point of quartz or quartzite—were later found at Yuanmou. Published drawings (Zhang, S. 1985, p. 141) show the Yuanmou tools to be much like the European eoliths and the Oldowan industry of East Africa. Layers of cinders, containing mammalian fossils, were also found with the tools and hominid incisors. According to Jia (1980, p. 8), "The cinders were in heaps at some spots while sparse and scattered elsewhere." The strata yielding the incisors gave a probable paleomagnetic date of 1.7 million years within a range of 1.6–1.8 million years (Jia 1980, p. 9).

There are problems with this Early Pleistocene age for Yuanmou *Homo erectus. Homo erectus* is thought to have evolved from *Homo habilis* in Africa about 1.5 million years ago, and then migrated elsewhere about 1.0 million years ago. *Homo habilis* is not thought to have left Africa. Implicit in Jia's age estimate for the Yuanmou hominid is a separate origin for *Homo erectus* in China. Jia seems to require the presence in China about 2.0 million years ago of *Australopithecus* or *Homo habilis,* something forbidden by current theory.

In this regard, Lewis R. Binford and Nancy M. Stone (1986, p. 15) stated: "It should be noted that many Chinese scholars are still wedded to the idea that man evolved in Asia. This view contributes to the willingness of many to uncritically accept very early dates for Chinese sites and to explore the possibility of stone tools being found in Pliocene deposits." One could also say that because Western scholars are wedded to the idea that humans evolved in Africa they uncritically reject very early dates for hominid fossils and artifacts around the world.

As previously mentioned, one need not suppose that either Africa or Asia was a center of evolution. There is, as shown in preceding chapters, voluminous evidence, much found by professional scientists, suggesting that humans of the modern type have lived on various continents, including South America, for tens of millions of years. And, during this same period, there is also evidence for various apelike creatures, some resembling humans more than others.

A question encountered in our discussions of anomalous cultural remains (Chapters 2–5) once more arises: What justification does one have for attributing the stone tools and signs of fire at Yuanmou to primitive *Homo erectus*?

The tools and signs of fire were not found close to the *Homo erectus* teeth (Jia 1985, p. 140). Two of the three tools lay 1.5 meters (5 feet) below the level of the teeth, and the third 1 meter (3 feet) above. The closest tool was 5 meters (about 16.4 feet) from the teeth. The others were up to 20 meters (65.6 feet) away.

Furthermore, as seen in this chapter, there is much evidence that *Homo sapiens* may have existed in China far earlier than is presently admitted. And we have already examined evidence from other parts of the world demonstrating the presence of *Homo sapiens* in the Early Pleistocene and earlier.

Aigner, representing mainline anthropological thought, reacted predictably to Jia Lanpo's suggested early dating for the Yuanmou hominid. She stated: "The hominid and faunal remains, as well as contemporaneous artifacts [occur] in level 25 at the base of the fourth stratigraphic unit, equivalent to earliest Middle Pleistocene times. A paleomagnetic age of 1.7 million years would place the strata and *H. erectus yuanmoensis* equivalent in age to the Olduvai Event and some 1 million years earlier than *H. erectus lantianensis*. Based on current reports, the remains are earliest Middle Pleistocene faunistically and stratigra- phically [about 1 million years old]. The paleomagnetic date could be applicable to another stratigraphic unit (the first or second). I am reluctant at this point to accept the date as valid for the hominid teeth" (Aigner 1981, pp. 52–54). This shows how dating procedures are far from exact. Dates are subject to extensive postexperimental revision and interpretation.

In 1983, the original paleomagnetic dating of 1.7 million years for the Yuanmou site, reported by Li Pu and his associates in 1976 and reconfirmed in 1977 by Cheng and his associates, was challenged by Liu Dongsheng and Ding Menglin (Wu, X. and Wang, L. 1985, p. 35). They proposed a different explanation of the magnetostratigraphic sequence at Yuanmou.

According to Wu Xinzhi and Wang Linghong: "Liu and Ding prefer to correlate the normal polarity member at Yuanmou with the Brunhes Epoch rather than with an event of normal polarity within the Matuyama. Furthermore, they have concluded that the layer yielding the fossils of *H. erectus* is situated at the base of the Brunhes Normal Epoch strata [Figure 9.8] and therefore might not be older than 0.73 million years B.P., and possibly only 0.5–0.6 million years old" (Wu, X. and Wang, L. 1985, pp. 35–36).

Paleomagnetic dating is based on the assumption that the earth's magnetic field undergoes shifts in polarity, which are recorded in the magnetic properties of the strata at a site. Upon conducting the required measurements on the strata, one obtains a sequence of normal and reversed polarities, which are grouped into various epochs, such as those named in the above passage (Brunhes Normal, Matuyama Reversed, etc.). The sequence of polarities is typically displayed in a column, with periods of normal polarity shown in black and reversed polarity in white. As can be seen, in Figure 9.8, there may be brief episodes of normal polarity in a reversed epoch and vice versa. One thus obtains a column of many black and white bars, of various thickness, representing the time of each polarity period. One may then compare the polarity sequence from a particular site, which may be quite complex, with the standard polarity sequence and its known chronometric dates. When the site polarity sequence is properly aligned with the

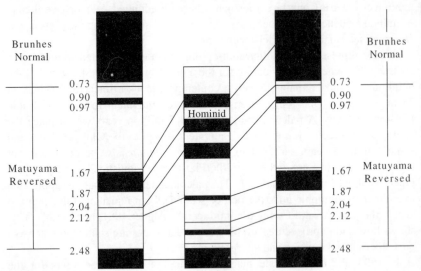

Figure 9.8. Cheng *et al.* (1977) and Li *et al.* (1976) established a correlation between the polarity sequence at Yuanmou (center) and the standard polarity sequence (left) that gave the hominid-bearing stratum an age of about 1.7 million years. But in 1983, Liu and Ding established a correlation between Yuanmou and the standard sequence (right) that gave the hominid-bearing stratum an age of approximately .73 million years (Wu, R. and Wang, L. 1985, p. 36).

standard sequence, one can then assign dates to the various strata at the site by comparison with the standard sequence. The problem is this: matching the polarity sequence obtained at a particular site with the standard sequence is not always easy—there is much room for interpretation, as we can see from the conflicting interpretations of the polarity sequence at Yuanmou.

In 1979, Chinese scientists using amino acid racemization methods dated animal fossils from Yuanmou to 0.80 million years. But Wu and Wang warned: "Fluctuations in average temperature on the enclosing sediments during burial may have had a significant influence on these determinations, and, as such fluctuations are not at present quantifiable, the resulting chronometric dates must be viewed with caution" (Wu, X. and Wang, L. 1985, pp. 36–37). The inventor of amino acid racemization dating, Bada, admitted that in certain cases the technique did not give reliable dates (Appendix 1.3.5). Wu and Wang stated that the fauna at the hominid site is Early Pleistocene, supporting the paleomagnetic date of 1.7 million years. They thus differed from Aigner, who said the fauna favored an early Middle Pleistocene age for Yuanmou.

Replying to the challenges to his proposed date of 1.7 million years, Jia Lanpo wrote: "Today, although the Lower Pleistocene age of the Yuanmou deposits has been called into question (Liu and Ding 1983), the associated fauna still reflects a Lower Pleistocene antiquity" (Jia 1985, p. 140). Reacting specifically to the more recent paleomagnetic date suggested by Liu and Ding, Jia (1985, p. 141) said: "Because of the apparent contradiction between these data and the extremely archaic mammalian fossil assemblage associated with the Yuanmou incisors, and due to the primitive morphological features of the teeth themselves, I do not share the opinions of Liu and Ding in this matter." So it appears there may exist at Yuanmou tools and signs of fire dating back to 1.7 million years ago.

9.2.11 Stone Tools at Xihoudu (Early Early Pleistocene)

In 1960, Jia Lanpo investigated Early Pleistocene sand and gravel deposits at Xihoudu in northern Shanxi province. He found 3 stones with signs of percussion, and more artifacts turned up in 1961 and 1962. Jia (1980, pp. 10–11) said: "In processing the stone artefacts collected at Xihoudu, we took extra care against the possibility of misjudgement by carefully analysing the scars on the stones to see if they were caused by natural, physical, or biological factors. But none could explain them. Later, we asked the late well-known geologist Li Siguang (also known as J. S. Lee) to examine the specimens from the angle of geomechanics to see if they could have been caused by natural agencies. His answer was: 'Unlikely. Shaped by man would be more appropriate.'" Because of Early Pleistocene faunal remains, the site was dated over a million years old. According to Jia (1985, p. 139), "Preliminary palaeomagnetic data indicate an absolute age for the Xihoudu site of about 1.8 million years B.P."

Among the dozens of stone tools were cores, flakes, choppers, scrapers, and heavy triangular points. Jia (1980, p. 11) wrote: "The choppers are single or double faced, and bear marks resulting from use. . . . a fragmented deer skull with both antler stumps attached was unearthed here, and on one stump transverse cuts could be observed. The cuts were most likely made by a sharp tool instead of being caused by erosion or other natural forces. Another antler bears marks of being scraped. These mark-bearing antlers lead us to believe . . . that over a million years ago, hominids at Xihoudu were using antler and bone tools."

Jia also found what appeared to be charred bones. This was established by visual comparison with burned bones from Zhoukoudian and by laboratory testing. But Jia (1980, p. 12) admitted: "Who the creators of the Xihoudu culture were has not yet been identified, for no human fossils have been found, not even so much as a tooth. From the antiquity of the site we may infer that the genus probably belongs to the Australopithecinae."

Aigner (1981, pp. 183–184), as one might well imagine, disagreed: "Despite the strong support for Lower Pleistocene human activity in north China claimed for Hsihoutu [Xihoudu], I am reluctant to accept unequivocally the materials at this time. . . . if Hsihoutu is verified, then humans occupied the north of China some 1,000,000 years ago and utilized fire. This would call into question some of our current assumptions about both the course of human evolution and the adaptational capabilities of early hominids."

If one could, however, become detached from current assumptions, interesting possibilities open up. Considering all the evidence gathered around the world over the past century or so, and not just the carefully selected evidence used to support present evolutionary views of human origins, one may infer something different than did Jia regarding Xihoudu. It is possible that some other hominid, perhaps *Homo sapiens,* might have been responsible for the cultural remains at Xihoudu. In fact, *Homo sapiens* is a more likely explanation for the presence of stone tools, bone tools, and fire than an australopithecine.

9.2.12 Concluding Words on China

This ends our review of finds in China. It appears that age determinations of fossil hominids have been distorted by morphological dating. When these ages are adjusted to reflect reasonable faunal date ranges, the total evidence fails to exclusively support an evolutionary hypothesis. Rather, the evidence appears also consistent with the proposal that anatomically modern human beings have coexisted with a variety of humanlike creatures throughout the Pleistocene.

10

Living Ape-Men?

In examining the fossil hominids of China (Chapter 9), we found signs that humans may have coexisted with more apelike hominids throughout the Pleistocene. In this chapter, we suggest that humans and ape-man-like creatures continue to coexist.

Over the past hundred or so years, researchers have accumulated substantial evidence that creatures resembling Neanderthals, *Homo erectus,* and the australopithecines even now roam wilderness areas of the world.

The existence of living ape-men, if admitted by scientists, offers new ways to interpret ambiguous paleoanthropological evidence. Hominid fossils once thought to be from the Middle Pleistocene or older periods may in fact be quite recent. The existence of living ape-men also calls into question the reliability of the scientific information processing system in zoology and anthropology.

10.1 HARD EVIDENCE IS HARD TO FIND

In 1775, Carl Linnaeus, the founder of the modern system of biological classification, listed three existing human species: *Homo sapiens, Homo troglodytes* (cave man), and *Homo ferus* (wild man). Although Linnaeus knew the latter two species only from travelers' reports and other secondary sources of information, he still included them within his *Systema Naturae* (Shackley 1983, p. 10).

Since 1775, much more evidence for the existence of living apelike wildmen has come to light. Professional scientists have (1) observed wildmen in natural surroundings, (2) observed live captured specimens, (3) observed dead specimens, and (4) collected physical evidence for wildmen, including hundreds of footprints. They have also interviewed nonscientist informants and investigated the vast amount of wildman lore contained in ancient literatures and traditions.

Despite this, no zoo or museum in a civilized nation has in its collection a wildman specimen, alive or dead. Many will say that all the wildman evidence mentioned above exists simply in reports, and that reports alone, even those

591

given by scientists, are not sufficient to establish the existence of wildmen. Hard evidence, available now, to anyone who wants to see it and touch it, is required.

But what is the real status of hard evidence? Can a physical object in and of itself confirm a certain idea about some aspect of human origins? The answer to that question is no. In paleoanthropology, as in many areas of science, evidence exists primarily in the form of reports.

The most important feature of an artifact or a fossil hominid bone, as far as paleoanthropologists are concerned, is its age. As we show in Appendix 1, radiometric and chemical methods are not very reliable age indicators. Therefore the best age indicator is stratigraphic position. Once a bone or artifact is taken out of the ground, evidence of its original stratigraphic position lies principally in the reports of its discovery.

For example, one afternoon in the early 1970s, Donald Johanson, the discoverer of "Lucy" (the most famous specimen of *Australopithecus afarensis*), found some fossilized bones lying on the surface, near his base camp in the Afar region of Ethiopia. At the very moment his fingers grasped one of those bones, lying upon Pliocene sediments, Johanson was, one might say, in touch with some hard evidence. But Johanson's discovery of those bones had no real scientific meaning until it was reported to other scientists. And from that time on, the discovery has existed, as far as the world of science is concerned, only in reports.

Were the bones discovered in the exact manner described in the reports? The answer to that question depends upon how much faith one places in the reporter and the reporting process.

One might take some comfort in the fact that the "actual" bones—the real hard evidence—are still present in a museum in Ethiopia. Of course, it is not so easy to obtain direct access to rare fossil specimens. But say you did have the proper scientific credentials and were able to go to Ethiopia and inspect the actual bones of Lucy. How would you know for certain that those were the bones picked up years ago by Johanson? You might compare them with the photographs and descriptions in the published reports. But here we go again—everything depends upon how much faith one places in the reports.

In some cases, the bones themselves are not available for inspection. For example, during World War II almost the whole collection of Beijing man (*Homo erectus*) fossils was lost during the Japanese occupation of China. The Beijing man fossils now exist only in the form of old written reports, photographs, and casts. And no one doubts that the originals did in fact exist.

But what about reports by scientists who claim they saw and examined dead specimens of wildmen, the corpses of which were not preserved? Most scientists will grant no credibility at all to such reports.

Here and in the case of Beijing man, the actual physical evidence is no longer available for inspection. Yet in one case the reports are believed, and in the other they are not. Why? We propose that reports about evidence conforming to the standard view of human evolution generally receive greater credibility than reports about nonconforming evidence. Thus deeply held beliefs, rather than purely objective standards, may become the determining factor in the acceptance and rejection of reports about controversial evidence.

10.2 CRYPTOZOOLOGY

For some researchers, the study of creatures such as wildmen comes under the heading of a genuine branch of science called cryptozoology. Cryptozoology, a term coined by the French zoologist Bernard Heuvelmans, refers to the scientific investigation of species whose existence has been reported but not fully documented. The Greek word *kryptos* means "hidden," so cryptozoology literally means "the study of hidden animals." There exists an International Society of Cryptozoology, the board of directors of which includes professional biologists, zoologists, and paleontologists from universities and museums around the world. The purpose of the society, as stated in its journal *Cryptozoology,* is "the investigation, analysis, publication, and discussion of all matters related to animals of unexpected form or size, or unexpected occurrence in time or space." A typical issue of *Cryptozoology* usually contains one or more articles by scientists on the topic of wildmen.

Is it really possible that there could be an unknown species of hominid on this planet? Many will find this hard to believe for two reasons. They suppose that every inch of the earth has been quite thoroughly explored. And they also suppose that scientists possess a complete inventory of the earth's living animal species. Both suppositions are incorrect.

First, even in countries such as the United States, there remain vast unpopulated and little-traveled areas. In particular, the northwestern United States still has large regions of densely forested, mountainous terrain which, although mapped from the air, are rarely penetrated by humans on the ground.

Second, a surprising number of new species of animals are still being found each year—about 5,000 according to a conservative estimate (Heuvelmans 1983, pp. 19–20). As might be suspected, the great majority of these, some 4,000, are insects. Yet Heuvelmans (1983, p. 21) noted: "Quite recently, in the mid 1970's, there were discovered each year, around 112 new species of fish, 18 new species of reptiles, about ten new species of amphibians, the same number of mammals, and 3 or 4 new species of birds."

Most of the mammals were small, and this might lead one to doubt that a large mammal, such as a wildman, might someday enter the list of living species. But the twentieth century has seen the discovery of many large species, some "known from native reports which were initially disbelieved" (Shackley 1983, p. 166).

The largest of the bears, the Kodiak bear, was unknown to science until 1899. The largest rhinoceros, Cotton's white rhino, was discovered in 1900. The mountain gorilla, the largest member of the ape family, turned up in 1901. The largest lizard, the Komodo dragon, was first captured in 1912. In 1975, the largest known peccary, or wild hog, *Catagonus wagneri,* was discovered in Paraguay. This animal was previously known only by Pleistocene fossils (Wetzel *et al.* 1975; Heuvelmans 1983, p. 12). In 1976, a large and entirely new species of shark, 4.5 meters (almost 15 feet) long and weighing over 700 kilograms (over 1,500 pounds), was caught by a U.S. Navy ship in the ocean waters off Hawaii (L. Taylor *et al.* 1983). So it is not completely outside the realm of possibility that science might someday come to fully accept the existence of wildmen, which may prove to be previously unknown types of hominids or primates, or surviving representatives of fossil hominids such as the australopithecines, *Homo erectus,* or the Neanderthals. It would not be the first time that science has found examples of "living fossils."

10.3 EUROPEAN WILDMEN

Many art objects of the Greeks, Romans, Carthaginians, and Etruscans bear images of semi-human creatures resembling wildmen. For example, in the Museum of Prehistory in Rome, there is an Etruscan silver bowl on which may be seen, among human hunters on horses, the figure of a large, ape-man-like creature (Wendt 1972, p. 15). Such imagery is, of course, subject to varying interpretations. The Russian scientist Boris Porshnev believed the humanlike creatures represented survivals of prehuman hominids. But British anthropologist Myra Shackley, who said wildmen may in fact exist in some parts of the world, asserted that the figures on classical Graeco-Roman art objects represent purely mythological beings such as satyrs (1983, pp. 18–19).

The satyr is a stylized and very recognizable figure, part human and part animal, occurring mainly on Greek vases. Typically, satyrs have horselike tails and are shown engaged in some kind of sporting or licentious behavior, perhaps connected with the cult of Dionysus. The hairy humanlike figure depicted on the Etruscan silver bowl, however, is shown not with revelers but in the midst of a hunting party of well-armed humans mounted on horses. The creature has no satyr's tail and appears to be carrying a crude club in one hand and a large stone, raised threateningly above his head, in the other.

During the Middle Ages, wildmen continued to be depicted in European art and architecture. A page from *Queen Mary's Psalter,* composed in the fourteenth century, shows a very realistically depicted hairy wildman being attacked by a pack of dogs (Shackley 1983, p. 25). Wildmen were thought to live in caves and forests, where they subsisted on berries and roots. They were not considered ordinary humans. Instead, they were said to be members of the animal kingdom, unable to speak or comprehend the existence of God.

10.4 NORTHWESTERN NORTH AMERICA

For centuries, the Indians of the northwestern United States and western Canada have believed in the reality of wildmen, known by various names, the most familiar of these being Sasquatch. In 1792, the Spanish botanist-naturalist José Mariano Moziño, in describing the Indians of Nootka Sound on Vancouver Island, Canada, stated (1970, pp. 27–28): "I do not know what to say about Matlox, inhabitant of the mountainous district, of whom all have an unbelievable terror. They imagine his body as very monstrous, all covered with stiff black bristles; a head similar to a human one, but with much greater, sharper and stronger fangs than those of the bear; extremely long arms; and toes and fingers armed with long curved claws. His shouts alone (they say) force those who hear them to the ground, and any unfortunate body he slaps is broken into a thousand pieces."

In 1784, the London *Times* printed a report that Indians at Lake of the Woods, Manitoba, had captured a "huge, manlike, hair-covered" creature (Shackley 1983, p. 35).

Describing the Spokane Indians of the Pacific Northwest, Elkanah Walker, a missionary who lived among them for 9 years, wrote in 1840: "They believe in the existence of a race of giants which inhabit a certain mountain, off to the west of us. This mountain is covered with perpetual snow. They inhabit its top. . . . They hunt and do all their work in the night. They are men stealers. They come to people's lodges in the night, when the people are asleep and take them and put them under their skins and take them to their place of abode without their even awakening. . . . They say their track is about a foot and a half long. . . . They frequently come in the night and steal their salmon from their nets and eat them raw. If the people are awake they always know when they are coming very near by the smell which is most intolerable" (Drury 1976, pp. 122–123).

Indians from the Columbia River region of the northwestern United States produced rock carvings that resembled the heads of apes. Anthropologist Grover Krantz (1982, p. 97) showed photographs of the heads to a number of scientists and noted: "Zoologists who did *not* know their source unanimously declared them to be representative of nonhuman, higher primates; those who *knew* the source insisted they must be something else!" Whatever the carvings may actually represent, Krantz's findings are significant. Preconceptions seem to determine what scientists are prepared to see, and one thing most scientists are definitely not prepared to see is apelike creatures in the American Northwest.

U.S. President Theodore Roosevelt included an intriguing wildman report in his book *The Wilderness Hunter* (1906, pp. 255–261). The incident took place in the Bitterroot Mountains, between Idaho and Montana. Wildman reports still come out of this region.

In the early to middle 1800s, a trapper named Bauman and his partner were exploring a particularly wild and lonely pass, through which ran a stream said to have many beaver. The two trappers set up camp late one afternoon and went out to explore for a couple of hours. Returning at dusk, they found that something had scattered their belongings around and had in "sheer wantonness" destroyed their lean-to. They rebuilt their lean-to, made supper, and then studied the footprints left by the beast. They noticed, quite to their surprise, that the malicious intruder had apparently walked off on two feet (bears usually go on all fours). This was a bit unsettling, but at last they managed to fall asleep under the lean-to.

Around midnight, they were awakened by some noise and saw a huge body standing at the opening of the lean-to. Their nostrils were assailed by a "strong wild-beast odor." Bauman fired a couple of shots at the creature. They figured he did not hit it, because they heard it move away through the woods.

The next day, the creature again ravaged the camp while Bauman and his partner were checking their traps. They found a trail of prints in the soft dirt, and these confirmed once more that their assailant, unlike a bear, had walked off on just two feet. That evening, they set up a roaring fire, which they kept going all night. Around midnight, the creature was heard moving through the woods, and it several times "uttered a harsh grating, long-drawn moan."

The following morning, Bauman and his partner decided to leave, but first they wanted to check their traps. As they moved through the forest, they sensed they were being followed. Roosevelt (1906, p. 259) said, "In the high, bright sunlight their fears seemed absurd to the two armed men, accustomed as they were, through long years of lonely wandering in the wilderness, to face every kind of danger from man, brute, or element." Bauman's partner returned to the camp before he did. When Bauman finally arrived, he found his partner dead. Said Roosevelt (1906, p. 260): "The footprints of the unknown beast-creature, printed deep in the soft soil, told the whole story."

Roosevelt had some thoughts about the episode. He wrote of Bauman: "he was of German ancestry, and in childhood had doubtless been saturated with all kinds of ghost and goblin lore, so that many fearsome superstitions were latent in his mind; besides he knew well the stories told by Indian medicine men in their winter camps, of the snow-walkers, and the spectres, and the formless evil beings that haunt the forest depths, and dog and waylay the lonely wanderer who after nightfall passes through the regions where they lurk; and it may be that when overcome by the horror of the fate that befell his friend, and when oppressed by the awful dread of the unknown, he grew to attribute, both at the time and still more in remembrance, weird and elfin traits to what was merely some ab-normally wicked and cunning wild beast; but whether this was so or not, no man can say" (Roosevelt 1906, pp. 254–255).

Roosevelt's psychological explanation of Bauman's tale is typical of the reasoning presently applied by those who have no desire to add wildmen to the North American faunal list. In this case, because of the vagueness of the account, it is not easy to offer counterarguments. Bauman did not get a clear look at the creature. But one might wonder what known large North America mammal typically prowls about on two feet rather than four? Bears will stand for a short time on two legs, but are not known to move any great distance in bipedal fashion. If the creature really was a bear, Bauman, an experienced backwoodsman, should have been able to identify it as such from the footprints, which he closely inspected. But he did not. What sort of animal could have made the footprints? Roosevelt (1906, p. 261) said that Bauman believed "the creature with which he had to deal was something either half-human or half devil, some great goblin-beast."

Taken on its own, the Bauman story is not very impressive as evidence for the existence of wildmen in North America, but when considered along with the more substantive reports it acquires greater significance.

On July 4, 1884, the *Colonist,* a newspaper published in Victoria, British Columbia, carried a story titled: "What is it? A strange creature captured above Yale. A British Columbian Gorilla." According to the article, Ned Austin, a railway engineer, spotted a humanlike creature ahead of him on the tracks, blew the whistle, and stopped. The creature darted up the side of a hill, with several railway employees in pursuit. After capturing the animal, described as "half man and half beast" (Shackley 1983, p. 35), the railway employees turned him over to Mr. George Tilbury.

The *Colonist* reported: "'Jacko,' as the creature has been called by his captors, is something of the gorilla type, standing about four feet seven inches in height and weighing 127 pounds. He has long, black, strong hair and resembles a human being with one exception, his entire body, excepting his hands (or paws) and feet is covered with glossy hair about one inch long. His forearm is much longer than a man's forearm, and he possesses extraordinary strength" (Shackley 1983, p. 35).

The paper added (Shackley 1983, p. 36): "Mr. Thos. White and Mr. Gouin, C.E., as well as Mr. Major, who kept a small store about half a mile west of the tunnel during the past two years, have mentioned seeing a curious creature at different points between Camps 13 and 17, but no attention was paid to their remarks as people came to the conclusion that they had seen either a bear or a stray Indian dog. Who can unravel the mystery that surrounds Jacko? Does he belong to a species hitherto unknown in this part of the continent?"

That the creature was not a gorilla seems clear—its weight was too small. Some might suppose that Jacko was a chimpanzee. But this idea was apparently considered and rejected by persons who were familiar with Jacko. Sanderson (1961, p. 27) mentioned "a comment made in another paper shortly after the original story was published, and which asked . . . how anybody could suggest that this 'Jacko' could have been a chimpanzee that had escaped from a circus."

Was the whole story perhaps a hoax? Myra Shackley thought not. She noted: "The newspaper account of Jacko was subsequently confirmed by an old man, August Castle, who was a child in the town at the time. The fate of the captive is not known, although some said that he (accompanied by Mr. Tilbury) was shipped east by rail in a cage on the way to be exhibited in a sideshow, but died in transit" (Shackley 1983, p. 36).

Furthermore, there were additional reports of creatures like Jacko from the same region. Zoologist Ivan Sanderson (1961, p. 29) said about Jacko in one of his collections of wildman evidence: "one of his species had been reported from the same area by Mr. Alexander Caulfield Anderson, a well-known explorer and an executive of the Hudson's Bay Company, who was doing a 'survey' of the newly opened territory and seeking a feasible trade route through it for his company. He reported just such hairy humanoids as having hurled rocks down upon him and his surveying party from more than one slope. That was in 1864."

In 1901, Mike King, a well-known lumberman, was working in an isolated region in northern Vancouver Island. He had to work alone. His native American employees refused to accompany him, fearing that the dreaded wildman of the woods lived there. Once, as King came over a ridge, he spotted a large humanlike creature covered with reddish brown fur. On the bank of a creek, the creature was washing some roots and placing them in two orderly piles beside him. The creature then left, running like a human being. King said: "His arms were peculiarly long and used freely in climbing and bush-running." Footprints observed by King were distinctly human, except for the "phenomenally long and spreading toes" (Sanderson 1961, pp. 34–35).

In 1941, several members of the Chapman family encountered a wildman at Ruby Creek, British Columbia. In 1959, Ivan Sanderson interviewed the Chapmans, who were native Americans, about what happened. On a sunny summer afternoon, Mrs. Chapman's oldest son alerted her to the presence of a large animal coming down out of the woods near their home. At first, she thought it was a large bear. But then, much to her horror, she saw that it was a gigantic man covered all over with yellow-brown hair. The hair was about 4 inches long. The creature moved directly towards the house, and Mrs. Chapman rounded up her three children and fled downstream to the village.

She estimated that the creature was about 7.5 feet tall. It had a relatively small head and a short, thick neck—practically no neck at all. Its body was completely human in shape, except the chest was immensely thick and the arms unusually long. Its shoulders were extremely wide. The naked regions of its face and its hands were much darker than the hair and appeared to be nearly black.

When Mr. Chapman returned home, a couple of hours after his wife had fled, he saw huge humanlike footprints all around the house. He was greatly alarmed, because, like almost all the native Americans of the Pacific Northwest, he had heard from childhood about the "big wild men of the mountains" (Sanderson 1961, p. 68). For the next week, giant humanlike footprints were found every day.

Moreover, said Sanderson, the Chapmans described the "strange gurgling whistle" emitted by the creature. According to Sanderson, this cry seemed identical to that heard by other persons in connection with similar creatures elsewhere in the Pacific Northwest.

In October of 1955, Mr. William Roe, who had spent much of his life hunting wild animals and observing their habits, encountered a wildman (Green 1978, pp. 53–56). The incident took place near a little town called Tete Jaune Cache in British Columbia. One day, said Roe in a sworn statement, he climbed up Mica Mountain to an old deserted mine and saw, at a distance of about 75 yards, what he first took to be a bear. When the creature stepped out into a clearing, Roe realized that it was something different: "My first impression was of a huge man, about six feet tall, almost three feet wide, and probably weighing somewhere near three hundred pounds. It was covered from head to foot with dark brown silver-tipped hair. But as it came closer I saw by its breasts that it was female. And yet, its torso was not curved like a female's. Its broad frame was straight from shoulder to hip. Its arms were much thicker than a man's arms, and longer, reaching almost to its knees. Its feet were broader proportionately than a man's, about five inches wide at the front and tapering to much thinner heels. When it walked it placed the heel of its foot down first, and I could see the grey-brown skin or hide on the soles of its feet" (Green 1978, pp. 53–55).

"It came to the edge of the bush I was hiding in, within twenty feet of me, and squatted down on its haunches," said Roe. "Reaching out its hands it pulled the branches of bushes toward it and stripped the leaves with its teeth. Its lips curled flexibly around the leaves as it ate. I was close enough to see that its teeth were white and even. . . . The head was higher at the back than at the front. The nose was broad and flat. The lips and chin protruded farther than its nose. But the hair that covered it, leaving bare only the parts of its face around the mouth, nose and ears, made it resemble an animal as much as a human. None of this hair, even on the back of its head, was longer than an inch, and that on its face was much shorter. Its ears were shaped like a human's ears. But its eyes were small and black like a bear's. And its neck also was unhuman. Thicker and shorter than any man's I had ever seen."

After a few minutes the creature became aware of Roe's presence in the bushes and departed. It was not exactly afraid but was apparently unwilling to have contact with a human being (Green 1978, p. 55).

In 1967, in the Bluff Creek region of Northern California, Roger Patterson and Bob Gimlin managed to shoot a short color film of a female Sasquatch. They also made casts of her footprints. These prints, which were 14 inches long, were 5.5 inches wide at the ball and 4 inches wide at the heel (Green 1978, p. 118).

Several opinions have been expressed about the film. While some authorities have said it is an outright fake, others have said they think it provides good evidence in favor of the reality of the Sasquatch. Mixed opinions have also been put forward. Dr. D. W. Grieve, an anatomist specializing in human walking, studied the film and had this to say: "My subjective impressions have oscillated between total acceptance of the Sasquatch on the grounds that the film would be difficult to fake, to one of irrational rejection based on an emotional response to the possibility that the Sasquatch actually exists. This seems worth stating because others have reacted similarly to the film. The possibility of a very clever fake cannot be ruled out on the evidence of the film. A man could have sufficient height and suitable proportions to mimic the longitudinal dimensions of the Sasquatch. The shoulder breadth however would be difficult to achieve without giving an unnatural appearance to the arm swing and shoulder contours" (Napier 1973, p. 220).

From his study of the film, Grieve estimated the length of the Sasquatch's foot to be 13.3 inches, which is consistent with the length of 14 inches reported for the footprints. John R. Napier (1973), however, believed that a 14-inch foot length was not consistent with the estimated body height of 6 feet 5 inches. In his computations, Napier, a respected British anatomist, used the ratio of foot length to body height in modern humans. He did not, however, explain why the physical proportions of the Sasquatch must be the same as those of modern humans.

Anthropologist Myra Shackley of the University of Leicester observed (1983, p. 43) that the majority view seems to be "that the film could be a hoax, but if so an incredibly clever one." Reacting similarly, Napier (1973, p. 95) stated: "Perhaps it was a man dressed up in a monkey-skin; if so it was a brilliantly executed hoax and the unknown perpetrator will take his place with the great hoaxers of the world." But then he added: "Perhaps it was the first film of a new type of hominid, quite unknown to science" (Napier 1973, p. 95). Concerning the charge of incredibly clever hoaxing, this explanation could be used to dismiss almost any kind of scientific evidence whatsoever. All one has to do is posit a sufficiently expert hoaxer. Therefore the hoax hypothesis should be applied only when there is actual evidence of hoaxing, as at Piltdown, for example. Ideally, one should be able to produce the hoaxer. Futhermore, even a demonstrated case of hoaxing cannot be used to dismiss entire categories of similar evidence.

10.5 MORE FOOTPRINTS

As far as Sasquatch footprints are concerned, independent witnesses have examined and reported hundreds of sets, and of these more than 100 have been preserved in photographs and casts (Green 1978, p. 348). Napier stated: "if any of them is real then as scientists we have a lot to explain. Among other things we shall have to re-write the story of human evolution . . . and we shall have to admit that there are still major mysteries to be solved in the world we thought we knew so well" (1973, p. 204).

Critics, however, assert that all these footprints have been faked. Undoubtedly, some footprints have been faked, a fact the staunchest supporters of the Sasquatch will readily admit. But could every single one of them be a hoax? Napier (1973, p. 124) stated that if all the prints are fakes "then we must be prepared to accept the existence of a conspiracy of Mafia-like ramifications with cells in practically every major township from San Francisco to Vancouver."

Grover S. Krantz, an anthropologist at Washington State University, was initially skeptical of Sasquatch reports. In order to determine whether or not the creature really existed, Krantz studied in detail some prints found in 1970 in northeast Washington State. In reconstructing the skeletal structure of the foot from the print, he noted that the ankle was positioned more forward than in a human foot. Taking into consideration the reported height and weight of an adult Sasquatch, Krantz, using his knowledge of physical anthropology, calculated just how far forward the ankle would have to be set. Returning to the prints, he found that the position of the ankle exactly matched his theoretical calculations. "That's when I decided the thing is real," said Krantz. "There is no way a faker could have known how far forward to set that ankle. It took me a couple of months to work it out with the casts in hand, so you have to figure how much smarter a faker would've had to be" (Huyghe 1984, p. 94).

Krantz (1983) and wildman expert John Green (1978, pp. 349–356) have written extensive reports on the North American footprint evidence. Typically the prints are 14 to 18 inches long and 5 to 9 inches wide, giving a surface roughly 3 to 4 times larger than that of an average human foot. Hence the popular name Bigfoot. To make a Sasquatch footprint as deep as an average human footprint would require a weight 3 to 4 times greater than that of an average-sized man. In all cases, however, whether the prints are in snow, mud, dirt, or wet sand, the Sasquatch prints are much deeper than those made by a man walking right next to them in the same material. Thus a weight of more than 3 or 4 times that of a man is required to make the Sasquatch prints. Green, wearing large fake feet and carrying 250 pounds on his back (for a total of 450 pounds), was unable to make a deep enough impression in firm wet sand. Moreover, Green's fake feet were only 14.5 inches long, small for a Sasquatch. Larger feet would have produced impressions of even smaller depth in the sand. Krantz (1983) estimated that to make typical Sasquatch prints a total weight of at least 700 pounds is required. Thus a 200-pound man would have to be carrying at least 500 pounds to make a good print.

But that is only the beginning. There are reports of series of prints extending from three-quarters of a mile up to several miles, in deserted regions far away from the nearest roads. The stride length of a Sasquatch varies from 4 to 6 feet (the stride length of an average man is about 3 feet). Try walking a mile with at least 500 pounds on your back and taking strides 5 feet long.

"A footprint machine, a kind of mechanical stamp, has been suggested," stated Napier (1973, p. 125), "but an apparatus capable of delivering a thrust of approximately 800 lb per square foot that can be manhandled over rough and mountainous country puts a strain on one's credulity." In addition, said Napier (1973, p. 125), careful studies of Sasquatch prints by Dr. Maurice Tripp, a geologist, revealed that impact ridges, which a footprint machine would be expected to leave, were not present.

Some of the reported series of tracks were in fresh snow, enabling observers to verify that no other marks were made at the ground level by some machine paralleling the prints. In several cases, the Sasquatch footprints indicated the maker strode over large logs, which a human of normal size could not have gotten over without disturbing the fresh snow clearly visible on their tops. Sometimes the Sasquatch prints went up or down embankments. In some cases, the distance between the toes of the footprints varied from one print to the next in a single series of prints. This means that besides all the other problems facing a hoaxer, he would have had to incorporate moving parts into his artificial feet.

Furthermore, in order to insure that some of his fake prints would be found, any hoaxer would probably have had to make more trails of footprints than were actually discovered—and that means a lot of work.

What about a device operated from a hovering craft? Such a device would undoubtedly be very expensive. A helicopter alone is not a cheap item, and a custom device for making the footprints would also cost a bit. Also, footprints have been found at the same time that a Sasquatch was actually seen or soon thereafter, as, for example, in the Patterson sighting in 1967 and the Chapman sighting in 1941. In other cases, people sleeping at campsites or work sites have gotten up in the morning and found newly made footprints nearby. In one case, the footprints went right alongside a man's camper truck (Green 1978, p. 352). If the prints had been made by a stamping machine, operated on the ground or from a helicopter, the people reporting the prints almost certainly would have been awakened.

In conclusion, critics have failed to explain all the footprints as the work of hoaxers. It would seem, therefore, that the footprints argue strongly for the reality of the Sasquatch, as demonstrated by the following case.

On June 10, 1982, Paul Freeman, a U.S. Forest Service patrolman tracking elk in the Walla Walla district of Washington State, observed a hairy biped around 8 feet tall, standing about 60 yards from him. After 30 seconds, the large animal walked away (Huyghe 1984, p. 94). Krantz (1983) studied casts of the creature's footprints and found dermal ridges, sweat pores, and other features in the proper places for large primate feet. Detailed skin impressions on the side walls of the prints indicated the presence of a flexible sole pad.

Krantz solicited opinions from other scholars and fingerprint experts. Tatyana Gladkova, a specialist in dermatoglyphics from the USSR Institute of Anthropology, said: "I see dermal ridges of the arch type distally directed. I see sweat pores. If it's a fake, it's a brilliant fake, on the level of counterfeiting, and by someone well versed in dermatoglyphics" (Krantz 1983, p. 78).

Douglas M. Monsoor, a master police fingerprint examiner from Lakewood, Colorado, stated: "I see the presence of ridge structure in these casts, which, in my examination, appears consistent with that type of ridge structure you would find in a human. Under magnification, they evidence all the minute characteristics similar to human dermal ridges. . . . If hoaxing were involved, I can conceive of no way in which it could have been done. They appear to be casts of impressions of a primate foot—of a creature different from any of which I am aware" (Krantz 1983, p. 79).

Ten years earlier, John R. Napier (1973, p. 125) declared that he found the prints he himself studied "biologically convincing." Napier (1973, pp. 204–205) stated: "The evidence that I have examined persuades me that some of the tracks are real, and that they are manlike in form. . . . But when the *size* of the tracks is taken into account, and the conclusion is reached that the man-like creature in question has a stature of at least 8 ft. and weighs upward of 800 lb., the mind starts to boggle at such a preposterous idea. The vision of such creatures stomping barefoot through the forests of north-west America, unknown to science, is beyond common sense. Yet reason argues this is the case. . . . Thomas Huxley's aphorism that 'logical consequences are the scarecrows of fools and the beacons of wise men' puts steel into my soul. . . . I am convinced that the Sasquatch exists." Coming from a scientist who headed the primate program at the Smithsonian Institution, this is a very strong statement. Napier is also one of Great Britain's leading experts in the field of primate anatomy. His name will come up often in our discussion of hominid fossil discoveries in Africa.

In the face of much good evidence, why do almost all anthropologists and zoologists remain silent about Sasquatch? Krantz observed, "They are scared for their reputations and their jobs" (Huyghe 1984, p. 96). Napier similarly noted: "One of the problems, perhaps the greatest problem, in investigating Sasquatch sightings is the suspicion with which people who claim to have seen a Sasquatch are treated by their neighbours and employers. To admit such an experience is, in some areas, to risk personal reputation, social status and professional credibility" (1973, p. 88). In particular, he told of "the case of a highly qualified oil company geologist who told his story but insisted that his name should not be mentioned for fear of dismissal by his company" (Napier 1973, p. 88). In this regard, Roderick Sprague, an anthropologist from the University of Idaho, said of Krantz: "It is Krantz's willingness to openly investigate the unknown that has cost him the respect of many colleagues as well as timely academic promotion" (1986, p. 103).

The majority of the Sasquatch reports come from the northwestern United States and British Columbia. However, there are also numerous reports from the eastern parts of the United States and Canada. For example, Green (1978) stated that there were, as of 1977, 11 reports from New York, more than 24 reports from Pennsylvania, 19 reports from Ohio, 18 from Michigan, 9 from Tennessee, more than 36 from New Jersey, 19 from Arkansas, 23 from Illinois, 30 from Texas, and 104 (maybe more) from Florida. Moving out west, Green recorded 74 reports from Montana, 32 from Idaho, 176 from Oregon, 281 from Washington, 225 from British Columbia, and 343 from California.

The volume of reports from the Pacific Northwest caused John R. Napier (1973, p. 96) to state: "The North American Bigfoot or Sasquatch has a lot going for it. It is impossible on the evidence . . . to say that it does not exist. Too many people claim to have seen it or at least to have seen footprints to dismiss its reality out of hand. To suggest that hundreds of people at worst are lying or, at best, deluding themselves is neither proper nor realistic."

"One is forced to conclude," said Napier, "that a man-like life-form of gigantic proportions is living at the present time in the wild areas of the north-western United States and British Columbia. . . . That such a creature should be alive and kicking in our midst, unrecognized and unclassifiable, is a profound blow to the credibility of modern anthropology" (Green 1978, p. 12). It might also be said that the existence of living ape-men in North America, from Washington and Oregon to Florida and New Jersey, is a blow not only to anthropology, but to biology, zoology, and science in general.

10.6 CENTRAL AND SOUTH AMERICA

Apelike wildmen are also reported in southern Mexico and throughout Central America. In *White Indians of Darien,* Richard Oglesby Marsh said a man told him that in 1920 he had killed a humanlike animal in Central America. In *Buckskin Joe,* Edward Jonathan Hoyt reported an encounter he had in 1898 in Honduras. A large, apelike creature, about 5 feet tall, crawled over the end of his bunk. Hoyt killed the animal, which resembled a human (Green 1978, p. 133).

From southern Mexico's tropical forests come accounts of beings called the Sisimite. Wendell Skousen, a geologist, said the people of Cubulco in Baja Verapaz reported: "There live in the mountains very big, wild men, completely clothed in short, thick, brown, hairy fur, with no necks, small eyes, long arms and huge hands. They leave footprints twice the length of a man's." Several persons said that they had been chased down mountainsides by the Sisimite. Skousen thought the creatures, which were said to travel sometimes on two legs and sometimes on all four, may have been bears. However, upon questioning the natives carefully, he wrote: "it looked like a bear, but it *wasn't* from the description that they gave—no conspicuous ears, no 'snout'" (Sanderson 1961, p. 159). Similar creatures are reported in Guatemala, where, it has been said, they kidnap women and children (Sanderson 1961, pp. 161–162).

People in Belize (formerly British Honduras) speak of semi-human creatures called Dwendis, which inhabit the jungles in the southern part of their country. The name Dwendi comes from the Spanish word *Duende,* meaning "goblin." Ivan Sanderson, who conducted research in Belize, wrote (1961, pp. 164–165): "Dozens told me of having seen them, and these were mostly men of substance who had worked for responsible organizations like the Forestry Department and who had, in several cases, been schooled or trained either in Europe or the United States. One, a junior forestry officer born locally, described in great detail two of these little creatures that he had suddenly noticed quietly watching him on several occasions at the edge of the forestry reserve near the foot of the Maya Mountains. . . . These little folk were described as being between three foot six and four foot six, well proportioned but with very heavy shoulders and rather long arms, clothed in thick, tight, close brown hair looking like that of a short-coated dog; having very flat yellowish faces but head-hair no longer than the body hair except down the back of the neck and midback." The Dwendis appear to represent a species different from the large Sasquatch of the Pacific Northwest of North America.

Most of Sanderson's informants told him that the Dwendis carried what appeared to be dried palm leaves or some kind of large hatlike object over their heads. Sanderson (1961, p. 165) observed: "This at first sounds like the silliest thing, but when one has heard it from highly educated men as well as from simple peasants, and all over an area as great as that from Peten [southern Mexico] to Nicaragua, one begins to wonder." He then pointed out: "There are many Mayan bas-reliefs that show pairs of tiny little men with big hats but no clothes, standing among trees and amid the vast legs of demi-gods, priests, and warriors. They are also much smaller than the peasants bearing gifts to the temples" (Sanderson 1961, p. 166).

From the Guianas region of South America come accounts of wildmen called Didis. Early explorers heard reports about them from the Indians, who said they were about five feet tall, walked erect, and were covered with thick black hair.

In 1931, Nelloc Beccari, an anthropologist from Italy, heard an account of the Didi from Mr. Haines, the Resident Magistrate in British Guiana. Heuvelmans gave this summary of what Haines related to Beccari: "Haines told him that he had come upon a couple of *di-di* many years before when he was prospecting for gold. In 1910 he was going through the forest along the Konawaruk, a tributary which joins the Essequibo just above its junction with the Potaro, when he suddenly came upon two strange creatures, which stood up on their hind feet when they saw him. They had human features but were entirely covered with reddish brown fur. . . . the two creatures retreated slowly and disappeared into the forest" (Sanderson 1961, pp. 179–180).

After giving many similar accounts in his book about wildmen, Sanderson (1961, p. 181) stated: "The most significant single fact about these reports from Guiana is that never once has any local person—nor any person reporting what a local person says—so much as indicated that these creatures are just 'monkeys.' In all cases they have specified that they are tailless, erect, and have human attributes."

From the eastern slopes of the Andes in Ecuador come reports of the Shiru, a small fur-covered hominidlike creature, about 4 to 5 feet tall (Sanderson 1961, p. 166). In Brazil, people tell of the large apelike Mapinguary, which leaves giant humanlike footprints and is said to kill cattle (Sanderson 1961, p. 174).

10.7 YETI: WILDMEN OF THE HIMALAYAS

Diaries and other papers of British officials residing in the Himalayan region of the Indian subcontinent during the nineteenth century contain sporadic references to sightings and footprints of wildmen called Yeti. The Yeti were first mentioned by B. H. Hodgson, who from 1820 to 1843 served as British resident at the Nepalese court. Hodgson reported that in the course of a journey through northern Nepal his bearers were frightened by the sight of a hairy, tailless, humanlike creature.

Many will suggest, on hearing a report like this (and hundreds have been recorded since Hodgson's time), that the Nepalese mistook an ordinary animal for a Yeti. The usual candidates for mistaken identity are bears and the langur monkey. But it is hard to imagine that lifelong residents of the Himalayas, intimately familiar with the wildlife, would have made such mistakes. Myra Shackley observed that Yeti are found in Nepalese and Tibetan religious paintings depicting hierarchies of living beings. "Here," said Shackley (1983, p. 60), "bears, apes, and langurs are depicted separate from the wildman, suggesting there is no confusion (at least in the minds of the artists) between these forms."

After reviewing the available reports, Ivan Sanderson (1961, p. 358) compiled the following composite description of the Yeti: "Somewhat larger than man-sized and much more sturdy, with short legs and long arms; clothed in long rather shaggy fur or hair, same length all over and not differentiated. Naked face and other parts jet black; bull-neck and small conical head with heavy brow-ridges; fanged canine teeth; can drop hands to ground and stand on knuckles like gorilla. . . . heel very wide and foot almost square and very large, second toe longer and larger than first, and both these separated and semi-opposed to the remaining three which are very small and webbed."

During the nineteenth century, at least one European reported personally seeing a captured animal that resembled a Yeti. A South African man told Myra Shackley (1983, p. 67): "Many years ago in India, my late wife's mother told me how her mother had actually seen what might have been one of these creatures at Mussorie, in the Himalayan foothills. This semi-human was walking upright, but was obviously more animal than human with hair covering its whole body. It was reportedly caught up in the snows. . . . his captors had it in chains."

During the twentieth century, sightings by Europeans of wildmen and their footprints continued, increasing during the Himalayan mountain-climbing expeditions of the 1930s. In 1938, H. W. Tilman followed a trail of footprints for a mile on a glacier, at an elevation of 19,000 feet. Speaking of one of his Sherpa guides, Tilman stated: "Sen Tensing, who had no doubt whatever that the creatures . . . that made the tracks were 'Yetis' or wild men, told me that two years before, he and a number of other Sherpas had seen one of them at a distance of about 25 yards at Thyangbochi. He described it as half man and half beast, standing about five feet six inches, with a tall pointed head, its body covered with reddish brown hair, but with a hairless face. . . . Whatever it was that he had seen, he was convinced that it was neither a bear nor a monkey, with both of which animals he was, of course, very familiar" (Heuvelmans 1962, pp. 136–137).

During the Second World War, a man named Slavomir Rawicz escaped from a Siberian prisoner-of-war camp and made his way by foot to India. In *The Long March* (1956), a book describing his experiences, Rawicz stated that while traveling across the Himalayas, he encountered two wildmen, six feet in height and covered with long reddish hair. However, explorers familiar with the region traversed by Rawicz have pointed out some inconsistencies in his account of his journey. For example, he took an inordinate amount of time to travel a certain section of the route—even at his own stated rate of progress. Critics thus insinuated that Rawicz's book, including the story of the wildmen, was largely if not completely fictional (Shackley 1983, pp. 54–55).

In November of 1951, Eric Shipton, while reconnoitering the approaches to Mt. Everest, found footprints on the Menlung glacier, near the border between Tibet and Nepal, at an elevation of 18,000 feet. Shipton followed the trail for a mile. Already well known as a mountaineer, Shipton could not easily be accused of publicity-seeking. A close-up photograph of one of the prints has proved convincing to many. Myra Shackley (1983, pp. 55–56) wrote: "Indeed, even the doubters admit that Shipton's famous footprints, seen on the Menlung Glacier in 1951, cannot readily be explained away."

The footprints were quite large. John R. Napier considered the possibility that the particular size and shape of the best Shipton footprint could have been caused by melting of the snow. Napier, however, noted (1973, p. 140): "Eric Shipton agrees that melting and sublimation might be responsible for the appearance, but he points out quite correctly that it would be reasonable to expect the narrow ridges behind and between the little toes to be the first features to disappear in these circumstances." For Napier, Shipton's observation appeared to rule out the snow-melting explanation, or at least make it far less likely. Napier proposed another possibility: "that the footprint is double—two tracks superimposed. But a double—what? I don't know."

Napier (1973, p. 141) concluded: "Something must have made the Shipton footprint. Like Mount Everest, it is there, and needs explaining. I only wish I could solve the puzzle; it would help me sleep better at night. Of course, it would settle a lot of problems if one could simply assume that the Yeti is alive. . . . The trouble is that such an assumption conflicts with the principles of biology as we know them." In the end, Napier suggested that the Shipton footprint was the result of superimposed human feet, one shod and the other unshod. In general, Napier, who was fully convinced of the existence of the North American Sasquatch, was highly skeptical of the evidence for the Yeti. But, as we shall see later in this section, new evidence would cause Napier to become more inclined to accept the Himalayan wildmen.

In the course of his expeditions to the Himalaya Mountains in the 1950s and 1960s, Sir Edmund Hillary gave attention to evidence for the Yeti, including footprints in snow. He concluded that in every case the large footprints attributed to the Yeti had been produced by the merging of smaller tracks of known animals, by superimposition and melting. To this Napier (1973, pp. 57–58), himself a skeptic, replied: "The signs of melting are so obvious that no one with any experience would confuse a melted footprint with a fresh one. Not all the prints seen over the years by reputable observers can be explained away in these terms; there must be other explanations for footprints, including, of course, the possibility that they were made by an animal unknown to science."

But although Napier was unwilling to completely reject the existence of an unknown hominid, he was nevertheless inclined to regard this as the least probable or desirable alternative. In 1956, Professor E. S. Williams photographed some prints on the Biafo glacier in the Karakoram mountains. Napier, who thought it likely that they were the superimposed prints of the front and rear paws of a bear, said (1973, p. 130): "It is impossible to state categorically that Williams's prints are those of a bear and not of a Yeti, but in the spirit of Bishop of Ockham it seems more reasonable to explain a phenomenon in terms of the known rather than the unknown."

Of course, in avoiding the relatively straightforward explanation that a peculiar set of tracks in snow was made by an unknown animal, one is forced to come up with all kinds of speculative hypotheses about the superimposition of prints of various animals and humans, or the transformation of such prints by melting, in a manner not clearly understood. And this would also appear to be a violation of a key aspect of Ockham's razor—namely, that the simplest of competing theories is preferable to the more complex.

In addition to Westerners, native informants also gave a continuous stream of reports on the Yeti. Lord Hunt, who headed a Mount Everest expedition in 1953, told of an incident recounted by the Tibetan Buddhist abbot of the Thyangboche monastery: "he gave a most graphic description of how a Yeti had appeared from the surrounding thickets, a few years back in the winter when the snow lay on the ground. This beast, loping along sometimes on its hind legs and sometimes on all fours, stood about five feet high and was covered with gray hair" (Shackley 1983, p. 62).

In 1958, Tibetan villagers from Tharbaleh, near the Rongbuk glacier, came upon a drowned Yeti, said Myra Shackley in her book on wildmen. The villagers described the creature as being like a small man with a pointed head and covered with reddish-brown fur (Shackley 1983, p. 1983).

Some Buddhist monasteries claim to have physical remains of the Yeti. One category of such relics is Yeti scalps, but the ones studied by Western scientists are thought to have been made from the skins of known animals (Shackley 1983, pp. 65–66). In 1960, Sir Edmund Hillary mounted an expedition to collect and evaluate evidence for the Yeti and sent a Yeti scalp from the Khumjung monastery to the West for testing. The results indicated that the scalp had been manufactured from the skin of the serow, a goatlike Himalayan antelope. But some disagreed with this analysis. Shackley (1983, p. 66) said they "pointed out that hairs from the scalp look distinctly monkey-like, and that it contains parasitic mites of a species different from that recovered from the serow."

In the 1950s, Western explorers sponsored by American businessman Tom Slick obtained samples from a mummified Yeti hand kept at Pangboche. Shackley (1983, p. 66) stated: "detailed investigation of small skin samples back in European laboratories failed to reach a diagnosis. Local rumour maintains that the hand comes from a rather poorly mummified lama, but it has some curiously anthropoid features."

In May of 1957, the *Kathmandu Commoner* carried a story about a Yeti head that had been kept for 25 years in the village of Chilunka, about 50 miles northeast of Kathmandu. The head reportedly had been severed from the corpse of a Yeti slain by Nepalese soldiers, who had hunted down the creature after it had killed many of their comrades (Shackley 1983, p. 66). Concerning another specimen, Shackley noted that Chemed Rigdzin Dorje, a Tibetan lama, spoke of the existence of a complete mummified Yeti.

Over the years, sightings continued. In 1970, mountaineer Don Willans was researching an approach to Annapurna, a high peak in northern Nepal. He found some tracks and at night saw an apelike creature bounding across the snow. Napier (1973, p. 135), still skeptical, said it could have been a langur monkey.

In 1978, Lord Hunt, who headed the British Mt. Everest expedition of 1953, saw Yeti tracks and heard the high-pitched cry the Yeti is said to make. Lord Hunt, described by Shackley as "a vigorous champion of the Yeti," had come upon similar tracks in 1953. In both 1953 and 1978, the tracks were found at altitudes of 15,000 to 20,000 feet, too high for the either the black or red bears of the Himalayas. Shackley (1983, p. 56) stated: "The tracks seen by Lord Hunt in 1978 were very fresh, and it was possible to see the impression of the toes, convincing him that the footprint represented the actual size and shape of the feet, about 13¾ in. long and 6¾ in. broad. . . . This is especially interesting since it has, of course, been frequently contended that such tracks are made either by other animals (bears or langurs being the most favoured), or by the impressions of human feet which have become exaggerated in the melting snow."

It is interesting to note that science has recognized the existence of many fossil species on the strength of their footprints alone. Heuvelmans (1982, p. 3) stated: "The hypotheses and reconstructions of cryptozoology (regarding animals actually alive) are no more daring, questionable, fantastic, or illegitimate than those upon which paleontology has based its reconstructions of the fauna of past ages. . . . It seemed perfectly legitimate to give the scientific name *Chirotherium* to a fossil genus known only by its tracks, found in Germany, England, France, Spain, Italy, and the United States, and of which some 20 species have been described. Yet, at the same time, it seemed ridiculous, premature, and absurd to describe scientifically the Himalayan Yeti, known *not only* by many tracks not identifiable with any known animal, but also by morphology and behavior as related by numerous eyewitnesses."

In 1986, Marc E. Miller and William Caccioli, of the New World Explorers Society, retraced the route of Hillary's 1960 Yeti expedition, visiting the Buddhist monasteries at Khumjung, Thyangboche, and Pangboche. At Khumjung, Miller and Caccioli interviewed Khonjo Khumbi, the village elder who accompanied Hillary to the United States with the famous Yeti scalp. Khonjo told Miller and Caccioli that in the course of his travels through Tibet he had seen whole Yeti furs. The High Lama of the Thyangboche monastery also said he had seen such furs in the homes of great hunters.

Miller and Caccioli (1986, p. 82) reported that they received possible Yeti chest hairs from an elderly woman of Khumjung village in Tibet: "We were told that her son was carrying potatoes along a trail in 1978, and was allegedly attacked by a Yeti. The Yeti was described as a large male, nearly 7 feet tall, and covered with dark and reddish hair. During the course of the attack, the young man took his potato hoe and struck the Yeti across the chest. The Yeti fled into the higher mountain region. The young man struggled back to Khumjung village to his mother, and described his encounter with the Yeti. His wounds were serious, and he later died."

Figure 10.1. Areas where Yetis have been sighted in Central Asia and the Himalayas are shaded with vertical black bars (after Shackley 1983, pp. 78–79).

Figure 10.2. Drawing of a Mongolian Almas from a 19th-century Tibetan book (Shackley 1983, p. 97).

10.8 THE ALMAS OF CENTRAL ASIA

The Sasquatch and the Yeti, from the descriptions available, are large and very apelike. But there is another wildman, the Almas, which seems smaller and more human. Reports of the Almas are concentrated in an area extending from Mongolia in the north, south through the Pamirs, and then westward into the Caucasus region. Similar reports come from Siberia and the far northeast parts of the Russian republic.

Early in the fifteenth century, Hans Schiltenberger was captured by the Turks and sent to the court of Tamerlane, who placed him in the retinue of a Mongol prince named Egidi. After returning to Europe in 1427, Schiltenberger wrote about his experiences. In his book, he described some mountains, apparently the Tien Shan range in Mongolia: "The inhabitants say that beyond the mountains is the beginning of a wasteland which lies at the edge of the earth. No one can survive there because the desert is populated by so many snakes and tigers. In the mountains themselves live wild people, who have nothing in common with other human beings. A pelt covers the entire body of these creatures. Only the hands and face are free of hair. They run around in the hills like animals and eat foliage and grass and whatever else they can find. The lord of the territory made Egidi a present of a couple of forest people, a man and a woman. They had been caught in the wilderness, together with three untamed horses the size of asses and all sorts of other animals which are not found in German lands and which I cannot therefore put a name to" (Shackley 1983, p. 93).

Myra Shackley (1983, pp. 93–94) found Schiltenberger's account especially credible for two reasons: "First, Schiltenberger reports that he saw the creatures *with his own eyes*. Secondly, he refers to Przewalski horses, which were only rediscovered by Nicholai Przewalski in 1881. . . . Przewalski himself saw 'wildmen' in Mongolia in 1871."

A drawing of an Almas is found in a nineteenth-century Mongol compendium of medicines derived from various plants and animals. The text next to the picture reads: "The wildman lives in the mountains, his origins close to that of the bear, his body resembles that of man, and he has enormous strength. His meat may be eaten to treat mental diseases and his gall cures jaundice" (Shackley 1983, p. 98).

Shackley (1983, p. 98) noted: "The book contains thousands of illustrations of various classes of animals (reptiles, mammals and amphibia), but not one single mythological animal such as are known from similar medieval European books. All the creatures are living and observable today. There seems no reason at all to suggest that the Almas did not exist also and illustrations seem to suggest that it was found among rocky habitats, in the mountains."

In 1937, Dordji Meiren, a member of the Mongolian Academy of Sciences, saw the skin of an Almas in a monastery in the Gobi desert. The lamas were using it as a carpet in some of their rituals. Shackley (1983, pp. 103–104) stated: "The hairs on the skin were reddish and curly. . . . The features [of the face] were hairless, the face had eyebrows, and the head still had long disordered hair. Fingers and toes were in a good state of preservation and the nails were similar to human nails."

A report of a more recent sighting of live wildmen was related to Myra Shackley by Dmitri Bayanov, of the Darwin Museum in Moscow. In 1963, Ivan Ivlov, a Russian pediatrician, was traveling through the Altai mountains in the southern part of Mongolia. Ivlov saw several humanlike creatures standing on a mountain slope. They appeared to be a family group, composed of a male, female, and child. Ivlov observed the creatures through his binoculars from a distance of half a mile until they moved out of his field of vision. His Mongolian driver also saw them and said they were common in that area. Shackley (1983, p. 91) stated: "So we are not dealing with folktales or local legends, but with an event that was recorded by a trained scientist and transmitted to the proper authorities. There is no reason to doubt Ivlov's word, partly because of his impeccable scientific reputation and partly because, although he had heard local stories about these creatures he had remained sceptical about their existence."

After his encounter with the Almas family, Ivlov interviewed many Mongolian children, believing they would be more candid than adults. The children provided many additional reports about the Almas. For example, one child told Ivlov that while he and some other children were swimming in a stream, he saw a male Almas carry a child Almas across it (Shackley 1983, pp. 91–92).

In 1980, a worker at an experimental agricultural station, operated by the Mongolian Academy of Sciences at Bulgan, encountered the dead body of a wildman: "I approached and saw a hairy corpse of a robust humanlike creature dried and half-buried by sand. I had never seen such a humanlike being before covered by camel-colour brownish-yellow short hairs and I recoiled, although in my native land in Sinkiang I had seen many dead men killed in battle. . . . The dead thing was not a bear or ape and at the same time it was not a man like Mongol or Kazakh or Chinese and Russian. The hairs of its head were longer than on its body" (Shackley 1983, p. 107).

The Pamir mountains, lying in a remote region where the borders of Tadzhikistan, China, Kashmir, and Afghanistan meet, have been the scene of many Almas sightings. In 1925, Mikhail Stephanovitch Topilski, a major-general in the Soviet army, led his unit in an assault on an anti-Soviet guerilla force hiding in a cave in the Pamirs. One of the surviving guerillas said that while in the cave he and his comrades were attacked by several apelike creatures. Topilski ordered the rubble of the cave searched, and the body of one such creature was found. Topilski reported (Shackley 1983, pp. 118–119): "At first glance I thought the body was that of an ape. It was covered with hair all over. But I knew there were no apes in the Pamirs. Also, the body itself looked very much like that of a man. We tried pulling the hair, to see if it was just a hide used for disguise, but found that it was the creature's own natural hair. We turned the body over several times on its back and its front, and measured it. Our doctor made a long and thorough inspection of the body, and it was clear that it was not a human being."

"The body," continued Topilski, "belonged to a male creature 165–170 cm [about 5½ feet] tall, elderly or even old, judging by the greyish colour of the hair in several places. The chest was covered with brownish hair and the belly with greyish hair. The hair was longer but sparser on the chest and close-cropped and thick on the belly. In general the hair was very thick, without any underfur. There was least hair on the buttocks, from which fact our doctor deduced that the creature sat like a human being. There was most hair on the hips. The knees were completely bare of hair and had callous growths on them. The whole foot including the sole was quite hairless and was covered by hard brown skin. The hair got thinner near the hand, and the palms had none at all but only callous skin."

Topilski added: "The colour of the face was dark, and the creature had neither beard nor moustache. The temples were bald and the back of the head was covered by thick, matted hair. The dead creature lay with its eyes open and its teeth bared. The eyes were dark and the teeth were large and even and shaped like human teeth. The forehead was slanting and the eyebrows were very powerful. The protruding jawbones made the face resemble the Mongol type of face. The nose was flat, with a deeply sunk bridge. The ears were hairless and looked a little more pointed than a human being's with a longer lobe. The lower jaw was very massive. The creature had a very powerful chest and well developed muscles. . . . The arms were of normal length, the hands were slightly wider and the feet much wider and shorter than man's."

In 1957, Alexander Georgievitch Pronin, a hydrologist at the Geographical Research Institute of Leningrad University, participated in an expedition to the Pamirs, for the purpose of mapping glaciers. On August 2, 1957, while his team was investigating the Fedchenko glacier, Pronin hiked into the valley of the Balyandkiik River. Shackley (1983, p. 120) stated: "at noon he noticed a figure standing on a rocky cliff about 500 yards above him and the same distance away. His first reaction was surprise, since this area was known to be uninhabited, and his second was that the creature was not human. It resembled a man but was very stooped. He watched the stocky figure move across the snow, keeping its feet wide apart, and he noted that its forearms were longer than a human's and it was covered with reddish grey hair." Pronin saw the creature again three days later, walking upright. Since this incident, there have been numerous wildman sightings in the Pamirs, and members of various expeditions have photographed and taken casts of footprints (Shackley 1983, pp. 122–126).

We shall now consider reports about the Almas from the Caucasus region. According to testimony from villagers of Tkhina, on the Mokvi River, a female Almas was captured there during the nineteenth century, in the forests of Mt. Zaadan. For three years, she was kept imprisoned, but then became domesticated and was allowed to live in a house. She was called Zana. Shackley (1983, p. 112) stated: "Her skin was a greyish-black colour, covered with reddish hair, longer on her head than elsewhere. She was capable of inarticulate cries but never developed a language. She had a large face with big cheek bones, muzzle-like

prognathous jaw and large eyebrows, big white teeth and a 'fierce expression.'"
Eventually Zana, through sexual relations with a villager, had children. Some of
Zana's grandchildren were seen by Boris Porshnev in 1964. In her account of
Porshnev's investigations, Shackley (1983, p. 113) noted: "The grandchildren,
Chalikoua and Taia, had darkish skin of rather negroid appearance, with very
prominent chewing muscles and extra strong jaws." Porshnev also interviewed
villagers who as children had been present at Zana's funeral in the 1880s.

In the Caucasus region, the Almas is sometimes called Biaban-guli. In 1899, K.
A. Satunin, a Russian zoologist, spotted a female Biaban-guli in the Talysh hills of
the southern Caucasus. He stated that the creature had "fully human movements"
(Shackley 1983, p. 109). The fact that Satunin was a well-known zoologist makes his
report particularly significant.

In 1941, V. S. Karapetyan, a lieutenant colonel of the medical service of the
Soviet army, performed a direct physical examination of a living wildman
captured in the Dagestan autonomous republic, just north of the Caucasus
mountains. Karapetyan said: "I entered a shed with two members of the local
authorities. When I asked why I had to examine the man in a cold shed and not
in a warm room, I was told that the prisoner could not be kept in a warm room.
He had sweated in the house so profusely that they had had to keep him in the
shed. I can still see the creature as it stood before me, a male, naked and bare-
footed. And it was doubtlessly a man, because its entire shape was human. The
chest, back, and shoulders, however, were covered with shaggy hair of a dark
brown colour. This fur of his was much like that of a bear, and 2 to 3 centimeters
[1 inch] long. The fur was thinner and softer below the chest. His wrists were
crude and sparsely covered with hair. The palms of his hands and soles of his feet
were free of hair. But the hair on his head reached to his shoulders partly covering
his forehead. The hair on his head, moreover, felt very rough to the hand. He had
no beard or moustache, though his face was completely covered with a light
growth of hair. The hair around his mouth was also short and sparse. The man
stood absolutely straight with his arms hanging, and his height was above the
average—about 180 cm [almost 5 feet 11 inches]. He stood before me like a
giant, his mighty chest thrust forward. His fingers were thick, strong and
exceptionally large. On the whole, he was considerably bigger than any of the
local inhabitants. His eyes told me nothing. They were dull and empty—the eyes
of an animal. And he seemed to me like an animal and nothing more" (Sanderson
1961, pp. 295–296). Significantly, the creature had lice of a kind different from
those that infect humans. It is reports like this that have led scientists such as
British anthropologist Myra Shackley and Soviet anatomist Dr. Zh. I. Kofman to
conclude that the Almas may represent a relict population of Neanderthals or
perhaps even *Homo erectus* (Shackley 1983, p. 114). What happened to the
wildman of Dagestan? According to published accounts, he was shot by his

Soviet military captors as they retreated before the advancing German army.

In the 1950s, Yu. I. Merezhinski, senior lecturer in the department of ethnography and anthropology at Kiev University, was doing research in Azerbaijan, in the northern part of the Caucasus region. From local people, Merezhinski heard reports of an Almas-like wildman called the Kaptar. Khadzi Magoma, an expert hunter, told Merezhinski that he would take him to a stream where the Kaptar sometimes bathed at night. In exchange, the hunter asked Merezhinski to take a flash photo of the creature for him. Merezhinski agreed, and they went to the stream, near which a few albino Kaptars were said to live. Shackley (1983, p. 110) stated: "sure enough Merezhinski saw one from a distance of only a few yards, clearly discernible on the river bank through the bushes. It was damp, lean and covered from head to foot with white hair. Unfortunately the reality of the creature was too much for Merezhinski, who instead of photographing it shot at it with his revolver but missed in his excitement. The old hunter, furious at the deception, refused to repeat the experiment."

Here once more we have a report by a professional scientist who directly observed a wildman. As an anthropologist, Merezhinski was particularly well qualified to evaluate what he saw. It is reports like this that tend to dispel the charge that the Almas is a creature that exists only in folklore.

And as far as folklore is concerned, accounts of the Almas and other wildmen are not necessarily a sign that the Almas is imaginary. Dmitri Bayanov, of the Darwin Museum in Moscow, asked (1982, p. 47): "Is the abundant folklore, say, about the wolf or the bear not a consequence of the existence of these animals and man's knowledge of them?" Bayanov (1982, p. 47) added: "Therefore we say that, if relic hominoids were *not* reflected in folklore and mythology, then their reality can be called into question."

10.9 WILDMEN OF CHINA

"Chinese historical documents, and many city and town annals, contain abundant records of Wildman, which are given various names," states Zhou Guoxing of the Beijing Museum of Natural History (Zhou, G. 1982, p. 13).

Two thousand years ago, the poet-statesman Qu Yuan made many references to Shangui (mountain ogres) in his verses. Li Yanshow, a historian who lived during the T'Ang Dynasty (A.D. 618–907), stated that the forests of Hubei province sheltered a band of wildmen. Wildmen also appeared in the writings of Li Shizhen, a pharmacologist of the Ming Dynasty (A.D. 1368–1644). In the fifty-first volume of his massive work on medical ingredients, he described several species of humanoid creatures, including one named Fei-fei.

Li wrote: "'Feifei,' which are called 'manbear,' are also found in the mountainous areas in west Shu (part of Sichuan Province today) and Chu division, where people skin them and eat their palms. The You mountain of Sha county, Fujian province, sees the same ones, standing about one zhang (equal to 3.1 meters [just over 10 feet]) in height and smiling to the people they come across, and are called 'shandaren' (men as big as mountains), 'wildmen,' or 'shanxiao'" (Zhou, G. 1982, p. 13).

In the eighteenth century, the Chinese poet Yuan Mei made reference to strange creatures inhabiting the wild regions of Shanxi province, calling them "monkeylike, yet not monkeylike" (Yuan and Huang 1979, p. 57).

According to Zhou: "Even today, in the area of Fang County, Hubei Province, there are still legends about 'maoren' (hairy men) or 'wildmen.' A local chronicle, about 200 years old, says that 'the Fang mountain lying 40 li (2 li equals one kilometer [.62 mile]) south to the county town is precipitous and full of holes, where live many maoren, about one zhang high and hair-coated. They often come down to eat human beings and chickens and dogs, and seize those who fight with them.' A lantern on which there is an ornament of a 'maoren' figure was unearthed in this area during an archaeological excavation. It has been dated at 2,000 years" (Zhou, G. 1982, pp. 13–14).

There have been many other reports of wildmen from the Hubei province in central China. In 1922, a militiaman is said to have captured a wildman, but there are no further records of this incident (Poirier *et al.* 1983, p. 32).

In 1940, Wang Zelin, a graduate of the biology department of Northwestern University in Chicago, was able to directly see a wildman shortly after it was shot to death by hunters. Wang was driving from Baoji, in Shanxi Province, to Tianshui, in Gansu Province, when he heard gunfire ahead of him. He got out of the car to satisfy his curiosity and saw a corpse. It was a female creature, six and a half feet tall and covered with a coat of thick greyish-red hair about one and a quarter inches long. The hair on its face was shorter. The cheek bones were prominent, and the lips jutted out. The hair on the head was about one foot long. According to Wang, the creature looked like a reconstruction of the Chinese *Homo erectus* (Yuan and Huang 1979, p. 57; Shackley 1983, pp. 79–82).

Ten years later, another scientist, Fun Jinquan, a geologist, saw some living wildmen. Zhou Guoxing stated: "With the help of local guides, he watched, at a safe distance, two local Wildmen in the mountain forest near Baoji County, Shanxi Province, in the spring of 1950. They were mother and son, the smaller one being 1.6 meters [5.25 feet] in height. Both looked human" (Zhou, G. 1982; p. 14).

In 1957, a middle-school teacher of biology in Zhejiang province obtained the hands and feet of a "manbear" killed by local peasants. Zhou Guoxing wrote: "In December 1980, I went to Sui Chang to study these hand and foot specimens. I concluded beyond any doubt, that they belong to a higher primate, and have morphological traits of both ape and monkey. The eyewitnesses thought that they had belonged to a Wildman, or of a manlike 'strange animal,' but after examining the specimens, I determined that they were not the hands and feet of a Wildman. They might possibly belong to an enormous monkey (perhaps of a species of macaque not previously recorded in this area). . . . There is no denying the possibility that they came from an unknown primate in the Jiolong Mountain area" (Zhou, G. 1982, p. 18).

Talk of the existence of an enormous monkey, previously unknown, raises interesting questions about the Beijing *Homo erectus* finds. Beijing man, as generally portrayed in textbooks and films, was quite human and almost civilized, a Middle Pleistocene hunter, fire maker, and cave dweller. But as several dissenting scientists noted, most of the Beijing man fossils were thick, big-browed partial skulls with smashed braincases. They appeared to represent not a relatively advanced protohuman but rather the unfortunate animallike prey of some more intelligent hominid. Perhaps the large, hitherto unknown species of macaque posited by Zhou in order to explain away some modern wildman evidence also inhabited China during the Middle Pleistocene, and the smashed skulls in the Zhoukoudian cave belonged to it. Or perhaps the broken skulls of Zhoukoudian belonged to the *Homo erectus*-like creature described above by Wang Zelin.

In 1961, workers building a road through the heavily forested Xishuang Banna region of Yunnan province in southernmost China reported killing a humanlike female primate. The creature was 1.2–1.3 meters (about 4 feet) tall and covered with hair. It walked upright, and according to the eyewitness reports, its hands, ears, and breasts were like those of a female human. The Chinese Academy of Sciences sent a team to investigate, but they were not able to obtain any physical evidence. Some suggested that the workers had come upon a gibbon. But Zhou Guoxing stated: "The present author recently visited a newsman who took part in that investigation. He stated that the animal which had been killed was not a gibbon, but an unknown animal of human shape. It is worth noting that, over the past 2 years or so, some people in the western border areas of Yunnan Province say that the above-mentioned kind of Wildman still move about, and that another one has since been killed" (Zhou, G. 1982, pp. 15–16).

In 1976, six cadres from the Shennongjia forestry region in Hubei province were driving at night down the highway near the village of Chunshuya, between Fangxian county and Shennongjia. On the way, they encountered a "strange tailless creature with reddish fur" (Yuan and Huang 1979, p. 56). Fortunately, it stood still long enough for five of the people to get out of the car and look at it from a distance of only a few feet, while the driver kept his headlights trained on it. The observers were certain that it was not a bear or any other creature with which they were familiar. They reported the incident in a telegram to the Chinese Academy of Sciences in Peking.

Over the years, Academy officials had received many similar reports from the same region of Hubei province. So when they heard about this incident, they decided to thoroughly investigate the matter. A scientific expedition consisting of more than 100 members proceeded to Hubei province. They collected physical evidence, in the form of hair, footprints, and feces, and recorded sightings by the local inhabitants (Yuan and Huang 1979). Subsequent research has added to these results.

Altogether, more than a thousand footprints have been found in Hubei province, some more than 19 inches long (Poirier *et al.* 1983, p. 34). Over 100 hairs have been collected, the longest measuring 21 inches. Some of the hairs were supplied by persons who claimed to have seen wildmen; others were taken from trees against which wildmen were said to have rubbed. Frank E. Poirier, an anthropologist at Ohio State University, reported (Poirier *et al.* 1983, p. 33): "The hair was studied by the Hubei Provincial Medical College and the Institute of Vertebrate Paleontology and Paleoanthropology in Beijing. The general consensus is that the hair belongs to a higher primate (monkey, ape, or human)."

Some have sought to explain sightings of wildmen in the Shennongjia region of Hubei province as encounters with the rare golden monkey, which inhabits the same area. The golden monkey might very well account for reports of creatures glimpsed for a moment at a great distance. But consider the case of Pang Gensheng, a local commune leader, who was confronted in the forest by a wildman.

Pang, who stood face to face with the creature, at a distance of five feet for about an hour, said: "He was about seven feet tall, with shoulders wider than a man's, a sloping forehead, deep-set eyes and a bulbous nose with slightly upturned nostrils. He had sunken cheeks, ears like a man's but bigger, and round eyes, also bigger than a man's. His jaw jutted out and he had protruding lips. His front teeth were as broad as a horse's. His eyes were black. His hair was dark brown, more than a foot long and hung loosely over his shoulders. His whole face, except for the nose and ears, was covered with short hairs. His arms hung down to below his knees. He had big hands with fingers about six inches long and thumbs only slightly separated from the fingers. He didn't have a tail and the hair on his body was short. He had thick thighs, shorter than the lower part of his leg. He walked upright with his legs apart. His feet were each about 12 inches long and half that broad—broader in front and narrow behind, with splayed toes" (Yuan and Huang 1979, pp. 58–59).

Zhou Guoxing has suggested that the wildman of Hubei province might be a relict population of *Gigantopithecus,* a large apelike hominid that inhabited southern China during the Middle Pleistocene. Zhou noted that in the forests of Hubei province some types of trees from the Tertiary have survived, as have the panda and other mammals from the Middle Pleistocene (Zhou, G. 1982, p. 22).

10.10 WILDMEN OF MALAYSIA AND INDONESIA

In 1969, John McKinnon, who journeyed to Borneo to observe orangutans, came across some humanlike footprints. McKinnon asked his Malay boatman what made them. "Without a moment's hesitation he replied 'Batutut,'" wrote McKinnon, "but when I asked him to describe the beast he said it was not an animal but a type of ghost. . . . Batutut, he told me, is about four feet tall, walks upright like a man and has a long black mane. . . . Like other spirits of the forest the creature is very shy of light and fire" (Green 1978, p. 134).

Later, in Malaya, McKinnon saw some casts of footprints even bigger than those he had seen in Borneo, but he recognized them as definitely having been made by the same kind of creature. The Malayans called it Orangpendek (short fellow). McKinnon stated: "Again natives spoke of a creature with long hair, who walks upright like a man. Drawings and even photographs of similar footprints found in Sumatra are attributed to the Sedapa or Umang, a small, shy, long-haired, bipedal being living deep in the forest" (Green 1978, pp. 134–135). According to Ivan Sanderson, these footprints differ from those of the anthropoid apes inhabiting the Indonesian forests (the gibbon, siamang, and orangutan). They are also distinct from those of the sun bear (Sanderson 1961, p. 219).

Early in the twentieth century, L. C. Westenek, a governor of Sumatra, received a written report about an encounter with a Sedapa wildman. The overseer of an estate in the Barisan Mountains, along with some workers, observed the Sedapa from a distance of 15 yards. The overseer said he saw "a large creature, low on its feet, which ran like a man, and was about to cross my path; it was very hairy and it was not an orang-utan; but its face was not like an ordinary man's. It silently and gravely gave the men a disagreeable stare and then ran calmly away" (Sanderson 1961, pp. 216–217).

In a journal article about wildmen published in 1918, Westenek recorded a report from a Mr. Oostingh, who lived in Sumatra. Once while proceeding through the forest, he came upon a man sitting on a log and facing away from him. Oostingh stated: "I saw that he had short hair, cut short, I thought; and I suddenly realised that his neck was oddly leathery and extremely filthy. 'That chap's got a very dirty and wrinkled neck!' I said to myself. His body was as large as a medium-sized native's and he had thick square shoulders, not sloping at all. . . . he seemed to be quite as tall as I (about 5 feet 9 inches). Then I saw that it was not a man."

"It was not an orang-utan," declared Oostingh. "I had seen one of these large apes a short time before." What was the creature if not an orangutan? Oostingh could not say for sure: "It was more like a monstrously large siamang, but a siamang has long hair, and there was no doubt that it had short hair" (Sanderson 1961, p. 220).

In 1918, Mr. Van Heerwarden, a hunter, began finding tracks of the Sedapa in Sumatra. The footprints he saw were shaped like those of a small human being. Van Heerwarden also heard reports about the Sedapa from natives. In October of 1923, he himself spotted one in a tree: "I discovered a dark and hairy creature on a branch. . . . The *sedapa* was also hairy on the front of its body; the colour there was a little lighter than on the back. The very dark hair on its head fell to just below the shoulder-blades or even almost to the waist. . . . Had it been standing, its arms would have reached to a little above its knees; they were therefore long, but its legs seemed to me rather short. I did not see its feet, but I did see some toes which were shaped in a very normal manner. . . . There was nothing repulsive or ugly about its face, nor was it at all apelike" (Sanderson 1961, pp. 222–223). After observing it for a while, Van Heerwarden allowed the creature to run away.

The presence of large humanlike creatures in the forests of the Indonesian archipelago is relevant to the dating of fossils of *Homo erectus* found in Java (Chapter 7). Paleoanthropologists assume that fossils displaying *Homo erectus* morphology must be 800,000 or more years old, even when they are found on the surface. In fact, almost all the fossils of *Homo erectus* from Java have been surface finds. But if creatures resembling *Homo erectus* are still roaming the forests of Indonesia, then the general practice of dating a fossil by its morphology is not secure. On January 20, 1986, Bernard Heuvelmans wrote in response to a letter from our researcher, Stephen Bernath: "I am convinced myself that fossils, especially remains of Hominoids, are dated not after the strata they have been found in but after a prejudiced idea of the strata they should have been found in according to the classical scheme of human evolution, which is completely wrong."

In practice, morphological dating has worked like this. During the 1930s, many fossil skull fragments were found lying on the surface at various locations in Java. The formations in the region have since been dated by the potassium-argon method to the Middle Pleistocene. The potassium-argon method, it may be recalled, is used to date volcanic materials, not the bones themselves. Ideally, for a bone to be assigned a Middle Pleistocene date using this method, it should have been found lying beneath an undisturbed layer of volcanic material. But in Java this was not the case, for almost without exception the fossils labeled *Homo erectus* were found lying on the surface or in unspecified locations. Scientists have simply assumed that the bones eroded from Middle Pleistocene formations, in which they had supposedly been deposited hundreds of thousands of years ago. The reason scientists feel comfortable in making this assumption is that they are certain that hominids with *erectus* morphology have been extinct since the Middle Pleistocene. But perhaps not. Heuvelmans stated in his letter of January 20, 1986: "small hairy hominoids, with long straight hair on the head, the *Nittaewo,* were exterminated at the end of the 18th century in Sri Lanka. According to British leading primatologist Osman Hill (1945), these dwarfs could be modern representatives of *Homo erectus.*"

Such evidence makes possible another explanation for *erectus*-like fossils found on the surface of Middle Pleistocene formations, in Java and elsewhere. As far as Java is concerned, perhaps as little as 10,000 years ago, a Sedapalike creature died by a stream bed or lake shore, and its bones became fossilized in the sediments. In very recent times, a piece of the fragmented skull reappeared on the surface, where it was discovered by native collectors, who turned it over to a paleoanthropologist. Upon seeing its primitive *erectus*-like morphology, the paleoanthropologist assigned it to the Middle Pleistocene, giving it a date of 800,000 years or more. The fossil was then described in textbooks and cited as more proof for the hypothesis that modern human beings evolved over the past several hundred thousand years from more apelike ancestors. But the fossil may not actually belong to the Middle Pleistocene. It could in fact be much more recent.

Whether the Java *Homo erectus* fossils are recent or ancient, the existence of living *erectus*-like creatures (or recently living ones) in Java shows the coexistence of such creatures with humans in modern times. And, as we have seen in previous chapters, there is much evidence such creatures coexisted with humans in the distant past. This throws the accepted pattern of human evolution into complete confusion.

Paleoanthropologists, in the face of such evidence, or in ignorance of it, will insist that human beings of modern type could not have existed any earlier than one hundred thousand years ago, and certainly not in the Early Pleistocene, the Pliocene, or the Miocene.

But if there is uncertainty about what kinds of hominids may be around today, how can we be so sure about what kinds of hominids may or may not have been around in the distant past?

Empiric investigation of the fossil record may not be a sure guide. As Heuvelmans stated in a letter (April 15, 1986) to our researcher Stephen Bernath: "do not overestimate the importance of the fossil record. Fossilization is a very rare, exceptional phenomenon, and the fossil record cannot thus give us an exact image of life on earth during the past geological periods. The fossil record of primates is particularly poor because very intelligent and cautious animals can avoid more easily the very conditions of fossilization—such as sinking in mud or peat, for instance."

The empiric method undoubtedly has its limitations, and the fossil record is incomplete and imperfect. But when all the fossil evidence, including that for very ancient humans and living ape-men, is objectively evaluated, the pattern that emerges is one of coexistence rather than sequential evolution.

10.11 AFRICA

Native informants from several countries in the western part of the African continent, such as the Ivory Coast, have given accounts of a race of pygmylike creatures covered with reddish hair. Europeans have also encountered them: "During one of his expeditions in the course of 1947, the great elephant-hunter Dunckel killed a peculiar primate unknown to him; it was small with reddish-brown hair and was shot in the great forest . . . between the Sassandra and Cavally rivers" (Sanderson 1961, p. 189). Natives are said to have bartered with these red-haired pygmies, called Sehites, leaving various trinkets in exchange for fruits (Sanderson 1961, p. 190).

Wildman reports also come from East Africa. Capt. William Hitchens reported in the December 1937 issue of *Discovery:* "Some years ago I was sent on an official lion-hunt in this area (the Ussure and Simibit forests on the western side of the Wembare plains) and, while waiting in a forest glade for a man-eater, I saw two small, brown, furry creatures come from dense forest on one side of the glade and disappear into the thickets on the other. They were like little men, about 4 feet high, walking upright, but clad in russet hair. The native hunter with me gazed in mingled fear and amazement. They were, he said, *agogwe,* the little furry men whom one does not see once in a lifetime" (Sanderson 1961, p. 191). Were they just apes or monkeys? It does not seem that either Hitchens or the native hunter accompanying him would have been unable to recognize an ape or monkey. Many reports of the Agogwe emanate from Tanzania and Mozambique (Green 1978, p. 133).

From the Congo region come reports of the Kakundakari and Kilomba. About 5.5 feet tall and covered with hair, they are said to walk upright like humans. Charles Cordier, a professional animal collector who worked for many zoos and museums, followed tracks of the Kakundakari in Zaire in the late 1950s and early 1960s. Once, said Cordier, a Kakundakari had become entangled in one of his bird snares. "It fell on its face," said Cordier, "turned over, sat up, took the noose off its feet, and walked away before the nearby African could do anything" (Green 1978, p. 133).

Reports of such creatures also come from southern Africa. Pascal Tassy (1983, pp. 132–133), of the Laboratory of Vertebrate and Human Paleontology, wrote in a review of Heuvelmans's *Les Bêtes Humaines d'Afrique* (which has a chapter on relict australopithecines): "Philip V. Tobias, now on the Board of Directors of the International Society of Cryptozoology, once told Heuvelmans that one of his colleagues had set traps to capture living australopithecines." Tobias, from South Africa, is a recognized authority on *Australopithecus.*

According to standard views, the last australopithecines perished approximately 750,000 years ago, and *Homo erectus* died out around 200,000 years ago.

The Neanderthals, it is said, vanished about 35,000 years ago, and since then fully modern humans alone have existed throughout the entire world. Yet many sightings of different kinds of wildmen in various parts of the world strongly challenge the standard view. Also, recent fossil skulls reportedly display anomalously primitive features. For example, *Nature* (1908, vol. 77, p. 587) published a report by Dr. K. Stolyhwo on a recent Neanderthal skull, found as part of a skeleton in a tomb that also contained a suit of chain armor and iron spearheads. Stolyhwo said the skull was similar to the Spy Neanderthal skull.

10.12 MAINSTREAM SCIENCE AND WILDMAN REPORTS

Despite all the evidence we have presented, most recognized authorities in anthropology and zoology decline to discuss the existence of wildmen. If they mention wildmen at all, they rarely present the really strong evidence for their existence, focusing instead on the reports least likely to challenge their disbelief.

Skeptical scientists say that no one has found any bones of wildmen; nor, they say, has anyone produced a single body, dead or alive. But as we have seen, hand and foot bones of wildmen, and even a head, have been collected. Competent persons report having examined bodies of wildmen. And there are also a number of accounts of capture. That none of this physical evidence has made its way into museums and other scientific institutions may be taken as a failure of the process for gathering and preserving evidence. The operation of what we could call a knowledge filter tends to keep evidence tinged with disrepute outside official channels.

However, some scientists with solid reputations, such as Krantz, Napier, Shackley, Porshnev, and others, have found in the available evidence enough reason to conclude that wildmen do in fact exist, or, at least, that the question of their existence is worthy of serious study.

Myra Shackley wrote to our researcher Steve Bernath on December 4, 1984: "As you know, this whole question is highly topical, and there has been an awful lot of correspondence and publication flying around on the scene. Opinions vary, but I guess that the commonest would be that there is indeed sufficient evidence to suggest at least the possibility of the existence of various unclassified manlike creatures, but that in the present state of our knowledge it is impossible to comment on their significance in any more detail. The position is further complicated by misquotes, hoaxing, and lunatic fringe activities, but a surprising number of hardcore anthropologists seem to be of the opinion that the matter is very worthwhile investigating."

So there is some scientific recognition of the wildman evidence, but it seems to be largely a matter of privately expressed views, with little or no official recognition.

11

Always Something New Out of Africa

The controversies surrounding Java man and Beijing man, what to speak of Castenedolo man and the European eoliths, have long since subsided. As for the disputing scientists, most of them are in their graves, their bones on the way to disintegration or fossilization. But today Africa, the land of *Australopithecus* and *Homo habilis,* remains an active battlefield, with scientists skirmishing to establish their views on human origins.

Only in the later decades of the twentieth century did paleoanthropologists shift the main focus of their discipline from Europe and Asia to Africa. But the importance of Africa was foreseen by Darwin (1871, p. 199), who wrote in *The Descent of Man*: "In each great region of the world the living mammals are closely related to the extinct species of the same region. It is, therefore, probable that Africa was formerly inhabited by extinct apes closely allied to the gorilla and chimpanzee; and as these two species are man's nearest allies, it is somewhat more probable that our early progenitors lived on the African continent than elsewhere."

In this chapter, we survey the history of paleoanthropological discoveries in Africa. The finds from the early part of the twentieth century, such as Reck's skeleton (Section 11.1) and the Kanjera skulls and Kanam jaw (Section 11.2), were controversial. According to their discoverers, these fossils represented evidence for anatomically modern humans in the Early Pleistocene. Anomalous finds continued to occur even in the latter part of the twentieth century. Among these we may number the Kanapoi humerus (Section 11.5.1), the ER 1481 femur from Lake Turkana (Section 11.6.3), and the Laetoli footprints (Section 11.10). Scientists have said all of them are morphologically within the modern human range. But instead of taking these fossils as evidence for anatomically modern humans in unexpectedly ancient times, scientists have generally said they show that protohuman creatures such as *Australopithecus* and *Homo habilis* had skeletal features resembling those of modern humans. Indeed, most scientists

627

have consistently depicted *Australopithecus* and *Homo habilis* as essentially human below their apelike heads. They also say that these creatures were exclusively terrestrial and bipedal in the human fashion. But there is much evidence that this view is mistaken, and that the australopithecines and habilines were very well adapted for life in the trees (Section 11.8).

11.1 RECK'S SKELETON

The first significant African discovery related to human origins and antiquity occurred in 1913. In that year, Professor Hans Reck, of Berlin University, conducted investigations at Olduvai Gorge in Tanzania, then German East Africa. During his stay at Olduvai Gorge, Reck found a human skeleton that would remain a source of controversy for decades.

11.1.1 The Discovery

While one of Reck's African collectors was searching for fossils on the northern slope of Olduvai Gorge, he spotted a piece of bone sticking up from the earth near a bush (Wendt 1955, p. 418). After clearing away the surface rubble, the collector saw parts of a complete and fully human skeleton embedded in the rock. He summoned Reck, who then had the skeleton taken out in a solid block of hardened sediment. The human skeletal remains, including a complete skull (Figure 11.1), were strongly cemented in the surrounding matrix, which had to be chipped with hammers and chisels (MacCurdy 1924a, p. 423).

Reck identified a sequence of five beds at Olduvai Gorge. The first four beds are water-laid volcanic tuffs of various colors. Bed I is grey and yellow. Bed II is generally of a buff color, although the upper portion has a reddish tint. Bed III is bright red, while Bed IV is grey, or brownish. Bed V, a loesslike deposit, is brownish (Hopwood 1932, p. 192). At the top and base of Bed V are hard whitish layers of a limestonelike deposit of calcrete, or steppe-lime. The sequence of beds (Table 11.1) outlined by Reck is still in use today, except that upper Bed IV is now referred to as the Masek formation and Bed V has been divided into several distinct formations (M. Leakey 1978, p. 3). From oldest to youngest they are the Lower Ndutu, Upper Ndutu, and Naisiusiu formations (Oakley *et al.* 1977, p. 169).

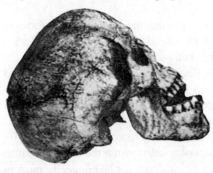

Figure 11.1. This skull is from a fully human skeleton found in 1913 by H. Reck at Olduvai Gorge, Tanzania (Reck 1933, plate 31).

TABLE 11.1

Stratigraphy of Olduvai Gorge, Tanzania

Est. Age (Years B.P.)	
	Calcrete layer
	Naisiusiu
32,000	
	Upper Ndutu
60,000	
	Lower Ndutu
400,000	Calcrete layer
	Masek (formerly part of IV)
600,000 2,000,000	
700,000	——— III/IV ———
1,150,000	
	II
1,700,000	
	I
2,000,000	

After Oakley *et al.* (1977, pp. 166–169).

The skeleton was from the upper part of Bed II. Just below the skeleton were fossils of *Elephas antiquus recki* (Hopwood 1932, p. 192). To Reck, the faunal evidence indicated the human skeleton was of Middle Pleistocene age, roughly contemporary with Dubois's Java man, now thought to be about 800,000 years old. Modern dating methods, however, give the uppermost part of Bed II a late Early Pleistocene date of around 1.15 million years (Oakley *et al.* 1977, p. 166).

The overlying layers were not, however, intact. The skeleton had been found on the side of Olduvai Gorge, about 3 or 4 meters (10 to 13 feet) below the level of the plain (Protsch 1974, p. 379). Here (Figure 11.2) the overlying layers (Beds III, IV, and V) had been worn by erosion. Bed II was, however, still covered by rubble from bright red Bed III and from Bed V (Hopwood 1932, p. 194). It was clear to Louis Leakey (1932b) that perhaps as little as 50 years ago, the site would have been covered by "a small relic of Bed 3 overlain by Bed 5," the latter containing hard layers of calcrete. Beds III and V were present on the slope just above the spot where the skeleton was found. Bed IV was missing in the immediate area, apparently removed by erosion before the deposition of Bed V.

Reck, understanding the significance of his find, carefully considered the possibility that the human skeleton had arrived in Bed II through burial or earth movements. He determined this was not the case. Reck (1914b) said: "The bed in which the human remains were found, without any accompanying cultural objects, showed no sign of disturbance. The spot appeared exactly like any other

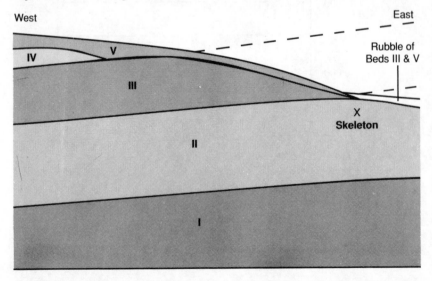

Figure 11.2. This section of the northern slope of Olduvai Gorge (after Hopwood 1932, p. 192) shows the location where H. Reck found a fully human skeleton in 1913 in upper Bed II. Bed II is 1.15–1.7 million years old (Oakley *et al.* 1977, p. 166).

in the horizon. There was no evidence of any refilled hole or grave" (Hopwood 1932, p. 193).

Later, Reck (1926) provided this account: "In some graves, the existence of refilling can be obscured by tamping down the ground and so forth. But artificial disturbance of the strata also results in the mixing together of different kinds of excavated earth. This should be quite evident in the present case. . . . But there was no sign of mixing of earth of different colors, nor were there any fragments of calcrete found mixed into the earth by the skeleton. Nothing of the sort was observed during the inspection of the original site and its surroundings at Olduvai nor in later examination of the matrix in which the skeleton was encased during transport to Germany" (Reck 1926, pp. 85–86; Hopwood 1932, p. 193). According to Reck, the strata at the site had not undergone any geological resorting by which, as Wendt (1955, p. 420) suggested, a recent layer, containing the skeleton, might have been forced into an older one.

In an unpublished manuscript, Reck observed: "The sediment . . . is so constituted that the artificial breaking of the bed with its visible layering by the digging of a grave would necessarily be recognizable. The wall of the grave would have a definite border, an edge that would show in profile a division from the undisturbed stone. The grave filling would show an abnormal structure and heterogeneous mixture of excavated materials, including easily recognizable pieces of calcrete. Neither of these signs were to be found despite the most attentive inspection. Rather the stone directly around the skeleton was not distinguishable from the neighboring stone in terms of color, hardness, thickness of layers, structure, or order" (Hopwood 1932, pp. 193–194).

In his first report, Reck (1914a) observed: "The skeleton in the grave was complete, though somewhat shifted and compressed [Figure 11.3]. It lay in horizontal position, exactly parallel to the layers of stone in which it was embedded, just as were all the faunal remains" (Hopwood 1932, p. 193). This sheds light on the question of burial. Beds I through IV at Olduvai were laid down by water in a lake bed, producing a distinct sequence of thin layers. Before the deposition of the overlying Ndutu Beds, starting 400,000 years ago (Oakley *et*

Figure 11.3. The skull found by H. Reck in Bed II of Olduvai Gorge was distorted (Reck 1933, plate 30). W. O. Dietrich (1933) believed this distortion argued against the skeleton being a recent shallow burial.

al. 1977, p. 166), faulting tilted the Olduvai strata in an east-west direction. If the skeleton had been buried in Bed II fairly recently, it probably would have intersected at an angle the layers of Bed II, here tilted about 7 degrees (Hopwood 1932, p. 192).

The skeleton's distortion by compression was also significant (Figure 11.3, p. 629). W. O. Dietrich, writing in 1933, stated that this feature of the skeleton argued against its being a recent, shallow burial in the top of Bed II. Its condition indicated a substantial accumulation of sediments had once covered it (Dietrich 1933, pp. 299–303). According to Reck, the deposition of the skeleton took place during the formation of Bed II. Later, the full weight of Beds III and IV would have covered the skeleton. Still later, after the erosion of Bed IV, Beds III and V would have covered the skeleton. All in all, the skeleton's condition and stratagraphic position appeared to rule out recent burial.

Reck returned to Germany, carrying Olduvai man's skull with him. He left the rest of the skeleton, encased in a block of Bed II sediment, to be shipped with the expedition's baggage.

Upon Reck's return to Germany, his African discovery attracted immediate attention, both in the popular press and in scientific circles. A leading American anthropologist, George Grant MacCurdy (1924a, p. 423) of Yale University, considered Reck's discovery to be genuine: "The human skeleton . . . came from the next to the lowest horizon (No. 2). . . . The skeleton was found some 3 or 4 meters (10 to 13 feet) below the rim of the Oldoway gorge, which here is about 40 meters [131 feet] deep. The skeleton bore the same relation to the stratified beds as did the other mammalian remains, and was dug out of the hard clay tufa with hammer and chisel just as these were. In other words, the conditions of the find were such as to exclude the possibility of an interment. The human bones are therefore as old as the deposit (No. 2)." He also agreed that the skeleton was of modern type: "Judging from the photograph of the skeleton still *in situ*, the man of Oldoway gorge did not belong to the Neandertal, but rather to the Aurignacian type" (MacCurdy 1924a, p. 423). Aurignacian refers to Cro-Magnon man, the first representative of *Homo sapiens sapiens* in Europe.

11.1.2 Leakey's Conversion

Louis Leakey (1928, p. 499) examined Reck's skeleton in Berlin, but he initially judged it more recent than Reck had claimed. Other scientists agreed.

In 1931, Leakey and Reck, attempting to settle the issue, visited the site where the skeleton had been found. Along with them were A. T. Hopwood of the British Museum of Natural History, Donald MacInnes, and geologist E. V. Fuchs. After studying the geology, Leakey and Hopwood were won over to Reck's point of view. Leakey was also influenced by new discoveries of stone implements in

Beds I and II of Olduvai Gorge. As we have seen, Reck originally reported that no cultural remains were found in Bed II, a fact that had caused Leakey to judge the skeleton not very old (Goodman 1983, p. 107).

In a letter published in *Nature,* the prestigious British science review, Leakey, Hopwood, and Reck confirmed that the skeleton was not buried from Bed IV, as Leakey had suggested in his book *The Stone Age Cultures of Kenya Colony* (1931), but was native to Bed II, as originally reported by Reck. They concluded that the skeletal remains belonged to an anatomically modern *Homo sapiens* who had lived during Africa's Upper Kamasian pluvial (rainy) period (L. Leakey *et al.* 1931), equivalent to the Mindel glacial period of the European middle Middle Pleistocene. This made Reck's skeleton roughly contemporary with Beijing man and Java man, both from the Middle Pleistocene. But, as previously mentioned, uppermost Bed II at Olduvai Gorge is now given a late Early Pleistocene age of 1.15 million years. By modern accounts, *Homo sapiens sapiens* is not thought to be more than 100,000 years old, although some specimens regarded as early *Homo sapiens* are dated at around 300,000 years.

In an article published in the *Times* of London, Leakey stated that his firsthand research in Africa had established "almost beyond question that the skeleton of a human being found by Professor Reck in 1913 is the oldest authentic skeleton of *Homo sapiens*" (Goodman 1983, p. 107). This led Leakey to announce that Beijing man and Java man were not direct human ancestors. How could they be, when Reck's skeleton, fully human, was just as old as they were?

Hopwood later published his own account of the 1931 expedition to Olduvai. Hopwood (1932, p. 193) stated: "Examination of the site in 1931 confirmed the observation that the bed in which the skeleton lay was undoubtedly, Bed II." Hopwood (1932, p. 194) added: "The slope is covered by rubble from Beds III and V in such a manner that it is difficult to see how a shallow grave could be dug and filled again without including some of this rubble."

From his study of the stratigraphy and the rate of erosion, Hopwood concluded that as little as 250 years ago "the place where the skeleton lay would certainly have been covered by the lower hard layer [of Bed V calcrete], which is ten to twelve inches thick." Hopwood pointed out that the calcrete layers at the site were extremely hard. He once saw laborers working with heavy crow bars take two full days to dig a hole just 2 feet square and 3 feet deep through similar material. The nearly impenetrable character of the calcrete appeared to rule out burial (Hopwood 1932, p. 194). Furthermore, the Bed II sediments themselves were quite hard at that point. The skeleton found by Reck in 1913 had to be extracted with hammers and chisels.

After reproducing statements from Reck's original reports, Hopwood (1932, p. 194) stated: "It is clear that Professor Reck, when he found the skeleton, thought it possible that he might be dealing with an intrusive burial, that he was careful to look for evidence for this, and that he failed to find it."

Hopwood (1932, p. 195) concluded: "it seems to follow from the original evidence of Professor Reck that the skeleton lay in undisturbed sediment without trace of foreign matter. The ethnological evidence appears to show, that despite physical resemblances, the skeleton is not of the Masai, who inhabit the country today, and that in pre-Masai days the actual part of the bed was in such a position that it was inaccessible to a tribe only with native tools. Hence the conclusion of my colleagues and myself that the skeleton was enclosed in Bed II before that bed was covered by later deposits; and in that sense we regard the skeleton as contemporary with Bed II."

Around this time, Sir Arthur Keith, who initially thought Reck's skeleton recent, also adopted the Bed II date. But not everyone agreed with the conclusion that Leakey and Hopwood reached after their 1931 expedition.

11.1.3 Cooper and Watson Launch Their Attack

In February of 1932, *Nature* printed a letter by zoologists C. Forster Cooper of Cambridge and D. M. S. Watson of the University of London. They suggested that the completeness of the skeleton found by Reck clearly indicated it was a recent burial (Cooper and Watson 1932a, p. 312).

Cooper and Watson (1932a, p. 312) stated: "Complete mammalian skeletons of any age are, as field palaeontologists know, of great rarity. When they occur, their perfection can usually be explained as the result of sudden death and immediate covering by volcanic dust." Even here, Cooper and Watson admitted that examples of complete, naturally-deposited skeletons, although rare, do in fact occur. They gave one circumstance for such an occurrence and indicated there might be others.

Cooper and Watson, casting further doubt on the claimed age for Reck's skeleton, contended that no one had yet found anatomically modern human skeletal remains anywhere near as old. They dismissed the Galley Hill skeleton, claiming it was "never seen *in situ* by any trained observer" (Cooper and Watson 1932a, p. 312). This is not unreasonable. As far as we can tell, the Galley Hill skeleton could be recent, but there is also evidence suggesting it could be from the middle Middle Pleistocene (Section 6.1.2.1). Cooper and Watson then mentioned the Ipswich skeleton, observing that it had "been withdrawn by its discoverer." While it is true that J. Reid Moir did change his mind about the age of the Ipswich skeleton, our own study (Section 6.1.3) shows that there is still reason to think it might be from the middle Middle Pleistocene.

Cooper and Watson (1932a, p. 312) then referred obliquely to "other fragments, found long ago . . . entirely without satisfactory evidence as to their mode of occurrence." They ignored (or were ignorant of) the finds at Castenedolo, Italy (Section 6.2.2). There G. Ragazzoni, a professional geologist, found *in situ*, in a Pliocene formation, a fairly complete and anatomically modern human skeleton, as well as parts of others.

In May 1932, Leakey replied to Cooper and Watson. In a letter to *Nature*, he argued that no more than 50 years ago the reddish-yellow upper part of Bed II would have been covered by an intact layer of bright red Bed III. If the skeleton had been buried in recent times (50 or more years ago), there should have been a mixture of bright red and reddish-yellow sediments in the grave filling. Such was not the case. "I was lucky enough personally to examine the skeleton at Munich while it was still intact in its original matrix," wrote Leakey, "and could detect no trace whatever of such admixture or disturbance." He added: "The bones of the skeleton . . . are, as far as I know, every bit as mineralized as most of the bones from Bed No. 2 itself" (L. Leakey 1932a, p. 721). This would argue against their being very recent.

Leakey, however, agreed with Cooper and Watson that Reck's skeleton had arrived in its position in Bed II by burial, but he did not think the burial was recent. "My own personal belief," wrote Leakey, "is that contemporary man, living on the edge of the then existing Oldoway lake, buried the skeleton into the muddy, clayey edge of the lake whilst Bed No. 2 was in the process of being deposited, for Bed No. 2 is essentially a shallow water deposit at the place where the skeleton was found" (L. Leakey 1932a, p. 721). Reck, on the other hand, believed that the individual had drowned and been covered by sedimentation.

Some scientists had called attention to apparent filing of the teeth of Reck's skeleton, suggesting this was characteristic of the tribal people inhabiting the region during recent historical times. To this Leakey replied: "I have personally examined the so-called 'filing' of the teeth of the Oldoway man on the original specimen at Munich, and this 'filing' has no resemblance to any filing done by native tribes to-day, and it is, to my mind, exceedingly doubtful if it can be called filing at all" (L. Leakey 1932a, p. 721).

Leakey then referred to his own finds at Kanam and Kanjera (Section 11.2), which he believed supported the Middle Pleistocene antiquity of Reck's skeleton. "Actually *in situ* at a place called Kanam," stated Leakey, "in the same horizon as the Pre-Chellean tools and the *Deinotherium,* we found a fragment of a mandible of *Homo sapiens* type, thus putting *Homo sapiens* in East Africa back one stage further than Oldoway Man—in fact, in deposits of the same age as Bed No. 1 at Oldoway" (L. Leakey 1932a, p. 722). The upper part of Bed I at Olduvai is now thought to be about 1.7–1.8 million years old (Oakley *et al.* 1977, p. 166).

About the Kanjera finds, Leakey reported: "We have . . . found fragments of the skulls of three different individuals of *Homo sapiens* type completely mineralised and just washed out of the exposures by the rains. They are in the same state of complete mineralisation as the remains of *Elephas antiquus,*

Hipparion, etc., from the same beds, and I have personally no doubt whatever that they were *in situ* a month or two ago, before the beginning of the present rainy season. These later remains are probably, then, the contemporary of the Oldoway skeleton, and since we have fragments which make up the greater part of the skull cap of one of the [Kanjera] individuals, an interesting comparison will be possible later on" (L. Leakey 1932a, p. 722).

C. Forster Cooper and D. M. S. Watson were still not satisfied. In June 1932, they said in a letter to *Nature* that red pebbles from Bed III may perhaps have been discolored. "Mere proximity to a large decaying body often alters the character of a matrix," said Cooper and Watson (1932b). This would explain why Reck and Leakey did not see the Bed III pebbles in the matrix surrounding the skeleton. Hopwood, however, disagreed that Bed III pebbles would have lost their bright red color. He pointed out that the top of Bed II, in which the skeleton was found, was also reddish and stated: "The reddish colour of the matrix is against the theory that any inclusions of Bed III would have been decolorised by decomposition products" (Hopwood 1932, p. 194).

In support of their post-Bed II burial hypothesis, Cooper and Watson offered additional explanations for the absence of Bed III materials in the supposed grave filling. According to Cooper and Watson (1932b), the grave diggers would have taken the red Bed III materials out first and thrown them back in last, on the top. This would explain why no Bed III materials were present in the matrix immediately surrounding the skeleton in Bed II. But this hypothesis depends on a fairly deep grave, with lots of Bed II materials being thrown out of the grave upon the previously removed Bed III materials. This would insure little mixing when the materials were placed back into the grave. But the hardness of the Bed II materials argues against a deep burial. When Reck found the skeleton, it had to be removed with chisels. So if there were a burial, it would most likely have been a shallow one. And in a shallow grave, dug through the rubble of Beds III and V a short distance into Bed II, mixing of materials from Beds II, III, and V would have been hard to avoid in the grave refilling. Since no mixing was visible, there was, all things considered, probably no post-Bed II burial.

Another suggestion—the skeleton was buried horizontally into Bed II, from the side of Olduvai Gorge. Therefore, no Bed III materials were found in the skeleton's matrix. But Hopwood (1932, p. 194) said: "It would appear that the onus of proof lies on those who might wish to make such a suggestion." The hardness of Bed II poses a substantial obstacle to horizontal burial.

Furthermore, in an October 1932 letter to *Nature,* Leakey (1932b) pointed out that the side of the cliff had receded about 2 feet since 1913. At this rate, a few centuries ago the side of the cliff would have been many yards past the present position of the skeleton. So any burial by horizontal tunneling must have taken place fairly recently. And Hopwood (1932, p. 194) noted: "The present inhabitants of the country, the Masai, rarely bury their dead." And if they did, they did not dig tunnels. Hopwood, describing current Masai burial practices, said: "the shallow grave (about one metre [3 feet] deep) is filled with stones and earth." The stones are meant "to keep hyaenas from abstracting the body." Reck's skeleton was not surrounded by stones.

Leakey's measurements also showed that since 1913 erosion had lowered the land surface near the skeleton's resting place by about 6 inches. Repeating a conclusion he had expressed in his May 1932 letter, Leakey said: "my own estimate is that a time less than fifty years before Prof. Reck came to Oldoway, the site where he found the skeleton was covered by a deposit consisting of a very small relic of Bed 3 overlain by Bed 5 and the steppe lime" (L. Leakey 1932b). Therefore, if a burial took place 50 or more years ago, workers would have had to dig through bright red layers of Bed III materials and the hard calcrete layers of Bed V. And neither Leaky nor Reck had seen any materials from Bed III or Bed V present in the skeleton's matrix.

But Cooper and Watson (1932b) called Leakey's 50-year estimate "a guess." They thought it was possible that Bed II could have been exposed for a very much longer time, allowing the skeleton to be buried without the difficulty of digging through the bright red Bed III materials or the hard calcrete layers of Bed V. The longer period of time would also allow for the subsequent fossilization of the skeleton. But the high rate of erosion observed by Leakey did not support the view maintained by Cooper and Watson.

Also, Hopwood (1932, p. 194) observed that Bed II was, at the time the skeleton was excavated, covered with a rubble of Bed III and Bed V materials, along with pieces of steppe lime, or calcrete. So even if the overlying beds were not intact, a very recent burial should nevertheless have caused their loose materials to be mixed in the grave filling.

In his October letter, Leakey responded to criticism of his proposal that Reck's skeleton had been buried during the formation of Bed II, a shallow Middle Pleistocene lake bottom. Leakey suggested that the deposit might have been dry during parts of the year, as often occurs with African lakes. He also remarked that burial in shallow water is not unknown. "Even to-day in certain circumstances," he wrote, "some native tribes dispose of the bodies of undesirables, such as suicides, in just such a way, 'so as to prevent the spirit from escaping'" (L. Leakey 1932b).

Leakey also replied to a suggestion by Cooper and Watson that the Kanam and Kanjera discoveries were irrelevant to the solution of the question of the age of Reck's skeleton. "I must, however, add," he wrote, "that I do regard the discovery of the Kanam mandible and Kanjira skulls as relevant to the Oldoway problem, in that they at least show that *Homo sapiens* was in existence at the time when Bed 2 at Oldoway was being formed" (L. Leakey 1932b, p. 578).

11.1.4 Reck and Leakey Change Their Minds

Despite the broadsides from Cooper and Watson, Reck and Leakey seemed to be holding their own. But in August 1932, P. G. H. Boswell, a geologist from the Imperial College in England, gave a perplexing report in the pages of *Nature*.

Professor Mollison had sent to Boswell from Munich a sample of what Mollison said was the matrix surrounding Reck's skeleton. Mollison, it may be noted, was not a completely neutral party. As early as 1929, he had expressed his belief that the skeleton was that of a Masai tribesman, buried in the not too distant past (Protsch 1974, p. 380).

Boswell (1932, p. 237) stated that the sample supplied by Mollison contained "(*a*) pea-sized bright red pebbles like those of Bed 3, and (*b*) chips of concretionary limestone indistinguishable from that of Bed 5 and enclosing at least one mineral (an amphibole), in relative abundance, not found in Beds 2 and 3, but present in Bed 4." Boswell took all this to mean that the skeleton had been buried after the deposition of Bed V, which is topped by a hard layer of steppe-lime, or calcrete. At the time he wrote his report, he was unaware that there was also a layer of calcrete at the bottom of Bed V.

The presence of the bright red Bed III pebbles and Bed V limestone chips in the sample sent by Mollison certainly calls for some explanation. Reck and Leakey had both carefully examined the matrix at different times over a period of 20 years. They did not report any mixture of Bed III materials or chips of limestonelike calcrete, even though they were specifically looking for such evidence. So it is remarkable that the presence of red pebbles and limestone chips should suddenly become apparent.

In short, we are faced with contradictory testimony. It would appear that at least one of the participants in the discovery and the subsequent polemics was guilty of extremely careless observation—or cheating.

Reck had studied the matrix at the site. And both Reck and Leakey had studied the matrix directly in contact with the skeleton in Munich. Did they fail to see the red pebbles and chips of limestone, or make false statements about their absence in the matrix? Neither possibility seems likely.

Later, Boswell and other scientists in England studied a sample sent from Munich, in isolation from any of the bones. Mollison, we have already noted, had for years expressed his own view that the skeleton was a recent burial. His statement assuring Boswell that the sample was part "of the material in which the Oldoway skeleton had been embedded" is thus open to question.

Cooper and Watson (1932b) had pointed out in one of their letters to *Nature:* "The photographs published by Prof. Reck show that the whole of the upper and a good deal of the lateral surfaces of the skeleton were exposed during the excavation made for its removal. . . . It need scarcely be pointed out that the only material certainly of the grave infilling carried to Munich in this way is that which is contained within the ribs and between the limbs and the trunk." Did Mollison carefully take his sample from within the ribs or between the legs of Reck's skeleton? Or did he take it from matrix materials that may have come from elsewhere on the block of sediment that contained the skeleton? None of the reports we have seen give any information that would allow these questions to be answered.

Even if the matrix sample supplied by Mollison was suitable for analysis, the presence of limestone chips (containing amphibole) is of ambiguous significance. E. J. Wayland (1932), head of the Geological Survey of Uganda, wrote in a letter to *Nature:* "The fact that the matrix . . . contained bits of concretionary limestone containing a mineral characteristic of Bed 4 does not prove the burial to be post-Bed 5, for Bed 4 contains concretionary limestone, and for that matter so do the other beds, not excluding Bed 2."

It seems that Boswell's mineral test, if accepted at face value, most strongly supports a Bed IV burial. During such a burial, Bed IV limestone chips and bright red Bed III pebbles could have been mixed into Bed II sediments. But a Bed IV burial would still give the anatomically modern skeleton an unexpectedly great age (Table 11.1, p. 629) of 400,000 to 700,000 years.

Keep in mind, however, that Reck, who examined the skeleton *in situ*, saw no signs of limestone chips or bright red pebbles, although he looked carefully for them. This suggests that no burial activity disturbed any layers of limestonelike calcrete in Beds II, III, IV, or V.

The debate about the age of Reck's skeleton became more complicated when Leakey brought new soil samples from Olduvai. Boswell and J. D. Solomon studied them at the Imperial College of Science and Technology. They reported their findings in the March 18, 1933 issue of *Nature,* in a letter signed also by Leakey, Reck, and Hopwood.

The letter contained this very intriguing statement: "Samples of Bed II, actually collected at the 'man site,' at the same level and in the immediate vicinity of the place where the skeleton was found consist of pure and wholly typical Bed II material, and differ very markedly from the samples of matrix of the skeleton which were supplied by Prof. Mollison from Munich" (L. Leakey *et al.* 1933, p. 397). This adds to our suspicion that the matrix sample supplied by Mollison to Boswell may not have been representative of the material closely surrounding Reck's skeleton.

Reck and Leakey, however, apparently concluded from the new observations that the matrix sample from Reck's skeleton was in fact some kind of grave filling, different from pure Bed II material. As far as we can tell, they offered no satisfactory explanation for their previous opinion that the skeleton had been found in pure, unmistakable Bed II materials.

Instead, both Reck and Leakey joined Boswell, Hopwood, and Solomon in concluding that "it seems highly probable that the skeleton was intrusive into Bed II and that the date of the intrusion is not earlier than the great unconformity which separates Bed V from the lower series" (L. Leakey *et al.* 1933, p. 397).

It remains somewhat of a mystery why both Reck and Leakey changed their minds about a Bed II date for Reck's skeleton. Perhaps Reck was simply tired of fighting an old battle against odds that seemed more and more overwhelming. At the time Reck had discovered his skeleton, many scientists were still somewhat uncertain about the evolutionary status of Dubois's Middle Pleistocene Java man. This left some room for controversial discoveries such as Reck's. But by the 1930s, after Black's discovery of Beijing man, the scientific community had become more uniformly committed to the idea that a transitional ape-man was the only proper inhabitant of the Middle Pleistocene. An anatomically modern *Homo sapiens* skeleton in Bed II of Olduvai Gorge did not make sense except as a fairly recent burial.

Leakey, almost alone, remained very much opposed to the idea that Java man (*Pithecanthropus*) and Beijing man (*Sinanthropus*) were human ancestors. In his discoveries at Kanam and Kanjera, he believed he had indisputable evidence for the presence of *Homo sapiens* in the same period as *Pithecanthropus* and *Sinanthropus* (and Reck's skeleton). So perhaps he abandoned the fight over Reck's highly controversial skeleton in order to strengthen support for his own recent finds at Kanam and Kanjera.

There is substantial circumstantial evidence in support of this hypothesis. In the issue of *Nature* (March 25, 1933) immediately following the one carrying Leakey's reversal on Reck's skeleton (March 18, 1933), there appeared the following notice in *Nature*'s "News and Views" section: "On March 18–19 a conference summoned by the Royal Anthropological Institute met at St. John's College, Cambridge, under the presidency of Sir Arthur Smith Woodward, to receive reports on the human skeletal remains discovered by Dr. Leakey's archaeological expedition to East Africa in the autumn of last year." We shall discuss the conference report later in this chapter (Section 11.2.3). For now, we simply note that Leakey's statement abandoning his previous position on the antiquity of Reck's skeleton appeared in *Nature* on the same day as the opening of a conference that would bear heavily on his reputation as a scientist.

In the March 18 issue of *Nature,* C. Stanton Hicks, of the University of Adelaide, Australia, complained about the disadvantages of practicing science in the outlying regions of the British Empire: "The old-established scientific societies with all their tradition and prestige, their facilities for publication and criticism of original work, and their influence in paving the way to higher posts, are in Great Britain." Leakey was a colonial, born and raised in British East Africa. In the conference convened to review his discoveries at Kanam and

Kanjera, this promising young scientist's fate hung in the balance. He would perhaps be accepted into the elite circles of British science, and given a post at Cambridge, or perhaps be banished into obscurity, lucky to occupy a professorship in an outlying university. It is quite possible Leakey thought it best to withdraw his reputation from somebody else's controversial fossil and thus pave the way for acceptance of his own better-dated finds at Kanam and Kanjera. After all, some of the most vocal opponents of Reck's skeleton, such as Boswell, Solomon, Cooper, Watson, and Mollison, would be sitting on the committee that would review the Kanam jaw and Kanjera skulls. As we shall see, the committee accepted the Kanam and Kanjera finds.

In his memoirs, Louis Leakey (1972, pp. 37–38) gave a brief and somewhat confusing review of his involvement with Reck's skeleton. He said the debate about the skeleton's age was resolved by a mineral analysis conducted by Boswell after a 1935 visit to Olduvai Gorge. This version is repeated, almost verbatim, in Cole's 1975 biography of Leakey. But, as far as we can tell, Boswell's mineral analysis was performed in 1933, and the results were reported in the March 18, 1933 letter to *Nature,* signed by Boswell, Leakey, Reck, Hopwood, and Solomon.

According to the new view outlined in the March 1933 letter, intact Beds III and IV at Olduvai Gorge were stripped away by erosion at the location of the skeleton. Bed II, thus exposed, would have probably still been covered by some remnants of Bed III and perhaps a thin layer of calcrete, or steppe-lime. The burial supposedly took place at this time. Subsequent to the burial, the layers of Bed V, including thick, hard layers of calcrete, were deposited. The authors of the March 1933 letter said: "it seems certain that the skeleton was deposited where it was found before the main mass of Bed V, and the overlying steppe-lime were formed, that is, the skeleton appears to have been buried at the time of the existence of the old land surface connected with the steppe-lime at the base of Bed V" (L. Leakey *et al.* 1933, pp. 397–398).

This still gives a potentially anomalous age for the fully human Reck's skeleton. The base of Bed V is about 400,000 years old, according to current estimates (Table 11.1, p. 629). Therefore, even according to the revised position taken by Reck and Leakey, the skeleton could be at least 400,000 years old. This is true even if, as Boswell claimed in his August 1932 letter, the matrix sample supplied by Mollison contained deep red pebbles like those of Bed III and pieces of steppe-lime with a mineral characteristic of Bed IV. Today, however, most scientists believe that *Homo sapiens sapiens* first appeared about 100,000 years ago, as shown by the Border Cave discoveries in South Africa.

The March 1933 letter to *Nature* concluded with some interesting observations about stone tools found "in the basal deposits of Bed V" and on an "old land surface" at the same level as the steppe-lime just below Bed V. These tools, said the authors, had "very close affinities with the phase C of the Upper Kenyan Aurignacian" (Leakey *et al.* 1933, p. 398). Archeologists first used the term Aurignacian in connection with the finely-made artifacts of Cro-Magnon man (*Homo sapiens sapiens*) found at Aurignac, France. According to standard opinion, tools of the Aurignacian type did not appear before 30,000 years ago.

The Kenyan Aurignacian is now called the Kenyan Capsian, and the industry referred to above is called Upper Kenyan Capsian C. An Upper Kenyan Capsian C industry is found at Gamble's Cave, Kenya. Gamble's Cave is considered Holocene, or less than 10,000 years old (Oakley *et al.* 1977, pp. 36–37).

The presence of tools characteristic of anatomically modern humans just below Bed V and in the basal layers of Bed V at Olduvai Gorge is significant. The tools lend support to the idea that anatomically modern humans, as represented by Reck's skeleton, were present in this part of Africa at least 400,000 years ago. Alternatively, one could attribute the tools to *Homo erectus.* But this would mean granting to *Homo erectus* toolmaking abilities substantially greater than scientists currently accept.

In *The Stone Age Races of Kenya* (1935), Leakey repeated his view that Reck's skeleton had been buried into Bed II from a land surface that existed during the formation of Bed V. But now he favored a time much later in that period, contemporary with the Upper Kenyan Capsian C industry at Gamble's Cave. Rainer Protsch (1974, p. 382) wrote: "The contemporaneity was not based on association of the hominids in both localities with that culture, but on the association of one with that culture [at Gamble's Cave] and similar physical types of the hominids in both sites." In our discussion of discoveries made in China, we examined the practice of morphological dating. Here again we see the primary role that the morphology of a skeleton plays in assigning it a date. And Leakey was not alone. Concerning the dating of Reck's skeleton, Protsch (1974, p. 382) noted: "Weinert [1934] argued against an early age of these *Homo sapiens* remains from a purely theoretical evolutionary point of view."

In 1971, Mary Leakey repeated the position taken by her husband in *The Stone Age Races of Kenya:* "The skull is of *Homo sapiens* type and resembles those of the Kenya Capsian from Gamble's Cave II and Naivasha Railway Rockshelter in Kenya. A living site with a microlithic industry dated about 10,000 B.P. is known to exist within a short distance of the Olduvai burial and it is possible that the two are associated" (M. Leakey 1971, p. 225).

But even if the hypothesis that the skeleton was buried into Bed II during the deposition of Bed V is accepted, the skeleton could still be up to 400,000 years old. As mentioned above, that is when the post-Bed IV sediments began to accumulate at Olduvai. Other than its anatomically modern character, the Leakeys had little justification for assigning Reck's skeleton to recent rather than earlier Bed V times.

11.1.5 The Radiocarbon Dating of Reck's Skeleton

Reiner Protsch later attempted to remedy this situation by dating Reck's skeleton itself. Without such a determination, all that could truthfully be said (granting the Bed V burial hypothesis) was that the skeleton could be anywhere from 400,000 to perhaps a few thousand years old.

In 1929, Mollison had measured the organic content of Reck's skeleton under ultraviolet light, hoping to gain insight into its age. Sonia Cole (1975, p. 93), Leakey's biographer, said: "he found a great contrast between it and very recent bones on the one hand and the fossil fauna from Olduvai Bed II on the other." According to Cole, the differences in organic content indicated to Mollison that the bones were of different ages. In particular, Reck's skeleton would have to be younger than the other fossils found in Bed II. Mollison's ultraviolet measurements are said to have substantially influenced Leakey to change his mind about the antiquity of Reck's skeleton (Cole 1975, p. 93).

But Protsch contradicted Cole's statement, quoted above, that Mollison had found the organic content of Reck's skeleton to be different from that of the fauna from Bed II. Protsch (1974, p. 380) said Mollison had found "identical results for the organic content of the hominid and the fauna of Bed II." This demonstrates the difficulties one encounters in trying to unravel the truth about a case like this.

According to Protsch (1974, p. 380), Mollison obtained an organic content of 4.8 percent for the Olduvai human skeleton and 5.3 percent for a skull, only a few thousand years old, from the Ofnet Cave in Bavaria. Mollison used this determination to assign a date of approximately five thousand years to the Olduvai skeleton. Protsch later ran his own tests, using modern microanalytical methods to measure the amount of collagen, the main organic constituent of bone. He obtained an organic content of 2.7 percent for Reck's skeleton and 16.56 percent for the Bavarian skull. This invalidated the earlier determination by Mollison. Not much can be read into either set of results, because bones from different locations can lose their organic content at greatly different rates (Appendix 1.1).

Eventually, fragments of bone thought to belong to the original skeleton were dated by the radiocarbon method. Protsch (1974) obtained for his sample an age of 16,920 years.

The skull was considered too valuable to use for testing, and the rest of the skeleton had disappeared from a Munich museum during the Second World War. Protsch (1974, p. 383) stated: "Through the courtesy of G. Glowatzki, director of the Staatssammlung, some very fragmentary postcranial material still imbedded in earth was found and used for radiocarbon dating. This material consisted mainly of rib fragments, long bone fragments, and pieces of vertebrae. Some of the bones were covered by the preservative Sapon, a lacquer, which was easily flaked off and removed. Many parts were not covered by this preservative, but nevertheless the same chemical pretreatment was given to all bone material. This postcranial material most likely belongs to the Olduvai Hominid I since it was marked as such."

But if the bones were "still embedded in earth," how did they become fragmented? The original skeleton was said to be intact. Also, the hard Bed II material in which the skeleton was encased upon arrival in Germany was not exactly "earth." Reports of the discovery say the skeleton had to be removed with hammers and chisels, indicating the matrix was stonelike in hardness.

From the bone fragments available to him, Protsch was able to gather a sample of only 224 grams, about one third the normal size of a test sample for the method he used. Although he obtained an age of 16,920 years for the human bone, he apparently got very much different dates from other materials from the same site. "Several other radiocarbon dates were run, but could be contaminated by either recent or old radiocarbon, since these sample materials were mostly calcrete or fresh water shells," said Protsch (1974, p. 384). But the human fossil material may also have been contaminated by recent radiocarbon.

In Appendix 1, we discuss in detail the difficulties involved in radiocarbon dating of bones that have been exposed to contamination. By 1974, the remaining bone fragments from Reck's skeleton, if they in fact belonged to Reck's skeleton, had been lying around in a museum for over 60 years and had been soaked in an organic preservative (Sapon).

Protsch did not describe what chemical treatment he used to eliminate recent carbon 14 contributed by the Sapon. Thus we have no way of knowing to what degree the contamination from this source was eliminated.

In Appendix 1.3.2, we also describe other sources of contamination, including: (1) saprophytes growing in and feeding on bone, (2) humic and fulvic acids, (3) exogenous amino acids, (4) improper collection procedures. Protsch (1974) did not discuss any of these sources of contamination or what procedures he used to try to eliminate them. All of these sources of contamination, if not properly dealt with, would cause the carbon 14 test to yield a falsely young age.

The procedures employed today (Appendix 1.3.2.1) are much more exacting than those used by the radiocarbon laboratories that dated Reck's skeleton in the early 1970s.

The radiocarbon method is applied only to the collagen, or protein, fraction of the bone. This protein must be extracted from the rest of the bone by an extremely rigorous purification process.

Scientists then determine whether a sample's amino acids (the building blocks of proteins) correspond to those found in collagen. If they do not correspond, this suggests that amino acids may have entered the bone from outside. According to

Jeffrey Bada (1985a, pp. 256–257), who conducted extensive research at Olduvai Gorge, bones can absorb amino acids from groundwater. These amino acids, being of a different age than the bone, could yield a falsely young radiocarbon date.

Even though a bone has a noncollagen profile, the amino acids could still be original to the bone (the collagen could have decayed, leaving only some of its constituent amino acids). In all cases, one should therefore date each amino acid separately. If any of the amino acids yield dates different from any of the others, this suggests the bone is contaminated and not suitable for carbon 14 dating.

Concerning the radiocarbon tests on Reck's skeleton reported by Protsch, the laboratories that performed them could not have dated each amino acid separately. This requires a dating technique (accelerator mass spectrometry) that was not in use in the early 1970s. Neither could these labs have been aware of the stringent protein purification techniques now deemed necessary.

Is it fair to subject Protsch's dating of Reck's skeleton to such retrospective criticism? After all, the requirements we are talking about were not in effect then. But if modern authorities are correct, and the rigorous purification and dating procedures outlined in Appendix 1.3.2.1 are actually necessary, then it is not unfair to measure Protsch's study against these standards. And when we do so, we can only conclude that the radiocarbon date Protsch gave for Reck's skeleton is unreliable. In particular, the date could very well be falsely young.

There are documented cases of bones from Olduvai Gorge giving falsely young radiocarbon dates. For example, a bone from the Upper Ndutu beds yielded an age of 3,340 years. The Upper Ndutu beds, part of what used to be called Bed V, are from 32,000 to 60,000 years old. A date of 3,340 years would thus be too young by at least a factor of ten. Bada (1985a, p. 255) attributed the unexpectedly young radiocarbon date to deterioration of the bone's original collagen and contamination by secondary carbon compounds from the ground. The radiocarbon dating of Reck's skeleton is thus questionable.

From his radiocarbon date of 16,920 years, Protsch came to the conclusion that the skeleton had been buried in Bed II during the deposition of the upper part of Bed V, which formed after Beds III and IV had been eroded (Protsch 1974, p. 384). Uppermost Bed V, now called the Naisiusiu formation, also yielded fossil material (an ostrich egg shell) with a radiocarbon date of approximately 17,000 years.

Nevertheless, burial from upper Bed V times still seems somewhat problematic. From the reports of Leakey, Hopwood, and others, it is apparent that as little as a few hundred years ago, the spot where Reck's skeleton was found would have been covered by intact Bed V. How much of Bed V is hard to tell, but Louis Leakey (1932b) reported that between 1913 and 1931 the land surface at the site had eroded 6 inches. And at the base Bed V there would have been a hard layer of calcrete, 10 to 12 inches thick.

Furthermore, Bed II itself was quite hard. According to the original reports of the excavation, Reck's skeleton had to be taken out with the aid of hammers and chisels. It hardly seems likely that primitive tribal people would have engaged in the arduous efforts necessary to dig a grave in such resistant rock. One way around this difficulty is to suppose, contrary to the geological evidence, that 17,000 years ago the Bed V calcrete was not present and that the Bed II sediments were still soft a million years after they were deposited—a highly improbable scenario.

Protsch also reported uranium content test results of 3 parts per million for both some middle Bed V faunal remains and Reck's skeleton (now called Olduvai Hominid 1). But he correctly pointed out that the specimens were from different localities, which reduces the value of the comparison. Uranium isotopes may accumulate at vastly different rates in different localities (Protsch 1974, p. 384).

Protsch (1974, pp. 382–383) said stone tools had been discovered in the Naisiusiu Beds, corresponding to the upper part of the old Bed V (Table 11.1, p. 629). These were somewhat like those found at Gamble's Cave (Upper Kenyan Capsian C), some distance away from Olduvai. At Gamble's Cave, skeletons of *Homo sapiens sapiens* had been found, like Reck's skeleton, in a contracted burial position (Protsch 1974, p. 381). Protsch thought these facts lent support to a Naisiusiu Bed origin for Reck's skeleton. But in 1933, Louis Leakey had said that stone tools of the Upper Kenyan Capsian C type had been found just below and in the basal layers of Bed V (L. Leakey *et al.* 1933, p. 398). This could be taken as evidence that men like those of Gamble's Cave (*Homo sapiens sapiens*) existed at least 400,000 years ago in this part of Africa.

But Protsch (1974, p. 382) said about Reck's skeleton: "Theoretically, several facts speak against an early age of the hominid, such as its morphology." This suggests that the skeleton's modern morphology was one of the main reasons Protsch doubted it was as old as Bed II or even the base of Bed V.

If Reck's skeleton were classified as *Homo erectus*, it is hard to imagine that anyone would now be raising any serious objections to its presence in Bed II. In 1960, a *Homo erectus* cranium (OH 9) was found on the surface at Olduvai Gorge (Poirier 1977, p. 223). It was nevertheless assigned to upper Bed II, giving it an age of over 1 million years. Adhering to the base of the cranium was a matrix matching that of Bed II. Yet a determined debunker could always attribute this to "secondary cementation."

What about the observations by Leakey (1932a, p. 721) and Mollison (Protsch 1974, p. 380) that the human bones are fossilized to the same degree as the animal bones found nearby in Bed II? Protsch (1974, p. 382) said that "a relative age determined by the state of fossilization of bones is invalid for a positive chronological diagnosis." Here, we agree with Protsch.

Yet scientists have used such determinations of relative age, as measured by differences in fluorine, uranium, or nitrogen content, to discredit many of the anomalously old *Homo sapiens* fossils we have discussed (such as Galley Hill). They also have used relative age determinations to date many accepted finds. But in the case of Reck's skeleton, Protsch said such results have no value.

Curiously enough, Protsch himself used a relative age determination to confirm that the bone fragments he had tested were actually from Reck's skeleton. Protsch (1974, p. 383) reported: "to check whether the skull and the fragmentary bones belonged together, two separate microanalytical tests were made on the skull and some post cranial primary bone. The [nitrogen] values of 0.45% (British Museum) and 0.43% (UCLA) are remarkably similar and give support to a positive association of the bones." So to be fair, if the Bed II faunal remains and Reck's skeleton were fossilized to a similar degree (Leakey 1932a, p. 721), could not this also be taken as supporting (although not proving) a "positive association of the bones?"

All in all, Protsch appears to have done a needed service—the cleaning up of a problem discovery, fitting it nicely into the accepted evolutionary sequence. By 1974, it is clear, no one in the mainstream of human evolutionary thought was prepared to accept a fully modern human being existing at least 400,000 years ago, contemporary with *Homo erectus*. Protsch himself admitted that his theoretical expectations ruled this out. By giving a plausible carbon 14 date and identifying the skeleton with tools found nearby in upper Bed V and with skeletal remains found at Gamble's Cave along with similar tools, Protsch put Reck's skeleton in an appropriate paleoanthropological niche. The case was closed.

But the case made by Protsch in favor of a Late Pleistocene burial was very weak. First of all, it is not at all certain that the bone sample he tested actually belonged to the original Reck's skeleton, which, except for the skull, disappeared during the Second World War. Furthermore, the carbon 14 method is not infallible, especially when applied to bones that were exposed to contamination for over 60 years. It is also possible that the bones were contaminated with recent carbon while they were buried in the ground at Olduvai Gorge. And, as we have seen, the radiocarbon dating methods employed by Protsch have been superseded by more rigorous procedures.

11.1.6 Probable Date Range of Reck's Skeleton

We are now left with several alternative explanations, which we shall now summarize, for the age of Reck's skeleton. First we have the original determination by Reck that it was deposited naturally during the formation of Bed II. Reck carefully searched for signs of intrusive burial (especially chips of limestone and other materials from the overlying beds) and found none "despite the most attentive inspection" (Hopwood 1932, pp. 193–194). This gives a date of over 1.15 million years for the skeleton, which is fully human. Second, we have Leakey's view that the skeleton was deliberately buried during the deposition of upper Bed II, which also gives a date of over 1.15 million years. Third, we have the revised position, taken by Reck, Leakey, and others, that the skeleton was buried into Bed II during the time Bed V was being deposited. In adopting their revised position, Reck and Leakey in effect reversed their previous statements that they had observed no mixture of materials from overlying beds in the matrix of the skeleton. It is significant that Leakey recanted his position on Reck's skeleton just before a commission of scientists, including the critics of his prior views on Reck's skeleton, was to pass judgement on his own discoveries at Kanam and Kanjera. The new position adopted by Reck, Leakey, and others yields a date range of from 400,000 to perhaps 10,000 years for the skeleton. Primarily on the basis of its modern morphology, the skeleton was assigned a very recent date within this range. During the Second World War, much of the skeleton was lost. Finally, in 1974, in an attempt to confirm an uppermost Bed V date, Protsch published a radiocarbon test result of about 17,000 years for a bone sample that may not have been from the original Reck skeleton. Even if the sample was from Reck's skeleton, the dating techniques that were used are now considered unreliable.

In our discussion of China, we introduced the concept of a probable date range (Section 9.2.1) as the fairest age indicator for controversial discoveries. The available evidence suggests that Reck's skeleton (OH 1) should be assigned a probable date range extending from the late Early Pleistocene (1.15 million years) to the late Upper Pleistocene (10,000 years). There is much evidence that argues in favor of the original Bed II date proposed by Reck. Particularly strong is Reck's observation that the thin layers of Bed II sediment directly around the skeleton were undisturbed. Also arguing against later burial is the rocklike hardness of Bed II. Reports favoring a Bed V date seem to be founded upon purely theoretical objections, dubious testimony, inconclusive test results, and highly speculative geological reasoning. But even these reports yield dates of up to 400,000 years for the skeleton.

A skeleton of *Homo sapiens sapiens* type with an age of 1.15 million years, or even .4 million years, does not fit the current evolutionary scenario. But Reck's skeleton does not seem out of place when seen in the context of the evidence documented in this book. This evidence demonstrates the presence of anatomically modern humans throughout the Early Pleistocene, Pliocene, Miocene, and even earlier. Only the radiocarbon date reported by Protsch suggests Reck's skeleton might be fairly recent, but as we have seen, this date has its problems.

It would undoubtedly take a time-traveling detective with supersensory powers to give us the real story of Reck's skeleton and its age. And Reck's skeleton is not exceptional. Most of the discoveries scientists have used to build up their picture of human evolution are similarly ambiguous, their significance obscured by professional rivalries and imperfect investigative methods.

11.2 THE KANJERA SKULLS AND KANAM JAW

In 1932, Louis Leakey announced discoveries at Kanam and Kanjera, near Lake Victoria in western Kenya. The Kanam jaw and Kanjera skulls, he believed, provided good evidence of *Homo sapiens* in the Early and Middle Pleistocene.

11.2.1 Discovery of the Kanjera Skulls

Kanjera lies on the south shore of Lake Victoria's Kavirondo Gulf. When Leakey visited Kanjera in 1932 with Donald MacInnes, they found stone hand axes and fragments of five human skulls, designated Kanjera 1–5. Leakey (1960d, p. 204) said: "I found part of No. 3 specimen *in situ* myself, and I have no doubt about its genuineness." The expedition also found a human femur.

According to Leakey, the fossil-bearing beds at Kanjera were equivalent to Bed IV at Olduvai Gorge. The faunal studies of H. B. S. Cooke (1963) confirmed this, which means the Kanjera beds range from 400,000 to 700,000 years old (Table 11.1, p. 629). But the morphology of the Kanjera skull pieces was quite modern. Leakey (1960d, p. 203) wrote: "The front part of the skull is preserved, in a damaged condition, in two of the specimens, and from this we can see that there was no trace of a bony brow-ridge above the eyes. Instead we find a very small and simple form much as in a child, but certainly of *Homo sapiens* type." Scientists now think modern *Homo sapiens* appeared about 100,000 years ago in Africa. An age of 400,000 years for the Kanjera skulls would, however, be acceptable for the oldest African early *Homo sapiens* (Bräuer 1984, p. 394). But the author of a recent survey attributed the Kanjera skulls to *Homo sapiens* cf. *sapiens* (Groves 1989, p. 291), indicating they are anatomically modern.

11.2.2 Discovery of the Kanam Jaw

At Kanam, Leakey initially found teeth of *Mastodon* and a single tooth of *Deinotherium* (an extinct elephantlike mammal), as well as some crude stone implements. Because *Deinotherium* was the marker fossil for Bed I at Olduvai Gorge, Leakey believed the Kanam formations were of the same Early Pleistocene age—about 1.7–2.0 million years old, according to current estimates (Oakley *et al.* 1977, pp. 166, 169).

On March 29, 1932, Leakey's collector, Juma Gitau, brought him a second

Deinotherium tooth from a gully at Kanam. Leakey told Gitau to keep digging in the same spot. Working a few yards from Leakey, Gitau hacked out a block of travertine (a hard calcium carbonate deposit) and broke it open with a pick. He saw a tooth protruding from a piece of travertine and showed it to MacInnes, who identified the tooth as human. MacInnes summoned Leakey. Together they searched for more human fossils, but none turned up (L. Leakey 1960d, p. 202).

Upon chipping away the travertine surrounding Gitau's find, they saw the front part of a human lower jaw with two premolars. Leakey thought the jaw from the Early Pleistocene Kanam formation was much like that of *Homo sapiens,* and he announced its discovery in a letter to *Nature* (Cole 1975, p. 91). According to Cooke (1963), the Kanam fauna is older than that of Bed I at Olduvai Gorge, making the Kanam beds at least 2.0 million years old.

Almost without exception, today's scientists believe that the human lineage extends from *Australopithecus* in the Late Pliocene and Early Pleistocene, to *Homo erectus* in the Early and Middle Pleistocene, and thence to *Homo sapiens sapiens* in the Late Pleistocene. Against this background, a *Homo sapiens* jaw in the earliest Pleistocene seems strangely out of place. But in the early 1930s, the now dominant view of human origins, although held by some, was still a somewhat tentative hypothesis. A good many British scientists regarded *Australopithecus,* discovered in 1925 by Raymond Dart in South Africa, as a variety of ape, with no direct connection to the line of human descent. Similarly, many scientists never accepted Java man, and Beijing man had only newly arrived on the scene. Although some scientists did believe Java man and Beijing man, now classified as *Homo erectus,* were genuine human ancestors, the picture was somewhat clouded by Piltdown man. Whereas Java man and Beijing man had humanlike jaws and apelike skulls, Piltdown man, roughly the same age, featured an apelike jaw and humanlike skull. The Neanderthals also had to be fit into the picture. Some thought they were direct human ancestors, others thought they were on an evolutionary side branch. In short, scientists of the 1930s held varying views about the progress of human evolution and the antiquity of the modern human type. Therefore, not all of them would automatically reject finds such as Leakey made at Kanam and Kanjera.

For Leakey, the Kanam and Kanjera fossils showed that a hominid close to the modern human type had existed at the time of Java man and Beijing man, or even earlier. If he was correct, Java man and Beijing man (now *Homo erectus*) could not be direct human ancestors, nor could Piltdown man with his apelike jaw.

11.2.3 A Commission of Scientists Decides
on Kanam and Kanjera

On March 18 and March 19, 1933, the human biology section of the Royal Anthropological Institute met to consider Leakey's discoveries at

Kanam and Kanjera. Chaired by Sir Arthur Smith Woodward, 28 scientists issued reports on four categories of evidence: geological, paleontological, anatomical, and archeological (Woodward *et al.* 1933, pp. 477–478). The geology committee concluded that the Kanjera and Kanam human fossils were native to the beds in which they were found. The paleontology committee said the Kanam beds were Early Pleistocene, while the Kanjera beds were no more recent than Middle Pleistocene. The archeology committee noted the presence at both Kanam and Kanjera of stone tools in the same beds where the human fossils had been found.

The anatomical committee said the Kanjera skulls exhibited "no characteristics inconsistent with the reference to the type *Homo sapiens* (Woodward *et al.* 1933, p. 477). The same was true of the Kanjera femur.

About the Kanam jaw, the anatomy experts said: "With the possible exceptions of the thickness of the symphysis, the conformation of the anterior internal surface, and what seems to be a large pulp-cavity of the first right molar tooth, the Committee is not able to point to any detail of the specimen that is incompatible with its inclusion in the type of the *Homo sapiens*" (Woodward *et al.* 1933, p. 478). The symphysis, the joint between the two halves of the lower jaw, runs down the middle of front part of the jaw.

The species designation *Homo sapiens,* as employed today by most, although not all, paleoanthropologists, includes early *Homo sapiens, Homo sapiens neanderthalensis,* and *Homo sapiens sapiens* (fully modern humans). But in 1933, the Neanderthals were generally considered distinct from *Homo sapiens,* and the first representatives of early *Homo sapiens,* as presently conceived by many workers, either had not been discovered or had not been reported to the scientific world. The first report on the Steinheim skull, discovered in 1933, came out in 1935. And the Swanscombe skull fragments were not found until 1935 and 1936. So when the members of the anatomical committee classed the Kanjera skulls and Kanam jaw as *Homo sapiens,* they presumably meant they were within the range of anatomically modern humans.

Although the committee stated that the remains could be classified as *Homo sapiens,* Leakey assigned the jaw to a new species, *Homo kanamensis,* which he considered the immediate ancestor of *Homo sapiens.* According to Cole (1975, pp. 103–104), Leakey later dropped the name *kanamensis* in favor of *sapiens.*

11.2.4 Boswell Strikes Again

Shortly after the 1933 conference gave Leakey its vote of confidence, geologist Percy Boswell began to question the age of the Kanam and Kanjera fossils. Leakey, who had experienced Boswell's attacks on the age of Reck's skeleton, decided to bring Boswell to Africa, hoping this would resolve his doubts. But all did not go well.

Upon returning to England, Boswell (1935) submitted to *Nature* a negative report on Kanam and Kanjera: "Unfortunately, it has not proved possible to find the exact site of either discovery, since the earlier expedition (of 1931–32) neither marked the localities on the ground nor recorded the sites on a map. Moreover, the photograph of the site where the mandible was found, exhibited with the jaw fragment at the Royal College of Surgeons, was, through some error, that of a different locality." Having examined Leakey's original field notes, Boswell (1935) said "it is regrettable that the records are not more precise."

Boswell found the geological conditions at the sites confused. He said that "the clayey beds found there had frequently suffered much disturbance by slumping." From this Boswell (1935) concluded: "The date of entombment of human remains found in such beds would be inherently doubtful."

But what about the committee that had given Leakey its endorsement? "It seems likely," said Boswell (1935) "that if the facts now brought forward had been available to the Committee, a different report would have been submitted." Boswell concluded that the "uncertain conditions of discovery . . . force me to place Kanam and Kanjera man in a 'suspense account.'"

11.2.5 Leakey Responds

Replying to Boswell's charge that he had not properly marked the sites, Leakey (1936) stated in a letter to *Nature* that he had in fact done so. Unfortunately, the iron pegs he used had disappeared, perhaps taken by natives for spearheads or fishhooks. He had not marked the sites on a map, but only because no maps of sufficient detail existed. He had considered hiring a surveyor to make maps, but had not done so because of lack of money. Instead, he had taken photographs to identify the sites, but these had been spoiled by a malfunction in his camera.

His own photographs of Kanam and Kanjera ruined, Leakey had selected some by Miss Kendrick, a member of his expedition, to display with his fossils in England. In his letter to *Nature,* Leakey explained he had misinterpreted the label on one of Kendrick's photographs and had mistakenly used it to show the site where the Kanam jaw had been found. But he pointed out: "I carefully refrained from using any photographs *as evidence* in connexion with my claim for the antiquity of the Kanam mandible, and only used them to show the general nature of the sites" (L. Leakey 1936).

Furthermore, Leakey felt he had been able to show Boswell the locations where he had found his fossils. Leakey (1936) wrote: "At Kanjera I showed him the exact spot where the residual mound of deposits had stood which yielded the

Kanjera No. 3 skull *in situ*. . . . the fact that I did show Prof. Boswell the site is proved by a small fragment of bone picked up there in 1935 which fits one of the 1932 pieces."

Regarding the Kanam jaw, Leakey stated in his memoirs: "It had been found in direct association with Lower Pleistocene fossils such as *Deinotherium* and *Mastodon,* and the matrix adhering to it was entirely similar to that which Boswell had now seen in the Kanam West gullies" (L. Leakey 1972, p. 35). Boswell did not mention the matrix adhering to the jaw in his letter to *Nature*.

Leakey (1972, p. 35) added: "Boswell, however, remained doubtful because no scientist had seen the jaw *in situ*. He would not agree to accept Juma's statement that it had been dug out while he was working on the *Deinotherium* tooth." Of course, if this standard were to be applied across the board, then many thoroughly accepted discoveries would also have to be thrown out. The Heidelberg (Mauer) jaw, for example, was discovered by a German sand pit worker. And almost all of the Java man discoveries reported by von Koenigswald were found by native collectors.

Regarding the location of the Kanam jaw, Leakey (1972, p. 35) said: "we had originally taken a level section right across the Kanam West gullies, using a Zeiss-Watts level, and could therefore locate the position to within a very few feet—and, in fact, we did so. I had brought with me a copy of the cross section, taken from a tree that could still be located on one side of the gully to another tree on the other side. On this cross section was a mark showing the point where the jaw had been recovered. I had, therefore, no doubt at all that I was showing Boswell and Wayland the right place within a few feet."

Boswell suggested that even if the jaw was found in the Early Pleistocene formation at Kanam, it had entered somehow from above—by "slumping" of the strata or through a fissure. To this Leakey (1960d, pp. 202–203) later replied: "I cannot accept this interpretation, for which there is no evidence. The state of preservation of the fossil is in every respect identical to that of the Lower Pleistocene fossils found with it. Had the Kanam mandible been a specimen representing some specialized extinct type of man (such as used to be called 'primitive') no one would have suggested that it was not contemporary with the other fossils of the same horizon. . . . the fact that the Kanam mandible has a distinct chin eminence certainly influenced some people against accepting its authenticity."

Boswell's preconceptions about the morphology of hominids in the Early Pleistocene apparently motivated his attacks on the age of the Kanam jaw, and of Reck's skeleton (Section 11.1.4). Leakey (1972, pp. 35–36) said in his memoirs: "he actually told us that were it not for the counterindication provided by the Piltdown jaw, which showed that man in the Lower Pleistocene had a simian shelf and extremely apelike characteristics, he would be inclined to accept the Kanam evidence, since the mineralization of the specimen compared closely with that of other fossils from the same deposits." Of course, British scientists later declared the Piltdown jaw to be a fake (Chapter 8).

11.2.6 Kanam and Kanjera After Boswell

Despite Boswell's attacks on the Kanam and Kanjera finds, a few well-known scientists continued to keep open minds about Leakey's original claims. Robert Broom, who in the 1930s found the first adult specimens of *Australopithecus,* wrote (1951, p. 13): "I have looked into this controversy very carefully and have no hesitation in saying that I have the fullest confidence in Leakey's work; I am quite satisfied that Leakey found these remains where he says he found them, and that they prove modern man is far older than a few English scientists had thought—perhaps even as old as the Lowest Pleistocene." Broom's use of the words "modern man" to describe the Kanam and Kanjera fossils suggests he regarded them as similar to *Homo sapiens sapiens.* Broom (1951, pp. 11–12) characterized the Kanjera fossils as "skulls of early man with a large brain, and without any of the characters of Neanderthal man." Standard texts give a lot of attention to Broom's *Australopithecus* finds, but usually fail to mention his unorthodox views on Kanam and Kanjera.

Philip V. Tobias of South Africa said about Kanjera (1968, p. 182): "Boswell did *not* disprove the claim that human fragments were found in a Middle Pleistocene deposit; he only failed to find additional evidence confirmatory of Leakey's claim. Thus there is a good prima facie case to re-open the question of Kanjera."

And the Kanjera case was in fact reopened. Leakey's biographer Sonia Cole (1975, p. 358) wrote: "In September 1969 Louis attended a conference in Paris sponsored by UNESCO on the theme of the origins of *Homo sapiens.* . . . the 300 or so delegates unanimously accepted that the Kanjera skulls were Middle Pleistocene."

Leakey originally suggested that the fossil-bearing formation at Kanjera was equivalent to Olduvai Bed IV, which is approximately 400,000 to 700,000 years old (early to middle Middle Pleistocene). By 1960, however, Leakey had modified his position. He said the Kanjera skulls were the same age as the Swanscombe skull (L. Leakey 1960d, p. 204), which is about 300,000 years old. In the paper Leakey presented at the UNESCO conference, he maintained his view that the deposits at Kanjera and Swanscombe were "of comparable age." But as we have seen, H. B. S. Cooke (1963), a leading authority on African mammals, confirmed Leakey's original view that the Kanjera beds were the same age as Olduvai Bed IV. In his Paris paper, Leakey (1971, p. 26) also asserted that the Kanjera skulls had "brow-ridges of modern *Homo sapiens* appearance."

Tobias (1962, p. 344) said about the Kanam jaw: "Nothing that Boswell said really discredited or even weakened the claim of Leakey that the mandible belonged to the stratum in question, nor did Boswell deny the faunal and cultural associations previously attributed to this stratum. . . . a number of subsequent writers have gratuitously assumed that Boswell's report invalidated all Leakey's claims. Although Leakey answered some of Boswell's specific criticisms, the reply has seldom been quoted and little cognizance has seemingly been taken of it." But, as we shall see below, Tobias had his own ideas about the age and evolutionary status of the Kanam jaw.

11.2.7 Morphology of the Kanam Jaw

Scientists have described the Kanam jaw in a multiplicity of ways. In 1932, a committee of English anatomists proclaimed it *Homo sapiens* (Woodward *et al.* 1933). Louis Leakey initially attributed the jaw to a new species, *Homo kanamensis,* a direct ancestor of *Homo sapiens.* But his biographer Sonia Cole (1975, pp. 103–104) said he soon gave up that designation in favor of *Homo sapiens.* Sir Arthur Keith (1935, p. 163), the dean of British anthropologists, also considered the Kanam jaw *Homo sapiens.* But in the 1940s Keith decided the jaw was most likely from an australopithecine (Tobias 1968, p. 180).

Tobias, an expert on the Australopithecinae, disagreed. After comparing the Kanam jaw with available *Australopithecus* jaws, Tobias (1968, p. 181) found it was, among other things, "much less robust" and different in the "general conformation and orientation" of the front part of the jaw.

Tobias (1962, p. 341) suggested that some of the *sapiens*-like features of the front part of the Kanam jaw might be, at least partially, the result of bone growth in response to a tumor on the inner surface of the front part of the jaw. Tobias was, however, not the first to notice the tumor.

Almost 20 years earlier, Sir Arthur Keith (1935, p. 163) wrote: "the chin of this representative of early humanity was the seat of a bony tumour of an exceedingly rare kind. The tumour, which grew from the deep aspect of the jaw, just behind the chin, has spread over and obscured the normal features of this region. Enough remains, however, to make quite certain that in dimensions and in its features, the chin region of this early being was shaped as in primitive types of living humanity—such as the aborigines of Australia." In other words, Keith, at that time, took the chin features to be within the range of anatomically modern humans, *Homo sapiens sapiens.*

Despite the effects of the bone tumor on the inner surface of the chin, Tobias (1962, p. 349) thought the lower front part of the Kanam jaw had some features like that of the modern human chin—although not as well developed. For example, the Kanam jaw, like the human jaw, has a pronounced incurvation below the level of the teeth and an outward swelling of the bone at the base of the front part of the jaw (Figures 11.4g–h).

But Tobias also called attention to the depth and thickness of the jaw, the relatively large size of some of the teeth, and other features that he regarded as primitive. Tobias (1962, p. 355) observed: "Several, though not all, of these features might be encountered individually as exceptional variants among modern African mandibles." He thought the Kanam jaw most closely resembled the late Middle Pleistocene mandible from Rabat in Morocco, and Upper Pleistocene mandibles such as those from the Cave of Hearths in South Africa and Dire-Dawa in Ethiopia (Tobias 1968, p. 181).

Recent workers class Rabat and Cave of Hearths as "early archaic *Homo sapiens*" (Bräuer 1984, pp. 380, 394). The Rabat mandible is said to have no true chin (Howell 1978, pp. 196, 204), while the Cave of Hearths mandible is said to have "a slight to moderate chin" (Tobias 1971, p. 338). The Dire-Dawa mandible, said to have no true chin, is nevertheless listed as *Homo sapiens sapiens* (Howell 1978, p. 214).

According to Tobias (1968, pp. 190–191), all of these mandibles displayed "neanderthaloid" features. He placed them, along with other neanderthaloid fossils, in the subspecies *Homo sapiens rhodesiensis*, which he regarded as transitional between *Homo erectus* and more developed African Neanderthals.

In 1960, Louis Leakey (1960d, p. *xix*), retreating from his earlier view that the Kanam jaw was *sapiens*-like, wrote: "it becomes highly probable that the Kanam mandible represents, in fact, a female of *Zinjanthropus*."

Leakey had found *Zinjanthropus* in 1959, at Olduvai Gorge (Section 11.4.1). He briefly promoted this apelike creature as the first toolmaker, and thus the first truly humanlike being. Shortly thereafter, fossils of *Homo habilis* were found at Olduvai. Leakey quickly demoted *Zinjanthropus* from his status as toolmaker, placing him among the robust australopithecines (*A. boisei*).

In the early 1970s, Leakey's son Richard, working at Lake Turkana, Kenya, discovered fossil jaws of *Homo habilis* that resembled the Kanam jaw. Since the Lake Turkana *Homo habilis* jaws were discovered with a fauna similar to that at Kanam, the elder Leakey changed his mind once more, suggesting that the Kanam jaw could be assigned to *Homo habilis* (L. Leakey 1972, p. 36; Cole 1975, p. 362).

That scientists have attributed the Kanam jaw to almost every known hominid (*Australopithecus, Australopithecus boisei, Homo habilis,* Neanderthal man, Early *Homo sapiens,* anatomically modern *Homo sapiens*) shows the difficulties involved in properly classifying hominid fossil remains.

Tobias's suggestion that the Kanam jaw came from a variety of early *Homo sapiens*, with neanderthaloid features, has won wide acceptance. Yet as can be seen in Figure 11.4, which shows outlines of the Kanam mandible and other

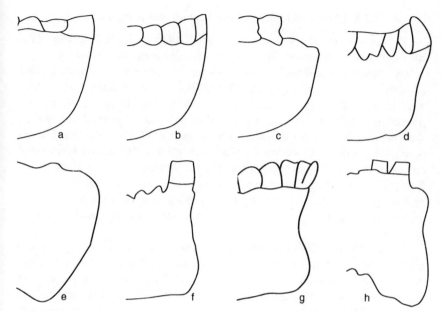

Figure 11.4 The outlines of the mandibles shown here (not to scale) were traced from published photographs, except for (a) and (g), which were traced from a drawing. (a) *Australopithecus,* Omo, Ethiopia (Eckhardt 1972, p. 103); (b) *Homo erectus,* Heidelberg (Mauer), Germany (Osborn 1916, p. 98); (c) Early *Homo sapiens,* Arago, France (Stringer *et al.* 1984, p. 64); (d) Neanderthal, Shanidar, Iraq (Gowlett 1984, p. 104); (e) *Homo sapiens rhodesiensis* ("neanderthaloid" according to P. V. Tobias), Cave of Hearths, South Africa (Tobias 1971, p. 338); (f) *Homo sapiens sapiens,* Border Cave, South Africa (Bräuer 1984, p. 381); (g) *Homo sapiens sapiens,* modern South African native (Zuckerman 1954, p. 308); (h) the Kanam mandible (Tobias 1962, p. 345).

hominid mandibles, the contour of the Kanam mandible's chin region is similar to that of the Border Cave specimen (f), recognized as *Homo sapiens sapiens,* and to that of a modern South African native (g). All three share two key features of the modern human chin, namely, an incurvation toward the top and a swelling outward at the base.

But even if one were to accept Tobias's view that the Kanam jaw was neanderthaloid, one would still not expect to discover Neanderthals in the Early Pleistocene, over 1.9 million years ago. Neanderthaloid hominids came into existence at most 400,000 years ago (Bräuer 1984, p. 394) and persisted until about 30,000 or 40,000 years ago, according to most accounts. We note that some workers (Bräuer 1984) confine the Neanderthal line to Eurasia and a small area of North Africa adjacent to Europe. These workers would not expect to find Neanderthals at Kanam in East Africa.

11.2.8 Chemical Testing of the Kanam and Kanjera Fossils

To ascertain the age of the Kanam jaw and Kanjera skulls, K. P. Oakley of the British Museum performed fluorine, nitrogen, and uranium content tests.

The Kanam jaw and the Kanjera skulls had about the same fluorine content as other bones from the Early and Middle Pleistocene formations where they were found (Oakley 1974, p. 257; 1975, p. 151). But Oakley (1974, p. 257) pointed out that "in volcanic areas (particularly under tropical conditions) fluorine analysis does not provide a reliable method of relative dating." If this is so, one wonders why he ran the tests. Nevertheless, the results he reported are consistent with the hypothesis that the human bones at Kanam and Kanjera are as old as the faunal remains at those sites. Of course, the agreement in fluorine content might, as Oakley suggested, be the result of uneven fluorine absorption in a volcanic, tropical environment. But then again, it might not.

Oakley (1974, p. 257) found that a Kanjera 4 skull fragment showed just a trace of nitrogen (0.01 percent), while a Kanjera 3 skull fragment showed none. Neither of the two animal fossils tested showed any nitrogen. The presence of "measurable traces" of nitrogen in the Kanjera 4 skull fragment meant, said Oakley (1974, p. 258), that all the human fossils were "considerably younger" than the Kanjeran fauna.

But certain deposits, such as clay, preserve nitrogen, sometimes for millions of years (Appendix 1.1.2). So perhaps the Kanjera 4 fragment was protected from nitrogen loss by clay. In any case, the Kanjera 3 fragment, like the animal samples, had no nitrogen. It is possible that this human bone was younger than the animal bones, and lost its nitrogen fairly quickly. But the test results do not dictate this interpretation—the bones could be the same age.

As shown in Table 11.2, the uranium content values for the Kanjera human fossils (8–47 parts per million) overlapped the values for the Kanjeran fauna (26–216 parts per million). This could mean they were of the same age.

But the human bones averaged 22 parts per million while the mammalian fauna averaged 136 parts per million. To Oakley (1974, p. 257), the substantial difference between the averages meant that "the Kanjera hominids, although fossilized (Upper Pleistocene?), are considerably younger than the Kanjeran faunal stage (Middle Pleistocene)."

Similar uranium contents results were obtained at Kanam. The Kanam mandible had 4–12 parts per million eU_3O_8, while the Kanam fauna had 60–214 parts per million (Oakley 1975, p. 151). "The low radiometric values of the Kanam jaw fragment strongly suggest that it is younger than the Kanam fauna," said Oakley (1975, p. 151).

TABLE 13.2

Uranium Content of Kanjera Hominid Fossils

Fossil Identification	Description of Fragment Tested	Uranium (eU$_3$O$_8$) Content (parts per million)
Kanjera 3	orbital fragment, *in situ*	15
	right parietal fragment, *in situ*	21
	cranial fragments from surface	16, 27, 27, 30, 42
	femoral fragment from surface	8, 14
Kanjera 4	frontal fragments from surface	11, 21, 35
Fauna	Kanjeran mammal fragments	26, 131, 146, 159, 216

The data in this table are from Oakley (1974, p. 257).

While the uranium content values, as reported, are consistent with the Kanjeran and Kanam faunas being older than the human bones, there are reasons for caution. The values reported for the Kanjeran fauna—26, 131, 146, 159, and 216 parts eU$_3$O$_8$ per million—vary widely. The highest value is 8.3 times greater than the lowest, although the bones are supposedly of the same general age. Also, the uranium content of the Kanjera 3 human fossils ranged from 8 to 42 parts per million, differing by a factor of 5 in a single individual. The high and low values for the Kanam fauna vary by a factor of 3.5, and for the Kanam jaw itself by a factor of 3. This reinforces our observation (Appendix 1.2.4) that the rate at which a bone absorbs uranium depends on many highly variable conditions—such as the concentration of uranium in the groundwater, the rate of groundwater flow, and the nature of the surrounding sediment. Also, different kinds of bone (and, apparently, even different parts of the same bone) may absorb uranium at greatly different rates. All of this tends to reduce the value of uranium content as a relative age indicator.

Oakley himself pointed out: "the distribution of uranyl ions in ground-water, like that of fluorine ions is subject to very considerable variation from place to place . . . it appears that fossil bones of Upper Pleistocene or early Holocene age in Kugata near Mount Homa [close to Kanam] not only contain *more* fluorine than bones of Lower Pleistocene age at Kanam, but are *more* radioactive on account of adsorbed uranium" (Tobias 1968, p. 181).

Significantly, the uranium content values that Oakley reported in 1974 were apparently not the first he had obtained. In a paper published in 1958, Oakley said, immediately after discussing the uranium content testing of the Kanam jaw: "Applied to the Kanjera bones our tests did not show any discrepancy between the human skulls and the associated fauna" (1958, p. 53). It would appear that Oakley was not satisfied with these early tests and later performed additional tests on the Kanjera bones, obtaining results that were more to his liking.

Leakey reported that some of the Kanjera human skull fragments, now classified *Homo sapiens* cf. *sapiens* (Groves 1989, p. 291), were found *in situ* in the Middle Pleistocene Kanjeran deposits. Oakley (1974, p. 257), however, explained: "When deposits such as the Kanjera beds become waterlogged during the wet season, bones lying on the surface become readily incorporated, so that when subsequently discovered they can easily have the appearance of occurring *in situ*." Is this what actually happened? Maybe yes, maybe no.

Oakley had to stretch even further to account for the Kanam jaw. After pointing out that Tobias had said the Kanam jaw was comparable to the Middle Pleistocene Rabat jaw, Oakley (1975, p. 152) said: "I suggest that during some interval in Middle Pleistocene times the jaw lay on a surface littered with fossils weathered out from the Kanam beds and it became embedded with these derived fossils in a block of surface limestone which was eventually down-faulted or trapped in a fissure penetrating the Kanam Beds. This would explain the low uranium content and the high degree of calcification, and at the same time take into account L. S. B. Leakey's statement in his memoirs . . . that Juma Gitau discovered the Kanam jaw fragment while engaged on extracting a molar of tooth of *Deinotherium*. As he expressed it to me: 'the jaw was in the same block as an *undoubted* Lower Pleistocene fossil.'"

Oakley had no trouble inventing special geological scenarios to explain away the stratigraphic evidence. But he offered no proof, such as positive signs of faulting, that these scenarios were correct. Operating as Oakley did, one can easily dispose of any unwanted stratigraphic evidence whatsoever.

But even if we do grant stratigraphic resorting, this does not necessarily show that the hominid fossils at Kanam and Kanjera were younger than the mammalian fossils at these sites. For example, Tobias (1968, p. 181) said: "The low radiometric values of the Kanam mandible do not necessarily bespeak a *recent* age for the jaw, but only a *different* history and probably a *different* age as compared with the other Kanam fauna." In fact, if the Kanam jaw had been washed in from a Late Pliocene deposit with a low uranium content, it could be *older* than the Early Pleistocene animal fossils in the Kanam bed.

Tobias, however, chose a more comfortable alternative. "Nothing in these results," he said "would rule out the possibility that the Kanam mandible was derived from *Middle* Pleistocene beds in the vicinity, such as those of Rawe close to Kanam West" (Tobias 1968, p. 181). A late Middle Pleistocene date would be favorable for his view that the jaw is neanderthaloid.

Our review of the chemical testing of the Kanam and Kanjera fossils leads us to the following conclusions. The fluorine and nitrogen content tests gave results consistent with the human bones being as old as their accompanying faunas. This interpretation can nevertheless be challenged. The uranium content test gave results consistent with the human bones being younger than their accompanying faunas. But here again, if one chooses to challenge this interpretation, one will find ample grounds to do so.

All in all, the results of chemical and radiometric tests do not eliminate the possibility that the Kanam and Kanjera human fossils are contemporary with their accompanying faunas. The Kanjera skulls, said to be anatomically modern (Groves 1989, p. 291), would thus be equivalent in age to Olduvai Bed IV, which is 400,000 to 700,000 years old. The taxonomic status of the Kanam jaw is uncertain. Recent workers hesitate to call it anatomically modern, although this designation cannot be ruled out completely. If it is as old as the Kanam fauna, which is older than Olduvai Gorge Bed I, then the Kanam mandible would be over 1.9 million years old. Also, crude pebble tools were found at Kanam, and more advanced Chellean tools were found at Kanjera.

11.3 THE BIRTH OF AUSTRALOPITHECUS

In 1924, Josephine Salmons noticed a fossil baboon skull sitting above the fireplace in a friend's home. Salmons, a student of anatomy at the University of the Witwatersrand in Johannesburg, South Africa, took the specimen to her professor, Dr. Raymond A. Dart. She thus set off a train of events that would win Dart worldwide fame.

The baboon skull given to Dart by Salmons was from a limestone quarry at Buxton, near a town called Taung, about 200 miles southwest of Johannesburg. Upon learning this, Dart asked his friend Dr. R. B. Young, a geologist, to visit the quarry and see what else might be found. At the Buxton quarry, Young found a limestone wall, the surface of which showed signs of old caves, filled in with a hard mixture of sand and travertine (a deposit of calcium carbonate). It was this old cave filling that contained the fossils, including many baboon bones. In fact, baboons still inhabited caves on nearby cliffs. When the sections of the wall containing the ancient cave deposits were blasted, Young collected some fossil-bearing chunks and sent them to Dart (Keith 1931, pp. 39–46).

11.3.1 The Taung Child

Two crates of fossils arrived at Dart's home on the very day a friend's wedding was to be held there. Dart's wife pleaded with him to leave the fossils alone until after the wedding, but Dart opened the crates. In the second crate,

Dart saw something that astonished him: "I found the virtually complete cast of the interior of a skull among them. This brain cast was as big as that of a large gorilla" (Wendt 1972, p. 208). Dart then found another piece of rock that appeared to contain the facial bones.

After the wedding guests departed, Dart began the arduous task of detaching the bones from their stony matrix. Without proper instruments, he used his wife's knitting needles to carefully chip away the stone without damaging the fossil remains. Dart wrote: "No diamond cutter ever worked more lovingly or with such care on a priceless jewel—nor, I am sure, with such inadequate tools. But on the seventy-third day, December 23, the rock parted. I could view the face from the front. . . . The creature which had contained this massive brain was no giant anthropoid such as a gorilla. What emerged was a baby's face, an infant with a full set of milk teeth and its permanent molars just in the process of erupting. I doubt if there was any parent prouder of his offspring than I was of my Taung baby on that Christmas" (Fisher 1988, p. 27).

After freeing the bones, Dart reconstructed the skull (Figure 11.5). He characterized the Taung baby's brain as unexpectedly large, about 500 cubic centimeters. The average brain capacity of a large male adult gorilla is only about 600 cubic centimeters. Dart noted the absence of a brow ridge and suggested that the teeth displayed some humanlike features (Boule and Vallois 1957, pp. 87–88). The front teeth were smaller in relation to the back teeth than in the apes, the canines were not as pointed, and there was no diastema. The diastema is a gap

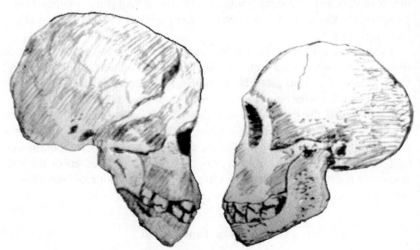

Figure 11.5. Left: The infant *Australopithecus* skull from a quarry near Taung, South Africa, after a photograph by A. R. Hughes (Day 1989, p. 14). Right: The skull of an immature gorilla, after Eckhardt (1972, p. 95).

between the teeth of the lower jaw in apes. The gap accommodates the tips of the large canines protruding downward from the upper jaw. The teeth of apes tend to be arranged in a U-shaped fashion, with the rows of back teeth on either side of the jaw running straight and parallel to each other. The teeth of the Taung specimen, like those of human beings, were arranged in a curved, parabolic dental arcade. The youthful age of the creature could be determined from the fact that among the 24 teeth, 20 were milk teeth and 4 were permanent molars (Keith 1931, p. 52).

Dart also noted that the foramen magnum, the opening for the spinal cord, was set toward the center of the base of the skull, as in human beings, rather than toward the rear, as in adult apes. Dart took this to indicate the creature had walked upright, which meant the Taung specimen was, in his eyes, clearly a human ancestor.

Dart prepared a report for *Nature,* the prestigious British science journal, and sent it off to England. He also told B. G. Paver, news editor for the *Johannesburg Star:* "Perhaps I may shortly have news for you that will not be merely a good local lead. I may have something of worldwide significance connected with man's origin to announce shortly" (Dart 1959, p. 23). Dart gave Paver, who was interested in anthropology, sufficient information to put together an article but made Paver promise not to print it until after his scientific report was published. *Nature,* however, held Dart's article for review by other scientists. Paver became impatient and jumped the gun. He published his own story on February 3, 1925. The *Nature* article appeared four days later (Dart 1959, p. 34). Despite the plan going somewhat awry, Dart's intuition was correct. He became an overnight celebrity, and letters of praise (and blame) began to pour in.

In his *Nature* article, Dart reported: "The specimen is of importance because it exhibits an extinct race of apes intermediate between living anthropoids and man" (Wendt 1972, p. 209). From the accompanying fossils, he estimated his find's age at 1 million years, and called it *Australopithecus africanus*—the southern ape of Africa. *Australopithecus,* he believed, was ancestral to all other hominid forms.

Although public interest and adulation flared up quickly, reaction from the scientific community was substantially more reserved. In England, Sir Arthur Keith and Sir Arthur Smith Woodward received the report from Dart with utmost caution.

Sir Arthur Keith's initial reaction was to give Dart the benefit of the doubt. Keith said: "Professor Dart is not likely to be led astray. If he has thoroughly examined the skull we are prepared to accept his decision" (Johanson and Edey 1981, p. 45). But Keith's later pronouncements were negative: "one is inclined to place *Australopithecus* in the same group or sub-family as the chimpanzee and gorilla. . . . It seems to be akin to both" (*Nature,* Feb. 14, 1924). Some German scientists, such as Hans Weinert, also thought the Taung specimen was nothing more than an anthropoid ape.

The dating of the find also figured into Keith's disapproval. "The Taung ape is much too late in the scale of time to have any place in man's ancestry," he wrote (Johanson and Edey 1981, p. 45). Dart had estimated that the Taung specimen was about 1 million years old. Keith had consistently held that human beings of modern type had existed for well over 100,000 years. But Keith's ideas about the pace of evolution would not allow a transformation from a creature as apelike as the Taung specimen to modern *Homo sapiens* in so short a period of time.

Grafton Elliot Smith was even more critical. In his response to Dart's article in *Nature,* he noted: "Many of the features cited by Professor Dart as evidence of human affinity, especially the features of the jaw and teeth mentioned by him, are not unknown in the young of the giant anthropoids and even in the adult gibbon" (Dart 1959, p. 36).

As time went by, Smith became increasingly unfavorable. In May 1925, in a lecture delivered at University College, Smith stated, in remarks reported in the *Times* of London: "It is unfortunate that Dart had no access to skulls of infant chimpanzees, gorillas, or orangs of an age corresponding to that of the Taung skull, for had such material been available he would have realized that the posture and poise of the head, the shape of the jaws, and many details of the nose, face, and cranium upon which he relied for proof of his contention that *Australopithecus* was nearly akin to man, were essentially identical with the conditions met in the infant gorilla and chimpanzee" (Dart 1959, p. 38).

Grafton Elliot Smith's critique remains valid even today. As we shall see, despite the enshrinement of *Australopithecus* as an ancestor of human beings, several scientists remain doubtful. Anatomical features that to some scientists suggest incipient humanity fall for others within the ape family's range of variation.

The popular press, initially favorable, also began to adopt a different attitude toward *Australopithecus,* making Dart's baby, as it came to be called, a subject of jokes and ridicule. A popular journal, *The Spectator,* asked readers to submit epitaphs for *Australopithecus.* One entry selected for publication read (Dart 1959, p. 38):

> Here lies a man, who was an ape.
> Nature, grown weary of his shape,
> Conceived and carried out the plan
> *By which the ape is now the man.*

Dart also received trouble from another quarter—Biblical creationists angry with him for proposing the forbidden missing link between ape and man. Dart (1959, p. 40) wrote: "Letters from religious people all over the world poured into my office, warning me that I was 'sitting on the brink of the eternal abyss of flame' and would later 'roast in the general fires of Hell.'"

In 1931, Dart was invited to London to give a report about his *Australopithecus* find before the Zoological Society of London. At the same meeting, Davidson Black gave his report introducing Beijing man. Black's presentation was consummately professional, delivered confidently with well-prepared visual aids. Dart, Taung fossil in hand, apparently stumbled through a weak presentation, simply restating his old case, first made in 1924. He failed to change any minds (Dart 1959, pp. 57–58). Dart later submitted a lengthy monograph on *Australopithecus* to the Royal Society, which refused to publish the work in full. Dart therefore withdrew it.

11.3.2 Dart Retreats

Dart was dismayed by the cool reception he received from the British scientific establishment. "Perhaps like Davidson Black," he said, "I should have traveled overseas with my specimens to evoke support for my beliefs" (Dart 1959, p. 51). Instead, Dart remained quietly in South Africa, teaching comparative anatomy at the University of the Witwatersrand in Johannesburg. For many years, he stopped hunting for fossils.

British scientists, led by Sir Arthur Keith, maintained their opposition to Dart's *Australopithecus* throughout the 1930s. Keith (1931, p. 82) said that he found the brain markings on the endocranial cast of the Taung specimen to be like those of the gorilla or chimpanzee, and not at all human. He recognized some differences between the brain of the Taung specimen and apes but concluded that "the difference is not such as to lead us to separate *Australopithecus* from the category of anthropoid apes and place it in a separate group—one intermediate to the highest ape and lowest form of humanity" (Keith 1931, p. 86).

The facial skeleton also appeared quite apelike to Keith. He wrote: "Our comparison of the profile and full-face of the Taung specimen with corresponding views of human and anthropoid skulls leaves no doubt as to the true status of *Australopithecus,* viz. that in all its essential characters it is a true anthropoid ape" (Keith 1931, p. 103).

Here Piltdown man, believed to be similar in geological age to the Taung specimen, entered Keith's calculations. The skull of Piltdown man, as we saw in Chapter 10, was like that of *Homo sapiens.* This fact argued against *Australopithecus,* with its apelike skull, being in the line of human ancestry.

Keith (1931, p. 109) also held that the Taung specimen's teeth, relatively bigger than those of a human child, were apelike.

What about the position of the foramen magnum, the opening through which the spinal cord enters the bottom of the skull? Keith pointed out that Dart had wrongly compared the position of the foramen magnum in the juvenile Taung skull with that of adult human beings and adult chimpanzees.

In adult humans, the foramen magnum is located toward the center of the bottom of the skull. This indicates erect posture. In adult chimpanzees, the foramen magnum is located toward the back of the skull, indicating a quadrupedal posture. The Taung skull's foramen magnum was located in the adult human position, so Dart thought this was good evidence for erect posture in *Australopithecus*.

Keith, however, pointed out that Dart should have compared the position of the foramen magnum in the infant Taung specimen with that of an infant chimpanzee instead of that of an adult. The foramen magnum of a baby chimp lies in around the same position as that of either the Taung specimen (Keith 1931, p. 110).

Also, the foramen magnum of a human child is situated more forward than the foramen magnum of the Taung baby or a baby chimp.

That the foramen magnum in the Taung specimen was located near the base of the skull, rather than the rear, did not, therefore, allow scientists to draw any conclusions about the posture of an adult *Australopithecus*. For that, they required an adult specimen of *Australopithecus,* complete with lower limbs, and no such specimen had yet been found.

Keith (1931, p. 115) concluded: "A close examination of all the features of the Taung skull—the size and configuration of the brain, the composition of the cranial walls, the features of face, the characters of jaws and teeth and the manner in which the head was hafted to the neck—leave me in no doubt as to the nature of the animal to which the skull formed part; *Australopithecus* was an anthropoid ape."

As for the few humanlike characteristics of the specimen, Keith (1931, p. 53) said: "The features wherein *Australopithecus* departs from living African anthropoids and makes an approach towards man cannot be permitted to outweigh the predominance of its anthropoid affinities." For Keith, the total evidence ruled out the possibility that *Australopithecus* was, as most modern paleoanthropologists firmly believe, a human ancestor.

11.3.3 Broom and Australopithecus

When Dart retired from the world stage, his friend Dr. Robert Broom took up the battle to establish *Australopithecus* as a human ancestor. From the beginning, Broom displayed keen interest in Dart's discovery. Soon after the Taung baby made his appearance, Broom rushed to Dart's laboratory. According to Dart (1959, p. 35): "he strode over to the bench on which the skull reposed and dropped on his knees 'in adoration of our ancestor,' as he put it." British science, however, demanded an adult specimen of *Australopithecus* before it would kneel in adoration. Early in 1936, Broom vowed to find one.

On August 17, 1936, G. W. Barlow, the supervisor of the Sterkfontein limestone quarry, gave Broom a brain cast of an adult australopithecine. Broom (1951, p. 44) later went to the spot where the brain cast had turned up and recovered several skull fragments. From these he reconstructed the skull of *Plesianthropus transvaalensis.* The deposits in which the fossil was discovered are thought to be between 2.2 and 3.0 million years old (Groves 1989, p. 198).

More discoveries followed, including the lower part of a femur (TM 1513). Broom and G. W. H. Schepers (1946) described this femur as essentially human (Zuckerman 1954, p. 310). W. E. Le Gros Clark, initially skeptical of this description, later admitted that the femur "shows a resemblance to the femur of *Homo* which is so close as to amount to practical identity." In 1949, W. L. Straus, Jr. (1949) said that the femur "resembles man and cercopithecid monkey in about equal degree" (Zuckerman 1954, p. 311). But according to a modern worker, the key diagnostic features of the Sterkfontein femur (TM 1513) are distinct from those of cercopithecid monkeys and African apes and are "characteristic of modern Man" (Tardieu 1981, pp. 77–79). Since the TM 1513 femur was found by itself, it is not clear that it belongs to a *Plesianthropus* individual. It is possible, therefore, that it could belong to a more advanced hominid, perhaps one resembling anatomically modern humans.

On June 8, 1938, Barlow gave Broom a fragment of a palate with a single molar attached. Broom, as usual, paid Barlow for the fossil, but when Broom asked from where it had come, Barlow was evasive. Broom noticed that the matrix was different from that in which the fossils from Sterkfontein were usually embedded. Some days later, he again visited Barlow and this time insisted that he reveal the source of the fossil.

Barlow told Broom that Gert Terblanche, a local schoolboy, had given him the fossil palate. Broom obtained some teeth from Gert, and together they went to the nearby Kromdraai farm, where the boy had gotten the teeth by pounding them from a fossil skull. Broom collected the skull fragments, and Gert also gave Broom a piece of lower jaw and more teeth. After reconstructing the partial skull, Broom saw it was different from the Sterkfontein type. He called the new creature *Paranthropus robustus.* As the name *robustus* indicates, this australopithecine hominid had a larger jaw and bigger teeth than *Australopithecus africanus,* represented by the Taung baby, and the gracile *Plesianthropus* specimens from Sterkfontein. The Kromdraai site is now considered to be approximately 1.0 to 1.2 million years old (Groves 1989, p. 198), although some have suggested an age of up to 1.8 million years (Tobias 1978, p. 67).

Broom also found at Kromdraai a fragment of humerus (the bone of the upper arm) and a fragment of ulna (one of the bones of the lower arm). He said: "had they been found isolated probably every anatomist in the world would say that they were undoubtedly human" (Broom 1950, p. 57).

In 1947, Le Gros Clark wrote that the humerus fragment from *Paranthropus* (TM 1517) displayed "a very close resemblance to the humerus of *Homo sapiens* and none of the distinctive features found in the recent anthropoid apes" (Zuckerman 1954, p. 310).

As might be expected, not everyone accepted this assessment of the TM 1517 humerus. In 1949, Straus said that "it is in general more like the average chimpanzee than like the average man." But he added that "this probably should not be stressed since it consistently falls within the ranges of variation of both species" (Zuckerman 1954, p. 311). A subsequent morphometric analysis done by H. M. McHenry (1972, p. 95) puts the TM 1517 humerus from Kromdraai "within the human range." As we have seen, scientists attribute the TM 1517 humerus to *Paranthropus robustus,* a robust australopithecine. Significantly, a robust australopithecine humerus from Koobi Fora, Kenya (ER 739), fell outside the human range in McHenry's study (1972, p. 95). So perhaps the TM 1517 humerus belonged to something other than a robust australopithecine. It is not impossible that the Kromdraai humerus and ulna, like the Sterkfontein femur, belonged to more advanced hominids, perhaps resembling anatomically modern humans.

World War II interrupted Broom's excavation work in South Africa. During this interval, he began the task of fully describing his *Australopithecus* discoveries, including Dart's Taung specimen.

After the war, Broom found another australopithecine skull (St 5) at Sterkfontein (Figure 11.6). Later he discovered further remains of an adult female australopithecine (St 14)—including parts of the pelvis, vertebral column, and legs. Their morphology, along with certain features of the Sterkfontein skulls,

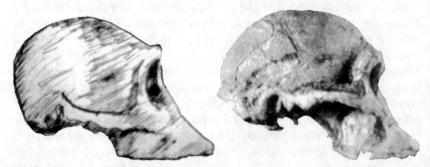

Figure 11.6. Left: The skull of a female chimpanzee (after Zuckerman 1954, p. 308). Right: The St 5 *Plesianthropus (Australopithecus) transvaalensis* skull discovered by Robert Broom at Sterkfontein, South Africa (Broom *et al.* 1950, plate 1).

demonstrated, in Broom's opinion, that the australopithecines had walked erect (Zuckerman 1954, p. 310).

11.3.4 Paranthropus and Telanthropus

At Swartkrans, near Sterkfontein, Robert Broom and J. T. Robinson found, beginning in 1947, fossils of a robust australopithecine called *Paranthropus crassidens* (large-toothed near-man). This creature had large strong teeth and a bony crest on top of the skull. The crest served as the point of attachment for big jaw muscles.

In addition to the fossils of *Paranthropus crassidens*, Broom and Robinson found the jaw of another kind of hominid in the Swartkrans cave. They attributed the jaw (SK 15), smaller and more humanlike than that of *Paranthropus crassidens*, to a new hominid called *Telanthropus capensis*.

Member 1 at Swartkrans, where all of the *Paranthropus* bones were found, is now said to be 1.2 to 1.4 million years old (Groves 1989, p. 198) or 1.8 million years old (Susman 1988, p. 782). But ages of 2.0 million and 2.6 million years have also been proposed (Tobias 1978, p. 65). Member 2, where the SK 15 *Telanthropus* mandible was found, is said to be 300,000 to 500,000 years old. Member 2 is said to represent an erosion channel. This makes it hard to tell how old the SK 15 jaw really is. It could have been washed in with other bones in the Middle Pleistocene. Or perhaps it could have been eroded from Early Pleistocene Member 1. In general, dating fossils found in the South African caves is quite difficult. The caves have been periodically filled and refilled over the course of 1 to 2 million years, resulting in an exceedingly confused stratigraphy. Those who have a degree of faith in chemical dating methods may take note that K. P. Oakley tested the fluorine content of the *Telanthropus* jaw and found it to be the same as the *Paranthropus* fossils (Broom and Robinson 1952, p. 113).

In 1961, Robinson "sank" the genus *Telanthropus* and reclassified the Swartkrans jaw as *Homo erectus* (Brain 1978, p. 140). Broom and Robinson (1952), however, had previously noted several differences between the SK 15 teeth and those of Beijing man and Java man, both of which are now classified as *Homo erectus*. In terms of these differences, the SK 15 teeth were more like those of modern humans. Broom and Robinson also described other ways in which the SK 15 teeth were similar to those of modern humans. But the lower front part of the jaw was damaged, making it "impossible to be sure whether there was a trace of a chin or not" (Broom and Robinson 1952, p. 110). The affinities of this apparently somewhat humanlike jaw remain a mystery.

Broom and Robinson found another humanlike lower jaw at Swartkrans. This fragmentary mandible (SK 45) came not from an erosion channel but from the main deposit containing the *Paranthropus* fossils. Broom and Robinson (1952,

p. 112) said: "In shape it is more easily matched or approached by many modern *Homo* jaws than by that of *Telanthropus.*" Robinson later referred the SK 45 jaw to *Telanthropus* and then to *Homo erectus* (Brain 1978, p. 140). But there are reasons, admittedly not unclouded, to consider other possibilities. Emphasizing the ambiguous nature of the *Telanthropus* fossils, a recent worker (Groves 1989, p. 275) assigned them to an unnamed species of *Homo.*

11.3.5 Paranthropus a Toolmaker?

In the years 1979–1983, C. K. Brain of the Transvaal Museum recovered fossil bones of 130 hominid individuals, 30 crude bone tools, and some crude stone tools. The newly discovered Swartkrans fossils included a relatively small number of well-preserved hand and foot bones.

Speaking of the 8 hand bones from Member 1 at Swartkrans, Randall L. Susman (1988, p. 783) said they indicated "that the robust australopithecines had much the same morphological potential for refined precision grasping and for tool-behavior as do modern humans." Susman (1988, pp. 782–783) noted, however, that the hand bones retained an apelike overall morphology.

The bone tools found at Swartkrans, according to Susman (1988, p. 783), have wear patterns indicating they were used for digging. Susman (1988, p. 783) therefore proposed that *Australopithecus (Paranthropus) robustus* had used stone and bone implements "for vegetable procurement and processing."

Most workers believe the making of tools is an exclusive trait of the genus *Homo,* starting with *Homo habilis.* According to this view, big-jawed *Paranthropus,* a robust australopithecine unconnected to the *Homo* line, munched vegetable matter like the modern gorilla, without the aid of tools.

In a *New York Times* report (1988), Donald C. Johanson, discoverer of Lucy, the most famous representative of *Australopithecus afarensis,* said about the Swartkrans hand bones: "The big question is, how can we be 100 percent sure these hands are not from a *Homo* individual."

Susman admitted that "the attribution of individual fossils to *Paranthropus* is complicated by the presence of a second hominid taxon (*Homo* c.f. *erectus*) at Swartkrans." But he pointed out: "In Member 1, however, more than 95% of the cranio-dental remains are attributed to *Paranthropus.* This fact suggests that there is an overwhelming probability that any one specimen recovered from Member 1 samples [represents] *Paranthropus"* (Susman 1988 p. 782). But even Susman (1988, p. 782) admitted that a thumb metacarpal (SK 84) found in Member 1 at Swartkrans in 1949 probably belonged to a *Homo* individual rather than *Paranthropus.*

So any matching of hand bones with hominid species at Swartkrans is still uncertain. The only thing that could end this uncertainty would be the discovery of hand bones in undisputed connection with other *Paranthropus* fossils.

But even if the new hand bones do belong to *Paranthropus,* there is no guarantee that *Paranthropus,* rather than *Homo,* made any of the stone and bone tools found at Swartkrans. "Did the two species live side by side?" asks anthropologist Eric Delson of the City University of New York. "Did [*P.*] *robustus* use leftovers of *Homo erectus* tool kits? There is no way to test these questions adequately" (Bower 1988, p. 345).

11.3.6 Makapansgat and Final Victory

In 1925, Raymond A. Dart investigated a tunnel at Makapansgat, South Africa. Noting the presence of blackened bones, Dart (1925) concluded hominids had used fire there. In 1945, Philip V. Tobias, then Dart's graduate student at the University of the Witwatersrand, found the skull of an extinct baboon in the cave deposits of Makapansgat and called it to Dart's attention. In 1947, Dart himself went back out into the field, after a lapse of two decades, to hunt for *Australopithecus* bones at Makapansgat.

At Makapansgat, Dart (1948) found australopithecine skull fragments (including an occipital) and other bones, along with more signs of fire. Dart therefore called the creature who lived there *Australopithecus prometheus,* after the Titan who stole fire from the gods. Today, *Australopithecus prometheus* is classified, along with the Taung and Sterkfontein specimens, as *Australopithecus africanus,* distinct from the robust australopithecines of Kromdraai and Swartkrans.

Most of the Makapansgat fossils came from dumps of broken rock in front of the quarry there. From the matrix surrounding the fossils, Dart said he was able to correlate them with identifiable fossil-bearing strata nearby. Had anatomically modern human fossils been recovered in such fashion, any claims for their great age would have been subjected to merciless criticism. This is because the main hominid layers at Makapansgat have been dated at about 3 million years by paleomagnetic methods (K. Weaver 1985, p. 596).

Dart discovered 42 baboon skulls at Makapansgat, 27 of which had smashed fronts. Seven more showed blows on the left front side (Dart 1959, p. 106). Dart, suspecting that australopithecines had been the cause of the damage, requested that R. H. Mackintosh, a specialist in forensic medicine at the University of the Witwatersrand, examine the skulls. Dart and Mackintosh concluded that the skulls showed signs of having been struck by a "powerful downward, forward, and inward blow, delivered from the rear upon the right parietal bone by a double-headed object" (Wendt 1972, p. 224). They believed the weapon was an antelope's humerus (the bone of the upper forelimb). The joint of an antelope humerus, noted Dart, exactly fit the double impressions on several broken baboon skulls.

From the evidence he gathered at Makapansgat, Dart created a lurid portrait of *Australopithecus prometheus* as a killer ape-man, bashing in the heads of baboons with primitive bone tools and cooking their flesh over fires   in the Makapansgat cave. While his robust cousins had remained in the forest, peacefully munching vegetables, and becoming extinct, this more advanced hominid had, according to Dart, ventured into the dry savannahs, to survive by his ruthless wits, and begin the long journey to humanity.

Dart said: "Man's predecessors differed from living apes in being confirmed killers; carnivorous creatures, that seized living quarries by violence, battered them to death, tore apart their broken bodies, dismembered them limb from limb, slaking their ravenous thirst with the hot blood of victims and greedily devouring their writhing flesh" (Johanson and Edey 1981, p. 40).

Dart and Mackintosh also ascertained that australopithecines were killed in the same way as the baboons (Wendt 1972, pp. 226–227). "*Australopithecus* lived a grim life," wrote Dart (1959, p. 191). "He ruthlessly killed fellow australopithecines and fed upon them as he would upon any other beast,   young or old."

Today, however, paleoanthropologists characterize Dart's portrait of *Australopithecus* as somewhat exaggerated. Johanson and Edey (1981, p. 65) call the vision of the killer ape-man "something of an embarrassment to anthropologists, who honor Dart for his dazzling recognition of the first australopithecine, but shake their heads over this later aberration."

In addition to antelope bones, Dart collected at Makapansgat many other animal remains that he believed had been used as daggers, choppers, saws, clubs, and so forth. He grouped these into what he called an "osteodontokeratic" industry, comprising tools made from bones, teeth, and horns (Dart 1957). In 1954, C. K. Brain found pebble tools at Makapansgat, 25 feet above the main layers in which the australopithecine fossils were found. One possible conclusion: a hominid more advanced than *Australopithecus* was the maker of the tools. But Dart (1959, pp. 159–160) pointed out that *Australopithecus* skeletal fragments were also to be found in the   same layer as the pebble tools.

Dart's views about *Australopithecus* hunting activity at Makapansgat aroused heavy opposition. Some scientists said that the combination of *Australopithecus* fossils, mammalian bones, and broken baboon skulls represented not hominid occupation sites but the lairs of hyenas or leopards.

To this Dart (1959, pp. 120–131) replied that hyenas, in particular, do not tend to leave such accumulations of bones in their lairs. However, C. K. Brain replied with a more sophisticated version of the carnivore hypothesis that eventually won the day. "Over a period of years," wrote Richard Leakey and Roger Lewin (1977, p. 96), "Brain observed that a combination of scavenging habits of local carnivores, and the differential resistance to weathering of various types of bone, produces a bone collection virtually identical to the one Dart found in the cave: the osteodontokeratic culture is apparently no more than the left overs from many leopard and hyena meals!"

Nevertheless, this version does not seem to account for some of the evidence reported by Dart. For example, Dart (1959, p. 166) told of finding a gazelle horn wedged solidly into the core of an antelope femur, clear evidence of an intentional act. Dart also noted that the bones of birds, turtles, and porcupines, not the normal prey of hyenas and leopards, were among those found in the cave.

Concerning the evidence for fire at Makapansgat, some researchers said the black deposits were not ash (Oakley 1954, 1956). Others claimed that although there might be signs of fire, the australopithecines were not the cause of them (Broom 1950, p. 74; Johanson and Edey 1981, p. 69).

But even though Dart's views were discredited, there was a positive result. According to Herbert Wendt (1972, p. 222), the controversy over the Makapansgat discoveries "brought the australopithecines into the news, and enhanced their status even in the eyes of their original critics."

Another key event was the publication, in 1946, of a monograph on the australopithecines by Broom and Schepers. The National Academy of Sciences of the United States gave Broom and his coauthor the Daniel Giraud Medal for the most important biological work published in that year.

Sir Arthur Keith wrote in 1947: "When Professor Dart of the University of the Witwatersrand, Johannesburg, announced in *Nature* the discovery of a juvenile *Australopithecus* and claimed for it a human kinship, I was one of those who took the point of view that when the adult form was discovered it would prove to be nearer akin to the living African anthropoids—the gorilla and the chimpanzee. Like Professor Le Gros Clark I am now convinced on the evidence submitted by Dr. Robert Broom that Professor Dart was right and I was wrong. The Australopithecinae are in or near the line which culminated in the human form" (Dart 1959, pp. 80–81). At long last, *Australopithecus* had won recognition in the power centers of British paleoanthropology.

11.3.7 Controversy Continues

With the new status of *Australopithecus* came a change in perception. Increasingly, the vast majority of scientists began to see *Australopithecus* as less and less apelike and more and more humanlike. Right up to the present, the place of *Australopithecus* in the direct line of human descent is taken as an indisputable fact by most paleoanthropologists. Pictures of australopithecines generally show them as essentially human from the neck down. Furthermore, the types of behavior displayed by the australopithecines in these pictures are such that figures of humans could be easily substituted. But even after mainstream English science changed its mind about *Australopithecus*, some scientists resisted. To these recalcitrant renegades, the undistorted facts continued to reveal a starkly apelike portrait of *Australopithecus*. According to their view, a picture of an *Australopithecus* individual should show it hanging by its arms from the branch of a tree rather than walking erect and humanlike on the ground.

The primary dissenter, in the early aftermath of English acceptance of *Australopithecus,* was Sir Solly Zuckerman, secretary of the Zoological Society of London and later a science adviser to the British government. In a comprehensive study, Zuckerman (1954) found that the teeth, skull, jaws, brain, and limbs of *Australopithecus* were essentially apelike. He therefore believed that attempts to identify australopithecines as human ancestors were misguided. Today, a new generation of dissident researchers is raising and sustaining the same objections to overly humanlike characterizations of *Australopithecus*. We shall give detailed attention to their views, and those of Zuckerman, in Section 11.8.

11.4 LEAKEY AND HIS LUCK

After the professional and personal disappointments he encountered in the late 1930s, Louis Leakey continued his work in East Africa, assisted by his second wife, Mary. They searched for fossils of Early Pleistocene human ancestors, which Leakey believed would be quite different from *Australopithecus* and *Homo erectus*. Eventually, the Leakeys would get lucky and make a series of important finds. But for decades they had to be content with stone tools.

11.4.1 Zinjanthropus

A site of particular interest was Olduvai Gorge in Tanzania. There the Leakeys found crude pebble choppers in Bed I, said to be 1.7 to 2.0 million years old (Oakley *et al.* 1977, p. 169). They also found round stones that appeared to have been used as bolas (Section 5.3.2). Leakey even found a bone implement he believed had been used for working leather. The standard image of Early Pleistocene hominids is one of ape-men scavenging carcasses of lion kills, not of protohumans working leather and hunting with bolas.

The stone tools Leakey found at Olduvai were not enough to satisfy him. "The remains of the men themselves still elude us," he said (Goodman 1983, p. 111). Finally, on July 17, 1959, Mary Leakey came across the shattered skull of a young male hominid in Bed I at site FLK. The skull was designated OH 5.

By one account, Leakey came out, looked at the OH 5 skull, and instead of rejoicing said: "Why, it's nothing but a goddamned robust australopithecine" (Johanson and Edey 1981, pp. 91–92). "When he saw the teeth he was disappointed since he had hoped we would find a *Homo* and not an *Australopithecus,*" said Mary Leakey (Johanson and Edey 1981, p. 92).

Mary Leakey eventually pieced together hundreds of fragments, comprising the facial region and the rear part of the hominid's braincase. The creature had a saggital crest, a bony ridge running lengthwise along the top of the skull. In this respect, it was very much like *Australopithecus robustus.* Leakey nevertheless created a new species for OH 5, partly because its teeth were bigger than the South African *robustus* specimens. Leakey called the new find *Zinjanthropus boisei.* Zinj is a name for East Africa and *boisei* refers to Mr. Charles Boise, one of the Leakeys' early financial backers (Wendt 1972, p. 232).

Along with the skull, the cranial capacity of which was about 530 cc, Leakey (1960a, pp. 1050–1051) found bones of mammals, including antelope and pig: "An extensive and rich living floor . . . has been uncovered. . . . All the larger animal bones have been broken open to obtain the marrow; all jaws and skulls of animals are smashed. A high proportion of the bones represent immature animals. Many more stone tools of the Oldowan culture have also been found." This assemblage apparently caused Leakey to give up his initial reserve and proudly declare to the world that he had found the remains of the first stone tool maker, and hence the first "true man."

Why Leakey decided to attribute the tools found at the FLK site to *Zinjanthropus* is somewhat puzzling. Similar tools had been found along with australopithecine remains at Sterkfontein. But Leakey had then said this proved only that the australopithecines of Sterkfontein "were contemporary with a type of early man who made these stone tools, and that the australopithecines were probably the victims which he killed and ate" (Goodman 1983, p. 113).

The FLK site presented a similar situation, calling for a similar explanation. But in a *Nature* article on *Zinjanthropus* Leakey said: "There is no reason whatsoever, in this case, to believe that the skull represents the victim of a cannibalistic feast by some hypothetical more advanced type of man" (Goodman 1983, p. 113).

Leakey became the first superstar that paleoanthropology had seen in a while. Along with *Zinjanthropus,* Leakey flew from Africa to the University of Chicago late in 1959 to participate in the Darwin Centennial, marking the one hundredth anniversary of the publication of *The Origin of Species* (Goodman 1983, p. 115).

The National Geographic Society honored Leakey with funds, publication of lavishly illustrated articles, television specials, and worldwide speaking tours. In 1962, the Society awarded him its highest award, the gold Hubbard Medal, for "revolutionizing knowledge of prehistory by unearthing fossils of earliest man . . . in East Africa" (Goodman 1983, p. 117).

The National Geographic Society is somewhat different from the other foundations active in paleoanthropological research, such as the Carnegie and

Rockefeller foundations. Its funds did not represent the fortune of a single individual or family. The Society started out small and grew on the strength of individual membership contributions, in exchange for which donors received the Society's now famous journal.

Alexander Graham Bell did, however, play an instrumental role in getting the National Geographic Society started. Although Bell did not give large sums of money, he hired Gilbert Grosvenor to supervise the publication of the Society's magazine, and paid his salary from his own pocket for many years. When Grosvenor took over editorial duties in 1899, the magazine of the National Geographic Society was a dry technical journal, intended mainly for specialists in geography. He quickly transformed it into a pictorial magazine with vast popular appeal among the middle and upper classes.

Much of the considerable social influence enjoyed by the National Geographic Society has derived from its carefully cultivated relationships with America's social and political elites. Its board of trustees has consistently represented a cross section of the aristocracies of money and merit. Gilbert Grosvenor himself, from an old New England family, was a cousin of William Howard Taft, who served as President of the United States and Chief Justice of the Supreme Court. Grosvenor married Alexander Graham Bell's daughter, and his son Melville Bell Grosvenor followed him into leadership of the National Geographic Society.

Through its Committee for Research and Exploration, the National Geographic Society expends funds for scientific work in geography and related fields. Results are publicized not only through the magazine of the Society but through school bulletins, news releases, lecture series, films, and television specials.

Until the Society backed Louis Leakey, it had not, the record of its grants shows, supported any work directly related to evolution. Since then, however, the National Geographic Society has been one of the most influential forces in educating the general public, at least in the United States, about the story of human evolution. Exactly why the Society suddenly became so active in this field, starting in 1959, is not explained in any of the accounts of its history we have thus far seen. We would welcome information about this.

In the September 1960 issue of *National Geographic* magazine, Louis Leakey (1960b, p. 433) wrote, in a big photo article about *Zinjanthropus:* "In some respects this new Stone Age skull more closely resembles that of present day man than it does the skulls of the gorilla or of the South African near-men ... *Zinjanthropus* represents a stage of evolution nearer to man as we know him today than to the near-men of South Africa." The article, provocatively titled "Finding the World's Earliest Man," featured an artist's representation of *Zinjanthropus.* Notwithstanding his huge jowls and low forehead, *Zinjanthropus* was depicted as blatantly humanlike—a shameless propaganda move.

But despite an outpouring of publicity, the reign of *Zinjanthropus* was all too brief. F. Clark Howell said: "It obviously was not a man. It was even less manlike than the least manlike of those two South African types" (Johanson and Edey 1981, p. 92). The two South African types were *Australopithecus africanus* (from Taung, Sterkfontein, and Makapansgat) and *Australopithecus robustus* (from Kromdraai and Swartkrans). *Robustus* was considered the least manlike.

Leakey's biographer, Sonia Cole (1975, pp. 239–240), wrote: "He must have wished he could have eaten his words. . . . Granted that Louis had to persuade the National Geographic Society that in Zinj he had a likely candidate for 'the first man' in order to ensure their continued support—but need he have stuck out his neck quite so far? Even a layman looking at the skull could not be fooled: Zinj, with his gorilla-like crest on the top of the cranium and his low brow, was quite obviously far more like the robust australopithecines of South Africa than he was like modern man—to whom, quite frankly, he bears no resemblance at all."

11.4.2 Homo Habilis

In 1960, about a year after the discovery of *Zinjanthropus,* Leakey's son Jonathan found the skull of another hominid (OH 7) nearby in a slightly lower level of Bed I, judged to be about 2 million years old. In addition to the skull, the OH 7 individual included the bones of a hand. Also in 1960, the bones of a hominid foot (OH 8) were found. In succeeding years, more discoveries followed, mostly teeth and fragments of jaw and skull. The fossil individuals were given colorful nicknames: Johnny's Child, George, Cindy, and Twiggy. Some of the bones were found in the lower part of Bed II.

Philip Tobias, the South African anatomist, gave the first newly found skull a capacity of 680 cc, far larger than *Zinjanthropus* at 530 cc, and larger even than the biggest australopithecine skull, at roughly 600 cc. It was, however, around 100 cc less than the smallest *Homo erectus* skulls (Wendt 1972, pp. 245–246).

Leakey sent the OH 7 hand bones to Dr. John Napier of the Royal Free Hospital in England. The results of Napier's study were pleasing to Leakey. The bones, said Napier (1962, p. 409), were "strikingly human in one revealing and . . . critical character. The tips of the fingers and thumb were surmounted by broad, stout, flat, nail-bearing terminal phalanges, a condition that, as far as we know, is found only in man."

The Leakeys sent the OH 8 foot bones to Michael Day for reconstruction. Day, recalling his impressions on completing his work, later said: "My hair stood on end. The foot was completely human" (Cole 1975, p. 253).

Like *Zinjanthropus,* the fossils of the new creatures were found along with broken animal bones and stone tools, scattered across a so-called living floor.

Some distance away from one of the new sites, but at the same level, a circle of large stones was found. The Leakeys interpreted this as the foundation for a windbreak made of brush, giving rise to speculation that the large-brained Olduvai hominid had made use of base camps.

Louis Leakey decided he had now come upon the real toolmaker of the lower levels of Olduvai, the real first true human. His bigger brain confirmed his status, although it was Darwin himself who had said that "one cannot measure intelligence in cubic centimeters" (Wendt 1972, p. 246). A full report on the new Olduvai hominid was published in 1964 by Louis Leakey, John Napier, and Philip Tobias. In this paper (Leakey *et al.* 1964), they called the creature *Homo habilis.* The name, suggested by Raymond Dart, means, "handy man." The designation *Homo* signified a close family relation to modern humans. As we shall see, however, many scientists doubted whether the honor was merited.

After the discovery of *Homo habilis, Zinjanthropus,* no longer the first true human, was demoted to *Australopithecus boisei,* a somewhat more robust variety of *Australopithecus robustus.* Both of these robust australopithecines had saggital crests, and are regarded not as human ancestors but as evolutionary offshoots that eventually became extinct.

The whole business of saggital crests complicates matters somewhat. Male gorillas and some male chimpanzees also have saggital crests, whereas the females of these species do not (Fix 1984, p. 32). This leads to the possibility that creatures assigned to different australopithecine species, on the grounds that some have saggital crests and others do not, may simply represent sexual variants within a single species. For example, Mary Leakey (1971, p. 281) said: "The possibility that *A. robustus* and *A. africanus* represent the male and female of a single species deserves serious consideration." If the possibility raised by Mary Leakey were found to be correct, this would mean that generations of experts have been wildly mistaken about the australopithecines.

11.4.3 Leakey's Views on Human Evolution

With the discovery at Olduvai Gorge of *Homo habilis,* a creature contemporary with the early australopithecines but with a bigger brain, Louis Leakey believed he had excellent evidence supporting his view that neither *Australopithecus* nor *Homo erectus* were in the direct line of human ancestry (Figure 11.7). He later wrote: "For too long scientists have been confused by earlier theories and in particular by those which derived *Homo sapiens* from classical forms of Neanderthal man, which in turn was supposed to have been derived from *Homo erectus,* that in turn was said to have been originated in the Australopithecines. . . . Today the vast amount of evidence that has been accumulated shows us clearly that the stock which was leading to ourselves—as distinct from

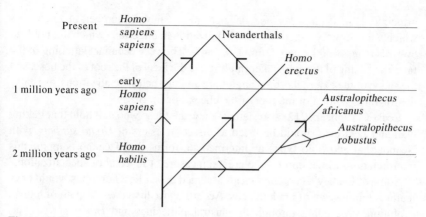

Figure 11.7. According to Louis Leakey (1960d, pp. 210–211; 1971, p. 27), neither *Australopithecus* nor *Homo erectus* was ancestral to modern humans. The Neanderthals, said Leakey (1971, p. 27), were probably the result of crossbreeding between *Homo erectus* and *Homo sapiens.* Today, the details of human evolution remain a subject of active debate. But most paleoanthropologists favor a progression from one of the australopithecines to *Homo habilis, Homo erectus,* early *Homo sapiens,* and then the Neanderthals and modern humans.

Homo erectus—was already present some 2 million years ago in East Africa and that, at that time, it was contemporary with *Australopithecus*. We should therefore expect to find evidence that true *Homo,* as well as primitive *Australopithecus,* was already present during the late stages of the Pliocene, about 4 million years ago" (L. Leakey 1971, p. 25).

Although Leakey was now willing to settle for somewhat primitive *Homo habilis* as the representative of true humanity in the Early Pleistocene, he had earlier believed that the fully modern human type extended that far back in geological time. As we have seen Leakey initially supported Reck's anatomically modern skeleton, found in Bed II of Olduvai Gorge (Section 11.1). He also campaigned on behalf of his own finds of *sapiens*-like human fossils at Kanam and Kanjera (Section 11.2). These finds, all of which Leakey originally thought to be from the Middle and Early Pleistocene, would have been the contemporaries of *Australopithecus* and *Homo erectus*. Later, Leakey withdrew his support of a Middle Pleistocene date for Reck's skeleton when challenged by Boswell, and soon thereafter saw his own finds at Kanam and Kanjera discredited in the eyes of most scientists by the same persistent critic. But in reviewing the controversies over these fossils, we have found, despite some ambiguity, sufficient reason to keep them as evidence for *sapiens*-like beings in Africa 1–2 million years ago.

In Leakey's opinion, the major problem with the standard view of human origins was that it resulted in a progression that appeared to violate evolutionary principles. "*Australopithecinae* or 'near-men' show a number of characters which very strongly suggest over-specialization in directions which did not lead towards man," said Leakey (1960d, p. 184). "The very peculiar flattening of the face, the raising of the eye sockets high above the level of the root of the nose, and the shape of the external orbital angles are among such specializations, as is also the forward position of the root of the cheek-bone process."

Leakey (1960c, p. 212) also stated: "there are those who still hold that Peking man and Java man should be listed as direct ancestors of *Homo sapiens,* with Neanderthal and Solo types as intermediate forms, but I cannot support this interpretation, which implies too great a measure of reversal of specialization."

Some of Leakey's contemporaries assumed the earliest hominids would have features reminiscent of modern apes. According to this view, the path of human evolution, proceeding through the australopithecines and *Homo erectus,* involves a progressive diminution of these primitive apelike features. According to Leakey (1960d), this idea is incorrect.

Certain features of modern apes, such as large brow ridges, are not primitive, said Leakey, but are instead fairly recent specializations. *Proconsul,* an Early Miocene African ape thought to be at the very root of the human line, did not have large brow ridges. "There is no trace whatsoever of a ridge of bone over the eyes, separating the brain-case from the face," wrote Leakey (1960d, p. 175).

Modern humans, with their small brow ridges, according to Leakey, preserve the primitive condition found in the Miocene apes. *Australopithecus, Homo erectus,* and the Neanderthals, with their large brow ridges, depart, like the modern apes, from this primitive condition. The now-dominant evolutionary progression thus involves an evolutionary reversal that Leakey thought unlikely. Miocene apes with no brow ridges give rise to early hominids with heavy brow ridges, and these hominids in turn give rise to modern humans, with small brow ridges. Furthermore, the Miocene apes like *Proconsul* have thin skulls, while the australopithecines, *Homo erectus,* and the Neanderthals have relatively thick skulls. Modern humans have thin skulls, implying another evolutionary reversal.

The advocates of punctuated equilibrium in evolution have a response to Leakey, namely that such reversals can be expected (Stanley 1981, p. 155). One of the great advantages of the punctuated equilibrium theory, which holds that speciation occurs not gradually over long periods of time but in rapid bursts, is that it allows advocates of evolution to easily explain away all kinds of contradictions found in the fossil record.

Apart from size, the physical structure of modern human brow ridges is different from that of other hominids. "The brow-ridge over each eye is made up of two component parts in *Homo sapiens,*" wrote Leakey (1960d, p. 164). "One part in each case starts just above the nose and extends sideways and slightly upwards to overlap the second part, which, on either side, starts at the extreme edge to right and left of the eye-socket respectively, and extends inwards and slightly downwards. Thus, above the center of each eye-socket, there is an overlap of the two elements." In Neanderthal, *Homo erectus,* and *Australopithecus,* the large brow ridges are most often composed of a single barlike mass of bone running horizontally over the eye sockets. To Leakey (1960d, p. 165), the presence of such barlike brow ridges "suggested not an ancestral stage in human evolution but a side branch that has become more specialized, in this respect, than any *Homo sapiens* type."

In addition to features found in the earliest presumed human ancestors (the Miocene apes such as *Proconsul*), modern humans also have, said Leakey, other specializations that distinguish them from *Homo erectus* and *Australopithecus.*

For example, the jaw of modern *Homo sapiens* has a chin eminence, which Leakey (1960d, p. 168) described as a "bony buttress on the front of the middle line of the jaw." Living apes do not have a true chin eminence, and neither do *Homo erectus* and *Australopithecus.*

According to Leakey, the purpose of the chin eminence is to strengthen the front portion of the jaw. In apes this is accomplished by the simian shelf, a ridge of bone running between the two sides of the forward part of the lower jaw. In Neanderthals, *Homo erectus, Homo habilis,* and *Australopithecus,* none of which have a simian shelf, the strengthening is accomplished by thickening the entire front portion of the jaw.

In making his case, Leakey also considered the presence of a feature of the facial skeleton called the canine fossa. Leakey (1960d, pp. 165–166) stated: "If we look at the facial region of different types of *Homo sapiens* we find that . . . there is always present a depression or hollow in the bone beneath each eye, which is called the 'canine fossa.' . . . In the great apes and in the skulls of human species other than *Homo sapiens* it is only very rarely seen and is more commonly replaced by a convexity or puffing out of the bone in that region."

Other anatomical differences between *Homo sapiens* and its presumed ancestors, as discussed by Leakey (1960d), involved the tympanic plate around the ear hole, the mastoid process, the articulation of the jaw, and the position of the foramen magnum.

Time, said Leakey, was another problem. Not only was *Homo habilis* contemporary with *Australopithecus,* thus eliminating the latter, in Leakey's mind, as a human ancestor—there was also trouble with the supposed transition from *Homo erectus* to *Homo sapiens.* Leakey (1971, p. 27) wrote: "The textbooks, on the whole, still suggest that *Homo sapiens* stems from *Homo erectus;* this view can no longer be sustained. The time interval between Java and Peking man in Asia, or the Olduvai form of *Homo erectus* in Tanzania, and the appearance of *Homo sapiens* over a wide area from Europe and east Africa is far too short." The later specimens of *Homo erectus* in Java and China, and in the upper levels of Olduvai Gorge, existed from 200,000 to 500,000 years ago in the Middle Pleistocene. Early *Homo sapiens,* is said to have appeared 300,000 to 400,000 years ago. In other words, *Homo erectus* and *Homo sapiens* were roughly contemporary, and this, to Leakey, seemed to eliminate *Homo erectus* as a human ancestor, although others might suggest that humans branched from *Homo erectus* far earlier.

Here we are, of course, restricting ourselves to conventionally accepted fossil evidence. In previous chapters, we have argued that the totality of evidence— including the fully modern human skeleton found in a Pliocene formation at Castenedolo in Italy, the advanced stone artifacts and human skeletal remains found in Eocene formations in the California gold country, and much else—does not support an evolutionary origin of the modern human type. If this is correct, then we should not expect the various hominid finds in Africa and elsewhere to line up neatly in an evolutionary sequence. And they do not.

If *Australopithecus, Homo erectus,* and the Neanderthals were not human ancestors, then how were they to be explained in terms of evolution? As far as the australopithecines were concerned, Leakey (1960d, p. 180) said it was likely that "they represent a very aberrant and specialized offshoot from the stock which gave rise to man."

Louis Leakey also had some iconoclastic opinions about the relationships among the various australopithecines. "Textbook views that . . . the robust *Australopithecus* is . . . a late specialized variant of the so-called gracile one . . . cannot any longer be regarded as valid," wrote Leakey (1971, p. 27). "A number of examples of the robust Australopithecines have now been found in deposits much older than Olduvai, while side by side with them are specimens that apparently represent ancestors of *Homo habilis.*" Leakey here seems to be referring to discoveries by his son Richard at sites near Lake Turkana, Kenya. As we shall see (Section 11.6), the dating of these Lake Turkana deposits was controversial. Richard Leakey originally favored an age of 2.9 million years, but he eventually agreed with critics that the deposits were about 2 million years old. Even so, the robust australopithecines fossils from Lake Turkana would be as old as the gracile australopithecine fossils found in South Africa. The elder Leakey therefore thought he had good reason to challenge the widely accepted belief that the robust australopithecines were derived from the gracile ones. Leakey's proposal was given additional support in 1986, with the discovery of the so-called Black Skull (Section 11.11), which pushed the robust australopithecines back to 2.5 million years ago. Mary Leakey, as we have seen, outdid her husband in boldness—she suggested the robust and gracile australopithecines might be the males and females of the same species.

As for Java man and Peking man, representing *Homo erectus,* Leakey (1960d, p. 186) also considered them "nothing but various aberrant and over-specialized branches that broke away *at different times* from the main stock leading to *Homo.*"

Leakey was not alone in his views about *Homo erectus.* In 1972, J. B. Birdsell, an anthropologist at the University of California at Los Angeles, wrote: "It is very difficult to visualize how any of the known forms of *Homo erectus* could have evolved into the grade of *Homo sapiens. . . .* nowhere can it be demonstrated that men of the *Homo erectus* grade did evolve into modern populations" (Goodman 1983, p. 121). Of course, there are many who would disagree with Leakey and Birdsell.

Some authorities have placed much emphasis on fossils such as Rhodesia man in Africa, Solo man in Java, and the European Neanderthals. These, they say, show clearly an evolutionary transition between *Homo erectus* and *Homo sapiens.* But Leakey (1971, p. 27) had another explanation: "Is it not possible that they are all variants of the result of crossbreeding between *Homo sapiens* and *Homo erectus?*" One might object that such crossbreeding would have yielded hybrids that were unable to reproduce. But Leakey pointed out that American bison cross fertilely with ordinary cattle.

So whereas some scientists would have *Homo erectus* evolving into the Neanderthals, who then give rise to modern *Homo sapiens,* Leakey would have all three coexisting. And as we have seen (Section 9.2.9), there is substantial evidence from the Chinese Middle Pleistocene that *Homo erectus* coexisted with varieties of *Homo sapiens,* including Neanderthals. In fact, there is evidence that *erectus*-like creatures may exist today in isolated wilderness regions, including China (Chapter 10). There are even reports that they have interbred with humans (Section 10.8). All of this agrees with our proposal that various humanlike and apelike creatures have coexisted in the distant past, just as at present.

11.4.4 Evidence for Bone Smashing in the Middle Miocene

During the late 1960s, Louis Leakey made some interesting discoveries at Fort Ternan, Kenya. The fossil-bearing formations at this site are said to be from 12.5 to 14.0 million years old (Butzer 1978, p. 198), which makes them Middle Miocene. After noting that hundreds of relatively undamaged fossil mammal bones had been found at Ft. Ternan, Leakey (1968, p. 528) said: "In striking contrast to this situation, there are in the same deposit, and at the same level, small areas of fossils where the bones have been broken up, and where the damage includes excellent examples of depressed fractures of the types usually associated with 'a blunt instrument.' . . . We also recovered a peculiar lump of lava exhibiting several battered edges, and with every appearance of having been used to smash bones." According to Glynn Isaac (1978, p. 229), Leakey believed that the lava was not of the kind found normally in the deposit; therefore it must have been transported to the site. Leakey concluded that an apelike Miocene hominid called *Kenyapithecus* had used the lava stone to crack bones for marrow. E. L. Simons (1978, pp. 548–549) and others considered *Kenyapithecus* to be an African variety of the Asian hominid *Ramapithecus*. Currently, however, scientists do not think the ramapithecines can be classified as hominids (Section 3.9). Because no one (as far as we know) now attributes tool behavior to Miocene apes, we are left wondering what hominid used stones to break bones for marrow at Ft. Ternan over 12.5 million years ago. We do not know, but, as we noted in Chapter 2, modern humans leave similar broken bone assemblages.

11.5 A TALE OF TWO HUMERI

In 1965, Bryan Patterson and W. W. Howells found a surprisingly modern-looking hominid humerus (upper arm bone) at Kanapoi, Kenya. In 1977, French workers found a similar humerus at Gombore, Ethiopia.

11.5.1 The Kanapoi Humerus

The Kanapoi humerus fragment, consisting of the intact lower (or distal) part of the bone, was found on the surface. But B. Patterson and Howells (1967, p. 64) noted: "Color, hardness, and degree of mineralization agree with those of numerous specimens collected *in situ* in the sediments." Potassium-argon tests on volcanic materials above the bone-bearing sediments yielded dates of 2.9 and 2.5 million years. Paleomagnetic tests showed the lava displayed reverse polarity. The Matuyama Reverse Epoch began 2.5 million years ago, consistent with the potassium-argon results (B. Patterson and Howells 1967, p. 64).

The Pliocene lake sediments also yielded a fauna earlier than that found in Bed I of Olduvai Gorge. Patterson and Howells said it corresponded to the early Villafranchian of Europe. Another researcher later commented on the faunal remains: "These are comparable to those found at the Mursi site in the Omo River valley, with an age of 4.0 to 4.5 million years" (Senut 1979, p. 113). Patterson accepted this as a reasonable date for the layer from which the humerus was thought to have eroded (Oakley *et al.* 1977, p. 59).

Could the bone have been intrusive in the deposit? B. Patterson and Howells (1967, p. 64) stated: "The excellent state of preservation—the fragment shows no significant postmortem damage other than the break that separated it from the remainder of the original bone—rules out the possibility of derivation from later deposits that may once have been present in the vicinity of the capping lava."

B. Patterson and Howells (1967, p. 65) said the Kanapoi humerus was "readily distinguishable . . . from gorilla and orangutan." They then made detailed "morphological and metrical comparisons" with human beings, chimpanzees, and *Australopithecus*.

Patterson and Howells measured 7 features on 40 human humeri, 40 chimpanzee humeri, and a cast of the distal humerus of *Paranthropus robustus* (Kromdraii TM 1517), the only australopithecine distal humerus then available. They concluded: "In these diagnostic measurements Kanapoi Hominoid 1 is strikingly close to the means of the human sample. It is larger than the individual of *Paranthropus robustus* represented by the corresponding humeral fragment from Kromdraai in each measurement" (B. Patterson and Howells 1967, p. 65). Patterson and Howells (1967, p. 65) added: "*Paranthropus* emerges from these morphological comparisons as rather less man-like than Kanapoi Hominoid 1." Further emphasizing the humanlike character of the Kanapoi humerus, they said: "there are individuals in our sample of man on whom measurements . . . of Kanapoi Hominoid I can be duplicated almost exactly" (B. Patterson and Howells 1967, p. 66).

Patterson and Howells would not have dreamed of suggesting that the Kanapoi humerus belonged to an anatomically modern human. Nevertheless, if an anatomically modern human had died at Kanapoi 4.0–4.5 million years ago, he or she might have left a humerus exactly like the one they found.

Further confirmation of the humanlike morphology of the Kanapoi humerus (KNM KP 271) came from anthropologists Henry M. McHenry and Robert S. Corruccini of the University of California. Using multivariate analysis techniques, they compared 16 different measurements of the Kanapoi humerus with those of the humeri of all species of anthropoid apes, three species of monkeys, and two fossil hominids—Kromdraai (TM 1517) and East Rudolf (ER 739). McHenry and Corruccini (1975, p. 227) concluded that "the hominid fossil from Kanapoi resembles *Homo sapiens* very closely." Elsewhere in the same study they noted: "The Kanapoi fossil is quite close to *Homo,* especially the Eskimo sample" (McHenry and Corruccini 1975, p. 235). Amplifying this, they stated that "the Kanapoi humerus is barely distinguishable from modern *Homo*" and "shows the early emergence of a *Homo*-like elbow in every subtle detail" (McHenry and Corruccini 1975, p. 240).

In an earlier study, McHenry (1973) wrote: "A humeral fragment has been found at Kanapoi that is almost five million years old yet almost indistinguishable in shape from many modern humeri. Geologically much younger australopithecine humeri at one or two million years are vastly different from those of modern man."

In his Harvard doctoral thesis, McHenry (1972, p. 95) stated that the Kanapoi humerus fell "within the human range." We have employed a simple multivariate analysis technique to evaluate the raw data supplied by McHenry in his thesis. We calculated the 16-dimensional vectors represented by his 16 measurements for each humerus, and took the size of the angles between any two vectors as indicators of the degree of similarity between the two humeri. A smaller angle means a greater similarity. This method, it should be noted, is size-independent.

In other words, bones of the same conformation, though being of different size, will show a difference of zero degrees. Confirming McHenry, we found that at 2.75 degrees the Kanapoi humerus vector was closest to *Homo sapiens* . For comparison, the angle of Kanapoi with chimpanzee was 4.40 degrees. With *Australopithecus robustus* (Kromdraai TM 1517) the angle was 4.51 degrees, and with *Australopithecus boisei* (East Rudolf) it was 4.83 degrees. In other words, the Kanapoi humerus differed from those of the australopithecines.

C. E. Oxnard (1975a, p. 97) agreed with McHenry's analysis. He stated: "we can confirm clearly that the fossil from Kanapoi is very humanlike." In his discussion, Oxnard pointed out that the Kanapoi humerus, although 4 millions years old, was quite modern in form, while the australopithecine humeri from later periods were much less so. This led Oxnard (1975a, p. 121) to suggest, as did Louis Leakey, that the australopithecines were not in the main line of human evolution. Keeping *Australopithecus* as a human ancestor would result in a very unlikely progression from the humanlike Kanapoi humerus, to the markedly less humanlike humerus of *Australopithecus,* and then to one more humanlike again.

Michael A. Day (1978, p. 315) said about the Kanapoi humerus: "it is hard to point to a single anatomical feature or group of features that is not well known in modern man. Functionally it must be nearly identical with the modern human condition."

A dissenting view may be found in a study by Marc R. Feldesman, of Portland State University in Oregon. From his own multivariate analysis of 15 fossil humeri and humeri of 22 species of monkey and apes, Feldesman (1982a, p. 73) concluded: "The Kanapoi distal humerus (KP 271), far from being more 'human-like' than *Australopithecus,* clearly associates with the hyperrobust Australo-pithecines from Lake Turkana." The Lake Turkana specimen closest to KP 271, according to Feldesman, was ER 739, now thought to represent *Australopithecus boisei.* This is exactly the reverse of McHenry's conclusion. McHenry found that KP 271 was close to *Homo sapiens* and distant from ER 739. Because Feldesman did not supply his raw data in his report, we could not evaluate his results.

In our discussion of fossil discoveries in China (Section 9.2.1), we made extensive use of the concept of possible date ranges. That is to say, when confronted with reports giving different ages for certain fossils, we established a range of possibilities that included all likely ages. Here we want to introduce a similar concept—that of possible morphology ranges. Concerning the Kanapoi humerus, we can say, on the basis of the reports we have cited, that its morphology range extends to the modern human end of the spectrum.

11.5.2 The Gombore Humerus

In 1977, French researchers (Chavaillon *et al.* 1977) reported finding a humanlike humerus at the Gombore site in Ethiopia, about 55 kilometers south of the capital, Addis Ababa.

The Gombore humerus was, however, more recent than the Kanapoi humerus. Noting that stone tools were found near the Gombore humerus, Brigitte Senut (1979, pp. 112–113) stated: "The stone industry of Gombore IB is like that of the upper part of Bed I and the base of Bed II at Olduvai (Tanzania), which have been dated at 1.7 million years by the potassium-argon method. The same radiometric method applied to basalt at the Ethiopian site gives the layers in which the Oldowan tools were found a date older than 1.5 million years." The first excavators (Chavaillon *et al.* 1977, p. 961) also noted: "The site is an Oldowan encampment, with a shelter and organized zones containing different types of tools."

Senut (1979, p. 111) said, in an English summary of one of her French papers, that the Gombore humerus could, along with the Kanapoi humerus, "be attributed to the genus *Homo.*" Concerning the Kanapoi humerus, Senut was in agreement with B. Patterson and Howells (1967), McHenry and Corruccini (1975), McHenry (1972, 1973), Oxnard (1975a), and Day (1978), who all thought the Kanapoi humerus to be unlike that of *Australopithecus.* Senut differed from Feldesman (1982a), who thought the Kanapoi humerus to be like that of *Australopithecus boisei* (ER 739).

Like Senut (1979), the original discoverers of the Gombore humerus hesitated to designate it as anything more than *Homo* (Chavaillon *et al.* 1977). Similarly, Feldesman (1982a, p. 92), who thought the Kanapoi humerus to be like those of australopithecines, said: "The Gombore specimen appears to be closer to *Homo* than to anything else." But Chavaillon and his coworkers (1977, p. 962) noted: "in the lateral view, the bone very much resembles *Homo sapiens sapiens.*"

Senut later found other features that were humanlike. "Gombore IB 7594, which was primitively [first] attributed to the genus *Homo* (Chavaillon *et al.* 1977, Senut 1979), cannot be differentiated from a typical modern human," she wrote (Senut 1981b, p. 91).

So now we seem to have two very ancient and humanlike humeri to add to our list of evidence challenging the currently accepted scenario of human evolution. These are the Kanapoi humerus at 4.0–4.5 million years in Kenya and the Gombore humerus at more than 1.5 million years in Ethiopia. At the very least, the Early Pliocene Kanapoi humerus "could challenge the new phylogenies tending to show that only one genus and one species (*Australopithecus afarensis*) was living at this date" (Senut 1979, p. 111). The Kanapoi and Gombore humeri also support the nonevolutionary view that human beings of modern type have coexisted with other humanlike and apelike creatures for a very long time.

11.6 RICHARD, SON OF LEAKEY

Louis Leakey's son Richard at first avoided fossil hunting, working instead as a safari organizer for clients including the National Geographic Society.

Eventually, however, Richard took up the family profession. Although he had no university training, he began to develop his own reputation as a competent paleoanthropologist.

In 1967, Richard Leakey, then just 23 years old, led the Kenya section of an international paleoanthropological expedition to the Omo region of southern Ethiopia. Unhappy at having to turn over fossils he discovered to professional scientists, Leakey suddenly left the Omo site. He flew by helicopter to Koobi Fora, on the crocodile-infested eastern shores of Kenya's Lake Rudolf, now called Lake Turkana. On his very first walk around Koobi Fora, Leakey found a stone tool and fossil pig jaw. The site was promising, but he needed funding in order to systematically develop it.

In January of 1968, Richard Leakey journeyed to Washington, D.C., where he got a grant of 25,000 dollars from the National Geographic Society's Committee for Research and Exploration. Returning to Kenya, Leakey set up a permanent camp at Koobi Fora.

That first year saw no major discoveries, but in 1969 Richard and his wife Meave found an australopithecine skull. Over the next few years, fossils of three more *Australopithecus* individuals turned up (R. Leakey 1973b, p. 820). Also, Glynn Isaac found hundreds of crude stone tools at several Early Pleistocene sites near Koobi Fora (R. Leakey 1973b, p. 820). *Australopithecus* was not known to have been a toolmaker. So who had made the tools?

11.6.1 Skull ER 1470

In August of 1972, Bernard Ngeneo, a member of Leakey's team, found at Lake Turkana a shattered skull that appeared to give an answer. Richard's wife Meave, a zoologist, reconstructed the skull, designated ER 1470. Alan Walker of the University of Nairobi estimated that its cranial capacity was over 810 cc (R. Leakey 1973a, p. 449), bigger than the robust australopithecines. For example, the robust OH 5 *Australopithecus boisei* specimen from Olduvai, formerly called *Zinjanthropus,* had a cranial capacity of just 530 cc (R. Leakey 1973a, p. 450). The ER 1470 skull was in fact as large as some smaller *Homo erectus* skulls, which range between 750 and 1100 cc. The average human skull is about 1400 cc. Among adult humans, the very lowest cranial capacities are in the low 800s (Brodrick 1971, p. 84).

Viewed from the rear, the sides of the reconstructed ER 1470 skull were nearly vertical, as in *Homo sapiens.* In *Australopithecus* and *Homo erectus*, the sides of the skull, seen from the rear, slope noticeably towards each other at the top (Figure 9.1, p. 556). Furthermore, the domed forehead of ER 1470 was not as receding as that of *Australopithecus* or *Homo erectus,* and the brow ridges were smaller. The skull walls of ER 1470 were thinner than those of *Australopithecus* or

Homo erectus. Also, the foramen magnum, the opening in the base of the skull for the spinal cord, was located farther forward than in *Australopithecus.* In other words, several features of the somewhat primitive ER 1470 skull were characteristic of advanced species of the genus *Homo* (Fix 1984, pp. 50–51; R. Leakey 1973a, p. 448).

Richard Leakey initially hesitated to designate a species for the ER 1470 skull, but eventually decided to call it *Homo habilis.* This strengthened the evidence for *Homo habilis* from Olduvai Gorge, announced by Louis Leakey in the 1960s.

What made the ER 1470 skull so unusual was its age. The stratum yielding the skull lay below the KBS Tuff, a volcanic deposit with a potassium-argon age of 2.6 million years. The skull itself was given an age of 2.9 million years, as old as the oldest australopithecines. The KBS Tuff's age was later challenged, with critics favoring an age of less than 2 million years (Section 11.6.5).

11.6.2 Evolutionary Significance of the ER 1470 Skull

Louis Leakey was pleased with his son's discovery. ER 1470 vindicated his long-held view that a line of human ancestors, separate from *Australopithecus* and *Homo erectus,* extended far into the past.

Richard Leakey also believed his find had revolutionary implications for human evolution. "Either we toss out this skull or we toss out our theories of early man," he wrote in *National Geographic* (R. Leakey 1973b, p. 819). "It simply fits no previous models of human beginnings." The model most widely accepted involved three steps. *Australopithecus africanus,* with some specimens as much as 3 million years old (Groves 1989, p. 198), gave rise to early *Homo* (*H. habilis* and then *H. erectus*), which in turn gave rise to *Homo sapiens.* But Leakey (1973b, p. 819) believed that the ER 1470 skull, larger and more humanlike than that of *Australopithecus africanus,* "leaves in ruins the notion that all early fossils can be arranged in an orderly sequence of evolutionary change."

J. B. Birdsell (1975) of UCLA agreed this was true, even if the ER 1470 skull proved to be 2 million rather than 2.9 million years old. "From the very nature of its characteristics cranium 1470 does not seem to fit the standard scheme of the three grades of human evolution," he wrote in the second edition of his textbook *Human Evolution* (Fix 1984, p. 60).

In a *National Geographic* article, Richard Leakey included a chart showing two separate lines of hominid development. On one line, at about 3 million years ago, Leakey placed the ER 1470 hominid. Next on this line came *Homo habilis* at roughly 2 million years ago. At 1 million years ago, *Homo habilis* gave way to *Homo erectus,* which was followed at the very top of the chart by *Homo sapiens.*

The second (completely separate) line in Richard Leakey's chart showed *Australopithecus* starting at 3 million years ago and finishing at 1 million years ago. Leakey (1973b, p. 819) commented: "Probably a relative rather than a forebear of mankind, apelike *Australopithecus* existed for at least 2 million years before it reached an evolutionary dead end." Leakey believed, however, that further research would turn up a common ancestor for *Australopithecus* and the *Homo* line at around 4 million years ago.

Richard Leakey differed from his father by keeping *Homo erectus* in the direct line of human ancestry, "Most people would now agree that '1470' should be called *Homo habilis* and that it is a direct ancestor of *Homo erectus*," he wrote (R. Leakey 1984, p. 154).

But the transition from ER 1470 to *Homo erectus* troubled Birdsell (1975), who wrote: "Anatomically in some ways such an evolutionary stage would seem retrogressive, for in a real sense it postulates that more archaic forms of men evolved out of a surprisingly advanced form, ER-1470" (Fix 1984, p. 137). Birdsell's statement is of interest because the progression from *Homo habilis* to *Homo erectus* is one of the cardinal doctrines of recent evolutionary thought. If this progression turns out to be improbable, that would present severe problems for the conventional account of human evolution. The progression is arguably improbable because it involves, for example, going from skull ER 1470, with moderate brow ridges, to *Homo erectus,* with massive barlike brow ridges, back to *Homo sapiens,* with small brow ridges.

Such difficulties did not, however, trouble Richard Leakey. Recently, he said he considers *Homo habilis* and *Homo erectus* to be nothing more than early stages of one species—*Homo sapiens* (Willis 1989, pp. 154–155).

Richard Leakey has made other interesting statements about human beginnings. For instance, he wrote in his book *Origins*: "If we are honest we have to admit we will never fully know what happened to our ancestors in their journey towards modern humanity: the evidence is simply too sparse" (R. Leakey and Lewin 1977, pp. 11–12).

And in *People of the Lake* (R. Leakey and Lewin 1978, p. 17), Leakey said: "If someone went to the trouble of collecting together in one room all the fossil remains so far discovered of our ancestors (and their biological relatives) who lived, say, between five and one million years ago, he would need only a couple of large trestle tables on which to spread them out. . . . Yet with a confidence that may strike the uninitiated as something close to supernatural—if not to plain madness—prehistorians can now construct a view of human origins that is anything but crude, and may even bear some resemblance to the truth." The evidence on the trestle tables would not, of course, be complete. Much has been suppressed or forgotten, and if it were placed back on the tables, it would be harder for confident prehistorians to construct plausible evolutionary lineages.

11.6.3 Humanlike Femurs from Koobi Fora

Some distance from where the ER 1470 skull had been found, but at the same level, John Harris, a paleontologist from the Kenya National Museum, discovered a quite humanlike upper leg bone. Harris summoned Richard Leakey (1973b, pp. 823, 828), who later reported: "Amid a mass of shattered elephant bone lay both ends of the femur of a remarkably advanced hominid. Further search turned up the missing pieces, parts of the tibia and a fragment of the fibula. . . . John also discovered another femur. All these leg bones lay in deposits older than 2.6 million years. Do they belong to our new-found '1470 man?' Frustratingly, we cannot be sure. It is quite clear, however, that these femurs are unlike those of *Australopithecus,* and astonishingly similar to those of modern man." The femurs would later be attributed to *Homo habilis.*

The first femur, with associated fragments of tibia and fibula, was designated ER 1481 and the other ER 1472. An additional fragment of femur was designated ER 1475. Like the ER 1470 skull, the femurs were found on the surface. But Richard Leakey (1973a, p. 448) wrote in *Nature:* "The unrolled condition of the specimens and the nature of the sites rules out the possibility of secondary deposition—there is no doubt in the minds of the geologists that the provenance is as reported. All the specimens are heavily mineralized and the adhering matrix is similar to the matrix seen on other fossils from the same sites." In other words, Leakey was certain the bones had recently weathered out of the fossil-bearing deposits from below the KBS Tuff.

Leakey (1973a, p. 450) stated in a scientific journal that these leg bones "cannot be readily distinguished from *H. sapiens* if one considers the range of variation known for this species." In a *National Geographic* article, Leakey (1973b, p. 821) repeated this view, saying the leg bones were "almost indistinguishable from those of *Homo sapiens.*"

Comparing the newly found ER 1481 femur with a femur of *Australopithecus,* Leakey (1973b, p. 828) said: "The more ovoid, less robust shaft neck of *Australopithecus* implies that the latter, though capable of walking upright, did so only for short periods." The "stronger neck shaft" of the new femur, Leakey (1973b, p. 828) added, "suggests its owner probably walked upright as his normal mode of locomotion."

Concerning ER 1481, Richard Leakey (1973a, p. 450) wrote: "When the femur is compared with a restricted sample of modern African bones, there are marked similarities in those morphological features that are widely considered characteristic of modern *H. sapiens.* The fragments of tibia and fibula also resemble *H. sapiens.*" He further stated: "The head of the femur is large and set on a robust cylindrical neck which takes off from the shaft at a more obtuse angle than in known *Australopithecus* femurs" (R. Leakey 1973a, pp. 449–450).

Other scientists agreed with Leakey's analysis. In 1976, B. A. Wood, anatomist at the Charing Cross Hospital Medical School in London, showed that in terms of three critical variables (femur neck length, femur head size, and femur neck shape), the ER 1472 and ER 1481 femurs always fell within a single standard deviation from the modern human mean. Wood (1976, p. 502) wrote: "The data . . . clearly show that femurs 1472 and 1481 from East Rudolf belong to the 'modern human walking' locomotor group." Christine Tardieu (1981), also identified several humanlike features of the lower parts of the ER 1481 and ER 1472 femurs. Other workers found the femurs different from those of *Homo erectus* (Section 11.7.1).

Although most scientists would never dream of it, one could consider attributing the Koobi Fora femurs to a hominid very much like modern *Homo sapiens,* living in Africa about 2.9 million years ago (about 2.0 million years ago if you choose to believe the revised date of 1.9 million years for the KBS Tuff).

The ER 1472 and ER 1481 femurs show that distinctly anomalous discoveries are not confined to the nineteenth century. They have continued to occur with astonishing regularity up to the present day, right under our very noses, so to speak, although hardly anyone recognizes them for what they are. In Africa alone, we are building up quite a catalog: Reck's skeleton, the Kanam jaw, the Kanjera skulls, the Kanapoi humerus, the Gombore humerus, and now the Lake Turkana femurs. All have been either attributed to *Homo sapiens* or described as being very humanlike. Except for the Middle Pleistocene Kanjera skulls, all were discovered in Early Pleistocene or Pliocene contexts.

11.6.4 The ER 813 Talus

In 1974, B. A. Wood (1974a, p. 135) described a talus (ankle bone) found between the KBS Tuff, then given an age of 2.6 million years, and the overlying Koobi Fora Tuff, with an age of 1.57 million years. Wood compared the fossil talus, designated ER 813, with hundreds of others, including those of modern humans, gorillas, chimpanzees, and other arboreal primates.

Using multivariate statistical techniques, Wood analyzed the ankle bones in terms of 3 angular and 5 linear measurements. He concluded: "In all the variates, the fossil aligned with the modern human tali" (Wood 1974a, p. 135). Wood further stated: "the functional implications of the canonical analysis results, combined with the close morphological affinity of the fossil talus with the modern human bones, make it possible that the locomotor pattern of this early hominid was like that of modern man" (1974a, p. 136).

If we accept the younger date for the KBS Tuff, the humanlike ER 813 talus would be 1.5 to 1.9 million years old, roughly contemporary with creatures designated as *Australopithecus robustus, Homo erectus,* and *Homo habilis.*

In a subsequent report, Wood (1976, pp. 500–501) said his tests confirmed "the similarity of KNM-ER 813 with modern human bones," showing it to be "not significantly different from the tali of modern bushmen." One could therefore consider the possibility that the KNM-ER 813 talus belonged to an anatomically modern human in the Early Pleistocene or Late Pliocene.

C. E. Oxnard (1975a, p. 121) wrote of ER 813: "description and examination using canonical analysis by Wood (1974) confirms that it is indeed very similar to modern man and is thus unlike the australopithecine specimens." Challenging the ancestral status of *Australopithecus,* Oxnard (1975a, p. 121) added: "Unless evolution took the talus through a stage where it was much like man (as at East Rudolf), then through a stage where it was uniquely different from man (as at Olduvai and possibly Kromdraai), and back again to a stage like man (modern man), then australopithecine fossils had to have been unrelated to any direct human line."

Of course, if the KNM-ER 813 talus really did belong to a creature very much like modern human beings, it fits, like the ER 1481 and ER 1472 femurs, into a continuum of such finds reaching back millions of years. In this case, any talk of an evolving human line, to which hominid fossils different from those of modern humans may be related, directly or indirectly, becomes irrelevant.

11.6.5 The Age of the KBS Tuff

The KBS Tuff was named after Kay Behrensmeyer, the Yale geologist who first identified it. Such volcanic tuffs can be dated by the potassium-argon method. If the dated tuff can be properly traced over difficult terrain, it can be used to determine a minimum age for fossils found below it. Over the years, workers obtained differing potassium-argon ages for the KBS Tuff, with substantial impact on the dating of fossil hominids at Lake Turkana.

11.6.5.1 The Potassium-Argon Dating Method

The potassium-argon (K/Ar) method relies on the decay of radioactive potassium 40 into argon, a stable gas. In principle, one can, by correctly measuring the amounts of potassium and argon in a sample, calculate its age. The more argon, the older the sample.

In practice, there are many difficulties in using this method. For the age range in question (2–4 million years), the accumulation of argon is very small. The measurements are thus extremely sensitive to any artificial loss or gain of argon.

Exposed to weathering, a sample may lose some of its argon. In this case, the measured age would be younger than the sample's true age If materials from older deposits get mixed into a sample, thus adding argon, the measured age would then be older than the true age of the sample.

In testing a sample, this question always arises: Has there been any argon loss or gain? In making such judgements, the investigator has wide latitude for personal interpretation.

Illustrating the difficulties inherent in the potassium-argon method, scientists have obtained ages ranging from 160 million to 2.96 billion years for Hawaiian lava flows that occurred in the year 1800. A report in the *Journal of Geophysical Research* stated: "It is possible that some of the abnormally high potassium-argon ages . . . may be caused by the presence of excess argon contained in fluid and gaseous inclusions" (Funkhouser and Naughton 1968, p. 4606).

Potassium-argon tests often yield such unexpected results, far older or younger than the generally accepted ages for the formations being dated. One researcher (Woodmorappe 1979) compiled a list of 275 discrepant potassium-argon dates. From his tables, we have selected a few representative dates from the geological era most relevant to our study, the later Tertiary (Table 11.3).

In a potassium-argon study of formations in the western United States, geologist R. L. Mauger (1977, p. 37) stated: "In general, dates in the 'correct ball park' are assumed to be correct and are published, but those in disagreement with other data are seldom published nor are discrepancies fully explained." And

TABLE 11.3

Discrepant Potassium-Argon Dates (Later Tertiary)

Age Expected	Age Obtained	Formation/ Location	Original Reference
5 myr	1–10.6	Bailey Ash/California, U.S.A.	Bandy and Ingle 1970
9.5	13–31	tuff/Nevada, U.S.A.	Marvin and Cole 1978
10	95	basalt/Nigeria	D. Fisher 1971
10	153	Nogales Formation (tuff)/ Arizona, U.S.A.	Marvin *et al.* 1973
23	3.4	Suta Volcanics/ Solomon Islands	Chivas and MacDougall 1978
27	31–43	volcanics/Kamchatka, Russian Republic	Pozdeyev 1973

Data in this table are from J. Woodmorappe (1979, Table 1).

geologist J. B. Waterhouse (1978, p. 316) noted: "It is, of course, all too facile to 'correct' various values by explanations of leakage, or initially high concentrates of strontium or argon. These explanations may be correct, but they must first be related to a time line or 'cline of values' itself subject to similar adjustments and corrections on a nonstatistical and nonexperimental basis."

This raises an important issue. E. T. Hall (1974, p. 15), director of Oxford's Research Laboratory for Archaeology and the History of Art, warned: "the greatest temptation is the one which leads an archaeologist selectively to believe evidence which seems to confirm the theories upon which he thinks his professional reputation rests. When the evidence comes from complex scientific techniques which are error prone and involve principles not wholly understood even by the scientists themselves, the dangers are great indeed."

Potassium-argon dating is such an error-prone technique. When, however, ordinary persons, or even scientists in disciplines other than those directly connected with the paleoanthropological enterprise, hear that a fossil has been dated by the potassium-argon method, they think the matter has been settled by science in a thoroughly reliable fashion. But when one gets beyond the screen of footnotes and suitably restrained phrasing in paleoanthropological reports, one frequently discovers that the dating is quite nebulous. The strongest argument in favor of a particular date is often the personal commitment of a scientist whose ideas are supported by the date.

Radiometric dates, said E. T. Hall (1974, p. 15), "tend to acquire a spurious infallibility for the layman or for quasi-scientists like archeologists. They believe because they want to believe."

11.6.5.2 The Potassium-Argon Dating of the KBS Tuff

In 1969, Richard Leakey sent samples of the KBS tuff to England for potassium-argon testing. According to E. T. Hall (1974, p. 15), the age obtained was a seemingly impossible 220 million years.

In 1970, F. T. Fitch and J. A. Miller, having received new samples, ran more potassium-argon tests and decided that the KBS Tuff was 2.6 million years old.

In 1972, Richard Leakey discovered the ER 1470 *Homo habilis* skull. Because the skull came from well below the KBS Tuff, he assigned it an age of 2.9 million years. This was controversial, because it made *Homo* as old as the oldest *Australopithecus*.

Results of paleomagnetic studies by Dr. A. Brock of the University of Nairobi confirmed the potassium-argon date given by Fitch and Miller for the KBS Tuff. (For an explanation of the paleomagnetic dating method, see Section 9.2.10.)

Brock found that samples in and near the KBS Tuff were of normal polarity (Brock and Isaac 1974, p. 346). This was consistent with the potassium-argon

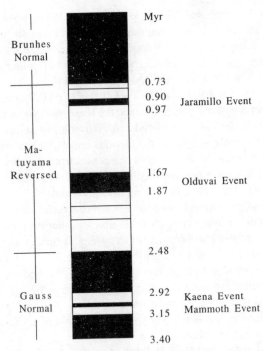

Figure 11.8. This is the standard paleomagnetic polarity scale (after Wu, X. and Wang, L. 1985, p. 36). Brock and Isaac (1974) believed the polarity sequence at Koobi Fora supported an age of 2.6 million years, in the Gauss Normal Epoch, for the KBS Tuff.

date of 2.6 million years obtained by Fitch and Miller, which, if correct, falls in the upper part of the Gauss Normal Epoch (Figure 11.8). Below the KBS Tuff, the samples were predominantly of normal polarity. But Brock also found in this region two short intervals of reversed polarity, which he identified with the Kaena and Mammoth Events (Brock and Isaac 1974, p. 346). This tended to confirm that the KBS Tuff was in the Gauss Normal Epoch and was somewhere between 2.5 and 2.9 million years old.

Brock stated that skull ER 1470 came from "a level equivalent to that in which the Kaena and Mammoth events have been identified" (Brock and Isaac 1974, p. 347). He added: "An age of 2.7 to 3.0 Myr . . . is strongly indicated" (Brock and Isaac 1974, p. 347). Referring to the potassium-argon dates by Fitch and Miller, Brock concluded that "in every case the isotopic and paleomagnetic dates are consistent" (Brock and Isaac 1974, p. 347). Brock also found his version consistent with the faunal chronology prepared by Vincent J. Maglio.

Maglio identified marker fossils in the hominid-bearing sediments of the Lake Turkana sites. Skull ER 1470 was found in the zone containing *Mesochoerus limnetes,* an extinct pig. Maglio compared the pig teeth found at Lake Turkana with those found at the Shungura site in Ethiopia, where it had been demonstrated that the pig teeth increased in size with the passage of time. At Shungura, teeth of the size found at Lake Turkana fell in a time period extending from 1.8 million to 2.7 million years ago.

Maglio (1972, p. 383) noted: "The indicated age range includes the K/Ar date of 2.6 m.y. for the KBS tuff located within the sedimentary unit containing this fauna." Richard Leakey believed that Maglio's report supported his position on the ages of the KBS Tuff and the important ER 1470 skull.

But other scientists, who had different ideas about the relative antiquity of *Australopithecus* and *Homo,* were not happy about the potassium-argon age of 2.6 million years for the KBS Tuff (Johanson and Edey 1981). They pointed to new faunal studies that seemed to make the KBS Tuff much younger.

Basil Cooke (1976), for example, said the size range of pig teeth from below the KBS Tuff matched that of Ethiopian pig teeth with an age of 2 million years. If this correlation were accepted, the KBS Tuff would have to be less than 2 million years old.

Richard Leakey proposed differential rates of evolution as a possible explanation. Perhaps the pigs at Lake Turkana and their teeth got bigger earlier than those in Ethiopia because of a more favorable environment.

The dispute over the pig teeth and the KBS Tuff was a major topic at a February 1975 conference in London. Fitch and Miller presented the results of new potassium-argon tests, which yielded an age for the KBS tuff of 2.4 million rather than 2.6 million years (R. Leakey 1984, p. 167).

Another group of scientists showed uranium fission track evidence confirming the new potassium-argon date of 2.4 million years (R. Leakey 1984, p. 168). When uranium 238, a radioactive element, decays into lead, particles released during fission leave tracks in zircon crystals. By measuring the amount of uranium and counting the tracks in the crystals, one can estimate the crystals' age.

At the London meeting, Basil Cooke used his fossil pig evidence to dispute the potassium-argon and fission track dates. But Richard Leakey (1984, p. 168), who did not regard Cooke's results as conclusive, strongly defended the potassium-argon age of 2.4 million years for the KBS Tuff.

Not long afterward, Garniss Curtis of the University of California published his own potassium-argon test results (Curtis *et al.* 1975, p. 395). He obtained ages of 1.6 million and 1.8 million years for the KBS Tuff.

Curtis said the samples tested by Fitch and Miller were probably contaminated with argon from older inclusions. Fitch and Miller said Curtis's samples possibly suffered argon loss, giving a date younger than the actual date of the tuff. Who was right? From the information provided in the published reports, it is hard to tell.

In one of their reports, Fitch and Miller (1976) did, however, give an interesting insight into potassium-argon dating procedures. They arranged some of their dated samples from the KBS Tuff into 4 groups, having average ages of 221 million years, 3.02 million years, 8.43 million years, and 17.5 million years. They also listed over a dozen other individual samples with ages ranging from 0.52 to 2.54 million years. This bewildering array of dates comprises the actual results of the potassium-argon testing of samples from the KBS Tuff.

All dates older and younger than the ones finally published were thrown out, mainly because the researchers assumed the samples had been in some way contaminated or degassed. They proposed, for example, that flowing water could have mixed new and old volcanic materials or that water from hot springs could have released argon originally trapped in the sampled material (Fitch and Miller 1976, p. 125).

When Anthony J. Hurford and his associates published the conclusions of their fission track test, presented in preliminary form at the 1975 London meeting, they, like Fitch and Miller, disputed the 1.8 million year date for the KBS tuff obtained by Curtis. They stated: "Fission-track dating of zircon separated from two pumice samples from the KBS Tuff in the Koobi Fora Formation, in Area 131, East Rudolf, Kenya, gives an age of 2.44±0.08 Myr for the eruption of the pumice. This result is compatible with the previously published K-Ar and 40Ar/39Ar age spectrum estimate of 2.61±0.26 Myr for the KBS Tuff in Area 105, but differs from the more recently published K-Ar date of 1.82±0.04 Myr for the KBS Tuff in Area 131. This study does not support the suggestion that pumice cobbles of different ages occur in the KBS Tuff" (Hurford *et al.* 1976, p. 738).

Curtis had suggested that the Fitch and Miller dates of 2.61 and 2.42 million years were the result of older pumice included in the KBS tuff. Hurford also pointed out that his results were compatible with the paleomagnetic results obtained earlier by Brock and Isaac (Hurford *et al.* 1976, p. 740).

In another development, it turned out that Curtis's potassium-argon age for the KBS Tuff was "flawed by an improperly adjusted weighing balance" (Johanson and Shreeve 1989, p. 99).

Meanwhile, Richard Leakey commissioned John Harris and Tim White to study the faunal conclusions reached by Basil Cooke. As it turned out, their investigation confirmed Cooke's results. Leakey, as leader of the Koobi Fora project, prevailed upon Harris to remove any mention of how this faunal evidence related to the hominids of Lake Turkana. White, in protest, asked to have his name removed from the paper before it was published. Harris did not remove White's name. The paper was rejected by *Nature,* but a revised version was published by the American journal *Science* in 1977 (Johanson and Edey 1981, pp. 240–242).

The controversy dragged on for several years. The younger age for the KBS Tuff was very much favored by Don Johanson and Tim White, who promoted *Australopithecus afarensis* (including "Lucy") as the ultimate ancestor of both *Homo habilis* and *Australopithecus africanus*. *Afarensis* was around 3 million years old. The skull of *afarensis* was typically australopithecine, small-brained with heavy brow ridges. Having the much bigger, smooth-browed ER 1470 cranium at around 2.9 million years, as Richard Leakey originally suggested, would have made *afarensis* an unlikely ancestor of ER 1470, classified as *Homo habilis*.

In order to put an end to the controversy, Richard Leakey decided to call in additional researchers. "It was only in 1980," wrote Leakey, "that a broad consensus was finally achieved. . . . Glynn [Isaac] and I decided we should invite other geophysicists to work on the KBS date. Eventually we managed to arrange for several different laboratories to evaluate the same material from split samples, using two methods: fission-track dating, as well as conventional potassium-argon. This was done quietly and with little fanfare. As a result, it became quite clear that the KBS tuff is no more than 1.9 million years old . . . it would be prudent to think of the skull KNM-ER 1470 as being about two million years old" (R. Leakey 1984, p. 170).

The case of the KBS Tuff is intriguing. Initially, Leakey had potassium-argon dates, faunal evidence, paleomagnetic dates, and fission track dates supporting an age of 2.6 million years. Then, a few years later, he said new potassium-argon dates, faunal evidence, and fission track dates favored an age of 1.9 million years.

Richard Leakey's allusion to consensus is instructive. Researchers party to such an agreement may announce that their consensus must be correct because it is supported by dating methods A, B, and C. But as we have seen, various dating methods tend to give age ranges broad enough to support a number of age determinations.

Many place excessive, even unquestioning, faith in published age determinations, unaware of the many sources of error inherent in current dating methods. They do not adequately appreciate the crucial role that the judgements of individual researchers play in arriving at a published date from among the spread of dates often obtained from a series of tests. These complex judgements can easily be influenced by the researcher's expectations and preconceptions.

11.7 OH 62: WILL THE REAL HOMO HABILIS PLEASE STAND UP?

Artists, working from fossils and reports supplied by paleoanthropologists, have typically depicted *Homo habilis* as having an essentially humanlike body

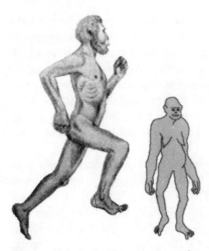

Figure 11.9. Left: This drawing (after Johnson and Edey 1981, p. 286) shows *Homo habilis,* as generally depicted before 1987. Below the head, the anatomy is essentially human. Right: After OH 62 was found at Olduvai Gorge in 1987, a new picture of *Homo habilis* (after Budiansky 1987, p. 10) emerged, far smaller and more apelike than before.

except for its apelike head (Figure 11.9).

Occasionally, scientists have raised questions about such depictions. "Were the australopithecines hairy? Was *Homo habilis* slightly less hairy, just to give it a hint of human respectability?" asked Richard Leakey. "Certainly, all the portraits ever painted of our ancestors show this kind of pattern. But as no artist has ever seen a living hominid, and as we have no way of knowing whether they were naked or not, it will remain a favorite topic of after-dinner speculation and fantasy forevermore" (Leakey and Lewin 1978, p. 66).

In any case, a very humanlike portrait of *Homo habilis* persisted until 1987. In that year, Tim White and Don Johanson reported they had found in lower Bed I at Olduvai the first *Homo habilis* individual (OH 62) with postcranial bones clearly associated with a cranium.

Johanson and his coworkers (1987, p. 205) stated: "This specimen's craniodental anatomy indicates attribution to *Homo habilis,* but its postcranial anatomy, including small body size [less than 3.5 feet] and relatively long arms, is striking similar to that of some early *Australopithecus* individuals." Drawings of the new *Homo habilis* (Figure 11.9) were decidedly more apelike than those of the past.

Wood (1987, p. 188) noted: "The shape and size of the proximal femur, and the anatomy and relative lengths of the limb bones, both run counter to the view which sees *H. habilis* as a biped with a postcranial skeleton that is essentially modern human in its morphology, proportions and, by inference, function."

Johanson and his coworkers (1987, p. 209) concluded it was likely that scientists had incorrectly attributed to *Homo habilis* many postcranial bones discovered prior to 1987.

11.7.1 Implications for the ER 1481 and ER 1472 Femurs

The OH 62 find supports our suggestion that the ER 1481 and ER 1472 femurs from Koobi Fora, described as very much like those of modern *Homo sapiens*

(Section 11.6.3), might have belonged to anatomically modern humans living in Africa during the Late Pliocene. These femurs have been attributed by some workers to *Homo habilis* and by others to *Homo erectus.* But these attributions are questionable. Showing this will, however, take a few paragraphs of unavoidably obscure and intricate analysis of bone morphology.

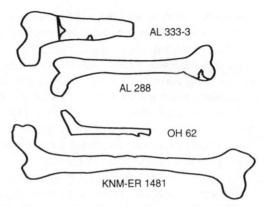

Figure 11.10. (Traced from Johanson and Shreeve 1989, photo section.) According to D. Johanson, the KNM-ER 1481 femur from Koobi Fora, Kenya, is from a male *Homo habilis.* It is, however, much larger than the OH 62 female *Homo habilis* femur from Olduvai Gorge. Attributing both femurs to the same species implies an unusual degree of sexual dimorphism. They display a greater size difference than the male (AL 333-3) and female (AL 288) *Australopithecus afarensis* femurs from Hadar, Ethiopia. Some workers have said that the degree of sexual dimorphism in the Hadar sample is too great to be accommodated within a single species. The same may be true of the KNM-ER 1481 and OH 62 femurs.

In his book *Lucy's Child* (Johanson and Shreeve 1989, photo section), D. Johanson suggested that the comparatively large ER 1481 femur was the *Homo habilis* male counterpart to the smaller OH 62 female *Homo habilis* femur.

But the attribution of the ER 1481 femur to the same species as OH 62 involves a remarkable degree of sexual dimorphism for *Homo habilis.* The ER 1481 femur is much bigger than the OH 62 female femur (Figure 11.10).

Johanson believed, however, that the femurs of Lucy (AL 288) and a male hominid (AL 333-3) from the Hadar, Ethiopia, site displayed a degree of sexual dimorphism similar to that of the OH 62 and ER 1481 femurs. This made it conceivable, to Johanson, that the OH 62 and ER 1481 femurs might belong to a single species of hominid. To us, however, the degree of sexual dimorphism in OH 62 and ER 1481 appears much greater than that in the Hadar femurs.

Furthermore, although Johanson thought that both Hadar femurs belonged to one species (*Australopithecus afarensis*), other paleoanthropologists have said that the AL 333-3 femur, along with many other fossils attributed by Johanson to *Australopithecus afarensis,* actually belonged to *Homo* individuals (Groves 1989, pp. 260–263).

One of these workers (Zihlman 1985, pp. 216–217) demonstrated that putting all the Hadar hominids in one species would involve sexual dimorphism more extreme than that encountered in the most sexually dimorphic anthropoid apes (Section 11.9.8). Zihlman therefore believed Johanson was not justified in assigning all the Hadar fossils to a single species.

If the Hadar fossils were too sexually dimorphic to be included in one species, we believe the same would be true of the ER 1481 and OH 62 femurs, which seem to manifest an even greater degree of sexual dimorphism than the Hadar femurs.

Johanson's attribution of the ER 1481 and OH 62 femurs to a single species appears to be a consequence of his belief that only one hominid species other than *Australopithecus boisei* (namely, *Homo habilis*) existed around 2 million years ago in East Africa (Willis 1989, p. 263).

As we shall see in Section 11.7.5, some workers have suggested that *Homo habilis* represents at least two species, including, perhaps, an australopithecine. Wood (1987), for example, proposed that small, apelike OH 62 might represent an East African gracile australopithecine rather than *Homo habilis*.

Accepting this, one might try to keep the traditional picture of *Homo habilis*. One could then, as previously, attribute the ER 1481 and ER 1472 femurs to *Homo habilis*, as represented by the somewhat humanlike ER 1470 skull. But the ER 1481 and ER 1472 femurs were found some distance from the ER 1470 skull, which means there is no solid reason to connect them. Attribution of the ER 1481 and ER 1472 femurs to *Homo habilis* is therefore questionable.

Some workers have suggested that the ER 1481 and ER 1472 femurs, and other bones attributed to *Homo habilis*, should be attributed to *Homo erectus* (Wood 1987, p. 188).

Even before the discovery of OH 62, Kennedy (1983) assigned the ER 1481 femur to *Homo erectus*. Kennedy's view would involve extending the age of African *Homo erectus* from about 1.6 million to at least 2 million years, since femur ER 1481 was found below the KBS Tuff at Koobi Fora.

In coming to her conclusion, Kennedy relied on comparative analysis of several femoral shaft measurements. But Trinkaus (1984, p. 137) noted that out of these measurements only one, the midshaft diameter, showed a "significant difference" (more than two standard deviations from the mean) from a sample of early anatomically modern human femurs. Trinkaus's early anatomically modern human sample included 24 fossil femurs from Cro-Magnon, Predmost, and other early *Homo sapiens sapiens* sites. We suspect, however, that if the midshaft diameter of ER 1481 were compared with a sample that represented the total variation among living humans, it would fall closer to the mean. The other femoral shaft measurements of ER 1481 reported by Kennedy all fell within the range of early anatomically modern humans. This suggests that ER 1481 might be assigned to *Homo sapiens* rather than *Homo erectus*.

There are other reasons why attribution of the ER 1481 and ER 1472 femurs to *Homo erectus* is questionable. Since the discovery of Java man in the 1890s, scientists have written numerous books and articles describing femurs said to be those of *Homo erectus*. But until recently, no femurs, or other postcranial bones, have ever been found in direct connection with a cranium of *Homo erectus*. There-fore, it is not absolutely certain that any of the femurs scientists had previously described actually belonged to *Homo erectus* individuals.

In 1984, however, members of Richard Leakey's team found a *Homo erectus* boy (KNM-WT 15000) at Lake Turkana. KNM-WT 15000 was assigned an age of 1.6 million years. The skeleton comprised associated cranial and postcranial ele-ments, including the femur (Brown *et al.* 1985, p. 788).

According to the discoverers (Brown *et al.* 1985, p. 791), several features of the KNM-WT 15000 *Homo erectus* femur were different from those normally encoun-tered in *Homo sapiens*. Other workers (Johanson *et al.* 1987, p. 209) also called attention to "*Australopithecus*-like aspects of . . . proximal femoral anatomy in early *Homo erectus* (KNM-WT 15000)." On the other hand, several workers have found the KNM-ER 1481 femur to be very much like modern human femurs and unlike those of australopithecines (Section 11.6.3).

Furthermore, Day and Molleson (1973, p. 128) said that most of the hominid femurs generally attributed to *Homo erectus* (such as the OH 28 femur from Olduvai Gorge and the *Sinanthropus* femurs from Zhoukoudian) were unlike those of modern human beings.

But Day and Molleson found the Java man femurs from Trinil, generally classi-fied as *Homo erectus*, to be distinct from the OH 28 and Chinese *Homo erectus* femurs and almost identical to those of modern humans. They thus concluded that the Trinil femurs belonged not to *Homo erectus* but to anatomically modern humans (Section 7.1.8). This may put *Homo sapiens sapiens* at the Trinil site about 800,000 years ago.

In a study by Wood (1976), the ER 1481 and ER 1472 femurs came closer to the human mean in several key features than the Trinil I femur, which Day and Molleson said was anatomically modern and distinct from that of *Homo erectus*.

All of this suggests that it would not be correct to assign the anatomically modern ER 1481 and ER 1472 femurs to either *Homo erectus* or *Homo habilis*.

11.7.2 The Leap from OH 62 to KNM-WT 15000

The discoverers of OH 62 had to grapple with the evolutionary link between the new, more apelike *Homo habilis* and *Homo erectus*. "The juxtaposition of an otherwise relatively derived *H. erectus* postcranium at ~ 1.6 Myr (KNM-WT 15000) and a postcranially primitive *H. habilis* at ~ 1.8 Myr (OH 62) may imply an abrupt transition between these taxa in eastern Africa," they stated (Johanson *et al.* 1987, p. 209). In paleoanthropology, the term "derived" is applied to a skeletal element that has supposedly undergone a significant and progressive morphological change relative to the same element in a supposedly ancestral form.

The *H. habilis-H. erectus* transition proposed by Johanson involves some rather extreme morphological changes, including a big change in size. Richard Leakey, applying normal human growth patterns, said that the *Homo erectus* boy, who was 5.6 feet tall, would probably have grown to over 6 feet tall as an adult. The female OH 62, on the other hand, was only about 3.25 feet tall, smaller than Lucy, who was about 3.5 feet tall.

How tall were the OH 62-type males? That is hard to say. Some presumably male *Australopithecus afarensis* individuals from the same Hadar, Ethiopia, site as Lucy may have been as much as 5 feet tall. On this basis, one might propose that an OH 62-type male might have been almost 5 feet tall. But, as we have seen, some workers say the size difference between the large and small Hadar specimens is too great to be accommodated within a single sexually dimorphic species. It seems likely, therefore, that the male companion to the 3.25-foot-tall OH 62 adult female was not much more than 4 feet tall. Altogether, an evolutionary leap from small, apelike OH 62 to big, humanlike KNM-WT 15000 in less than 200,000 years seems implausible.

Advocates of the much-debated punctuational model of evolution, however, can easily accept the transition. Unlike the traditional gradualists, punctuationalists assert that evolution proceeds by rapid episodes of change interrupted by long periods of stasis. The periods of change are so brief, say the punctuationalists, that intermediate forms are rarely preserved in the fossil record. Punctuationalism can, therefore, accommodate a variety of troublesome evolutionary anomalies, such as the *habilis* to *erectus* transition proposed by Johanson.

"The very small body size of the OH 62 individual," said its discoverers, "suggests that views of human evolution positing incremental body size increase through time may be rooted in gradualistic preconceptions rather than fact" (Johanson *et al.* 1987, p. 209). But punctuational views may also be rooted in preconception rather than fact. The paleontological facts, considered in their entirety, suggest that various ape-man-like and humanlike beings, including some resembling modern humans, coexisted throughout the Pleistocene, and earlier.

In summary, the OH 62 specimen, seen as *Homo habilis,* delivers a triple blow to conventional ideas about human evolution. (1) OH 62 shatters the prevailing humanlike portrayal of *Homo habilis,* as presented in book and magazine illustrations, television shows, and museum exhibits. (2) The primitive morphology of OH 62 raises questions about the taxonomic status of very humanlike postcranial bones, such as the ER 1481 femur, which have been attributed to *Homo habilis.* To what kind of hominid should they now be assigned? It is possible they belonged to an anatomically modern human species that coexisted with *Homo habilis,* the australopithecines, and *Homo erectus* around 2 million years ago in Africa. (3) The size and geological age of OH 62 make the convention-ally accepted evolutionary transition from *Homo habilis* to *Homo erectus* less plausible. Of course, if one were to classify OH 62 as an australopithecine that would resolve some of these difficulties.

11.7.3 Conflicting Assessments of Other Homo Habilis Fossils

It was not only new evidence such as OH 62 that challenged the long-ac-cepted picture of *Homo habilis.* Previously discovered fossil evidence relating to *Homo habilis,* originally interpreted by some authorities as very humanlike, was later characterized by others as quite apelike.

11.7.3.1 The OH 8 Foot

As mentioned earlier (Section 11.4.2), a fairly complete foot skeleton, labeled OH 8, was found in Bed I at Olduvai Gorge. Dated at 1.7 million years, the OH 8 foot was associated with other fossils classified by L. Leakey as *Homo habilis* (OH 7) and was also attributed to this species (Lewis 1980, pp. 275, 290).

M. H. Day and J. R. Napier (1964) said the OH 8 foot very much resembled that of *Homo sapiens,* thus contributing to the overall humanlike picture of *Homo habilis.* According to Day and Napier, the OH 8 foot showed that *Homo habilis* walked upright.

But O. J. Lewis (1980, p. 291), anatomist at St. Bartholomew's Hospital Medical College in London, wrote: "The attribution of these remains to the taxon *Homo* has been a source of controversy." He showed the functional morphology of the OH 8 foot was more like that of chimpanzees and gorillas (Table 11.4).

TABLE 11.4

Apelike Features of the OH 8 Foot Reported by O. J. Lewis (1980)

1. Articulations between the metatarsals "are like the chimpanzee" (p. 294).
2. Ankle joint surfaces "retain the apelike form" (p. 291).
3. Form of the talus (ankle bone) is like that "seen in the extant African apes" (p. 291).
4. Disposition of the heel similar to that of gorillas and chimpanzees (p. 291).
5. Hallux (large toe) capable of being extended sideways, with some "residual grasping functions" (p. 293).

Commenting on the 1964 study by Day and Napier, Lewis (1980, p. 294) noted that "conservative arboreal features of the tarsus [ankle] . . . escaped comment." The suggestion that the OH 8 ankle manifested arboreal features is intriguing. It certainly does not serve the propaganda purposes of evolutionists to have the public visualizing a supposed human ancestor like *Homo habilis* climbing trees with an aboreally adapted foot rather than walking tall and brave across the African savannahs. When the owner of the OH 8 foot did walk on the ground, it probably did so in a chimpanzeelike manner, said Lewis (1980, p. 296).

From Lewis's study of the OH 8 foot, one could therefore conclude that *Homo habilis* was much more apelike than most scientists have tended to believe. The OH 62 discovery supports this view. Another possible conclusion: the OH 8 foot did not belong to *Homo habilis* but to an australopithecine. This view was favored by Wood (1974b) and Lewis (1980, p. 295). A related conclusion is that *Homo habilis* itself was, as Oxnard (1975b) proposed, simply a variant of *Australopithecus.* Oxnard, said Lewis (1980, p. 295), thought "the australopithecines (including OH 8) were at least partially arboreal primates retaining efficient climbing capabilities associated with a bipedal capacity probably of a type no longer seen." Of course, the proposal that *Australopithecus* was even partially arboreal defies the conventional view that this creature was humanlike from the neck down and walked fully upright on the ground. In Section 11.8, we give a detailed discussion of this issue.

Over the years, scientists have described the OH 8 foot skeleton as humanlike (Day and Napier 1964), apelike (Lewis 1980), intermediate between human and ape (Day and Wood 1968), distinct from both human and ape (Oxnard 1972), and orangutanlike (Lisowski *et al.* 1974). This demonstrates once more an important characteristic of paleoanthropological evidence—it is often subject to multiple, contradictory interpretations. Partisan considerations often determine which view prevails at any given point in time

11.7.3.2 The OH 7 Hand

The OH 7 hand was also found at Olduvai Gorge (Section 11.4.2), as part of the type specimen of *Homo habilis.* Napier (1962, p. 409) described the hand as quite human in some of its features, especially the finger tips. As in the case of the OH 8 foot, subsequent studies showed the OH 7 hand to be very apelike, calling into question either its attribution to *Homo habilis* or the generally accepted humanlike picture of *Homo habilis,* which the original interpretation of the OH 7 hand helped create.

C. E. Oxnard (1984, p. 334-ii) was highly critical of Napier's original study of the *Homo habilis* hand: "being convinced that he was looking at a pre-human hand that made tools, he interpreted three features in which that hand was similar to a human hand as more weighty than ten in which he found it similar to those of apes." Oxnard identified evolutionary bias (seeing a fossil as "pre-human") as the key factor in Napier's attempt to characterize an essentially apelike structure as human.

Randall L. Susman and Jack T. Stern noted that the OH 7 finger bones had large areas for the insertion of a muscle (the flexor digitorum superficialis) that apes use when hanging from branches. "The impressions for this muscle are greater in relative area than in any living ape or modern humans," they said (Susman and Stern 1979, p. 572).

Susman and Stern (1979, p. 565) therefore concluded: "Prominent markings for insertions of these muscles in a fossil hand (such as O.H. 7) suggest use of the forelimb in suspensory climbing behavior."

Susman and Stern (1979, p. 572) noted in addition that the finger bones of the OH 7 hand were thick and curved like those of chimpanzees, indicating, like the flexor digitorum superficialis muscle, a degree of arboreal suspensory behavior.

In others words, *Homo habilis,* or whatever creature owned the OH 7 hand, may have spent much of its time hanging by its arms from tree limbs. This apelike image differs from the very humanlike portrait of *Homo habilis* and other supposed human ancestors one usually encounters in Time-Life picture books and National Geographic Society television specials.

11.7.4 Cultural Level of Homo Habilis

A reevaluation of the cultural evidence at *Homo habilis* sites also casts doubt on the conventional humanlike interpretation of *Homo habilis.*

Louis and Mary Leakey designated the *Homo habilis* sites at Olduvai as "living floors." They viewed particular combinations of hominid and animal fossils, along with stone tools, as signs of permanent or semipermanent habitation. From such interpretations of the evidence came detailed paintings, showing *Homo habilis* families living in base camps, with hunting parties returning with animal carcasses to be butchered with stone tools.

But according to Binford (1981, p. 252), the Leakeys' characterization of *Homo habilis* sites as "living floors" was the result of wishful thinking: "the researchers have a generalized idea as to what the past was like and they have then *accommodated* all the archaeological-geological facts to this idea. This is not exactly science." Binford went on to criticize the notion of living floors in terms of their "integrity" and "resolution."

Binford believed the *Homo habilis* sites were of low integrity. By this he meant there was no certainty that *Homo habilis* was in fact responsible for the animal bones found at the sites. The bones could very well have been the result of natural deaths, which would have occurred fairly often on the shores of the ancient lake that deposited the sediments at Olduvai. The bones might also have been brought to their resting places by carnivorous animals rather than hominids.

For Binford, the term "resolution" meant the time during which the faunal remains and artifacts were deposited. For the concept of a "living floor" to be meaningful, the resolution should be quite high—that is to say, the faunal remains and artifacts should have been deposited over a relatively short period of time. But Binford believed that the resolution at the *Homo habilis* sites at Olduvai Gorge was low, and that the faunal remains and artifacts were deposited over very long periods of time. This would decrease the certainty that hominid behavior was responsible for the association of a particular assemblage of bones and artifacts.

If, for example, one interprets a scatter of stones and bones as having been deposited simultaneously, one might talk of a habitation site. But if the bones and artifacts were deposited one by one over the course of hundreds or thousands of years, as animals chanced to die, and scavenging hominids chanced to drop stone tools, the supposition that one has found a habitation site becomes far less likely.

About the reputed living floor at the famous *Zinjanthropus* site at Olduvai, where remains of *Homo habilis* were also found nearby, Binford (1981, p. 282) said: "given its demonstrably low integrity and resolution, arguments about base camps, hominid hunting, sharing of food, and so forth are certainly premature and most likely wildly inaccurate. The only clear picture obtained is that of a hominid scavenging the kills and death sites of other predator-scavengers for abandoned anatomical parts of low food utility, primarily for purposes of extracting bone marrow. Some removal of marrow bones from kills is indicated, but there is no evidence of 'carrying food home.' Transport of the scavenged parts away from the kill site to more protected locations in a manner identical to that of all other scavengers is all that one need imagine to account for the unambiguous facts preserved in Olduvai."

Thus, according to Binford, *Homo habilis* was definitely not a hunter. In fact, Binford has concluded that hunting is an activity exclusively characteristic of modern *Homo sapiens*. "There are many people," he said, "who are just outraged because I've suggested that early men, including the Neanderthals, weren't hunters" (A. Fisher 1988a, p. 37).

There are some scientists, such as Henry Bunn of the University of Wisconsin (A. Fisher 1988a, p. 38), who have disputed Binford's conclusions about the Olduvai sites. Nevertheless, Binford's analysis provides a refreshing alternative to the usual overly humanized presentation of "*Homo*" *habilis*.

"There were all these wonderful renderings in popular magazines and books of little bands of bushmanlike people sitting around with daddy off hunting and momma gathering plant foods and grandma teaching the baby. But that was just a projection of modern man onto ancient man," declared Binford in an interview (A. Fisher 1988a, p. 37). "We have had far too much of what I tend to think of as the *National Geographic* approach to research," said Binford (1981, p. 297).

Binford's revised view of the cultural evidence at *Homo habilis* sites, together with the revised view of *Homo habilis* anatomy, raises many questions about how humanlike *Homo habilis* really was.

Finally, we should remember that *Homo habilis* is not the only creature that could have been responsible for the stone tools found at sites yielding *Homo habilis* fossils. The same is true of the circle of stones found at Olduvai site DK, interpreted by some as part of a shelter. Mary Leakey said that living African tribal people make and use the same kinds of tools and erect the same kinds of shelters (Section 3.7.3). This suggests that beings like modern *Homo sapiens,* rather than *Homo habilis,* could have made both the tools and the shelter about 1.5 million to 2.0 million years ago in the Early Pleistocene.

11.7.5 Does Homo Habilis Deserve to Exist?

In light of the contradictory evidence connected with *Homo habilis,* some researchers have proposed that there was no justification for "creating" this species in the first place.

Doubts about the taxonomic reality of *Homo habilis* arose right from the start. Even Tobias and Napier, who had joined Louis Leakey in proposing the new species in April of 1964, expressed caution. Tobias and Napier wrote in a letter to the *Times* of London on June 5, 1964 that "anatomy alone could not tell us whether the creature was a very advanced australopithecine or the lowliest hominine" (Cole 1975, p. 256). In making this statement, Tobias and Napier presumably meant that stone tools and broken animal bones associated with the creature's ambiguous skeletal remains justified designating it the earliest representative of the genus *Homo.*

The dental evidence was a cause of concern among some researchers, including T. J. Robinson. Johanson wrote: "He said that one could find greater shape differences in a population of modern humans than Leakey had found between *habilis* and the australopithecines—or, in fact, between *habilis* and *Homo erectus.* Robinson's point was that on dental evidence alone there was too narrow a slot between *Australopithecus* and *Homo erectus* to yield room for another species" (Johanson and Edey 1981, p. 102). As we have seen, however, there are, aside from the teeth, significant differences between *Homo habilis,* as represented by the small OH 62 individual, and *Homo erectus.*

Wilfred Le Gros Clark said: "'Homo habilis' has received a good deal of publicity since his sudden appearance was announced, and it is particularly unfortunate that he should have been announced before a full and detailed study of all the relevant fossils can be complete. . . . From the brief accounts that have been published, one is led to hope that he will disappear as rapidly as he came" (Fix 1984, p. 143). Le Gros Clark consistently maintained his early opposition to *Homo habilis.*

And C. Loring Brace wrote: "*Homo habilis* is an empty taxon inadequately proposed and should be formally sunk" (Fix 1984, p. 143).

If the bones attributed by some workers to *Homo habilis* were not to be interpreted as a new species, then what did they represent? T. J. Robinson argued that *Homo habilis* had been mistakenly derived from a mixture of skeletal elements belonging to *Australopithecus africanus* and *Homo erectus.* Even Louis Leakey suggested that *Homo habilis* might actually have embraced two *Homo* species, one giving rise to *Homo sapiens* and the other to *Homo erectus* (Wood 1987, p. 187).

Concerning the new OH 62 discovery, Wood pointed out that this hominid individual had been classified as *Homo habilis* by Johanson and his coworkers primarily because its craniodental remains resembled those of the *habilis*-like Stw 53 skull from the Sterkfontein site in South Africa. But Wood (1987, p. 188) observed: "The logical 'trail' becomes tenuous because Stw 53 has merely been likened to *H. habilis,* and not formally attributed to it, even though more than a decade has elapsed since its discovery." Wood appeared to suggest, though somewhat indirectly, that OH 62 might in fact be attributed to *Australopithecus africanus,* which he said was "the most likely alternative taxonomic attribution for Stw 53" (1987, p. 188).

According to Wood (1987, p. 187), one interpretation of the OH 62 find is that it "confirms that the range of variation within material from the early Pleistocene of East Africa assigned to early *Homo* is now too great to be sensibly encompassed within one taxon." Wood himself favored this view.

So in the end, we find that *Homo habilis* is about as substantial as a desert mirage, appearing now humanlike, now apelike, now real, now unreal, according to the tendency of the viewer. Taking the many conflicting views into consideration, we find it most likely that the *Homo habilis* material belongs to more than one species, including a small, apelike, arboreal australopithecine (OH 62 and some of the Olduvai specimens), an early species of *Homo* (ER 1470 skull), and anatomically modern humans (ER 1481 and ER 1472 femurs).

11.8 OXNARD'S CRITIQUE OF AUSTRALOPITHECUS

According to most paleoanthropologists, *Australopithecus* was a direct human ancestor, with a very humanlike postcranial anatomy. Advocates of this view have also asserted that *Australopithecus* walked erect, in a manner practically identical to modern human beings. But right from the very start, some researchers objected to this depiction of *Australopithecus*. Influential English scientists, including Sir Arthur Keith (1931), said that the *Australopithecus* was not a hominid but a variety of ape (Sections 11.3.1–3).

This negative view persisted until the early 1950s, when the combined effect of further *Australopithecus* finds and the fall of Piltdown man created a niche in mainstream paleoanthropological thought for a humanlike *Australopithecus.*

But even after *Australopithecus* won mainstream acceptance as a hominid and direct human ancestor, opposition continued. Louis Leakey (1960d, 1971) held that *Australopithecus* was an early and very apelike offshoot from the main line of human evolution (Section 11.4.3). Later, his son Richard Leakey (1973b) took much the same stance (Section 11.6.2).

In the early 1950s, Sir Solly Zuckerman (1954) published extensive biometric studies showing *Australopithecus* was not as humanlike as imagined by those who favored putting this creature in the lineage of *Homo sapiens.* From the late 1960s through the 1980s, Charles E. Oxnard of the University of Chicago, employing multivariate statistical analysis, renewed and amplified the line of attack begun by Zuckerman.

In this section, we shall focus on Oxnard's studies of *Australopithecus,* except those dealing specifically with *Australopithecus afarensis* (Lucy). The latter are included in our general discussion of *Australopithecus afarensis* (Section 11.9).

11.8.1 A Different Picture of Australopithecus

In *Uniqueness and Diversity in Human Evolution,* Oxnard (1975a, p. vii) wrote: "Whereas the conventional wisdom about human evolution depends upon the (apparent) marked similarity between modern man and the various australopithecine fossils, the studies here indicate that these fossils are uniquely different from modern man in many respects."

Oxnard's interpretation of the fossil evidence profoundly unsettles the evolutionary status of *Australopithecus.* According to Oxnard (1975b, p. 394), "it is rather unlikely that any of the Australopithecines . . . can have any direct phylogenetic link with the genus *Homo."*

In Table 11.5, we review the observations that led Oxnard to this conclusion. The table also includes material from Zuckerman's studies.

Oxnard believed there is much that remains to be known about *Australopithecus,* and that what we do know does not conform to the customary image of this creature. Oxnard (1975a, p. 123) observed: "All of this makes us wonder about the usual presentation of human evolution in encyclopedias and popular publications, where not only are the australopithecines described as being of known bodily size and shape, but where, in addition, such characteristics as bipedality . . . and even facial features are happily reconstructed."

TABLE 11.5

Anatomical Features Cited by S. Zuckerman and C. E. Oxnard Indicating That Australopithecines Were Not Human Ancestors

Brain:
"endocranial casts of the Australopithecinae . . . do not appear to diverge in any material way from existing apes" (Zuckerman 1954, p. 305).

"estimates of endocranial volume do not depart from the range of size met with in the great apes" (Zuckerman 1954, p. 304).

"suggestions that the Australopithecinae may have had higher relative brain weights than, say, chimpanzees" have not been substantiated (Zuckerman 1954, p. 304).

Teeth and Jaws:
"with the exception of their incisors and canines, the size and general shape of the [australopithecine] jaws and teeth . . . were very much more like those of the living apes than like acknowledged members of the Hominidae, either living or extinct" (Zuckerman 1954, pp. 306–307).

Shape of Skull:
"resembles. . . . the ape—so much so that only detailed and close studies can reveal the difference between them" (Zuckerman 1954, p. 307).

Shoulder Bone (Sterkfontein Sts 7 scapula):
"does not resemble that of man to any degree. . . . almost as well-adapted structurally for suspension of the body by the limbs as is the corresponding part of the present-day gibbon. . . . more specialized in this respect than in even the highly specialized chimpanzee" (Oxnard 1968, p. 215). Oxnard dismissed suggestions that the Sterkfontein scapula was too distorted to yield accurate measurements. He also rejected accusations that the Sterkfontein scapula was nonhominid.

Has an abnormally large area for attachment of the biceps muscle, which must have been extraordinarily well developed, as it is in the gibbons (Oxnard 1968, p. 215).

Collar Bone (Olduvai OH 48 *Homo habilis* clavicle):
"whereas in humans the clavicle is scarcely twisted at all, in the various apes, as in the Olduvai clavicle, it is heavily twisted. This particular feature does not fit with the idea that the fossils are functionally close to man" (Oxnard 1984, p. 323). Oxnard, like others (Section 11.7.5), considered *Homo habilis* to be an australopithecine.

TABLE 11.5—*Continued*

Hand Bones:

"quite different from those of humans. . . . evidence seems to relate to abilities for grasping with power reminiscent of what we find in the orang-utan. . . . some are curved enough that they must have operated in this arboreal-grasping mode" (Oxnard 1984, p. 311, citing Susman 1979, Susman and Creel 1979, Susman and Stern 1979).

Engineering stress analysis showed *Australopithecus* fingers were inefficient in the chimpanzee knuckle-walking mode but "*efficient* in the hanging-climbing mode as also is the orang-utan" (Oxnard 1984, p. 313). Human finger structure was "inefficient in both modes" (Oxnard 1984, p. 314).

Pelvis (including Sterkfontein Sts 14):

"although there is no doubt about the similarity in shape of the iliac bones of man and Sterkfontein pelvis . . . it is also clear that this blade is positioned quite differently in man and the fossil" (Oxnard 1975a, p. 52).

Joint structure in the australopithecine hip "apparently not inconsistent with quadrupedalism" (Zuckerman *et al.* 1973, p. 152).

Muscle attachments not "inconsistent with . . . an occasional or habitual quadrupedal gait" (Zuckerman *et al.* 1973, p. 152).

Pelvic structure points to hindlimb capable of "an 'acrobatic' function" (Zuckerman *et al.* 1973, p. 156).

Pubis and ischium (bones of the lower part of the pelvis) chimpanzeelike (Zuckerman 1954, p. 313).

Femurs:

"show the small heads and inclined femoral necks that might be expected in animals capable of quadrupedal activities" (Oxnard 1975b, p. 394).

Talus (ankle bone):

"the general morphological similarity . . . is with the aboreal ape *Pongo*" (Oxnard 1975a, pp. 86–87). *Pongo* is the orangutan.

"in the shape of their talus, the . . . fossils may be reflecting functions of the foot that may relate to acrobatic aboreal climbing such as is reminiscent of the extant species *Pongo*" (Oxnard 1975a, p. 89).

Conclusion:

"Pending further evidence we are left with the vision of intermediately sized animals, at home in the trees, capable of climbing, performing degrees of acrobatics and perhaps of arm suspension" (Oxnard 1975a, p. 89). See our Figure 11.11, p. 714.

Figure 11.11. Most scientists describe *Australopithecus* as an exclusively terrestrial biped, humanlike from the head down. But according to some studies by S. Zuckerman and C. E. Oxnard, *Australopithecus* was more apelike. Although capable of walking on the ground bipedally (left), *Australopithecus* was also "at home in the trees, capable of climbing, performing degrees of acrobatics [right] and perhaps of arm suspension" (Oxnard 1975a, p. 89). The unique functional morphology of *Australopithecus* led Zuckerman and Oxnard to doubt it is a human ancestor. Illustrations by Miles Tripplett.

11.8.2 The Pelvis of Australopithecus

Of particular interest is the *Australopithecus* pelvis. Scientists who believe humans evolved from australopithecines often assert that the *Australopithecus* pelvis is similar to that of modern *Homo sapiens*. In both humans and australopithecines, the ilium, the broad upper part of the pelvis, is of roughly the same shape. The ilium of the chimp is more narrow (Figure 11.12). Some researchers have taken the visual resemblance between the human ilium and that of *Australopithecus* as proof that *Australopithecus* stood upright and walked very much like modern human beings.

But the impact of this demonstration is reduced when one considers the orientation of the ilium to the rest of the pelvis in apes, humans, and australopithecines. The comparison can best be made when the hip sockets are turned toward the viewer (Figure 11.13).

As can be seen, the ape ilium is situated in a manner different from that of a human being. The ape's iliac blade is oriented so that only the edge is visible.

In *Australopithecus*, the ilium is oriented like that of apes rather than humans (Oxnard 1984, p. 311). To Oxnard and Zuckerman, this suggested apelike or uniquely nonhuman elements in the musculature, posture, and locomotor pattern of *Australopithecus.*

The typical visual presentation of the human ilium and that of *Australopithecus,* showing both to be of the same shape, is therefore somewhat deceptive, in that their different orientations are usually not mentioned.

Even the claimed similarity in shape of the ilium in *Australopithecus* and human beings is not complete in all respects. Zuckerman (1954, p. 345) observed: "When the least breadth of the ilium is expressed as a percentage of the greatest breadth. . . . the [australopithecine] fossils are pongid [apelike]."

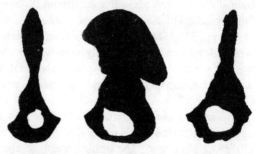

Figure 11.12. In *Australopithecus* (right) and a modern human (center), the broad iliac blade of the pelvis is of similar shape. Some have taken this as proof that australopithecines walked upright in human fashion. A chimpanzee ilium (left) is shaped differently. After Oxnard (1975a, p. 53), not to scale.

Figure 11.13. When the pelvis is viewed with the hip socket toward the observer, the ilium of *Australopithecus* (left) is oriented like that of the chimpanzee (right) and not like that of the human being (center). This, and other features of the australopithecine pelvis, indicated to Oxnard and Zuckerman that *Australopithecus* was capable of quadrupedal and tree-climbing behavior similar to that of the orangutan. After Oxnard (1975a, p. 55), not to scale.

There are other significant differences. Zuckerman (1954, pp. 344–345) said about the specimens of *Australopithecus* pelvis that he studied: (1) "in their maximum iliac breadth they were smaller than in man, but of the size usual in apes"; (2) "the extent of the gluteal [muscle] areas was significantly smaller than in the chimpanzee and man, but of the size found in the gorilla, and . . . in the orang"; (3) "The size of the auricular surface, the area with which the sacrum [tail bone] articulates, was significantly smaller than in man, but similar to that in apes."

Regarding the size of the auricular surface and that of the iliac tuberosity (the large rounded prominence for the attachment of muscles and ligaments on the upper part of the ilium), Zuckerman (1954, p. 346) stated: "Schultz (1930) has shown that the great relative size of these two areas in man is related to the erect attitude, and to the transmission of the weight of the trunk, head, and upper limbs on the sacroiliac articulations. Their smaller size in the great apes can be related to the more quadrupedal posture and gait of these animals. In view of their equally small size in the fossil specimens, it is difficult not to believe that the Australopithecines walked in the same way as do apes." Modern proponents of a more humanlike view of *Australopithecus* consistently and vehemently deny this possibility.

According to Zuckerman, features of the *Australopithecus* pelvis identified by some as decidedly human were subject to alternate interpretations. One of these humanlike features, according to Broom, Robinson, and Schepers (1950), was "the presence of a well-developed anterior inferior iliac spine." Zuckerman (1954, pp. 343–344), however, observed: "Such a spine may imply a ligament whose development is normally associated with the maintenance of the erect posture. On the other hand the spine is also well developed in many quadrupedal animals, e.g. the menotyphlous insectivores, and many carnivores and rodents (Straus 1929)."

In a set of drawings, Oxnard showed the hips and lower limbs of a human, an ape, and an australopithecine placed as if all three were quadrupedal. Oxnard (1975a, p. 57) noted: "The similarities of the ape and *Australopithecus* are most evident." This could be taken to indicate that *Australopithecus* was well adapted for quadrupedal locomotor behavior.

In 1973, Oxnard assisted Zuckerman and other researchers in conducting an extensive multivariate statistical analysis comparing the pelvis of *Australopithecus* with the pelvises of 430 primates, representing 41 genera.

The pelvis study considered 4 measurements relating to joints and 5 relating to muscular attachments. When all 9 features of the pelvis were considered together, *Australopithecus* proved to be unique, differing from both human beings and the nonhuman primates.

Zuckerman and Oxnard therefore concluded that it was "conceivable that the habitual posture and gait of *Australopithecus* might have been unique by displaying a combination of quadrupedalism and bipedalism" (Zuckerman *et al.* 1973, p. 153).

Amplifying this suggestion, Zuckerman and Oxnard further stated: "the locomotor use of the hindlimb might have been composite, involving possibly quadrupedalism, bipedalism, and maybe other types of activity, such as an 'acrobatic' function" (Zuckerman *et al.* 1973, p. 156). Their comparative studies demonstrated that among sub-human primates "the group approximating most closely to *Australopithecus* comprises genera in which the hindlimb sometimes supports, sometimes suspends the animal, and generally operates in many planes of space" (Zuckerman *et al.* 1973, p. 159). It is difficult to overstate how strongly this contradicts the conventional picture of *Australopithecus,* which is never shown hanging from a tree limb by its legs.

11.8.3 Zuckerman and Oxnard on Suppression of Evidence

The paper by Zuckerman and Oxnard on the pelvic study was originally presented at a symposium of the Zoological Society of London in 1973. At the conclusion of the symposium, Zuckerman made some important remarks. He said: "for more than 25 years anatomists and anthropologists—I am talking about physical anthropologists now—have been turning themselves inside out, persuading themselves and others that the obviously simian characteristics of the australopithecine fossils could be reconciled with the model of some assumed proto-human type. Over the years I have been almost alone in challenging the conventional wisdom about the australopithecines—alone, that is to say, in conjunction with my colleagues in the school I built up in Birmingham—but I fear to little effect. The voice of higher authority had spoken, and its message in due course became incorporated in text books all over the world" (Zuckerman 1973, pp. 450–451).

The situation has not changed since Zuckerman spoke in 1973. The voices of authority in paleoanthropology and the scientific community in general have managed to keep the humanlike view of *Australopithecus* intact. The extensive and well-documented evidence contradicting this favored view remains confined to the pages of professional journals, where it has little or no influence on the public in general, even the educated public.

Zuckerman (1973, p. 451) also stated: "in my view what above all has denied the study of the palaeontology of the higher Primates the right to be regarded as a serious science is the fact that over the years *ex cathedra* pronouncements about what constitutes a unique human characteristic in a bone have usually proved nonsense. My belief is that they will always do so."

Zuckerman (1973, p. 451) explained: "It could well be that some feature or group of features in a fossil bone—maybe those having some definable mechanical significance—proves to be more like the corresponding features in man than in the living apes. Almost invariably other features in the same region would be likely to turn out far more ape-like than human. In combination, we end up with something that differs from both men and apes, and which would thus be unique. What conclusion does one then draw, one might well ask. Are we to suppose that the fossils are ancestral to one group, or to the other, or neither? This is the kind of question people try to answer, but we have to recognize that it is at the same time the sort of question which is not amenable to any answer which would be scientifically final."

Oxnard believed that much of the evidence required to find an answer had dropped out of sight. Reviewing the decades-long controversy about the nature of *Australopithecus,* Oxnard (1984, pp. 317–318) said: "In the uproar, at the time, as to whether or not these creatures were near ape or human, the *opinion* that they were human won the day. This may well have resulted not only in the defeat of the contrary *opinion* but also in the burying of *that part of the evidence* upon which the contrary opinion was based. If this is so, it should be possible to unearth this *other part of the evidence.* This evidence may actually be more compatible with the new view; it may help open the possibility that these particular australopithecines are neither like African apes nor humans, and certainly not intermediate, but something markedly different from either."

Of course, this is exactly the point we have been making throughout this book. Evidence has been buried. We ourselves have uncovered considerable amounts of such buried evidence relating to the antiquity of the modern human type.

11.8.4 Opposition to Statistical Studies

Some have claimed that the statistical approach employed by Oxnard and Zuckerman is inappropriate and misleading.

For example Robert Broom said: "I regard all biometricians in the field of morphology as fools" (Johanson and Edey 1981, p. 76). Donald Johanson, discoverer and defender of Lucy, ridiculed Zuckerman, accusing him of "kicking up more and more biometric dust" and firing off "statistical salvos" (Johanson and Edey 1981, p. 76).

Johanson noted: "To give Zuckerman his due, there *were* resemblances between ape skull and australopithecine skulls. The brains were approximately the same size, both had prognathous (long, jutting) jaws, and so on. What Zuckerman missed was the importance of some traits that australopithecines had in common with men" (Johanson and Edey 1981, p. 76).

In this regard, Johanson cited Charles A. Reed, of the University of Illinois, who said: "No matter that Zuckerman wrote of such characters as being 'often inconspicuous'; the important point was the presence of several such incipient characters in functional combinations. This latter point of view was one which, in my opinion, Zuckerman and his co-workers failed to grasp, even while they stated that they did. Their approach . . . was extremely static in that they essentially demanded that a fossil to be considered by them to show any evidence of evolving toward living humans, must have essentially arrived at the latter status before they would regard it as having begun the evolutionary journey" (Johanson and Edey 1981, p. 76).

In citing Reed against Zuckerman in this way, Johanson was being somewhat hypocritical. Johanson and others sharing his views certainly did not characterize *Australopithecus* as an apelike creature with "incipient" human features. Rather they said *Australopithecus* was practically human from the neck down, especially in terms of humanlike bipedal locomotion. In other words, Johanson and others were themselves guilty of insisting that a distant ancestor of living humans had "essentially arrived at the latter status." Reacting to this exaggerated claim, Zuckerman, and later Oxnard, were just saying it was wrong, and that the anatomy and locomotor behavior of *Australopithecus* were essentially apelike.

Johanson, Reed, and others have also ignored the implications of findings by Oxnard and Zuckerman that *Australopithecus* had anatomical features that were uniquely different from those of apes and modern humans (Section 11.8.5). Contrary to the usual view, *Australopithecus* was not, according to Oxnard and Zuckerman, morphologically intermediate between humans and apes. Thus it is unlikely that *Australopithecus* was a human ancestor, unless one wants to invoke an evolutionary path that took the human line on a big australopithecine detour.

One point that Oxnard made in response to critics of his somewhat complicated mathematical approach was that simple visual evidence also established his conclusion that *Australopithecus* had a significant degree of quadrupedal, acrobatic, and suspensory capability.

For example, Oxnard observed that the articular, or joint, surfaces of the lower limbs of human beings are large relative to the articular surfaces of the upper limbs. Oxnard (1984, p. 316) stated: "This befits their bipedal status in which the lower limb takes all the body weight."

Simple visual inspection also revealed that in African apes the articular surfaces of the upper and lower limbs are more equal in size. According to Oxnard (1984, p. 316), this indicates a pattern of behavior "in which both limbs participate in bearing the body weight (and the upper limbs somewhat more than the lower, however that may be, whether through quadrupedal knuckle-walking on the ground or through quadrumanal climbing in the trees)." Quadrumanal (four-handed) climbing involves use of grasping hands and handlike feet by arboreal primates such as the gibbon and orangutan. In fact, in the gibbon and orangutan, which move through the trees mainly by using their arms, the articular surfaces of the upper limbs are larger than those of the lower limbs.

Oxnard (1984, p. 316) noted that as far as *Australopithecus* is concerned, "the fossils . . . resemble most, among living primates, the equivalent parts from apes (and among the apes, the orang-utan) more closely than they do humans." Like orangutans, *Australopithecus* has larger articular surfaces in the upper limbs than the lower (Oxnard 1975a, pp. 117–119). "These facts should be set alongside the comment of Richard Leakey (1973c), who reports that preliminary indications point to a relatively short lower limb and a long upper limb for the australopithecines," said Oxnard (1984, p. 316). Such proportions are decidely apelike and, along with the proportions of the articular surfaces, suggest a component of orangutanlike forelimb suspension in the locomotor repertoire.

Oxnard did not deny that *Australopithecus* manifested bipedal behavior. After all, apes can also walk on two legs in some fashion. Nevertheless, Oxnard (1984, p. 316) concluded about the australopithecines: "however able these creatures were at walking on two legs, they were also convincing quadrupeds and perhaps excellent climbers, feats denied to man today." Oxnard (1984, p. 316) warned: "Such findings must make us wonder whether the australopithecine pattern of bipedal adaptation really reflects a transitional phase to man." In other words, he doubted the common belief that *Australopithecus* is a human ancestor.

11.8.5 Implications of Uniqueness

Summarizing his findings, Oxnard (1975b, p. 393) stated: "Between the very early Miocene apes and ancient man is the tantalizing set of fossils known as *Australopithecus*. . . . most workers feel that the overall position of these fossils is adequately fixed, with a taxonomic label as clearly *Hominidae,* an evolutionary label as on the line to man or very close to it, and a functional label as a human type of biped. . . . But our current studies are providing very different ideas. In the multivariate investigations reported here, the various australopithecine fossils are usually quite different from both man and the African apes. . . . Viewed as a genus, they are a mosaic of features unique to themselves and features bearing some resemblance to those of the orang-utan."

Let us consider one example of uniqueness in the australopithecine anatomy— the talus, or ankle bone. The multivariate statistical technique employed by Oxnard involves measuring a fixed number of features on a bone, in this case the talus. The results of such a study can be visually represented, for each bone, as a point in multidimensional space. For example, if one is measuring three features of a bone, the combination of these features can be displayed as a point in a three-dimensional space. Four features would require a four-dimensional space, and so on. The relationships between bones or sets of bones can thus be examined. Points clustered together represent bones that are morphologically similar. Figure 11.14 shows the morphological relationships of the ankle bones of modern humans, African apes, orangutans, and *Australopithecus*. As can be seen, the point representing the talus of *Australopithecus* lies in its own domain, distant from modern humans and African apes, and close to orangutans. Oxnard found the same to be true of other parts of the australopithecine anatomy.

According to modern theory, the African apes, particularly the chimpanzees, are the closest relatives of modern humans. Scientists hypothesize that the hominids split from the ancestors of modern chimpanzees several million years ago.

Since, according to this view, modern humans and chimpanzees share a common (though as yet undiscovered) ancestor, then *Australopithecus,* as a hominid predecessor of modern humans, should be morphologically intermediate between humans and chimpanzees. Oxnard's finding that the morphology of the australopithecines is uniquely different from that of modern humans and chimpanzees calls into question their supposed evolutionary relationship.

That the anatomy of *Australopithecus,* although unique, resembles that of *Pongo* (the orangutans) is particularly troubling. Accepting this, evolutionists would have to say that the hominids developed an orangutanlike functional morphology in the australopithecine stage (independently, however, from the orangutans) and then veered back toward the modern human condition. Of course, given the flexibility of evolutionary theorizing, anything is possible. But the view of *Australopithecus* emerging from the studies of Oxnard and Zuckerman introduces vexing complications.

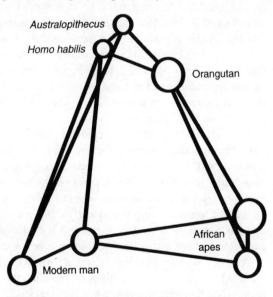

Considering the anatomical uniqueness of the australopithecines, Oxnard (1975b, p. 394) said: "If these estimates are true, then the possibility that any of the australopithecines is a direct part of human ancestry recedes."

Groves (1989, p. 307), after reviewing studies by Oxnard and others, agreed that "the locomotor system of *Australopithecus africanus* was unique— not simply an intermediate stage between us and apes." He found the same to be true of other species of *Australopithecus.* This

Figure 11.14. This display (after Oxnard 1975a, p. 82) depicts the results of a multivariate statistical analysis of the talus (ankle bone) in various hominids and apes. The talus of *Australopithecus* and that of *Homo habilis* (considered by Oxnard to be an australopithecine) are morphologically distant from those of modern humans and African apes. Given the view that humans and African apes such as the chimpanzee share a common ancestor, one would expect the australopithecine talus to occupy an intermediate position. Instead, it occupies a unique position, close to that of the orangutan. The same is true of other australopithecine bones. To Oxnard, this raised doubts about the status of *Australopithecus* as a human ancestor.

fact, along with other aspects of the hominid fossil record, caused him to suggest that "non-Darwinian" principles were required to explain an evolutionary progression from *Australopithecus* to modern human beings (Groves 1989, p. 316).

11.8.6 Oxnard on the Antiquity of Homo

Like Louis and Richard Leakey, Oxnard believed that the *Homo* line was far more ancient than the standard evolutionary scenario allows. In this connection, Oxnard called attention to some of the fossils we have previously discussed, such as the humanlike ER 813 talus, over 1.5 million years old (Section 11.6.4). "Description and examination using multivariate methods [Wood 1974a] confirms that it is indeed similar to modern man and unlike the australopithecine specimens," said Oxnard (1975b, p. 394). He also mentioned the Kanapoi humerus, perhaps 4 or more million years old. Citing research by B. Patterson and W. W. Howells (1967), Oxnard (1975b, p. 394) said the Kanapoi humerus had been "shown to be very similar to that of modern man." His own research backed up that judgement (Oxnard 1975a). From such evidence, Oxnard (1984, p. 332) concluded that the genus *Homo* was 5 or more million years old.

Oxnard (1975b, p. 395) predicted that "more evidence of earlier forms that are more like man than australopithecines will be found." He held that such "fossil remnants will be discovered outside Africa because . . . human or prehuman populations must have existed in other places, with migrations, and with multiple evolutionary lines." But as we have seen in Part I, much evidence for completely humanlike forms existing at very early times has already been found, in Europe and the Americas as well as Africa. Such evidence is so extensive that talk of evolutionary lines, either single or multiple, becomes problematic.

All one can say with certainty is that various humanlike and apelike creatures seem to have coexisted for millions of years into the past. Oxnard (1975b, p. 395) approached this interpretation when he suggested: "We may have to accept that the australopithecine form (or forms) of locomotion, tool using, and tool making may be merely one (or more) unsuccessful evolutionary experiments existing in parallel with those of man." Here the mention of tool using and making refers to *Homo habilis,* which Oxnard regarded as an australopithecine.

Elsewhere, Oxnard (1984, p. 1) gave this succinct statement of his principal conviction: "the conventional notion of human evolution must now be heavily modified or even rejected . . . new concepts must be explored."

11.9 LUCY IN THE SAND WITH DIATRIBES

Donald Johanson studied anthropology at the University of Chicago, under F. Clark Howell. As a young graduate student, eager to learn the romantic business of hominid fossil hunting, Johanson accompanied Howell to Africa, working at the Omo site in Ethiopia.

After two seasons work at Omo, Johanson found himself in Paris. There he met Maurice Taieb, a French geologist, who told him about Hadar, a promising Plio-Pleistocene site in the Afar desert, in northeastern Ethiopia. In 1972, Johanson surveyed the region with Taieb, and after returning to the United States received a National Science Foundation grant to explore it more thoroughly. Johanson hoped to find hominid fossils.

In 1973, Johanson returned to Africa, but before going to Hadar he attended a conference of paleoanthropologists in Nairobi. There he met Richard Leakey, who had captured everyone's attention with skull ER 1470 (Section 11.6.1), said to be 2.9 million years old. Leakey, by then quite famous, asked Johanson, still an unknown, what he was up to. Johanson replied that he would soon be hunting for hominids at Hadar in northern Ethiopia. "Do you really expect to find hominids there?" asked Leakey. Johanson replied yes, adding "older than yours." He bet Leakey a bottle of wine he would do it. "Done!" said Leakey (Johanson and Edey 1981, pp. 134–135). Right from the start, it seems, Johanson was motivated by glamor. Finding hominids is special. It gets one headlines, interviews, and foundation grants, as well as recognition from one's colleagues.

11.9.1 The Hadar Knee (AL 129)

By the end of his first season at Hadar, Johanson was in trouble. His National Science Foundation grant money, which was supposed to have lasted two years, was almost gone. Johanson worried he would be labeled incompetent. Furthermore, he had not found any of those glamorous hominid fossils. Johanson noted: "I had not *exactly* promised hominids when I put in my request for funds from the National Science Foundation, but I knew when I wrote up my grant proposal that if I did not include a strong pitch for hominids I would get no money at all; the likelihood of being sent to Ethiopia to collect pig's teeth was remote" (Johanson and Edey 1981, p. 154).

Despite his financial problems, Johanson continued scouting for fossils. One afternoon, he found the upper portion of a tibia, a long bone between the knee and the ankle. The bone was obviously from some kind of primate. Nearby, Johanson found a distal femur, the lower end of a thighbone. From the way the femur and tibia fit together, Johanson believed he had found the complete knee joint not of some ancient monkey but of a hominid, an ancestor of modern humans. The deposits yielding the fossils were over 3 million years old, making this one of the oldest hominid finds ever made (Johanson and Edey 1981, p. 155).

Johanson felt that "his whole reason for being there, the core of his own most secret motivation" had been justified (Johanson and Edey 1981, p. 159).

In scientific publications that followed, Johanson reported that the Hadar knee (AL 129) was 4 million years old and belonged to a primitive australopithecine with a fully human bipedal gait (Johanson and Coppens 1976).

In support of his contention that AL 129 was characteristically human in structure, Johanson cited the presence of a valgus knee. A valgus knee is one in which the femur slants outward from the knee to the hip, at an angle from the lower part of the leg. Humans have a valgus knee. In African apes, the femur rises straight from the knee to the hip, in line with the lower part of the leg.

At 15 degrees from vertical, the angle of valgus in AL 129 was, however, much higher than the adult human mean of 9 degrees (Stern and Susman 1983, p. 296). This suggests that the locomotor behavior of AL 129, even if terrestrial and bipedal, might have been quite different from that of adult humans. In human children 3 to 4 years of age, the degree of valgus is as great as that in the AL 129 femur (Stern and Susman 1983, p. 296). The high angle is reflected in a child's knock-kneed stance and somewhat awkward gait. The creature with the AL 129 knee may have stood and walked in similar fashion.

Furthermore, Jack T. Stern and Randall L. Susman of the State University of New York at Stony Brook noted that the presence of a valgus knee is not exclusively associated with terrestrial bipeds. Orangutans and spider monkeys, both of which spend most of their time in trees, have valgus knees (Stern and Susman 1983, p. 298).

As we have seen (Section 11.8.5), C. E. Oxnard and others have found that the functional morphology of the australopithecines has orangutan affinities. The valgus knee in AL 129 could thus represent yet another orangutanlike feature in *Australopithecus*. The totality of orangutan resemblances suggests arboreal behavior in *Australopithecus*, which Oxnard, Zuckerman, and the Leakeys did not consider ancestral to modern humans.

In his account of the discovery of AL 129, Johanson did not mention that primates other than humans have a valgus knee. It seems there are two possible explanations why he did not. Either he was unaware that orangutans and spider monkeys have the same outward slanting femur as humans, or he was aware of this but deliberately neglected to mention it because it would have complicated the case he was trying to make.

According to Brigitte Tardieu (1979), key features of the AL 129 femur and tibia, other than the degree of valgus, fell outside the modern human range. "These traits . . . led her to conclude that despite clear adaptations to terrestrial bipedality in the small Hadar hominid, the precise mechanism of this bipedality could not be specified and that it must have occurred along with some degree of arboreal behavior," said Stern and Susman (1983, p. 298).

Stern and Susman (1983, pp. 298–299) themselves concluded: "Since, aside from the degree of valgus, the knee of the small Hadar hominid possesses no modern trait to a pronounced degree, and since many of these traits may not serve to specify the precise nature of the bipedality that was practiced, we must agree with Tardieu that the overall structure of the knee is compatible with a significant degree of arboreal locomotion."

It is intriguing that the views of scientists like Tardieu, Stern, and Susman, though appearing in scientific journals, are rarely encountered in popular presentations or general textbooks. This points to the existence of a pattern of knowledge filtration in the scientific community that tends, consciously or unconsciously, to suppress information that would complicate the relatively simple picture of human evolution presented to the public in general and to students at all levels of the educational system, except, perhaps, graduate students working directly in the field of physical anthropology.

Be that as it may, Johanson's lucky find saved the day for him, sparing him the embarrassment of leaving Ethiopia fundless and fossilless. Johanson said of the Hadar knee find: "It had brought me up a step; in my dealings with other scientists I was standing taller. I now had a unique hominid fossil of my own" (Johanson and Edey 1981, p. 165).

The glamor factor won Johanson 25,000 dollars from supporters in Cleveland, where Johanson held a post at the Museum of Natural History. The new funds allowed Johanson to return for a second year of work at Hadar.

11.9.2 Alemayehu's Jaws

Alemayehu Asfaw was an employee of the Ethiopian Ministry of Culture, and by the terms of Johanson's agreement with the Ethiopian government, he was working at the Hadar site. In October of 1974, Alemayehu found a fossil jaw that he thought belonged to a baboon, but Johanson said it was hominid. Other similar jaws soon turned up. Classifying them proved difficult. Johanson asked Richard Leakey to come and have a look at them. Leakey took up the invitation, and arrived accompanied by his mother Mary Leakey and wife Meave. Together with Johanson they examined the jaws and judged them to be *Homo,* making them the oldest *Homo* fossils yet found (Johanson and Edey 1981, pp. 172–176).

11.9.3 Lucy

On November 30, 1974, Donald Johanson and Tom Gray were searching Locality 162 at the Hadar site, collecting bits of mammalian bone. After some time, Gray was ready to call it quits and go back to the camp. Johanson, however, suggested they check out a nearby gully. Other members of the expedition had already thoroughly searched it, but Johanson, who had been feeling "lucky" all day, decided to have one more look. Gray and Johanson did not find much. But as they were about to leave, Johanson spotted a piece of arm bone lying exposed on the surface. He thought it was hominid. Gray disagreed, saying it was probably from a monkey. Then Gray found a piece of skull and a part of a femur. They seemed definitely hominid. As they looked around, they could see scattered on the surface other bones—apparently from the same hominid individual. Johanson and Gray started jumping and howling in the 110-degree heat, celebrating what was obviously an extremely significant find. Finally they calmed down, realizing their boots were probably smashing some of the precious bones. After collecting a few hominid fossils, they headed back to camp. That evening Johanson and his coworkers partied while a Beatles song, "Lucy in the Sky with Diamonds," blared repeatedly from the camp sound system. From the lyrics of that song, the female hominid received her name, Lucy (Johanson and Edey 1981, pp. 16–18).

By a combination of potassium-argon, fission track, and paleomagnetic dating methods, Johanson determined that Lucy was 3.5 million years old (Johanson and Edey 1981, pp. 200–203).

11.9.4 The First Family

In 1975, Johanson was back at Hadar, this time with a *National Geographic* photographer, who recorded another important discovery. On the side of a hill, Johanson and his team found the fossil remains of 13 hominids, including males, females, and children. The group was called the First Family. They were the same geological age as Lucy, about 3.5 million years old.

Stone tools were also found at the First Family site. They were made of basalt, and Johanson said they were "of somewhat better workmanship" than tools from the lower levels of Olduvai Gorge (Johanson and Edey 1981, p. 231).

How old were the tools? The fact that they were found on the surface made dating them somewhat difficult. In his book *Lucy,* Johanson reported the views of John Harris, a tool expert, who had worked at Lake Turkana: "He said that it was really impossible to date a surface-found tool at Lake Turkana because modern humans who needed rough blades to chop animals were making similar implements in profusion as recently as a thousand years ago, and that there were even a few people who were making them there today" (Johanson and Edey 1981, pp. 229–230). In Harris's opinion, the surface-found tools at Hadar could also have been recent.

To remove their doubts about the age of the stone tools found on the surface at the First Family site, Harris and Johanson conducted some excavations and were rewarded by discovering a number of tools *in situ.* They judged the level at which they were found to be 2.5 million years old (Johanson and Edey 1981, p. 231). No hominid fossils were found along with these tools. Because

Australopithecus was not known to have manufactured stone tools, Johanson speculated that *Homo habilis* was the toolmaker. But the oldest *Homo habilis* fossils were only about 2 million years old. Johanson simply proposed that *habilis* remains of the same age as the tools would eventually be found. As we have seen, there are, however, fossil remains resembling the modern human type from Early Pleistocene and Pliocene contexts in Africa (Sections 11.1, 11.2, 11.5, 11.6.3, and 11.6.4) and elsewhere (Section 6.2). It is thus possible that anatomically modern humans could have made the Hadar tools.

With the First Family, the major discoveries at Hadar, which also included the Hadar knee, Alemayehu's jaws, and Lucy, were completed. We shall now examine how these fossils were interpreted and reinterpreted by various parties.

11.9.5 Two Hominids at Hadar?

In classifying his finds, Johanson initially relied heavily upon the judgement of Richard and Mary Leakey that the Alemayehu jaws and First Family specimens were *Homo* (Johanson and Edey 1981, p. 217). If Lucy and the AL 129 femur and tibia were australopithecine, as Johanson believed, then there were two kinds of hominids at Hadar.

In a December 1976 *National Geographic* article, Johanson made a clear distinction between the First Family, which he thought represented *Homo,* and Lucy, which he thought represented an early *Australopithecus* (Fix 1984, p. 70). This two-species view was reflected in a number of scientific papers published by Johanson and various coauthors.

Richard Leakey later said that Lucy, with her V-shaped jaw and other primitive features represented "a late *Ramapithecus*" (Johanson and Edey 1981, p. 279). *Ramapithecus,* as previously noted (Section 3.9), was an extremely primitive apelike creature living in the Miocene and Pliocene. It may be recalled that *Ramapithecus,* originally considered the root of the hominid line, was later reclassified as nonhominid and ancestral to the orangutans.

Given the orangutan affinities of *Australopithecus,* as detailed by Oxnard (Section 11.8.5), maybe Leakey's idea that Lucy was a ramapithecine was right.

11.9.6 Johanson and White Decide on a Single Hadar Species

Johanson was later influenced to change his mind about the number of species at Hadar. The person who convinced him to do so was Timothy D. White, a paleontologist who had worked at Lake Turkana with Richard Leakey. White, on faunal grounds, disputed Leakey's dating of the KBS tuff (Section 11.6.5.2). Eventually, he left Lake Turkana and for a time worked at Laetoli, Kenya, where Mary Leakey had found hominid jaws similar to those at Hadar.

Johanson and White first met briefly in Africa. In the summer of 1977, when Johanson was back at the Cleveland Museum of Natural History studying his Hadar fossils, he asked White to bring samples of the Laetoli fossils.

White came and convinced Johanson to accept the following points: (1) the U-shaped jaws discovered at Hadar by Johanson and those discovered at Laetoli by Mary Leakey were of the same species; (2) the species was not *Homo,* as Johanson and the Leakeys had originally thought, but a new kind of australopithecine; (3) the V-shaped jaw of Lucy was also of the same species, being a female sexual variant of the other U-shaped jaws. Referring to a scientific paper in which he had advocated the two-species concept, Johanson said: "I would withdraw that paper today if I could" (Johanson and Edey 1981, p. 209).

Johanson and White (1979) soon announced their new species, calling it *Australopithecus afarensis,* after the Afar region of Ethiopia where most of the specimens were found.

According to Johanson and White, *Australopithecus afarensis* gave rise to two lineages. The first led by way of *Australopithecus africanus* to the robust australopithecines. The second lineage led by way of *Homo habilis* to *Homo erectus* and thence to *Homo sapiens.* In constructing this phylogenetic hypothesis, Johanson relied primarily upon dental evidence. The molars of *A. afarensis* were the smallest of all the australopithecines. The molars of *A. africanus* were larger, and those of the robust australopithecines larger still. This, to Johanson, indicated an evolutionary development. In Lucy's *Homo* offspring, the molars grew progressively smaller, representing a separate, parallel line of evolutionary development. It all seemed to fit together quite nicely.

11.9.7 A. Afarensis: Overly Humanized?

Johanson said that *Australopithecus afarensis* individuals had "smallish, essentially human bodies" (Johanson and Edey 1981, p. 275). But several scientists have strongly disagreed with Johanson's picture of *Australopithecus afarensis.* These dissenters have painted a far more apelike portrait of Lucy and her relatives. In most cases, their views on Lucy parallel the earlier work of Oxnard, Zuckerman, and others on *Australopithecus.* If the dissenting view is correct, as it appears to be, then Johanson's description of *Australopithecus afarensis* can only be considered as misleading.

It seems that Johanson imposed a humanlike interpretation upon Lucy's essentially apelike anatomy for the propaganda purpose of enhancing her evolutionary status as a human ancestor. Johanson himself said: "There is no such thing as a total lack of bias. I have it; everybody has it. The fossil hunter in the field has it. . . . In everybody who is looking for hominids there is a strong urge to learn more about where the human line started. If you are working back at around three million, as I was, that is very seductive, because you begin to get an idea that that is where *Homo* did start. You begin straining your eyes to find *Homo* traits in fossils of that age" (Johanson and Edey 1981, p. 257). Johanson gave this confession to explain why he originally characterized the First Family fossils and the Alemayehu jaws as *Homo,* but it also applies to his insistence on seeing in Lucy traits of a creature well on the way to becoming human.

11.9.7.1 The Skull of Australopithecus Afarensis

The Hadar fossils did not include a complete skull of an *A. afarensis* individual, but Tim White managed to pull together a partial reconstruction, using cranial fragments, pieces of upper and lower jaw, and some facial bones from several First Family individuals. According to Johanson, the reconstructed skull "looked very much like a small female gorilla" (Johanson and Edey 1981, p. 351). The forehead was low, the large jaw projected far beyond the upper part of the face, and there was no chin. The general apelike appearance was also reflected in anatomical details such as the mandibular fossa (the place where the lower jaw attaches to the skull), the tympanic plate, and the mastoid process. All of these were apelike, not humanlike (Johanson and Edey 1981, pp. 272–273). Furthermore, the cranial capacity of *A. afarensis* (380–450 cc) overlapped that of chimpanzees (330–400 cc) and other apes. Here there was no dispute between Johanson and his critics. Both agreed that the *afarensis* head was apelike.

Johanson and White believed the skull was, however, different from that of previously known australopithecines. But W. W. Ferguson (1984) and P. Schmid (1983) pointed out that White's reconstruction of the *Australopithecus afarensis* skull was incorrect. Correcting the mistake "makes the resulting construction a great deal more like *A. africanus,*" said Groves (1989, p. 263). P. V. Tobias (1980) said all the Hadar and Laetoli fossils were not a new species but were just subspecies of *Australopithecus africanus.* According to Tobias, *Australopithecus africanus* was the ancestor of *Homo,* while for Johanson and White *Australopithecus africanus* was the ancestor of only the robust australopithecines.

Originally, Johanson thought the *A. afarensis* U-shaped jaws were humanlike and like the Leakeys assigned them to the genus *Homo.* Later Johanson said they were "distinct from apes and from any of the later hominids" (Johanson and Edey 1981, p. 271). But his detailed descriptions showed the Hadar jaws to be in fact quite apelike.

In humans, the teeth in the jaw are arrayed in a parabolic curve. In the Hadar jaws, such as AL 200, the teeth on either side of the jaw are set in straight, parallel rows, as in the apes, although the rearmost molars are sometimes slightly displaced (Johanson and Edey 1981, pp. 267–268). Both in apes and the Hadar fossils the palate is flat (Johanson and Edey 1981, p. 270). In humans it is arched.

TABLE 11.6

Evidence for Arboreality in Postcranial Anatomy of A. Afarensis

1. General anatomy of Lucy's shoulder blade was characterized as "virtually identical to that of a great ape and had a probability of less than 0.001 of coming from the population represented by our modern human sample" (Susman *et al.* 1984, pp. 120–121).

2. Lucy's shoulder blade has a shoulder joint which points upward (Oxnard 1984, p. 334-i; Stern and Susman 1983, p. 284). This would allow "use of the upper limb in elevated positions as would be common during climbing behavior" (Stern and Susman 1983, p. 284).

3. *A. afarensis* wrist bones are apelike. "Thus we may conclude that *A. afarensis* possessed large and mechanically advantageous wrist flexors, as might be useful in an arboreal setting" (Stern and Susman 1983, p. 282).

4. *A. afarensis* metacarpals (the bones in the palm region of the hand) "have large heads and bases relative to their parallel-sided and somewhat curved shafts—an overall pattern shared by chimpanzees." This "might be interpreted as evidence of developed grasping capabilities to be used in suspensory behavior" (Stern and Susman 1983, pp. 282, 283).

5. The finger bones are even more curved than in chimpanzees and are morphologically chimpanzeelike (Stern and Susman 1983, pp. 282–284; Susman *et al.* 1984, p. 117; Marzke 1983, p. 198).

6. *A. afarensis* humerus (upper arm bone) has features that are "most likely related to some form of arboreal locomotion" (Oxnard 1984, p. 334-i; see also Senut 1981, p. 282).

7. One of the long bones in the forearm, the ulna, resembles that of the pygmy chimpanzee (Feldesman 1982b, p. 187).

8. Vertebrae show points of attachment for shoulder and back muscles "massive relative to their size in modern humans" (Cook *et al.* 1983, p. 86). These would be very useful for arboreal activity (Oxnard 1984, p. 334-i).

9. "Recently Schmid (1983) has reconstructed the A.L. 288-1 rib cage as being chimpanzee-like" (Susman *et al.* 1984, p. 131).

10. Blades of hip oriented as in chimpanzee (Stern and Susman 1983, p. 292). Features of *afarensis* hip therefore "enable proficient climbing" (Stern and Susman 1983, p. 290).

TABLE 11.6—*Continued*

11. The thighbone of Lucy "probably comes from an individual with the ability to abduct the hip in the manner of pongids," allowing for "movement in the trees" (Stern and Susman 1983, p. 296).

12. Knee joint is loose, as in gibbon. "The mobility and prehensility of the foot are greatly complemented" (Tardieu 1981, p. 76), making it good for climbing.

13. Lucy had valgus knee, as do humans. But "the orang-utan and the spider monkey . . . are extremely able arborealists that have similar valgus angles as humans" (Oxnard 1984, p. 334-ii; see also Prost 1980).

14. Lucy had "a relatively short hindlimb . . . comparable to that seen in apes of similar body size." This "would clearly facilitate climbing" (Susman *et al.* 1984, pp. 115, 116).

15. Feet have long, curved toes and a mobile ankle joint, making them well suited for grasping limbs and climbing in trees (Susman *et al.* 1984, p. 125). Also, the big toe is divergent, as in the apes (Susman *et al.* 1984, pp. 137–138).

As in the apes, the canines of the Hadar jaws were conical. In humans, the inner surface of the canine is flattened. In order to accommodate the projecting lower canine of *A. afarensis,* the upper jaw has a noticeable gap between the incisor and the canine. Other australopithecines also have the same gap. This gap, called a diastema, is also present in apes but not in humans.

Departures from the ape condition were minor. In an ape, the first premolar has a single cusp. In humans, the first premolar has a prominent second cusp. In all of the Hadar specimens except Lucy, the first premolar has a slightly developed second cusp (Johanson and Edey 1981, p. 270).

All in all, the apelike condition of the Hadar jaws is so pronounced that even Johanson admitted: "If David Pilbeam were to find any of them in Miocene deposits without any associated long bones, he would surely say it was an ape" (Johanson and Edey 1981, p. 376).

11.9.7.2 Postcranial Anatomy

Now we move on to the postcranial anatomy of *A. afarensis,* particularly Lucy. Several workers have found *A. afarensis* to be rather apelike, thus challenging Johanson's view that Lucy was terrestrially bipedal in human fashion. Table 11.6 summarizes the evidence for arboreality in the postcranial anatomy of *A. afarensis,* and we shall amplify some of the points in this section.

Oxnard (1968) called attention to features of the Sterkfontein scapula suggesting that australopithecines probably engaged in holding the arms over the head in hanging behavior (Section 11.8.1, Table 11.5).

A. afarensis has the same kind of scapula. Stern and Susman (1983, p. 284) concluded that the shoulder joint of *A. afarensis* was "directed far more cranially than is typical of modern humans and that this trait was an adaptation to use of the upper limb in elevated positions as would be common during climbing behavior."

Johanson (1976, p. 808) had said that the Hadar hands bore "an uncanny resemblance to our own—in size, shape, and function." But this appears to be incorrect.

Stern and Susman (1983, p. 284) concluded: "A summary of the morphologic and functional affinities of the Hadar hand fossils leads inexorably to an image of a suspensory adapted hand, surprisingly similar to hands found in the small end of the pygmy chimpanzee–common chimpanzee range." M. W. Marzke (1983, p. 198), sharing this view, stated that the curved bones of the *A. afarensis* hand "recall the the bony apparatus which accommodates the well developed flexor musculature in living apes and positions it for efficient hook-like grip of the branches by the flexed fingers during arboreal climbing and feeding."

So thus far we have in *A. afarensis* a gorillalike head, an upward-pointing shoulder joint indicating that the arm was used for suspensory behavior, and a hand with a powerful wrist and curved fingers, suitable for climbing. One can just imagine the effects of a painting or model of Lucy engaged in suspensory or other arboreal behavior. This would surely detract from her image as a creature well on the way to human status. Even if one believes Lucy could have evolved into a human being, one still has to admit that her anatomical features appear to have been misrepresented for propaganda purposes.

The distal humerus, the elbow region of the upper arm bone, fits the apelike pattern already established. Brigitte Senut, a physical anthropologist at the French Museum of Natural History, conducted a study of the outlines of cross sections of the distal humerus in living primates, including human beings, and fossil hominids. Senut (1981a, p. 282) discovered that the distal humerus of Lucy (AL 288-1M) was "pongid-like." Pongids are the anthropoid apes, such as chimpanzees, gorillas, and orangutans. Senut (1981a, p. 282) concluded: "The scheme in the Afar specimen [Lucy] would suggest . . . its apelike pattern might be a result of a kind of suspension."

Senut (1981a, p. 282) went on to say: "From our point of view, we would say that this specimen may be too pongid-like (i.e. specialized) to be in our ancestry." From the standpoint of mainstream paleoanthropological thought, this is an extremely heretical view.

Feldesman (1982a, p. 91) found Lucy's humerus to be most closely related to the pygmy chimpanzee, *Pan paniscus.*

As far as the bones of the lower arm are concerned, Feldesman (1982b, p. 187) found that "'Lucy' (AL 288) clearly resembles *Pan paniscus* in proximal ulnar morphology." The ulna is the innermost of the two bones making up the forearm (the radius is the other). The proximal, or upper, part of the ulna joins the humerus at the elbow.

In 1985, Della Collins Cook, an anthropologist, and three coauthors, among them Donald Johanson, published a study of the vertebral column of *Australopithecus afarensis.*

Cook and Johanson claimed: "The AL-288 vertebrae correspond to those of modern humans in remarkable detail" (Cook *et al.* 1983, p. 84). They noted, however, that the "Hadar vertebrae depart from the morphological pattern found in modern humans in a few details that may have functional significance" (Cook *et al.* 1983, p. 86). These "few details" were not trivial. For example, according to Cook and Johanson, the spinous processes of the *A. afarensis* neck and upper back vertebrae were quite long. The spinous process, a bony projection on the back side of the vertebrae, serves as a point of attachment for muscles. According to Cook and Johanson, the length and surface features of the spinous processes indicated that in *A. afarensis* the back and shoulder muscles were "massive relative to their size in modern humans" (Cook *et al.* 1983, p. 86).

Oxnard (1984, p. 334-i) stated that the features of the *A. afarensis* vertebrae reported by Cook and Johanson "are likely to have provided the stress bearing structures necessary to support the actions of very powerful shoulder muscles in climbing and arboreal activities suggested by our prior studies of the scapula and clavicle of other australopithecines."

C. Owen Lovejoy, a supporter of Johanson, claimed that the *afarensis* hip was suitable only for upright walking (Johanson and Edey 1981, pp. 347–348). But the *afarensis* hip structure is significantly different from that in human beings. In particular, Lucy's iliac blade, like that of other australopithecines, is positioned as in apes (Section 11.8.2, Figure 11.13). Susman said: "Therefore, we are of the opinion that the orientation of the iliac blades in the Hadar species is well-suited for a part-time climber" (Susman *et al.* 1984, p. 132).

In *Scientific American,* Lovejoy (1988) reasserted his familiar claims that Lucy's pelvic structure and musculature were very similar to those of humans. We will not here repeat the detailed demonstrations by Zuckerman, Oxnard, Stern, Susman, and others that the pelvic morphology of *A. afarensis* had quite a bit in common with arboreal primates, and was better suited for climbing than walking (Section 11.8.2).

What is perhaps most significant about Lovejoy's presentation is that he does not once directly mention his opponents and their arguments. This adds to our suspicions that the views of Zuckerman, Oxnard, Stern, Susman, Prost, and others are being suppressed for propaganda purposes on the level of secondary presentations for the wider scientific community, educational institutions, and the public in general. The views of the advocates of arboreality for *A. afarensis* are represented almost solely in the primary level of publication, in the obscure pages of scientific journals intended for specialists. They are, however, not at all well represented in publications like *Scientific American*, college textbooks on anthropology, and popular books and television programs dealing with the topic of human evolution. Arboreal habits would not look well in the hominid advertised as the oldest known creature directly ancestral to modern humanity.

Femurs from Lucy and the First Family group challenge claims by Johanson and Lovejoy that the lower limb of *A. afarensis* was distinctly human in morphology and function. Stern and Susman (1983, p. 296) concluded that the proximal (upper) part of Lucy's femur "probably comes from an individual with the ability to abduct the hip in the manner of pongids," allowing for "movement in the trees."

Measurements of several features of the lower (distal) end of the AL 333-4 First Family femur showed it to be outside the human range and within the ranges of chimpanzees, gibbons, and several species of monkeys. In fact "the distal end of the AL 333-4 femur actually appears less human-like than that of a woolly monkey" (Stern and Susman 1983, p. 297).

Christine Tardieu, an anthropologist at the Museum of Natural History in Paris, gave a slightly different assessment of the AL 333-4 distal femur, finding it barely within the modern human range, at "the extreme end closest to the apes" (Stern and Susman 1983, p. 299). Thus, as often happens, we find ourselves confronted with contradictory interpretations of the same fossil material, but on the whole, the femurs in question appear to be apelike.

Tardieu, in addition to measuring the AL 333-4 femur of the First Family group, also conducted studies of the distal femur of Lucy. She gave special attention to the notch in the femur that holds the upper end of the tibia, the larger of the two bones of the lower leg. In humans, the spine of the tibia fits tightly into the notch of the femur. In apes, the fit is looser. In this regard, Lucy is in the range of the gibbon. Tardieu (1981, p. 76) stated: "The loose fit of the articular surfaces . . . and the consequent laxity of the knee joint signify that the leg and the foot can be placed on the substrate in a much freer fashion than in Man." This would be good for climbing, but unsatisfactory for extensive walking on the ground.

Commenting on Tardieu's study of Lucy's knee, Oxnard (1984, p. 334-ii) said she was led to "conclude that . . . its locking mechanism was not developed, implying that full extension of the leg in walking, a key point of human bipedality, was lacking." Such features "suggested to Tardieu that 'Lucy' spent a considerable period of time climbing in the trees" (Oxnard 1984, p. 334-ii).

One can just imagine Lucy, hanging lazily from a tree limb by one of her arms, bending a small, dangling foot back from the ankle, while rotating her lower leg from the knee to bring the backward reaching foot in contact with a nearby limb.

The knee of Lucy (AL 288), like the original Hadar knee complex (AL 129), had a significant degree of valgus. Johanson, Lovejoy, and others held this to be an indication of humanlike posture and terrestrial bipedal gait. But, as we have seen, the orangutan and spider monkey have similar valgus angles, and they are arboreal.

In our anatomical survey, we have now progessed to the controversial feet of *A. afarensis*. Even Johanson had a difficult time disguising the manifestly apelike condition of Lucy's foot. He wrote: "The *afarensis* phalanges are arched, and proportionally a good deal longer than those in modern feet. They might almost be mistaken for finger bones" (Johanson and Edey 1981, p. 345). Johanson also noted that the *A. afarensis* foot had "very large muscles whose presence is betrayed by markings along the sides of the phalanges" (Johanson and Edey 1981, p. 345). Such muscles would have been useful in hindlimb grasping.

It is amazing that Johanson could so candidly acknowledge the very apelike morphology of the *afarensis* foot and yet refuse to draw the obvious conclusion that it was used in arboreal behavior. Instead, Johanson stated: "Although similarly curved phalanges and muscle markings are found in the chimpanzee—reflecting the chimp's ability to climb trees—Latimer warns that this does not mean that *afarensis* was a tree climber too" (Johanson and Edey 1981, p. 346). Bruce Latimer was one of Johanson's graduate students and worked with him quite closely in Ethiopia on the Hadar finds, so his impartiality is suspect. He was later employed by Johanson to help with the reconstruction of *A. afarensis*. It is not unexpected that Latimer would agree with his professor, mentor, and employer that *afarensis* was a fully terrestrial biped. But researchers operating from more detached and independent standpoints have reached totally different conclusions, which seem to be more in harmony with the evidence.

In studying the most complete *A. afarensis* foot, AL 333-115 from the First Family group, Stern and Susman (1983, p. 306) found that the proximal phalanges (the bones at the base of each toe) had a "strikingly pongid morphology." This was true in terms of both their length and curvature.

Susman, reporting the conclusions of an investigation into the curvature of proximal phalanges in a variety of apes, stated that the chimpanzee and bonobo, or pygmy chimpanzee, had "the most curved toe bones of any ape plotted" (Susman *et al.* 1984, p. 125). And the proximal phalanges of AL 333-115 were "more curved than in the average bonobo" (Susman *et al.* 1984, p. 125). In other words, *A. afarensis* was apparently more apelike, in this respect, than any of the living apes. Human proximal phalanges are nearly straight.

Like the proximal phalanges, the other toes bones of *A. afarensis* also displayed apelike features. Altogether, the long, curved toes of *A. afarensis,* accompanied by powerful grasping muscles, would have been well suited for arboreal behavior.

Susman concluded: "at the very least the small individuals should have been able to grab with their toes as well as 2-year old children grab with their fingers. The large Hadar individuals probably could use their toes for simple grasping as effectively as considerably older human children use their fingers. . . . the strength of the grip may have well exceeded the strength of hand grip in young humans" (Susman *et al.* 1984, p. 124). Lending support to this conclusion, the *A. afarensis* fibula (the smaller of the two bones of the lower leg) was quite robust, indicating the presence of powerful muscles for flexing the foot (Susman *et al.* 1984, p. 124).

According to Johanson, Latimer, who was opposed to arboreality, concluded that "*afarensis* was an exceptionally strong walker, and that its elongated toes may have been of service to it when moving over rough stony ground, or in mud, where some slight gripping ability would have been useful" (Johanson and Edey 1981, pp. 345–346).

Stern and Susman (1983, p. 308) found this notion "untenable," observing that "curved toes are found only in species that engage in arboreal behavior."

Stern and Susman (1983, p. 308) further stated: "There is no evidence that any extant primate has long, curved, heavily muscled hands and feet for any purpose other than to meet the demands of full or part-time arboreal life."

Another apelike feature of the *A. afarensis* foot can be found in the hallux, or big toe. Studies by Susman showed that the *A. afarensis* hallux could be extended sideways, like the human thumb (Susman *et al.* 1984, pp. 137–138).

The hallux of *A. afarensis* was relatively smaller than that of some arboreal primates, causing Latimer to suggest that *A. afarensis* was not well suited for climbing trees. But Susman pointed out that the highly arboreal gibbon also has a small hallux (Susman *et al.* 1984, p. 137). Altogether, the picture that emerges of the *afarensis* foot is extremely apelike—a foot with long, curved, fingerlike toes and a highly mobile, thumblike big toe.

Tim White, one of the promoters of *A. afarensis,* has responded negatively to attempts to characterize Lucy as fully, or even partially, arboreal. White stated: "We are wary of this approach which makes the interpretive leap from curved phalanges into the trees" (White and Suwa 1987, p. 512). As we have seen, wariness is always required in approaching empirical treatments of human origins and antiquity. But we should perhaps be more wary of the interpretive leap from curved phalanges *out* of the trees, since greatly curved phalanges in extant primates are an exclusively arboreal adaptation. This is especially true of curved phalanges existing in combination with an upward pointing shoulder joint and other signs of arboreal capability.

From the toes, let us now move on to the A. afarensis ankle, including its articulations with the tibia and fibula, the bones of the lower leg.

Regarding the articular surfaces of the fibula of A. afarensis, Stern and Susman (1983, p. 305) wrote that they provide "evidence for a significant component of arboreality in the behavior of A. afarensis." Johanson's supporters such as Latimer disagreed with this analysis (Latimer et al. 1987). In overall appearance, however, the lower part of Lucy's fibula, the part that connects with the ankle, is different from that of a human being and almost identical to that of the pygmy chimpanzee (Susman et al. 1984, p. 130).

Stern and Susman (1983, p. 302) argued that Lucy's foot could be bent back further than in humans. "This trait would seem to be useful in reaching for branches with the feet and in hindlimb suspension," they noted (Stern and Susman 1983, p. 299). According to Stern and Susman (1983, p. 300), Lucy's ankle was structured so that she would have "had difficulty in maintaining a vertical orientation of the trunk and might have progressed bipedally in a manner unlike that of humans and more like that of an African ape."

Johanson's supporters took a completely opposite position, namely, that the ankle of A. afarensis was almost totally adapted for a humanlike, terrestrial bipedal gait, making impossible any substantial arboreal behavior. Latimer and Lovejoy in particular have published several articles micro-analyzing every curve of the A. afarensis foot and ankle bones as proof of exclusive terrestrial bipedalism (Gomberg and Latimer 1984, Latimer et al. 1987, Latimer and Lovejoy 1990a, Latimer and Lovejoy 1990b).

We note, however, that an author of a recent survey (Groves 1989) takes the side of Stern, Susman, Tardieu, Oxnard, and others who have argued for a substantial component of arboreality in Australopithecus afarensis and the australopithecines generally. Groves (1989, p. 310) said that in the australopithecines "bipedal locomotion was only part of a pattern which also incorporated sophisticated climbing ability."

J. H. Prost (1980) of the University of Chicago concluded that the australopithecines, including Lucy, were primarily quadrupedal vertical climbers. "Quadrupedal vertical climbing produced a large number of . . . traits which have incorrectly assumed to have been bipedal adaptations," stated Prost (1980, p. 186).

According to Prost (1980, p. 175), the australopithecines, including A. afarensis, would have possessed, in addition to their aboreal capabilities, the capacity for "facultative terrestrial bipedalism." The word facultative means "optional" or "taking place under some conditions but not under others." In other words, the predominantly arboreal australopithecines, if the situation demanded, would have been able to move bipedally on the ground, perhaps in running from one tree to another some distance away. This type of behavior is observed in many primates, including chimpanzees, orangutans, and gibbons. So the fossil evidence in no way obligates one to attribute to *A. afarensis* any specifically human locomotor behavior. According Prost (1980, p. 188), the first true terrestrial bipedal hominid was *Homo habilis* (as understood before the discovery of the apelike OH 62 individual) or early *Homo erectus*.

R. H. Tuttle posited the existence of preaustralopithecine hominids displaying a kind of arboreal bipedalism. He called them hylobatians, after the genus *Hylobates,* which includes the modern gibbon. Tuttle (1981, p. 90) stated: "Vertical climbing on tree trunks and vines and bipedalism on horizontal boughs were conspicuous components of their locomotor repertoire. They commonly stood bipedally while foraging in trees . . . Short bursts of bipedal running and hindlimb-propelled leaps may have been important for the manual capture of insects and small vertebrates with which they supplemented their vegetable fare."

According to Tuttle (1981, p. 89), the Hadar hominids "had curved fingers and toes, strong great toes and thumbs, and other features that suggest they were rather recently derived from arboreal hominids [his hylobatians] and that they probably continued to enter trees, perhaps for night rest and some foraging."

Studies of primate behavior apparently support the arboreal implications of the fossil morphology of *A. afarensis*. Susman stated: "We feel, based on extensive literature on free-ranging primates, that creatures such as represented by A.L. 288-1 could not have survived full-time on the ground. Today, all primates from common chimpanzees (which range from 27 up to 70 kg [59 to 154 lb.]), to vervet monkeys and baboons (which range from less than 3 to over 40 kg [7 to >88 lb.]), are obliged at least to sleep in trees (or on rocky cliff-faces). They all feed in trees" (Susman *et al.* 1984, pp. 150–151). Susman pointed out that pollen studies showed the presence of trees at the Hadar site (Susman *et al.* 1984, p. 151).

Having completed our review of the anatomy of *Australopithecus afarensis,* we conclude that Johanson was incorrect in stating that Lucy and her relatives were predominantly terrestrial bipeds and had "essentially human bodies" (Johanson and Edey 1981, p. 275). The picture that emerges is one of an arboreally adapted creature with long, curved toes and fingers, a long, heavily muscled arm equipped with an upward-pointing shoulder joint, a pelvis structured like that of apes, and a knee complex resembling that of the orangutan.

This view is not, however, very well represented in popular presentations. In order to maintain a believable human evolutionary sequence, the scientific community apparently requires, for propaganda purposes, a credible human ancestor in the Late Pliocene and Early Pleistocene. The erect, bipedal, non-arboreal hominid, with apelike head and humanlike body, as portrayed by Johanson and his disciples, satisfies this requirement far better than the almost totally apelike and wholly or partially arboreal creature that emerges from the studies of Stern, Susman, Oxnard, and others. This judgement is supported by the fact that the views of Johanson, Latimer, and Lovejoy make their way into college textbooks, popular books on evolution, televisions specials, and so on, with hardly a hint of any serious opposing conception. This, we believe, is not an accident. The informal gatekeepers and guardians of scientific orthodoxy are apparently quite careful about what reaches the public.

11.9.8 Opposition to the Single Species Hypothesis

The idea that the large and small hominid individuals from Hadar and Laetoli represent a single sexually dimorphic species (*Australopithecus afarensis*) has not won universal acceptance among scientists.

Adrienne Zihlman (1985, p. 214) of the University of California (Santa Cruz) stated: "The interpretation of extreme sexual dimorphism for these fossils has been a mere assertion from the beginning . . . and has continued to be so."

In one of her reports, Zihlman (1985, pp. 216–217) supplied some data on sexual dimorphism in human beings, various apes, and *A. afarensis*. She found: "The Hadar fossils suggest even greater dimorphism than exists in orangutans, a species where males may be more than three times the body weight of females. This means that 'A. afarensis' is more sexually dimorphic than any living hominoid. From the point of view of size, more than one species is strongly implied."

In the human species, males average only about 20 percent heavier than females. So even if, for the sake of argument, one accepts that *A. afarensis,* with males more than three times heavier than females, did represent one species, the extreme degree of dimorphism argues strongly in support of apelike rather than humanlike morphology and behavior. And if Zihlman is right, and there were two species, not one, at Hadar, then Tim White sold Donald Johanson an illusion.

Todd Olson, an anthropologist at the City College of New York, concluded from cranial evidence that more than one species was present at the Hadar site. Olson discovered that the mastoid process in the larger Hadar individuals (such as AL 333–45) was "pneumatized" with small air pockets. The mastoid process is a bony projection behind the ear. A pneumatized mastoid is characteristic of *Australopithecus robustus.* The mastoid in the small Hadar individuals (Lucy), *A. africanus,* and *Homo sapiens* is nonpneumatized. The difference in mastoid structure between the large and small Hadar individuals, along with dental evidence, convinced Olson that two species rather than one were found at Hadar (Herbert 1983, pp. 10–11). The larger individuals were, according to

Olson, a population related to *Australopithecus robustus,* and the smaller individuals, including Lucy, were the earliest members of the *Homo* line. This is an interesting variation of the original two-species interpretation of the Hadar fossils, as proposed by Richard Leakey, who placed the larger individuals in the *Homo* line and characterized the smaller Lucy as a surviving *Ramapithecus.* Johanson and his supporters "took great exception to Olson's analysis, showing that the AL 333-45 basicranium is distorted and, if anything, is *Homo*-like" (Groves 1989, p. 262).

Dental evidence has also caused some workers to question the the view that a single species was present at Hadar. In Lucy, the first premolar has a single cusp, but in the other Hadar jaws, the premolars, like those of modern humans, have a double cusp. *Science News* reported: "Yves Coppens, director of the Musee de l'Homme in Paris . . . and an original cosigner on the paper identifying *A. afarensis* as a species has now reversed himself based on the dental evidence—specifically the existence of both single-cusp and bicuspid premolars in the sample—he says there must have been two species coexisting at Hadar" (Herbert 1983, p. 11). Johanson and White, however, said that in an evolving line, some individuals would have the single cusp and others the bicuspid tooth.

Stern and Susman, like Johanson, originally believed the Hadar fossils represented the males and females of a single species exhibiting a high degree of sexual dimorphism. According to their view, the small females, including Lucy, would have been quite arboreal, the larger males less so.

Stern, however, eventually backed down from the sexual dimorphism concept. *Science News* reported in 1983: "he argues that the finger bones clearly sort themselves into two groups; one group [the small individuals] has strongly curved fingers—exactly like African apes—and the other [the large individuals] has less curved . . . fingers, halfway between gorillas and humans" (Herbert 1983, p. 9).

Stern said: "The finger bones pushed me over the edge. Taken in conjunction with the differences in the ankles and leg bones, I had to ask myself: Do you ever see such difference in living animals? And the answer is no—never. It's just too big a difference to be sexual dimorphism" (Herbert 1983, p. 9). Apparently, both species would have manifested arboreal behavior. Even the large First Family specimens had finger bones curved more than those of humans. They also had, as we have seen, long curved toes and a femoral anatomy similar to that of apes.

Where does all this leave us regarding our understanding of *Australopithecus afarensis*? Johanson and White and their supporters say *A. afarensis,* a terrestrial biped, was ancestral to *A. africanus* and the robust australopithecines, a line that finished in extinction. They also said *A. afarensis* was ancestral to the line leading from *Homo habilis* to *Homo sapiens.* Others say *A. afarensis* was a variety of

A. africanus, which gave rise to the *Homo* line. Still others take a two-species approach. Tardieu (1981), studying the postcranial evidence, particularly the femurs, concluded that the larger individuals at Hadar represented the *Homo* line and the smaller individuals, like Lucy, something else. Y. Coppens, from studies of the dental evidence, reached a similar conclusion (Weaver 1985, pp. 592, 595). Richard Leakey also took the multiple-species approach, claiming that Lucy was a surviving *Ramapithecus* whereas the larger Hadar specimens represented the *Homo* line. Olson, studying features of the cranial anatomy, concluded that the larger Hadar individuals were like *Australopithecus robustus,* whereas Lucy was the first species in the *Homo* line (Herbert 1983, pp. 10–11). Susman felt the large and small Hadar types represented a single, partly arboreal species. Stern originally agreed with this, but later adopted a two-species view, as a result of his studies of the finger anatomy. Finally, Oxnard and others believed *A. afarensis* to be an apelike arboreal creature with no direct relation to the human line.

This brief review does not, however, exhaust the various opinions about the phylogenetic status of *A. afarensis.* "For Ferguson (1983, 1984) the Hadar sample contains three different taxa: *Sivapithecus* sp., *Australopithecus africanus,* and *Homo antiquus* (new species)," noted Groves (1989, p. 262). Groves himself (1989, p. 263), in his comprehensive taxonomic survey of the hominids, said: "Certainly the post-cranial data are absolutely clear, and split the Hadar sample into two divisions." Groves (1989, p. 263) classified one Hadar group as early *Homo* and the other as an unnamed new hominid genus. Under the species designation *Australopithecus afarensis,* he kept only the Laetoli jaws. So Groves, like Ferguson, found three species instead of one in the *A. afarensis* fossils of Johanson and White.

Within the scientific community there is as of yet no unanimous picture of what the australopithecines, including *A. afarensis,* were really like, both in terms of their morphology and their phylogenetic relation with modern humans. The field is still wide open and full of conflicting views.

Nevertheless, we find the argument for a substantial component of arboreality in the locomotor behavior of *A. afarensis* more credible than that for exclusive terrestrial bipedalism. There also appears to be good reason to suppose the Hadar hominid fossils represent more than one species. Furthermore, we favor the view, espoused by Louis and Richard Leakey, that no australopithecine, including *A. afarensis,* warrants being labeled a human ancestor.

Just as today we find true humans coexisting with various categories of apes, some more humanlike than others, the same was true in the past, as far back as our research can carry us. In fact, an objective review of the evidence yields signs of anatomically modern human beings tens of millions of years ago, a fact distinctly incompatible with any current evolutionary model.

11.10 THE LAETOLI FOOTPRINTS

The Laetoli site is located in northern Tanzania, about 30 miles south of Olduvai Gorge. *Laetoli* is the Masai word for red lily. The area was first explored by the Leakeys in 1935. Later, Mary Leakey returned to Laetoli and discovered some hominid jaws, which she regarded as early *Homo.*

One day in 1979, Dr. Andrew Hill of the Kenya National Museum and several other members of Mary Leakey's expedition were playing around, throwing pieces of elephant dung at each other. In the course of this sport, Hill noticed some marks on the ground. They proved to be fossil footprints of animals. Subsequently, Peter Jones and Philip Leakey, the youngest son of Louis and Mary Leakey, discovered among the footprints some that appeared to have been made by hominids. The prints had been impressed in layers of volcanic ash, dated by Garniss Curtis, using the potassium-argon method, at from 3.6 to 3.8 million years old.

National Geographic magazine featured an article by Mary Leakey titled "Footprints in the Ashes of Time." A caption to a photo of some hominid prints read: "The best-preserved print shows the raised arch, rounded heel, pronounced ball, and forward-pointing big toe necessary for walking erect. Pressures exerted along the foot attest to a striding gait" (M. Leakey 1979, p. 452). Dr. Louise Robbins, a footprint expert from the University of North Carolina, observed: "They looked so human, so modern, to be found in tuffs so old" (M. Leakey 1979, p. 452).

Readers who have accompanied us this far in our intellectual journey will have little difficulty in recognizing the Laetoli footprints as potential evidence for the presence of anatomically modern human beings over 3.6 million years ago in Africa. We were, however, somewhat astonished to encounter such a striking anomaly in the unexpected setting of the more recent annals of standard paleoanthropological research. What amazed us most was that scientists of worldwide reputation, the best in their profession, could look at these footprints, describe their humanlike features, and remain completely oblivious to the possibility that the creatures that made them might have been as humanlike as ourselves.

Their mental currents were running in the usual fixed channels. Mary Leakey (1979, p. 453) wrote: "at least 3,600,000 years ago, in Pliocene times, what I believe to be man's direct ancestor walked fully upright with a bipedal, free-striding gait. . . . the form of his foot was exactly the same as ours."

Who was the ancestor? Here we once more confront the debate, between the Leakeys on one hand and Johanson and White on the other, about the number and type of species represented by the fossil materials from Hadar and Laetoli.

Taking the Leakeys' point of view, the Laetoli footprints would have been made by a nonaustralopithecine ancestor of *Homo habilis.* Taking the Johanson-White point of view, the Laetoli footprints would have been made by *Australopithecus afarensis.* In either case, the creature who made the prints would have had an apelike head and other primitive features.

But why not a creature with fully modern feet and fully modern body? There is nothing in the footprints that rules this out. Furthermore, we have compiled in this book quite a bit of fossil evidence, some of it from Africa, that is consistent with the presence of anatomically modern human beings in the Early Pleistocene and the Late Pliocene.

The most prominent set of tracks at Laetoli represented the footprints of three hominids, one larger than the others. Applying an anthropological rule of thumb that a hominid's foot length represents 15 percent of the creature's height, Mary Leakey (1979, p. 453) calculated that the largest hominid stood 4 feet, 8 inches tall, whereas the next largest stood 4 feet tall. The smallest would have been still shorter. Leakey hypothesized that the largest individual was an adult male, the next largest an adult female, and the smallest a child. Admitting this was only a guess, she suggested the alternative possibility that the second largest set of prints might represent a juvenile male (M. Leakey 1979, p. 453). One cannot, however, be certain that the largest tracks represent a fully adult form either. Even so, the heights of the creatures that made the two larger sets of tracks, as estimated by Mary Leakey, fall within the modern human adult range.

Are we perhaps exaggerating the humanlike features of the Laetoli footprints? Let us see what various researchers have said. Louise M. Robbins, who provided an initial evaluation of the Laetoli prints to Mary Leakey in 1979, later published a more detailed report. Several sets of tracks, identified by letters, were found at Laetoli. In examining the "G" trails, representing the three individuals described by Mary Leakey as a possible family group, Robbins (1987, p. 501) found that the prints "share many features that are characteristic of the human foot structure."

Robbins (1987, p. 501) noted: "Each hominid has a non-divergent great toe, or toe 1, and that toe is about twice as large as toe 2 beside it." She found the spacing between toes 1 and 2 "no greater than one finds in many people today, including individuals who habitually wear shoes" (1987, p. 501). Robbins also found "the ball region of the hominids' feet is of human form" and added that the feet displayed "a functionally stable longitudinal arch structure" (1987, p. 501). Finally, she observed that "the heel impressions in the hominids' footprints appear human in their form and in their locomotory performance" (Robbins 1987, p. 501).

Robbins (1987, p. 501) therefore concluded that "the four functional regions—heel, arch, ball, and toes—of the hominids' feet imprinted the ash in a typically human manner" and that "the hominids walked across the ash surface in characteristic human bipedal fashion."

Concerning the size of the prints, Robbins (1987, p. 502) stated: "The assumed dimensions of the G-2 footprints do indeed fall well within the adult male range of a sample of American subjects, and the measurements of G-3's footprints fall in the lower portion of the range for adult females in the American sample. The dimensions of the G-1 footprints, however, are well below dimensional ranges for American adults but within foot length and width ranges for a small sample of immature individuals. . . . Nonetheless, it is mere conjecture at this stage of hominid footprint investigation to suggest that the Site G hominids may have been a male, a female, and an offspring who were walking from an area of falling volcanic ash."

M. H. Day studied the prints using photogrammetric methods. Photogrammetry is the science of obtaining exact measurements through the use of photography. Photogrammetric methods are extensively used by cartographers in making accurate contour maps from aerial photographs. Day (1985, p. 121), having found the same techniques useful on the miniature geography of footprints, stated: "What these footprints, and their photogrammetric analysis, show is that bipedalism of an apparently human kind was established 3.6 million years ago. The mechanism of weight and force transmission through the foot is extraordinarily close to that of modern man." His study showed the prints had "close similarities with the anatomy of the feet of the modern human habitually unshod; arguably the normal human condition" (Day 1985, p. 121).

Typically, Day (1985, p. 125) concluded: "There is now no serious dispute as to the upright stance and bipedal gait of the australopithecines."

But what proof did he have that an australopithecine made the Laetoli footprints? There is no reason to rule out the possibility that some unknown creature, perhaps very much like modern *Homo sapiens,* was the cause of them.

R. H. Tuttle (1981, p. 91) stated: "The shapes of the prints are indistinguishable from those of striding, habitually barefoot humans."

Tuttle (1987, p. 517) concluded: "Strictly on the basis of the morphology of the G prints, their makers could be classified as *Homo* sp. because they are so similar to those of *Homo sapiens,* but their early date would probably deter many palaeoanthropologists from accepting this assignment. I suspect that if the prints were undated, or if they had been given younger dates, most experts would probably accept them as having been made by *Homo.*" Tuttle (1987, p. 517) also stated: "They are like small barefoot *Homo sapiens.*"

Furthermore, Tuttle held that the *A. afarensis* foot could not have made the prints. Of the AL 333-115 foot, he said: "The shafts of the proximal phalanges are markedly curved ventrally. This feature is characteristic of certain full-time and part-time arboreal apes and monkeys. . . . It is difficult to imagine a foot with such markedly curved phalanges fitting neatly into the footprints at Laetoli"

(Tuttle 1981, p. 91). The same would be true of any australopithecine foot.

Stern and Susman (1983) objected to this. Convinced that the apelike *A. afarensis* foot had made the Laetoli footprints, they proposed that the ancient hominids had walked across the volcanic ash with their long toes curled under their feet, as chimpanzees have sometimes been observed to do. Curled-under toes would explain why the *A. afarensis* footprints at Laetoli so much resembled those made by the relatively short-toed human foot.

Could an australopithecine walking with curled toes have made the humanlike prints? Tuttle (1985) found this extremely unlikely. If the Laetoli hominid had long toes, then, said Tuttle, one would expect to find two patterns of toe impressions—long extended toes and short curled toes, with extra-deep knuckle marks. Tuttle (1985, p. 132) observed: "Neither pattern exists at Laetoli G so we can infer that their lateral toes were quite short." This meant the long-toed *afarensis* foot could not have made the prints.

Even Tim White, who believed *Australopithecus afarensis* made the footprints, stated: "The Stern and Susman (1983) model of toe curling 'as in the chimpanzee' predicts substantial variation in lateral toe lengths seen on the Laetoli prints. This prediction is not borne out by the fossil prints" (White and Suwa 1987, p. 495).

Stern and Susman did in fact claim that a few of the Laetoli footprints gave signs of toes longer than in humans. Tuttle (1985, p. 132) admitted that "the right foot of G-1 sometimes left peculiar marks distal to the toe tips." To Stern and Susman, the marks forward of the "toe tips" represented the actual toe tips of uncurled toes. But Tuttle had another explanation for the marks. He wrote: "These are best explained by . . . the tendency for G-1 to drag its foot on lift off probably due to pathology of the lower limb" (Tuttle 1985, p. 132). The fact that the peculiar markings appeared only on one foot of one individual, and then only sometimes, lends support to Tuttle's explanation.

Stern and Susman (1983) also suggested that the Laetoli prints did not have a deep rounded impression at the base of the big toe, representing the ball of the foot in humans. They regarded this as evidence that the foot that made the prints was not human. But Tuttle (1985, p. 132) observed that "humans commonly leave prints devoid of these features as may be seen in prints on the beach." And, as we have seen, Robbins (1987, p. 501) said the prints she studied did have a "humanlike" ball region.

Directly challenging Johanson, White, Latimer, and Lovejoy, who asserted *Australopithecus afarensis* made the Laetoli prints, Tuttle (1985, p. 130) said: "Because of digital curvature and elongation and other skeletal features that evidence aboreal habits . . . it is unlikely that *Australopithecus afarensis* from Hadar, Ethiopia, could make footprints like those at Laetoli." Such statements have provoked elaborate counterattacks from Johanson and his followers, who have continued to promote the idea that *A. afarensis* could have made the tracks.

Tim White, for example, published a study (White and Suwa 1987) of the Laetoli prints in which he disputed Tuttle's contention that their maker was a hominid more advanced than *A. afarensis*.

White asserted: "there is not a single shred of evidence among the 26 hominid individuals in the collection of over 5,000 vertebrate remains from Laetoli that would suggest the presence of a more advanced Pliocene hominid at this site" (White and Suwa 1987, p. 496). But, as we have seen in our review of African hominid fossils, there are in fact a few "shreds" of evidence for the presence of *sapiens*-like creatures in the Pliocene, some not far from Laetoli. Also, it is well known that human skeletal remains are quite rare, even at sites where there are other unmistakable signs of a human presence.

Like Tuttle, White rejected the curled-toe hypothesis of Stern and Susman. Instead, White tried to fit the foot of *A. afarensis* to the Laetoli prints. This was very difficult because no complete foot skeleton of *A. afarensis* had been found at the Hadar site. A partial foot skeleton, however, had been recovered. This was the AL 333-115 foot skeleton, which included only bones from the front part of the foot—phalanges and metatarsal heads.

According to White, the best tracks at Laetoli were in the G-1 trail, representing the smallest of the three individuals of the G group. Even White admitted that the phalanges of AL 333-115 were "obviously incompatible with the G-1 tracks" (White and Suwa 1987, p. 497). Stern and Susman, and Tuttle, found them incompatible with any of the tracks. White, however, pointed out that the AL 333-115 individual represented one of the larger, presumably male, members of the First Family group and proposed that the foot of Lucy, one of the smaller, female individuals, might have fitted the G-1 Laetoli prints.

But the only bones recovered from Lucy's foot were an ankle bone and two toe bones. White therefore decided to use a partial *Homo habilis* foot skeleton (OH 8) from Olduvai Gorge to reconstruct the rear part of Lucy's foot. White reduced the OH 8 foot by 10 percent to bring it down to the size of Lucy's ankle bone (talus). He then scaled the large AL 333-115 toes bones down to the size of Lucy's few toe bones, and used them to make up the rest of the foot (White and Suwa 1987, p. 502). According to White, this speculatively reconstructed foot matched the prints.

White predicted that "the discovery of a complete foot skeleton at Hadar or Laetoli will conform in its basic proportions with the reconstruction described in this paper" (White and Suwa 1987, p. 512). But this prediction remains to be fulfilled. It is interesting that the most complete *afarensis* foot skeleton now available (AL 333-115) definitely does not fit any of the prints.

White also predicted that "the Laetoli prints will eventually be shown to be subtly distinct from those left under analogous conditions by anatomically modern humans" (White and Suwa 1987, pp. 510, 512). But as far as anyone can see now, they are indistinguishable from those of modern humans. Even White himself once said: "Make no mistake about it. They are like modern human footprints. If one were left in the sand of a California beach today, and a four-year-old were asked what it was, he would instantly say that somebody had walked there. He wouldn't be able to tell it from a hundred other prints on the beach, nor would you. The external morphology is the same. There is a well-shaped modern heel with a strong arch and a good ball of the foot in front of it. The big toe is in a straight line. It doesn't stick out to the side like an ape toe" (Johanson and Edey 1981, p. 250).

And Tuttle (1985, p. 130) noted: "in all discernible morphological features, the feet of the individuals that made the G trails are indistinguishable from those of modern humans."

11.11 BLACK SKULL, BLACK THOUGHTS

In 1985, Alan Walker of Johns Hopkins University discovered west of Lake Turkana a fossil hominid skull stained dark by minerals. Called the Black Skull, it raised questions about Donald Johanson's view of hominid evolution.

According to Johanson, *Australopithecus afarensis* gave rise to two lines of hominids. This arrangement can be visualized as a tree with two branches. The trunk is *Australopithecus afarensis*. On one branch is the *Homo* line, proceeding from *Homo habilis* to *Homo erectus* to *Homo sapiens*. On the second branch are the australopithecines arising from *Australopithecus afarensis*.

Johanson and White claimed that *Australopithecus afarensis* gave rise to *Australopithecus africanus*, which in turn gave rise to *Australopithecus robustus*. The trend was toward larger teeth and jaws, and a larger skull with a ridge of bone, the saggital crest, running lengthwise along the top. The saggital crest served as a point of attachment for the powerful jaw muscles of robust australopithecines. *Australopithecus robustus* then supposedly gave rise to the superrobust *Australopithecus boisei*, which manifested all the above-mentioned features in an extreme form.

In an article titled "Baffling Limb on the Family Tree," Walker's wife Pat Shipman, also of Johns Hopkins University, explained the evolutionary significance of the Black Skull, designated KNM-WT 17000.

The first specimens of *Australopithecus robustus* were, it was thought, about 2 million years old (Johanson and Edey 1981, p. 283). But the Black Skull, with its *Australopithecus boisei* features, including the largest cranial crest of any hominid (Shipman 1986, p. 91), was 2.5 million years old. Shipman believed this meant that *Australopithecus boisei* and the *boisei*-like Black Skull could not be descended from *Australopithecus robustus*, as believed by Johanson and others.

So where does that leave us? Here is one possibility suggested by Shipman. On our hominid family tree, we could now go from *Australopithecus afarensis* up one branch to *Australopithecus africanus*. Then from *Australopithecus africanus* could come two separate branches. On one branch is *Australopithecus robustus* and on the other *Australopithecus boisei* and the *boisei*-like Black Skull. In other words, instead of deriving *Australopithecus boisei* from *Australopithecus robustus*, both originate from *Australopithecus africanus*.

But perhaps not. "All known *africanus* skulls share many features that are derived, i.e, advanced, relative to those of the new skull, such as a moderate flexion or angling of the base of the cranium and a deep jaw joint with a bony lump in front of it," said Shipman (1986, p. 91).

So, according to Shipman, another possibility now emerges—that *Australopithecus africanus*, although ancestral to *Australopithecus robustus*, might not have been ancestral to *Australopithecus boisei* and the *boisei*-like Black Skull.

This leaves us with a three-branched family tree. Down at the bottom we still have *Australopithecus afarensis*. Above are three branches—the *Homo* line on the first, *Australopithecus boisei* and the Black Skull on the second, and then *Australopithecus africanus* on the third, leading to *Australopithecus robustus*.

But Shipman pointed out that it then becomes difficult to account for the fact that *Australopithecus boisei* and *Australopithecus robustus* are so similar. If *Australopithecus robustus* came from *Australopithecus africanus* and *Australopithecus boisei* from *Australopithecus afarensis*, then *Australopithecus boisei* and *Australopithecus robustus* would have had to develop their robust similarities independently by parallel evolution, something that is possible but unlikely.

According to Shipman, another way to explain the similarities between *Australopithecus boisei* and *Australopithecus robustus* is to propose that *Australopithecus robustus* was not descended from *Australopithecus africanus* and that *Australopithecus robustus* and *Australopithecus boisei* had a common ancestor besides *Australopithecus africanus*—perhaps *Australopithecus afarensis*.

So now we have a four-branched tree, with *Australopithecus afarensis* at the bottom. Above are the *Homo* line, *Australopithecus africanus*, *Australopithecus robustus*, and *Australopithecus boisei*, all separate from each other.

Shipman found it very hard to believe that a single species, *Australopithecus afarensis*, could have given rise to four separate lineages. So where did the four new species come from?

Shipman suggested that one should take a very hard look at the idea that *Australopithecus afarensis* represents just one sexually dimorphic species. She pointed out, as we have discussed in Section 11.9.8, that some scientists have concluded that "at least two species of *Australopithecus* and possibly *Homo* are mistakenly lumped together into *afarensis*" (Shipman 1986, p. 90).

Walker said it is likely that "the specimens identified as *Australopithecus afarensis* include two species, one of which directly gives rise to *Australopithecus boisei*" (Walker *et al.* 1986, p. 522).

How did Johanson respond to the discovery of the *boisei*-like Black Skull? He admitted that the Black Skull complicated things, making it impossible to arrange *Australopithecus africanus, Australopithecus robustus,* and *Australopithecus boisei* in a single line of succession coming from *Australopithecus afarensis.* Johanson proposed 4 possible arrangements of these species, along the lines we have been discussing, without suggesting which one was correct (Johanson and Shreeve 1989, p. 126). There was, he said, not yet enough evidence to decide among them.

The uncertainty about the number of species at Hadar, combined with the confused relationships among the successor species (*Australopithecus africanus, Australopithecus robustus, Australopithecus boisei,* and *Homo habilis*), create problems for evolutionists attempting to construct a phylogenetic tree for these hominids. Shipman (1986, p. 92) stated: "the best answer we can give right now is that we no longer have a very clear idea of who gave rise to whom." Walker warned that the discovery of KNM 17000 suggested "that early hominid phylogeny has not yet been finally established and that it will prove to be more complex than has been stated" (Walker *et al.* 1986, p. 522).

In the midst of the new complexity, one question is especially important—the origin of the *Homo* line. Shipman told of seeing Bill Kimbel, an associate of Johanson, attempt to deal with the phylogenetic implications of the Black Skull. "At the end of a lecture on Australopithecine evolution, he erased all the tidy, alternative diagrams and stared at the blackboard for a moment. Then he turned to the class and threw up his hands," wrote Shipman (1986, p. 93). Kimbel eventually decided the *Homo* line came from *Australopithecus africanus* (Willis 1989). Johanson and White continued to maintain that *Homo* came directly from *Australopithecus afarensis.*

After she considered various phylogenetic alternatives and found the evidence for all of them inconclusive, Shipman (1986, p. 93) stated: "we could assert that we have no evidence whatsoever of where *Homo* arises from and remove all members of the genus *Australopithecus* from the hominid family. . . . I've such a visceral negative reaction to this idea that I suspect I am unable to evaluate it rationally. I was brought up on the notion that *Australopithecus* is a hominid." This is one of the more honest statements we have heard from a mainstream scientist involved in paleoanthropological research.

In the foregoing discussion, we have considered only the evidence that is generally accepted by most scientists. Needless, to say, if we were to also consider the evidence for anatomically modern humans in very ancient times that would complicate the matter even further.

Having reviewed the history of African discoveries related to human evolution, we can make the following summary observations. (1) There is a significant amount of evidence from Africa suggesting that beings resembling anatomically modern humans were present in the Early Pleistocene and Pliocene. (2) The conventional image of *Australopithecus* as a very humanlike terrestrial biped appears to be false. (3) The status of *Australopithecus* and *Homo erectus* as human ancestors is questionable. (4) The status of *Homo habilis* as a distinct species is questionable. (5) Even confining ourselves to conventionally accepted evidence, the multiplicity of proposed evolutionary linkages among the hominids in Africa presents a very confusing picture. Combining these findings with those from the preceding chapters, we conclude that the total evidence, including fossil bones and artifacts, is most consistent with the view that anatomically modern humans have coexisted with other primates for tens of millions of years.

Appendices

Appendices

Appendix 1

Chemical and Radiometric Testing
Of Anomalous Human Skeletal Remains

Paleoanthropologists attempt to comprehend human origins by studying the morphology of hominid bones. According to evolutionary thinking, hominid bones more robust or apelike than modern human bones are more primitive, and thus likely to be older. Hominid bones that closely resemble modern human bones are judged recent. But problems arise when bones displaying the features of anatomically modern human bones are found in unexpectedly old geological contexts. In recent times, scientists have employed a variety of chemical and radiometric tests to resolve controversies about the ages of some of the discoveries we have discussed—such as Castenedolo in Italy (Section 6.2.2) and Galley Hill in England (Section 6.1.2.1).

In dating hominid bones, scientists have traditionally emphasized the relationship of the bones to the stratigraphic layer in which they were found. Are they contemporaneous with the layer or not? By showing that a bone is contemporaneous with a definite stratum in an excavation or exposed section, one can ascertain its age with the maximum degree of certainty. This certainty will not, of course, be absolute. But in general, if a bone is found lying *in situ,* in a well-defined stratum, rather than on the surface of the ground or in a cave, one can be reasonably certain of its age.

Most of the modern African discoveries of *Australopithecus, Homo habilis,* and *Homo erectus* have occurred on the surface. The same is true of most of the *Homo erectus* finds in Java. On the other hand, almost all of the discoveries of anomalously old human bones mentioned in Chapter 6 occurred *in situ,* in well-defined strata. In this respect, these discoveries, largely forgotten, are superior to many now fully accepted.

In some cases, anomalously old human bones were not uncovered by professional scientists. The skeleton found in Pliocene clay at Savona, Italy, is one example. Some think that only observations by professional scientists, present at the moment of discovery, are sufficient to insure that bones did not arrive in their position in the earth by intrusive burial or some kind of geological disturbance. Hence, it is said, discoveries not conforming to this standard must be rejected. The unfairness of this procedure becomes apparent when one considers that a good number of accepted discoveries were also made by nonscientists. Workmen were responsible for many of the hominid finds made in China during and since the 1950s. The same is true of the vast majority of *Homo erectus* finds in Java. Some were made by paid collectors, operating without scientists present, and others were made by farmers. The famous Heidelberg jaw was uncovered by German workmen, and some of the early *Australopithecus* finds in South Africa were made by limestone quarry workers, and, in one case, by a schoolboy. Yet these are not rejected.

Also, even though professional scientists were not present when some anomalously old human bones were discovered, other educated observers were present, and they testified that they detected no signs of intrusive burial. At Savona, for example, educated observers noticed that there was no white sand from the upper recent layers in the bluish Pliocene clay surrounding the skeletal remains. At Galley Hill, educated persons who saw the bones *in situ,* although not themselves scientists, checked for signs of burial and found none.

Furthermore, at Castenedolo, a professional geologist, Ragazzoni, was present when some of the human bones were uncovered there in Pliocene clay. And after careful inspection, he found no signs of intrusive burial.

Even though scientists have almost universally rejected cases such as Castenedolo, there has sometimes remained among them a lingering worry that the discoveries might be genuine. It is perhaps for this reason that scientists have welcomed newly developed chemical and radiometric tests that promise objective age determinations with a high degree of certainty. Results from these tests have been employed to discredit the great ages originally attributed to several of the skeletal remains discussed in previous chapters.

The tests give two kinds of age determinations—relative and absolute. The fluorine content test, for example, gives relative ages, determined by comparing a bone's fluorine content with that of bones from the same site or other sites of known ages. The carbon 14 test is said to give an absolute age, calculated by the rate at which this radioactive element decays.

The advantage of the new tests is apparent. It is often hard for skeptics to prove from old reports that an anomalously old bone was not deposited in a given stratum at the time of that stratum's formation. A test that dates the bone itself provides an ideal solution to this problem. Either the bone's date corresponds with that of the stratum in which it was found, or it does not. If the test date is quite recent, then it does not matter if the original reports say the bone was found in a Pliocene stratum overlain by undisturbed layers of sediment. The original reports can be safely dismissed, with appropriate explanations.

Or can they? The procedure outlined above depends upon the assumption that the results of chemical and radiometric tests are more reliable than conclusions derived from stratigraphic analysis. This assumption is widely held. Radiocarbon dating, in particular, has been generally accepted as a source of "absolute dates." Some call it, jokingly, the "word of God" (*Science News* 1977b, p. 197). But these radiometric and chemical dating methods are not infallible.

It is difficult to quantify the degree of fallibility of a scientific procedure, but available evidence suggests that the conclusions provided by the modern dating techniques are not intrinsically more reliable than those based upon the professional analysis of stratigraphy. It is therefore appropriate to weigh the findings obtained by chemical, radiometric, and stratigraphic methods in a balanced way to arrive at a final conclusion about any given paleoanthropological discovery.

A1.1 THE NITROGEN CONTENT TEST

The antiquity of a bone can be measured in terms of its nitrogen content. This principle was understood by nineteenth-century scientists who measured the organic content of bone, calling it "animal matter."

Fresh bone contains about 5 percent nitrogen (Fleming 1976, p. 185), concentrated in its proteins. Most of the bone protein is in the form of collagen fibers. After an animal dies, its bone proteins are broken down into peptides and ultimately into amino acids. As time passes, these soluble nitrogenous breakdown products can be dissolved from the bone, and its nitrogen content drops. Thus, the less nitrogen in a bone, the older it should be.

A1.1.1 Nitrogen Content of the Castenedolo, Galley Hill, and La Denise Human Skeletal Remains

Oakley (1980, p. 40) said that in 1965 one or two bones from each of three individuals found in Pliocene clay at Castenedolo (the man, woman, and a child) were tested for nitrogen content. These tests yielded values ranging from 1.2 to 2.6 percent nitrogen.

Middle Pleistocene bones from other Italian sites had nitrogen contents varying from under 0.1 percent to 0.3 percent nitrogen, far lower than the Castenedolo bones. This suggested the Castenedolo bones were more recent. Bones from Italian Late Pleistocene sites ranged from under 0.1 percent to 3.7 percent nitrogen, bracketing the Castenedolo values. Bones from Italian Holocene sites were in the range of 0.5 to 4.4 percent nitrogen, also bracketing the Castenedolo values.

The discoverer of the Castenedolo skeletons, Ragazzoni, said the fossils had been found in undisturbed Pliocene clay. But the nitrogen content test cast doubt on a Pliocene date for the bones. Oakley concluded that the Castenedolo bones must be Holocene, or Late Pleistocene at the oldest. Oakley also stated that a radiocarbon test showed the Castenedolo bones to be only about 1,000 years old (Section A1.3.8).

In the case of Galley Hill, a similar procedure was used to test the skeletal remains found there in a Middle Pleistocene stratum. Oakley (1980, p. 17) reported that the femur and humerus from Galley Hill had nitrogen contents of 1.61 percent and 2.04 percent, respectively. These results were compared with those for bones from other sites of known age in England and elsewhere (Fleming 1976, p. 188). The results suggested a recent Holocene date for the Galley Hill skeleton. This was confirmed by a radiocarbon date of approximately 3,300 years.

The human skull fragment from La Denise, France, yielded a nitrogen content of 1.68 percent (Oakley 1980, p. 32). The formation in which it was found was Pleistocene, but comparison with bones from other sites in France suggested a Holocene date.

At first glance, the evidence from the nitrogen content test seems very conclusive. But scientists have determined that the rate of nitrogen decrease in a bone depends heavily on the physical and chemical properties of the surrounding geological materials, the levels of moisture and temperature at the site, and the size of the bone fragment itself (Fleming 1976, p. 186). This means that if two bones were buried in the same material for the same time, but at different sites that had been subjected to different moisture and temperature levels, then the two bones would have different nitrogen levels. Typically, not only do the temperature and moisture histories vary at different sites, but the type of material (sand, clay, etc.) and the type of bone itself also vary. All this means that comparing nitrogen contents of bones from different sites to determine their relative ages is of questionable value.

A1.1.2   Remarkable Instances of Nitrogen Preservation

There is extensive evidence that some geological materials, particularly clay, help bones to retain their nitrogen longer than other materials.

Fleming (1976, p. 186) reported: "A tusk of the early elephant genus, *Elephas primigenius,* found in the frozen mud of Siberia contained collagen fibrils with perfect periodicity in its banded structure, even though it was some 15,000 years old." Collagen contains nitrogen.

Oakley (1980, p. 20) reported that in London the ulna of a woolly rhinoceros was discovered in clay at a depth of 13 meters (about 43 feet). The ulna was carbon 14 dated at 30,000 years, yet had a nitrogen content of 3.42 percent (almost as much as a modern bone). A mammoth femur found 6 meters (about

20 feet) higher in a layer of sand had a nitrogen content of only 0.1 percent. Oakley (1980, p. 20) said: "The reason for the preservation of so much of the protein in the Lloyds rhinoceros bone is that it was embedded in an unoxidized clay—an environment in which collagen decays very much more slowly than in sand or gravel."

In April 1990, Michael T. Clegg, a paleobotanist at the University of California at Riverside, reported that the Miocene Clarkia deposit in Idaho had yielded well-preserved plant and animal remains 17 million years old. The New York Times News Service (1990, p. a-2) reported: "The Idaho site consists of layers of sediments that encase the remains of flowers, stems and leaves, insects and fish, that are exceptionally well preserved because of the low oxygen content and cold temperature of the water in which they were deposited." Strands of DNA, by far the oldest ever seen, were taken from a magnolia leaf, which was still green. DNA contains nitrogen. It appears the other organic remains from the Clarkia deposit also would have retained nitrogen in their proteins.

Newell (1959, pp. 496–497) reported several instances of still more remarkable preservation of proteins, which contain nitrogen. For example, in the Eocene lignite deposits of the Geiseltal in Germany (38–55 million years old) "more than six thousand remains of vertebrate animals and a great number of insects, molluscs and plants were found." Newell pointed out that "the compressed remains of soft tissues of many of these animals showed details of cellular structure and some of the specimens had undergone but little chemical modification."

An even more striking case is provided by some exoskeletons of the eurypterid *Megalograptus,* discovered in a clay deposit of Ordovician age (438–505 million years old). Eurypterids are large aquatic arthropods related to the king crabs. Newell stated: "The original chitin is but little altered in these fossils and still retains a scorpionid color pattern. Chitin usually is quickly destroyed by bacterial decomposition under aerobic conditions. The preservation of these eurypterids illustrates the influence of finegrained matrix on preservation of fossils."

Technically speaking, chitin is a polysaccharide, not a protein. But in living insects and crustaceans, chitin is found combined with proteins. Edward Atkins, a biophysics researcher at the University of Bristol, England, said (1985, p. 382): "Examination of insect cuticle . . . reveals the polysaccharide chitin . . . embedded in a globular protein matrix." The molecules interlock to form the hard exoskeleton. If either the polysaccharide (chitin) or the proteins were missing, the exoskeleton would quickly disintegrate.

Newell's use of the word "chitin" therefore seems to indicate the exoskeleton, consisting of chitin in combination with its associated proteins. If by "chitin" he meant only the polysaccharide, it is difficult to explain his observation that "the original chitin is but little altered in these fossils and still retains a scorpionid color pattern." From this statement it appears that proteins were present and that they were almost unaltered, despite being over 400 million years old.

Newell (1959) also said traces of amino acids, which contain nitrogen, were still present in 1.7-billion-year-old spores of algae and fungi from the Gunflint formation of Ontario. He added that there are "innumerable reports of examples of highly improbable preservation." In this regard, Oakley (1957, p. 76) reported: "the rate of disappearance of the protein (collagen) in bone is much slower than has generally been supposed. The amino-acids of bone protein have recently been identified in marine fish remains some 300 million years old."

It is thus clear that nitrogenous proteins may be preserved for extremely long periods of time, particularly when bones are embedded in a fine-grained matrix such as clay. Significantly, the Castenedolo bones were found in clay, judged to be Pliocene in age. Oakley (1980, p. 17) said that the Galley Hill bones tested for nitrogen content were the femur and humerus, and he noted (Oakley and Montagu 1949, p. 36) that these bones were found in a stratum of silty (some say 'clayey') loam of Middle Pleistocene age. De Mortillet (1883, p. 241) reported that the frontal skull fragment from a Pleistocene formation at La Denise, France, was originally situated in a clayey limonite layer. It is thus possible that the high nitrogen contents of the Castenedolo, Galley Hill, and La Denise bones, relative to other bones of Pliocene and Pleistocene age, may have been the result of their preservation in a clayey or fine-grained matrix.

A1.1.3 Nitrogen Content of the Florisbad and Olmo Skulls with Reference to Castenedolo and Galley Hill

Keeping in mind the possibility that nitrogen content does not vary directly with age, let us now consider the nitrogen content of a human skullcap found in 1932 at Florisbad, in southern Africa, and of a human skull found in 1863 at Olmo, Italy. These two cases add significantly to our understanding of how best to interpret the nitrogen content of the Castenedolo and Galley Hill skeletons.

The Florisbad skullcap, said to display primitive features, has been described by some as Neanderthaloid and by others as early *Homo sapiens* (Rightmire 1984, p. 310). According to Bräuer (1984, p. 347), the Florisbad skull is around 80,000 to 150,000 years old, an age thought appropriate for its morphology.

At Florisbad there were alternating layers of peat (peat I being the lowest) and sand. Oakley (1957, p. 77) reported the nitrogen contents of bones from the different levels (Table A1.1). Oakley said that at first glance this data showed the Florisbad skull, with its high nitrogen content, was a recent intrusive burial.

But then Oakley (1957, p. 77) pointed out that the material in which the Florisbad skull was discovered is "really a peaty clay" whereas the other bones

TABLE A1.1

Nitrogen Content of Bones from Florisbad, South Africa

Specimen	Percent Nitrogen
Bone from uppermost sand	0.3
Bone from sand between peats II and III	0.2
Florisbad skull from peat I	1.7
Bone from sand below peat I	0.1

in Table A1.1 were found in sand. He said that this difference in materials was sufficient to explain the huge difference in nitrogen values, because bones in clay retain their protein much longer than bones in sand.

Oakley used the fact that the Florisbad skull was found in clay to explain its high nitrogen content without recourse to the hypothesis of intrusive burial. But in the case of the Castenedolo bones, also found in clay and showing a high nitrogen content relative to bones from other Italian sites of similar and lesser age, Oakley suggested intrusive burial. This demonstrates the wide latitude an investigator has in interpreting nitrogen content test results.

The Florisbad skull has a nitrogen content (1.7 percent) comparable to that of the Castenedolo bones (1.2 to 2.6 percent); yet, if we accept the radiocarbon age of about 1,000 years for the Castenedolo bones and Bräuer's age of 80,000 to 150,000 years for the Florisbad skull, the Florisbad skull would be 80 to 150 times older than the Castenedolo bones. The nitrogen content of the Florisbad skull is, furthermore, almost identical to that of the Galley Hill skeleton (1.6 to 2.0 percent); yet, if we accept the radiocarbon age of about 3,300 years for Galley Hill, the Florisbad skull would be about 24 to 45 times older. The nitrogen contents of the Castenedolo and Galley Hill skeletons are, therefore, consistent with ages far greater than their radiocarbon ages of 1,000 years and 3,500 years.

The Olmo skull from central Italy is also relevant to our discussion of the nitrogen content test, particularly regarding Castenedolo. Professor Azzaroli of the University of Florence reported (personal communication, 1988) that the Olmo skull is more than 40,000 years old—the associated fauna had no measurable carbon 14 activity—but is probably less than 70,000 years old. Oakley (1980, p. 42) said the Olmo skull has a nitrogen content of 3.32 percent and probably came from a layer of late Pleistocene gravel, although the discoverer, Professor Cocchi, originally assigned it to a layer of Early Pleistocene clay (Keith 1925). The Olmo skull has a nitrogen content higher than that of Castenedolo (1.2 to 2.6 percent); yet the Olmo skull is considered 40 to 70 times older than the Castenedolo bones. The high nitrogen content of the Olmo skull is even more interesting if the skull came from the gravel, instead of the clay, because gravel is not conducive to protein retention.

It thus becomes apparent that there is no certain relationship between the ages of bones and their nitrogen contents. The fact that the Castenedolo bones have a relatively high nitrogen content does not prove they are very young.

A1.2 THE FLUORINE AND URANIUM CONTENT TESTS

The nitrogen content test measures the decay of organic compounds and consequent loss of nitrogen in the bone over time. Another type of test measures the accumulation of mineral elements in bone. Small quantities of fluorine and uranium are present in groundwaters, and buried bones absorb them. In general, the greater the quantity of fluorine and uranium in a bone, the greater its age.

In the case of fluorine, the process of absorption begins when fluorine ions in water come in contact with buried bones. The fluorine ions tend to displace the hydroxyl ions in hydroxyapatite, the main constituent of ordinary bone. Thus the hydroxyapatite is gradually transformed into fluorapatite. Once the fluorine is incorporated into the chemical structure of the bone, it tends to remain there—so this is an irreversible process (Oakley 1980, p. 4).

A1.2.1 Fluorine Content of the Galley Hill Skeleton

The stratum containing the anatomically modern Galley Hill skeleton was Middle Pleistocene, about 300,000 years old. Oakley and Montagu (1949) compared the fluorine contents of bones from the Galley Hill human skeleton and bones from other sites. These sites were the Barnfield pit at Swanscombe (a half mile away from the Galley Hill pit), Baker's Hole (three-quarters of a mile away), Crayford (several miles away), and Northfleet (one mile away).

The fluorine content of Middle Pleistocene bones from the Barnfield pit at Swanscombe was 1.7 to 2.8 percent, that of Late Pleistocene bones from Baker's Hole was 0.9 to 1.4 percent, and that of Holocene bones from the Barnfield pit at Swanscombe and from the Northfleet site was 0.05 to 0.3 percent. Oakley and Montagu reported an average value of .34 percent fluorine for several bones of the Galley Hill skeleton. In addition, Oakley (1980, p. 17) reported .56 percent fluorine for a femur from this skeleton. Oakley and Montagu (1949) concluded that the Galley Hill skeleton was of late Late Pleistocene or Holocene age.

Fleming (1976, p. 188) reached the same conclusion. Fleming expressed his results as the ratio of fluorine to phosphate ($100F/P_2O_5$) in a bone sample.

Expressing results as simply the percentage of fluorine in a sample is sometimes misleading, because a sample may contain fine silt or other inorganic contaminants in addition to bone. Although Oakley and Fleming were in basic agreement about the recent age of the Galley Hill skeleton, they published different fluorine to phosphate ratios. Oakley (1980, p. 17) reported a value of $100F/P_2O_5 = 2.0$, whereas Fleming (1976, p. 188) reported $100F/P_2O_5 = 0.5$, or 4 times less fluorine.

In our discussion of the nitrogen content test, we learned there are many factors that can influence nitrogen preservation. Similarly, there are a variety of factors that influence the rate at which fluorine is absorbed by buried bones. Fleming (1976, p. 181) said: "The rate of fluorine uptake by fossil bone varies from site to site, dependent upon climate, amount of the element in circulation and hydrological conditions of the burial medium."

Unfortunately, Oakley and Montagu (1949), Oakley (1980), and Fleming (1976) did not report the fluorine content of the groundwater at Galley Hill or the sites used for comparison. Therefore it is possible that the reported differences between the fluorine content of the Galley Hill skeleton and the fluorine content of bones from the other sites could be caused by differences in the fluorine content of the groundwater. Oakley and Montagu (1949, p. 36) claimed that the soil acidity at the Galley Hill pit was substantially higher than that at the Barnfield pit. If the acidity differs at these two sites, then it is quite possible that the concentration of fluorine and other elements also differs.

Furthermore, even if the concentration of fluorine in the groundwater was determined at the sites in question, were any measurements made proving that this concentration was constant since the Middle Pleistocene? How much might it have varied? What effect did the surrounding matrix and various chemicals in the environment have on the rate of fluorine incorporation? Oakley and Montagu (1949), Oakley (1980), and Fleming (1976) did not answer these questions. But without answering them one cannot draw any firm conclusions about the ages of bones from different sites by comparing their fluorine contents.

A1.2.2 Fluorine Content of the La Denise Bones

Oakley (1980, p. 32) reported fluorine contents of 0.06–0.07 percent for the human bones from a Pleistocene volcanic formation at La Denise, France. A skull fragment of *Dicerorhinus*, a Pleistocene rhinoceros, was found in the same locality, but at a good distance from the place where the human bones were found. It yielded a fluorine content of 1.53 percent. The fluorine content of the La Denise human bones was also compared with that of bones from elsewhere in France (Oakley 1980, p. 34). But Oakley did not report the concentration of fluorine in the groundwater at any of the comparison locations or at the location of the

La Denise human bones. The above-mentioned cautions are therefore applicable in this case as well.

Another factor is the effect of volcanic deposits, such as exist at La Denise. Oakley (1980, p. 3) reported that the fluorine method is "not applicable in regions where fluorine is excessively abundant in the ground-water, as it is in most volcanic areas with tropical weathering, where fluorination of vertebrate materials occurs rapidly and sometimes in a random fashion." La Denise is not at present in a tropical climate, but at some time in the past, during Pleistocene interglacial periods, the weather was much warmer in Europe than at present (hippopotamuses swam in English rivers).

If volcanic deposits introduce large amounts of fluorine into the groundwater, it seems the La Denise human frontal bone, if it is Middle or Early Pleistocene, should have a much higher fluorine content than .06 percent. One might expect the fluorine content to be at least as high as that of the mammalian bone found in the same volcanic deposits. Oakley's suggestion of rapid and random fluorination in such an environment may explain the difference in fluorine contents. Here again we see how important it is to determine the fluorine content of the groundwater at the exact locations of bones being compared, not only for the present but for hundreds of thousands of years in the past. This essential knowledge is practically impossible to ascertain, thus complicating attempts to assign ages to fossil bones by comparisons of fluorine content.

A1.2.3 Fluorine Content of the Castenedolo Skeletons

We are not alone in pointing out the complicating effects of past variations in groundwater fluorine content. In his discussion of the fluorine dating of the Castenedolo human remains, found in a layer of Pliocene clay, Oakley adopted an approach to this problem that bears close attention.

Oakley (1980, p. 40) believed the Castenedolo bones were introduced into the Pliocene layer in which they were found during the Holocene, about 1,000 years ago, as determined by radiocarbon dating. Nevertheless, he reported the fluorine content of these bones was 0.4–0.6 percent, a level he considered high for bones of late Holocene age (Oakley 1980, p. 42). The high fluorine content was difficult to explain because "the Laboratory of the Government Chemist, London, could not detect fluorine in a sample of well water from Castenedolo" (Oakley 1980, p. 42). Oakley was forced to conclude that "in this tectonically rather unstable region the composition of ground-water is probably liable to considerable variation in the course of a millennium" (1980, p. 42).

Yet if one assumes the well water represents the groundwater at the actual site where the Castenedolo bones were discovered, and that its negligible fluorine content has been stable for a long period of time, the relatively high fluorine content of the bones could be seen as confirming their Pliocene age.

We have called attention to possible variations in concentrations of elements in groundwater with the passing of time, pointing out how this casts doubt on dating techniques dependent on measuring accumulations of these elements in bone. This would seem to be an obvious point, but scientists rarely if ever mention it in discussions of the fluorine content of human skeletal remains. This is especially true when scientists like Oakley use a low fluorine content test result to confirm a recent date for a find thought to be anomalously old, as in the case of Galley Hill. In the past, the fluorine content of groundwater at the site could have been quite low, and this would explain the small accumulation of fluorine in a bone genuinely old. But this possibility is not given due consideration.

And now suddenly we find Oakley bringing up the variation argument—but only when the fluorine test suggested a considerable antiquity for bones that he preferred to regard as recent. It is highly unlikely that Oakley would have spoken of variation in the groundwater's fluorine content if the fluorine content of the Castenedolo bones had been acceptably low. In all of his other reports on fluorine dating of anomalously old human bones he never mentioned such variation.

In the case of Castenedolo, Oakley raised tectonic instability as the cause of variation in the fluorine content of groundwater there. But he provided no factual evidence to back up his assertion that the region has been tectonically unstable over the last 1,000 years, and if unstable, that such instability could give rise to a change in the fluorine content of the groundwater sufficient to produce the observed fluorine content in the bones. The fluorine content of the groundwater would have had to have been very high for several hundred years and then decreased down to nothing. The whole case rests on this scenario, but no convincing, detailed evidence was presented in favor of it.

A1.2.4 Uranium Content of the Castenedolo Bones

Oakley also reported the results of uranium content tests on the Castenedolo bones. Uranium, like fluorine, is present in varying amounts in groundwater, and as the water percolates through buried bones, they can absorb this element from it. The rate of accumulation of uranium in bone depends on the amount of this element in the water and also on the "hydrological conditions prevailing at the site and on the nature of the mineral matrix" (Oakley 1980, p. 8). The amount of uranium (actually several elements of the uranium family) in a bone is determined by counting beta decays with an appropriately screened Geiger counter.

Oakley (1980, p. 40) stated that "radiometric assays of the Castenedolo bones indicated an unexpectedly high uranium content" varying from 9 to 32 parts per million of uranium oxide. He then said: "The most likely explanation of the radioactivity of the bones is that the local ground-water is rich in uranyl ions.

Professor E. Anati of the Centro Camuno di Studi Preistorici, Brescia, sent a sample of well water from Castenedolo to the Centro di Studi Nucleari E. Fermi in Milan, where Professor Terrani determined 180 ml contained 1.58±.39 micrograms of uranium. As this is an unusually high proportion of uranium for ground-water, it probably accounts for accumulation of this element in the bones to the levels recorded after an exposure of less than 1000 years" (Oakley 1980, pp. 40–42).

But does it? Before Oakley's statement can be accepted, one would have to experimentally determine the rate at which bones, in the same condition as the Castenedolo bones and buried in the same matrix, accumulate uranium when exposed to groundwater containing the above-mentioned concentration of uranium. Moreover, is there any guarantee that the concentration of uranium in the well water is the same as that of the groundwater that actually flowed through the matrix surrounding the bones? How far away from the site of the discovery of the Castenedolo human bones was the well water sample taken? It should be kept in mind that at Castenedolo the bones were found on the side of a hill, at a depth of only a few feet, and therefore they would probably have been exposed only to rainwater run-off, not water like that from a well dug down to the local water table.

Also, we have seen how Oakley suggested that the fluorine concentration of the groundwater at Castenedolo may have varied considerably over the last thousand years. Yet he does not suggest that the uranium concentration may also have varied, and perhaps was very low for a long time. Of course, if that were the case, the high uranium content of the Castenedolo human bones would indicate a great antiquity for them.

The chemical tests we have considered so far have all turned out to be plagued with uncertainty. It is apparent that their results cannot stand alone. They are ambiguous and thus subject to multiple interpretation, leaving lots of room for different investigators to "prove" different favored theses. Therefore, conclusions based on such tests cannot automatically overrule stratigraphic evidence, which in some cases may be more sound and compelling.

A1.3 THE CARBON 14 DATING OF BONES

The carbon 14 test is different from the nitrogen, fluorine, and uranium content tests in that it gives absolute rather than relative ages. Here, the word "absolute" means that the date is a number in years B.P. (before present), rather than a statement of the form "A is older than B." Carbon 14 dates enjoy a high degree of confidence in the scientific community, and have been used to assign recent dates for a number of anomalous human skeletal remains. The absolute radiocarbon date thus connotes "the final truth" regarding such discoveries.

Much of the early radiocarbon dating was done at the University of Chicago. A key part of the apparatus is the counter sleeve, containing the sample to be dated. R. E. Taylor (1987, p. 160) wrote: "At one time the Chicago laboratory used four counter sleeves which were given the names of 'Matthew, Mark, Luke, and John'—so named because they gave the 'gospel truth.' . . . Later two other sleeves named 'Peter' and 'Paul' were used." Let us see if such faith is warranted.

A1.3.1 Principles of Carbon 14 Dating

Most of the carbon in the atmosphere is in the form of stable, nonradioactive carbon 12. The rare radioactive isotope carbon 14 is also present. It decays spontaneously with a half-life of approximately 5,730 years. This means that if 1 gram of carbon 14 is initially present, then after 5,730 years have elapsed only 0.5 gram will remain. After another 5,730 years, only 0.25 gram will remain, and after yet another 5,730 years have passed by only 0.125 gram will be present.

The ratio of rare carbon 14 to abundant carbon 12 in a living organism is assumed to be the same as that in the atmosphere. The carbon atoms in the body are being constantly replaced by new carbon 14 and carbon 12, in the normal atmospheric proportions. But when an organism dies, it no longer equilibrates its carbon compounds as does the living world. Its carbon 14 and carbon 12 are not replenished from the environment, and the ratio of carbon 14 to carbon 12 falls as the radioactive carbon 14 decays and the carbon 12 remains constant.

By measuring how much of the total carbon is carbon 14 at a given time, we can compute how much time has passed since the organism died. The smaller the amount of carbon 14, as a percentage of the total carbon, the older the sample. There are, however, several problems with this method.

(1) The oldest objects dateable by the carbon 14 method are only around 40,000 years old. By the time the carbon 14 in a bone has gone through 7 or 8 half-life reductions, the remaining amount is hardly measurable. All that can be said is that the object is older than 40,000 years. Some scientists hope that the new accelerator mass spectrometry (AMS) method of radiocarbon dating may be good for up to 100,000 years—if problems involving the chemical processing of extremely small amounts of residual radiocarbon can be overcome.

(2) Scientists can only assume that the percentage of carbon 14 in the atmosphere at large has been approximately the same for many thousands of years.

(3) Scientists can only assume that the half-life of carbon 14 has been constant over many thousands of years.

(4) Most importantly, scientists have to insure that no carbon compounds from the environment pollute the sample to be dated. Such contamination can occur when the sample is buried in the ground, when it is being collected, when it is being stored prior to the dating test, and during the test procedure itself.

A1.3.2 The Problem of Contamination

We shall now examine in detail the various ways that a bone can become contaminated with carbon compounds. First, however, we need to understand the changes a bone undergoes after the organism it belongs to dies.

According to Berger *et al.* (1964, p. 999), dry modern bone has a composition of roughly 50 percent calcium phosphate, 10 percent calcium carbonate (containing inorganic carbon), 25 percent collagen and other proteins (containing organic carbon), and 5 to 10 percent bone fat (also containing organic carbon). Organic carbon is that found in living tissues. Over the long period of time a bone is in the ground, foreign elements enter into it owing to the fact that bones are porous and contain many small canals. The process called fossilization occurs as the collagen and bone fat, which contain organic carbon, are gradually leached away while minerals gradually build up inside the canals and other available openings throughout the porous bone structure. The bone fat breaks down quickly whereas the collagen can remain for much longer, depending on the physical and chemical properties of the matrix in which the bone is buried.

A1.3.2.1 Contamination in the Ground

We are now in a position to understand the ways a bone can be contaminated when it is buried in the ground. The calcium carbonate portion can be replaced with groundwater carbonates. This carbonate can be from a very old source or a modern one. This exchange results in the carbon 14 age determination test giving a falsely old or young age when the calcium carbonate portion of the bone is used (Berger *et al.* 1964; R. Taylor 1987, p. 55; Nelson *et al.* 1986, p. 750).

Most radiocarbon laboratories are aware of this potential source of error, and hence they avoid using the calcium carbonate portion for age determination. The inorganic parts of the bone are typically removed by dissolving the bone in weak hydrochloric acid (HCl), which leaves the residual collagen and other proteins. The age is then determined on only the protein portion of the bone. Hassan and Ortner (1977) and Berger *et al.* (1964) have reported that there is no exchange of carbon in the environment with the carbon in the collagen—a fact that has led many paleoanthropologists to place confidence in the ages determined on the collagen fraction of the bone.

But other scientists have asserted that there are many types of organic inclusions in bones. These are not dissolved away by the hydrochloric acid pretreatment that removes inorganic carbon. Organic inclusions are extremely difficult to detect and remove physically (Hassan and Ortner 1977, Tamers and Pearson 1965). These foreign organic substances can thus contribute varying amounts of carbon 14, causing a radiocarbon test to yield a false age.

Hassan and Ortner (1977, p. 134) reported the presence in bone of a light brown, fibrous material, which they described as "hyphae (algae) which may grow and feed on bone." In most cases, this fibrous material was coated with calcite, a carbonate. Hassan and Ortner also suggested that, in general, rootlets could be present. These contaminants would introduce modern carbon and hence tend to cause a falsely young date. Hassan and Ortner (1977, p. 135) concluded: "Radiocarbon dating of the organic fraction of bone could be erroneous if hyphae, rootlets and charcoal or wood have not been removed. Most of these inclusions would be extremely difficult to hand pick. The standard methods of decalcification of bone do not remove such inclusions which could be a major source of error in radiocarbon dating of bone collagen."

Berger *et al.* (1964) reported that saprophytes ("possibly fungi or algae which feed on the organic material and perhaps also on the mineral matter") may also be present. They added: "Moreover, bones should be checked microscopically for excessive bore canals with associated possible foreign protein."

In an attempt to remove these kinds of contaminants, Longin (1971) devised a technique of dissolving the collagen (after decalcification of the bone) in hot water. Some organic inclusions, such as saprophytes, are said to separate out in an insoluble residue, leaving a gelatin for dating. Longin checked his results by comparing the dates obtained on the gelatin with dates obtained from charcoal from the same level at which the bones were found. Charcoal is generally considered to be a more reliable substance for dating than bone. The dates from the gelatin matched those from the charcoal, whereas untreated collagen gave younger dates. So Longin's method appears to have successfully removed contaminants from collagen, but we shall see that some very important carbon 14 dates on anomalous skeletons have not used this technique.

In addition to the problems discussed above, there are reports that significant amounts of modern amino acids can enter buried bones from the environment, thus giving a falsely young carbon 14 age. Amino acids native to the bone are called indigenous, while those entering from the environment are called exogenous or secondary. Bada (1985a, p. 256) reported that bones older than approximately 100,000 years from the Olduvai Gorge region "contain only trace quantities of indigenous aspartic acid, and that secondary aspartic acid introduced from surrounding soils and percolating ground waters is present in significant quantities."

Another example he reported is a bone from the Upper Ndutu Beds. This bone yielded a carbon 14 age on its collagen fraction of only 3,340 years, although other radiocarbon dates and geological considerations indicated that the bone should be more than 29,000 years old. Bada (1985a, p. 255) concluded that this sample "was badly contaminated with secondary carbon components having a relatively recent radiocarbon age."

If the bone really is over 29,000 years old and yet the carbon 14 test yielded an age of only 3,340 years, then this indicates something very significant about carbon 14 dating, namely that the method can be off by a factor of 8 or more. It is not clear whether the error was due to contamination by exogenous amino acids or other organic compounds, as Bada thought, or whether it resulted from some mistake in the complicated laboratory procedure—but it appears that a huge error was introduced somewhere.

If a bone over 40,000 years old (the effective limit for radiocarbon dating) had yielded such an incorrect date, the magnitude of the error could have been far greater. For example, a bone 3 million years old, contaminated with recent carbon, could have yielded a date of 3,000 years, instead of its proper radiocarbon date of over 40,000 years. The amount of contamination required to reduce a date of 3 million years to 3,000 years is just slightly greater than that required to reduce a date from 29,000 years to 3,000 years, as in the case above reported by Bada. In the hypothetical case we are discussing, the contaminated bone would actually be 1000 times older than its presumed radiocarbon date.

Stafford *et al.* (1987) reported that humic and fulvic acids can enter into bones and cause a falsely young carbon 14 age. Humic and fulvic acids are products of the decomposition of organic materials (such as plants) and contain carbon. Stafford *et al.* analyzed previous attempts to free samples of these contaminants. Berger and Libby (1966) advised dissolving away inorganic bone materials with hydrochloric acid (HCl) and then treating the insoluble collagen residue with sodium hydroxide (NaOH). This last step was supposed to remove humic contaminants, but Stafford *et al.* found this procedure insufficient. Even the Longin technique, which purifies the HCl-insoluble collagen in a warm water bath, was found to be inadequate for this purpose.

Stafford *et al.* therefore recommended the following procedure. Ideally, a bone selected for dating should contain over 0.2 percent nitrogen. Then one should ascertain that the protein in a bone sample is collagenlike in composition. Then the collagen, or gelatin obtained by the Longin method, should be treated with XAD resin to remove the humates. Stafford *et al.* (1987, p. 31) warned: "HCl insoluble residues, untreated gelatin, and acid-soluble phases may occasionally yield accurate dates, but there are no known chemical criteria for predicting when dates will be spurious on these fractions." They said that, ideally, one should individually date each amino acid that makes up the collagen.

Stafford *et al.* (1987, pp. 31–32) advised: "Bones with non-collagen amino acid compositions and <0.2% N do not date as accurately as bones with substantial amounts of collagen. Even XAD treatment may not be effective in yielding accurate ages on bones that are diagenetically altered."

An example of a bone that did not yield a good date even after XAD resin treatment is the Del Mar tibia (see Sections A1.3.4 and A1.3.5 for more on the

Del Mar finds). This bone gave a uranium series test result of about 11,000 years. It was then subjected to radiocarbon testing. Stafford and Tyson (1989, p. 391) reported that the "XAD-2 purified, weak HCl insoluble hydrolyzate" from this tibia yielded an age of 5,380 years, but the single amino acid glycine fraction of this hydrolyzate gave an age of only 1,150 years. They said (1989, p. 393): "The discordance between the total-amino-acid and glycine c-14 dates for the tibia are an example of a bone that cannot be dated accurately by radiocarbon methods." All the amino acids present in the dated sample should have been of the same age.

Stafford and Tyson (1989, p. 393) added: "Because radiocarbon dates on the tibia cannot be used to prove or disprove age, the 11,000 year uranium series date . . . cannot be proven to be invalid and may be closer to the tibia's true age than ages obtained from radiocarbon dating." They concluded that the tibia was "contaminated by exogenous amino acids, thereby making its age uncertain." This shows that a bone can yield a falsely young age, even after the inorganic portions of the bone have been removed and the remaining protein subjected to rigorous purification techniques.

A1.3.2.2 Contamination During Collection and Storage

It is possible a radiocarbon test sample can become contaminated during collection and storage. J. Gordon Ogden of the Biology Department of Dalhousie University in Halifax, Canada, reported (1977, p. 172): "Samples collected and stored in cloth or plastic bags are almost useless, as are samples that have been wrapped in paper or tissue, particularly if the sample has been allowed to remain moist. Fungi may and probably will grow upon it. Although it is not universally the case, many fungi are capable of utilizing some atmospheric CO_2 in their carbon metabolism." Ogden, the director of a radiocarbon laboratory, jokingly remarked to his audience at the New York Academy of Sciences: "many of you may resolve never to trust a radiocarbon date again." The addition of 5 percent modern carbon to a sample 100,000 years old reduces its age to an apparent 24,000 years (R. Taylor 1987, p. 117).

Another scientist, J. W. Michels of Pennsylvania State University, reported (1973, pp. 163–164) that special care must be exercised in removing the sample from its original environment. It should not be touched with the hand. Michels recommended that only metal or glass should come in contact with the object to be dated and that the sample should be wrapped tightly in new aluminum foil. Samples dropped on the ground should be discarded. Other experts advise that samples be placed immediately in airtight containers, to avoid contamination from atmospheric carbon 14 produced by nuclear weapons tests.

A less stringent view was expressed by R. E. Taylor (1987, p. 68): "Ordinary cleanliness, caution, and common sense are usually sufficient. One does not want to touch a sample if, for example, one's hand is covered with oil, handcreams, or powdered coal. Routine precautions will avoid placing samples in contact with any modern organic materials that might mix with the sample matrix. As an obvious example, this means that samples should not be packed in such materials as cotton or paper cuttings. Clean metal containers with screw tops and metal foil are the best materials in which to package samples intended for c-14 analysis . . . some types of plastics used to fabricate containers may 'outgas' organic compounds and should be avoided." This is still quite a formidable list of precautions. How, then, can we tell whether or not such precautions were in fact observed? Ideally, the reporting process should include exact descriptions of how each sample was handled from the time it was removed from the ground to the time it was dated in a radiocarbon laboratory. Obviously this is impossible in many of the cases with which we are concerned, for the bones lay in museums for decades before being dated.

Owing to the various problems with carbon 14 dating, Ogden (1977, p. 173) said: "It may come as a shock to some, but fewer than 50 percent of the radiocarbon dates from geological and archaeological samples in northeastern North America have been adopted as 'acceptable' by investigators."

A1.3.3 The Dating of Laguna Man and Los Angeles Man

Now that we understand the basics of carbon 14 dating, let us see how this method has been applied to the dating of important human skeletal remains. We will start with some controversial finds from Southern California, the radiocarbon dating of which sheds considerable light on the reliability of this method. The story begins with the radiocarbon dating of two human skulls from Laguna Beach and Los Angeles, proceeds to a consideration of another dating technique called amino acid racemization dating, and ends with a discussion of a new method of radiocarbon dating known as AMS, or accelerator mass spectrometry.

Before trying to date some potentially ancient, and therefore highly significant, human remains from North America, Rainer Berger of the University of California at Los Angeles (UCLA) wanted to see whether or not radiocarbon dates on collagen were accurate. He did so by comparing radiocarbon dates on collagen with radiocarbon dates on charcoal or other items from the same level at the same site (Table A1.2). It is important to note that some of the dates in the left column of this table were determined by carbon 14 labs other than UCLA, in fact five different labs, and yet they agree closely with the UCLA dates on collagen in the right column.

At a certain point in his investigation, Berger concluded that with sufficient purification, the collagen portion can yield reliable dating results. He then used this technique to date the Laguna skull at 17,150±1,470 years old. Since this skull

TABLE A1.2

Radiocarbon Dates for Collagen
Compared with Dates for Charcoal and Other Substances

Location	Age on Charcoal or Other Item	Age on Collagen
Mexico	1,830 ± 55	1,720 ± 130
Mentohotep, Egypt	2,030 B.C. (historic)	2050 B.C.
Sakkara, Egypt	3,980 ± 60	3,845 ± 60
	4,000 ± 60	
Anzabegovo, Yugoslavia	6,950 ± 80	6,700 ± 80
South Africa	7,050 ± 45	7,380 ± 120
	9,580 ± 85	9,230 ± 160
	9,450 ± 55	
	10,030 ± 55	10,100 ± 190
	9,780 ± 60	10,120 ± 200
	10,500 ± 400	
	35,000 ± 2,400	35,630 ± 2,500
	38,550 ± 3,800	38,680 ± 2,000
	35,700 ± 1,100	32,400 ± 2,500
	36,100 ± 900	34,800 ± 2,500
California	14,400 ± 300	14,500 ± 190
	14,640 ± 115	
Nevada	2,510 ± 80	2,500 ± 80
	2,590 ± 80	
Olduvai Gorge	17,000	17,550 ± 1,000

Charcoal and other carbon-bearing substances are believed to yield more reliable radiocarbon dates than bone collagen. To test the reliability of UCLA procedures for dating bone collagen, Rainer Berger compared UCLA bone collagen dates with dates for charcoal and other substances from the same sites. The results in this table (after Berger 1975, p. 176) led Berger to conclude that the UCLA radiocarbon dating procedures for bone collagen were satisfactory.

was unearthed in 1933 and was later examined many times by anthropologists, a pretreatment with ether was employed to remove organic substances introduced by the handling. The collagen portion was isolated by dilute HCl treatment followed by "sodium hydroxide extraction" to remove humic acid contaminants.

A long bone fragment associated with the Laguna skull yielded a date of over 14,800 years, which agrees fairly well with the date for the skull.

Berger also dated the Los Angeles skull. The Los Angeles skull was found in 1936 in Late Pleistocene sandy clay at a depth of 4 meters (13 feet) below the bed of an ancient river (Berger 1975, p. 180). A couple of months later, below the same river bed and at the same depth, the remains of a mammoth were found. According to Berger, the most reliable bone dates are obtained by isolating the amino acids native to the skull. To accomplish this, he employed the "liquid chromatographic separation technique." In particular, Berger (1975, p. 181) was trying to eliminate the possibility of contamination from petroleum deposits, although there was no visible evidence of such contamination on the skull. An age of over 23,600 years was obtained for the skull. According to Berger, this age is in agreement with the age of the Pleistocene stratum in which the skull had been discovered.

As of 1989, Dr. Berger still considered the dates of 17,150 and 23,600 years for the Laguna and Los Angeles skulls to be completely correct and based on excellent and meticulous laboratory techniques. Yet we may recall that according to the dominant view, human beings did not enter North America until about 12,000 years ago. Archeologists holding this view contend that Berger's radiocarbon dates are the result of error. And as we will see, these dates became the focus of an intense controversy, culminating in their total rejection in 1984–85.

A1.3.4 Amino Acid Racemization Dating of Bone

Berger's dates for the Laguna and Los Angeles skulls were controversial, but other North American human remains were thought by some researchers to be still older—over 40,000 years old. This placed them beyond the effective range of radiocarbon dating. But during the 1970s, Jeffrey Bada did extensive work on a new chemical technique that measured the rate of amino acid racemization. Bada believed the amino acid racemization (AAR) technique gave good age determinations for protein material that might be older than 40,000 years. In sections A1.3.4–6, we shall review some of the history of AAR dating and its bearing on the reliability of radiocarbon dating. Initially, the accuracy of the AAR method was thought to be quite good, because AAR dates obtained on materials less than 40,000 years old agreed with carbon 14 dates. Later, researchers obtained a different set of carbon 14 results, using a new method, accelerator mass spectrometry (AMS). Faced with this inconsistency, researchers decided to throw out the results of both the standard carbon 14 and AAR dating methods. But such radical shifts in consensus cast doubt on the whole enterprise. At the very least, we are confronted with many examples of discordant radiocarbon dates.

Amino acids can exist in two different forms—L and D. In the proteins of most living organisms, the amino acids are in the L state. But this situation is thermodynamically unstable because in the state of chemical equilibrium, the L and D forms should be existing in equal amounts. After a certain time, specific to each amino acid, the originally predominant L form in a dead organism will have transformed into a mixture containing nearly equal amounts of the L and D forms. This process, which begins when the organism dies, is called racemization. Since racemization is a chemical reaction, its speed is strongly dependent on temperature.

Under certain conditions, if one knows the temperature history of a bone and can measure the ratio of L to D forms of a particular amino acid, one can then calculate the age of the bone (Bada 1985a). Since it is hard to know the temperature history of a bone that may have been around for many thousands of years, the extent of racemization (the ratio of D to L forms) is measured in a bone that is from the same region and has been dated by some other means. Using the known age and the D/L ratio from this bone, one can calculate a calibration constant k, which can be used to calculate the age of other bones from the same environment—at least in principle.

The amino acid generally used for racemization dating is aspartic acid, and the corresponding calibration constant is called k_{asp}, where "asp" stands for aspartic acid.

For the Southern California region, the Laguna skull was chosen as the calibration specimen. A value of $k_{asp} = 1.08 \times 10^{-5}$ was obtained, and this value of k_{asp} was used to calculate the age of the Los Angeles skull. The Los Angeles skull gave a date of 26,000 years old, a value agreeing nicely with Berger's carbon 14 date of 23,600 years old.

In addition to the Los Angeles skull, Bada and coworkers also dated other human skeletal remains from California. Several specimens from the San Diego area were tested because geomorphological evidence suggested they might be of considerable age (Bada *et al.* 1974). La Jolla SDM16755 gave an AAR date of 28,000 years, La Jolla SDM16742 gave a date of 44,000 years, and Del Mar SDM16704 gave a date of 48,000 years. SDM is an abbreviation for San Diego Museum of Man.

A year later, Bada and Helfman (1975) also dated a skeleton from Sunnyvale (near San Jose) at around 70,000 years old, and three new samples from various parts of the Del Mar skeleton (SDM16704) yielded dates of 46,000±3,000 years old, thus confirming the earlier AAR age assignment of 48,000 years. In addition, Bada and Helfman dated three new skeletons from San Diego County. La Jolla SDM16740 gave an AAR date of 39,000 years old, La Jolla SDM16724 gave a date of 27,000 years old, and Batiquitos Lagoon SDM16706 gave a date of 45,000 years old.

TABLE A1.3

**Late Pleistocene Amino Acid Racemization Dates
Obtained for Human Skeletal Remains from California**

Location	AAR age (years)
Los Angeles skull	26,000
La Jolla SDM16755	28,000
La Jolla SDM16742	44,000
Del Mar SDM16704	48,000
	46,000
Sunnyvale SV-B	70,000
La Jolla SDM16740	39,000
La Jolla SDM16724	27,000
Batiquitos Lagoon SDM16706	45,000

All of these AAR ages (Bada *et al.* 1974, Bada and Helfman 1975) contradict the dominant view that humans entered North America around 12,000 years ago. Many of the ages also contradict the view, accepted by a minority of researchers, that the peopling of the New World began about 30,000 years ago. Only the Los Angeles skull was dated previously using the carbon 14 method, yielding an age of 23,600 years or more (Berger 1975).

As a further check on the Del Mar and Sunnyvale skeletons, another lab (NASA), using "independent processing and analytical procedures," verified the D/L ratios obtained by Bada.

All of these dates (Table A1.3) contradict the standard view that human beings entered North America not much more than 12,000 years ago. Many of them also surpass the age limit of about 30,000 years that is gaining support among a minority of researchers. As such, it is perhaps not surprising that a decade after Bada's findings were published, a paper with the title "New World colonized in Holocene" appeared in the pages of the prestigious journal *Nature* (Burleigh 1984). The paper declared that the new radiocarbon dating technique of accelerator mass spectrometry had pushed all of Bada's dates back into the Holocene, thus adding to "the increasing body of evidence placing the human colonization of the New World in this period." In an article in the same issue of *Nature*, Bada renounced his earlier findings (Bada *et al.* 1984).

What went wrong with Bada's dates? Initially they seemed quite plausible. One strong argument for the California dates was that the amino acid racemization test gave good results on bones from many other parts of the world. Bada and other scientists obtained amino acid racemization (AAR) dates for a variety of bones previously dated at several laboratories by geomorphological, carbon 14, thermoluminescence, or historical techniques (Table A1.4, pp. 776–777).

Dates in Table A1.4 were taken from Bada *et al.* (1974, 1979), Bada and Helfman (1975), Bada and Deems (1975), Bada and Masters (1982), Bada (1985a), Ike *et al.* (1979), Matsu'ura and Ueta (1980), and Belluomini (1981). Note: Where a k_{asp} value is listed in place of an AAR date, the sample in question was used to calibrate the AAR dates below it in the table.

On the whole, the results reported in Table A1.4 show good agreement between AAR and the other dating methods. Especially noteworthy is the strong agreement in California. In addition to the Los Angeles skull, the Pleistocene horse skeleton found at Scripps confirmed the Laguna k_{asp} of $1.08´10^{-5}$. The horse skeleton was assigned an age of over 30,000 years by radiocarbon dating of associated charcoal, and this was consistent with an AAR age of 50,000 years, calculated using the Laguna k_{asp} value (Bada and Masters 1982, pp. 175–177).

Similar confirmation was provided by dwarf mammoth bones from Santa Rosa Island. These remains had given carbon dates of 29,700±3,000 years on charred bone and 30,400±2,500 years on bone collagen. When the Laguna k_{asp} was adjusted by a formula to the average 13.7 degree Centigrade temperature of the coastal islands, an AAR date of 33,000 years was obtained for the bones, in good agreement with the radiocarbon dates (Bada *et al.* 1974, pp. 184–185).

Many bones of Holocene age were AAR dated using a modified k_{asp}, adjusted by a formula to reflect Holocene temperatures. Since the AAR dates all agreed well with their corresponding radiocarbon dates, they tended to confirm the validity of the Laguna k_{asp}. In particular, skeleton SDM16709 was AAR dated twice by Bada and coworkers, and was also AAR dated independently by two other labs. The reported results agreed with its radiocarbon age of around 8,000 years. However, there is one discrepancy in the reports: the D/L ratio of .142 reported in Bada and Masters (1982) should give an age of 4,900 years for SDM16709, rather than the reported 7,900 years.

There were a couple of cases of strong disagreement between AAR dates and carbon 14 dates. A bone from the Ndutu Beds of Olduvai Gorge is an important example. This was given an AAR date of 33,000 years old and a radiocarbon date of 3,340 years old. However, radiocarbon dates on other Ndutu bones were in the 29,000-year range, and Bada did not hesitate to attribute the 3,340-year date to massive contamination by new carbon (Section A1.3.2.1).

It is significant that the Ndutu bone with the radiocarbon date of 3,340 years old was extensively decomposed, and had only 0.05 percent nitrogen. Bada (1985a, p. 255) argued that its AAR date was still reliable. In several cases he cited evidence indicating that the degree of decomposition (or diagenesis) of a bone has no impact on its AAR date. He pointed out that Holocene Nasera rock shelter skeletons from Africa have the same D/L ratios as some nearby Holocene skeletons from Olduvai Gorge, even though the former have a collagenous amino acid composition and the latter have a noncollagenous amino acid

TABLE A1.4

Agreement Between Amino Acid Racemization Dates and Dates Obtained by Radiocarbon and Other Methods

Location	C-14 or Other Age (yrs.)	AAR Age (yrs.)
California		
Laguna skull	17,150±1,470	k_{asp}=1.08´10⁻⁵
Los Angeles Man	>23,600	26,000
Santa Rosa Island	30,400±2,500	at 13.7 C: 33,000
	29,700±3,000	
Scripps horse	>30,000	~50,000
		adjusted Holocene k_{asp}=1.50´10⁻⁵
SDM18402	6,250±1,250	~6,000
SDM19241	6,700±150	5,700
SDM16709	8,360±75	8,100
		7,900
		9,400
Ora64	6,960±140	6,800
Tranquillity	2,550±60	3,800
Stanford Man	5,130±70	4,700
SDA-66	9,040±210	5,400
	7,750±400	6,900
Marin	3,270±70	2,800
Arizona		
Murray Springs	5,640±60	k_{asp}=4.84´10⁻⁵
	11,230±340	10,500
	~300	~400
Double Adobe	10,420±100	9,900
Muleta Cave, Mallorca, Spain	16,850±200	k_{asp}=1.25´10⁻⁵
	18,890±200	18,600
	28,600±600	33,700
Egypt (dated historically)		
Tarkhan	3200–2700 B.C.	k_{asp}=9.31´10⁻⁵
		3360 B.C.
		3220 B.C.
		A.D. 760
Sakkara	2500–2400 B.C.	2030 B.C.
Tura	2100–1600 B.C.	
Israel, Mt. Carmel Caves		
Sefunim layer 8-9	~27,000	k_{asp}=1.64´10⁻⁵
Tabun, layer C	40,900±1,000	44,000
layer E	~80,000	54,000

TABLE A1.4—*Continued*

Location	C-14 or Other Age (yrs.)	AAR Age (yrs.)
Czechoslovakia		
Stranska Skala	690,000	$k_{asp}=7.70'10^{-7}$
Kulna cave, layer 7a	45,600±2,850	50,000
layer 11	last interglacial	80,000
Italy		
Palidoro, R9441	14,580±130	$k_{asp}=2.91'10^{-6}$
R946	15,340±140	15,650
R947	15,660±130	16,350
R948	15,900±150	15,000
Grotto Maritza, R1271	10,420±150	$k_{asp}=4.76'10^{-6}$
Piana del Fucino, R1270R	10,420±60	11,500
Grotta Polesini, R1265	10,090±80	$k_{asp}=5.21'10^{-6}$
Grotta Paglicci, R1334R	15,270±220	16,150
Gargano Promontory, R1323	15,320±250	23,900
Rignano Garganico, R1324	17,200±150	$k_{asp}=3.95'10^{-6}$
R1269	24,210±410	22,900
R1269R	24,720±420	20,300
Cayonu, Turkey	8,340	$k_{asp}=1.21'10^{-5}$
	7,620	8,700
Taishaku, Japan	>20,000	23,300±400
South Africa		
Nelson Bay Cave	16,700±240	$k_{asp}=4.92'10^{-5}$
	18,660±110	20,000
	18,100±550	
Klasies River Mouth Cave I	~80,000–120,000	110,000
		90,000
		89,000
		65,000
Swartklip I	~80,000	~110,000
Heuningsneskrans	20,500±300	$k_{asp}=2.64'10^{-5}$
	23,900±800	24,800
		31,000
Olduvai Gorge, Tanzania		
Middle Naisiusiu Beds	17,550±1,000	$k_{asp}=1.48'10^{-5}$
Upper Ndutu Beds	3,340±800	33,000
	>29,000	56,000
Nasera	21,600±400	$k_{asp}=1.80'10^{-5}$
	22,900±400	26,000

composition (Bada 1985a, p. 254). The collagenous composition is typical of unmodified or slightly modified bone, and the noncollagenous composition develops as the bone decomposes. Bada previously noted that three different Egyptian skeletons of the same historical age yielded nearly identical D/L ratios: "even though the bones have differed greatly in their state of preservation and amino acid content" (Bada *et al.* 1979, p. 745).

Bada compiled some of the information presented in Table A1.4 to support his original contention that properly calibrated AAR ages of fossil bones "agree closely with ages determined by radiocarbon or other chronological information" (Ike *et al.* 1979, p. 525). Indeed, in 1975 Bada pointed out that in his extensive studies up to that date: "the largest difference between the aspartic acid racemization and radiocarbon deduced ages is 18%. The average difference is 7%" (Bada and Helfman 1975, p. 163).

But after the AMS dating of the California skeletons, Bada declared the new data "reveal no clear relationship between the radiocarbon ages of the various skeletons and the extent of aspartic acid racemization" (Bada *et al.* 1984, p. 443). Before discussing these AMS studies in detail, we will first give an independent example of evidence that contradicts Table A1.4, and shows a total lack of correlation between AAR ages and radiocarbon ages.

Lajoie *et al.* (1980) measured aspartic acid D/L ratios (and from these calculated ages) for six human bones from the San Francisco region. The aspartic acid ages were compared with carbon 14 dates (Table A1.5). As we can see, the ratios between the AAR dates and the corresponding carbon 14 dates vary wildly, especially in the case of SV-B, the Sunnyvale human skeleton. We note, however, that the carbon 14 age for SV-B reported here is based on indirect association with other samples dated by the carbon 14 method.

The rate of aspartic acid racemization depends on temperature. One might ask if any of the bones in Table A1.5 had been exposed to unusual extremes of heat. Such heat would have speeded up the racemization process, giving an age that was far too old, as in the case of the Sunnyvale skeleton. But Lajoie *et al.* said that none of the bones in Table A1.5, or their immediate environments, gave any indication of heating by fire. Also, all the sites are within a few miles of each other, so it is hard to conceive of vast differences in their climatic temperature histories. Thus it would seem that temperature variations cannot account for the variation in AAR ages shown in the table.

How are we to explain the radical disagreement between carbon 14 and AAR dates reported by Lajoie *et al.*? These researchers observed (1980, p. 480): "Theoretically, D/L values should increase with age and effective temperature. However, there is virtually no correlation between D/L values for aspartic acid, the amino acid data most widely used for bone dating, and age in our samples. Therefore some factor other than age is expressed."

TABLE A1.5

**Comparison of Carbon 14 and Amino Acid Racemization Dates
for Six Human Skeletons from the San Francisco Region**

Bone Sample	C-14 age	D/L	AAR age	Age Ratio
S-III	2,270	.10	6,880	3.0
C	2,710	.13	11,700	4.3
UV-B	3,170	.125	10,900	3.4
S-II	4,375	.092	5,600	1.3
SV-B	4,500	.42	62,000	13.8
B	4,900	.069	1,900	.4

Here the ratios of AAR ages to corresponding carbon 14 ages vary widely, demonstrating no correlation between the two (after Lajoie *et al.* 1980). The AAR ages were calculated from the aspartic acid D/L ratios.

Here it is taken for granted that the carbon 14 dates measure actual age, and that the differing AAR dates must be erroneous. Actually, however, we are simply faced with an inconsistency between two methods of computing dates, and either or both may be yielding spurious results. We will gain greater insight into this when considering in detail the events that led to Bada's radical change of opinion about his own AAR dates. To discuss this, we must first introduce the AMS technique for radiocarbon dating.

A1.3.5 Ages in Chaos: AMS vs. AAR Dating

In the late 1970s, researchers began to develop a new radiocarbon dating method known as accelerator mass spectrometry (AMS). Conventional radiocarbon dating relies on counting the decay of radioactive carbon 14 atoms in a sample with Geiger counters. The AMS method uses a particle accelerator to separate carbon 14 atoms, which are slightly heavier than carbon 12 atoms. These are then directly counted.

In the 1980s, the AMS method was used to redate organic fractions of some of the controversial California skeletal remains discussed in the previous section, and consequently, their Late Pleistocene amino acid racemization dates were uniformly discarded in favor of Holocene AMS dates. Some of the results obtained by the AMS radiocarbon testing of skeletons that had previously been

TABLE A1.6

Comparison of Ages for Human Skeletal Remains Found by Amino Acid Racemization, Carbon 14, and Accelerator Mass Spectrometry Methods

Skeleton	AAR Age	Old C-14 Age	New AMS Age Ratio
Sunnyvale	70,000		3,600–4,850
			6,300±400
Del Mar SDM16704	48,000		4,900
			5,400±120
Otavalo, Ecuador	28,000	28,000	2,300–2,670
		2,500	
La Jolla SDM16755	28,000		(I) 1,700–1,930
			(II) 4,820–6,330
San Jacinto	37,000		3,020
Yuha	23,600	21,500–22,125	1,750–3,930
			>26,600
Truckhaven	23,600	4,990±250	<500
Haverty	>50,000		4,050–7,900
(Angeles Mesa)			
Los Angeles Man	26,000	>23,600	3,560
Laguna skull	calibration	17,150	5,100±500

In the 1980s, the newly developed AMS method gave Holocene ages for human skeletal remains previously dated to the Late Pleistocene by AAR and carbon 14 methods. Data for this table was compiled from Bada *et al.* (1984), R. E. Taylor *et al.* (1985), Bada and Finkel (1983), and Goodman (1982).

assigned high AAR ages are summarized in Table A1.6.

We note that the dates (I) and (II) for SDM16755 were thought to be for skeletons at different depths at the same site. We also note that the Otavalo skull previously had conflicting radiocarbon dates of 28,000 and 2,500 years old. This skull also had a thermoluminescence age of 25,000 years, thus confirming its 28,000-year radiocarbon and AAR dates (Goodman 1982, p. 70).

The Yuha skeleton also deserves mention. This skeleton had previously been dated at 21,500±400 years old and 21,500±1,000 years old by applying the conventional radiocarbon method to caliche (a carbonate deposit) adhering to the bones (Stafford *et al.* 1987, pp. 33–35). Nine AMS dates, six on the bones and three on the caliche, gave dates in the range 1,750 to 3,930 years old. But two

AMS dates on "petrocalcic horizon caliche" associated with the bones were 3,030±270 years old and over 26,600 years old. An AAR date of 23,600 years old was obtained on the bones, and a thorium 230 date of 19,000±3,000 years old was obtained for the caliche. Here the explanation offered is that both Holocene and Pleistocene caliche had somehow gotten cemented onto the bones. The AAR date is thought to be wrong, as usual, but it happened to agree with the misleading dates from the Pleistocene caliche.

Of particular importance in Table A1.6 is the fact that the new AMS dates reported by R. E. Taylor *et al.* disagreed with the previous carbon 14 dates for the Laguna and Los Angeles skulls. Yet we have previously given reasons why the original ages of 17,150 years and over 23,600 years respectively for these skulls can be considered accurate (Section A1.3.3). According to Berger, who reported the original carbon 14 dates for the Laguna and Los Angeles skulls, the new AMS date for the Los Angeles skull conflicts not only with the previous carbon 14 date but also with the site stratigraphy. The Los Angeles skull was found in a Pleistocene stratum. How can we resolve this conflict?

One possible solution was offered by Berger (personal communication, 1989). He suggested that there might have been something wrong with the samples used for the AMS dating of the Laguna and Los Angeles skulls. Specifically, the Oxford AMS dating of the Laguna skull, which yielded an age of 5,100 years (Bada *et al.* 1984), was done on amino acids prepared a decade earlier by Bada for racemization dating, and these may have been contaminated somewhere along the line. Berger said that the samples he himself took (which resulted in the decay count carbon 14 ages of 17,150 years and over 23,600 years) were good, but he is not so sure about the samples used for the recent AMS dates. To confirm this, Berger planned to take new samples and perform a new AMS test using the University of California AMS facility.

In the case of Laguna and Los Angeles, radiocarbon tests (AMS carbon 14 vs. decay count carbon 14) on the same bones gave radically different results. The earlier decay count carbon 14 age for Laguna man is 3 times its AMS age, and the earlier decay count age for Los Angeles man is a staggering 6 (or more) times its AMS age. Scientists have arguments in favor of both the old and new dates, but one thing is clear—they cannot both be right. Somebody made a serious mistake somewhere, but it is not easy to figure out who and where. The conflicting dates do not increase our confidence in carbon 14 dating in general.

If the AMS Laguna skull date of 5,100 years is correct, it invalidates the value of $k_{asp}= 1.08´10^{-5}$, which Bada calculated using Berger's decay count date of 17,150 years. The new AMS date indicates that the Laguna k_{asp} should be about 3.36 times (17,150 years/5,100 years) bigger. Because the Laguna k_{asp} was the racemization constant Bada used to obtain ages for the other skeletons from California, which ranged from 26,000 to 70,000 years old, the AAR ages computed for these other skeletons must in turn be about 3.36 times too large.

And if the new Los Angeles skull AMS date is right, then its D/L ratio of 0.35 yields a k_{asp} of 8.3×10^{-5}. This is about 7.7 times Bada's original value (and 2.3 times the revised Laguna k_{asp}). AAR dates calculated with this k_{asp} would be even smaller than those obtained from the revised k_{asp} for the Laguna skull.

If Bada's original Laguna k_{asp} is wrong, how were Bada and other researchers able to use it to obtain AAR ages for eleven skeletons in California (Table A1.4) that agreed so well with carbon 14 or geomorphological dates for these bones? If one were to use the larger, presumably more correct values for k_{asp}, then in many cases one would obtain AAR ages which are much smaller than the ages obtained by carbon 14 and other methods.

For example, the Scripps Pleistocene horse and the Santa Rosa Island mammoth would be projected into the Holocene by the revised AAR calibrations. The many California Holocene dates of Table A1.4 would also be reduced far below their carbon 14 dates.

Bada (1985a, pp. 245–248) thought that one cause of anomalously high AAR and carbon 14 dates was contamination by amino acids from the environment, and he cited studies of fossil dinosaur bones that indicate such contamination does take place. The extraneous amino acids could increase the measured AAR date if they had a high D/L ratio, and they could also produce a high radiocarbon date if they were depleted in carbon 14 (Bada *et al.* 1984, p. 443). Of course, such contamination could also produce a low AAR or radiocarbon date if the amino acids were from recent living organisms. Since amino acids in the environment break down chemically with the passage of time (or are eaten by micro-organisms) it seems that any contamination would most likely be from recent amino acids, which would result in unusually low dates.

Bada believed that the discordant California AAR dates could also be attributed, at least in part, to the decomposition of proteins in the bone. Regarding the highly racemized California skeletons, Bada *et al.* (1984, p. 443) said that "the collagenous matter in these bones had undergone severe diagenesis [decomposition], which may have altered the D/L-Asp ratio."

Bada (1985a, p. 261) offered the following explanation of how bone diagenesis could account for enhanced racemization of aspartic acid: "During the process of collagen hydrolysis [breakdown], aspartic acid residues may end up preferentially at the N-terminal position of the peptide hydrolysis products. This in turn could give rise to a more rapid racemization rate, since amino acids at this position in peptides have been found to have racemization rates greatly exceeding those of amino acid residues at other positions."

A somewhat similar idea was offered by Bada's colleague, Patricia Masters (1987). According to her, the skeletons giving anomalously high AAR ages all had noncollagenous amino acid compositions, indicating that their collagen had largely decomposed, and that only noncollagenous proteins (NCPs) were left. Research by Masters showed that aspartic acid in NCPs tends to racemize about ten times faster than it does in collagen. Thus, highly decomposed bones will tend to give much greater AAR ages than undecomposed bones. According to Masters, this explains the anomalously high AAR dates of the California skeletons.

Here is what must have happened, according to Masters's scenario (which we do not regard as proven):

(1) Due to an unfortunate fluke, the Laguna skull received an erroneous radiocarbon date, which happened to be falsely old.

(2) The protein in the skull was noncollagenous, resulting in an abnormally high degree of racemization for the aspartic amino acid.

(3) Using the falsely old radiocarbon date and an abnormally high level for aspartic acid in noncollagenous protein, Bada obtained an incorrect k_{asp} for the Laguna skull.

(4) By chance, this incorrect k_{asp} for the noncollagenous Laguna skull was such that it would give correct AAR dates for collagenous bone.

(5) Bada, using the Laguna k_{asp}, was therefore able to get correct AAR dates for the Holocene bones from California listed in Table A1.4 (SDM18402 through Marin 152). All of these bones had previously yielded radiocarbon ages of less than 10,000 years. These bones were also, by chance, all collagenous. This means that their level of aspartic acid racemization was consistent with the radiocarbon age of the bones. If any of the bones had been noncollagenous, the aspartic acid would have been racemized to a much greater extent than normal. Therefore the age obtained by the AAR method would have been abnormally great, compared with the radiocarbon age. This would have tipped off the researchers, who were unaware of the effect of noncollagenous protein on amino acid racemization, that there was a problem with their method. But this did not happen. Instead, the researchers took the results as confirmation that the AAR method was working properly.

(6) The AAR test was also applied to another set of bones (Table A1.3). By chance, they were, except for the Los Angeles skull, all noncollagenous. This means that their aspartic acid was racemized at a rate much higher than normal. When they were AAR dated using the incorrect Laguna skull k_{asp}, they therefore yielded incorrect ages ranging from 28,000 to 70,000 years. But at the time the AAR dates were obtained, these bones had not yet been radiocarbon dated. Thus their Pleistocene AAR ages were not initially suspect. Of course, the Los Angeles skull did have a radiocarbon date, but by chance this was erroneous, and it happened to match its erroneous AAR date. Later radiocarbon dating yielded Holocene dates for all of these bones, including the Los Angeles skull.

(7) Moreover, the agreement between AAR dates on decomposed and undecomposed bone of the same age in Egypt and Kenya (mentioned above)

must also be a fluke. According to Masters, the decomposed bone, presumably not collagenous, should have given a higher AAR age than the undecomposed, presumably collagenous, bone. Perhaps the apparently decomposed bone was still collagenous, or the apparently undecomposed bone was actually noncollagenous.

(8) And what about the Santa Rosa Island mammoth? As it stands now, it has a Pleistocene radiocarbon age obtained from charred bone. It also has a Pleistocene AAR age (Table A1.4), obtained by Bada. Since this AAR date agreed with the radiocarbon date, it seemed to confirm Bada's dating method. But why did it agree? According to the ideas of Masters, there are two possible explanations for the agreement between the radiocarbon date and the AAR date. Either the mammoth skeleton is collagenous, and is thus as old as its AAR and radiocarbon dates indicate. Or it is noncollagenous and thus much younger than the Pleistocene AAR age calculated using the Laguna k_{asp} (perhaps Holocene). The agreement in dates would therefore have resulted from an erroneously high radiocarbon date as well as an erroneously high AAR date. The same thing can be said for the Scripps horse. This skeleton was buried in sand, and it is thus quite possible that it has a low nitrogen content and is noncollagenous. Bones buried in sand tend to lose their protein content rather quickly.

Accepting Masters's view, we see that the original AAR dates obtained by Bada and others involve a remarkable sequence of coincidences, and it may be that things really happened as outlined above. If so, however, we should be very cautious about the results of all dating schemes that depend on the complex behavior of atoms and molecules in varying environmental circumstances over thousands of years. This includes both chemical and radiometric methods.

Alternatively, perhaps things did not actually happen according to Masters's bone diagenesis theory. Consider the following. Bada originally used the Laguna skull k_{asp} to get an AAR date of 48,000 years old for the Del Mar skeleton. The k_{asp} represents the rate at which the L form of an amino acid converts to the D form. A smaller k_{asp} represents a slower rate of racemization, a bigger k_{asp} a faster rate of racemization. The faster the rate of racemization, the less time it takes for a given amount of racemization to take place. Thus for the same D/L ratio, age calculations made using a bigger k_{asp} will result in a younger age for a sample than calculations made using a smaller k_{asp}.

Subsequent to Bada's AAR date of 48,000 years for the Del Mar skeleton, Stafford and Tyson (1989, p. 391) cited five concordant AMS dates of about 5,000 years for the sphenoid bone of the Del Mar skull—making the skull about ten times younger than Bada thought.

If one calculated a k_{asp} for the Del Mar skeleton using the new AMS ages of 5,000 years, this k_{asp} would be about 10 times bigger than the Laguna skull's k_{asp}, which Bada used to compute his AAR age of 48,000 years for the Del Mar skeleton. A k_{asp} 10 times bigger gives an age 10 times smaller than 48,000 years.

A k_{asp} 10 times bigger for the Del Mar skeleton makes sense according to Masters (1987), who maintained that the Del Mar skeleton was poorly preserved and had a noncollagenous profile. Masters held that amino acids in noncollagenous bone protein racemize at a rate approximately 10 times faster than amino acids in collagenous bone protein. But Stafford and Tyson (1989, p. 393) list the percentages of each amino acid in the Del Mar sphenoid. The amino acid profile is definitely collagenous.

If the bone diagenesis theory of Masters is true, then the possibilities here are as follows.

(1) The samples for the aspartic acid D/L measurement and the amino acid profile evaluations were all taken from the same location on the skull (the sphenoid). This means either Masters, who said the skeleton was noncollagenous, or Stafford and Tyson, who said the sphenoid was collagenous, made a big mistake in their evaluations of amino acid profiles. If Stafford and Tyson were right, and the bone is collagenous, this calls into question the very young AMS dates of 5,000 years and lends credibility to the original AAR date of 48,000 years obtained by Bada, whose methodology worked well on bone that happened to be collagenous.

(2) The samples were taken from different parts of the skull. Masters said that the whole skeleton, including the skull, is noncollagenous, but the fact that Stafford and Tyson reported a collagenous profile for the sphenoid indicates that at least one part of the skull is collagenous whereas other parts are noncollagenous. If this is true, then one can get a very different result depending on which part of a bone one samples. This could clearly produce havoc in all attempts to evaluate bone by chemical or radiometric techniques.

Another example illustrating these two alternatives involves the nitrogen content of the skull from Del Mar. Bada and Helfman (1975, table 7) give a nitrogen content of 0.10 percent for the Del Mar skull, but part of the sphenoid bone from this same skull is given a nitrogen content of 0.69 percent by Stafford and Tyson (1989, p. 393). There are several possibilities: (1) it is a typographical error; (2) both samples were taken from the same part of the skull (the sphenoid), in which case someone made a huge error in the nitrogen determination; (3) the samples were taken from different parts of the skull, which means that different parts of the same skull can have vastly different states of preservation (as judged by nitrogen content). If this is true, then nitrogen content values of bones must be used with great caution.

Unfortunately, errors that may be typographical abound in the literature on the dating of bones. Here we cannot resist mentioning two irksome examples. Table 1 in Masters (1987) cites an AMS date of 5,400±120, called OxA-188, for an amino acid extract from a cranial fragment from the Del Mar skeleton. But in Masters's table 2, this date is said to apply to the femur from this skeleton. This is important, since, the original excavator of the Del Mar skeleton said that two excavation sites were involved: an upper midden (W-34), where a femur was found, and a lower midden (W-34-A), where a skull, mandible, and a few ribs are the only bones mentioned (Rogers 1963, p. 5). The upper midden was thought to date between 5,000 and 7,500 years, and the lower midden was thought to be much older (Bada *et al.* 1974, p. 792). In Masters's table 1, the AMS date of 5,400±120 is said to refer to W-34, the upper midden.

The second example also comes from Masters (1987, table 2). There an AMS date of 5,270±100, identified as OxA-774, is given for dentine from the Del Mar skeleton. But Stafford and Tyson (1989, p. 391) assign this date, with the same identification, to the Del Mar "femur" (which they identify as a tibia).

In summary, we have seen in the last two sections many examples of coincidences between AAR dates and dates based on radiocarbon and other methods. According to current views on dating, these coincidences do not come about because the different dating methods have correctly measured the ages of various sample bones. What then is responsible for these coincidences?

One possibility is that they are due to chance. For example, in the Yuha case, one person measured the radiocarbon age of caliche adhering to a skeleton now regarded as Holocene, hoping to thereby find the age of the skeleton. Another researcher measured the skeleton's aspartic D/L ratio and perhaps applied an incorrect k_{asp}. By chance both got nearly identical Late Pleistocene dates.

Another possibility is that of fraud. On a conscious or semiconscious level, researchers may "fudge" their data in order to bring it into line with their expectations. Though we have no definite evidence of such fraud in the cases discussed in the last two sections, fraud is, unfortunately, a well-known phenomenon in highly competitive scientific fields (Broad and Wade 1982).

A1.3.6 A Case from Israel

In all the confusion and error that marks the AAR controversy, one thing practically never brought into question is the validity of the AMS dates. AAR dates may fall by the wayside, and conventional carbon 14 dates may be repudiated, but AMS dates are presented as a voice of absolute authority. Yet it is quite possible that AMS dating can also go astray. In this regard, we may consider an interesting case reported by A. J. Legge (1986, pp. 13–21).

Oxford AMS dating of charred *Triticum* (wheat) grains from Nahal Oren, Israel, revealed a curious fact. Although all the grains were excavated from the same layer (layer 6), one of the grains was dated at over 33,000 years old, another was dated at 2,940 years old, and a third was dated at 3,100 years old (the charred part). A humic extract for this last specimen was said to yield an AMS carbon 14 date of 6,650 years old. The standard decay counting carbon 14 method yielded dates on bone of S1=15,800 years old and S2=16,880 years old for two different levels in layer 6. The grain dated 33,000 years old was found just beneath S2, the grain dated 2,940 years old came from the same level as S2, and the grain dated at 3,100 years old came from just above S1. Layer 6 was described as "extremely hard, stony and brecciated."

Legge interpreted this data to mean that a combination of rodent and insect burrowing had brought the younger grains down to the 16,000-year level, even though this would mean that the grains must have been carried down more than 3 meters (about 10 feet) and into a very hard layer. Legge cited evidence that ants store uncharred grains underground, but he admitted that it is unlikely that they would store grains as heavily charred as the three in question. Furthermore, why would ants bother to dig so far into a very hard, stony layer? Legge attributed the presence of the oldest grain, found in a layer half as old as it was, to "re-deposition of material from older archaeological levels."

Regarding the reliability of AMS testing, there are two ways to look at what occurred with the dating of the Nahal Oren charred grains. One is the way advocated by Legge (and many other scientists in his field). According to them, the AMS dates are to be considered correct, and some means must be sought to explain the positions of the grains. The other view is that the AMS dates are wrong and the "younger" and "older" grains are in reality the same age as the stratum in which they were found. If this is true, then we have a case in which the AMS carbon 14 test can give dates much younger or older than expected. Another option is that the decay count carbon 14 ages for layer 6 might be wrong and the AMS ages for the two younger grains might be correct, although Legge believed the decay count dates were in fact correct.

An important insight can be gained from this case. In general, AMS dates are taken by scientists as correct, and certainly more reliable than stratigraphic evidence, even in a case such as this, where there are arguments suggesting that the AMS dates are wrong. But such confidence may not be warranted.

One problem with AMS dating, as well as with the earlier decay counting carbon 14 methods, is that many different chemical processing stages are required to produce a datable sample from a piece of bone or other organic material. Exacting precision is required, and small mistakes can introduce errors into the final age determination. The extremely small size of the AMS samples contributes to the chances for contamination. R. E. Taylor (1987, p. 94) stated: "The collection and pretreatment of microsamples intended for AMS analysis must operate under constraints much more stringent than those routinely employed in most decay counting laboratories. In some cases 'clean room' environments may increasingly be required for AMS sample processing." Furthermore, the dating apparatus itself is highly complex, offering a large potential for error.

A1.3.7 The Carbon 14 Dating of the Galley Hill Skeleton

Having reviewed the development of the techniques of radiocarbon dating, we can see that at first researchers were not aware of various sources of contamination. As time passed, more and more elaborate techniques for identifying and removing contaminants were introduced, along with improved methods of measuring the carbon 14 in a sample. This raises an important question: If a radiocarbon date was obtained in earlier times using procedures that were considered adequate then, but that do not meet current dating standards, can that date still be regarded as reliable? The answer must be no. If the more rigorous recent standards are indeed required, then they must apply to all radiocarbon dates, whenever they were determined.

With this in mind, we are now prepared to analyze the radiocarbon dating of two important finds that are relevant to our understanding of human origins and antiquity. These are the Galley Hill and Castenedolo skeletons. Both are fully human. On stratigraphic grounds, the Galley Hill skeleton was originally referred to the Middle Pleistocene, at about 300,000 years ago (Section 6.1.2.1). The Castenedolo human skeletons were found in Pliocene clay (Section 6.2.2).

In the case of Galley Hill, there is evidence of extreme contamination. We have already mentioned Oakley and Montagu's statement (1949, p. 37) that the Galley Hill skeleton was "treated with 'gelatine' and later dipped in preservative solution." Gelatin is usually made from cow hooves, skin, and connective tissue, which contain a great amount of modern collagen. This collagen is chemically very similar to bone collagen, the portion of the bone normally used in radiocarbon testing. Thus the gelatin could not be eliminated by the standard decalcification process meant to keep the original bone collagen.

Was any attempt made to eliminate the gelatin and other contaminants? The published report on the dating issued by the British Museum Research Laboratory (the lab that did the test) simply said (Barker and Mackey 1961, p. 39): "In all cases, only the organic fraction of the sample was used as a source of carbon, and the procedure adopted was as follows: Sample was broken into small pieces, either by coarse grinding or pounding in a mortar, and was treated with cold dilute hydrochloric acid in order to remove carbonates and to decalcify the material. Resulting granular gel was washed thoroughly by repeated soaking with cold water and was finally dried, prior to combustion."

Cold, dilute hydrochloric acid is used to eliminate only the inorganic parts of the bone. Thus this procedure in and of itself would not have been sufficient to remove the gelatin or preservatives (unfortunately the name of the preservative was not given, but most preservatives contain carbon). Oakley (1980, p. 17) mentioned the 1961 carbon 14 test, but he did not explicitly state whether or not the sample that was actually used in the carbon 14 test was decontaminated beforehand. Thus, there is no indication from the published reports on the test that a decontamination procedure capable of eliminating the known organic contaminants was used.

If no such decontamination procedure was employed, then the test result of 3,310 years old obtained by the British Museum must be considered invalid. In particular, the gelatin would have introduced large amounts of modern carbon, causing the test to yield an age far younger than the true age of the skeleton. Considering the importance of removing the contaminants, we wrote to the British Museum in 1989 for information on what decontamination procedures, if any, were used. We also inquired about other details of the radiocarbon test procedure. But thus far we have not received a reply.

Besides the gelatin and preservatives, there are other sources of contamination that were apparently not considered. For example, in the accounts of the radiocarbon testing of the Galley Hill skeleton, written by Barker and Mackey (1961) and Oakley (1980), we find no mention that sodium hydroxide (NaOH) or XAD resin was used to remove humic and fulvic acids, which Stafford *et al.* (1987, p. 30) called "the predominant contaminants in fossil bone."

Similarly, there is no mention that contamination by saprophytes—fungi that grow and feed on bone (Hassan and Ortner 1977, Berger *et al.* 1964)—was properly handled. Ogden (1977) said that fungi probably grow on samples, especially in a moist environment. The Galley Hill bones were exhumed in 1888 but were not carbon 14 dated until 1961. They were thus exposed to the damp English climate for many years. It is quite possible that fungi entered and spread throughout the porous bone structure. The extent to which preservatives may have inhibited this growth is not at all certain.

The metabolism of fungus cells can contribute new carbon from atmospheric carbon dioxide (Ogden 1977, p. 172). Over the course of the 73 years between the discovery of the Galley Hill skeleton and its radiocarbon dating, much of its old carbon could have been replaced by modern carbon, resulting in a carbon 14 age far younger than the skeleton's true age.

Oakley and Montagu (1949, p. 38) reported that the Galley Hill human bones "showed scoring by rootlets." Yet there is no mention in Barker and Mackey (1961) or Oakley (1980) that the bones were inspected to see to what extent rootlets were present. If present, rootlets would have introduced modern carbon and thus caused a falsely young date for the bones.

Ogden (1977, p. 172) said that samples wrapped in paper are "almost useless," but Newton (1895, p. 518) quoted a letter from Robert Elliott, who reported that after he removed the Galley Hill skeleton from its matrix he carefully wrapped the bones in "soft paper and brought them home."

Finally, the Galley Hill skeleton apparently had a noncollagenous amino acid profile. Oakley (1980, p. 12) stated: "A paper-chromatogram prepared from a sample of the ulna of a woolly rhinoceros preserved in Upper Pleistocene clay at the Lloyd's site, London, showed strongly the main amino-acids composing collagen. In marked contrast the Galley Hill skeleton . . . gave a chromatogram which showed only a few of these amino acids and those in reduced strength." This means that collagen was not a major constituent of the Galley Hill bone.

This is an interesting observation. Even if the Galley Hill skeleton was depleted in collagen, surely the gelatin preservative would have contained this protein. Also, the fact that the Galley Hill skeleton had a nitrogen content of 1.6 to 2.0 percent suggests that it should have had a collagen profile. If the collagen was there, where did it go? And if the skeleton was noncollagenous, what nitrogenous compounds were responsible for its high nitrogen content?

Could it be that an unmentioned decontamination procedure was employed that removed the bone collagen as well as the contaminants? Was the bone nitrogen largely from proteins of organisms invading the bone from outside? Did it derive from amino acids absorbed from the environment, as Bada (1985a) said can happen? Or did the collagen belonging to both the bone and the gelatin preservative decay through the action of saprophytes during the 73 years from 1888 to 1961? We certainly do not know the answers to these questions. However, it is clear that bones with a noncollagenous amino acid profile are presently considered doubtful candidates for radiocarbon dating (Stafford *et al.* 1987, Stafford and Tyson 1989).

Altogether, the evidence surrounding the radiocarbon testing of the Galley Hill skeleton seems to cast severe doubt on the validity of the frequently cited carbon 14 date of 3,310 years. Is it possible a radiocarbon date could be so far off the mark? Scientists accept that errors of this magnitude have in fact occurred in radiocarbon dating. For example, the Los Angeles skull, dated with meticulous attention by Berger to over 23,600 years, was later dated by AMS to 3,560 years. One of these dates must be wrong. If Berger's date is wrong, it is wrong by a factor of 7. A somewhat larger error in Berger's date for the Los Angeles skull would give a range of 3,560 years to over 40,000 years—the equivalent of infinity in conventional radiocarbon dating. Note here that radiocarbon dating is very sensitive to error at the high end of the date range.

A1.3.8 The Carbon 14 Dating of the Castenedolo Skeleton

Ragazzoni (1880) and Sergi (1884) reported that the bones of two adults (male and female) and two children were discovered in Pliocene clay at Castenedolo, Italy, in 1880. In 1969, the British Museum Research Laboratory did a carbon 14 test on the vertebrae and ribs of the female skeleton and got an age of 958±116 years. Concerning removal of contaminants, the report on the test published by the British Museum only said: "Samples were pretreated for removal of contaminants, with dilute hydrochloric acid and, where appropriate, with dilute alkali also. Bone and antler samples were demineralized in low vacuum with 0.75 N hydrochloric acid at ambient temperature, leaving only the protein fraction (collagen) which was washed and dried before combustion" (Barker, Burleigh, and Meeks 1971, pp. 157–188).

It is clear that the carbon dating of the Castenedolo bones was not carried out with proper attention to standard decontamination procedures known in 1969 and today. As in the case of Galley Hill, modern procedures to remove humic and fulvic acids, saprophytes, and rootlets were not employed. In 1986 we wrote to Richard Burleigh, one of the scientists who actually did the dating in 1969. After describing the various sources of carbon 14 contamination, we asked him what procedures he used to eliminate them.

Dr. Burleigh replied by letter (December 3, 1986): "In fact it is very difficult to give definite answers to your questions. All that you have said about the possible contamination of the organic fraction of these bones is undeniable and is a risk taken with many radiocarbon analyses, particularly where, as in this instance, there are arguments in favour of a high antiquity. I can only say that I do not *think* that the material dated was significantly contaminated in any of the ways you suggest (and I also happen to believe that the results were reliable and do represent the true age of these particular human remains). As I recall these bones had been buried in a matrix consisting of Tertiary calcareous marine clay which probably would have protected them from contamination *in situ* and should not of itself have affected the results of age measurement."

This does not, however, address the problem of contamination after excavation. Burleigh did not state that the actual bones he tested (vertebrae and ribs from the female skeleton) were still covered with the clay matrix. Sergi (1884, pp. 314–315) stated that when he saw the female skeleton in 1883 the ribs and vertebral column were then embedded in a mass of clay. But in order to have been identified, some of these bones must have been sticking out of the clay, or else the clay must have been very thin. When clay dries it often cracks. Thus it is likely that at least some parts of the skeletal remains were exposed to contamination during the 89 years they were in museums before being dated.

The young dates obtained on the Castenedolo ribs and vertebrae could thus have resulted from extensive contamination with modern carbon. Let us consider a model in which a Pliocene Castenedolo bone remains uncontaminated for 3 million years, due perhaps to protection by a clay matrix. Then, after the bone's excavation, the metabolism of invading organisms causes 88 percent of the bone's original organic carbon to be replaced with modern carbon. In this case, the radiocarbon age of the bone will come out to 1,000 years. Since the Castenedolo bone was disinterred 89 years before it was dated, organisms would have had to replace about 1 percent of its organic carbon per year to achieve this result. This is certainly not inconceivable.

Furthermore, Bada (1985a, p. 251) reported: "Only dense, compact bone pieces (e.g. tibia, femur) are generally processed [for dating], because porous bones (such as ribs and vertebrae) generally contain large quantities of extraneous material and thus are difficult to clean." The bones dated from the Castenedolo finds were the ribs and vertebrae, which would be highly vulnerable to contamination.

Critical to the question of contamination would be information about the amino acid profile of the bone protein. A noncollagenous profile, indicating amino acids had entered from the environment, would cast doubt on the recent date obtained by the British Museum. In 1989, we asked the British Museum if the amino acid composition of the Castenedolo bones had ever been determined, but we have not yet received a reply.

A further possibility, of course, is that an error in the radiocarbon dating of the Castenedolo bones might have been caused not by contamination but by some mistake in the laboratory procedure, which is highly complex.

We have already given examples of radiocarbon dates that are considered "way off." For instance, the original carbon 14 dates obtained by Berger for the Los Angeles and Laguna skulls (Sections A1.3.3–5) are now considered by dating authorities to be incorrect; but we have come across no explanation of exactly what Berger did wrong.

Based on the published reports, the procedure Berger used is far superior to that used in the Galley Hill and Castenedolo cases. From an objective point of view, there is, therefore, more reason to accept Berger's dates than the dates for Galley Hill and Castenedolo, yet the establishment has done just the opposite. One can always assert that something must have gone wrong with Berger's laboratory procedures, but why raise this objection in the case of Laguna and Los Angeles and not in the case of Galley Hill and Castenedolo? It seems that carbon 14 dates are accepted when they agree with preconceived notions and rejected otherwise.

Let us consider some implications of a date of only 1,000 years old for the Castenedolo bones. In medieval times it would be unlikely that a body would be buried without a simple coffin or at least some sheets. The sediment in which the bones were found (clay) is known to preserve organic materials such as wood (wooden ships older than 1,000 years old have been excavated), yet no signs of coffins, burial shrouds, or personal effects were discovered. Also, how did the female skeleton get underneath the red clay and yellow sand without disturbing these layers? From his observations at the Castenedolo site, Sergi (1884, p. 315) said that such disturbance would not have escaped the notice of even an ordinary person, what to speak of a trained geologist like Ragazzoni.

Significantly, the bones of the two children and man found at Castenedolo were dispersed horizontally over several square meters (Sergi 1884, p. 315), with the bones of the two children apparently mixed with one another (Ragazzoni 1880, p. 122). Furthermore, these bones did not represent a complete skeleton for either the man or the children. There was no evidence of major earth movements—and extremely unusual earth movements would have been required to widely disperse bones of buried corpses horizontally and mix bones from different individuals together.

This constitutes strong evidence that the Castenedolo bones are not the result of recent intrusive burial. We note that the radiocarbon method was not used to date the bones of the man or children, and the significance of the dispersed position of these skeletons in the strata was ignored by most scientists writing about them.

In short, there is conflicting evidence about the age of the Castenedolo bones—a carbon 14 date and a nitrogen content test (Section A1.1.1) in favor of a recent age, an ambiguous uranium content test (Section A1.2.4), and a fluorine content test (Section A1.2.3) and stratigraphic observations (Section 6.2.2) in favor of high antiquity. In almost all cases of anomalously old human bones, scientists choose to accept carbon 14 dates even when they radically contradict the stratigraphic evidence. But is it really fair that all weight should be given to the former and none to the latter? The stratigraphic evidence is unusually strong in favor of a Pliocene age for the Castenedolo bones, whereas we have observed that the carbon 14 dating is far from perfect.

A1.3.9 Radiocarbon Dates vs. Stratigraphy: Which Is More Reliable?

We have examined two important cases (Castenedolo and Galley Hill) in which radiocarbon dates conflict radically with stratigraphy. In both of these cases the radiocarbon dating was far from perfect, yet scientists accept the carbon 14 dates and reject the stratigraphic evidence. It seems strange that carbon 14 dating is given what amounts to absolute status.

First of all, it is widely admitted that bone is not a very suitable material for radiocarbon dating. Tamers and Pearson stated (1965, p. 1053): "Bone dates are generally regarded as considerably less reliable than determinations using charcoal, wood, or other materials with large carbon contents (except shell). Most laboratories are in agreement that, whenever possible, bone should not be used for radiocarbon dating." More recently, Stafford *et al.* (1987, p. 24) said: "Bone is not usually recommended for ^{14}C dating . . . because its ^{14}C ages are either discordant with associated charcoal dates or ages for different fractions of the bone are discordant with each other."

Even using the most up-to-date methods of accelerator mass spectrometry there are still problems. J. A. J. Gowlett, of the Oxford Research Laboratory for Archaeology, outlined a program for using AMS to redate many sites already dated by the decay-counting radiocarbon method. But he admitted: "In many cases dating the finds will not remove the ambiguities from the situation. Lists of radiocarbon dates on their own do little to tell us about quality of association. There can be no substitute for excavation, conducted to exacting standards, coupled with appropriate use of chronological methods" (Gowlett 1986, p. 56). In other words, site stratigraphy must remain an important component in assigning a date to a find. In fact, Gowlett held there is no substitute for it. This judgment is particularly relevant to the Castenedolo skeletons, which were excavated by a professional geologist.

Another problem with radiocarbon dates is that they are sometimes adjusted or go unreported if they lie outside the the accepted time frames of human origins and antiquity. For example, P. A. Colinvaux (1964, p. 314) in one of his studies rejected one third of his own radiocarbon dates in order to "erect a chronology which is compatible with the known late Pleistocene and Recent history."

Within the scientific community, there exist institutions that watch over the variegated developments in the field of paleoanthropology, seeking to maintain in reported results a consistency with an accepted chronology. Of the Oxford radiocarbon laboratory, Gowlett and Hedges (1986, preface) said: "On occasion, a laboratory as 'producer' will take a slightly different view of problems from the archaeologist as 'consumer'. Our approach in the laboratory has been to vet [evaluate] the output of dates, to try to pin down 'unacceptable' or 'rogue' dates as early as possible, and to redate material where questions remain."

This process of eliminating and adjusting "unacceptable" and "rogue" evidence has been going on in a systematic fashion for well over a century. Indeed, those responsible have been quite successful in their endeavors. They have, in fact, been able to eliminate or adjust almost all evidence that contradicts dominant views about the progress of human evolution—in most cases simply for the reason that the evidence contradicts these dominant views. Of course, other reasons can always be found, among them radiocarbon dates.

But when all of the difficulties surrounding radiocarbon dating are taken into account, we find that the radiocarbon technique is not deserving of any more confidence than the other dating techniques we have examined in this chapter, namely the nitrogen content, fluorine content, uranium content, and amino acid racemization tests. Potassium-argon and uranium series dating, the results of which are similarly suspect, are discussed elsewhere (Sections 11.6.5.1 and 5.4.4.1 respectively). Therefore, results obtained when these methods are applied to controversial discoveries need not take precedence over the stratigraphical evidence originally reported in such cases.

Appendix 2
Evidence for Advanced Culture
In Distant Ages

Up to this point, most of the evidence we have considered gives the impression that even if humans did exist in the distant past, they remained at a somewhat primitive level of cultural and technological achievement. One might well ask the following question. If humans had a long time to perfect their skills, then why do we not find ancient artifacts indicative of an advancing civilization?

Charles Lyell (1863, p. 379) expressed this doubt in his *Antiquity of Man:* "instead of the rudest pottery or flint tools . . . we should now be finding sculptured forms, surpassing in beauty the master-pieces of Phidias or Praxiteles; lines of buried railways or electric telegraphs, from which the best engineers of our day might gain invaluable hints; astronomical instruments and microscopes of more advanced construction than any known in Europe, and other indications of perfection in the arts and sciences." The following reports do not quite measure up to this standard, but some of them do give hints of unexpected accomplishments.

These reports are, however, somewhat disconcerting. Not only are some of the objects decidedly more advanced than stone tools, but many also occur in geological contexts far older than we have thus far considered.

The reports of this extraordinary evidence emanate, with some exceptions, from nonscientific sources. And often the artifacts themselves, not having been preserved in standard natural history museums, are impossible to locate.

We ourselves are not sure how much importance should be given to this highly anomalous evidence. But we include it for the sake of completeness and to encourage further study.

In this appendix, we have included only a sample of the published material available to us. And given the spotty reporting and infrequent preservation of these highly anomalous discoveries, it is likely that the entire body of reports now existing represents only a small fraction of the total number of such discoveries made over the past few centuries.

A2.1 ARTIFACTS FROM AIX EN PROVENCE, FRANCE (AGE UNCERTAIN)

In his book *Mineralogy,* Count Bournon recorded an intriguing discovery that had been made by French workmen in the latter part of the eighteenth century. In his description of the details about the discovery, Bournon wrote: "During the years 1786, 1787, and 1788, they were occupied near Aix en Provence, in France, in quarrying stone for the rebuilding, upon a vast scale, of the Palace of Justice. The stone was a limestone of deep grey, and of that kind which are tender when they come out of the quarry, but harden by exposure to the air. The strata were separated from one another by a bed of sand mixed with clay, more or less calcareous. The first which were wrought presented no appearance of any foreign bodies, but, after the workmen had removed the ten first beds, they were astonished, when taking away the eleventh, to find its inferior surface, at the depth of forty or fifty feet, covered with shells. The stone of this bed having been removed, as they were taking away a stratum of argillaceous sand, which separated the eleventh bed from the twelfth, they found stumps of columns and fragments of stone half wrought, and the stone was exactly similar to that of the quarry: they found moreover coins, handles of hammers, and other tools or fragments of tools in wood. But that which principally commanded their attention, was a board about one inch thick and seven or eight feet long; it was broken into many pieces, of which none were missing, and it was possible to join them again one to another, and to restore to the board or plate its original form, which was that of the boards of the same kind used by the masons and quarry men: it was worn in the same manner, rounded and waving upon the edges."

Count Bournon, continuing his description, stated: "The stones which were completely or partly wrought, had not at all changed in their nature, but the fragments of the board, and the instruments, and pieces of instruments of wood, had been changed into agate, which was very fine and agreeably colored. Here then, we have the traces of a work executed by the hand of man, placed at a depth of fifty feet, and covered with eleven beds of compact limestone: every thing tended to prove that this work had been executed upon the spot where the traces existed. The presence of man had then preceded the formation of this stone, and that very considerably since he was already arrived at such a degree of civilization that the arts were known to him, and that he wrought the stone and formed columns out of it."

These passages appeared in the *American Journal of Science* in 1820 (vol. 2, pp. 145–146); today, however, it is unlikely such a report would be found in the pages of a scientific journal. Scientists simply do not take such discoveries seriously.

A2.2 LETTERS IN MARBLE BLOCK, PHILADELPHIA (AGE UNCERTAIN)

In 1830, letterlike shapes were discovered within a solid block of marble from a quarry 12 miles northwest of Philadelphia. The marble block was taken from a depth of 60–70 feet. This was reported in the *American Journal of Science* (vol. 19, 1831, p. 361). The quarry workers removed layers of gneiss, mica slate, hornblende, talcose slate, and primitive clay slate before coming to the layer from which the block containing the letterlike shapes was cut.

While they were sawing through the block, the workmen happened to notice a rectangular indentation, about 1.5 inches wide by .625 inches high, displaying two raised characters (Figure A2.1). Several respectable gentlemen from nearby Norris-town, Pennsylvania, were called to the scene and inspected the ob-ject. It is hard to explain the formation of the characters as products of natural physical processes. This suggests the characters were made by intelligent humans from the distant past.

Figure A2.1. Raised letterlike shapes found inside a block of marble from a quarry near Philadelphia, Pennsylvania (Corliss 1978, p. 657; *American Journal of Science* 1831, vol. 19, p. 361). The block of marble came from a depth of 60–70 feet.

A2.3 NAIL IN DEVONIAN SANDSTONE, NORTH BRITAIN

In 1844, Sir David Brewster reported that a nail had been discovered firmly embedded in a block of sandstone from the Kingoodie (Mylnfield) Quarry in North Britain. Dr. A. W. Medd of the British Geological Survey wrote to us in 1985 that this sandstone is of "Lower Old Red Sandstone age" (Devonian, between 360 and 408 million years old). Brewster was a famous Scottish physicist. He was a founder of the British Association for the Advancement of Science and made important discoveries in the field of optics.

In his report to the British Association for the Advancement of Science, Brewster (1844) stated: "The stone in Kingoodie quarry consists of alternate layers of hard stone and a soft clayey substance called 'till'; the courses of stone vary from six inches to upwards of six feet in thickness. The particular block in which the nail was found, was nine inches thick, and in proceeding to clear the rough block for dressing, the point of the nail was found projecting about half an inch (quite eaten with rust) into the 'till,' the rest of the nail lying along the surface of the stone to within an inch of the head, which went right down into the body of the stone." The fact that the head of the nail was buried in the sandstone block would seem to rule out the possibility the nail had been pounded into the block after it was quarried.

A2.4 GOLD THREAD IN CARBONIFEROUS STONE, ENGLAND

On June 22, 1844, this curious report appeared in the London *Times* (p. 8): "A few days ago, as some workmen were employed in quarrying a rock close to the Tweed about a quarter of a mile below Rutherford-mill, a gold thread was discovered embedded in the stone at a depth of eight feet." Dr. A. W. Medd of the British Geological Survey wrote to us in 1985 that this stone is of Early Carboniferous age (between 320 and 360 million years old).

A2.5 METALLIC VASE FROM PRECAMBRIAN ROCK AT DORCHESTER, MASSACHUSETTS

The following report, titled "A Relic of a Bygone Age," appeared in the magazine *Scientific American* (June 5, 1852): "A few days ago a powerful blast was made in the rock at Meeting House Hill, in Dorchester, a few rods south of Rev. Mr. Hall's meeting house. The blast threw out an immense mass of rock, some of the pieces weighing several tons, and scattered fragments in all directions. Among them was picked up a metallic vessel in two parts, rent asunder by the explosion. On putting the two parts together it formed a bell-shaped vessel, 4-1/2 inches high, 6-1/2 inches at the base, 2-1/2 inches at the top, and about an eighth of an inch in thickness. The body of this vessel resembles zinc in color, or a composition metal, in which there is a considerable portion of silver. On the side there are six figures or a flower, or bouquet, beautifully inlaid with pure silver, and around the lower part of the vessel a vine, or wreath, also inlaid with silver. The chasing, carving, and inlaying are exquisitely done by the art of some cunning workman. This curious and unknown vessel was blown out of the solid pudding stone, fifteen feet below the surface. It is now in the possession of Mr. John Kettell. Dr. J. V. C. Smith, who has recently travelled in the East, and examined hundreds of curious domestic utensils, and has drawings of them, has never seen anything resembling this. He has taken a drawing and accurate dimensions of it, to be submitted to the scientific. There is not doubt but that this curiosity was blown out of the rock, as above stated; but will Professor Agassiz, or some other scientific man please to tell us how it came there? The matter is worthy of investigation, as there is no deception in the case."

The editors of *Scientific American* ironically remarked: "The above is from the Boston *Transcript* and the wonder is to us, how the *Transcript* can suppose Prof. Agassiz qualified to tell how it got there any more than John Doyle, the blacksmith. This is not a question of zoology, botany, or geology, but one relating to an antique metal vessel perhaps made by Tubal Cain, the first inhabitant of Dorchester."

According to a recent U.S. Geological Survey map of the Boston-Dorchester area, the pudding stone, now called the Roxbury conglomerate, is of Precambrian age, over 600 million years old. By standard accounts, life was just beginning to form on this planet during the Precambrian. But in the Dorchester vessel we have evidence indicating the presence of artistic metal workers in North America over 600 million years before Leif Erikson.

A2.6 A TERTIARY CHALK BALL FROM LAON, FRANCE

The April 1862 edition of *The Geologist* included an English translation of an intriguing report by Maximilien Melleville, the vice president of the Société Académique of Laon, France. In his report, Melleville described a round chalk ball (Figure A2.2) discovered 75 meters (about 246 feet) below the surface in early Tertiary lignite beds near Laon.

Lignite (sometimes called ash) is a soft brown coal. The lignite beds at Montaigu, near Laon, lie at the base of a hill and were mined by horizontal shafts. The main shaft ran 600 meters (about 1,969 feet) into a bed of lignite 2.3 meters (about 7.5 feet) thick, above which lay sandy clay with identifiable fossil shells.

In August of 1861, workmen digging at the far end of the shaft, 225 feet below the surface of the hill, saw a round object fall down from the top of the excavation. The object was about 6 centimeters (2.36 inches) in diameter and weighed 310 grams (about 11 ounces).

Figure A2.2. This chalk ball (Melleville 1862b, plate 4) was discovered in an Early Eocene lignite bed near Laon, France. On the basis of its stratigraphic position, it can be assigned a date of 45–55 million years ago.

Melleville (1862a, p. 146) stated: "They looked to see exactly what place in the strata it had occupied, and they are able to state that it did not come from the interior of the 'ash,' but that it was imbedded at its point of contact with the roof of the quarry, where it had left its impression indented." The workmen carried the chalk ball to a Dr. Lejeune, who informed Melleville.

Melleville (1862a, p. 147) then stated: "long before this discovery, the workmen of the quarry had told me they had many times found pieces of wood changed into stone . . . bearing the marks of human work. I regret greatly now not having asked to see these, but I did not hitherto believe in the possibility of such a fact."

According to Melleville (1862a, p. 147), there was no possibility that the chalk ball was a forgery: "It really is penetrated over four-fifths of its height by a black bituminous colour that merges toward the top into a yellow circle, and which is evidently due to the contact of the lignite in which it had been for so long a time plunged. The upper part, which was in contact with the shell bed, on the contrary has preserved its natural colour—the dull white of the chalk. . . . As to the rock in which it was found, I can affirm that it is perfectly virgin, and presents no trace whatever of any ancient exploitation. The roof of the quarry was equally intact in this place, and one could see there neither fissure nor any other cavity by which we might suppose this ball could have dropped down from above."

Regarding human manufacture of the chalk object, Melleville was cautious. He wrote: "from one fact, even so well established, I do not pretend to draw the extreme conclusion that man was contemporary with the lignites of the Paris basin. . . . My sole object in writing this notice is to make known a discovery as curious as strange, whatever may be its bearing, without pretending to any mode of explanation. I content myself with giving it to science, and I shall wait before forming an opinion in this respect, for further discoveries to furnish me with the means of appreciating the value of this at Montaigu" (Melleville 1862a, p. 148).

Geology's editors wrote: "we consider his resolution wise in hesitating to date back the age of man to the lower Tertiary period of the Paris basin without further confirmatory evidence" (Melleville 1862a, p. 148). Gabriel de Mortillet (1883, p. 76) suggested that a piece of white chalk was rolled in the waves of the incoming Tertiary seas and after it became round was left where it was found.

This does not, however, seem to be a likely explanation. First of all, the ball had features inconsistent with the action of waves. Melleville (1862a, p. 147) reported: "Three great splinters with sharp angles, announce also that it had remained during the working attached to the block of stone out of which it was made, and that it had been separated only after it was finished, by a blow, to which this kind of fracture is due." If wave action is accepted as the explanation of the general roundness of the object, this action should also have smoothed the sharp edges described by Melleville. Furthermore, it is likely that sustained exposure to waves would have disintegrated a piece of chalk.

De Mortillet (1883, p. 28) stated that the ball was found in an Early Eocene stratum. If humans made the ball, they must have been in France 45–55 million years ago. As extraordinary as this might seem to those attached to the standard evolutionary views, it is in keeping with the evidence considered in this book.

A2.7 OBJECTS FROM ILLINOIS WELL BORINGS (MIDDLE PLEISTOCENE)

In 1871, William E. Dubois of the Smithsonian Institution reported on several man-made objects found at deep levels in Illinois. The first object was a copper quasi coin (Figure A2.3) from Lawn Ridge, in Marshall County, Illinois. In a letter to the Smithsonian Institution, J. W. Moffit stated that in August 1870 he was drilling a well using a "common ground auger" (W. Dubois 1871, p. 224). When Moffit brought the auger up from a depth of 125 feet, he discovered the coinlike object "on the auger."

To get down to 125 feet, Moffit drilled through the following strata: 3 feet of soil; 10 feet of yellow clay; 44 feet of blue clay; 4 feet of clay, sand, and gravel; 19 feet of purple clay; 10 feet of brown hard pan; 8.5 feet of green clay; 2 feet of vegetable mould; 2.5 feet of yellow clay; 2 feet of yellow hard pan; and 20.5 feet of mixed clay.

A. Winchell (1881, p. 170) also described the coinlike object. Winchell quoted a letter by W. H. Wilmot, who listed a sequence of strata slightly different from that given by Moffit. Wilmot reported that the quasi coin had been discovered in the well boring at a depth of 114 feet rather than 125 feet.

Using the sequence of strata given by Winchell (1881, p. 170), the Illinois State Geological Survey (personal communication, September 1984) gave an estimate for the age of the deposits at the 114-foot level. They would have formed during the Yarmouthian Interglacial "sometime

Figure A2.3. This coinlike object, from a well boring near Lawn Ridge, Illinois, was reportedly found at a depth of about 114 feet below the surface (Winchell 1881, p. 170). According to information supplied by the Illinois State Geological Survey, the deposits containing the coin are between 200,000 and 400,000 years old.

between 200,000 and 400,000 years ago."

W. E. Dubois (1871, p. 225) said that the shape of the quasi coin was "polygonal approaching to circular," and that it had crudely portrayed figures and inscriptions on both sides. The inscriptions were in a language that Dubois could not recognize, and the quasi coin's appearance differed from any known coin.

Dubois concluded that the coin must have been made in a machine shop. Noting its uniform thickness, he said the coin must have "passed through a rolling-mill; and if the ancient Indians had such a contrivance, it must have been pre-historic" (W. Dubois 1871, p. 225). Furthermore, Dubois reported that the coin must have been cut with shears or a chisel and the sharp edges filed down.

The quasi coin described above suggests the existence of a civilization at least 200,000 years ago in North America. Yet beings intelligent enough to make and use coins (*Homo sapiens sapiens*) are generally not thought to have lived much earlier than 100,000 years ago. According to standard views, metal coins were first used in Asia Minor during the eighth century B.C.

Moffit also reported that other artifacts were found in nearby Whiteside County, Illinois (W. Dubois 1871, p. 224). At a depth of 120 feet, workmen discovered "a large copper ring or ferrule, similar to those used on ship spars at the present time. . . . They also found something fashioned like a boat-hook." Mr. Moffit added: "There are numerous instances of relics found at lesser depths. A spear-shaped hatchet, made of iron, was found imbedded in clay at 40 feet; and stone pipes and pottery have been unearthed at depths varying from 10 to 50 feet in many localities." In September 1984, the Illinois State Geological Survey wrote to us that the age of deposits at 120 feet in Whiteside County varies greatly. In some places, one would find at 120 feet deposits only 50,000 years old, while in other places one would find Silurian bedrock 410 million years old.

A2.8 A CLAY IMAGE FROM NAMPA, IDAHO (PLIO-PLEISTOCENE)

A small human image, skillfully formed in clay, was found in 1889 at Nampa, Idaho (Figure A2.4). The figurine came from the 300-foot level of a well boring. G. F. Wright (1912, pp. 266–267) wrote: "The record of the well shows that in reaching the stratum from which the image was brought up they had penetrated first about fifty feet of soil, then about fifteen feet of basalt, and afterwards passed through alternate beds of clay and quicksand . . . down to a depth of about three hundred feet, when the sand pump began to bring up numerous clay balls, some of them more than two inches in diameter, densely coated with iron oxide. In the lower portion of this stratum there were evidences of a buried land surface, over which there had been a slight accumulation of vegetable mould. It was from this point that the image in question was brought up at a depth of three hundred and twenty feet. A few feet farther down, sand rock was reached."

As for the figurine, Wright (1912, p. 267) noted: "The image in question is made of the same material as that of the clay balls mentioned, and is about an inch and a half long; and remarkable for the perfection with which it represents the human form. . . . It was a female figure, and had the lifelike lineaments in the parts which were finished that would do credit to the classic centers of art."

"Upon showing the object to Professor F. W. Putnam," wrote Wright (1912, pp. 267-269), "he at once directed attention to the character of the incrustations of iron upon the surface as indicative of a relic of considerable antiquity. There were patches of anhydrous red oxide of iron in protected places upon it, such as could not have been formed upon any fraudulent object. In visiting the locality in 1890 I took special pains, while on the ground, to compare the discoloration of the oxide upon the image with that upon the clay balls still found among the debris which has come from the well, and ascertained it to be as nearly identical as it is possible to be. These confirmatory evidences, in connection with the very satisfactory character of the evidence furnished by the parties who made the discovery, and confirmed by Mr. G. M. Cumming, of Boston (at that time superintendent of that division of the Oregon Short Line Railroad, and who knew all the parties, and was upon the ground a day or two after the discovery) placed the genuineness of the discovery beyond reasonable doubt. To this evidence is to be added, also, the general conformity of the object to other relics of man which have been found beneath the lava deposits on the Pacific coast. In comparing the figurine one cannot help being struck with its resemblance to numerous 'Aurignacian figurines' found in prehistoric caverns in France, Belgium, and Moravia. Especially is the resemblance striking to that of 'The Venus impudica' from Laugerie-Basse." The Nampa image is also similar to the famous Willendorf Venus, thought to be about 30,000 years old (Figure A2.5).

Figure A2.4. Figurine (Wright 1912, p. 268) from a well at Nampa, Idaho. This object is of Plio-Pleistocene age, about 2 million years old.

Figure A2.5. The Willendorf Venus, from Europe, dated at 30,000 years old (MacCurdy 1924a, p. 260).

Wright also examined the bore hole to see if the figurine could have slipped down from a higher level. He stated: "To answer objections it will be well to give the facts more fully. The well was six inches in diameter and was tubed with heavy iron tubing, which was driven down, from the top, and screwed together, section by section, as progress was made. Thus it was impossible for anything to work in from the sides. The drill was not used after penetrating the lava deposit near the surface, but the tube was driven down, and the included material brought out from time to time by use of a sand pump" (Wright 1912, p. 270).

Responding to our inquiries, the United States Geological Survey stated in a letter (February 25, 1985) that the the clay layer at a depth of over 300 feet is "probably of the Glenns Ferry Formation, upper Idaho Group, which is generally considered to be of Plio-Pleistocene age." The basalt above the Glenns Ferry formation is considered Middle Pleistocene.

Other than *Homo sapiens sapiens,* no hominid is known to have fashioned works of art like the Nampa figurine. The evidence therefore suggests that humans of the modern type were living in America around 2 million years ago, at the Plio-Pleistocene boundary.

That the Nampa figurine strongly challenges the evolutionary scenario was noted by W. H. Holmes of the Smithsonian Institution. Holmes (1919, p. 70) wrote in his *Handbook of Aboriginal American Antiquities:* "According to Emmons, the formation in which the pump was operating is of late Tertiary or early Quaternary age; and the apparent improbability of the occurrence of a well-modeled human figure in deposits of such great antiquity has led to grave doubt about its authenticity. It is interesting to note that the age of this object, supposing it to be authentic, corresponds with that of the incipient man whose bones were, in 1892, recovered by Dubois from the late Tertiary or early Quaternary formations of Java."

Here we find the Java man discovery, itself questionable, once more being used to dismiss evidence for humans of modern abilities in the early Quaternary or Tertiary (see Section 5.5.13 for another example). The evolutionary hypothesis was apparently so privileged that any evidence contradicting it could be almost automatically rejected. But although Holmes doubted that beings capable of making the Nampa image could have existed at the same time as the primitive Java ape-man, we find today that humans, of various levels of technological expertise, coexist in Africa with gorillas and chimpanzees.

Holmes (1919, p. 70) went on to say: "Like the auriferous gravel finds of California, if taken at its face value the specimen establishes an antiquity for Neolithic culture in America so great that we hesitate to accept it without further confirmation. While it may have been brought up as reported, there remains the possibility that it was not an original inclusion under the lava. It is not impossible that an object of this character could have descended from the surface through some crevice or water course penetrating the lava beds and have been carried through deposits of creeping quicksand aided by underground waters to the spot tapped by the drill." It is instructive to note how far a scientist like Holmes will go to explain away evidence he does not favor. One should keep in mind, however, that any evidence, including evidence currently used to buttress the theory of evolution, could be explained away in this fashion.

A barrier to the supposition that the Nampa image was recently manufactured by Indians and somehow worked its way down from the surface may be found in this statement by Holmes (1919, p. 70): "It should be remarked, however, that forms of art closely analogous to this figure are far to seek, neither the Pacific slope on the west nor the Pueblo region on the south furnishing modeled images of the human figure of like character or of equal artistic merit."

A2.9 GOLD CHAIN IN CARBONIFEROUS COAL FROM MORRISONVILLE, ILLINOIS

On June 11, 1891, *The Morrisonville Times* reported: "A curious find was brought to light by Mrs. S. W. Culp last Tuesday morning. As she was breaking a lump of coal preparatory to putting it in the scuttle, she discovered, as the lump fell apart, embedded in a circular shape a small gold chain about ten inches in length of antique and quaint workmanship. At first Mrs. Culp thought the chain had been dropped accidentally in the coal, but as she undertook to lift the chain up, the idea of its having been recently dropped was at once made fallacious, for as the lump of coal broke it separated almost in the middle, and the circular position of the chain placed the two ends near to each other, and as the lump separated, the middle of the chain became loosened while each end remained fastened to the coal. This is a study for the students of archaeology who love to puzzle their brains over the geological construction of the earth from whose ancient depth the curious is always dropping out. The lump of coal from which this chain was taken is supposed to come from the Taylorville or Pana mines [southern Illinois] and almost hushes one's breath with mystery when it is thought for how many long ages the earth has been forming strata after strata which hid the golden links from view. The chain was an eight-carat gold and weighed eight penny-weights."

In a letter to Ron Calais, Mrs. Vernon W. Lauer, the present publisher of the *Morrisonville Times*, stated: "Mr. Culp was editor and publisher of the Times in 1891. Mrs. Culp, who made the discovery, moved to Taylorville after his death— remarried and her death occurred on February 3, 1959." Calais told our research assistant (Stephen Bernath) that he had information the chain was given to one of Mrs. Culp's relatives after her death, but Calais could not trace the chain further.

The Illinois State Geological Survey has said the coal in which the gold chain was found is 260–320 million years old. This raises the possibility that culturally advanced human beings were present in North America during that time.

A2.10 CARVED STONE FROM LEHIGH COAL MINE NEAR WEBSTER, IOWA (CARBONIFEROUS)

The April 2, 1897 edition of the *Daily News* of Omaha, Nebraska, carried an article titled "Carved Stone Buried in a Mine," which described an object from a mine near Webster City, Iowa. The article stated: "While mining coal today in the Lehigh coal mine, at a depth of 130 feet, one of the miners came upon a piece of rock which puzzles him and he was unable to account for its presence at the bottom of the coal mine. The stone is of a dark grey color and about two feet long, one foot wide and four inches in thickness. Over the surface of the stone, which is very hard, lines are drawn at angles forming perfect diamonds. The center of each diamond is a fairly good face of an old man having a peculiar indentation in the forehead that appears in each of the pictures, all of them being remarkably alike. Of the faces, all but two are looking to the right. How the stone reached its position under the strata of sandstone at a depth of 130 feet is a question the miners are not attempting to answer. Where the stone was found the miners are sure the earth had never before been disturbed." Inquiries to the Iowa State Historical Preservation and Office of State Archaeology at the University of Iowa revealed nothing new. The Lehigh coal is probably from the Carboniferous.

A2.11 IRON CUP FROM OKLAHOMA COAL MINE (CARBONIFEROUS)

On January 10, 1949, Robert Nordling sent a photograph of an iron cup to Frank L. Marsh of Andrews University, in Berrien Springs, Michigan. Nordling wrote: "I visited a friend's museum in southern Missouri. Among his curios, he had the iron cup pictured on the enclosed snapshot" (Rusch 1971, p. 201).

At the private museum, the iron cup had been displayed along with the following affidavit, made by Frank J. Kenwood in Sulphur Springs, Arkansas, on November 27, 1948: "While I was working in the Municipal Electric Plant in Thomas, Okla. in 1912, I came upon a solid chunk of coal which was too large to use. I broke it with a sledge hammer. This iron pot fell from the center, leaving the impression or mould of the pot in the piece of coal. Jim Stall (an employee of the company) witnessed the breaking of the coal, and saw the pot fall out. I traced the source of the coal, and found that it came from the Wilburton, Oklahoma, Mines" (Rusch 1971, p. 201). According to Robert O. Fay of the Oklahoma Geological Survey, the Wilburton mine coal is about 312 million years old. In 1966, Marsh sent the photo of the cup and the correspondence relating to it to Wilbert H. Rusch, a professor of biology at Concordia College, in Ann Arbor, Michigan. Marsh stated: "Enclosed is the letter and snap sent me by Robert Nordling some 17 years ago. When I got interested enough in this 'pot' (the size of which can be gotten at somewhat by comparing it with the seat of the straight chair it is resting on) a year or two later I learned that this 'friend' of Nordling's had died and his little museum was scattered. Nordling knew nothing of the whereabouts of the iron cup. It would challenge the most alert sleuth to see if he could run it down. . . . If this cup is what it is sworn to be, it is truly a most significant artifact" (Rusch 1971, p. 201). It is an unfortunate fact that evidence such as this iron cup tends to get lost as it passes from hand to hand among people not fully aware of its significance.

A2.12 A SHOE SOLE FROM NEVADA (TRIASSIC)

On October 8, 1922, the *American Weekly* section of the *New York Sunday American* ran a prominent feature titled "Mystery of the Petrified 'Shoe Sole' 5,000,000 Years Old," by Dr. W. H. Ballou. Ballou (1922, p. 2) wrote: "Some time ago, while he was prospecting for fossils in Nevada, John T. Reid, a distinguished mining engineer and geologist, stopped suddenly and looked down in utter bewilderment and amazement at a rock near his feet. For there, a part of the rock itself, was what seemed to be a human footprint! [Figure A2.6] Closer inspection showed that it was not a mark of a naked foot, but was, apparently, a shoe sole which had been turned into stone. The forepart was missing. But there was the out-line of at least two-thirds of it, and around this outline ran a well-defined sewn thread which had, it appeared, attached the welt to the sole. Further on was another line of sewing, and in the center, where the foot would have rested had the object been really a shoe sole, there was an indentation, exactly such as would have been made by the bone of the heel rubbing upon and wearing down the material of which the sole had been made. Thus was found a fossil which is

Figure A2.6. Partial shoe sole in Triassic rock from Nevada (Ballou 1922). The Triassic is dated at 213–248 million years ago.

the foremost mystery of science today. For the rock in which it was found is at least 5,000,000 years old."

Reid brought the specimen to New York, where he tried to bring it to the attention of other scientists. Reid reported: "On arrival at New York, I showed this fossil to Dr. James F. Kemp, geologist of Columbia University; Professors H. F. Osborn, W. D. Matthew and E. O. Hovey, of the American Museum of Natural History. All of these men reached the same conclusion, in effect that 'it was the most remarkable natural imitation of an artificial object they had ever seen.' These experts agreed, however, that the rock formation was Triassic, and manufacturers of shoes agreed that originally the specimen was a hand-welted sole. Dr. W. D. Matthew wrote a brief report on the find declaring that while all the semblances of a shoe were present, including the threads with which it had been sewn, it was only a remarkable imitation, a *lusus naturae*, or 'freak of nature'" (Ballou 1922, p. 2). Curiously enough, an inquiry to the American Museum of Natural History resulted in a reply that the report by Matthew is not in their files.

Reid, despite Matthew's dismissal, nevertheless persisted: "I next got hold of a microphotographer and an analytical chemist of the Rockefeller Institute, who, on the outside, so as not to make it an institute matter, made photos and analyses of the specimen. The analyses proved up [removed] any doubt of the shoe sole having been subjected to Triassic fossilization. . . . The microphoto magnifications are twenty times larger than the specimen itself, showing the minutest detail of thread twist and warp, proving conclusively that the shoe sole is not a resemblance, but is strictly the handiwork of man. Even to the naked eye the threads can be seen distinctly, and the definitely symmetrical outlines of the shoe sole. Inside this rim and running parallel to it is a line which appears to be regularly perforated as if for stitches. I may add that at least two geologists whose names will develop some day have admitted that the shoe sole is valid, a genuine fossilization in Triassic rocks" (Ballou 1922, p. 2). The Triassic rock bearing the fossil shoe sole is now recognized as being far more than 5 million years old. The Triassic period is now generally dated at 213–248 million years ago.

A2.13 BLOCK WALL IN AN OKLAHOMA MINE (CARBONIFEROUS?)

W. W. McCormick of Abilene, Texas, reported his grandfather's account of a stone block wall that was found deep within a coal mine: "In the year 1928, I, Atlas Almon Mathis, was working in coal mine No. 5., located two miles north of Heavener, Oklahoma. This was a shaft mine, and they told us it was two miles deep. The mine was so deep that they let us down into it on an elevator. . . . They pumped air down to us, it was so deep" (Steiger 1979, p. 27).

One evening, Mathis was blasting coal loose by explosives in "room 24" of this mine. "The next morning," said Mathis (Steiger 1979, p. 27), "there were several concrete blocks laying in the room. These blocks were 12-inch cubes and were so smooth and polished on the outside that all six sides could serve as mirrors. Yet they were full of gravel, because I chipped one of them open with my pick, and it was plain concrete inside." Mathis added: "As I started to timber the room up, it caved in; and I barely escaped. When I came back after the cave-in, a solid wall of these polished blocks was left exposed. About 100 to 150 yards farther down our air core, another miner struck this same wall, or one very similar" (Steiger 1979, p. 27). The coal in the mine was probably Carboniferous, which would mean the wall was at least 286 million years old.

According to Mathis, the mining company officers immediately pulled the men out of the mine and forbade them to speak about what they had seen. This mine was closed in the fall of 1928, and the crew went to mine number 24, near Wilburton, Oklahoma.

Mathis said the Wilburton miners told of finding "a solid block of silver in the shape of a barrel . . . with the prints of the staves on it" (Steiger 1979, p. 28). The coal from Wilburton was formed between 280 and 320 million years ago.

Admittedly, these are very bizarre stories, accompanied by very little in the way of proof. But such stories are told, and we wonder how many of them there are and if any of them are true.

We recently ran across the following wall-in-coal-mine story: "It is . . . reported that James Parsons, and his two sons, exhumed a slate wall in a coal mine at Hammondville, Ohio, in 1868. It was a large, smooth wall, disclosed when a great mass of coal fell away from it, and on its surface, carved in bold relief, were several lines of hieroglyphics" (Jessup 1973, p. 65). Of course, such stories could be tall tales, but they might also be leads for interesting research.

A2.14 LATE TWENTIETH-CENTURY DISCOVERIES

The foregoing sampling of discoveries indicating a relatively high level of civilization in very distant ages was compiled from reports published in the nineteenth and early twentieth centuries, but similar reports continue up to the present day. We shall now review some of them.

A2.14.1 Metallic Tubes from Chalk in France (Cretaceous)

Y. Druet and H. Salfati announced in 1968 the discovery of semi-ovoid metallic tubes of identical shape but varying size in Cretaceous chalk (Corliss 1978, pp. 652–653). The chalk bed, exposed in a quarry at Saint-Jean de Livet, France, is estimated to be least 65 million years old. Having considered and

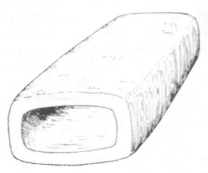

Figure A2.7. Metallic tube found at Saint-Jean de Livet, France, in a 65-million-year-old chalk bed (Corliss 1978, p. 652).

eliminated several hypotheses, Druet and Salfati concluded that intelligent beings had lived 65 million years ago.

Desiring more information, we wrote to the geomorphology laboratory at the University of Caen, to which Druet and Salfati reportedly turned over their specimens (Figure A2.7), but we have not received a reply. We invite readers to communicate to us any information they might have about this case or similar cases, for inclusion in future editions of this book.

A2.14.2 Shoe Print in Shale from Utah (Cambrian)

In 1968, William J. Meister, a draftsman and amateur trilobite collector, reported finding a shoe print in the Wheeler Shale near Antelope Spring, Utah. This shoelike indentation (Figure A2.8) and its cast were revealed when Meister split open a block of shale. Clearly visible within the imprint were the remains of trilobites, extinct marine arthropods. The shale holding the print and the trilobite fossils is from the Cambrian, and would thus be 505 to 590 million years old.

Meister (1968, p. 98) described the ancient shoelike impression in an article that appeared in the *Creation Research Society Quarterly:* "The heel print was indented in the rock about an eighth of an inch more than the sole. The footprint was clearly that of the right foot because the sandal was well worn on the right side of the heel in characteristic fashion."

Meister (1968, p. 99) supplied the following important piece of additional information: "On July 4, I accompanied Dr. Clarence Coombs, Columbia Union College, Tacoma, Maryland, and Maurice Carlisle, graduate geologist, University of Colorado at Boulder, to the site of the discovery. After a couple of hours of digging, Mr. Carlisle found a mudslab, which he said convinced him that the discovery of fossil tracks in the location was a distinct possibility, since this discovery showed that the formation had at one time been at the surface."

Scientists who were made aware of the Meister discovery were sometimes contemptuous in their dismissals. This is evident from private correspondence supplied to us by George F. Howe of Los Angeles Baptist College, who requested that we quote from it anonymously. A geologist from Brigham Young University, quite familiar with the Antelope Springs region, wrote in 1981 that the track represented "an oddity of weathering which uninformed people mistakenly interpret for fossil forms."

A professor of evolutionary biology from a Michigan university stated, when asked about the Meister print: "I am not familiar with the trilobite case . . . but I would be greatly surprised if this isn't another case of fabrication or willful misrepresentation. There is not one case where a juxtaposition of this type has ever been confirmed. So far the fossil record is one of the best tests that evolution has occurred. I put the creationists and those that believe in a flat earth in the same category. They simply do not want to believe in facts and hard evidence. There is not much you can do with such people. . . . Nothing has emerged in recent years to refute the fact that evolution has, and continues to occur, irrespective of what the self-proclaimed

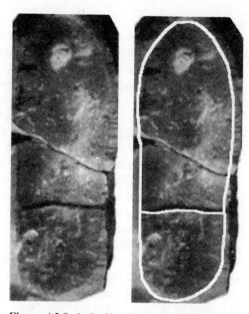

Figure A2.8. Left: Shoelike print discovered by William Meister in Cambrian shale near Antelope Spring, Utah (our photograph). If genuine, the shoe print would be over 505 million years old. Right: Outlined in white, the Meister print does not deviate from the shape of a modern shoe.

'scientific' creationists claim. The ability of individuals in our society to be duped and brainwashed, either intentionally or unknowingly, by our mass media and certain leaders never ceases to amaze me."

The evolutionary biologist admitted he had not familiarized himself with the "facts and hard evidence" relating to the Meister sandal print before passing judgment. He was thus guilty of the same sin he accused the creationists of committing. We do not necessarily accept the Meister print as genuine, but we believe it should be evaluated on its own merits, rather than on the basis of inflexible preconceptions.

William Lee Stokes, a biologist and geologist at the University of Utah, examined the Meister print shortly after it was discovered. Stokes (1974, p. 139) stated: "After seeing the specimen I explained to Mr. Meister why I could not accept it as a footprint and why geologists in general would not accept it. At the very least, we would expect a true footprint to be one of a sequence showing right and left prints somewhat evenly spaced, of the same size and progressing regularly in one direction. . . . It is most significant that no other matching prints were obtained. I know of no instance where a solitary one-of-a-kind impression has been accepted and reported in a scientific journal as a genuine footprint no matter how well-preserved it might be." But in an article that appeared in *Scientific American,* a French scientist (de Lumley 1969) reported a single humanlike footprint from the Middle Pleistocene habitation site at Terra Amata in southern France (Section 6.1.4).

Stokes (1974, p. 139) further stated: "A true footprint should also show displacement or squeezing aside of the soft material into which the foot was pressed. . . . From my examination of this specimen I can say that there is no evidence of squeezing or pushing aside of the matrix."

In 1984, one of us (Thompson) visited Meister in Utah. Close inspection of the print revealed no obvious reason why it could not be accepted as genuine. Concerning squeezing aside of the matrix, much depends on the consistency of the matrix and the nature of the object making the imprint. The rounded contours of a bare foot result in more pushing aside of the matrix than the sharp edges of the soles of footwear. We have observed that shoes and sandals can leave very sharp impressions in relatively compact, moist beach sand, with very little sign of pushing aside of the matrix. Shale, the rock in which the Meister print was made, is formed by the consolidation of clay, mud, or silt. One could microscopically examine the grain structure of the shale within the region of the print in order to determine whether or not there is any evidence suggesting that the print was not caused by pressure from above.

Stokes (1974, pp. 139–140) concluded that the Meister specimen was the result of spalling, a natural fracturing of the rock, and stated that the geology department of the University of Utah had in its collection several products of spalling, some resembling footprints. One would have to see these specimens to judge if they really resemble footprints to the extent the Meister specimen does. The shape of the Meister print, as shown by our visual inspection and computer analysis, almost exactly matches that of a modern shoe print.

Furthermore, spalling normally occurs on the surfaces of rocks. The Meister print, however, was found in the interior of a block of shale that was split. Significantly, the shale in the region of the print is of a rougher texture than the shale on the other parts of the split block's surface. This suggests that the rock split where it did not accidentally but because of a line of weakness along the boundary of the two textures. One could, therefore, propose that an ancient shoe caused this shoe-shaped area of weakness. Alternatively, the area of weakness might have resulted from some other unknown cause, in which case the shoelike shape is entirely coincidental. This would be a rather remarkable freak of nature, for the print does not even slightly depart from the shape of a genuine shoe.

The Meister print, as evidence for a human presence in the distant past, is ambiguous. Some scientists have dismissed the print after only cursory examination.

Others have rejected it sight unseen, simply because its Cambrian age puts it outside the realm of what might be expected according to evolutionary theory. We suggest, however, that the resources of empirical investigation have not yet been exhausted and that the Meister print is worthy of further research.

A2.14.3 Grooved Sphere from South Africa (Precambrian)

Over the past several decades, South African miners have found hundreds of metallic spheres, at least one of which has three parallel grooves running around its equator (Figure A2.9). The spheres are of two types—"one of solid bluish metal with white flecks, and another which is a hollow ball filled with a white spongy center" (Jimison 1982). Roelf Marx, curator of the museum of Klerksdorp, South Africa, where some of the spheres are housed, said: "The spheres are a complete mystery. They look man-made, yet at the time in Earth's history when they came to rest in this rock no intelligent life existed. They're nothing like I have ever seen before" (Jimison 1982).

We wrote to Roelf Marx for further information about the spheres. He replied in a letter dated September 12, 1984: "There is nothing scientific published about the globes, but the facts are: They are found in pyrophyllite, which is mined near the little town of Ottosdal in the Western Transvaal. This pyrophyllite $(Al_2Si_4O_{10}(OH)_2)$ is a quite soft secondary mineral with a count of only 3 on the Mohs' scale and was formed by sedimentation about 2.8 billion years ago. On the other hand the globes, which have a fibrous structure on the inside with a shell around it, are very hard and cannot be scratched, even by steel." The Mohs' scale of hardness is named after Friedrich Mohs, who chose ten minerals as references points for comparative hardness, with talc the softest (1) and diamond the hardest (10).

In his letter to us, Marx said that A. Bisschoff, a professor of geology at the University of Potchefstroom, told him that the spheres were "limonite concretions." Limonite is a kind of iron ore. A concretion is a compact,

Figure A2.9. A metallic sphere from South Africa with three parallel grooves around its equator (photo courtesy of Roelf Marx). The sphere was found in a Precambrian mineral deposit, said to be 2.8 billion years old.

rounded rock mass formed by localized cementation around a nucleus.

One problem with the hypothesis that the objects are limonite concretions concerns their hardness. As noted above, the metallic spheres cannot be scratched with a steel point, indicating they are extremely hard. But standard references on minerals state that limonite registers only 4 to 5.5 on the Mohs' scale, indicating a relatively low degree of hardness (Kourmisky 1977). Furthermore, limonite concretions usually occur in groups, like masses of soap bubbles stuck together. They do not, it seems, normally appear isolated and perfectly round, as is the case with the objects in question. Neither do they normally appear with parallel grooves encircling them (Figure A2.9).

For the purposes of this study, it is the sphere with three parallel grooves around its equator that most concerns us. Even if it is conceded that the sphere itself is a limonite concretion, one still must account for the three parallel grooves. In the absence of a satisfactory natural explanation, the evidence is somewhat mysterious, leaving open the possibility that the South African grooved sphere—found in a mineral deposit 2.8 billion years old—was made by an intelligent being.

Appendix 3
Summary of Anomalous Evidence Related to Human Antiquity

In Table A3.1, sites mentioned in this book are listed in order of the published minimum ages we find most likely or otherwise worthy of consideration. The following is a glossary of terms used in the table.

eoliths = naturally broken stone with one or more edges intentionally modified or worn by use.

paleoliths = stones purposely fashioned by chipping into a recognizable tool type.

neoliths = the most advanced stone tools and utensils.

human = identified by at least some workers as anatomically modern human.

incised, broken, carved, or scraped bones = purposely modified animal bones.

TABLE A3.1

Summary of Anomalous Evidence Related to Human Antiquity (General)

Period / Myr	Site	Category	Reference	Section
Precambrian				
2800	Ottosdalin, South Africa	grooved metallic sphere	Jimison 1982	A2.14.3
>600	Dorchester, Mass.	metal vase	*Scientific Amer.,* June 5, 1852	A2.5
Cambrian				
505–590	Antelope Spring, Utah	shoe print	Meister 1968	A2.14.2

815

TABLE A3.1—*Continued*

Period / Myr	Site	Category	Reference	Section
Devonian 360–408	Kingoodie Quarry, Scotland	iron nail in stone	Brewster 1844	A2.3
Carboniferous 320–360	Tweed, England	gold thread in stone	*Times* (London) June 22, 1844	A2.4
312	Wilburton, Oklahoma	iron pot	Rusch 1971	A2.11
286–360	Webster, Iowa	carved stone	*Daily News,* Omaha, Neb., April 2, 1897	A2.10
286–320	Macoupin, Illinois	human skeleton	*The Geologist,* December 1862	6.3.1
286–320	Rockcastle County in Kentucky, and other sites	humanlike footprints	Burroughs 1938	6.3.2
280–320	Wilburton, Oklahoma	silver object	Steiger 1979	A2.13
260–320	Morrisonville, Illinois	gold chain	*Morrisonville Times,* June 11, 1891	A2.9
260–320	Heavener, Oklahoma	block wall in coal	Steiger 1979	A2.13
Triassic 213–248	Nevada	shoe print	Ballou 1922	A2.12
Jurassic 150	Turkmenian Republic	human footprint	*Moscow News* 1983, no. 24	6.3.3
Cretaceous 65–144	Saint-Jean de Livet, France	metal tubes in chalk	Corliss 1987a	A2.14.1

TABLE A3.1—*Continued*

Period / Myr	Site	Category	Reference	Section
Eocene				
50–55	Picardy, France	eoliths	Breuil 1910	3.4.1
50–55	Clermont, France	eoliths, paleoliths	Breuil 1910	3.4.1
45–55	Laon, France	chalk ball, cut wood	Melleville 1862	A2.6
38–55	Barton Cliff, England	carved stone	Fisher 1912	2.16
38–55	Essex, England	eoliths, paleoliths	Warren 1920	3.3.7
38–45	Delémont, Switzerland	human skeleton	de Mortillet 1883	6.2.7
Oligocene				
33–55	Boston Tunnel, Tuolumne Table Mt., Calif.	neolith, carved stone	Whitney 1880	5.5.8
33–55	Montezuma Tunnel, Tuolumne Table Mt., Calif.	neoliths	Whitney 1880	5.5.9
33–55	Tuolumne Table Mt., Calif.	human skeleton	Winslow 1873	6.2.6.2
26–54	Baraque Michel, Belgium	paleoliths	Rutot 1907	4.4
26–54	Bay Bonnet, Belgium	paleoliths	Rutot 1907	4.4
26–30	Boncelles, Belgium	paleoliths	Rutot 1907	4.4

TABLE A3.1—*Continued*

Period / Myr	Site	Category	Reference	Section
Early Miocene >23	Spring Valley Mine, Oroville, Calif.	neoliths	Whitney 1880	5.5.12
>23	Sugar Loaf, Oroville, Calif.	neoliths	Whitney 1880	5.5.12
20–25	Thenay, France	paleoliths	Bourgeois 1867	4.2
Middle Miocene 12.5–14	Ft. Ternan, Kenya	broken bones, eolith	L. Leakey 1968	11.4.4
12–25	Santacrucian Formation, Argentina	paleoliths, signs of fire, cut bones, broken bones, burned bones	F. Ameghino 1912	5.1.5
12–19	Billy, France	incised bone	Laussedat 1863	2.6
12–19	Sansan, France	broken bones	Garrigou 1871	2.7
12–19	Pouancé, France	incised bone	Bourgeois 1867	2.12
12–19	Clermont, France	incised bone	Pomel and de Mortillet 1876	2.14
Late Miocene 9–55	Tuolumne Table Mt., Calif.	Snell collection, neoliths, advanced paleoliths, human jaw	Whitney 1880	5.5.4
				6.2.6.4

TABLE A3.1—*Continued*

Period / Myr	Site	Category	Reference	Section
9–55	Valentine Mine, Tuolumne Table Mt., Calif.	neolith, human skull fragment	Whitney 1880	5.5.5 6.2.6.3
9–55	Stanislaus Co. Mine, Tuolumne Table Mt., Calif.	neolith	Whitney 1880	5.5.6
9–55	Sonora Tunnel, Tuolumne Table Mt., Calif.	stone bead	Whitney 1880	5.5.7
9–55	Tuolumne Table Mt., Calif.	neolith (King pestle)	Becker 1891	5.5.10
9–10	Harital-yangar, India	eolith	Prasad 1982	3.9
>8.7	Placer County, Calif.	human bones	Whitney 1880	6.2.6.5
7–9	Aurillac, France	paleoliths	Verworn 1905	4.3
5–25	Midi de France, France	human skeleton	de Mortillet 1883	6.2.7
5–25	TagusValley, Portugal	paleoliths	Ribeiro 1872	4.1.1
5–25	Dardanelles, Turkey	carved bone, broken bones, flint flake	Calvert 1874	2.10
5–12	Yenang-yaung, Burma	paleoliths	Noetling 1894	4.8

TABLE A3.1—*Continued*

Period / Myr	Site	Category	Reference	Section
5–12	Pikermi, Greece	broken bones	von Dücker 1872	2.8
5–12	Entrerrean Formation, Argentina	paleoliths, signs of fire, incised bones, broken bones, scraped bones, burned bones	F. Ameghino 1912	5.1.5
>5	Marshall Mine, San Andreas, Calif.	neoliths	Whitney 1880	5.5.11
>5	Smilow Mine, San Andreas, Calif.	neoliths	Whitney 1880	5.5.11
>5	Bald Hill, Calif.	human skull (hoax?)	Whitney 1880	6.2.6.1
>5	Clay Hill, Calif.	partial human skeleton (recent?)	Whitney 1880	6.2.6.6
Pliocene				
4–7	Antwerp, Belgium	cut shells, paleoliths, incised bones, human toe prints	Freudenberg 1919	4.5
4–4.5	Kanapoi, Kenya	human humerus	Patterson and Howells 1967	11.5.1
3.6–3.8	Laetoli, Kenya	human footprints	M. Leakey 1979	11.10
3–5	Monte Hermoso, Argentina	paleolith, hearths, slag, burned bones, burned earth, human vertebra	F. Ameghino 1888	5.1.1 6.2.4

TABLE A3.1—*Continued*

Period / Myr	Site	Category	Reference	Section
3–4	Castenedolo, Italy	partial human skeleton,	Ragazzoni 1880	6.2.2
		partial human skeletons (3),	Ragazzoni 1880	6.2.2
		human skeleton	Ragazzoni 1880	6.2.2
3–4	Savona, Italy	human skeleton	Issel 1867	6.2.3
2.5–144	Sub-Crag Detritus Beds, England	bone tools, sawed bone, eoliths, neolith	Moir 1917 Moir 1935 Moir 1929	2.16 2.18 3.3.3 5.3.1
2.5–3.0	According to standard opinion, the oldest stone tools are about 2.5–3.0 million years old at most, and occur only in Africa. One would not expect to find stone tools outside of Africa more than 1 million years ago—when *Homo erectus* is thought to have migrated from his African homeland.			
2.2–3	Sterkfontein, South Africa	human femur	Tardieu 1981	11.3.3
2–4	Kent Plateau, England	eoliths, paleoliths	Prestwich 1889	3.2
2–4	Rosart, Belgium	paleoliths	Rutot 1907	4.4
2–3	Harital-yangar, India	eoliths	Sankhyan 1981	3.6.4
2–3	San Valentino, Italy	pierced bone	Ferretti 1876	2.13
2–3	Monte Aperto, Italy	incised bones, flint blades	Capellini 1876	2.11
2–3	Acquatra-versa, Italy	paleolith	Ponzi 1871	4.6
2–3	Janicule, Italy	paleoliths	Ponzi 1871	4.6

TABLE A3.1—*Continued*

Period / Myr	Site	Category	Reference	Section
2–3	Miramar, Argentina	hearths, slag, burned earth	Hrdlicka 1912	5.1.7
2–3	Miramar, Argentina	paleoliths, neoliths	Roth *et al.* 1915, C. Ameghino 1914, Boman 1921	5.2
2–3	Miramar, Argentina	human jaw	Boman 1921	6.2.5
2.5	Hadar, Ethiopia	eoliths (attributed to *H. habilis*)	Johanson and Edey 1981	11.9.4
2–2.5	San Giovanni, Italy	incised bones	Ramorino 1865	2.5
2–2.5	Red Crag, England	pierced teeth	Charlesworth 1873	2.9
2–2.5	Red Crag, England	carved shell	Stopes 1881	2.15
2–2.5	Foxhall, England	paleoliths, signs of fire, human jaw	Moir 1927 / Collyer 1867	3.3.4 / 6.2.1
2	Soan Valley, Pakistan	eoliths	Bunney 1987	3.6.3
2	Nampa, Idaho	clay figurine	Wright 1912	A2.8
2	According to most scientists, the first toolmaking hominid was *Homo habilis*, the earliest fossils of which are just over 2 million years old and confined to Africa.			
Early Pleistocene				
1.8	Diring Yurlakh, Siberia	eoliths	Daniloff and Kopf 1986	3.6.4
1.8	Xihoudu, China	paleoliths, cut bones, charred bones	Jia 1980	9.2.12

TABLE A3.1—*Continued*

Period / Myr	Site	Category	Reference	Section
1.7–2	Olduvai, Tanzania	broken bone, polished bone, eoliths, paleoliths, bolas, bone tool (for leather work), stone circle (shelter base)	M. Leakey 1971 L. Leakey 1960	2.18 3.7.2 3.7.3 5.3.2 5.3.2 3.7.3
	All of the Olduvai material (above) is normally attributed to *Homo habilis,* but the bone leather-working tool, the shelter, and bolas suggest fully human capability.			
1.7–2	Kanam, Kenya	human jaw, eoliths	L. Leakey 1960	11.2.2
1.7	Yuanmou, China	paleoliths	Jia 1980	9.2.11
	According to the dominant view, the first hominid to leave Africa was *Homo erectus,* who did so about 1 million years ago. So who made the Yuanmou tools (above)?			
1.5–2.5	Ulalinka, Siberia	eoliths	Okladinov and Ragozin 1984	3.6.4
1.5–1.8	Koobi Fora, Kenya	human talus	Wood 1974	11.6.4
1.5	Gombore, Ethiopia	human humerus, eoliths	Senut 1981b	11.5.2
1.2–3.5	Dewlish, England	trench in chalk	Fisher 1912	2.17
1.2–2.5	Val d'Arno, Italy	incised bones	de Mortillet 1883	2.4
1.2–2	St. Prest, France	incised bones, eoliths	Desnoyers 1863 de Mortillet 1883	2.1 2.1

TABLE A3.1—*Continued*

Period / Myr	Site	Category	Reference	Section
1.15	Olduvai, Tanzania	human skeleton	Reck 1914a,b	11.1
1–2.5	Monte Hermoso, Argentina	eoliths	Hrdlicka 1912	5.1.2
1–1.9	Trinil, Java	human tooth	MacCurdy 1924a	7.1.5
1–1.8	Kromdraai, South Africa	human ulna, human humerus	Zuckerman 1954 McHenry 1973	11.3.3
1–1.5	Buenos Aires, Argentina	human skull	F. Ameghino 1909	6.1.5
1	According to most scientists, the first hominid to leave Africa was *Homo erectus*, who did so about 1 million years ago.			
Middle Pleistocene				
.83	Trinil, Java	human femurs	Day and Molleson 1973	7.1.8
.83	Trinil, Java	broken bones, charcoal, hearths	Keith 1911	7.1.5
.6	Gehe, China	neoliths (bolas, implying fully human capability)	Minshall 1989	5.3
.4–1.75	Cromer Forest Bed, England	bone tools, incised bone, sawn wood, paleoliths	Moir 1927 Moir 1924	2.19 2.20 3.3.5
.4–.7	Kanjera, Kenya	human skull fragments, paleoliths	L. Leakey 1960	11.2.1
.4	Olduvai, Tanzania	advanced paleoliths (modern human type)	L. Leakey 1933	11.1.4

TABLE A3.1—*Continued*

Period / Myr	Site	Category	Reference	Section
.33–.6	Ipswich, England	human skeleton	Keith 1928	6.1.3
.33	Galley Hill, England	human skeleton (burial?), paleoliths	Newton 1895	6.1.2.1
.33	Moulin Quignon, France	human jaw and paleoliths (forgeries?)	Keith 1928	6.1.2.2
.33	Clichy, France	partial human skeleton (hoax?)	Bertrand 1868	6.1.2.3
.3–.4	Terra Amata, France	shelters, hearths, bone tools, paleoliths, human footprint	de Lumley 1969	6.1.4
.3	Torralba, Spain	paleoliths	Binford 1981	6.1.4

Terra Amata and Torralba (above) are typical European Middle Pleistocene sites where stone tools and other artifacts are automatically attributed to *Homo erectus*. But anatomically modern humans could also be responsible for the artifacts.

Period / Myr	Site	Category	Reference	Section
.25–.45	Vértesszöllös, Hungary	human skull fragment	Pilbeam 1972	7.2
.25	Hueyatlaco, Mexico	advanced paleoliths	Steen-McIntyre 1981	5.4.4
.25	Sandia Cave, New Mexico	advanced paleoliths	*Smithsonian Misc. Coll.* v. 99, n. 23	5.4.5

The implements from Hueyatlaco and Sandia Cave (above) are of a type normally attributed only to *Homo sapiens sapiens* (maximum age 100,000 years in Africa).

TABLE A3.1—*Continued*

Period / Myr	Site	Category	Reference	Section
.2–.4	Lawn Ridge, Illinois	metal coin (oldest known coins 1000 B.C.)	Dubois 1871	A2.7
.1–1	Lantian, China	neoliths, (bolas, imply fully human capability)	Minshall 1989	5.3
.1–1	Tongzi, China	human teeth	Qiu 1985	9.2.2
.1–1	Liujiang, China	partial human skeleton	Han and Xu 1985	9.2.6
.1	Trenton, New Jersey	human femur, human skull fragments	Volk 1911	6.1.1

The Trenton fossils (above), with an age of 107,000 years, predate the oldest recognized anatomically modern human fossils (about 100,000 years old, from South Africa).

.1 — According to many scientists, anatomically modern humans first appeared about 100,000 (.1 million) years ago in Africa.

Late Pleistocene				
.08–.125	Piltdown, England	human cranium	Dawson and Woodward 1913	8
.03–2	La Denise, France	human skull fragments	de Mortillet 1883	6.1.2.4

La Denise and Piltdown fossils (above) are anomalous if they are over .1 million years old.

The following Pleistocene discoveries are anomalous only for North and South America (Table A3.2). According to most scientists, humans first entered North America not more than 12,000 (.012 million) years ago. Question marks after the dates of some of the following discoveries indicate they were later assigned AMS radiocarbon dates of less than 10,000 years.

TABLE A3.2

**Summary of Anomalous Evidence Related to Human Antiquity
(North and South America Only)**

Period / Myr	Site	Category	Reference	Section
Middle Pleistocene				
.3–.75	Anza-Borrego Desert, Calif.	incised bones	Graham 1988	2.3
.28–.35	El Horno, Mexico	paleoliths	Steen-McIntyre 1981	5.4.4
.2–.5	Calico, Calif.	eoliths	Simpson 1986	3.8.3
.2–.3	Toca da Esperança, Brazil	eoliths	de Lumley *et al.* 1988	3.8.4
.12–.19	Black's Fork River, Wyoming	paleoliths	Renaud 1940	4.9
Late Pleistocene				
.08–.09	Texas Street, San Diego, Calif.	eoliths	Carter 1957	3.8.2
.08	Old Crow River, Canada	incised bones	Morlan 1986	2.2
.07	Timlin, New York	paleoliths	Raemish 1977	5.4.3
.07?	Sunnyvale, Calif.	human bones	Bada and Helfman 1975	A1.3.4
.06–.12	Sheguiandah, Canada	paleoliths	T. E. Lee 1972	5.4.1
>.05	Whiteside County, Illinois	copper ring	W. E. Dubois 1871	A2.7
.048?	Del Mar, Calif.	human bones	Bada *et al.* 1974	A1.3.4

TABLE A3.2 — *Continued*

Period / Myr	Site	Category	Reference	Section
.045?	Bataquitos Lagoon,Calif.	human bones	Bada and Helfman 1975	A1.3.4
.044?	La Jolla, Calif.	human bones	Bada *et al.* 1974	A1.3.4
>.04	Santa Barbara Island, Calif.	hearth, eoliths, mammal bones	*Science News* 1977	3.8.1
.04	Lewisville, Texas	paleolith	Alexander 1978	5.4.2
.039	La Jolla, Calif.	human bones	Bada and Helfman 1975	A1.3.4
.03	El Cedral, Mexico	hearths, mammal bones	Lorenzo 1986	3.8.1
.03	Boq. do Sitio de P. Furada, Brazil	hearths, eoliths, painted rock	Guidon and Delibrias 1986	3.8.1
.028?	Otavalo, Ecuador	human skull	Goodman 1981	A1.3.5
.028?	La Jolla, Calif.	human bone	Bada *et al.* 1974	A1.3.4
.027?	La Jolla, Calif.	human bones	Bada and Helfman 1975	A1.3.4
.026?	Los Angeles, Calif.	human skull	Berger 1975	A1.3.3
.026?	Yuha, Calif.	human skeleton	Stafford *et al.* 1987	A1.3.5
.017?	Laguna, Calif.	human skull	Berger 1975	A1.3.3

Bibliography

Unless otherwise noted in the text, quotations from entries followed by (*) have been translated into English by us.

Aigner, J. S. (1978) Pleistocene faunal and cultural stations in south China. *In* Ikawa-Smith, F., ed. *Early Paleolithic in South and East Asia.* The Hague, Mouton, pp. 129–162.

Aigner, J. S. (1981) *Archaeological Remains in Pleistocene China.* Munich, C. H. Beck.

Aigner, J. S., and Laughton, W. S. (1973) The dating of Lantian man and his significance for analyzing trends in human evolution. *American Journal of Physical Anthropology, 39(1):* 97–110.

Alexander, H. L. (1978) The legalistic approach to early man studies. *In* Bryan, A. L., ed. *Early Man in America from a Circum-Pacific Perspective.* Edmonton, Archaeological Researches International, pp. 20–22.

Alsoszatai-Petheo, J. (1986) An alternative paradigm for the study of early man in the New World. *In* Bryan, A. L., ed. *New Evidence for the Pleistocene Peopling of the Americas.* Orono, Maine, Center for the Study of Early Man, pp. 15–26.

Ameghino, C. (1915) El femur de Miramar. *Anales del Museo nacional de historia natural de Buenos Aires, 26:* 433–450. (*)

Ameghino, F. (1908) Notas preliminares sobre el *Tetraprothomo argentinus,* un precursor del hombre del Mioceno superior de Monte Hermoso. *Anales del Museo nacional de historia natural de Buenos Aires, 16:* 105–242. (*)

Ameghino, F. (1909) Le *Diprothomo platensis,* un précurseur de l'homme du pliocène inférieur de Buenos Aires. *Anales del Museo nacional de historia natural de Buenos Aires, 19:* 107–209. (*)

Ameghino, F. (1910a) *Vestigios industriales en el eoceno superior de Patagonia.* Report to Congreso cientifico internacional americano, Buenos Aires, July 10–25, 1910, 8 pp.

Ameghino, F. (1910b) *Vestigios industriales en la formation entrerriana (oligoceno superior ó mioceno el más inferior).* Report to Congreso cientifico internacional americano, Buenos Aires, July 10–25, 1910, 8 pp.

Ameghino, F. (1911) Énumération chronologique et critique des notices sur les terres cuites et les scories anthropiques des terrains sédimentaires néogenes de l'Argentine parues jusqu'a la fin de l'année 1907. *Anales del Museo nacional de historia natural de Buenos Aires, 20:* 39–80. (*)

Ameghino, F. (1912) L'age des formations sedimentaires tertiaires de l'Argentine en relation avec l'antiquité de l'homme. *Anales del Museo nacional de historia natural de Buenos Aires, 22:* 45–75. (*)

Anderson, E. (1984) Who's who in the Pleistocene: a mammalian bestiary. *In* Martin, P. S., and Klein, R. G., eds. *Quaternary Extinctions.* Tucson, University of Arizona Press, pp. 40–90.

Antunes, M. T., Ferreira, M. P., Rocha, R. B., Soares, A. F., and Zybyszewski, G. (1980) Portugal: cycle alpin. *In* Dercourt, J., ed. *Géologie des Pays Européens*. Vol. 3. Paris, Bordas, pp. 103–149.

Atkins, E. (1985) Conformations in polysaccharides and complex carbohydrates. *Journal of Biosciences, 8:* 375–387.

Autran, A., and Peterlongo, J. M. (1980) Massif Central. *In* Dercourt, J., ed. *Géologie des Pays Européens*. Vol. 4. Paris, Bordas, pp. 4–123.

Ayres, W. O. (1882) The ancient man of Calaveras. *American Naturalist, 25(2):* 845–854.

Bada, J. L. (1985a) Amino acid racemization dating of fossil bones. *Annual Review of Earth and Planetary Sciences, 13:* 241–268.

Bada, J. L. (1985b) Aspartic acid racemization ages of California Paleoindian skeletons. *American Antiquity, 50:* 645–647.

Bada, J. L., and Deems, L. (1975) Accuracy of dates beyond the ^{14}C dating limit using the aspartic acid racemization reaction. *Nature, 255:* 218–219.

Bada, J. L., and Finkel, R. (1983) The Upper Pleistocene peopling of the New World: evidence derived from radiocarbon, amino acid, and uranium series dating. *In* Masters, P. M., and Flemming, N. C., eds. *Quaternary Coastlines and Marine Archaeology*. New York, Academic Press, pp. 463–479.

Bada, J. L., Gillespie, J. A., Gowlett, J. A. J., and Hedges, R. E. M. (1984) Accelerator mass spectrometry radiocarbon ages of amino acid extracts from California palaeoindian skeletons. *Nature, 312:* 442–444.

Bada, J. L., and Helfman, P. M. (1975) Amino acid racemization dating of fossil bones. *World Archaeology, 7:* 161–173.

Bada, J. L., and Masters, P. M. (1982) Evidence for a ~50,000 year antiquity of man in the Americas derived from amino-acid racemization in human skeletons. *In* Ericson, J. E., Taylor, R. E., and Berger, R., eds. *Peopling of the New World*. Los Altos, Ballena Press, pp. 171–179.

Bada, J. L., Masters, P. M., Hoopes, E., and Darling, D. (1979) The dating of fossil bones using amino acid racemization. *In* Berger, R., and Suess, H. E., eds. *Radiocarbon Dating*. Berkeley, University of California Press, pp. 740–756.

Bada, J. L., Schroeder, R. A., and Carter, G. F. (1974) New evidence for the antiquity of man in North America deduced from aspartic acid racemization. *Science, 184:* 791–793.

Ballou, W. H. (1922) Mystery of the petrified 'shoe sole' 5,000,000 years old. *American Weekly* section of the *New York Sunday American*, October 8, p. 2.

Bandy, O. L., and Ingle, J. C. (1970) Neogene planktonic events and radiometric scale, California. *Geological Society of America Special Paper, 124:* 143.

Barker, H., Burleigh, R., and Meeks, N. (1971) British Museum natural radiocarbon measurements VII. *Radiocarbon, 13:* 157–188.

Barker, H., and Mackey, J. (1961) British Museum natural radiocarbon measurements III. *Radiocarbon, 3:* 39–45.

Barnes, A. S. (1939) The differences between natural and human flaking on prehistoric flint implements. *American Anthropologist, N. S. 41:* 99–112.

Bartstra, G. J. (1978) The age of the Djetis beds in east and central Java. *Antiquity* 52: 30–37.

Bateman, P. C., and Wahrhaftig, C. (1966) Geology of the Sierra Nevada. *Bulletin of the California Division of Mines and Geology, 190:* 107–172.

Bayanov, D. (1982) A note on folklore in hominology. *Cryptozoology, 1:* 46–48.

Beaumont, P. B., de Villiers, H., and Vogel, J. C. (1978) Modern man in sub-Saharan Africa prior to 49,000 years B.P.: a review and evaluation with particular reference to Border Cave. *South African Journal of Science, 74:* 409–419.

Becker, G. F. (1891) Antiquities from under Tuolumne Table Mountain in California. *Bulletin of the Geological Society of America, 2:* 189–200.

Bellucci, G. and Capellini, G. (1884) L'Homme Tertiaire en Italie. *Congrès International d'Anthropologie et d'Archéologie Préhistoriques, Lisbon 1880, Compte Rendu,* p. 138.

Belluomini, G. (1981) Direct aspartic acid racemization dating of human bones from archaeological sites of central Italy. *Archaeometry 23:* 125–137.

Belyaeva, E. I., Dubrovo, I. A., and Alekseeva, L. J. (1962) Order *Proboscidea. In* Orlov, Y. A., ed. *Fundamentals of Paleontology,* Vol. 13. Series editor V. I. Gromova, Moscow. Translated from Russian by the Israel Program for Scientific Translations, 1968, pp. 349–372.

Berger, R. (1975) Advances and results in radiocarbon dating: early man in America. *World Archaeology, 7:* 175–183.

Berger, R., Horney, A. G., and Libby, W. F. (1964) Radiocarbon dating of bone and shell from their organic components. *Science, 144:* 999–1001.

Berger, R., and Libby, W. F. (1966) UCLA radiocarbon dates V. *Radiocarbon, 8:* 467–497.

Berggren, W. A., and Van Couvering, J. (1974) *The Late Neogene.* Amsterdam, Elsevier.

Bertrand, P. M. E. (1868) Crane et ossements trouves dans une carriere de l'avenue de Clichy. *Bulletins de la Societe d'Anthropologie de Paris (Series 2), 3:* 329–335. (*)

Binford, L. R. (1981) *Bones: Ancient Men and Modern Myths.* New York, Academic Press.

Binford, L. R., and Ho, C. K. (1985) Taphonomy at a distance: Zhoukoudian, 'the cave home of Beijing man?' *Current Anthropology, 26:* 413–430.

Binford, L. R., and Stone, N. M. (1986) The Chinese Paleolithic: an outsider's view. *Anthroquest: The Leakey Foundation News, 35:* 1, 14–21.

Birdsell, J. B. (1975) *Human Evolution,* 2nd edition. Chicago, Rand McNally.

Black, D. (1927) Further hominid remains of Lower Quaternary age from the Chou Kou Tien deposit. *Nature, 120:* 927–954.

Black, D. (1929) Preliminary notice of the discovery of an adult *Sinanthropus* skull at Chou Kou Tien. *Bulletin of the Geological Survey of China,* 8: 207–208.

Black, D. (1931) Evidence of the use of fire by *Sinanthropus*. *Bulletin of the Geological Survey of China, 11(2):* 107–108.

Black, D., Teilhard de Chardin, P., Yang, Z. [Young, C. C.], and Pei, W. [Pei, W. C.] (1933) Fossil man in China. *Memoirs of the Geological Survey of China, A.11:* 1–158.

Boman, E. (1921) Los vestigios de industria humana encontrados en Miramar (Republica Argentina) y atribuidos a la época terciaria. *Revista Chilena de Historia y Geografia, 49(43):* 330–352. (*)

Boswell, P. G. H. (1932) The Oldoway human skeleton. *Nature, 130:* 237–238.

Boswell, P. G. H. (1935) Human remains from Kanam and Kanjera, Kenya Colony. *Nature, 135:* 371.

Boule, M. (1923) *Fossil Men: Elements of Human Paleontology.* Edinburgh, Oliver and Boyd.

Boule, M. (1937) Le Sinanthrope. *L'Anthropologie, 47:* 1–22.

Boule, M., and Vallois, H. V. (1957) *Fossil Men.* London, Thames and Hudson.

Bourgeois, L. (1873) Sur les silex considérés comme portant les marques d'un travail humain et découverts dans le terrain miocène de Thenay. *Congrès International d'Anthropologie et d'Archéologie Préhistoriques, Bruxelles 1872, Compte Rendu,* pp. 81–92. (*)

Bowden, M. (1977) *Ape-Men, Fact or Fallacy?* Bromley, Sovereign Publications.

Bowen, D. Q. (1980) The Quaternary of the United Kingdom. *In* Dercourt, J., ed. *Geology of the European Countries.* Vol. 1. Paris, Bordas, pp. 418–421.

Bower, Bruce (1988) Retooled ancestors. *Science News, 133:* 344–345.

Brain, C. K. (1978) Some aspects of the South African australopithecine sites and their bone accumulations. *In* Jolly, C. J., ed. *Early Hominids of Africa.* London, Duckworth, pp. 130–161.

Bräuer, G. (1984) A craniological approach to the origin of anatomically modern *Homo sapiens* in Africa, and implications for the appearance of modern Europeans. *In* Smith, F. H., and Spencer, F., eds. *The Origin of Modern Humans: A World Survey of the Fossil Evidence.* New York, Alan R. Liss, pp. 327–410.

Bray, W. (1986) Finding the earliest Americans. *Nature, 321:* 726.

Breuil, H. (1910) Sur la présence d'éolithes a la base de l'Éocene Parisien. *L'Anthropologie, 21:* 385–408. (*)

Breuil, H. (1922) Les industries pliocenes de la region d'Ipswich. *Revue anthropologique, 32:* 226–229. (*)

Breuil, H. (1932) Le feu et l'industrie de pierre et d'os dans le gisement du 'Sinanthropus' à Choukoutien. *L'Anthropologie, 42:* 1–17. (*)

Breuil, H. (1935) L'état actuel de nos connaissances sur les industries paléothiques de Choukoutien. *L'Anthropologie, 45:* 740–746. (*)

Breuil, H., and Lantier, R. (1965) *The Men of the Old Stone Age.* New York, St. Martin's.

Brewster, D. (1844) Queries and statements concerning a nail found imbedded in a block of sandstone obtained from Kingoodie (Mylnfield) Quarry, North Britain. *Report of the British Association for the Advancement of Science, Notices and Abstracts of Communications*, p. 51.

Broad, W., and Wade, N. (1982) *Betrayers of the Truth*. New York, Simon and Schuster.

Brock, A., and Isaac, G. L. (1974) Paleomagnetic stratigraphy and chronology of hominid-bearing sediments east of Lake Rudolf, Kenya. *Nature, 247:* 344–348.

Brodrick, A. H. (1971) *Man and His Ancestors*. London, Hutchinson.

Broom, R. (1950) *Finding the Missing Link*. London, Watts.

Broom, R., and Robinson, J. T. (1952) Swartkrans ape-man. *Transvaal Museum Memoir, 6.*

Broom, R., Robinson, J. T., and Schepers, G. W. H. (1950) Sterkfontein ape-man *Pleisanthropus. Transvaal Museum Memoir, 4.*

Broom, R., and Schepers, G. W. H. (1946) The South African fossil ape-men, the Australopithecinae. *Transvaal Museum Memoir, 2.*

Brown, F., Harris, J., Leakey, R., and Walker, A. (1985) Early *Homo erectus* skeleton from west Lake Turkana, Kenya. *Nature, 316:* 788–793.

Brush, S. G. (1974) Should the history of science be rated X? *Science, 183:* 1164–1172.

Bryan, A. L. (1978) An overview of paleo-American prehistory from a circum-Pacific perspective. *In* Bryan, A. L., ed. *Early Man in America from a Circum-Pacific Perspective*. Edmonton, Archaeological Researches International, pp. 306–327.

Bryan, A. L. (1979) A preliminary look at the evidence for a standardized stone tool technology at Calico. *Quarterly of the San Bernardino County Museum Association, 26(4):* 75–79.

Bryan, A. L. (1986) Paleoamerican prehistory as seen from South America. *In* Bryan, A. L., ed. *New Evidence for the Pleistocene Peopling of the Americas*. Orono, Maine, Center for the Study of Early Man, pp. 1–14.

Budiansky, S. (1987) New light on when man came down from the trees. *U.S. News & World Report,* June 1, pp. 10–11.

Budinger, Jr., F. E. (1983) The Calico early man site. *California Geology, 66(4):* 75–82.

Bunney, S. (1987) First migrants will travel back in time. *New Scientist, 114(1565):* 36.

Burkitt, M. C. (1956) *The Old Stone Age*. New York, New York University.

Burleigh, R. (1984) New World colonized in Holocene. *Nature, 312:* 399.

Burroughs, W. G. (1938) Human-like footprints, 250 million years old. *The Berea Alumnus*. Berea College, Kentucky. November, pp. 46–47.

Butzer, K. W. (1978). Geological perspectives on early hominid evolution. *In* Jolly, C. J., ed. *Early Hominids of Africa*. London, Duckworth, pp. 191–217.

Calvert, F. (1874) On the probable existence of man during the Miocene period. *Journal of the Royal Anthropological Institute, 3:* 127.

Capellini, G. (1877) Les traces de l'homme pliocène en Toscane. *Congrès International d'Anthropologie et d'Archéologie Préhistoriques, Budapest 1876, Compte Rendu.* Vol. 1, pp. 46 – 62. (*)

Carrington, A. (1963) *A Million Years Before Man.* London, Weidenfeld & Nicholson.

Cartailhac, E. (1879) L'homme tertiaire. *Matériaux pour l'Histoire de l'Homme, 2nd series, 11:* 433 – 439. (*)

Carter, G. F. (1957) *Pleistocene Man at San Diego.* Baltimore, Johns Hopkins.

Carter, G. F. (1979) The blade and core stage at Calico. *Quarterly of the San Bernardino County Museum Association, 26(4):* 81–89.

Carter, G. F. (1980) *Earlier Than You Think: A Personal View of Man in America.* College Station, Texas A & M University.

Chang, K. (1962) New evidence on fossil man in China. *Science, 136:* 749 –759.

Chang, K. (1977) *The Archaeology of Ancient China.* 3rd edition. New Haven, Yale University.

Chang, K. (1986) *The Archaeology of Ancient China.* 4th edition. New Haven, Yale University.

Charlesworth, E. (1873) Objects in the Red Crag of Suffolk. *Journal of the Royal Anthropological Institute of Great Britain and Ireland, 2:* 91–94.

Chavaillon, J., Chavaillon, N., Coppens, Y., and Senut, B. (1977) Présence d'hominidé dans le site oldowayen de Gomboré I à Melka Kunturé, Éthiopie. *Comptes Rendus de l' Académie des Sciences, Series D, 285:* 961–963.

Cheng, G., Li, S., and Lin, J. (1977) Discussion of the age of *Homo erectus yuanmouensis* and the event of early Matuyama. *Scientia Geologica Sinica, 1:* 34 – 43.

Cheng, G., Lin, J., and Li, S. (1978) A research on the ages of the strata of Lantien man. *Gurenlie Lunwenji (Collected Papers of Palaeoanthropology).* Institute of Vertebrate Palaeontology and Palaeoanthropology, Chinese Academy of Sciences, Beijing, Science Press, pp. 151–157.

Chivas, A. R., and McDougall, I. (1978) Geochronology of the Koloula porphyry copper prospect, Guadalcanal, Solomon Islands. *Economic Geology, 73:* 682.

Choffat, P. (1884a) Excursion à Otta. *Congrès International d'Anthropologie et d'Archéologie Préhistoriques, Lisbon 1880, Compte Rendu,* pp. 61–67. (*)

Choffat, P. (1884b) Conclusions de la commission chargée de l'examen des silex trouvés à Otta. Followed by discussion. *Congrès International d'Anthropologie et d'Archaéologie Préhistoriques, Lisbon 1880, Compte Rendu,* pp. 92 –118. (*)

Cicha, I. (1970) *Stratigraphical Problems of the Miocene in Europe.* Prague, Czechoslovak Academy of Sciences.

Clark, W. B. (1979) Fossil river beds of the Sierra Nevada. *California Geology, 32:* 143–149.

Cole, S. (1975) *Leakey's Luck, The Life of Louis Leakey.* London, Collins.

Coles, J. M. (1968) Ancient man in Europe. *In* Coles, J. M., and Simpson, D., eds. *Studies in Ancient Europe.* Bristol, Leicester University, pp. 17–43.

Colinvaux, P. A. (1964) The environment of the Bering land bridge. *Ecological Monographs, 34:* 297–329.

Considine, D. M., ed. (1977) *Van Nostrand's Scientific Encyclopedia,* 5th edition. New York, Van Nostrand Reinhold.

Cook, D. C., Buikstra, J. E., DeRousseau, C. J., and Johanson, D. C. (1983) Vertebral pathology in the Afar australopithecines. *American Journal of Physical Anthropology, 60:* 83–101.

Cooke, H. B. S. (1963) Pleistocene mammal faunas of Africa, with particular reference to Southern Africa. *In* Howell, F. C., and Boulière, F., eds. *African Ecology and Human Evolution.* Chicago, Aldine, pp. 78–84.

Cooke, H. B. S. (1976) Suidae from Plio-Pleistocene strata of the Rudolf Basin. *In* Coppens, Y., Howell, F. C., Isaac, G., and Leakey, R. E., eds. *Earliest Man and Environments in the Lake Rudolf Basin.* Chicago, University of Chicago, pp. 251–263.

Coon, C. S. (1969) *Origin of Races.* New York, Alfred Knopf.

Cooper, C. F., and Watson, D. M. S. (1932a) The Oldoway human skeleton. *Nature, 129:* 312–313.

Cooper, C. F., and Watson, D. M. S. (1932b) The Oldoway human skeleton. *Nature, 129:* 903.

Corliss, W. R. (1978) *Ancient Man: A Handbook of Puzzling Artifacts.* Glen Arm, Sourcebook Project.

Cousins, F. W. (1971) *Fossil Man.* Emsworth, A. E. Norris.

Creely, R. S. (1965) Geology of the Oroville quadrangle, California. *Bulletin of the California Division of Mines and Geology, 184.*

Cuenot, C. (1958) *Teilhard de Chardin.* London, Burns & Oates.

Curtis, G. H., Drake, R. E., Cerling, T. E., and Hample, J. (1975) Age of KBS Tuff in Koobi Fora formation, East Rudolf, Kenya. *Nature, 258:* 395–398.

Daniloff, R., and Kopf, C. (1986) Digging up new theories of early man. *U.S. News & World Report,* September 1, pp. 62–63.

Dart, R. A. (1948) The Makapansgat proto-human *Australopithecus prometheus. American Journal of Physical Anthropology, New Series, 6:* 259–283.

Dart, R. A. (1957) The osteodontokeratic culture of *Australopithecus prometheus. Transvaal Museum Memoirs, 10:* 1–105.

Dart, R. A. (1959) *Adventures with the Missing Link.* New York, Viking Press.

Darwin, C. R. (1859) *The Origin of Species.* London, J. Murray.

Darwin, C. R. (1871) *The Descent of Man.* London, J. Murray.

Dawson, C., and Woodward, A. S. (1913) On the discovery of a Paleolithic human skull and mandible in a flint bearing gravel at Piltdown. *Quarterly Journal of the Geological Society, London, 69:* 117–151.

Dawson, C., and Woodward, A. S. (1914) Supplementary note on the discovery of a Palaeolithic human skull and mandible at Piltdown (Sussex). *Quarterly Journal of the Geological Society, London, 70:* 82–99.

Day, M. H. (1978) Functional interpretations of the morphology of postcranial remains of early African hominids. *In* Jolly, C. J., ed. *Early Hominids of Africa.* London, Duckworth, pp. 311–345.

Day, M. H. (1985) Hominid locomotion—from Taung to the Laetoli footprints. *In* Tobias, P. V., ed. *Hominid Evolution: Past, Present, and Future.* New York, Alan R. Liss, pp. 115–128.

Day, M. H. (1989) Fossil man: the hard evidence. *In* Durant, J. R., ed. *Human Origins.* Oxford, Clarendon, pp. 9–26.

Day, M. H., and Molleson, T. I. (1973) The Trinil femora. *Symposia of the Society for the Study of Human Biology, 2:* 127–154.

Day, M. H., and Napier, J. R. (1964) Hominid fossils from Bed I, Olduvai Gorge, Tanganyika: fossil foot bones. *Nature, 201:* 967–970.

Day, M. H., and Wood, B. A. (1968) Functional affinities of the Olduvai Hominid 8 talus. *Man, Second Series, 3:* 440–455.

Delson, E. (1977) Vertebrate paleontology, especially of nonhuman primates, in China. *In* Howells, W. W., and Tsuchitani, P. J., eds. *Palaeoanthropology in the People's Republic of China.* Washington, D. C., National Academy of Sciences, pp. 40–65.

De Lumley, H. (1969) A Palaeolithic camp at Nice. *Scientific American, 220(5):* 42–50.

De Lumley, H., de Lumley, M., Beltrao, M., Yokoyama, Y., Labeyrie, J., Delibrias, G., Falgueres, C., and Bischoff, J. L. (1988) Découverte d'outils taillés associés à des faunes du Pléistocene moyen dans la Toca da Esperança, État de Bahia, Brésil. *Comptes Rendus de l' Académie des Sciences, (Series II) 306:* 241–247. (*)

De Mortillet, G. (1883) *Le Préhistorique.* Paris, C. Reinwald. (*)

De Mortillet, G., and de Mortillet, A. (1881) *Musée Préhistorique.* Paris, C. Reinwald. (*)

De Quatrefages, A. (1884) *Hommes Fossiles et Hommes Sauvages.* Paris, B. Baillière. (*)

De Quatrefages, A. (1887) *Histoire Générale des Races Humaines.* Paris, A. Hennuyer. (*)

DeTerra, H. (1943) Pleistocene geology and early man in Java. *Transactions of the American Philosophical Society, New Series, 32:* 437–464.

Deméré, T. A., and Cerutti, R. A. (1982) A Pliocene shark attack on a cetotheriid whale. *Journal of Paleontology, 56:* 1480–1482.

Deo Gratias, Rev. [D. Perrando] (1873) Sur l'homme tertiaire de Savone. *Congrès International d'Anthropologie et d'Archéologie Préhistoriques, Bologna 1871, Compte Rendu,* pp. 417–420. (*)

Deperet, C. (1926) Fouilles prehistoriques dans le gisement des Hommes fossiles de la Denise, près le Puy-en-Velay. *Comptes Rendus de l' Académie des Sciences, 182:* 358–361. (*)

Desmond, A. (1976) *The Hot-Blooded Dinosaurs.* New York, Dial.

Desnoyers, M. J. (1863) Response à des objections faites au sujet d'incisions constatées sur des ossements de Mammiferes fossiles des environs de Chartres. *Comptes Rendus de l' Académie des Sciences, 56:* 1199–1204. (*)

Diamond, J. (1987) The American blitzkrieg: a mammoth undertaking. *Discover,* June, pp. 82 – 88.

Dietrich, W. O. (1933) Zur Altersfrage der Olduwaylagerstätte. *Centralblatt für Mineralogie, Geologie und Paläontologie, Geologie und Paläontologie Abteilung B, 5:* 299 –303.

Dreimanis, A., and Goldthwait, R. P. (1973) Wisconsin glaciation in the Huron, Erie, and Ontario lobes. *Geological Society of America Memoir, 136:* 71–106.

Drury, C. M., ed. (1976) *Nine Years with the Spokane Indians: The Diary, 1838 – 1848, of Elkanah Walker.* Glendale, California, Arthur H. Clark.

Dubois, E. (1932) The distinct organization of *Pithecanthropus* of which the femur bears evidence now confirmed from other individuals of the described species. *Proceedings of the Koninklijke Nederlandse Akademie van Wetenschappen Amsterdam, 35:* 716 –722.

Dubois, E. (1934) New evidence of the distinct organization of *Pithecanthropus. Proceedings of the Koninklijke Nederlandse Akademie van Wetenschappen Amsterdam, 37:* 139 –145.

Dubois, E. (1935) The sixth (fifth new) femur of *Pithecanthropus erectus. Proceedings of the Koninklijke Nederlandse Akademie van Wetenschappen Amsterdam, 38:* 850 –852.

Dubois, W. E. (1871) On a quasi coin reported found in a boring in Illinois. *Proceedings of the American Philosophical Society, 12(86):* 224 –228.

Durham, J. W. (1967) The incompleteness of our knowledge of the fossil record. *Journal of Paleontology, 41:* 559 –565.

Durrell, C. (1966) Tertiary and Quaternary geology of the northern Sierra Nevada. *Bulletin of the California Division of Mines and Geology, 190:* 185 –197.

Eckhardt, R. B. (1972) Population genetics and human origins. *Scientific American, 226(1):* 94 –103.

Edmunds, F. H. (1954) *British Regional Geology: The Wealden District.* London, Geological Survey.

Evans, P. (1971) Towards a Pleistocene time-scale. *In* Harland, W. B., et al., eds. *The Phanerozoic time-scale, a supplement. Part 2.* Geological Society of London, Special Publication No. 5, pp. 123 –356.

Feldesman, M. R. (1982a) Morphometric analysis of the distal humerus of some Cenozoic catarrhines; the late divergence hypothesis revisited. *American Journal of Physical Anthropology, 59:* 73 –95.

Feldesman, M. R. (1982b) Morphometrics of the ulna of some cenozoic 'hominoids.' *American Journal of Physical Anthropology, 57:* 187.

Ferguson, W. W. (1983) An alternative interpretation of *Australopithecus afarensis* fossil material. *Primates, 25:* 397 –409.

Ferguson, W. W. (1984). Revision of fossil hominid jaws from Plio/Pleistocene of Hadar, in Ethiopia including a new species of the genus *Homo* (Hominoidea:Homininae). *Primates, 25:* 519–529.

Fisher, A. (1988a) On the emergence of humanness. *Mosaic, 19(1):* 34–45.

Fisher, A. (1988b) The more things change. *Mosaic, 19(1):* 23–33.

Fisher, D. E. (1971) Excess rare gases in a subaerial basalt from Nigeria. *Nature, 232:* 60.

Fisher, O. (1905) On the occurrence of *Elephas meriodionalis* at Dewlish (Dorset). *Quarterly Journal of the Geological Society of London, 61:* 35–38.

Fisher, O. (1912) Some handiworks of early men of various ages. *The Geological Magazine, London, 9:* 218–222.

Fitch, F. J., and Miller, J. A. (1976) Conventional potassium-argon and argon-40/argon-39 dating of volcanic rocks from East Rudolf. *In* Coppens, Y., Howell, F. C., Isaac, G., and Leakey, R. E., eds. *Earliest Man and Environments in the Lake Rudolf Basin.* Chicago, University of Chicago, pp. 123–147.

Fix, W. R. (1984) *The Bone Peddlers.* New York, Macmillan.

Fleming, S. (1976) *Dating in Archaeology: A Guide to Scientific Techniques.* London, Dent.

Flint, R. F. (1971) *Glacial and Quaternary Geology.* New York, John Wiley.

Fosdick, R. D. (1952) *The Story of the Rockefeller Foundation.* New York, Harper.

Freudenberg,W.(1919)Die Entdeckung von menschlichen Fußspuren und Artefakten in den tertiären Gerölschichten und Muschelhaufen bei St. Gilles-Waes, westlich Antwerpen. *Praehistorische Zeitschrift, 11:* 1–56. (*)

Funkhouser, J. G., and Naughton, J. J. (1968) Radiogenic helium and argon in ultrabasic inclusions from Hawaii. *Journal of Geophysical Research, 73:* 4601–4607.

Garrigou, F. (1873) Sur l'etude des os cassés que l'on trouve dans divers gisements paleontologiques de l'époque Quaternaire et de l'époque Tertiaire. *Congrès International d'Anthropologie et d'Archéologie Préhistoriques, Bologna 1871, Compte Rendu,* pp. 130–148. (*)

Garrigou, F., and Filhol, H. (1868) M. Garrigou prie l'Académie de vouloir bien ouvrir un pli cacheté, déposé au nom de M. Filhol fils et au sien, le 16 mai 1864. *Comptes Rendus de l'Académie des Sciences, 66:* 819–820. (*)

The Geologist, London (1862) Fossil man. *5:* 470.

Gomberg, D. N., and Latimer, B. (1984) Observations on the transverse tarsal joint of *A. afarensis* and some comments on the interpretation of behaviour from morphology (abstract). *American Journal of Physical Anthropology, 61:* 164.

Goodman, J. (1982) *American Genesis.* New York, Berkley Books.

Goodman, J. (1983) *The Genesis Mystery.* New York, Times Books.

Gould, R. A., Koster, D. A., and Sontz, A. H. L. (1971) The lithic assemblage of the Western Desert aborigines of Australia. *American Antiquity 36(2):* 149–169.

Gould, S. J., and Eldredge, N. (1977) Punctuated equilibria: the tempo and mode of evolution reconsidered. *Paleobiology, 3:* 115–151.

Gowlett, J. A. J. (1984) *Ascent to Civilization.* London, Collins.

Gowlett, J. A. J. (1986) Problems in dating the early human settlement of the Americas. *In* Gowlett, J. A. J., and Hedges, R. E. M., eds. *Archaeological Results from Accelerator Dating.* Oxford, Oxford University Committee for Archaeology, pp. 51–59.

Gowlett, J. A. J., and Hedges, R. E. (1986) Editors' preface. *In* Gowlett, J. A. J, and Hedges, R. E. M., eds. *Archaeological Results from Accelerator Dating.* Oxford, Oxford University Committee for Archaeology.

Graham, D. (1988) Scientist sees an early mark of man. *San Diego Union,* October 31.

Green, J. (1978) *Sasquatch: The Apes Among Us.* Seattle, Hancock House.

Griffin, J. B. (1979) The origin and dispersion of American Indians in North America. *In* Laughlin, W. S., and Harper, A. B., eds. *The First Americans: Origins, Affinities, and Adaptations.* New York, Gustav Fischer, pp. 43–55.

Griffin, J. B. (1983) The Midlands. *In* Jennings, J. D., ed. *Ancient North Americans.* San Francisco, W. H. Freeman, pp. 243–302

Groves, C. P. (1989) *A Theory of Human and Primate Evolution.* Oxford, Clarendon.

Guidon, N., and Delibrias, G. (1986) Carbon-14 dates point to man in the Americas 32,000 years ago. *Nature, 321:* 769–771.

Guo, S., Zhou, S., Meng, W., Zhang, R., Shun, S., Hao, X., Liu S., Zhang, F., Hu, R., and Liu, J. (1980) The dating of Peking man by the fission track technique. *Kexue Tongbao 25(8):* 384.

Haeckel, E. (1905) *The Evolution of Man.* Vol. 1. New York, G. Putnam's Sons.

Hall, E. T. (1974) Article on radiometric dating. *Sunday Telegraph,* November 3, p. 15, cited by Bowden (1977, p. 54).

Han, D., and Xu, C. (1985) Pleistocene mammalian faunas of China. *In* Wu, R., and Olsen, J. W. eds., *Palaeoanthropology and Palaeolithic Archaeology of the People's Republic of China.* Orlando, Academic Press, pp. 267–289.

Harland, W. B., Cox, A. V., Llewellyn, P. G., Pickton, C. A. G., Smith, A. G., and Walters, R. (1982) *A Geologic Time Scale.* Cambridge, Cambridge University Press.

Harrison, E. R. (1928) *Harrison of Ightham.* London, Oxford University Press.

Harte, Bret (1912) *The Poetical Works of Bret Harte.* Boston, Houghton Mifflin.

Hassan, A. A., and Ortner, D. J. (1977) Inclusions in bone material as a source of error in radiocarbon dating. *Archaeometry, 19(2):* 131–135.

Haynes, C. V. (1973) The Calico site: artifacts or geofacts. *Science, 187:* 305–310.

Heizer, R. F., and Whipple, M. A. (1951) *The California Indians: A Source Book.* Berkeley, University of California Press.

Herbert, W. (1983) Lucy's family problems. *Science News, 124:* 8–11.

Heuvelmans, B. (1962) *On the Track of Unknown Animals.* London, Rupert Hart-Davis.

Heuvelmans, B. (1982) What is cryptozoology? *Cryptozoology, 1:* 1–12.

Heuvelmans, B. (1983) How many animal species remain to be discovered? *Cryptozoology, 2:* 1–24.

Hicks, C. S. (1933) Scientific centralisation in the British Empire. *Nature, 131:* 397.

Hill, O. (1945) Nittaewo, an unsolved problem of Ceylon. *Loris, 4:* 251–262.

Ho, T. Y., Marcus, L. F., and Berger, R. (1969) Radiocarbon dating of petroleum-impregnated bone from tar pits at Rancho La Brea, California. *Science, 164:* 1051–1052.

Holmes, W. H. (1899) Review of the evidence relating to auriferous gravel man in California. *Smithsonian Institution Annual Report 1898–1899,* pp. 419–472.

Holmes, W. H. (1919) Handbook of aboriginal American antiquities, Part I. *Smithsonian Institution, Bulletin 60.*

Hood, D. (1964) *Davidson Black.* Toronto, University of Toronto.

Hooijer, D. A. (1951) The age of *Pithecanthropus. American Journal of Physical Anthropology, 9:* 265–281.

Hooijer, D.A. (1956) The lower boundary of the Pleistocene in Java and the age of *Pithecanthropus. Quaternaria, 3:* 5–10.

Hopwood, A. T. (1932) The age of Oldoway man. *Man, 32:* 192–195.

Hough, J. L. (1958) *Geology of the Great Lakes.* Urbana, University of Illinois.

Howell, F. C. (1966) Observations on the earlier phases of the European Lower Paleolithic. *American Anthropologist, 68(2, part 2):* 89.

Howell, F. C. (1978) Hominidae. *In* Maglio, V. J., and Cooke, H. B. S., eds. *Evolution of African Mammals.* Cambridge, Harvard University.

Howells, W. W. (1977) Hominid fossils. *In* Howells, W. W., and Tsuchitani, P. J., eds. *Palaeoanthropology in the People's Republic of China.* Washington, D. C., National Academy of Sciences, pp. 66–77.

Hrdlicka, A. (1907) Skeletal remains suggesting or attributed to early man in North America. *Smithsonian Institution, Bureau of American Ethnology, Bulletin 33.*

Hrdlicka, A. (1912) Early man in South America. Washington, D. C., *Smithsonian Institution.*

Hurford, A. J., Gleadow, A. J. W., and Naeser, C. W. (1976) Fission-track dating of pumice from the KBS Tuff, East Rudolf, Kenya. *Nature, 263:* 738–740.

Huxley, T. H. (1911) *Man's Place in Nature.* London, Macmillan.

Huyghe, P. (1984) The search for Bigfoot. *Science Digest,* September, pp. 56–59, 94, 96.

Ike, D., Bada, J. L., Masters, P. M., Kennedy, G., and Vogel, J. C. (1979) Aspartic acid racemization and radiocarbon dating of an early milling stone horizon burial in California. *American Antiquity, 44:* 524–530.

Ingalls, A. G. (1940) The Carboniferous mystery. *Scientific American, 162:* 14.

Irving, W. N. (1971) Recent early man research in the north. *Arctic Anthropology, 8(2):* 68–82.

Irwin-Williams, C. (1978) Summary of archaeological evidence from the

Valsequillo region, Puebla, Mexico. *In* Bowman, D. L., ed. *Cultural Continuity in Mesoamerica.* London, Mouton, pp. 7–22.

Irwin-Williams, C. (1981) Comments on geologic evidence for age of deposits at Hueyatlaco archaeological site, Valsequillo, Mexico. *Quaternary Research, 16:* 258.

Isaac, G. L. (1978) The archaeological evidence for the activities of early African hominids. *In* Jolly, C. J., ed. *Early Hominids of Africa.* London, Duckworth, pp. 219–254.

Issel, A. (1868) Résumé des recherches concernant l'ancienneté de l'homme en Ligurie. *Congrès International d'Anthropologie et d'Archéologie Préhistoriques, Paris 1867, Compte Rendu,* pp. 75–89. (*)

Issel, A. (1889) Cenni sulla giacitura dello scheletro umano recentmente scoperto nel pliocene di Castenedolo. *Bullettino di Paletnologia Italiana, 15:* 89–109. (*)

Jacob, K., Jacob, C., and Shrivastava, R. N. (1953) Spores and tracheids of vascular plants from the Vindhyan System, India: the advent of vascular plants. *Nature, 172:* 166–167.

Jacob, T. (1964) A new hominid skull cap from Pleistocene Sangiran. *Anthropologica, New Series, 6:* 97–104.

Jacob, T. (1966) The sixth skull cap of *Pithecanthropus erectus. American Journal of Physical Anthropology, 25:* 243–260.

Jacob, T. (1972) The absolute age of the Djetis beds at Modjokerto. *Antiquity 46:* 148.

Jacob, T. (1973) Palaeoanthropological discoveries in Indonesia with special reference to finds of the last two decades. *Journal of Human Evolution, 2:* 473–485.

Jacob, T., and Curtis, G. H. (1971) Preliminary potassium-argon dating of early man in Java. *Contribution of the University of California Archaeological Research Facility, 12:* 50.

Jessup, M. K. (1973) *The Case for the UFO.* Garland, Texas, Varo Manufacturing Company.

Jia, L. (1975) *The Cave Home of Peking Man.* Beijing, Foreign Languages Press.

Jia, L. (1980) *Early Man in China.* Beijing, Foreign Languages Press.

Jia, L. (1985) China's earliest Palaeolithic assemblages. *In* Wu, R., and Olsen, J. W., eds. *Palaeoanthropology and Palaeolithic Archaeology of the People's Republic of China.* Orlando, Academic Press, pp. 135–145.

Jimison, S. (1982) Scientists baffled by space spheres. *Weekly World News,* July 27.

Johanson, D. C. (1976) Ethiopia yields first 'family' of man. *National Geographic, 150:* 790–811.

Johanson, D. C., and Coppens, Y. (1976) A preliminary anatomical description of the first Plio-pleistocene hominid discoveries in the Central Afar, Ethiopia. *American Journal of Physical Anthropology, 45:* 217–234.

Johanson, D. C., and Edey, M. A. (1981) *Lucy: The Beginnings of Humankind.* New York, Simon and Schuster.

Johanson, D.C., Masao, F.T., Eck, G.G., White, T.D., Walter, R.C., Kimbel, W.H., Asfaw, B., Manega, P., Ndessokia, P., and Suwa, G. (1987) New partial skeleton of *Homo habilis* from Olduvai Gorge, Tanzania. *Nature, 327:* 205–209.

Johanson, D. C., and Shreeve, J. (1989) *Lucy's Child.* New York, William Morrow.

Johanson, D. C., and White, T. D. (1979) A systematic assessment of the early African hominids. *Science, 203:* 321–330.

Jones, E. (1953) *The Life and Work of Freud.* Vol. 1. New York, Basic Books.

Josselyn, D. W. (1966) Announcing accepted American pebble tools: the Lively Complex of Alabama. *Anthropological Journal of Canada, 4(1):* 24–31.

Kahlke, H. (1961) On the complex *Stegodon-Ailuropoda* fauna of southern China and the chronological position of *Gigantopithecus blacki* von Koenigswald. *Vertebrata Palasiatica, 5(2):* 83–108.

Keith, A. (1928) *The Antiquity of Man.* Vol. 1. Philadelphia, J. B. Lippincott.

Keith, A. (1931) *New discoveries relating to the antiquity of man.* New York, W. W. Norton.

Keith, A. (1935) Review of *The Stone Age Races of Kenya,* by L. S. B. Leakey. *Nature, 135:* 163–164.

Kennedy, G. E. (1983) Femoral morphology in *Homo erectus. Journal of Human Evolution, 12:* 587–616.

Klaatsch, H. (1907) Review of *La question de l'homme tertiaire* by L. Mayet. *Zeitschrift für Ethnologie, 39:* 765–766. (*)

Klein, C. (1973) *Massif Armoricain et Bassin Parisien.* Strasbourg, Association des Publications près les Universités de Strasbourg. 2 vols.

Kourmisky, J., ed. (1977) *Illustrated Encylopedia of Minerals and Rocks.* London, Octopus.

Krantz, G. S. (1975) An explanation for the diastema of Javan *erectus* skull IV. *In* Tuttle, R. H., ed. *Paleoanthropology: Morphology and Paleoecology.* The Hague, Mouton, pp. 361–370.

Krantz, G. S. (1982) Review of Halpin, M., and Ames, M. M., eds. *Manlike Monsters on Trial: Early Records and Modern Evidence. Cryptzoology, 1:* 94–100.

Krantz, G. S. (1983) Anatomy and dermatoglyphics of three Sasquatch footprints. *Cryptozoology, 2:* 53–81.

Kurtén, B. (1968) *Pleistocene Mammals of Europe.* Chicago, Aldine.

Laing, S. (1893) *Problems of the Future.* London, Chapman and Hall.

Laing, S. (1894) *Human Origins.* London, Chapman and Hall.

Lajoie, K. R., Peterson, E., and Gerow, B. A. (1980) Amino acid bone dating. *In* Hare, P. E., ed. *Biogeochemistry of Amino Acids.* New York, John Wiley, pp. 477–489.

Latimer, B., and Lovejoy, C. O. (1990a) Hallucial metatarsal joint in *Australopithecus afarensis. American Journal of Physical Anthropology, 82:* 125–133.

Latimer, B., and Lovejoy, C. O. (1990b) Metatarsophalangeal joint of *Australopithecus afarensis. American Journal of Physical Anthropology, 83:* 13–23.

Latimer, B., Ohman, J. C., and Lovejoy, C. O. (1987) Talocrural joint in African hominoids: implications for *Australopithecus afarensis. American Journal of Physical Anthropology, 74:* 155–175.

Laussedat, A. (1868) Sur une mâchoire de Rhinoceros portant des entailles profondes trouvée à Billy (Allier), dans les formations calcaires d'eau douce de la Limagne. *Comptes Rendus de l'Académie des Sciences, 66:* 752–754. (*)

Le Gros Clark, W. E., and Campbell, B. G. (1978) *The Fossil Evidence for Human Evolution.* Chicago, University of Chicago.

Leakey, L. S. B. (1928) The Oldoway skull. *Nature, 121:* 499–500.

Leakey, L. S. B. (1931) *The Stone Age Cultures of Kenya Colony.* Cambridge, Cambridge University.

Leakey, L. S. B. (1932a) The Oldoway human skeleton. *Nature, 129:* 721–722.

Leakey, L. S. B. (1932b) The Oldoway human skeleton. *Nature, 130:* 578.

Leakey, L. S. B. (1935) *The Stone Age Races of Kenya.* London, Oxford University Press.

Leakey, L. S. B. (1936) Fossil human remains from Kanam and Kanjera, Kenya colony. *Nature, 138:* 643.

Leakey, L. S. B. (1960a) Recent discoveries at Olduvai Gorge. *Nature, 188:* 1050–1052.

Leakey, L. S. B. (1960b) Finding the world's earliest man. *National Geographic, 118:* 420–435.

Leakey, L. S. B. (1960c) The origin of the genus *Homo. In* Tax, S., ed. *Evolution after Darwin.* Vol. II. Chicago, Chicago University.

Leakey, L. S. B. (1960d) *Adam's Ancestors,* 4th edition. New York, Harper & Row.

Leakey, L. S. B. (1968) Bone smashing by Late Miocene Hominidae. *Nature, 218:* 528–530.

Leakey, L. S. B. (1971) *Homo sapiens* in the Middle Pleistocene and the evidence of *Homo sapiens'* evolution. *In* Bordes, F., ed. *The Origin of Homo sapiens.* Paris, Unesco, pp. 25–28.

Leakey, L. S. B. (1972) *By the Evidence: Memoirs, 1932–1951.* New York, Harcourt Brace Jovanovich.

Leakey, L. S. B. (1979) Calico and early man. *Quarterly of the San Bernardino County Museum Association 26(4):* 91–95.

Leakey, L. S. B, Hopwood, A. T., and Reck, H. (1931) Age of the Oldoway bone beds, Tanganyika Territory. *Nature, 128:* 724.

Leakey, L. S. B., Reck, H., Boswell, P. G. H., Hopwood, A. T., and Solomon, J. D. (1933) The Oldoway human skeleton. *Nature, 131:* 397–398.

Leakey, L. S. B., Tobias, P. V., and Napier, J. R. (1964) A new species of the genus *Homo* from Olduvai Gorge. *Nature, 202:* 7–9.

Leakey, M. D. (1971) *Olduvai Gorge.* Vol. 3. *Excavations in Beds I and II, 1960–1963.* Cambridge, Cambridge University.

Leakey, M.D. (1978) Olduvai fossil hominids: their stratigraphic positions and loca-tions. *In* Jolly, C. J., ed. *Early Hominids of Africa.* London, Duckworth, pp. 3–16.

Leakey, M. D. (1979) Footprints in the ashes of time. *National Geographic, 155:* 446–457.

Leakey, R. E. (1973a) Evidence for an advanced Plio-Pleistocene hominid from East Rudolf, Kenya. *Nature, 242:* 447–450.

Leakey, R. E. (1973b) Skull 1470. *National Geographic, 143:* 819–829.

Leakey, R. E. (1973c) Further evidence of Lower Pleistocene hominids from East Rudolf, North Kenya, 1972. *Nature, 242:* 170–173.

Leakey, R. E. (1984) *One Life.* Salem, New Hampshire, Salem House.

Leakey, R. E., and Lewin, R. (1977) *Origins.* New York, Dutton.

Leakey, R. E., and Lewin, R. (1978) *People of the Lake: Mankind and Its Beginnings.* Garden City, Anchor Press.

Lee, R. E. (1983) "For I have been a man, and that means to have been a fighter." *Anthropological Journal of Canada, 21:* 11–13.

Lee, T. E. (1964) Canada's national disgrace. *Anthropological Journal of Canada, 2(1):* 28–31.

Lee, T. E. (1966a) Untitled editorial note on the Sheguiandah site. *Anthropological Journal of Canada, 4(4):* 18–19.

Lee, T. E. (1966b) Untitled editorial note on the Sheguiandah site. *Anthropological Journal of Canada, 4(2):* 50.

Lee, T. E. (1968) The question of Indian origins, again. *Anthropological Journal of Canada, 6(4):* 22–32.

Lee, T. E. (1972) Sheguiandah in retrospect. *Anthropological Journal of Canada, 10(1):* 28–30.

Lee, T. E. (1977) Introduction to Carter, G. F., On the antiquity of man in America. *Anthropological Journal of Canada, 15(1):* 2–4.

Lee, T. E. (1981) A weasel in the woodpile. *Anthropological Journal of Canada, 19(2):* 18–19.

Lee, T. E. (1983) The antiquity of the Sheguiandah site. *Anthropological Journal of Canada, 21:* 46–73.

Legge, A. J. (1986) Seeds of discontent. *In* Gowlett, J. A. J., and Hedges, R. E. M., eds. *Archaeological Results from Accelerator Dating.* Oxford, Oxford University Committee for Archaeology, pp. 13–21.

Leriche, M. (1922) Les terrains tertiaires de la Belgique. *Congrès Géologique International (13e, Bruxelles), Livret-Guide des Excursions en Belgique, A4:* 1–46.

Lewis, O. J. (1980) The joints of the evolving foot, part III. *Journal of Anatomy, 131:* 275–298.

Li, P., Qian, F., Ma, X., Pu, Q., Xing, L., and Ju, S. (1976) A preliminary study of the age of Yuanmou man by paleomagnetic techniques. *Scientia Sinica, 6:* 579–591.

Li, R., and Lin, D. (1979) Geochemistry of amino acid of fossil bones from deposits of Peking man, Lantian man, and Yuanmou man in China. *Scientia Geologica Sinica, 1:* 56–61.

Liu, D., and Ding, M. (1983) Discussion on the age of "Yuanmou man." *Acta Anthropologica Sinica, 2(1):* 40–48.

Lisowski, F. P., Albrecht, G. H., and Oxnard, C. E. (1974). The form of the talus in some higher primates: a multivariate study. *American Journal of Physical Anthropology, 41:* 191–216.

Lohest, M., Fourmarier, P., Hamal-Nandrin, J., Fraipont, C., and Capitan, L. (1923) Les silex d'Ipswich: conclusions de l'enquéte de l'Institut International d'Anthropologie. *Revue Anthropologique, 33:* 44–67. (*)

Longin, R. (1971) New method of collagen extraction for radiocarbon dating. *Nature, 230:* 241–242.

Lorenzo, J. L. (1978) Early man research in the American hemisphere: appraisal and perspectives. *In* Bryan, A. L., ed. *Early Man in America From a Circum-Pacific Perspective.* Edmonton, Archaeological Researches International, pp. 1–9.

Lorenzo, J. L., and Mirambell, L. (1986) Preliminary report on archaeological and paleoenvironmental studies in the area of El Cedral, San Luis Potosi, Mexico 1977–1980. *In* Bryan, A. L., ed. *New Evidence for the Pleistocene Peopling of the Americas.* Orono, Maine, Center for the Study of Early Man, pp. 106–111.

Lovejoy, C. O. (1988) Evolution of human walking. *Scientific American, 259(5):* 118–125.

Lyell, Charles (1863) *Antiquity of Man.* London, John Murray.

Ma, X., Qian, F., Li, P., and Ju, S. (1978) Paleomagnetic dating of Lantian man. *Vertebrata PalAsiatica, 16(4):* 238–243.

MacCurdy, G. G. (1924a) *Human Origins: A Manual of Prehistory.* Vol. 3. *The Old Stone Age and the Dawn of Man and His Arts.* New York, D. Appleton.

MacCurdy, G. G. (1924b) What is an eolith? *Natural History, 24:* 656–658.

Macalister, R. A. S. (1921) *Textbook of European Archaeology.* Vol. 1. *Paleolithic Period.* Cambridge, Cambridge University.

Maglio, V. J. (1972) Vertebrate faunas and chronology of hominid-bearing sediments east of Lake Rudolf, Kenya. *Nature, 239:* 379–385.

Maglio, V. J. (1973) Origin and evolution of the Elephantidae. *American Philosophical Society Transactions, 63:* 1–149.

Malde, H. E., and Steen-McIntyre, V. (1981) Reply to comments by C. Irwin-Williams: archaeological site, Valsequillo, Mexico. *Quaternary Research, 16:* 418–421.

Mallery, A. H. (1951) *Lost America: The Story of Iron-Age Civilization Prior to Columbus.* Washington, D. C., Overlook.

Mammoth Trumpet (1984) Life in ice age Chile. *1(1):* 1.

Marks, P. (1953) Preliminary note on the discovery of a new jaw of *Meganthropus* von Koenigswald in the lower Middle Pleistocene of Sangiran, central Java. *Indonesian Journal of Natural Science, 109(1):* 26–33.

Marshall, L. G., Pascual, R., Curtis, G. H., and Drake, R. E. (1977) South American geochronology: radiometric time scale for Middle to Late Tertiary mammal-bearing horizons in Patagonia. *Science, 195:* 1325–1328.

Marshall, L. G., Webb, S. D., Sepkoski, Jr., J. J. and Raup, D. M. (1982)

Mammalian evolution and the great American interchange. *Science, 215:* 1351–1357.

Marvin, R. F., and Cole, J. C. (1978) Radiometric ages compilation A, U.S. Geological Survey. *Isochron/West: A Bulletin for Isotope Geochronology 22:* 9–10.

Marvin, R. F., Stern, T. W., Creasey, S. C., and Mehnert, H. H. (1973) Radiometric ages of igneous rocks from Pima, Santa Cruz, and Cochise Counties, southeastern Arizona. *United States Geological Survey Bulletin 1379:* 18.

Marzke, M. W. (1983) Joint function and grips of the *Australopithecus afarensis* hand, with special reference to the region of the capitate. *Journal of Human Evolution, 12:* 197–211.

Masters, P. M. (1987) Preferential preservation of noncollagenous protein during bone diagenesis: implications for chronometric and stable isotopic measurements. *Geochimica et Cosmochimica Acta, 51:* 3209–3214.

Matsu'ura, S., and Ueta, N. (1980) Fraction dependent variation of aspartic acid racemization of fossil bone. *Nature, 286:* 883–884.

Mauger, R. L. (1977) K-Ar ages of biotites from tuffs in Eocene rocks of the Green River, Washake, and Unita Basins, Utah, Wyoming, and Colorado. *University of Wyoming Contributions to Geology, 15:* 17–41.

McDougall, I., Compton, W., and Hawkes, D. D. (1963) Leakage of radiogenic argon and strontium from minerals in Proterozoic dolorites from British Guiana. *Nature, 198:* 564–567.

McHenry, H. M. (1972) Postcranial skeleton of Early Pleistocene hominids. Ph.D. thesis, Harvard University.

McHenry, H. M. (1973) Early hominid humerus from East Rudolf, Kenya. *Science, 180:* 739–741.

McHenry, H. M., and Corruccini, R. S. (1975) Distal humerus in hominoid evolution. *Folia Primatologica 23:* 227–244.

Meister, W. J. (1968) Discovery of trilobite fossils in shod footprint of human in "Trilobite Bed"— a Cambrian formation, Antelope Springs, Utah. *Creation Research Quarterly, 5(3):* 97–102.

Meldau, F. J. (1964) *Why We Believe in Creation, Not in Evolution.* Denver, Christian Victory.

Melleville, M. (1862a) Foreign intelligence. *The Geologist, 5:* 145–148.

Melleville, M. (1862b) Note sur un objet travaillé de main d'homme trouve dans les lignites du Laonnois. *Revue Archéologique, 5:* 181–186. (*)

Merriam, J. C. (1938) *The Published Papers of John Campbell Merriam.* Vol. IV. Washington, D. C., Carnegie Institution.

Michels, J. W. (1973) *Dating Methods in Archaeology.* New York, Seminar Press.

Millar, Ronald (1972) *The Piltdown Men.* London, Victor Gollancz.

Miller, M. E., and Caccioli, W. (1986) The results of the New World Explorers Society Himalayan Yeti Expedition. *Cryptozoology, 5:* 81–84.

Minshall, H. L. (1989) *Buchanan Canyon: Ancient Human Presence in the Americas.* San Marcos, Slawson Communications.

Moir, J. R. (1916) Pre-Boulder Clay man. *Nature, 98:* 109.

Moir, J. R. (1917a) A series of mineralized bone implements of a primitive type from below the base of the Red and Coralline Crags of Suffolk. *Proceedings of the Prehistoric Society of East Anglia, 2:* 116–131.

Moir, J. R. (1917b) A piece of humanly-shaped wood from the Cromer Forest Bed. *Man, 17:* 172–173.

Moir, J. R. (1919) A few notes on the sub-Crag flint implements. *Proceedings of the Prehistoric Society of East Anglia, 3:*158–161.

Moir, J. R. (1923) An early palaeolith from the glacial till at Sidestrand, Norfolk. *The Antiquaries Journal, 3:* 135–137.

Moir, J. R. (1924) Tertiary man in England. *Natural History, 24:* 637–654.

Moir, J. R. (1927) *The Antiquity of Man in East Anglia.* Cambridge, Cambridge University.

Moir, J. R. (1929) A remarkable object from beneath the Red Crag. *Man, 29:* 62–65.

Moir, J. R. (1935) The age of the pre-Crag flint implements. *Journal of the Royal Anthropological Institute, 65:* 343–364.

Mongait, A. (1959) *Archaeology in the U.S.S.R.* Moscow, Foreign Languages Publishing House.

Morlan, R. E. (1986) Pleistocene archaeology in Old Crow Basin: a critical reappraisal. *In* Bryan, A. L., ed. *New Evidence for the Pleistocene Peopling of the Americas.* Orono, Maine, Center for the Study of Early Man, pp. 27–48.

Moziño, J. M. (1970) *Noticias de Nutka: An Account of Nootka Sound in 1792.*Translated and edited by Iris Higbie Wilson. Seattle, University of Washington.

Napier, J. R. (1962) Fossil hand bones from Olduvai Gorge. *Nature, 196:* 400–411.

Napier, J. R. (1973) *Bigfoot: The Yeti and Sasquatch in Myth and Reality.* New York, Dutton.

Nelson, D. E., Vogel, J. S., Southon, J. R., and Brown, T. A. (1986) Accelerator radiocarbon dating at SFU. *Radiocarbon 28:* 215–222.

New York Times (1988) Fossil hands in S. African cave may upset ideas on evolution. May 6, p. A-12.

New York Times News Service (1990) 17-million-year-old leaf fossil yields strands of DNA. *San Diego Union,* April 12, p. A-2.

Newell, N. D. (1959) Symposium on fifty years of paleontology. Adequacy of the fossil record. *Journal of Paleontology, 33:* 488–499.

Newton, E. T. (1895) On a human skull and limb-bones found in the Paleolithic terrace-gravel at Galley Hill, Kent. *Quarterly Journal of the Geological Society of London, 51:* 505–526.

Nilsson, T. (1983) *The Pleistocene.* Dordrecht, D. Reidel.

Noetling, F. (1894) On the occurrence of chipped flints in the Upper Miocene of

Burma. *Records of the Geological Survey of India, 27:* 101–103.

Norris, R. M. (1976) *Geology of California.* New York, John Wiley.

O'Connell, P. (1969) *Science of Today and the Problems of Genesis.* Hawthorne, Christian Book Club of America.

Oakley, K. P. (1954) Evidence of fire in South African cave deposits. *Nature, 174:* 261–262.

Oakley, K. P. (1956) Fire as a Paleolithic tool and weapon. *Proceedings of the Prehistoric Society, New Series, 21:* 36–48.

Oakley, K. P. (1957) The dating of the Broken Hill, Florisbad, and Saldanha skulls. *In* Clark, J. D., ed. *Third Pan-African Congress on Prehistory.* London, Chatto and Windus, pp. 76–79.

Oakley, K. P. (1958) Physical Anthropology in the British Museum. *In* Roberts, D. F., ed. *The Scope of Physical Anthropology and Its Place in Academic Studies.* New York, Wenner Gren Foundation for Anthropological Research, pp. 51–54.

Oakley, K. P. (1961) *Man the Toolmaker.* London, British Museum (Natural History).

Oakley, K. P. (1974) Revised dating of the Kanjera hominids. *Journal of Human Evolution, 3:* 257–258.

Oakley, K. P. (1975) A reconsideration of the date of the Kanam jaw. *Journal of Archeological Science, 2:* 151–152.

Oakley, K. P. (1980) Relative dating of the fossil hominids of Europe. *Bulletin of the British Museum (Natural History), Geology Series, 34(1):* 1–63.

Oakley, K. P., Campbell, B. G., and Molleson, T. I. (1975) *Catalogue of Fossil Hominids.* Part III. *Americas, Asia, Australasia.* London, British Museum.

Oakley, K. P., Campbell, B. G., and Molleson, T. I. (1977) *Catalogue of Fossil Hominids.* Part I. *Africa,* 2nd edition. London, British Museum.

Oakley, K. P., and Hoskins, C. R. (1950) New evidence on the antiquity of Piltdown man. *Nature, 165:* 379–382.

Oakley, K. P., and Montagu, M. F. A. (1949) A re-consideration of the Galley Hill skeleton. *Bulletin of the British Museum (Natural History), Geology, 1(2):* 25–46.

Obermaier, H. (1924) *Fossil Man in Spain.* New Haven, Yale University.

Ogden, J. G. (1977) The use and abuse of radiocarbon dating. *Annals of the New York Academy of Sciences, 288:* 167–173.

Okladinov, A. P., and Ragozin, L. A. (1984) The riddle of Ulalinka. *Soviet Anthropology and Archaeology,* Summer 1984, pp. 3–20.

Osborn, H. F. (1910) *The Age of Mammals.* New York, Macmillan.

Osborn, H. F. (1916) *Men of the Old Stone Age.* New York, Charles Scribner's Sons.

Osborn, H. F. (1921) The Pliocene man of Foxhall in East Anglia. *Natural History, 21:* 565–576.

Osborn, H. F. (1927) *Man Rises to Parnassus.* London, Oxford University.

Osborn, H. F. (1928) *Man Rises to Parnassus,* 2nd edition. Princeton, Princeton University.

Oxnard, C. E. (1968) A note on the fragmentary Sterkfontein scapula. *American Journal of Physical Anthropology, 28:* 213–217.

Oxnard, C. E. (1972) Some African fossil foot bones: a note on the interpolation of fossils into a matrix of extant species. *American Journal of Physical Anthropology, 37:* 3–12.

Oxnard, C. E. (1975a) *Uniqueness and Diversity in Human Evolution.* Chicago, University of Chicago.

Oxnard, C. E. (1975b) The place of the australopithecines in human evolution: grounds for doubt? *Nature, 258:* 389–395.

Oxnard, C. E. (1984) *The Order of Man.* New Haven, Yale University.

Patterson, B., and Howells, W. W. (1967) Hominid humeral fragment from Early Pleistocene of northwestern Kenya. *Science, 156:* 64–66.

Patterson, L. W. (1983) Criteria for determining the attributes of man-made lithics. *Journal of Field Archaeology, 10:* 297–307.

Patterson, L. W., Hoffman, L. V., Higginbotham, R. M., and Simpson, R. D. (1987) Analysis of lithic flakes at the Calico site, California. *Journal of Field Archaeology, 14:* 91–106.

Payen, L. (1982) Artifacts or geofacts: application of the Barnes test. *In* Taylor, R. E., and Berger, R., eds. *Peopling of the New World.* Los Altos, Ballena Press, pp. 193–201.

Pei, J. (1980) An application of thermoluminescence dating to the cultural layers of Peking man site. *Quaternaria Sinica, 5(1):* 87–95.

Pei, W. (1939) The upper cave industry of Choukoutien. *Palaeontologica Sinica, New Series D, 9:* 1–41.

Peterlongo, J. M. (1972) *Guides Géologiques Régionaux: Massif Central.* Paris, Masson et Cie.

Phenice, T. W. (1972) *Hominid Fossils: An Illustrated Key.* Dubuque, William C. Brown.

Pilbeam, D. (1972) *The Ascent of Man, An Introduction to Human Evolution.* New York, Macmillan.

Poirier, F. E. (1977) *Fossil Evidence: The Human Evolutionary Journey,* 2nd edition. St. Louis, C. V. Mosby.

Poirier, F. E., Hu, H., and Chen, C. (1983) The evidence for wildman in Hubei province, People's Republic of China. *Cryptozoology, 2:* 25–39.

Pomerol, C. (1982) *The Cenozoic Era.* Chichester, Ellis Horwood.

Pomerol, C. and Feurgeur, L. (1974) *Guides Géologiques Régionaux: Bassin de Paris.* Paris, Masson et Cie.

Ponzi, G. (1873) Les relations de l'homme préhistorique avec les phénomènes géologiques de l'Italie centrale. *Congrès International d'Anthropologie et d'Archéologie Préhistoriques, Bologna 1871, Compte Rendu,* pp. 49–72. (*)

Pozdeyev, A. I. (1973) Late Paleogene terrestrial volcanism in Koryak Highland and its metallogenic characteristics. *International Geologic Review, 15:* 825–826.

Prasad, K. N. (1971) A note on the geology of the Bilaspur-Haritalyangar region.

Records of the Geological Survey of India, 96: 72–81.

Prasad, K. N. (1982) Was *Ramapithecus* a tool-user. *Journal of Human Evolution, 11:* 101–104.

Prest, V. K. (1969) Retreat of Wisconsin and recent ice in North America. *Geological Survey of Canada, Map 1257A.*

Prestwich, J. (1889) On the occurrence of Palaeolithic flint implements in the neighborhood of Ightham. *Quarterly Journal of the Geological Society of London, 45:* 270–297.

Prestwich, J. (1891) On the age, formation, and successive drift-stages of the Darent: with remarks on the Palaeolithic implements of the district and the origin of its chalk escarpment. *Quarterly Journal of the Geological Society of London, 47:* 126–163.

Prestwich, J. (1892) On the primitive character of the flint implements of the Chalk Plateau of Kent, with reference to the question of their glacial or pre-glacial age. *Journal of the Royal Anthropological Institute of Great Britain and Ireland, 21(3):* 246–262.

Prestwich, Sir John (1895) The greater antiquity of man. *Nineteenth Century, 37:* 617ff.

Previette, K. (1953) Who went there? *Courier-Journal Magazine,* Louisville, Kentucky, May 24.

Prost, J. (1980) The origin of bipedalism. *American Journal of Physical Anthropology, 52:* 175–190.

Protsch, R. (1974) The age and stratigraphic position of Olduvai hominid I. *Journal of Human Evolution, 3:* 379–385.

Puner, H. W. (1947) *Freud: His Life and His Mind.* New York, Grosset and Dunlap.

Qiu, Z. (1985) The Middle Palaeolithic of China. *In* Wu, R., and Olsen, J. W., eds. *Palaeoanthropology and Palaeolithic Archaeology of the People's Republic of China.* Orlando, Academic Press, pp. 187–210.

Raemsch, B. E., and Vernon, W. W. (1977) Some Paleolithic tools from northeast North America. *Current Anthropology, 18:* 97–99.

Ragazzoni, G. (1880) La collina di Castenedolo, solto il rapporto antropologico, geologico ed agronomico. *Commentari dell' Ateneo di Brescia,* April 4, pp. 120–128. (*)

Raup, D., and Stanley, S. (1971) *Principles of Paleontology.* San Francisco, W. H. Freeman.

Reck, H. (1914a) Erste vorläufige Mitteilungen über den Fund eines fossilen Menschenskeletts aus Zentral-afrika. *Sitzungsbericht der Gesellschaft der naturforschender Freunde Berlins, 3:* 81–95. (*)

Reck, H. (1914b) Zweite vorläufige Mitteilung über fossile Tiere- und Menschen-funde aus Oldoway in Zentral-afrika. *Sitzungsbericht der Gesellschaft der naturforschender Freunde Berlins, 7:* 305–318. (*)

Reck, H. (1926) Prähistorische Grab und Menschenfunde und ihre Beziehungen zur Pluvialzeit in Ostafrika. *Mitteilungen der Deutschen Schutzgebiete, 34:* 81–86. (*)

Reck, H. (1933) *Oldoway: Die Schlucht des Urmenschen.* Leipzig, F. A. Brockhaus.

Reeves, B., Pohl, J. M. D., and Smith, J. W. (1986) The Mission Ridge site and the Texas Street question. *In* Bryan, A. L., ed. *New Evidence for the Pleistocene Peopling of the Americas.* Orono, Maine, Center for the Study of Early Man, pp. 65–80.

Ribeiro, C. (1873a) Sur des silex taillés, découverts dans les terrains miocène du Portugal. *Congrès International d'Anthropologie et d'Archéologie Préhistoriques, Bruxelles 1872, Compte Rendu,* pp. 95–100. (*)

Ribeiro, C. (1873b) Sur la position géologique des couches miocènes et pliocènes du Portugal qui contiennent des silex taillés. *Congrès International d'Anthropologie et d'Archéologie Préhistoriques, Bruxelles 1872, Compte Rendu,* pp. 100–104. (*)

Ribeiro, C. (1884) L'homme tertiaire en Portugal. *Congrès International d'Anthropologie et d'Archaéologie Préhistoriques, Lisbon 1880, Compte Rendu,* pp. 81–91. (*)

Rightmire, G. P. (1984) *Homo sapiens* in Sub-Saharan Africa. *In* Smith, F. H., and Spencer, F., eds. *The Origin of Modern Humans: A World Survey of the Fossil Evidence.* New York, Alan R. Liss, pp. 327–410.

Robbins, L. M. (1987) Hominid footprints from Site G. *In* Leakey, M. D., and Harris, J., eds. *Laetoli: A Pliocene Site in Northern Tanzania.* Oxford, Clarendon Press, pp. 497–502.

Rogers, S. P. (1963) The physical characteristics of the aboriginal La Jollan population of Southern California. *San Diego Museum Papers, 4.*

Romer, A. S. (1966) *Vertebrate Paleontology.* Chicago, University of Chicago.

Romero, A. A. (1918) El *Homo pampaeus. Anales de la Sociedad Cientifica Argentina, 85:* 5–48. (*)

Roosevelt, T. (1906) *The Wilderness Hunter.* Vol. 2. New York, Charles Scribner's Sons.

Roth, S., Schiller, W., Witte, L., Kantor, M., Torres, L. M., and Ameghino, C. (1915) Acta de los hechos más importantes del descubrimento de objetos, instrumentos y armas de piedra, realizado en las barrancas de la costa de Miramar, partido de General Alvarado, provincia de Buenos Aires. *Anales del Museo de historia natural de Buenos Aires, 26:* 417–431. (*)

Roujou, A. (1870) Silex taillé découvert en Auvergne dans le miocène supérieur. *Matériaux pour l'Histoire de l'Homme 2:* 93–96.

Rusch, Sr., W. H. (1971) Human footprints in rocks. *Creation Research Society Quarterly, 7:* 201–202.

Rutot, A. (1906) Eolithes et pseudoéolithes. *Société d'Anthropologie de Bruxelles. Bulletin et Memoires. Memoires 25(1).* (*)

Rutot, A. (1907) Un grave problem: une industrie humaine datant de l'époque oligocène. Comparison des outils avec ceux des Tasmaniens actuels. *Bulletin de la Société Belge de Géologie de Paléontologie et d'Hydrologie, 21:* 439–482. (*)

Sanderson, I. T. (1961) *Abominable Snowmen: Legend Come to Life*. Philadelphia, Chilton.

Sanford, J. T. (1971) Sheguiandah reviewed. *Anthropological Journal of Canada, 9(1):* 2–15.

Sanford, J. T. (1983) Geologic observations at the Sheguiandah site. *Anthropological Journal of Canada, 21:* 74–87.

Sankhyan, A. R. (1981) First evidence of early man from Haritalyangar area, Himalchal Pradesh. *Science and Culture, 47:* 358–359.

Sankhyan, A. R. (1983) The first record of Early Stone Age tools of man from Ghummarwin, Himalchal Pradesh. *Current Science, 52:* 126–127.

Sartono, S. (1964) On a new find of another *Pithecanthropus* skull: an announcement. *Bulletin of the Geological Survey of Indonesia, 1(1):* 2–5.

Sartono, S. (1967) An additional skull cap of a *Pithecanthropus*. *Journal of the Anthropological Society of Japan (Nippon), 75:* 83–93.

Sartono, S. (1972) Discovery of another hominid skull at Sangiran, central Java. *Current Anthropology, 13(2):* 124–126.

Sartono, S. (1974) Observations on a newly discovered jaw of *Pithecanthropus modjokertensis* from the Lower Pleistocene of Sangiran, central Java. *Proceedings of the Koninklijke Nederlandse Akadamie van Wetenschappen, Amsterdam, Series B, 77:* 26–31.

Savage, D. E., and Russell, D. E. (1983) *Mammalian Paleofaunas of the World*. Reading, Addison-Wesley.

Schlosser, M. (1911) Beitrage zur Kenntnis der oligozänen Landsäugetiere aus dem Fayum. *Beitrage zur Paläontologie und Geologie, 24:* 51–167.

Schmid, P. (1983) Eine Rekonstruktion des Skelettes von A.L. 288 –1 (Hadar) und deren Konsequenzen. *Folia Primatologica, 40:* 283–306.

Schultz, A. H. (1930) The skeleton of the trunk and limbs of higher primates. *Human Biology, 2:* 303.

Schweinfurth, G. (1907) Über A. Rutots Entdeckung von Eolithen im belgischen Oligocän. *Zeitschrift für Ethnologie, 39:* 958–959.

Science News (1977a) Early man confirmed in America 40,000 years ago. *111:* 196.

Science News (1977b) Amino acid dating: now it has teeth. *111:* 196–197.

Science News (1988) Bone marks: tools vs. teeth. *134:* 14.

Science News Letter (1938a) Geology and ethnology disagree about rock prints. *34:* 372.

Science News Letter (1938b) Human-like tracks in stone are riddle to scientists. *34:* 278–279.

Senut, B. (1979) Comparaison des hominidés de Gombore IB et de Kanapoi: deux pièces du genre *Homo? Bulletin et Mémoires de la Société d'Anthropologie de Paris, 6(13):* 111–117.

Senut, B. (1981a) Humeral outlines in some hominoid primates and in Plio-pleistocene hominids. *American Journal of Physical Anthropology, 56:* 275–283.

Senut, B. (1981b) Outlines of the distal humerus in hominoid primates: application to some Plio-Pleistocene hominids. *In* Chiarelli, A. B., and Corrucini, R. S., eds. *Primate Evolutionary Biology.* Berlin, Springer Verlag, pp. 81–92.

Sergi, G. (1884) L'uomo terziario in Lombardia *Archivio per L'Antropologia e la Etnologia, 14:* 304–318. (*)

Sergi, G. (1912) Intorno all'uomo pliocenico in Italia. *Rivista Di Antropologia (Rome), 17:* 199–216. (*)

Shackley, M. (1983) *Wildmen: Yeti, Sasquatch and the Neanderthal Enigma.* London, Thames and Hudson.

Shipman, P. (1986) Baffling limb on the family tree. *Discover, 7(9):* 87–93.

Simons, E. L. (1978) Diversity among the early hominids: a vertebrate palaeontologist's viewpoint. In Jolly, C. J., ed. *Early Hominids of Africa.* London, Duckworth, pp. 543–566.

Simons, E. L., and Ettel, P. C. (1970) *Gigantopithecus. Scientific American, 22:* 76–85.

Simpson, R. D., Patterson, L. W., and Singer, C. A. (1981) Early lithic technology of the Calico Mountains site, southern California. *Calico Mountains Archeological Site, Occasional Paper.* Presented at the 10th Congress of the International Union of Prehistoric and Protohistoric Sciences, Mexico City.

Simpson, R. D., Patterson, L. W., and Singer, C. A. (1986) Lithic technology of the Calico Mountains site, southern California. *In* Bryan, A. L., ed. *New Evidence for the Pleistocene Peopling of the Americas.* Orono, Maine, Center for the Study of Early Man, pp. 89–105.

Singh, P. (1974) *Neolithic Cultures of Western Asia.* New York, Seminar.

Sinclair, W. J. (1908) Recent investigations bearing on the question of the occurrence of Neocene man in the auriferous gravels of the Sierra Nevada. *University of California Publications in American Archaeology and Ethnology, 7(2):* 107–131.

Slemmons, D. B. (1966) Cenozoic volcanism of the central Sierra Nevada, California. *Bulletin of the California Division of Mines and Geology, 190:* 199–208.

Smith, G. E. (1931) The discovery of primitive man in China. *Antiquity, 5:* 20–36.

Snelling, N. J. (1963) Age of the Roirama formation, British Guiana. *Nature, 198:* 1079–1080.

Sollas, W. J. (1911) *Ancient Hunters,* 1st edition. London, Macmillan.

Sollas, W. J. (1924) *Ancient Hunters,* 3rd edition, revised. London, Macmillan.

Southall, J. (1882) Pliocene man in America. *Journal of the Victoria Institute, 15:* 191–201.

Sparks, B. W., and West, R. G. (1972) *The Ice Age in Britain.* London, Methuen.

Spencer, F. (1984) The Neandertals and Their Evolutionary Significance: A Brief Historical Survey. *In* Smith, F. H., and Spencer, F., eds. *The Origin of Modern Humans: A World Survey of the Fossil Evidence.* New York, Alan R. Liss, pp. 1–49.

Spieker, E. M. (1956) Mountain-building chronology and nature of geologic time scale. *Journal of the American Association of Petroleum Geologists, 40:* 1769–1815.

Sprague, R. (1986) Review of *The Sasquatch and Other Unknown Hominoids,* V. Markotic, ed. *Cryptozoology, 5:* 99–108.

Stafford, T. W., Jull, A. J. T., Brendel, K., Duhamel, R. C., and Donahue, D. (1987) Study of bone radiocarbon dating accuracy at the University of Arizona NSF Accelerator Facility for Radioisotope Analysis. *Radiocarbon, 29:* 24–44.

Stafford, T. W., and Tyson, R. A. (1989) Accelerator radiocarbon dates on charcoal, shell, and human bone from the Del Mar site, California. *American Antiquity, 54:* 389–395.

Stainforth, R. M. (1966) Occurrence of pollen and spores in the Roraima Formation of Venezuela and British Guiana. *Nature, 210:* 292–294.

Stanley, S. M. (1981) *The New Evolutionary Timetable.* New York, Basic Books.

Steen-McIntyre, V., Fryxell, R., and Malde, H. E. (1981) Geologic evidence for age of deposits at Hueyatlaco archaeological site, Valsequillo, Mexico. *Quaternary Research 16:* 1–17.

Steiger, B. (1979) *Worlds Before Our Own.* New York, Berkley.

Stern, Jr., J. T., and Susman, R. L. (1983). The locomotor anatomy of *Australopithecus afarensis. American Journal of Physical Anthropology, 60:* 279–318.

Stokes, W. L. (1974) Geological specimen rejuvenates old controversy. *Dialogue, 8:* 138–141.

Stopes, H. (1881) Traces of man in the Crag. *British Association for the Advancement of Science, Report of the Fifty-first Meeting,* p. 700.

Stopes, M. C. (1912) The Red Crag portrait. *The Geological Magazine, 9:* 285–286.

Straus, Jr., W. L. (1929) Studies on the primate ilia. *American Journal of Anatomy, 43:* 403.

Stringer, C. B., Hublin, J. J., and Vandermeersch, B. (1984) The origin of anatomically modern humans in Western Europe. *In* Smith, F. H., and Spencer, F., eds. *The Origin of Modern Humans: A World Survey of the Fossil Evidence.* New York, Alan R. Liss, pp. 51–135.

Susman, R. L. (1979) Comparative and functional morphology of hominoid fingers. *American Journal of Physical Anthropology, 50:* 215–236.

Susman, R. L. (1988) Hand of *Paranthropus robustus* from Member I, Swartkrans: fossil evidence for tool behavior. *Science 240:* 781–783.

Susman, R. L., and Creel, N. (1979) Functional and morphological affinities of the subadult hand (O.H. 7) from Olduvai Gorge. *American Journal of Physical Anthropology, 51:* 311–332.

Susman, R. L., and Stern, Jr., J. T. (1979) Telemetered electromyography of the flexor digitorum profundus and flexor digitorum superficialis in *Pan troglodytes* and implications for interpretation of the O.H. 7 hand. *American Journal of Physical Anthropology, 50:* 565–574.

Susman, R. L., Stern, Jr., J. T., and Jungers, W. L. (1984) Arboreality and bipedality in the Hadar hominids. *Folia primatologica, 43:* 113–156.

Szabo, B. J., Malde, H. E., and Irwin-Williams, C. (1969) Dilemma posed by uranium–series dates on archaeologically significant bones from Valsequillo, Puebla, Mexico. *Earth and Planetary Science Letters, 6:* 237–244.

Tamers, M. A., and Pearson, F. J. (1965) Validity of radiocarbon dates on bone. *Nature, 208:* 1053–1055.

Tardieu, C. (1979) Analyse morpho-functionelle de l'articulation du genou chez les Primates. Application aux hominides fossiles. *Thesis, University of Pierre.*

Tardieu, C. (1981) Morpho-functional analysis of the articular surfaces of the knee-joint in primates. *In* Chiarelli, A. B., and Corrucini, R. S., eds. *Primate evolutionary biology.* Berlin, Springer Verlag, pp. 68–80.

Tassy, P. (1983) Review of *Les Bêtes Humaines d'Afrique,* by B. Heuvelmans. *Cryptozoology, 2:* 132–133.

Taylor, L. R., Compagno, L. J. V., and Struhsaker, P. J. (1983) Megamouth—a new species, genus, and family of lamnoid shark (*Megachasma pelagois,* family Megachasmidae) from the Hawaiian Islands. *Proceedings of the California Academy of Sciences, 43(8):* 87–110.

Taylor, R. E. (1987) *Radiocarbon Dating: An Archaeological Perspective.* Orlando, Academic Press.

Taylor, R. E., Payen, L. A., Prior, C. A., Slota, Jr., P. J. Gillespie, R., Gowlett, J. A. J., Hedges, R. E. M., Jull, A. J. T., Zabel, T. H., Donahue, D. J., and Berger, R. (1985) Major revisions in the Pleistocene age assignments for North American human skeletons by C-14 accelerator mass spectrometry: none older than 11,000 C-14 years B.P. *American Antiquity, 50:* 136–139.

Teilhard de Chardin, P. (1931) Le *Sinanthropus,* de Pekin. *L'Anthropologie, 41:* 1–11.

Teilhard de Chardin, P. (1965) *The Appearance of Man.* New York, Harper & Row.

Teilhard de Chardin, P., and Yang, Z. [Young, C. C.] (1929) Preliminary report on the Chou Kou Tien fossiliferous deposit. *Bulletin of the Geological Survey of China, 8:* 173–202.

Thorson, R. M., and Guthrie, R. D. (1984) River ice as a taphonomic agent: an alternative hypothesis for bone 'artifacts.' *Quaternary Research, 22:* 172–188.

Time-Life (1973) *Emergence of Man: The First Men.* New York, Time-Life Books.

Tobias, P. V. (1962) A re-examination of the Kanam mandible. *In* Mortelmans, G., and Nenquin, J., eds. *Actes du IVe Congres Panafricain de Prehistorie et de l'Etude du Quaternaire.* Tervuren, Belgium, Musee Royal de l'Afrique Centrale, pp. 341–360.

Tobias, P. V. (1968) Middle and early Upper Pleistocene members of the genus *Homo* in Africa. *In* Kurth, G., ed. *Evolution and Hominisation,* 2nd edition.

Stuttgart, Gustav Fischer, pp. 176–194.

Tobias, P. V. (1971) Human skeletal remains from the Cave of Hearths, Makapansgat, northern Transvaal. *American Journal of Physical Anthropology, 34:* 335–368.

Tobias, P. V. (1978) The South African australopithecines in time and hominid phylogeny, with special reference to dating and affinities of the Taung skull. *In* Jolly, C. J., ed. *Early Hominids of Africa.* London, Duckworth, pp. 44–84.

Tobias, P. V. (1979) Calico Mountains and early man in North America. *Quarterly of the San Bernardino County Museum Association, 26(4):* 97–98.

Tobias, P. V. (1980). 'Australopithecus afarensis' and *A. africanus:* critique and an alternative hypothesis. *Paleontologica Africane, 23:* 1–17.

Traill, D. A. (1986a) Schliemann's acquisition of the Helios Metope and his psychiatric tendencies. *In* Calder, W. M., and Traill, D. A., eds. *Myth, Scandal, and History: The Heinrich Schliemann Controversy and a First Edition of the Mycenaean Diary.* Detroit, Wayne State University, pp. 48–80.

Traill, D. A. (1986b) Priam's treasure: Schliemann's plan to make duplicates for illicit purposes. *In* Calder, W. M., and Traill, D. A., eds. *Myth, Scandal, and History: The Heinrich Schliemann Controversy and a First Edition of the Mycenaean Diary.* Detroit, Wayne State University, pp. 110–121.

Trinkaus, E. (1984) Does KNM-ER 1481A establish *Homo erectus* at 2.0 myr B.P.? *American Journal of Physical Anthropology, 64:* 137–139.

Tuttle, R. H., ed. (1975) *Paleoanthropology: Morphology and Paleoecology.* The Hague, Mouton, pp. 361–370.

Tuttle, R. H. (1981) Evolution of hominid bipedalism and prehensile capabilities. *Philosophical Transactions of the Royal Society of London, B, 292:* 89–94.

Tuttle, R. H. (1985) Ape footprints and Laetoli impressions: a response to the SUNY claims. *In* Tobias, P. V., ed. *Hominid Evolution: Past, Present, and Future.* New York, Alan R. Liss, pp. 129–133.

Tuttle, R. H. (1987) Kinesiological inferences and evolutionary implications from Laetoli biped trails G-1, G-2/3, and A. *In* Leakey, M. D., and Harris, J. eds. *Laetoli: A Pliocene Site in Northern Tanzania.* Oxford, Clarendon Press, pp. 508–517.

Van Andel, T. H. (1981) Consider the incompleteness of the geological record. *Nature, 294:* 397–398.

Vasishat, R. N. (1985) *Antecedents of Early Man in Northwestern India.* New Delhi, Inter-India Publications.

Vere, F. (1959) *Lessons of Piltdown.* Emsworth, A. E. Norris.

Verworn, M. (1905) Die archaeolithische Cultur in den Hipparionschichten von Aurillac (Cantal). *Abhandlungen der königlichen Gesellschaft der Wissenschaften zu Göttingen. Mathematisch-Physikalische Klasse, Neue Folge, 4(4):* 3–60. (*)

Volk, E. (1911) The archaeology of the Delaware Valley. *Papers of the Peabody Museum of American Archaeology and Ethnology, Harvard University, 5.*

Von Dücker, Baron (1873) Sur la cassure artificelle d'ossements recueillis dans le terrain miocène de Pikermi. *Congrès International d'Anthropologie et* *d'Archéologie Préhistoriques, Bruxelles 1872, Compte Rendu,* pp. 104–107.(*)

Von Koenigswald, G. H. R. (1937) Ein Unterkieferfragment des *Pithecanthropus* aus den Trinilschichten Mitteljavas. *Proceedings of the Koninklijke Nederlandse Akadamie van Wetenschappen Amsterdam, 40:* 883–893.

Von Koenigswald, G. H. R. (1940a) Neue *Pithecanthropus* Funde 1936–1938. *Wetenschappelijke Mededeelingen Dienst Mijnbouw Nederlandse Oost-Indie, 28:* 1–223.

Von Koenigswald, G. H. R. (1940b), Preliminary note on new remains of *Pithecanthropus* from central Java. *Proceedings of the Third Congress of Prehistorians of the Far East, Singapore, 1938,* pp. 91–95.

Von Koenigswald, G. H. R. (1947) Search for early man. *Natural History, 56:* 8–15.

Von Koenigswald, G. H. R. (1949a) The discovery of early man in Java and Southern China. *In* W. W. Howells, ed. *Early Man in the Far East.* Detroit, American Association of Physical Anthropologists, pp. 83–98.

Von Koenigswald, G. H. R. (1949b), The fossil hominids of Java. *In* van Bemmelen, R. W., ed. *The Geology of Indonesia.* Vol. IA. The Hague, Government Printing Office, pp. 106–111.

Von Koenigswald, G. H. R. (1956) *Meeting Prehistoric Man.* London, Thames and Hudson.

Von Koenigswald, G. H. R. (1968a) Observations upon two *Pithecanthropus* mandibles from Sangiran, central Java. *Proceedings of the Koninklijke Nederlandse Akadamie van Wetenschappen Amsterdam, Series B, 71:* 99–107.

Von Koenigswald, G. H. R. (1968b) Das absolute Alter des *Pithecanthropus Erectus* Dubois. *In* Kurth, G., ed. *Evolution and Hominisation,* 2nd edition. Stuttgart, Gustav Fischer Verlag, pp. 195–203.

Von Koenigswald, G. H. R., and Weidenreich, F. (1939) The relationship between *Pithecanthropus* and *Sinanthropus. Nature, 144:* 926–929.

Wadia, D. N. (1953) *The Geology of India,* 3rd edition. London, Macmillan.

Walker, A., Leakey, R. E., Harris, J. M., and Brown, F. H. (1986) 2.5-myr *Australopithecus boisei* from west of Lake Turkana, Kenya. *Nature, 322:* 517–522.

Wallace, A. R. (1869) *The Malay Archipelago.* New York, Dover.

Wallace, A. R. (1887) The antiquity of man in North America. *Nineteenth Century, 22:* 667–679.

Wallace, A. R. (1905) *My Life.* Vol. 2. London, Chapman & Hall.

Warren, S. H. (1920) A natural 'eolith' factory beneath the Thanet Sand. *Quarterly Journal of the Geological Society of London, 76:* 238–253.

Waterhouse, J. B. (1978) Chronostratigraphy for the world Permian. *In* Cohee, G. V., Glaessner, M. F., and Hedberg, H. D., eds. *Contributions to the Geologic Time Scale.* Tulsa, American Association of Petroleum Geologists Studies in Geology No. 6, pp. 299–322.

Wayland, E. J. (1932) The Oldoway human skeleton. *Nature, 130:* 578.

Weaver, K. F. (1985) The search for our ancestors. *National Geographic, 168:* 560–624.

Weaver, W. (1967) *U. S. Philanthropic Foundations.* New York, Harper & Row.

Weidenreich, F. (1935) The *Sinanthropus* population of Choukoutien (Locality 1) with a preliminary report on new discoveries. *Bulletin of the Geological Survey of China, 14(4):* 427–468.

Weidenreich, F. (1941) The extremity bones of *Sinanthropus pekinensis. Palaeontologia Sinica,* New Series, D, *5:* 1–150.

Weidenreich, F. (1943) The skull of *Sinanthropus pekinensis. Palaeontologia Sinica,* New Series D, *10:* 1–484.

Weidenreich, F. (1945) Giant early man from Java and South China. *Anthropological Papers of the American Museum of Natural History, 40:* 1–134.

Weiner, J. S. (1955) *The Piltdown Forgery.* Oxford, Oxford University.

Weiner, J. S., Oakley, K. P., and Le Gros Clark, W. E. (1953) The solution of the Piltdown problem. *Bulletin, British Museum (Natural History), Geology, 2(3):* 141–146.

Weiner, J. S., Oakley, K. P., and Le Gros Clark, W. E. (1955) Further contributions to the solution of the Piltdown problem. *Bulletin, British Museum (Natural History), Geology, 2(6):* 228–288.

Weinert, H. (1934) *Homo sapiens* im Altpaläolithischen Diluvium? *Zeitschrift für Morphologie und Anthropologie. Erb-und Rassenbiologie,* Stuttgart, pp. 459–468.

Wendt, H. (1955) *In Search of Adam.* Boston, Houghton Mifflin.

Wendt, H. (1972) *From Ape to Adam.* Indianapolis, Bobbs-Merrill.

West, R. G. (1968) *Pleistocene Geology and Biology.* New York, John Wiley.

West, R. G. (1980) *The Pre-glacial Pleistocene of the Norfolk and Suffolk Coasts.* Cambridge, Cambridge University.

Wetzel, R. M., Dubos, R. E., Martin, R. L., and Myers, P. (1975) *Catagonus:* an 'extinct' peccary, alive in Paraguay. *Science, 189:* 379–380.

White, T. D., and Suwa, G. (1987) Hominid footprints at Laetoli: facts and interpretations. *American Journal of Physcial Anthropology, 72:* 485–514.

Whitney, J. D. (1880) The auriferous gravels of the Sierra Nevada of California. *Harvard University, Museum of Comparative Zoology Memoir 6(1).*

Wilford, J. N. (1990) Mastermind of Piltdown hoax named. New York Times News Service story reprinted in *San Diego Union,* June 11, p. C-1.

Williams, S. (1986) Fantastic archaeology: alternate views of the past. *Epigraphic Society Occasional Papers, 15:* 41.

Willis, D. (1989) *The Hominid Gang.* New York, Viking.

Winchell, A. (1881) *Sparks from a Geologist's Hammer.* Chicago, S. C. Griggs.

Winslow, C. F. (1873) The President reads extracts from a letter from Dr. C. F. Winslow relating the discovery of human remains in Table Mountain, Cal. (Jan 1). *Proceedings of the Boston Society of Natural History, 15:* 257–259.

Witthoft, J. (1955) Texas Street artifacts, part I. *New World Antiquity, 2(9):* 132–134; part II, *2(12):* 179–184.

Wolpoff, M. H. (1980) *Paleoanthropology.* New York, Alfred A. Knopf.

Wood, B. A. (1974a) Evidence on the locomotor pattern of *Homo* from early Pleistocone of Kenya. *Nature, 251:* 135–136.

Wood, B. A. (1974b) Olduvai Bed I postcranial fossils: a reassessment. *Journal of Human Evolution, 3:* 373–378.

Wood, B. A. (1976) Remains attributable to *Homo* in the East Rudolf succession. *In* Coppens, Y., Howell, F. C., Isaacs, G. I., and Leakey, R. E., eds. *Earliest Man and Environments in the Lake Rudolf Basin.* Chicago, University of Chicago, pp. 490–506.

Wood, B. A. (1987) Who is the 'real' *Homo habilis? Nature, 327:* 187–188.

Woodmorappe, J. (1979) Radiometric geochronology reappraised. *Creation Research Quarterly, 16:* 102–129,147.

Woodward, A. S. (1917) Fourth note on the Piltdown gravel with evidence of a second skull of *Eoanthropus dawsoni. Quarterly Journal of the Geological Society of London, 73:* 1–8

Woodward, A. S. (1948) *The Earliest Englishman.* London, Watts.

Woodward, A. S., *et al.* (1933) Early man in East Africa. *Nature, 131:* 477–478.

Wright, G. F. (1912) *Origin and Antiquity of Man.* Oberlin, Bibliotheca Sacra.

Wu, R. [Woo, J.] (1965) Preliminary report on a skull of *Sinanthropus lantianensis* of Lantian, Shensi. *Scientia Sinica, 14(7):* 1032–1035.

Wu, R. [Woo, J.] (1966) The skull of Lantian man. *Current Anthropology, 7(1):* 83–86.

Wu, R. (1973) Lantian hominid. *Wenwu, Peking 6:* 41–44.

Wu, R., and Dong, X. (1985) *Homo erectus* in China. *In* Wu, R., and Olsen, J. W., eds. *Palaeoanthropology and Paleolithic Archaeology in the People's Republic of China.* Orlando, Academic Press, pp. 79–89.

Wu, R., and Lin, S. (1983) Peking man. *Scientific American, 248:* 86–94.

Wu, X., and Wang, L. (1985) Chronology in Chinese palaeoanthropology. *In* Wu, R., and Olsen, J. W., eds. *Palaeoanthropology and Palaeolithic Archaeology in the People's Republic of China.* Orlando, Academic Press, pp. 29–51.

Wu, X., and Wu, M. (1985) Early *Homo sapiens* in China. *In* Wu, R., and Olsen, J. W., eds. *Palaeoanthropology and Palaeolithic Archaeology in the People's Republic of China.* Orlando, Academic Press, pp. 91–106.

Wu, X., and Zhang, Z. (1985) Late Palaeolithic and Neolithic *Homo sapiens. In* Wu, R., and Olsen, J. W., eds. *Palaeoanthropology and Palaeolithic Archaeology of the People's Republic of China.* Orlando, Academic Press, pp. 107–134.

Yuan, Z., and Huang, W. (1979) 'Wild man' - fact or fiction? *China Reconstructs.* July, pp. 56–59.

Zaguin, W. H. (1974) The palaeogeographic evolution of the Netherlands during the Quaternary. *Geologie en Mijnbouw N.S. 53:* 369–385.

Zhang, S. (1985) The early Palaeolithic of China. *In* Wu, R., and Olsen, J. W., eds. *Palaeoanthropology and Palaeolithic Archaeology in the People's Republic of China.* Orlando, Academic Press, pp. 147–186.

Zhou, G. (1982) The status of wildman research in China. *Cryptozoology,* *1:* 13–23.

Zhou, M. [Chow, M.], Hu, C., and Lee, Y. (1965) Mammalian fossils associated with the hominid skull cap of Lantian, Shensi. *Scientia Sinica, 14:* 1037–1048.

Zihlman, A. L. (1985) *Australopithecus afarensis:* two sexes or two species? *In* Tobias, P.V., ed. *Hominid Evolution: Past, Present, and Future.* New York, Alan R. Liss, pp. 213–220.

Zuckerman, S. (1954) Correlation of change in the evolution of higher primates. *In* Huxley, J., Hardy, A. C., and Ford, E. B., eds. *Evolution as a Process.* London, Allen and Unwin, pp. 300–352.

•Zuckerman, S. (1973) Closing remarks to symposium. *The Concepts of Human Evolution. Symposia of the Zoological Society of London, 33:* 449 – 453.

Zuckerman, S., Ashton, E. H., Flinn, R. M., Oxnard, C. E., and Spence, T. F. (1973) Some locomotor features of the pelvic girdle in primates. *The Concepts of Human Evolution. Symposia of the Zoological Society of London, 33:* 71–165.

List of Tables

LIST OF ILLUSTRATIONS

Index

Order Form

☎ Telephone orders: Call 1-800-HIDDEN 1 (1-800-443-3361)
Have your VISA or MasterCard ready.

▤ FAX orders: 559-337-2354

✉ Postal orders: Torchlight Publishing, Inc.
P.O. Box 52, Badger
CA 93603, USA

Please send the following:

❑ *Forbidden Archeology: The Hidden History of the Human Race*
(Hardback, 952 pages, 143 illustrations, 24 tables, $44.95)

❑ *The Hidden History of the Human Race*
(Softbound, 352 pages, 69 illustrations, $15.95).
This is the condensed version of Forbidden Archeology.

❑ *Forbidden Archeology's Impact: How a controversial new book shocked the scientific community and became an underground classic.*
(Hardback, 592 pages, $35.00)

❑ Please send me more information on other books published by Torchlight Publishing.

❑ Please send me more information on the Bhaktivedanta Institute and add my name to your mailing list.

Company: _____

Name: _____

Address: _____

City: _____ State: _____ Zip: _____

(I understand that I may return any books for a full refund - for any reason, no questions asked.)

Sales tax: California residents add 7.25%.

Shipping: Book rate: USA—$4.00 for first book ($3.00 H.H.) and $3.00 for each additional book ($2.50 H.H.);
Canada—$6.00 for first book ($5.00 H.H.) and $3.50 for each additional book ($3.00 H.H.);
Foreign countries—$8.00 for first book, $4.00 for each additional book. (Surface shipping may take 3–4 weeks. Foreign orders please allow 6–8 weeks for delivery.)
Airmail per book (USA only)—$7.00 ($4.00 H.H.)

Payment: ❑ Check/Money order enclosed   Credit card: ❑ VISA ❑ MasterCard

Card number: _____

Name on card: _____Exp. date: _____

Signature: _____

Call 1 (800) 443-3361 and order now!